Lecture Notes in Computer Science 10484

Commenced Publication in 1973
Founding and Former Series Editors:
Gerhard Goos, Juris Hartmanis, and Jan van Leeuwen

More information about this series at http://www.springer.com/series/7412

Sebastiano Battiato · Giovanni Gallo
Raimondo Schettini · Filippo Stanco (Eds.)

Image Analysis and Processing - ICIAP 2017

19th International Conference
Catania, Italy, September 11–15, 2017
Proceedings, Part I

 Springer

Editors

Sebastiano Battiato 🆔
University of Catania
Catania
Italy

Giovanni Gallo 🆔
University of Catania
Catania
Italy

Raimondo Schettini 🆔
University of Milano-Bicocca
Milan
Italy

Filippo Stanco 🆔
University of Catania
Catania
Italy

ISSN 0302-9743   ISSN 1611-3349 (electronic)
Lecture Notes in Computer Science
ISBN 978-3-319-68559-5   ISBN 978-3-319-68560-1 (eBook)
https://doi.org/10.1007/978-3-319-68560-1

Library of Congress Control Number: 2017956081

LNCS Sublibrary: SL6 – Image Processing, Computer Vision, Pattern Recognition, and Graphics

This Springer imprint is published by Springer Nature
The registered company is Springer International Publishing AG
The registered company address is: Gewerbestrasse 11, 6330 Cham, Switzerland

Preface

The 2017 International Conference on Image Analysis and Processing, ICIAP 2017, was the 19th edition of a series of conferences promoted biennaly by the Italian Member Society (GIRPR) of the International Association for Pattern Recognition (IAPR). The conference traditionally covers both classic and the most recent trends in image processing, computer vision, and pattern recognition, addressing both theoretical and applicative aspects.

ICIAP 2017 (http://www.iciap2017.com) was held in Catania, during September 11–15, 2017, in the Benedictine Monastery of San Nicolò l'Arena. The monastery is a UNESCO World Heritage Site and today it hosts the Department of Humanities (DISUM) of the University of Catania. The conference was organized by Image Processing Laboratory, Department of Mathematics and Computer Science (DMI) of the University of Catania. Moreover, ICIAP 2017 was endorsed by the International Association for Pattern Recognition (IAPR), the Italian Member Society of IAPR (GIRPR), and received the institutional support of the University of Catania. Notable sponsorship came from several industrial partners such as STMicroelectronics, Micron, and iCTLab.

ICIAP is traditionally a venue for discussing image processing and analysis, pattern recognition, computer vision, and machine learning, from both theoretical and applicative perspectives, promoting connections and synergies among senior scholars and students, universities, research institutes, and companies. ICIAP 2017 followed this trend, and the program was subdivided into eight main topics, covering a broad range of scientific areas, which were managed by two area chairs per each topic. They were: Biomedical and Assistive Technology; Image Analysis, Detection and Recognition; Information Forensics and Security; Imaging for Cultural Heritage and Archaeology; Multimedia; Multiview Geometry and 3D Computer Vision; Pattern Recognition and Machine Learning; Video Analysis and Understanding.

Moreover, we hosted several prominent companies as well as start-ups to show their activities while assessing them with respect to the cutting-edge research in the respective areas.

ICIAP 2017 received 229 paper submissions coming from all over the world, including Australia, Austria, Brazil, Canada, China, Colombia, Cuba, France, Germany, Hungary, Iran, Ireland, Italy, Israel, Japan, Korea, Kuwait, Malaysia, Mexico, Poland, Portugal, Romania, Russia, Saudi Arabia, Serbia, Spain, South Africa, The Netherlands, Tunisia, Turkey, UK, USA. The paper review process was managed by the program chairs with the invaluable support of 15 area chairs, together with the Program Committee and a number of additional reviewers. The peer-review selection process was carried out by three distinct reviewers in most of the cases. This ultimately led to the selection of 138 high-quality manuscripts, 23 oral presentations, and 115 interactive papers/posters, with an overall acceptance rate of about 60%

(about 10% for oral presentations). The ICIAP 2017 proceedings are published as volumes of the *Lecture Notes in Computer Science* (LNCS) series by Springer.

The program also included five invited talks by distinguished scientists in computer vision pattern recognition and image analysis. We enjoyed the plenary lectures of Daniel Cremers, Technische Universität München, Irfan Essa, Georgia Institute of Technology, Fernando Peréz-Gonzalez, University of Vigo, Nicu Sebe, University of Trento, Roberto Scopigno, ISTI-CNR, and Alain Tremeau, University Jean Monnet, who addressed very interesting and recent research approaches and paradigms such as deep learning and semantic scene understanding in computer vision, multimedia forensics, and applications in the field of color retrieval and management and cultural heritage.

While the main conference was held during September 13–15, 2017, ICIAP 2017 also included five tutorials and seven workshops, held on Monday, September 11, and Tuesday, September 12, 2017, on a variety of topics.

The organized tutorials were: "Virtual Cell Imaging (Methods and Principles)" by David Svoboda; "Image Tag Assignment, Refinement, and Retrieval" by Xirong Li, Tiberio Uricchio, Lamberto Ballan, Marco Bertini, Cees Snoek, Alberto Del Bimbo; "Active Vision and Human Robot Collaboration" by Dimitri Ognibene, Fiora Pirri, Guido De Croon, Lucas Paletta, Mario Ceresa, Manuela Chessa, Fabio Solari; "Humans Through the Eyes of a Robot: How Human Social Cognition Could Shape Computer Vision" by Nicoletta Noceti and Alessandra Sciutti.

There was a special session, "Imaging Solutions for Improving the Quality of Life (I-LIFE'17)," organized by Dan Popescu and Loretta Ichim with eight interesting works selected by the organizers.

ICIAP 2017 also hosted seven half- or full-day satellite workshops: the "First International Workshop on Brain-Inspired Computer Vision (WBICV 2017)" organized by George Azzopardi, Laura Fernández-Robles, Antonio Rodríguez-Sánchez; "Third International Workshop on Multimedia Assisted Dietary Management (MADiMa 2017)" organized by Stavroula Mougiakakou, Giovanni Maria Farinella, Keiji Yanai; "Social Signal Processing and Beyond (SSPandBE 2017)" organized by Mariella Dimiccoli, Petia Ivanova Radeva, Marco Cristani; "Natural Human–Computer Interaction and Ecological Perception in Immersive Virtual and Augmented Reality (NIVAR 2017)" organized by Manuela Chessa, Fabio Solari, Jean-Pierre Bresciani; "Automatic Affect Analysis and Synthesis" organized by Nadia Berthouze, Simone Bianco, Giuseppe Boccignone, Paolo Napoletano; "International Workshop on Biometrics As-a-Service: Cloud-Based Technology, Systems, and Applications" organized by Silvio Barra, Arcangelo Castiglione, Kim-Kwang Raymond Choo, Fabio Narducci; "Background Learning for Detection and Tracking from RGBD Videos" organized by Massimo Camplani, Lucia Maddalena, Luis Salgado. The workshop papers were all collected in a separate volume of the LNCS series by Springer.

We thank all the workshop organizers and tutorial speakers who made possible such an interesting pre-conference program.

Several awards were conferred during ICIAP 2017. The "Eduardo Caianiello" Award was attributed to the best paper authored or co-authored by at least one young researcher (PhD student, postdoc, or similar); a Best Paper Award was also assigned after a careful selection made by an ad hoc appointed committee provided by Springer and IAPR.

The organization and the success of ICIAP 2017 were made possible thanks to the cooperation of many people. First of all, special thanks should be given to the area chairs, who made a big effort for the selection of the papers, together with all the members of the Program Committee. Second, we would also like to thank the industrial, special session, publicity, publication, and Asia and US liaison chairs, who, operating in their respective fields, made this event a successful forum of science.

Special thanks go to the workshop and tutorial chairs as well as all workshop organizers and tutorial lecturers for making richer the conference program with notable satellite events. Last but not least, we are indebted to the local Organizing Committee, mainly colleagues from IPLAB, who dealt with almost every aspects of the conference.

Thanks very much indeed to all the aforementioned people, since without their support we would have not made it.

We hope that ICIAP 2017 met its aim to serve as a basis and inspiration for future ICIAP editions.

September 2017

Sebastiano Battiato
Giovanni Gallo
Raimondo Schettini
Filippo Stanco

Organization

General Chairs

Sebastiano Battiato   University of Catania, Italy
Giovanni Gallo   University of Catania, Italy

Program Chairs

Raimondo Schettini   University of Milano-Bicocca, Italy
Filippo Stanco   University of Catania, Italy

Workshop Chairs

Giovanni Maria Farinella   University of Catania, Italy
Marco Leo   ISASI- CNR Lecce, Italy

Tutorial Chairs

Gian Luca Marcialis   University of Cagliari, Italy
Giovanni Puglisi   University of Cagliari, Italy

Special Session Chairs

Carlo Sansone   University of Naples Federico II, Italy
Cesare Valenti   University of Palermo, Italy

Industrial and Demo Chairs

Cosimo Distante   ISASI – CNR Lecce, Italy
Michele Nappi   University of Salerno, Italy

Publicity Chairs

Antonino Furnari   University of Catania, Italy
Orazio Gambino   University of Palermo, Italy

Video Proceedings Chair

Concetto Spampinato   University of Catania, Italy

US Liaison Chair

Francisco Imai   Canon US Inc., USA

Asia Liaison Chair

Lei Zhang   The Polytechnic University, Hong Kong, SAR China

Steering Committee

Virginio Cantoni   University of Pavia, Italy
Luigi Pietro Cordella   University of Naples Federico II, Italy
Rita Cucchiara   University of Modena and Reggio Emilia, Italy
Alberto Del Bimbo   University of Florence, Italy
Marco Ferretti   University of Pavia, Italy
Fabio Roli   University of Cagliari, Italy
Gabriella Sanniti di Baja   ICAR-CNR, Italy

Area Chairs

Biomedical and Assistive Technology

Domenico Tegolo   University of Palermo, Italy
Sotirios Tsaftaris   University of Edinburgh, UK

Image Analysis, Detection and Recognition

Edoardo Ardizzone   University of Palermo, Italy
M. Emre Celebi   University of Central Arkansas, USA

Imaging for Cultural Heritage and Archaeology

Matteo Dellepiane   ISTI-CNR, Italy
Herbert Maschner   University of South Florida, USA

Information Forensics and Security

Stefano Tubaro   Polytechnic University of Milan, Italy
Zeno Geradts   University of Amsterdam, The Netherlands

Multimedia

Costantino Grana   University of Modena and Reggio Emilia, Italy

Multiview Geometry and 3D Computer Vision

Andrea Fusiello   Università degli Studi di Udine, Italy
David Fofi   University of Burgundy, France

Pattern Recognition and Machine Learning

Dima Damen   University of Bristol, UK
Vittorio Murino   Italian Institute of Technology (IIT), Italy

Video Analysis and Understanding

François Brémond   Inria, France
Andrea Cavallaro   Queen Mary University of London, UK

Invited Speakers

Daniel Cremers   Technische Universität München, Germany
Irfan Essa   Georgia Institute of Technology, USA
Fernando Peréz-Gonzalez University of Vigo, Vigo, Spain
Nicu Sebe   University of Trento, Italy
Roberto Scopigno   ISTI-CNR, Italy
Alain Tremeau   Jean Monnet University, France

Program Committee

Lourdes Agapito   University College London, UK
Jake Aggarwal   University of Texas at Austin, USA
Irene Amerini   University of Florence, Italy
Djamila Aouada   University of Luxemburg, Luxemburg
Federica Arrigoni   University of Udine, Italy
Lamberto Ballan   University of Padova, Italy
Fabio Bellavia   University of Florence, Italy
Simone Bianco   University of Milan-Bicocca, Italy
Silvia Biasotti   CNR-IMATI, Italy
Manuele Bicego   University of Verona, Italy
Giulia Boato   University of Trento, Italy
Giuseppe Boccignone   University of Milan, Italy
Alex Bronstein   Israel Institute of Technology, Israel
Alfred Bruckstein   Israel Institute of Technology, Israel
Joachim Buhmann   ETH Zurich, Switzerland
Francesco Camastra   University of Naples Parthenope, Italy
Barbara Caputo   University of Rome La Sapienza, Italy
Modesto   University of Las Palmas de Gran Canaria, Spain
   Castrillon-Santana
Rama Chellappa   University of Maryland, USA
Aladine Chetouani   University of Orleans, France

Paolo Cignoni	CNR-ISTI, Italy
Gianluigi Ciocca	University of Milan-Bicocca, Italy
Carlo Colombo	University of Florence, Italy
Antonio Criminisi	University of Oxford, UK
Marco Cristani	University of Verona, Italy
Jin Dakai	National Institutes of Health, USA
Tanasi Davide	University of South Florida, USA
Maria De Marsico	University of Rome La Sapienza, Italy
Alessio Del Bue	Italian Institute of Technology, Italy
Cedric Demonceaux	Univ. Bourgogne Franche-Comte, France
Adrien Depeursinge	University of Applied Sciences Western Switzerland, Switzerland
Luigi Di Stefano	University of Bologna, Italy
Naveed Ejaz	Fraunhofer Institute of Integrated Circuits, Germany
Sabu Emmanuel	Kuwait University, Kuwait
Francisco Escolano	University of Alicante, Spain
Gianluca Foresti	University of Udine, Italy
Ana Fred	Technical University of Lisbon, Portugal
Carlo Gatta	Computer Vision Center, Spain
Andrea Giachetti	University of Verona, Italy
Giorgio Giacinto	University of Cagliari, Italy
Mehmet Gonen	School of Medicine, Koç University, Turkey
Marco Gori	University of Siena, Italy
Giorgio Grasso	University of Messina, Italy
Adlane Habed	University of Strasbourg, France
Edwin Hancock	University of York, UK
Anders Hast	Uppsala University, Sweden
Loretta Ichim	Polytechnic University of Bucharest, Romania
Sebastiano Impedovo	University of Bari, Italy
Ignazio Infantino	National Research Council of Italy, Italy
Federico Iuricich	University of Maryland, USA
Richard Jiang	Northumbria University, UK
Michal Kawulok	Silesian University of Technology, Poland
Marco La Cascia	University of Palermo, Italy
Michela Lecca	Fondazione Bruno Kessler, Italy
Ales Leonardis	University of Birmingham, UK
Salvatore Livatino	University of Hertfordshire, UK
Giosuè Lo Bosco	University of Palermo, Italy
Marco Loog	Delft University of Technology, The Netherlands
Carmen Alina Lupascu	Italian National Research Council, Italy
Lucia Maddalena	ICAR-CNR, Italy
Luca Magri	University of Verona, Italy
Simone Marinai	University of Florence, Italy
Eleonora Maset	University of Udine, Italy
Pier Luigi Mazzeo	Italian Research Council, Italy
Christian Micheloni	University of Udine, Italy

Additional Reviewers

Dario Allegra
Lorenzo Baraldi
Catarina Barata
Federico Bolelli
Rodu Nicolae Dobrescu
Amr Elkhouli
Recep Erol
Fausto Galvan
Messina Giuseppe
Francesco Gugliuzza
Sen Jia
Corneliu Lazar
Dario Lo Castro
Liliana Lo Presti

Giuseppe Mazzola
Filippo Luigi Maria Milotta
Marco Moltisanti
Vito Monteleone
Oliver Moolan-Feroze
Pietro Morerio
Alessandro Ortis
Toby Perrett
Roberto Pirrone
Giuseppa Sciortino
Diego Sona
Valeria Tomaselli
Roberto Vezzani

Endorsing Institutions

International Association for Pattern Recognition (IAPR)
Italian Group of Researchers in Pattern Recognition (GIRPR)
Springer

Institutional Patronage

University of Catania
Image Processing Laboratory IPLab

Sponsoring and Supporting Institutions

iCTLab
Micron
STMicroelectronics

Contents – Part I

Pattern Recognition and Machine Learning

Multiview Geometry and 3D Computer Vision

Image Analysis, Detection and Recognition

Contents – Part II

Multimedia

Biomedical and Assistive Technology

Information Forensics and Security

Imaging for Cultural Heritage and Archaeology

Imaging Solutions for Improving the Quality of Life

Video Analysis and Understanding

A Rank Aggregation Framework for Video Interestingness Prediction

Jurandy Almeida[1]([✉]), Lucas P. Valem[2], and Daniel C.G. Pedronette[2]

[1] Institute of Science and Technology, Federal University of São Paulo – UNIFESP,
São José dos Campos 12247-014, Brazil
jurandy.almeida@unifesp.br
[2] Department of Statistics, Applied Mathematics and Computing,
State University of São Paulo – UNESP, Rio Claro 13506-900, Brazil
{lucasvalem,daniel}@rc.unesp.br

Abstract. Often, different segments of a video may be more or less attractive for people depending on their experience in watching it. Due to this subjectiveness, the challenging task of automatically predicting whether a video segment is interesting or not has attracted a lot of attention. Current solutions are usually based on learning models trained with features from different modalities. In this paper, we propose a late fusion with rank aggregation methods for combining ranking models learned with features of different modalities and by different learning-to-rank algorithms. The experimental evaluation was conducted on a benchmarking dataset provided for the Predicting Media Interestingness Task at the MediaEval 2016. Two different modalities and four learning-to-rank algorithms are considered. The results are promising and show that the rank aggregation methods can be used to improve the overall performance, reaching gains of more than 10% over state-of-the-art solutions.

Keywords: Multimedia information retrieval · Predicting Media Interestingness · Learning-to-rank methods · Multimodal late fusion · Rank aggregation

1 Introduction

The production of multimedia data have been grown continuously and consistently. Supported by mobile devices, social networks and cloud environments, multimedia data can be generated, shared and stored everywhere. In this scenario, there is a growing demand for efficient systems able to manage large volumes of multimedia data and reduce the work and information overload when seeking a given content of interest [18].

However, several research challenges are involved, from content representation to its indexing and ranking according to user interests, specially considering different modalities. In many multimedia applications, the fusion of different

Thanks to Brazilian agencies FAPESP (grants 2013/08645-0, 2016/06441-7, and 2017/02091-4), CNPq (grant 423228/2016-1), and CAPES for funding.

modalities is essential for improving the overall performance [19,23]. The main motivation of fusion approaches consists in achieving a more precise representation of the data by combining features from distinct modalities, such as audio and visual content [20]. Additionally, different learning models capable of encoding user preferences can be also considered and fused as complementary information.

In this paper, a multimodal fusion framework based on rank aggregation is proposed for video interestingness prediction. Firstly, different audio and visual features are extracted for constructing a content-based representation. Subsequently, user preferences are encoded through learning-to-rank algorithms, used to construct rankers capable of predicting the interestingness degree of a video. Finally, rank aggregation methods are used for combining the multimodal information provided by different pairs of feature-rankers in order to improve the effectiveness of predictions. Experimental results demonstrate the potential of rank aggregation methods for combining multimodal information on interestingness prediction tasks, which can improve the state-of-the-art results [1] in more than 10%. In addition, the relevance of feature selection strategy is also discussed, providing useful guidance for future work.

This paper is organized as follows. Section 2 discusses related work. Section 3 presents the features, while Sect. 4 presents the learning-to-rank algorithms. Section 5 discusses the rank aggregation methods. Section 6 reports the results of our experiments. Finally, Sect. 7 states conclusions and presents future work.

2. Related Work

This section presents an overview of related work dedicated to video interestingness prediction. In this work, we are interested in multimodal approaches based on data fusion.

The pioneering work of Jiang et al. [16] introduced a new dataset for predicting the interestingness of videos, where a large number of features were evaluated and used to train prediction models with Ranking SVM [17]. According to their findings, audio and visual features are effective for approaching this task, and their fusion can improve the overall performance.

A lot of research on video interestingness prediction has been done for the MediaEval 2016 Predicting Media Interestingness Task [12]. This task aims to automatically select the most interesting video shots according to a common viewer by using features derived from audio-visual content or associated textual information. Ten groups submitted their results for the video subtask and six of them adopted a multimodal approach. The final ranking of these six groups based on the official results was: RECOD [1], UNIGECISA [26], RUC [8], NII-UIT [22], Technicolor [27], and BigVid [29].

Almeida [1] (RECOD team) extracted motion features from the video shots and used them to train four different ranking models, which were combined by a majority voting strategy. Here, we extend the work of Almeida by exploring data fusion (audio and visual data) to enhance video interestingness prediction.

Rayatdoost and Soleymani [26] (UNIGECISA team) used both audio and keyframe-based features provided for the task. Also, they extracted visual sentiment and emotional acoustic features. To obtain a single representation for each shot, they computed the mean and the standard deviation for all the keyframes. Then, principal component analysis (PCA) were applied to reduce the dimensionality of such features. Finally, three different regression models were trained based on the reduced features.

Chen et al. [8] (RUC team) used both audio and keyframe-based features provided for the task. In addition, they extracted statistical acoustic and deep learning features. A single representation for each shot was computed by applying mean pooling over all the keyframe-based features. Different features were combined by early fusion and used to train two different classification models.

Lan et al. [22] (NII-UIT team) used both audio and keyframe-based features provided for the task and also extracted deep learning features. A max pooling strategy was used to aggregate all the keyframe-based features into a single representation for each shot, which was used to train a SVM (Support Vector Machine) classifier. Classification models learned with different features are combined by late fusion using an average weighting scheme.

Shen et al. [27] (Technicolor team) used both audio and keyframe-based features provided for the task. They used such features to train two different deep neural network architectures.

Xu et al. [29] (BigVid team) used both audio and keyframe-based features provided for the task. Also, they extracted semantic features based on sentiment and style attributes. Average pooling over all the keyframe-based features was applied to compute a single representation for each shot. Such features were used to train three different learning models: a classification model using SVM, a ranking model using Ranking SVM, and a deep neural network. In addition, they also considered the combination between SVM and Ranking SVM using a score-level average late fusion.

In this work, we propose a late fusion with rank aggregation methods for combining ranking models learned with features of different modalities and by different learning-to-rank algorithms.

3 Feature Extraction

Two main approaches were used to encode video content. One of them encodes motion information by using *histogram of motion patterns* [2]. The other approach is based on audio information and considers the well-known *mel-frequency cepstral coefficients* [11].

3.1 Histogram of Motion Patterns

Instead of using any keyframe visual features, a simple and fast algorithm was adopted to encode visual properties, known as *histogram of motion patterns* (HMP) [2]. It considers the video movement by the transitions between frames. For each frame of an input video, motion features are extracted from the video stream. For that, 2×2 ordinal matrices are obtained by ranking the intensity values of the four luminance (Y) blocks of each macro block. This strategy is employed for computing both the spatial feature of the 4-blocks of a macro block and the temporal feature of corresponding blocks in three frames (previous, current, and next). Each possible combination of the ordinal measures is treated as an individual pattern of 16-bits (i.e., 2-bits for each element of the ordinal matrices). Finally, the spatio-temporal pattern of all the macro blocks of the video sequence are accumulated to form a normalized histogram.

3.2 Mel-Frequency Cepstral Coefficients

Besides encoding visual properties using HMP, we also used a representation very popular to encode audio information, called *mel-frequency cepstral coefficients* (MFCC) [11]. They are capable of representing the short-time power spectrum of a sound in an accurate and compact form. Initially, the audio signal is filtered with a Finite Impulse Response (FIR) filter to pre-amplify high frequencies. Then, the resulting signal is converted to frames of small duration (typically 20–40 ms). Next, such frames are weighted by a Hamming window aiming at removing any negative effects on its edges. After that, the power spectrum of each frame is computed by applying the Discrete Fourier Transform (DFT) and taking only the magnitude of the spectral coefficients. Thereafter, a filter bank of overlapping triangular filters, also known as Mel-scale filter bank, is used to smooth the spectrum and emphasize perceptually meaningful frequencies. Once the filterbank energies are computed, the logarithm of them is taken aiming at reducing large variations in energy, whose loudness is not perceived by humans. Finally, the Discrete Cosine Transform (DCT) is applied to the log Mel filterbank energies and then only the lower-order coefficients are used to form the feature vector.

4 Ranking Models

The interestingness of videos is a subjective concept that depends on judgments of different viewers on whether a video is interesting or not based on their experience in watching it. Due to this subjectiveness, the automatic prediction of the interestingness degree of a video is a challenging task.

To approach this task, we adopted the strategy proposed by Jiang et al. [16], where a machine learning model is trained aiming at comparing the interestingness between video pairs. In this way, given two videos to the system, it indicates the more interesting one. The basic idea is to use machine learning algorithms to learn a ranking function based on features extracted from training data, and then apply it to features extracted from testing data.

We have used four different learning-to-rank algorithms. The first three are based on pairwise comparisons: *Ranking SVM* [17], *RankNet* [6], and *Rank-Boost* [14]. The latter approach considers lists of objects by using *ListNet* [7].

Ranking SVM [17] is a pairwise ranking method that uses the Support Vector Machine (SVM) classifier to learn a ranking function. For that, each query and its possible results are mapped to a feature space. Next, a given rank is associated to each point in this space. Finally, a SVM classifier is used to find an optimal separating hyperplane between those points based on their ranks.

RankNet [6] is a pairwise ranking method that relies on a probabilistic model. For that, pairwise rankings are transformed into probability distributions, enabling the use of probability distribution metrics as cost functions. Thus, optimization algorithms can be used to minimize a cost function to perform pairwise rankings. The authors formulate this cost function using a neural network in which the learning rate is controlled with gradient descent steps.

RankBoost [14] is a pairwise ranking method that relies on boosting algorithms. Initially, each possible result for a given query is mapped to a feature space, in which each dimension indicates the relative ranking of individual pairs of results, i.e., whether one result is ranked below or above the other. Thus, the ranking problem is formulated as a binary classification problem. Next, a set of weak rankers are trained iteratively. At each iteration, the resulting pairs are re-weighted so that the weight of pairs ranked wrongly is increased whereas the weight of pairs ranked correctly is decreased. Finally, all the weak rankers are combined as a final ranking function.

ListNet [7] is an extension of RankNet that, instead of using pairwise rankings, considers all possible results for a given query as a single instance, enabling to capture and exploit the intrinsic structure of the data. Roughly speaking, it is a listwise ranking method that relies on the probability distribution of permutations. Initially, a given scoring function is used to define the permutation probability distribution for the predicted rankings. Then, another permutation probability distribution is defined for the ground truth. Next, the K-L divergence is used to compute the cross entropy between these two distributions, which is defined as the listwise ranking loss between them. Finally, a linear neural network model is trained through the gradient descent algorithm, which is used to minimize the listwise ranking loss.

5   Rank Aggregation Framework

Ranking has been established as a relevant task in many diverse domains, including information retrieval, natural language processing, and collaborative filtering [9]. However, in many situations, distinct ranking models produce different results. Additionally, the information provided by different ranking results is often complementary, and therefore, can be used for improving the effectiveness of the systems. This is the objective of rank aggregation methods, which aim at combining different rankings in order to obtain a more accurate one.

Rank aggregation approaches are often unsupervised, requiring no training data and can be seen as a way for obtaining a consensus ranking when multiple scores or ranked lists are provided for a set of objects. Different strategies have been used, considering mainly the information of the score computed for an object and the position (or rank) assigned to an object in a ranked list.

Formally, a rank aggregation method can be defined as follows. Let $\mathcal{C} = \{vs_1, vs_2, \ldots, vs_n\}$ be a collection of video shots, where n denotes the number of shots for the video being analyzed. Let $\mathcal{D} = \{D_1, D_2, \ldots, D_d\}$ be a set of rankers. Let the function $\rho_j(i)$ denotes the interestingness degree assigned by the ranker $D_j \in \mathcal{D}$ to the video shot $vs_i \in \mathcal{C}$.

Based on the score ρ_j, a ranked list $\tau_j = (vs_1, vs_2, \ldots, vs_n)$ can be computed. The ranked list τ_j can be defined as a permutation of the collection $\mathcal{C}$, which contains the most interesting video shots according to the ranker D_j. A permutation τ_j is a bijection from the set $\mathcal{C}$ onto the set $[n] = \{1, 2, \ldots, n\}$. For a permutation τ_j, we interpret $\tau_j(i)$ as the position (or rank) of the video shot vs_i in the ranked list τ_j. We can say that, if vs_i is ranked before vs_l in the ranked list τ_j, that is, $\tau_j(i) < \tau_j(l)$, then $\rho_j(i) \leq \rho_j(k)$.

Given the different scores ρ_j and their respective ranked lists τ_j computed by distinct rankers $D_j \in \mathcal{D}$, a rank aggregation method aims to compute a fused score $F(i)$ to each video shot vs_i. In this work, we used three different methods based on score and rank information, described in the following sections.

5.1   Borda Method

The Borda [30] method combines the rank information of each video shot in different ranked lists computed by different rankers. The Borda count method uses rank information in voting procedures. Rank scores are linearly assigned to video shots in ranked lists according to their positions and are summed directly.

More specifically, the distance is scored by the number of video shots not ranked higher than it in the different ranked lists [21]. The new score $F_B(i)$ is computed as follows:

$$F_B(i) = \sum_{j=0}^{d} \tau_j(i). \tag{1}$$

5.2   Reciprocal Rank Fusion

The Reciprocal Rank Fusion [10] uses the rank information for computing a new score according to a naive scoring formula:

$$F_R(i) = \sum_{j=0}^{d} \frac{1}{k + \tau_j(i)}, \tag{2}$$

The intuition behind the formula is based on the conjecture that highly-ranked shots are significantly more relevant than lower-ranked shots [10]. The constant k mitigates the impact of outlier rankers. For the experiments in this paper, $k = 16$ is used.

5.3   Multiplicative Rank Aggregation

A multiplicative approach [24] is used for the rank aggregation based on scores. The use of a multiplication approach is inspired by the Naïve Bayes classifiers. Given a set of scores computed by distinct rankers, such classifiers try to estimate the relevance probability assuming conditional independence among rankers. Considering the independence assumption, the scores of each ranker are multiplied. The fused score $F_M(i)$ for a given video shot vs_i is computed as:

$$F_M(i) = \prod_{j=1}^{d}(1 + \rho_j(i)). \tag{3}$$

6   Experiments and Results

Experiments were conducted on a benchmarking dataset provided by the MediaEval 2016 organizers for the Predicting Media Interestingness Task [12]. This dataset is composed of 78 Creative Commons licensed trailers of Hollywood-like movies. It is divided into a *development set* of 52 videos (67%) and a *test set* of 26 videos (33%). These videos were segmented by hand, producing a total of 7,396 video shots. After video segmentation, the development set has 5,054 shots and the test set has 2,342 shots.

Each video shot was represented by the HMP and MFCC features, as discussed in Sect. 3. For encoding visual properties, we extracted the HMP features directly from the video data. On other hand, for representing audio information, we used the MFCC features provided for the task [15]. Unlike HMP, MFCC produces multiple local features for a same video. To obtain a single representation, we built a Bag-of-Features (BoF) [5] model upon local MFCC features. In the BoF framework, visual words [28] are obtained by quantizing local features according to a pre-learned dictionary. Thus, a video sequence is represented as a normalized frequency histogram of visual words associated with each local feature. In this work, we construct a codebook of 4000 visual words using a random selection. For the dictionary creation, we used only the MFCC features extrated from the development set.

Once the features were extracted, they were used as input to train machine-learned rankers, as presented in Sect. 4. The SVM^{rank} package[1] [17] was used for running Ranking SVM. The RankLib package[2] was used for running RankNet, RankBoost, and ListNet. Ranking SVM was configured with a linear kernel. RankNet, RankBoost, and ListNet were configured with their default parameter settings. All those approaches were calibrated through a 4-fold cross validation on the development set. Next, the trained rankers were used to predict the rankings of test video shots. The rankings associated with the video shots of a same movie trailer were normalized using a z-score normalization. After that, the normalized rankings of all the rankers are combined using our proposed

---

[1] https://www.cs.cornell.edu/people/tj/svm_light/svm_rank.html.
[2] https://sourceforge.net/p/lemur/wiki/RankLib/.

framework, producing the final prediction scores. Finally, a thresholding method was applied to transform the prediction scores into binary decisions. It was found empirically that better results were obtained when a video shot is classified as interesting if its prediction score is greater than 0.7; otherwise, it is classified as non-interesting.

The effectiveness of our strategy was assessed using Mean Average Precision (MAP), which is the official evaluation metric adopted in the task. Our results were compared with those reported by Almeida[3] [1], which ranked 1st out of 10 groups in the MediaEval 2016 Predicting Media Interestingness Task.

Table 1 presents the results obtained by the HMP and MFCC features in isolation. On the development set, by analyzing the confidence intervals, it can be noticed that the performance of the different learning-to-rank algorithms is similar, with a small advantage to Ranking SVM. On the test set, however, Ranking SVM provided the best results for HMP whereas ListNet was the best for MFCC. These results indicate that the fusion of such learning-to-rank algorithms may be promising.

Table 1. Results obtained by HMP and MFCC on the development set using the machine-learned rankers in isolation.

Feature	Ranker	Development set			Test set
		Avg.	Conf. Interval (95%)		
			Min.	Max.	
HMP	Ranking SVM	15.19	13.99	16.38	18.15
	RankNet	13.82	12.09	15.55	16.17
	RankBoost	14.67	12.93	16.42	16.17
	ListNet	13.32	12.06	14.57	16.56
MFCC	Ranking SVM	14.19	12.27	16.12	15.87
	RankNet	13.33	11.49	15.17	17.10
	RankBoost	12.53	11.55	13.51	15.62
	ListNet	13.45	12.20	14.71	17.57

For combining the results provided by different features and machine-learned rankers, we adopted the strategy proposed by Almeida et al. [3]. Initially, we sorted the individual results obtained by each pair (feature & ranker) in a decreasing order of MAP. Then, each pair was selected according to its rank, i.e., the best was the first, the second best was the second, and so on. At each step, the next pair was combined with all the previous ones, as discussed in Sect. 5.

Figure 1 shows the MAP scores obtained by different rank aggregation methods on the development set. We show the behavior of such methods for combining

---

[3] The results reported by Almeida [1] refer to those obtained using only the HMP feature and are presented in Table 1.

the most effective pairs according to the average individual results achieved in the development set (see Table 1). The horizontal line denotes the MAP score for the best pair in isolation and forms a baseline for our proposed framework. The vertical line indicates the set of pairs which achieved the highest MAP score when combined with the rank aggregation methods. The error bars represent 95% confidence intervals computed from the 4 folds.

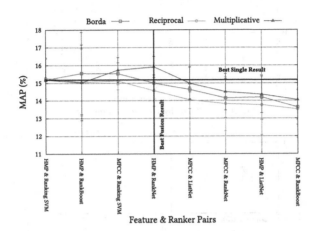

Fig. 1. MAP obtained by different rank aggregation methods on the development set.

We can see that, as more pairs are considered for late fusion, more effective results are obtained, until reach a peak. This is an expected behavior, because different features and machine-learned rankers may complement each other, which aggregates more information. From a certain point, however, non-relevant results from the less effective pairs exceed relevant results from the most effective ones and the gain decreases. By analyzing the confidence intervals, it is important to note that there is a high variance among the 4 folds. These results indicate that the ordering defined by such folds, i.e., from the most to the least effective pairs, is not consistent. This ordering is used for selecting the pairs to be combined by the rank aggregation methods.

Figure 2 shows the MAP scores obtained by different rank aggregation methods on the test set. In Fig. 2(a), features and machine-learned rankers were selected for late fusion with rank aggregation methods in a decreasing order of their average individual results on the development set (see Table 1). Notice that the rank aggregation methods did not improve the best individual result (i.e., HMP & Ranking SVM). The main reason for such results is the selection strategy adopted for defining the pairs to be used for combination.

In Fig. 2(b), we replicate the previous experiment, however a different selection strategy was adopted. In this figure, we show the MAP scores as the most effective pairs are used for combination. Unlike the previous experiment, instead of considering the decreasing order of average individual results from the development set, the ordering was defined based on the individual results achieved

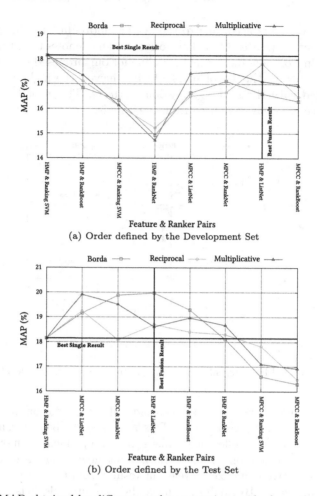

(a) Order defined by the Development Set

(b) Order defined by the Test Set

Fig. 2. MAP obtained by different rank aggregation methods on the test set.

in the test set (see Table 1). As we can see, the best fusion result was obtained by the Borda method in combining the four most effective pairs (i.e., HMP & Ranking SVM, MFCC & ListNet, MFCC & RankNet, HMP & ListNet), which achieved a MAP score equals to **19.97%**, yielding gains of more than **10%** with respect to the best single result (i.e., HMP & Ranking SVM).

Such positive results indicate the potential of rank aggregation methods for combining multimodal information and improving the interestingness prediction. At the same time, the importance of the selection strategy is also evident. The better results presented by the set of pairs defined by the effectiveness order on the test set indicate that unsupervised selection procedures can be exploited.

7   Conclusions

This paper presented a novel approach for predicting the interestingness of videos. Our method is based on combining the features of audio and visual modalities with rank aggregation methods. The proposed strategy relies on a late fusion of ranking models learned with different learning-to-rank algorithms.

Our approach was validated in the dataset of the MediaEval 2016 Predicting Media Interestingness Task. Conducted experiments demonstrate that our multimodal strategy yields better video interestingness prediction results when compared with those based on a single modality (either audio or visual information). Also, we show that, by using a proper selection strategy, the rank aggregation methods can be used to improve the overall performance, achieving significant gains in comparison with state-of-the-art solutions.

Future work includes the evaluation of other features (e.g., keyframe-based methods [13,25]), especially those encoding information from different modalities, as well as perform an extensive study on smarter selection strategies for combining learning-to-rank algorithms (e.g., genetic programming [4]).

References

1. Almeida, J.: UNIFESP at MediaEval 2016: predicting media interestingness task. In: Proceedings of the MediaEval 2016 Workshop (2016)
2. Almeida, J., Leite, N.J., Torres, R.S.: Comparison of video sequences with histograms of motion patterns. In: ICIP, pp. 3673–3676 (2011)
3. Almeida, J., Pedronette, D.C.G., Alberton, B., Morellato, L.P.C., Torres, R.S.: Unsupervised distance learning for plant species identification. IEEE J. Sel. Topics Appl. Earth Observ. Remote Sens. **9**(12), 5325–5338 (2016)
4. Andrade, F.S.P., Almeida, J., Pedrini, H., da Torres, R.S.: Fusion of local and global descriptors for content-based image and video retrieval. In: Alvarez, L., Mejail, M., Gomez, L., Jacobo, J. (eds.) CIARP 2012. LNCS, vol. 7441, pp. 845–853. Springer, Heidelberg (2012). doi:10.1007/978-3-642-33275-3_104
5. Boureau, Y.L., Bach, F., LeCun, Y., Ponce, J.: Learning mid-level features for recognition. In: CVPR, pp. 2559–2566 (2010)
6. Burges, C.J.C., Shaked, T., Renshaw, E., Lazier, A., Deeds, M., Hamilton, N., Hullender, G.N.: Learning to rank using gradient descent. In: ICML, pp. 89–96 (2005)
7. Cao, Z., Qin, T., Liu, T.Y., Tsai, M.F., Li, H.: Learning to rank: from pairwise approach to listwise approach. In: ICML, pp. 129–136 (2007)
8. Chen, S., Dian, Y., Jin, Q.: RUC at MediaEval 2016: predicting media interestingness task. In: Proceedings of the MediaEval 2016 Workshop (2016)
9. Chen, W., Liu, T.Y., Lan, Y., Ma, Z., Li, H.: Ranking measures and loss functions in learning to rank. In: NIPS, pp. 315–323 (2009)
10. Cormack, G.V., Clarke, C.L.A., Buettcher, S.: Reciprocal rank fusion outperforms condorcet and individual rank learning methods. In: SIGIR, pp. 758–759 (2009)
11. Davis, S., Mermelstein, P.: Comparison of parametric representations for monosyllabic word recognition in continuously spoken sentences. IEEE Trans. Acoust. Speech Sig. Process. **28**(4), 357–366 (1980)

12. Demarty, C.H., Sjöberg, M., Ionescu, B., Do, T.T., Wang, H., Duong, N.Q.K., Lefebvre, F.: Mediaeval 2016 predicting media interestingness task. In: Proceedings of the MediaEval 2016 Workshop (2016)
13. Duarte, L.A., Penatti, O.A.B., Almeida, J.: Bag of genres for video retrieval. In: SIBGRAPI, pp. 257–264 (2016)
14. Freund, Y., Iyer, R.D., Schapire, R.E., Singer, Y.: An efficient boosting algorithm for combining preferences. J. Mach. Learn. Res. **4**, 933–969 (2003)
15. Jiang, Y.G., Dai, Q., Mei, T., Rui, Y., Chang, S.F.: Super fast event recognition in internet videos. IEEE Trans. Multimedia **17**(8), 1174–1186 (2015)
16. Jiang, Y.G., Wang, Y., Feng, R., Xue, X., Zheng, Y., Yang, H.: Understanding and predicting interestingness of videos. In: AAAI, pp. 1113–1119 (2013)
17. Joachims, T.: Training linear SVMs in linear time. In: SIGKDD, pp. 217–226 (2006)
18. Kankanhalli, M.S., Lim, J.H.: Perspectives on Content-Based Multimedia Systems. Springer, Secaucus (2000). doi:10.1007/b116171
19. Kludas, J., Bruno, E., Marchand-Maillet, S.: Information fusion in multimedia information retrieval. In: Boujemaa, N., Detyniecki, M., Nürnberger, A. (eds.) AMR 2007. LNCS, vol. 4918, pp. 147–159. Springer, Heidelberg (2008). doi:10.1007/978-3-540-79860-6_12
20. Kokar, M.M., Tomasik, J.A., Weyman, J.: Formalizing classes of information fusion systems. Inf. Fusion **5**(3), 189–202 (2004)
21. Kozorovitsky, A.K., Kurland, O.: Cluster-based fusion of retrieved lists. In: SIGIR, pp. 893–902 (2011)
22. Lam, V., Do, T., Phan, S., Le, D.D., Satoh, S., Duong, D.A.: NII-UIT at MediaEval 2016 predicting media interestingness task. In: Proceedings of the MediaEval 2016 Workshop (2016)
23. Li, L.T., Pedronette, D.C.G., Almeida, J., Penatti, O.A.B., Calumby, R.T., Torres, R.S.: A rank aggregation framework for video multimodal geocoding. Multimedia Tools Appl. **73**(3), 1323–1359 (2014)
24. Pedronette, D.C.G., Torres, R.S.: Image re-ranking and rank aggregation based on similarity of ranked lists. Pattern Recogn. **46**(8), 2350–2360 (2013)
25. Penatti, O.A.B., Li, L.T., Almeida, J., da Torres, R.S.: A visual approach for video geocoding using bag-of-scenes. In: ICMR, pp. 1–8 (2012)
26. Rayatdoost, S., Soleymani, M.: Ranking images and videos on visual interestingness by visual sentiment features. In: Proceedings of the MediaEval 2016 Workshop (2016)
27. Shen, Y., Demarty, C.H., Duong, N.Q.K.: Technicolor at MediaEval 2016 predicting media interestingness task. In: Proceedings of the MediaEval 2016 Workshop (2016)
28. Sivic, J., Zisserman, A.: Video Google: a text retrieval approach to object matching in videos. In: ICCV, pp. 1470–1477 (2003)
29. Xu, B., Fu, Y., Jiang, Y.G.: BigVid at MediaEval 2016: predicting interestingness in images and videos. In: Proceedings of the MediaEval 2016 Workshop (2016)
30. Young, H.P.: An axiomatization of borda's rule. J. Econ. Theory **9**(1), 43–52 (1974)

Graph-Based Hierarchical Video Cosegmentation

Franciele Rodrigues[1], Pedro Leal[1], Yukiko Kenmochi[2], Jean Cousty[2],
Laurent Najman[2], Silvio Guimarães[1,2], and Zenilton Patrocínio Jr.[1(✉)]

[1] PUC Minas - ICEI - DCC - VIPLAB, Belo Horizonte, Brazil
francisbonfim@gmail.com, mr.pedro@outlook.com,
{sjamil,zenilton}@pucminas.br
[2] Université Paris-Est, LIGM, ESIEE Paris - CNRS, Champs-sur-Marne, France
{y.kenmochi,j.cousty,l.najman}@esiee.fr

Abstract. The goal of video cosegmentation is to jointly extract the
common foreground regions and/or objects from a set of videos. In this
paper, we present an approach for video cosegmentation that uses graph-
based hierarchical clustering as its basic component. Actually, in this
work, video cosegmentation problem is transformed into a graph-based
clustering problem in which a cluster represents a set of similar supervoxels belonging to the analyzed videos. Our graph-based Hierarchical Video
Cosegmentation method (or HVC) is divided in two main parts: (i) supervoxel generation and (ii) supervoxel correlation. The former explores only
intra-video similarities, while the latter seeks to determine relationships
between supervoxels belonging to the same video or to distinct videos.
Experimental results provide comparison between HVC and other methods from the literature on two well known datasets, showing that HVC
is a competitive one. HVC outperforms on average all the compared
methods for one dataset; and it was the second best for the other one.
Actually, HVC is able to produce good quality results without being too
computational expensive, taking less than 50% of the time spent by any
other approach.

Keywords: Graph-based segmentation · Video cosegmentation ·
Hierarchical clustering

1 Introduction

The goal of video cosegmentation is to jointly extract the common foreground
regions and/or objects from a set of videos. The video cosegmentation can be considered weakly supervised [7], since the presence of common foreground regions
and/or objects in multiple videos provides some indication that is not available
to the unsupervised problem of segmentation for a single video. That additional
information may help, but it may not be enough to reduce the ambiguity in video

F. Rodrigues, S. Guimarães and Z. Patrocínio—The authors are grateful to
FAPEMIG (PPM-00006-16), CNPq (Grant 421521/2016-3), PUC Minas and
CAPES for the financial support to this work.

S. Battiato et al. (Eds.): ICIAP 2017, Part I, LNCS 10484, pp. 15–26, 2017.
https://doi.org/10.1007/978-3-319-68560-1_2

cosegmentation of general content, due to the presence of multiple foreground regions and/or objects with low contrast to the background.

In this paper, we present a novel approach for video cosegmentation that uses graph-based hierarchical clustering as its basic component. Our graph-based **H**ierarchical **V**ideo **C**osegmentation method (HVC) presents two main technical contributions. The former is the adoption of a simple graph-based hierarchical clustering method as key component of the framework which respects two important principles of multi-scale set analysis, *i.e.*, *causality* and *location* principles [9]. Therefore, it is able to produce a set of video segments that are more homogeneous and whose borders are better defined using simple features to calculate dissimilarity measure between neighboring pixels and voxels (instead of several and expensive features which are very common in other approaches found in the literature). The second one is the removal of the need for parameter tuning and for the computation of a segmentation at finer levels, since it is possible to compute any level without computing the previous ones.

The few existing methods for video cosegmentation are all based on low-level features. In [11], the authors separated foreground and background regions through an iterative process based on feature matching among video frame regions and spatio-temporal tubes. The video cosegmentation method presented in [4] can extract multiple foreground objects by learning a global appearance model that connects segments of the same class. It also uses the Bag-of-Words (BoW) representation for multi-class video cosegmentation. While BoW provides more discriminative ability than basic color and texture features, they may be susceptible to appearance variations of foreground objects in different videos, due to factors such as pose change. In [15], the authors proposed a method which employs the object proposal [5] as the basic element, and uses the regulated maximum weight clique method to select the corresponding nodes for video multi-class segmentation. Finally, in [7], the authors proposed a multi-state selection graph in which a node representing a video frame can take multiple labels that correspond to different objects (also based on object proposal [5]). In addition, they used an indicator matrix to handle foreground objects that are missing in some videos, and they also presented an iterative procedure to optimize an energy function along with that indicator matrix.

The paper is organized as follows. Section 2 presents concepts about graph-based hierarchical clustering used in this work. While Sect. 3 describes our method to cope with video cosegmentation problem, Sect. 4 presents experimental results of our approach together with a comparative analysis with others methods from the literature. Finally, we draw some conclusions in Sect. 5.

2   Graph-Based Hierarchical Clustering

Following the seminal ideas proposed in [10], a hierarchy of partitions based on observation scales can be computed using a criterion for region-merging popularized by [6]. Moreover, it satisfies two important principles of multi-scale set analysis, *i.e.*, *causality* and *location* principles [9]. Namely, and in contrast with the approach presented in [6], the number of regions is decreasing when the scale parameter increases, and the contours do not move from one scale to another.

Thanks to that, one can compute the hierarchical observation scales for any graph, in which the adjacent graph regions are evaluated depending on the order of their merging in the fusion tree, *i.e.*, the order of merging between connected components on the minimum spanning tree (MST) of the original graph. Actually, one does not need to produce explicitly a hierarchy of partitions, since a weight map with observation scales can be used to infer the desired hierarchy, *e.g.*, by removing those edges whose weight is greater than a desired scale value. This map is a new edge-weighted tree created from MST in which each edge weight corresponds to the scale from which two adjacent regions connected by this edge are correctly merged, *i.e.*, there are no other sub-regions of these regions that might be merged before these two.

Following [10], for computing the weight map of observation scales, we consider the criterion for region-merging proposed in [6] which measures the evidence for a boundary between two regions by comparing two quantities: one based on intensity differences across the boundary, and the other based on intensity differences between neighboring pixels within each region. More precisely, in order to know whether two regions must be merged, two measures are considered. The *internal difference* $Int(X)$ of a region X is the highest edge weight among all the edges linking two vertices of X in MST. The *difference* $Diff(X,Y)$ between two neighboring regions X and Y is the smallest edge weight among all the edges that link X to Y. Then, two regions X and Y are merged when:

$$Diff(X,Y) \leq \min\left\{ Int(X) + \frac{\lambda}{|X|}, Int(Y) + \frac{\lambda}{|Y|} \right\} \qquad (1)$$

in which λ is a parameter used to prevent the merging of large regions, *i.e.*, larger λ forces smaller regions to be merged.

The merging criterion defined by Eq. (1) depends on the scale λ at which the regions X and Y are observed. More precisely, let us consider the *(observation) scale* $S_Y(X)$ *of* X *relative to* Y as a measure based on the difference between X and Y, on the internal difference of X and on the size of X:

$$S_Y(X) = (Diff(X,Y) - Int(X)) \times |X|. \qquad (2)$$

Then, the *scale* $S(X,Y)$ is simply defined as:

$$S(X,Y) = \max(S_Y(X), S_X(Y)). \qquad (3)$$

Thanks to this notion of a scale, Eq. (1) can be written as:

$$\lambda \geq S(X,Y). \qquad (4)$$

The core of [10] is the identification of the smallest scale value that can be used to merge the largest region to another one while guaranteeing that the internal differences of these merged regions are greater than the value calculated for smaller scales. The hierarchization of this principle has been successfully applied to several tasks: image segmentation [10], video segmentation [12–14], and video summarization [3]. In next section, we present our proposal to extend its application to the video cosegmentation problem.

3   Proposed Method

In this work, video cosegmentation problem is transformed into a graph-based clustering task in which a cluster (or connected component of the graph), computed from a graph partition, represents a set of similar supervoxels belonging to the analyzed videos. In order to do that, our proposed method, named HVC, is divided in two main parts: (i) supervoxel generation; and (ii) supervoxel correlation. The former explores only intra-video similarities, while the latter seeks to determine relationships between supervoxels belonging to the same video (intra-video similarity) or to distinct videos (inter-video similarity).

Figure 1 illustrates the steps of HVC method. First, each video is transformed into a video graph (step 1). Then, to explore the intra-video similarity, a hierarchy is computed from each video graph (step 2) and the identification of video segments (supervoxels) is made from each hierarchy (step 3). For each video, its set of supervoxels is described (step 4) and a single supervoxel graph is generated (step 5) containing all supervoxels from every video, in order to analyze both intra and inter-video similarities. Again, another hierarchy is computed from

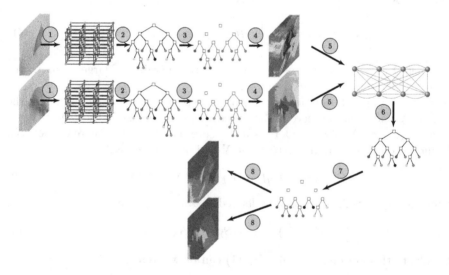

Fig. 1. Outline of our method: each video is transformed into a video graph (step 1); a hierarchy is computed from each video graph (step 2); the identification of video segments (supervoxels) is made from each hierarchy (step 3); each set of supervoxels is described (step 4) and a single supervoxel graph is generated (step 5); another hierarchy is computed from supervoxel graph (step 6); a partition of supervoxel graph is obtained (step 7); and, finally, the identification of connected components is made (step 8).

supervoxel graph (step 6) and a partition of supervoxel graph is obtained (step 7). And, finally, the identification of connected components (*i.e.*, "cosegments") is made (step 8).

An example of HVC results can be seen in Fig. 2 for both parts: supervoxels generation and correlation. The first part – supervoxel generation (steps 1 to 3) – adopts a hierarchical video segmentation (very similar to HOScale method proposed in [13,14]) that helps producing supervoxels that are more homogeneous and whose borders are better defined (HOScale exhibits high values for 3D segmentation accuracy and boundary recall and a low undersegmentation error [13,14]). The second part – supervoxels correlation (steps 4 to 8) – also utilizes a graph-based hierarchical clustering method based on [10], but applied to a complete graph generated from video segments obtained before. This removes the need for parameter tuning, resulting in a method that is not dependent on the hierarchical level, and consequently, making possible to compute any level without computing the previous ones [10]. Moreover, this is done using simple features to calculate dissimilarity measure between neighboring pixels and voxels (more details are given in Sect. 4).

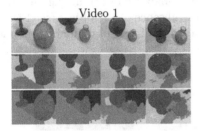 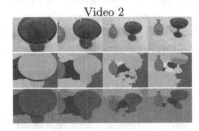

Video 1 Video 2

Fig. 2. HVC results for two videos with the same pair of vases. First row presents some samples of the original video frames. Video segments are illustrated at the second line (*i.e.*, pixels with the same color belong to the same supervoxel); and, finally, cosegmentation results are presented at the third line (*i.e.*, the same color is adopted to present pixels from common regions between videos).

The method HVC depends on: (i) the dissimilarity measure used in video graphs; (ii) the minimum size of a video segments (min_{vs}); (iii) the number of those segments (n_{vs}) per video; (iv) the dissimilarity measure used in supervoxel graph; (v) the minimum size of connected component (min_{cc}) for eliminating outliers during supervoxel clustering step; and (vi) the number of connected components (n_{cc}) used for obtaining a video cosegmentation.

4 Experiments

In order to evaluate our proposed method HVC, we used two well-known datasets: (i) ObMiC [7,8]; and (ii) MOViCS [4]. ObMiC dataset [7,8] is composed of four sets of video pairs each with two foreground objects in common, and the *ground truth* is manually labeled for each frame. MOViCS dataset [4] contains four video sets with 11 videos in total, and five frames of each video are labeled with *ground truth* at the pixel level.

During supervoxel generation, video graphs are the ones induced by the 26-adjacency pixel relationship, in which edge weights are calculated by a simple color gradient computed using the Euclidean distance in *Lab* color space, and we set n_{vs} to 100, 200, 300, 400, and 500. The supervoxel graph is generated as a complete graph, combining every possible number of video segments. In order to improve the strength of the relationship between supervoxels related to objects (or foreground regions) belonging to the same video (*i.e.*, intra-video similarity) an *objectness measure* (*i.e.*, a value which reflects how likely an image window covers an object of any category [1]) was used. The average value of *objectness* for every supervoxel was computed from the *objectness* values from its pixels. Following [1], to calculate the *objectness* value for a pixel p, the *objectness measure* was applied to 1,000 random windows for each video frame and the measure obtained for each window is added if it contains the pixel p. Actually, we adopted a normalized version of that *objectness measure* per pixel, called *heatmap*, in which pixels values are rescaled to $[0, 1]$ and used to produced a pseudo-colored image where areas with high probability of containing an object are shown in red, while dark blue indicates the absence of any object (see Fig. 3). Finally, n_{cc} is set to 5%, 10%, 15%, 20%, and 25% of the total number of nodes of the supervoxel graph.

Fig. 3. Examples of heatmaps generated from *objectness measure*. (Color figure online)

We have compared our method HVC against two cosegmentation methods from the literature[1]: (i) Regulated Maximum Weight Cliques (RMWC) [15]; and (ii) Multi-state Selection Graph (MSG) [7]. Differently from [7], the used MSG implementation does not have any post-processing, since the available code does not have any pixel-level refinement step in it. This allows a much fair comparison among different approaches because we can focus on the actual results generated by the cosegmentation methods (instead of considering improvements from post-processing steps that may be applied to the results of any approach).

To assess the quality of obtained cosegmentation results, we adopted two metrics (similar to [7]) to evaluate accuracy and error rate: (i) the *average Intersection-over-Union* (IoU); and (ii) the average per-frame pixel error (pFPE), respectively. We present IoU and pFPE scores that are optimal considering a constant scale parameter for the whole database (ODS) and a scale

---

[1] RMWC   is   available   at   http://www.dromston.com/projects/video_object_cosegmentation.php and MSG could be found at http://hzfu.github.io/proj_video_coseg.html.

parameter varying for each video (OVS) (analogously to [2]). Thus, HVC_D and HVC_V stand for the results of HVC with a constant scale parameter for the whole database (ODS) and a scale parameter varying for each video (OVS), respectively.

Table 1 presents accuracy results on both datasets. The method HVC_V outperforms on average RMWC for both datasets (for MOViCS dataset, the difference in average accuracy is only 1%). The performance of MSG is very poor on MOViCS dataset, but it has presented an average accuracy 5% greater than HVC_V on ObMiC dataset. As one can see in Fig. 4, good accuracy results are related to low values of pFPE. Actually, MSG method presented the lowest pFPE value on average for ObMiC dataset and the highest one for MOViCS dataset, which could explain its good results for the former and poor performance for the latter (e.g., see the results for video class *Tiger* on MOViCS dataset).

In order to assess qualitatively the obtained cosegmentation results, some examples for different approaches on ObMiC dataset are shown in Fig. 5. Results

Table 1. Accuracy results for different methods on ObMiC and MOViCS datasets.

(a) ObMiC dataset

Video class	RMWC	MSG	HVC_V	HVC_D
Dog	0.11	**0.62**	0.54	0.54
Monster	0.41	0.53	**0.65**	0.55
Skating	0.15	**0.59**	0.40	0.22
Person	0.23	**0.32**	0.26	0.22
Average	0.22	**0.51**	0.46	0.38

(b) MOViCS dataset

Video class	RMWC	MSG	HVC_V	HVC_D
Chicken	**0.58**	0.27	0.43	0.33
Giraffe	0.35	0.29	**0.44**	0.35
Lion	**0.55**	0.14	0.53	0.27
Tiger	0.40	0.10	**0.47**	0.36
Average	0.47	0.20	**0.48**	0.33

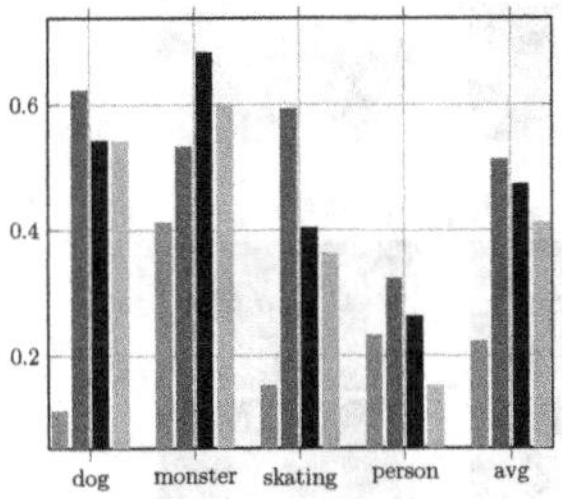

(a) IoU on ObMiC dataset. 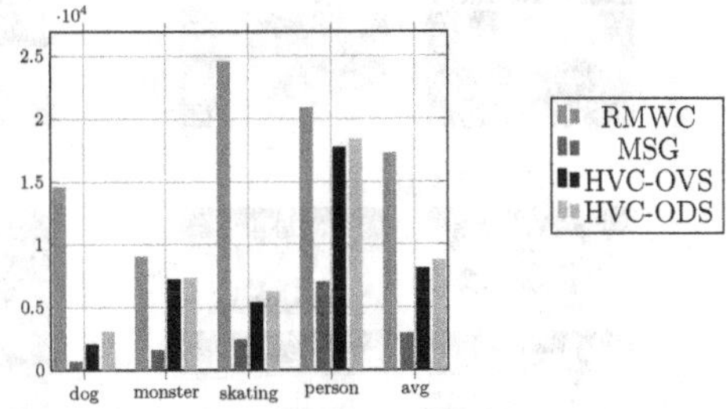 (b) pFPE on ObMiC dataset.

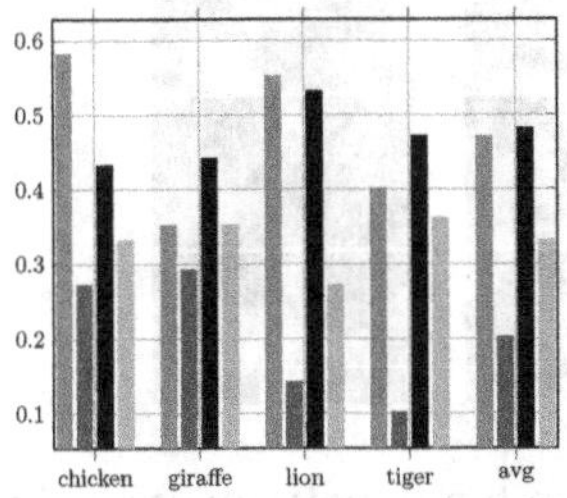

(c) IoU on MOViCS dataset. 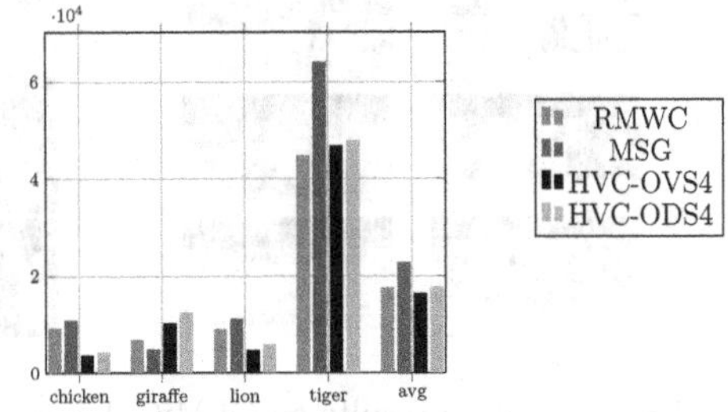 (d) pFPE on MOViCS dataset.

Fig. 4. Accuracy and error on the ObMiC and MOViCS dataset for different methods.

are presented for two videos from each class, along with the original video frames
and the expected results (*i.e.*, *ground truth*). For video class *Dog*, RMWC results
were very poor, while MSG and HVC produced similar results (with a little
advantage for MSG method). The same pattern can be observed for video class
Skating (but in this case MSG method was even better). For video class *Monster*, both RMWC and MSG methods have failed to identify one of the expected
objects. Moreover, MSG method has assigned an instance of those objects from
the first video to a different one in the second video. Finally, for video class *Person*, HVC was able to identify both persons (without the heads), while RMWC
and MSG have continued failing in identifying one of them. This is similar to
what happened for class *Monster*, except that in this case an object instance from
the first video was divided and assigned to distinct parts of the same object (by
RMWC) or to segments belonging to two different objects (by MSG).

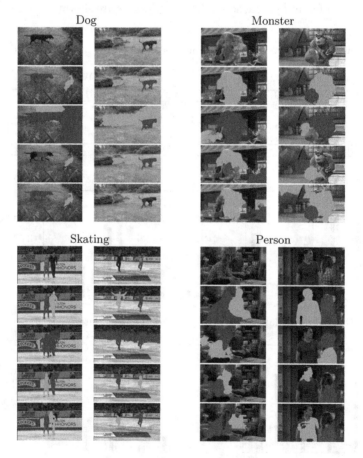

Fig. 5. Cosegmentation results on ObMiC dataset. From top to bottom: original video
frames, *ground truth*, RMWC [15], MSG [7], and our proposed method HVC.

Similarly, some results produced by different approaches on MOViCS dataset are shown in Fig. 6. As before, results are presented for each class, along with the original video frames and the expected results (*i.e.*, *ground truth*), but some classes have more than two results since they have more videos (03 for class *Tiger* and 04 for class *Lion*). For classes *Chicken* and *Lion*, RMWC has shown the best results followed closely by HVC method, while MSG results were very poor (it has divided some objects and has also considered some similar object instances as distinct). Finally, for classes *Giraffe* and *Tiger*, the opposite occurred: HVC presented best results followed by RMWC (while MSG showed some improvement only for class *Giraffe*).

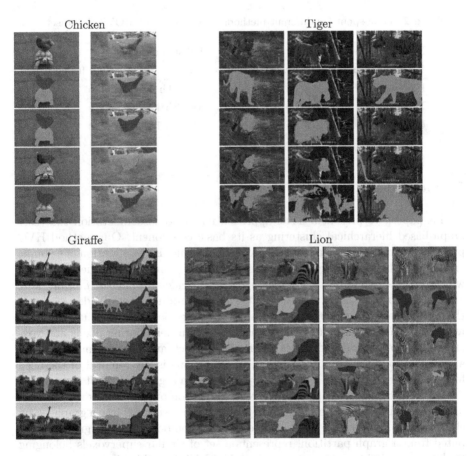

Fig. 6. Cosegmentation results on MOViCS dataset. From top to bottom: original video frames, *ground truth*, RMWC [15], MSG [7], and our proposed method HVC.

It is worth to mention that, for the class *Dog*, the proposed method HVC was not able to relate any segment of the second video to anyone belonging to the first one. This problem probably occurs due to the low differences between color averages of regions belonging to the same video. The same problem has also happened with RMWC (see the third video of the class *Tiger*).

Finally, HVC method is able to obtain very good results on both datasets using only a small amount of time. Table 2 presents total and average (per frame) time spent for tested methods on both datasets. For ObMiC dataset, HVC spent only 45.5% and 32.2% of the time spent on average by RMWC and MSG, respectively; while it spent on average 26.6% and 44.5% of the time spent by RMVC and MSG, respectively, for MOViCS dataset. The method MSG outperforms HVC on ObMiC dataset, but since it uses a great number of (computational expensive) features it took 211% more time to obtain the its results.

Table 2. Time spent for different methods on ObMiC and MOViCS datasets.

Method	ObMiC dataset		MOViCS dataset	
	Total	Avg. per frame	Total	Avg. per frame
RMWC	14 h 28 m 25 s	04 m 13 s	128 h 24 h 50	14 m 59 s
MSG	20 h 24 m 36 s	05 m 57 s	76 h 40 h 12	08 m 57 s
HVC	06 h 33 m 10 s	01 m 55 s	34 h 04 h 11	03 m 59 s

5   Conclusion

In this paper, we present a novel approach for video cosegmentation that uses graph-based hierarchical clustering as its basic component. Our method HVC presents two main technical contributions. The former is the adoption of a simple graph-based hierarchical clustering method as key component of the framework which respects two important principles of multi-scale set analysis, *i.e.*, *causality* and *location* principles [9]. Therefore, it is able to produce a set of video segments that are more homogeneous and whose borders are better defined using simple features to calculate dissimilarity measure between neighboring pixels and voxels (instead of several and expensive features which are very common in other approaches found in the literature). The second one is the removal of the need for parameter tuning and for the computation of a segmentation at finer levels, since it is possible to compute any level without computing the previous ones.

In this work, video cosegmentation problem is transformed into a graph-based clustering task in which a cluster (or connected component of the graph), computed from a graph partition, represents a set of similar supervoxels belonging to the analyzed videos. Our proposed method HVC is divided in two main parts: (i) supervoxel generation; and (ii) supervoxel correlation. The former explores only intra-video similarities, while the latter seeks to determine relationships between supervoxels belonging to the same video (intra-video similarity) or to distinct videos (inter-video similarity). Moreover, HVC uses simple features to calculate dissimilarity measure between neighboring pixels and voxels.

Experimental results provide quantitative and qualitative comparison involving new approach and other methods from the literature on two well known datasets, showing that HVC is a competitive approach. Concerning quality measures, HVC outperforms on average both tested methods for one dataset; and it presents on average an accuracy of 5% less than the best method for the other dataset. In spite of that, HVC method represents an attractive approach which is able to produce good quality results without being too computational expensive. When compared to the other methods, it took less than 50% of the time spent by any other approach.

In order to improve and better understand our results, further works involve inclusion of new features and automatic identification of the number of connected components; and also the application to another datasets.

References

1. Alexe, B., Deselaers, T., Ferrari, V.: Measuring the objectness of image windows. IEEE Trans. Pattern Anal. Mach. Intell. **34**(11), 2189–2202 (2012)
2. Arbelaez, P., Maire, M., Fowlkes, C., Malik, J.: Contour detection and hierarchical image segmentation. IEEE Trans. Pattern Anal. Mach. Intell. **33**, 898–916 (2011)
3. Belo, L., Caetano, C., Patrocínio Jr., Z.K.G., Guimarães, S.J.F.: Summarizing video sequence using a graph-based hierarchical approach. Neurocomputing **173**(3), 1001–1016 (2016)
4. Chiu, W.C., Fritz, M.: Multi-class video co-segmentation with a generative multi-video model. In: IEEE Conference on Computer Vision and Pattern Recognition (CVPR), pp. 321–328 (2013)
5. Endres, I., Hoiem, D.: Category independent object proposals. In: Daniilidis, K., Maragos, P., Paragios, N. (eds.) ECCV 2010. LNCS, vol. 6315, pp. 575–588. Springer, Heidelberg (2010). doi:10.1007/978-3-642-15555-0_42
6. Felzenszwalb, P.F., Huttenlocher, D.P.: Efficient graph-based image segmentation. Int. J. Comput. Vis. **59**(2), 167–181 (2004)
7. Fu, H., Xu, D., Zhang, B., Lin, S., Ward, R.K.: Object-based multiple foreground video co-segmentation via multi-state selection graph. IEEE Trans. Image Process. **24**(11), 3415–3424 (2015)
8. Fu, H., Xu, D., Zhang, B., Lin, S.: Object-based multiple foreground video co-segmentation. In: IEEE Conference on Computer Vision and Pattern Recognition (CVPR), pp. 3166–3173 (2014)
9. Guigues, L., Cocquerez, J.P., Le Men, H.: Scale-sets image analysis. Int. J. Comput. Vis. **68**(3), 289–317 (2006)
10. Guimarães, S.J.F., Cousty, J., Kenmochi, Y., Najman, L.: A hierarchical image segmentation algorithm based on an observation scale. In: Gimel'farb, G., et al. (eds.) SSPR/SPR 2012. LNCS, vol. 7626, pp. 116–125. Springer, Heidelberg (2012). doi:10.1007/978-3-642-34166-3_13
11. Rubio, J.C., Serrat, J., López, A.: Video co-segmentation. In: Lee, K.M., Matsushita, Y., Rehg, J.M., Hu, Z. (eds.) ACCV 2012. LNCS, vol. 7725, pp. 13–24. Springer, Heidelberg (2013). doi:10.1007/978-3-642-37444-9_2
12. Souza, K.J.F., Araújo, A.A., Guimarães, S.J.F., Patrocínio Jr., Z.K.G., Cord, M.: Streaming graph-based hierarchical video segmentation by a simple label propagation. In: 26th Conference on Graphics, Patterns and Images (SIBGRAPI), pp. 119–125 (2015)

13. Souza, K.J.F., Araújo, A.A., Patrocínio Jr., Z.K.G., Cousty, J., Najman, L., Kenmochi, Y., Guimarães, S.J.F.: Hierarchical video segmentation using an observation scale. In: 25th Conference on Graphics, Patterns and Images (SIBGRAPI), pp. 320–327 (2013)
14. Souza, K.J.F., Araújo Jr., A.A., Patrocínio, Z.K.G., Guimarães, S.J.F.: Graph-based hierarchical video segmentation based on a simple dissimilarity measure. Pattern Recogn. Lett. **47**, 85–92 (2014)
15. Zhang, D., Javed, O., Shah, M.: Video object co-segmentation by regulated maximum weight cliques. In: Fleet, D., Pajdla, T., Schiele, B., Tuytelaars, T. (eds.) ECCV 2014. LNCS, vol. 8695, pp. 551–566. Springer, Cham (2014). doi:10.1007/978-3-319-10584-0_36

Interest Region Based Motion Magnification

Manisha Verma$^{(\boxtimes)}$ and Shanmuganathan Raman

Electrical Engineering, Indian Institute of Technology Gandhinagar,
Gandhinagar, India
{manisha.verma,shanmuga}@iitgn.ac.in

Abstract. In this paper, we proposed a method known as interest region based motion magnification for amplification of invisible motions. This method enables one to magnify subtle motion in the video for specific objects of interest to the user. To achieve this task, we have used object extraction using kernel K-means approach, automatic scribble drawing using super pixels and Bezier curves, alpha matting, and Eulerian motion magnification. The proposed method is tested on previously used video sequences for motion magnification and our own new videos with large background motion. We show the effectiveness of the proposed method by comparing with Eulerian motion magnification technique. We have presented visual results and performed no-reference video quality assessment for original videos and motion magnified videos. We further discuss the future improvements for motion magnification applications.

Keywords: Eulerian motion magnification · Object segmentation · Image matting · Spatial-temporal analysis

1 Introduction

Human visual system understands the neighboring environment and processes the information through visible spectrum. The light reflected from the scene is sensed by the eyes and the brain performs complex processes through a network of neurons, receptors, and other specialized cells. Visualizing motion is the process of interpreting the speed and direction of small particles or objects in the nearby regions of a given scene. Human eyes can visualize the motion of objects which are significant. However, motion which can not be visualized by eyes might be important and might reveal invisible secrets actually present in the scene. Video motion magnification is an active research area over the past few years in which an imperceptible object motion is magnified and a synthetic video is generated where small motions are made perceptible to the eyes.

Wu *et al.* proposed temporal filtering based motion magnification and called it as Eulerian method [19] while improving the Lagrangian method that is based on motion tracking [7]. It follows the same fundamentals of fluid dynamics for Lagrangian and Eulerian approaches. Most of the work done in the motion magnification follows the uniform magnification for the entire scene. The proposed method tries to consider only the selected regions of interest from a video which

© Springer International Publishing AG 2017
S. Battiato et al. (Eds.): ICIAP 2017, Part I, LNCS 10484, pp. 27–39, 2017.
https://doi.org/10.1007/978-3-319-68560-1_3

exhibit imperceptible motion and therefore, need to be magnified. The proposed approach is also shown to reduce noise and remove outliers in the generated synthetic magnified video. Using this approach, the video is not only constrained to specific conditions, such as single object videos, and any given video could be processed in order to magnify the regions of interest.

It is not a completely automatic approach and requires user intervention to specify the object of interest. It is challenging as it follows many steps to complete and error incurred in one step can lead to erroneous output. Hence, it is needed to choose appropriate methods during each step. Recently, researchers have used Eulerian motion in tremor assessment [5] and endoscopic surgery [8]. Interest region based motion magnification can be very helpful in these types of applications to magnify a particular region or object.

The primary contributions of the proposed work are listed below.

1. Interest region based motion magnification has been performed for a given video with objects exhibiting imperceptible motion.
2. In addition, we show that the noise due to other sources present in the magnified region is reduced.
3. The approach is shown to work on videos of different natural scenes with objects exhibiting different kinds of motion and a video quality assessment is presented to check the video quality for noise.

The rest of the paper is organized as follows. In Sect. 2, related work is discussed. The framework of the proposed method including brief description of the techniques used is explained in Sect. 3. Experiments along with the results are discussed in Sect. 4. Finally, the paper is concluded in Sect. 5.

2   Related Work

Researchers have worked in artificial motion manipulation over the past decade and proposed different approaches using optical flow for many applications. Liu *et al.* proposed motion magnification for subtle changes in video [7]. They have used video registration to suppress camera shake motion, feature tracking to group correlated object motions, segmentation of motion trajectories, motion magnification followed by rendering of magnified video to fill the gaps. Wang *et al.* proposed "Cartoon Animation Filter" that produces motion exaggeration in artificial video of input video, which they claim to be more animated and alive. It subtracts the smoothed and time shifted version of second order derivative of signal from original signal [18]. These are Lagrangian approaches which make use of optical flow for motion magnification. First Eulerian motion magnification approach is proposed by Wu *et al.* [19]. Instead of calculating optical flow explicitly, they have magnified the temporal difference between the frames. This work is extended in phase-based magnification using complex steerable pyramids for noise reduction [16]. They magnified local phase variations in all the sub-bands of complex steerable pyramids. To improve the time complexity, Riesz pyramids are proposed for phase based motion magnification [17].

Motion magnification has been utilized in many applications. Deviation magnification of geometric structures is proposed by Wadhwa *et al.* [15]. Basic parametric shapes (e.g., lines and circles) are fitted in object of still images, and sampling and image matting are performed on particular object shape. Deviation is computed and magnified by a factor, and a rendered image is obtained in which deviation is magnified. Raja *et al.* proposed presentation attack detection scheme for iris recognition system by motion magnification of the phase information in the eye region [9]. Motion magnification has been used for face spoofing detection [2]. Subtle facial motions have been magnified, and texture based features have been used to detect the spoofing. Davis *et al.* used motion magnification to infer material properties by emphasizing small vibrations in the object [3].

Interest region based motion magnification is beneficial in terms of reducing outliers and noise. To the best of our knowledge, two papers have been proposed towards this direction. The first work is proposed by Kooij *et al.* in which they used depth maps to magnify objects of interest specified by depths [5]. However, it requires extra information of depth maps which is an additional task. On the other hand, our method works with only a given video for motion magnification. In the second work, Elgharib *et al.* proposed motion magnification in the presence of large motions and called it DVMAG (Dynamic video magnification) [4]. They calculated alpha matte of each frame by user specified scribbles and magnified motion in respective alpha mattes. They applied texture synthesis to fill the gaps in magnified videos. Our proposed method is different from [4] in the sense that DVMAG requires a large amount of user interaction to draw scribbles, and the proposed method required only two coordinates of the region of interest that can be easily automated in future using object proposals [10,11]. The proposed method, unlike [4], does not require texture synthesis to fill the detail gaps.

3   Interest Region Based Motion Magnification

Most of the motion magnification techniques which have been proposed in the past, are for whole video frame irrespective of the object of interest and restrain to record the video in such conditions that most of the frame area should contain the object of interest with minimum background. Hence, they can not be applied to standard video recorded in regular conditions, e.g., fast moving objects in the background. It will lead to more noise during the magnification. In the proposed work, motion magnification has been applied to specific objects in the video. Since the motion magnification is for imperceptible motions, the region of interest is assumed to be static in the video.

Challenges: Our method is based on the observation that large motions in background may affect the motion magnified video and bring extensive noise. Solving this issue bring new challenges in the work. The main challenge of this work is to get the object of interest and perform an automatic motion magnification. Previous two works [4,5] require additional information to get the object of interest. However, we made it possible with only two pixel locations marked by user on the first frame. The rest of the work is handled by an automated algorithm with no user intervention. The extracted object contains a sharp and distorted boundary, hence it can not be used as a mask. To get a fuzzy boundary mask, image matting is performed and for that automatic scribbles are drawn on background and foreground objects. From extraction of object of interest to scribble drawing, image matting, and motion magnification, all are performed automatically.

This algorithm has three main steps. In the first step, an object of interest is extracted from the first frame of video using kernel k-means approach [14], discussed in Sect. 3.1. In the second step, the image matting is performed on the input frame as explained in Sect. 3.2. To perform image matting, scribbles are required to be drawn on the foreground and background image parts which is done automatically. In the third step, video magnification is performed using Eulerian video magnification approach [19] as discussed in Sect. 3.3.

3.1   Object Segmentation

K-means segmentation approach is a partitioning method based on the sum of squared error in each cluster. In case of two segments C and $\bar{C}$, likelihood in energy function can be written as

$$\sum_{p \in C} ||I_p - \mu_C||^2 + \sum_{p \in \bar{C}} ||I_p - \mu_{\bar{C}}||^2 \qquad (1)$$

Kernel K-means (kKM) segmentation approach is adopted in the proposed work [14]. kKM is a well proven data clustering technique in machine learning, that makes use of kernel tricks to separate the complex structures which are non-linearly separable in the input space. Kernel K-means maps the data into a higher dimensional Hilbert space using a non-linear mapping ψ. The energy function of standard K-means segmentation was replaced by the following in kKM

$$E_k(C) = \sum_{p \in C} ||\psi(I_p) - \mu_C||^2 + \sum_{p \in \bar{C}} ||\psi(I_p) - \mu_{\bar{C}}||^2 \qquad (2)$$

where C and $\bar{C}$ are two segments, I_p are data points in clusters, μ_C and $\mu_{\bar{C}}$ are cluster means for C and $\bar{C}$ respectively. Detailed explanation regarding kKM, adaptive kKM, and kernel bandwidth can be found in [14]. Object segmentation using kKM is shown in Figs. 1(b) and 4.

3.2   Scribble Drawing and Alpha Matting

Scribbles are used to perform the image matting to extract foreground and background image regions. They specify the image regions which can be considered clearly as foreground (white scribbles) and clearly as background (black scribbles) as shown in Fig. 1(e). After the extraction of foreground object, black (on background) and white (on foreground) scribbles need to be drawn on the most

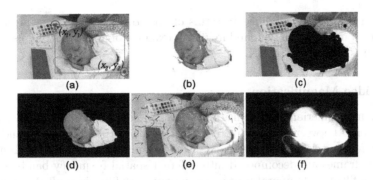

Fig. 1. (a) Original frame with manually drawn bounding box using coordinates (x_1, y_1) and (x_2, y_2), (b) Segmented object using kernel K-means, (c) Eroded background image, (d) Eroded foreground image, (e) Black and white scribbles drawn on background and foreground image using Bezier curves, and (f) Extracted alpha matte using [6]

feasible spatial locations in the image in different shapes so that diverse image regions can be covered through the scribbles. To achieve this, super pixel over segmentation [1] and Bezier curves are employed [13].

Initially, morphological operations are applied so that scribbles can be drawn only in the foreground or background part of image and not on the boundary as it may lead to a erroneous alpha map. Eroded background and foreground images are shown in Fig. 1(c) and (d). Bezier curves are drawn using six points chosen near the centroid of each superpixel as mentioned in [13]. Scribbled image is shown in Fig. 1(e).

Motion magnification in the interest region is a challenging problem, especially on the object boundary, and it should be applied to a finely segmented object. Otherwise, it may lead to false video magnification. Hence, it is required to perform matting to produce the best segmentation of the video frame with fuzzy boundaries. Matting is an approach which smoothens the boundaries of segmented objects and makes their appearance more natural while blending. Intensity of pixels in the image can be expressed as a linear combination of F (foreground) and B (background) pixels.

$$I(x, y) = \alpha(x, y)F(x, y) + (1 - \alpha(x, y))B(x, y) \tag{3}$$

where $\alpha(x, y)$ is foreground opacity. We have used closed form solution for extracting the alpha matte [6]. In this approach, F and B are assumed to be smooth in a local window around each pixel. Equation 3 can be rewritten as

$$\alpha(x, y) \approx aI(x, y) + b, \forall (x, y) \in w \tag{4}$$

where $a = \frac{1}{F-B}$, $b = \frac{B}{F-B}$ and w is a small image window. A cost function J, is minimized for α, a and b

$$J(\alpha, a, b) = \sum_{(p,q) \in I} \left(\sum_{(x,y) \in w} (\alpha(x, y) - a(p, q)I(x, y) - b(p, q))^2 + \varepsilon a(p, q)^2 \right) \tag{5}$$

which can be further modified in terms of only α. More details regarding the closed form matting can be found in [6]. An example of image matting using this approach is shown in Fig. 1(f).

3.3   Video Magnification

We have used Eulerian motion magnification approach that amplifies the temporal difference between consecutive frames [19]. It incorporates spatial and temporal processing to highlight the small motion present in the video. Initially, the video frames are decomposed into different spatial frequency bands using a Laplacian filter. In temporal processing, a band pass filter is applied to magnify few particular frequencies based on the application. Temporal filter is applied on all spatial sub-bands and all pixels uniformly. The extracted temporal filtered video frame is magnified by a factor γ_{mag}. The theory behind motion magnification using temporal filtering follows first-order Taylor series expansion of signal that is commonly used in optical flow estimation.

If $I(x,t)$ is the image signal of position x and time t, then the modified signal with γ_{mag} factor is given by

$$\hat{I}(x;t) = I(x;t) + \gamma_{mag}B(x;t) \qquad (6)$$

where $B(x;t)$ is the result of the temporal bandpass filter. The motion magnification factor γ_{mag} can be estimated using the following equation

$$(1 + \gamma_{mag})\delta(t) < \frac{\lambda_c}{8} \qquad (7)$$

where λ_c is the cut-off spatial frequency beyond which an attenuated version of γ_{mag} is used, and $\delta(t)$ is the video motion signal. Detailed mathematical explanation of the Eulerian motion magnification can be found in [19].

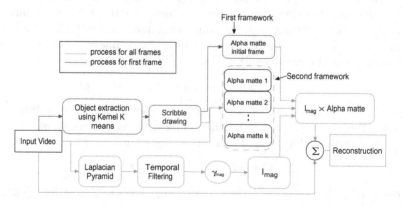

Fig. 2. Block diagram of the proposed framework. Blue lines shows the processes which need to be performed for only first frame. Red lines shows the processes which need to be performed for each frame of video. (Color figure online)

3.4   Pipeline

The proposed work follows a sequential approach using the three steps discussed above in order to achieve the task of motion magnification based on interest region in videos. A small amount of user input is required to process. In the first frame, user is asked to draw a box on the object of interest or to provide two coordinates for the same. In Fig. 1(a), coordinates $((x_1, y_1), (x_2, y_2))$ are shown in blue marker and bounding box in red. Next, kernel K-means approach is used to segment the object from the background. It gives an approximate object segment, which is used to get alpha matte. The eroded background image is fed to super pixel over segmentation, and scribbles are drawn on foreground and background image regions using Bezier curves near each centroid of all the superpixels. After this, image matting is performed on image and corresponding alpha matte is calculated.

An object of interest which needs to be magnified is assumed to be static in video with tiny motion. This assumption is valid as the motivation of the proposed work is to magnify small motions. It relaxes the algorithm, as it requires scribbles only in the first frame, and these scribbles are enough for other frames as the object is not exhibiting large movements. On the basis of this assumption, the following two types of frameworks are adopted in this work depending on the object of interest in the given video.

1. If the motion of object lies inside the object or it is extremely tiny, then alpha matte of only the first frame may work for all the frames. If this condition is satisfied, then calculating alpha matte for only one frame will be computationally very cheap as compared to the second framework.
2. In the second framework, alpha matte of each frame is calculated and further utilized on magnified video frames.

In both the above mentioned approaches, the alpha matte is calculated for the first frame or for all the frames. Other than that, video is magnified by a magnification factor using temporal filtering and alpha matte is multiplied with magnified temporal difference. In the first framework, same alpha matte of first frame is multiplied with temporal differences of all the frames. Besides, in the second framework, alpha matte of each frame is multiplied with corresponding temporal difference. Finally, the magnified temporal difference of only the foreground object (using alpha matte multiplication) is added to original frame. Block diagram of both the frameworks is illustrated in Fig. 2.

4   Experiments and Discussion

The proposed method is tested on videos with subtle motion. First frame of each video is shown in Fig. 3. We have used similar videos as [16, 19] and some videos are recorded in conditions with a moving background. For object extraction using kKM, hard constraint and smoothness parameters can be set according to the objects of interest. In most of the cases, hard constraints are set to 'on'

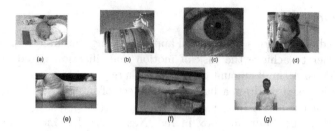

Fig. 3. (a) Baby, (b) Camera, (c) Eye, (d) Woman, (e) Wrist, (f) Hand, and (g) Person.

as objects of interest must lie inside the box provided by the user. Smoothness weight should be chosen more than zero to get a more flat image. An example of object extraction with and without smoothness constraint is shown in Fig. 4.

Fig. 4. (a) Original frame of *hand* sequence, (b) Extracted object with no smoothness, and (c) with .1 smoothness.

To remove the boundary of background and foreground image, erosion is performed with a disk of 10 or 20 radius. Next, the scribbles are drawn on background and foreground using superpixels and Bezier curves. Superpixel count can be placed from 50 to 100 depending on variability in the size of interest regions. Parameter used in magnification, i.e., band pass frequencies, sampling rate, magnification factor, and cut-off frequency are adopted from [19]. IIR, Butterworth, and ideal temporal filters are used in experiments to obtain temporal difference in frames. Results can be accessed online at: https://sites.google.com/site/manishaverma89/publications/int-reg-motion-mag.

Results in the form of spatial-temporal plots, are illustrated in Figs. 5, 6, 7 and 8. In Fig. 5, plots of *camera* sequence are shown. Pixels of a random column (shown in black line in Figs. 5, 6, 7 and 8) are plotted over time for each frame. Time and pixel intensities are plotted on x and y axis respectively. The first image in the Fig. 5 is the first frame of the *camera* sequence. Figure 5(a) is time-space plot of the original sequence, and there is no variation in pixel intensities over time. In Fig. 5(b), intensity variation of Eulerian method [19] is shown. It is clearly visible that motion magnification adds noise in the video as it magnifies the background. This problem will not appear if the background is ideally motionless, however, that is a very unlikely situation. In Fig. 5(b) and (c), only camera motion is visible as the background motion is not magnified. Figure 5(b) and (c) follow the first and the second framework respectively. It is noticeable that since the *camera* sequence has very tiny motion, there is no such difference in the first and second framework with respect to this example.

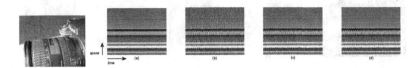

Fig. 5. Comparison of Wu *et al.* [19] and proposed method on *camera* video sequence. Original first frame and space-time plots of pixel intensities of (a) Original frame (no motion), (b) Wu et al. method [19] (uniform motion magnification), (c) Proposed method - first framework (interest region based motion magnification using first frame's alpha matte), and (d) Proposed method - second framework (interest region based motion magnification using each frame's alpha matte).

In a similar way, time-space plots are drawn for *eye* sequence. Iris is extracted as a foreground object and magnified throughout all the frames. Magnification using alpha matte (Fig. 6(c) and (d)) leads to noiseless magnification where only iris is magnified, and other portions are unchanged as they were in original sequence. However, Eulerian motion magnification magnifies the whole frame (Fig. 6(b)).

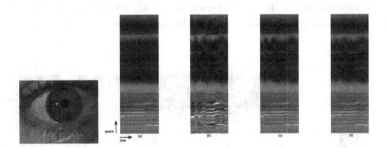

Fig. 6. Motion magnification comparison in *eye* video sequence. Original first frame and space-time plots of pixel intensities of (a) Original frame, (b) Wu et al. method [19], (c) Proposed method - first framework, and (d) Proposed method - second framework.

In the next two experiments, we have used videos with moving background. A video is recorded in such condition where a still hand is placed in front of a monitor displaying a video of waterfall. Hence the video has a subtle motion (of hand) with moving background (waterfall). First frame of video with a horizontal black line is shown for which the spatial intensity is plotted over time. Since the background is moving, Wu *et al.* [19] approach leads to high noise in background as shown in Fig. 7(b). On the other hand, our approaches (Fig. 7(c) and (d)) provide motion magnification with less noise.

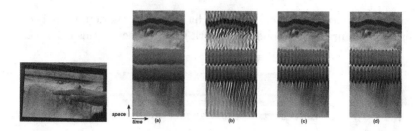

Fig. 7. Motion magnification comparison in *hand* video sequence. Original first frame and space-time plots of pixel intensities of (a) Original frame, (b) Wu et al. method [19], (c) Proposed method - first framework, and (d) Proposed method - second framework.

In the last experiment, a video is recorded where a person is sitting motionless, and another person is moving behind the first person. In Fig. 8, we have shown vertical and horizontal movements over time. In first column of Fig. 8(i) and (ii), first frame of video is shown with vertical and horizontal black lines respectively and corresponding motion graphs are shown in respective rows. Plot of space and time of original video is shown in Fig. 8(a) for both vertical and horizontal motions. The moving person is seen in the middle of all space-time plots, as in the middle of video sequence, the background person comes in the contact of foreground person. Extreme noise in the presence of the background is obtained by Wu *et al.* approach. Head of person is considered as foreground and extracted for motion magnification in the proposed approach and shown in Fig. 8(c) and (d). The minor difference between the first and the second framework can be

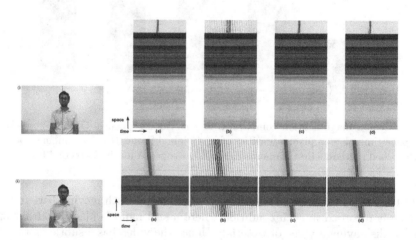

Fig. 8. Motion magnification comparison in *person* video sequence with (i) vertical and (ii) horizontal motions. Original first frame and space-time plots of pixel intensities of (a) Original frame, (b) Wu et al. method [19], (c) Proposed method - first framework, and (d) Proposed method - second framework.

seen at the boundary of the foreground object when it comes in contact with the background person.

We have presented a no-reference video quality assessment based on Video BLIINDS [12]. It computes video statistics and perceptual features, and feeds them to a learned support vector regressor for video quality prediction. Video quality of four videos, i.e., original video, motion magnified video produced by [19], motion magnified video using proposed framework 1, and proposed framework 2 are measured using Video BLIINDS [12] and shown in Table 1. The algorithm computes the differential mean opinion score (DMOS index), hence a low score implies better quality of the video. The DMOS index for the original video is less for almost all the videos. It is clearly visible that index of Wu *et al.* method is highly exceeding from both of the proposed frameworks in all the videos. There is a minor variation for proposed framework 1 and 2, and that depends on various factors, e.g., the movement of object, scribbles drawn in the first frame, and background motion. For two videos *hand* and *person*, the score for proposed magnified video is less than the original video, that could be possible due to training of Video BLIINDS. Other than that, for all the videos, the score using the proposed method is less than Wu *et al.* [19] method.

Table 1. Video quality assessment using video BLIINDS [12]

Video sequence	Original	Wu et al. [19]	Proposed framework 1	Proposed framework 2
Baby	−32.0891	77.62	54.12	53.99
Camera	9.21	109.44	62.46	76.42
Eye	−108.69	76.77	61.02	60.59
Woman	36.29	91.77	69.18	68.22
Wrist	−56.34	52.18	8.89	7.75
Person	41.36	110.63	26.57	30.19
Hand	82.51	114.70	62.88	69.27

5   Conclusion

In the proposed work, interest region based motion magnification is proposed which helps in reducing noise and removing the outliers in motion magnification. The proposed work makes use of object extraction, automatic scribble drawing, image matting and motion magnification to achieve the task. The proposed method would be very favourable for videos where the object of interest is not focused in camera and other motions (excluding object of interest) are present in the video. The proposed method is shown to work well on different videos as compared to uniform motion magnification.

In the future work, we will try to employ semantic object detection techniques and try to make a fully automatic system for magnification of specific objects with no user intervention. Any existing motion amplification methods (phase based complex steerable pyramids and Reisz pyramids) can then be employed to process the interest region. Region based motion magnification can be helpful in many applications.

Acknowledgement. The authors would like to thank Young Scientists Startup Research Grant, SERB-DST and Indian Institute of Technology Gandhinagar for support.

References

1. Achanta, R., Shaji, A., Smith, K., Lucchi, A., Fua, P., Süsstrunk, S.: SLIC super-pixels. Technical report (2010)
2. Bharadwaj, S., Dhamecha, T.I., Vatsa, M., Singh, R.: Computationally efficient face spoofing detection with motion magnification. In: Proceedings of the IEEE Conference on Computer Vision and Pattern Recognition Workshops, pp. 105–110 (2013)
3. Davis, A., Bouman, K.L., Chen, J.G., Rubinstein, M., Durand, F., Freeman, W.T.: Visual vibrometry: estimating material properties from small motion in video. In: Proceedings of the IEEE Conference on Computer Vision and Pattern Recognition, pp. 5335–5343 (2015)
4. Elgharib, M., Hefeeda, M., Durand, F., Freeman, W.T.: Video magnification in presence of large motions. In: Proceedings of the IEEE Conference on Computer Vision and Pattern Recognition, pp. 4119–4127 (2015)
5. Kooij, J.F.P., van Gemert, J.C.: Depth-aware motion magnification. In: Leibe, B., Matas, J., Sebe, N., Welling, M. (eds.) ECCV 2016. LNCS, vol. 9912, pp. 467–482. Springer, Cham (2016). doi:10.1007/978-3-319-46484-8_28
6. Levin, A., Lischinski, D., Weiss, Y.: A closed-form solution to natural image matting. IEEE Trans. Pattern Anal. Mach. Intell. **30**(2), 228–242 (2008)
7. Liu, C., Torralba, A., Freeman, W.T., Durand, F., Adelson, E.H.: Motion magnification. ACM Trans. Graph. **24**(3), 519–526 (2005)
8. McLeod, A.J., Baxter, J.S., de Ribaupierre, S., Peters, T.M.: Motion magnification for endoscopic surgery. In: SPIE Medical Imaging, p. 90360C. International Society for Optics and Photonics (2014)
9. Raja, K.B., Raghavendra, R., Busch, C.: Video presentation attack detection in visible spectrum iris recognition using magnified phase information. IEEE Trans. Inf. Forensics Secur. **10**(10), 2048–2056 (2015)
10. Redmon, J., Divvala, S., Girshick, R., Farhadi, A.: You only look once unified, real-time object detection. In: Proceedings of the IEEE Conference on Computer Vision and Pattern Recognition, pp. 779–788 (2016)
11. Ren, S., He, K., Girshick, R., Sun, J.: Faster R-CNN: towards real-time object detection with region proposal networks. In: Advances in Neural Information Processing Systems, pp. 91–99 (2015)
12. Saad, M.A., Bovik, A.C., Charrier, C.: Blind prediction of natural video quality. IEEE Trans. Image Process. **23**(3), 1352–1365 (2014)
13. Sonane, B., Ramakrishnan, S., Raman, S.: Automatic video matting through scribble propagation. In: Proceedings of the Tenth Indian Conference on Computer Vision, Graphics and Image Processing, pp. 87:1–87:8. ACM (2016)
14. Tang, M., Ben Ayed, I., Marin, D., Boykov, Y.: Secrets of grabcut and kernel k-means. In: Proceedings of the IEEE International Conference on Computer Vision, pp. 1555–1563 (2015)

15. Wadhwa, N., Dekel, T., Wei, D., Durand, F., Freeman, W.T.: Deviation magnification: revealing departure from ideal geometries. ACM Trans. Graph. **34**(6), 226 (2015)
16. Wadhwa, N., Rubinstein, M., Durand, F., Freeman, W.T.: Phase-based video motion processing. ACM Trans. Graph. **32**(4), 80 (2013)
17. Wadhwa, N., Rubinstein, M., Durand, F., Freeman, W.T.: Riesz pyramids for fast phase-based video magnification. In: Proceedings of IEEE International Conference on Computational Photography, pp. 1–10 (2014)
18. Wang, J., Drucker, S.M., Agrawala, M., Cohen, M.F.: The cartoon animation filter. ACM Trans. Graph. **25**(3), 1169–1173 (2006)
19. Wu, H.Y., Rubinstein, M., Shih, E., Guttag, J., Durand, F., Freeman, W.: Eulerian video magnification for revealing subtle changes in the world. ACM Trans. Graph. **31**(4), 1–8 (2012)

Investigating the Use of Space-Time Primitives to Understand Human Movements

Damiano Malafronte[1], Gaurvi Goyal[1], Alessia Vignolo[1,2], Francesca Odone[1], and Nicoletta Noceti[1(✉)]

[1] Università degli Studi di Genova, Genova, Italy
{damiano.malafronte,gaurvi.goyal}@dibris.unige.it,
alessia.vignolo@iit.it, {francesca.odone,nicoletta.noceti}@unige.it
[2] Istituto Italiano di Tecnologia, Genova, Italy

Abstract. In this work we start investigating the use of appropriately learnt space-time primitives for modeling upper body human actions. As a study case we consider cooking activities which may undergo large intra class variations and are characterized by subtle details, observed by different view points. With a BoK procedure we quantize each video frame with respect to a dictionary of meaningful space-time primitives, then we derive time series that measure how the presence of different primitives evolves over time. The preliminary experiments we report are very encouraging on the discriminative power of the representation, also speaking in favor of the tolerance to view point changes.

Keywords: Spatio-temporal interest points · Motion primitives · Multi-view motion analysis · Multi-view action analysis · Shearlet transform

1 Introduction

Understanding human motion and its regularities is a key research goal of Human-Machine Interaction, with a potential to unlock more refined abilities – such as the anticipation of action goals – and thus the design of intelligent machines able to proficiently and effectively collaborate with humans [1,2].

In this ongoing work we are interested in investigating HMI functionalities, where a machine (e.g. a robot) observes a human performing tasks and learns how to discriminate among the ones characterized by different dynamic properties [3]. We consider upper body human action primitives taking place in a specific setting, cooking in our case. For the time being, we restrict our attention to the actor, and do not exploit any contextual information which could be derived, for instance, by the presence of a tool or an object.

Since some time we have assisted to a growing interest towards the so-called space-time key-points. From the pioneering work of Laptev [4], who proposed an extension to the space-time of corner points, soon followed by alternative and possibly richer approaches [5,6], we have appreciated the power of these key-points as low level building blocks for motion analysis and action recognition.

S. Battiato et al. (Eds.): ICIAP 2017, Part I, LNCS 10484, pp. 40–50, 2017.
https://doi.org/10.1007/978-3-319-68560-1_4

Space-time key points mark special points where the signal undergoes a significant variation both in space and time, and for this reason they are quite rare. They carry meaningful information in particular when we analyze distinctive dynamic events, but they may be not as effective with more subtle actions or gestures.

In this work, instead of retaining the sole information provided by these hand-crafted space-time key-points, we learn *ad hoc* space-time local primitives for a given (class of) action(s). Given a dynamic event, different meaningful local primitives can be observed and associated with an appropriate meaning in space and time [7]. To achieve this goal we follow and unsupervised approach and consider a signal representation based on Shearlets [8,9]. Shearlets emerge among multi-resolution models by their ability to efficiently capture anisotropic features, to detect singularities [10,11] and to be stable against noise and blurring [12–14]. The effectiveness of Shearlets is supported by a well-established mathematical theory and confirmed by a variety of applications to image processing [9,14,15].

We propose a pipeline to represent the space-time information embedded in an image sequence. First, from the 2D + T shearlet coefficients we represent a space-time neighborhood by appropriately encoding the signal behavior in space and time. Then, we learn a dictionary of space-time local primitives or atoms meaningful for a specific action set. To do so, we follow a BoK approach [16], applying a clustering procedure to all the space-time points of a training set of image frames. The whole procedure is carried out in an unsupervised way, in the sense we do not use labels describing specific image features. Finally, we represent a video sequence as a set of time series depicting the evolution of the primitives frequency over time.

In the preliminary results we report, we analyze this information and evaluate whether it is meaningful and stable to multiple repetitions of the same action and discriminative among different but similar actions. We also evaluate its robustness to view point variations and investigate the descriptive power of dictionaries learnt by different datasets. Instead of addressing view-invariance as a general property we focus on a set of different view points that describe typical observation points in human-human interaction (ego-view, frontal view, lateral view) as they are meaningful to a natural HMI.

2   Shearlet Theory: An Overview

Here we briefly review the construction of the discrete shearlet transform of a $2D + T$ signal f by adapting the approach given in [17] for 3D signals.

Denoted by L^2 the Hilbert space of square-integrable functions $f : \mathbb{R}^2 \times \mathbb{R} \to \mathbb{C}$ with the usual scalar product $\langle f, f' \rangle$, the discrete shearlet transform $SH[f]$ of a signal $f \in L^2$ is the sequence of coefficients

$$SH[f](\ell, j, k, m) = \langle f, \Psi_{\ell,j,k,m} \rangle$$

where $\{\Psi_{\ell,j,k,m}\}$ is a family of filters parametrized by

1. A label $\ell = 0, \ldots, 3$ of 4 regions or pyramids $\mathcal{P}_\ell$ in the frequency domain;
2. The scale parameter $j \in \mathbb{N}$;
3. The shearing vector $k = (k_1, k_2)$ where $k_1, k_2 = -\lceil 2^{j/2} \rceil, \ldots, \lceil 2^{j/2} \rceil$;
4. The translation vector $m = (m_1, m_2, m_3) \in \mathbb{Z}^3$.

For $\ell = 0$ the filters, which do not depend on j and k, are

$$\Psi_{0,m}(x, y, t) = \varphi(x - cm_1)\varphi(y - cm_2)\varphi(t - cm_3), \tag{1}$$

where $c > 0$ is a step size and φ is a $1D$-scaling function. The system $\{\Psi_{0,m}\}_m$ takes care of the low frequency cube $\mathcal{P}_0 = \{(\xi_1, \xi_2, \xi_3) \in \widehat{\mathbb{R}}^3 \mid |\xi_1| \le 1, |\xi_2| \le 1, |\xi_3| \le 1\}$.

For $\ell = 1$ the filters are defined in terms of translations and two linear transformations (parabolic dilations and shearings)

$$A_{1,j} = \begin{pmatrix} 2^j & 0 & 0 \\ 0 & 2^{j/2} & 0 \\ 0 & 0 & 2^{j/2} \end{pmatrix} \qquad S_{1,k} = \begin{pmatrix} 1 & k_1 & k_2 \\ 0 & 1 & 0 \\ 0 & 0 & 1 \end{pmatrix}, \quad \text{so that}$$

$$\Psi_{1,j,k,m}(x, y, t) = 2^j \psi_1 \left(S_{1,k} A_{1,j} \begin{pmatrix} x \\ y \\ t \end{pmatrix} - \begin{pmatrix} cm_1 \\ \hat{c}m_2 \\ \hat{c}m_3 \end{pmatrix} \right), \tag{2}$$

where c is as in (1) and $\hat{c} > 0$ is another step size (in the rest of the paper we assume that $c = \hat{c} = 1$ for sake of simplicity). The system $\{\Psi_{1,j,k,m}\}$ takes care of the high frequencies in the pyramid along the x-axis: $\mathcal{P}_1 = \{(\xi_1, \xi_2, \xi_3) \in \widehat{\mathbb{R}}^3 \mid |\xi_1| \ge 1, |\frac{\xi_2}{\xi_1}| \le 1, |\frac{\xi_3}{\xi_1}| \le 1\}$. For $\ell = 2, 3$ we have a similar definition by interchanging the role of x and y (for $\ell = 2$) and of x and t (for $\ell = 3$).

Our algorithm is based on a nice property that allows us to associate with any shearing vector $k = (k_1, k_2)$ a direction (without orientation) parametrized by two angles, namely *latitude* and *longitude*, given by

$$(\cos \alpha \cos \beta, \cos \alpha \sin \beta, \sin \alpha) \qquad \alpha, \beta \in [-\frac{\pi}{2}, \frac{\pi}{2}]. \tag{3}$$

The correspondence depends on ℓ and, for the first pyramid, it is given by

$$\tan \alpha = \frac{2^{-j/2} k_2}{\sqrt{1 + 2^{-j} k_1^2}} \qquad \tan \beta = 2^{-j/2} k_1 \quad \alpha, \beta \in [-\frac{\pi}{4}, \frac{\pi}{4}].$$

The fact that Shearlets are sensitive to orientations allows us to discriminate among spatial-temporal features of different kinds [7, 18].

3   Building Dictionaries of Space-Time Primitives

1 - Space-Time Point Representation (Fig. 1). We start by considering a point $\hat{m}$ for the fixed scale $\hat{j}$ and the subset of shearings encoding different directions: $\mathbf{K} = \left\{ k = (k_1, k_2) \mid k_1, k_2 = -\lceil 2^{\hat{j}/2} \rceil, \ldots, \lceil 2^{\hat{j}/2} \rceil \right\}$. We perform the following steps:

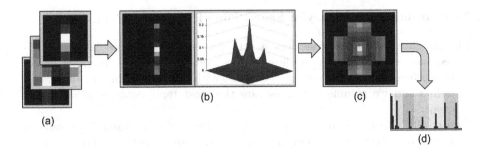

Fig. 1. $2D + T$ point representation: (a) Matrices $C_1(r,c)$, $C_2(r,c)$ and $C_3(r,c)$; (b) Object **C** both in gray-levels and 3D visualization; (c) Coefficients grouping; (d) The obtained representation **D**.

Figure 1a. We reorganize the information provided by $SH[f](\ell, \hat{j}, k, \hat{m})$ in three $M \times M$ matrices, each one associated with a pyramid ℓ, where each entry is related to a specific shearing: $C_\ell(r,c) = SH[f](\ell, \hat{j}, k_{rc}, \hat{m})$ with $\ell = 1, 2, 3$, where r and c, are discrete versions of k_1 and k_2.

Figure 1b. We merge the three matrices in a single one. The obtained overall representation **C** is centered on k_{max}, the shearing corresponding to the coefficient with the maximum value in the set $SH[f](\ell, \hat{j}, k, \hat{m})$, with $\ell \in \{1, 2, 3\}$ and $k \in \mathbf{K}$. The matrix **C** models how the shearlet coefficients vary in a neighborhood of the direction where there is the maximum variation, and it is built in a way so that the distance of every entry of **C** with respect to the center is proportional to the distance of the corresponding angles (as defined in (3)) from the angles associated with k_{max}. Different kinds of spatio-temporal elements can be associated with different kinds of local variations in **C** (see for instance Fig. 6).

Figure 1c. We now compute a compact rotation-invariant representation for point $\hat{m}$. We group the available shearings in subsets $\bar{s}_i$, according to the following rule: $\bar{s}_0 = \{k_{max}\}$ and $\bar{s}_i$ will contain the shearings in the i-th ring of values from k_{max} in **C**. We extract the values corresponding to the coefficients for $\bar{s}_1$ (by looking at the 8-neighborhood of k_{max}), then we consider the adjacent outer ring (that is, the 24-neighborhood without its 8-neighborhood) to have the coefficients corresponding to $\bar{s}_2$, and so on.

Figure 1d. We build a vector containing the values of the coefficients corresponding to each set: $\mathbf{D}(\hat{m}) = coeff_{\bar{s}_0} \frown coeff_{\bar{s}_1} \frown coeff_{\bar{s}_2} \frown \ldots;\ coeff_{\bar{s}_i}$ is the set of coefficients associated with each shearings subset $\bar{s}_i$:

$$coeff_{\bar{s}_0} = SH[f](\ell_{k_{max}}, \hat{j}, k_{max}, \hat{m})$$

$$coeff_{\bar{s}_i} = \left\{ SH[f](\ell_{\bar{s}_i}, \hat{j}, k_{\bar{s}_i}, \hat{m}), k_{\bar{s}_i} \in \bar{s}_i \right\},$$

where $\ell_{k_{max}}$ is the pyramid associated with the shearing k_{max} and where $\ell_{\bar{s}_i}$ represents the pyramid associated to each shearing $k_{\bar{s}_i}$.

2 - Learning a Dictionary of Space-Time Primitives (Figure 2).

Figure 2a. This phase considers a set of meaningful frames in a (set of) sequence(s). The frames are chosen automatically through a key-point detection process [18]. We select the N_f frames with the highest number of interest points and we assume that these are the most representative of an action event.

Figure 2b. We represent each point $\hat{m}$ of every selected frame by means of $\mathbf{D}(\hat{m})$, for a fixed scale $\hat{j}$. On each frame, we apply K-means and obtain a set of K cluster centroids, which we use as space-time primitives or atoms.

Figure 2c. We re-apply K-means on all the previously obtained atoms [7]. We end up with a dictionary $\mathcal{D}$ of N_a space-time primitives.

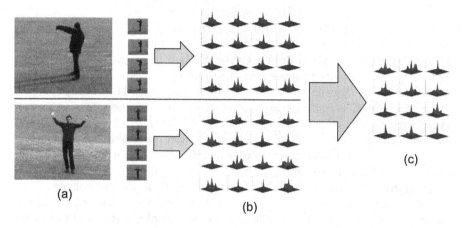

Fig. 2. Learning the dictionary. (a) Automatic selection of meaningful frames from the training set; (b) Atoms learnt by each sequence; (c) Dictionary summarization on the whole training set.

3 - Encoding a Video Sequence with Respect to a Dictionary (Figure 3). We now consider a sequence V of a given action.

Figure 3b. For each image frame $I_t \in V$ we follow a BoK approach and quantize points of I_t w.r.t the dictionary atoms, obtaining F_i^t frequency values (how many points in frame I_t can be associated with the $i - th$ atom).

Figure 3c. We filter out still primitives that are not useful to our purpose. To do this, we consider a point-wise index which we call *dynamism measure* (DM):

$$\text{DM}[\hat{m}] = SH[f](\ell_{k_{max}}, j, k_{max}, \hat{m}) \cdot cos(\Theta_{k_{max}}, \boldsymbol{n}) \qquad (4)$$

where for a given point $\hat{m}$ we consider the value corresponding to its maximum shearlet coefficient and its associated shearing parameter k_{max}; $\Theta_{k_{max}}$ is the associated direction obtained using (3) and $\boldsymbol{n}$ is the normal vector to the xy plane in our signal (i.e. aligned with the temporal axis). To discard still patterns we consider only the values of $DM[\hat{m}]$ which are above a given threshold τ. The angle $\Theta_{k_{max}}$ tells us whether a point belongs to a spatio-temporal structure which is moving or not[1], while the $SH[f](\ell_{k_{max}}, j, k_{max}, \hat{m})$ factor helps us to consider only points representing a *strong* spatio-temporal change. Finally, we compute temporal sequences of frequency values across time, obtaining N_a time series or profiles $\{P_j\}_{j=1}^{N_a}$, which summarize the content of the video sequence.

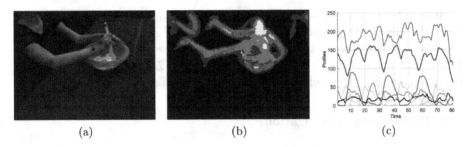

(a)　　　　　　　　(b)　　　　　　　　(c)

Fig. 3. Action encoding: (a) A sample frame; (b) The quantization w.r.t. the dictionary atoms; (c) Examples of temporal profiles (see text for details).

4 Experimental Analysis

4.1 Dataset and Experimental Protocol

The data we consider are drawn from a larger dataset of cooking actions that we will soon release to the research community. We have used three identical high resolution IP cameras, mounted on three tripods so that in all acquisitions we have a still uniform background and moving foreground objects. Figure 4 shows the setup and example video frames. The dataset includes repetitions of the same action observed from three different viewpoints: a frontal view (A), a lateral view (B), and an egocentric view, obtained by a camera mounted slightly above the subject's head (C). No specific constraints have been imposed to the volunteer.

For this preliminary analysis we are considering a subset of 3 actions. For each action and each view we consider 3 action instances. In the following experiments we consider dictionaries learnt from *Eating* actions only. For the detection phase (see [18]), we fix the number of selected frames N_f to 4 and consider only shearlet coefficients at scale 2. For the dictionary learning phase, the number of centroids per frame is $K = 8$, and the final dictionary size is $N_a = 12$.

---

[1] Points belonging to still spatio-temporal structure spawn surfaces over time, and the normal vector $\Theta_{k_{max}}$ for those points will belong to the xy plane, bringing the value for $cos(\Theta_{k_{max}}, n)$ to be 0.

Fig. 4. Acquisition setup

We evaluate the dissimilarity between action pairs by means of Dynamic Time Warping (DTW). Given two videos V^1 and V^2 depicting a certain action instance and described by two sets of temporal profiles $P^1 = \{P_i^1\}_{i=1}^{N_a}$ and $P^2 = \{P_i^2\}_{i=1}^{N_a}$ then $Dis(V^1, V^2) = \text{avg}_{i=1}^{N_a} DTW(P_i^1, P_i^2)$. Z-normalization is applied to the temporal profiles before computing the dissimilarity.

4.2 Preliminary Investigation

1. How informative are the learnt space-time dictionaries to discriminate among different actions of the same kind? In this experiment we consider comparisons between actions observed from a given viewpoint, described according to a dictionary obtained from the same view: we refer to such dictionaries as D_A, D_B, and D_C. In Fig. 5a we show the average DTW cost in aligning the instances of the action classes. We observe that on average the comparisons of actions from the same class have a lower cost. Among the 3, CAM_C appears to be the most challenging viewpoint. We may notice that *Eating* action is the best performing, as dictionaries are built on eating examples. At the same time we observe a good generalization to other actions.

2. What is the relationship between different dictionaries learnt from different viewpoint data? Is there any benefit in learning dictionaries from different views? To answer this question, we compare dictionaries specific to different views, and observe they encode similar spatio-temporal primitives. We build a dissimilarity matrix collecting the Euclidean distances between atoms of the two dictionaries. The atoms are then matched using the Hungarian algorithm, and their contributions are sorted in the dissimilarity matrix accordingly. As a consequence, on the main diagonal we may find agglomerations of atoms belonging to different dictionaries but encoding the same kind of spatio-temporal information. Figure 6 shows an example where dictionaries referring to CAM_A and CAM_B are considered, and where we highlighted groups of atoms carrying similar information. At the top of the diagonal a group of 3 atoms (Fig. 6a) describe moving edge-like structures, which correspond to surface in the space-time domain. Similarly, the primitives in Fig. 6b and c represent corner-like structures with a different amount of dynamic variations in the direction around the principal one.

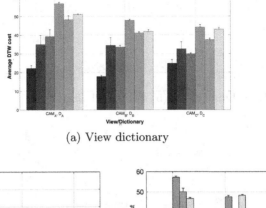

(a) View dictionary

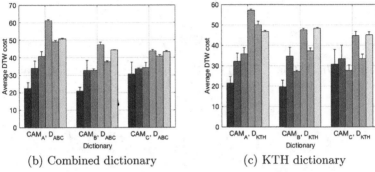

(b) Combined dictionary   (c) KTH dictionary

Fig. 5. Average DTW cost obtained when comparing actions of the same view using different dictionaries.

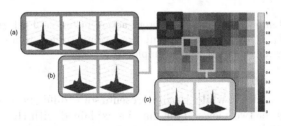

Fig. 6. An example of dissimilarity matrix between atoms of two different dictionaries (from CAM_A and CAM_B), with a selection of prototypes encoding different dynamic properties of the signal.

As we observe a large overlap between different dictionaries, we also consider the benefits of learning a joint dictionary from the 3 views, as this choice would simplify inter-view comparisons. Figure 5b shows how stable the performance is when adopting D_{ABC} for all the data.

3. To what extent the space-time representation is view-invariant?
Figure 7 provides a first qualitative answer to the question. The plots represent the average profiles of all actions instances. *Eating* is characterized by the high-

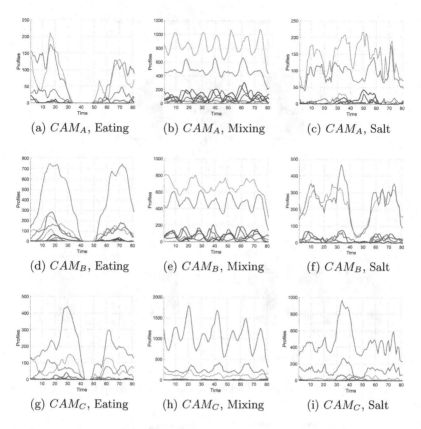

Fig. 7. Average temporal profiles of different action instances. Each row corresponds to a view (CAM_A, CAM_B, CAM_C), while each column refers to an action (Eating, Mixing, Salt). The dictionary D_{ABC} is employed.

est stability across views, while *Mixing* presents some differences in CAM_C with respect to the other two views. This may be explained with the fact the action is performed following a quasi-planar shape on the table, favouring a clear and regular apparent motion from the top view. *Salt* is a less constrained action characterized by a higher degree of instability over time and across views. Figure 8a reports the average DTW costs obtained from pairs of views. On the left (D_{ABC}) we confirm *Eating* is stable across views, while a higher intra-class variability is associated with *Mixing*. We also notice a similarity between *Eating* and *Salt*. A visual inspection of the corresponding profiles in Fig. 7 confirms the presence of common temporal patterns.

We observe that the different temporal profiles are characterized by an uneven amount of stability. This suggests that a selection of the profiles to be used in the comparison may be of benefit. This aspect is currently under investigation, as a proof of concept, in Fig. 8b we consider only one profile, the green one in Fig. 7. An improvement on the results may be appreciated.

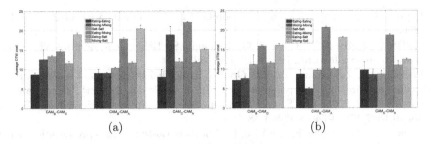

Fig. 8. Comparison between descriptions from different views.

4. Is it really useful to learn an ad hoc dictionary for a given set of data? As a final investigation, we reason on the necessity of using data of the considered scenario. To this purpose we consider an unrelated benchmark (KTH [19]) showing full body actions. Figure 5c shows the results obtained in this case. We notice a small degradation, but the overall performance is still acceptable. This speaks in favor of the potential of our space-time primitives to transfer knowledge between different settings.

5 Discussion

We presented an ongoing work on representing actions through space-time primitives learnt from data. The preliminary results on a small subset of data include useful insights on how to proceed: the representation is rich and incorporates not only space-time corners but also other local structures with a significant dynamic information; the learnt atoms are quite stable across views, with a strong discriminative power. The action representation is again quite stable across views, even if some actions seem to be intrinsically view-variant, and some views are more meaningful than others. Representations obtained from front and lateral views are very closely related, as expected.

Two main aspect are currently under investigation: *(i)* Capturing the temporal cross-correlation between different primitives, especially across views and *(ii)* Devising an action recognition module based on the proposed representation.

References

1. Tomasello, M., Carpenter, M., Call, J., Behne, T., Moll, H.: Understanding and sharing intentions: the origins of cultural cognition. Behav. Brain Sci. **28**(05), 675–691 (2005)
2. Elsner, C., Falck-Ytter, T., Gredebäck, G.: Humans anticipate the goal of other people's point-light actions. Front. Psychol. **3** (2012)
3. Vignolo, A., Rea, F., Noceti, N., Sciutti, A., Odone, F., Sandini, G.: Biological movement detector enhances the attentive skills of humanoid robot iCub. In: IEEE-RAS (Humanoids), pp. 338–344 (2016)

4. Laptev, I.: On space-time interest points. Int. J. Comput. Vis. **64**(2), 107–123 (2005)
5. Dollár, P., Rabaud, V., Cottrell, G., Belongie, S.: Behavior recognition via sparse spatio-temporal features. In: VS-PETS (2005)
6. Willems, G., Tuytelaars, T., Van Gool, L.: An efficient dense and scale-invariant spatio-temporal interest point detector. In: Forsyth, D., Torr, P., Zisserman, A. (eds.) ECCV 2008. LNCS, vol. 5303, pp. 650–663. Springer, Heidelberg (2008). doi:10.1007/978-3-540-88688-4_48
7. Malafronte, D., Odone, F., De Vito, E.: Local spatio-temporal representation using the 3D shearlet transform. In: SAMPTA (2017)
8. Labate, D., Lim, W.-Q., Kutyniok, G., Weiss, G.: Sparse multidimensional representation using shearlets. In: Optics & Photonics (2005)
9. Kutyniok, G., Labate, D.: Shearlets. Applied and Numerical Harmonic Analysis. Birkhäuser/Springer, New York (2012). doi:10.1007/978-0-8176-8316-0
10. Guo, K., Labate, D., Lim, W.-Q.: Edge analysis and identification using the continuous shearlet transform. Appl. Comput. Harmon. Anal. **27**(1), 24–46 (2009)
11. Kutyniok, G., Petersen, P.: Classification of edges using compactly supported shearlets. Appl. Comput. Harmon. Anal. **42**(2), 245–293 (2015)
12. Easley, G.R., Labate, D., Colonna, F.: Shearlet-based total variation diffusion for denoising. TIP **18**(2), 260–268 (2009)
13. Chen, Z., Hao, X., Sun, Z.: Image denoising in shearlet domain by adaptive thresholding. J. Inf. Comput. Sci. **10**(12), 3741–3749 (2013)
14. Duval-Poo, M.A., Noceti, N., Odone, F., De Vito, E.: Scale invariant and noise robust interest points with shearlets. IEEE Trans. Image Process. **26**(6), 2853–2867 (2017)
15. Duval-Poo, M.A., Odone, F., De Vito, E.: Edges and corners with shearlets. IEEE Trans. Image Process. **24**(11), 3768–3780 (2015)
16. Csurka, G., Dance, C., Fan, L., Willamowski, J., Bray, C.: Visual categorization with bags of keypoints. In: ECCV-W, vol. 1, no. 1–22 (2004)
17. Kutyniok, G., Lim, W., Reisenhofer, R.: Shearlab 3D: faithful digital shearlet transforms based on compactly supported shearlets. ACM Trans. Math. Softw. **42**, 5:1–5:42 (2016)
18. Malafronte, D., Odone, F., De Vito, E.: Detecting spatio-temporally interest points using the shearlet transform. In: Alexandre, L.A., Salvador Sánchez, J., Rodrigues, J.M.F. (eds.) IbPRIA 2017. LNCS, vol. 10255, pp. 501–510. Springer, Cham (2017). doi:10.1007/978-3-319-58838-4_55
19. Schuldt, C., Laptev, I., Caputo, B.: Recognizing human actions: a local SVM approach. In: ICPR, vol. 3 (2004)

Organizing Videos Streams for Clustering and Estimation of Popular Scenes

Sebastiano Battiato[1], Giovanni M. Farinella[1], Filippo L.M. Milotta[1,2(✉)],
Alessandro Ortis[1,2], Filippo Stanco[1], Valeria D'Amico[2], Luca Addesso[2],
and Giovanni Torrisi[2]

[1] Department of Mathematics and Computer Science,
University of Catania, Viale A. Doria, 6, 95125 Catania, Italy
{battiato,gfarinella,milotta,ortis,fstanco}@dmi.unict.it
[2] JOL WAVE, Telecom Italia, Viale A. Doria, 6, 95125 Catania, Italy
{valeria1.damico,luca.addesso,giovanni.torrisi}@telecomitalia.it

Abstract. The huge diffusion of mobile devices with embedded cameras has opened new challenges in the context of the automatic understanding of video streams acquired by multiple users during events, such as sport matches, expos, concerts. Among the other goals there is the interpretation of which visual contents are the most relevant and popular (i.e., where users look). The popularity of a visual content is an important cue exploitable in several fields that include the estimation of the mood of the crowds attending to an event, the estimation of the interest of parts of a cultural heritage, etc. In live social events people capture and share videos which are related to the event. The popularity of a visual content can be obtained through the "visual consensus" among multiple video streams acquired by the different users devices. In this paper we address the problem of detecting and summarizing the "popular scenes" captured by users with a mobile camera during events. For this purpose, we have developed a framework called RECfusion in which the key popular scenes of multiple streams are identified over time. The proposed system is able to generate a video which captures the interests of the crowd starting from a set of the videos by considering scene content popularity. The frames composing the final popular video are automatically selected from the different video streams by considering the scene recorded by the highest number of users' devices (i.e., the most popular scene).

Keywords: Video analysis · Clustering · Social cameras · Scene understanding

1 Introduction

During a social event, the audience typically uses its personal devices to record video clips related to the most interesting moments of the event. As a result, several videos will be related to the same visual contents, and this redundancy can be exploited to infer the most interesting moments of the event over time,

© Springer International Publishing AG 2017
S. Battiato et al. (Eds.): ICIAP 2017, Part I, LNCS 10484, pp. 51–61, 2017.
https://doi.org/10.1007/978-3-319-68560-1_5

according to the people interests on the observed scenes. The issue of crowd-popularity estimation through automatic video processing is not trivial due to the variability of the visual contents observed by multiple devices: different points of view, pose and scale of the objects, lighting conditions and occlusions. The differences between device models should be also taken into account, since they imply different characteristics of the lens, color filter arrays, resolution and so on. For instance, even using two devices with similar (or equal) sensors the colors recorded will not necessarily be the same because devices responses are processed with different non-linear transformations due to the differences on the Imaging Generation Pipelines (IGPs). They can vary from device to device and even on an per-image basis [1,2].

We propose a system called RECfusion to estimate the popularity of scenes related to multiple video streams. The streams are analyzed with the aim to create a continuous video flow, obtained by mixing the several input channels, taking into account the most popular scenes over time to reflect the interests of the crowd. Then, the clusters of the different scenes are tracked over time. This allows to have not only the most popular scene at each time, but also the other scenes of interest and give the possibility to introduce a scenes story log allowing the user to select the scene of interest among all the detected ones.

The reminder of the paper is structured as follows: in Sect. 2 we discuss related studies about crowd-saliency inference from multi-device videos. In Sect. 3 an overview of the RECfusion framework is given together with the description of its three main modules: intraflow analysis, interflow analysis and cluster tracking. In Sect. 4 a proper dataset is introduced, whereas in Sect. 5 we report the experimental settings and the results. We conclude the paper with a final discussion and hints for possible future works in Sect. 6.

2   Related Works

Different papers about crowd-saliency inference from multi-device videos have been proposed in literature in the past. The works in [3,4] exploit Structure from Motion (SfM) to estimate a 3D reconstruction of the scene and the pose of employed devices. Hoshen et al. [5] uses egocentric video streams considering a single camera model acquired by different participants to create a single popular video of an event. However, in [3–5] the number of the different popular scenes and the number of the devices are known a priori. Saini et al. [6] developed the framework MoViMash with the purpose of replicate the behavior of a movie director: the system learns from a labeled sets of video frames "how to" and "when" perform transitions between different views. However, this technique is hardly adaptable for a real-time context, since for each different recorded scene a proper learning phase should be tuned to. ViComp is another framework similar to MoViMash [7]. In ViComp the final output video consists in a combination of several video streams from multiple sources. The combination is obtained by selecting high quality video segments according to their audio-visual ranking scores. It selects the best video stream among a pool of available ones basing on degradation and noise caused by video compression [8] and estimated camera pan and tilt [9].

The aforementioned approaches achieve significant results but, compared to them, our approach (RECfusion) does not need any prior knowledge or training stage and is able to combine videos from an unknown number and types of recording devices. RECfusion is a framework with a popularity-based video selection approach: it clusters the video streams and selects the best video stream from each cluster exploiting clustering metrics.

3 RECfusion System Overview

RECfusion is a framework designed for automatic video curation driven by the popularity of the scenes acquired by multiple devices. Given a set of video streams as input, the framework can group these video streams according to the viewed similarity and popularity of the scenes over time, then it automatically suggests a video stream to be used as output acting like a "virtual director". With the aim to mitigate the aforementioned differences in the color representation of the devices, due their different IGPs, the video frames are pre-processed by an equalization algorithm. This step helps the further computations that compares frames captured by different devices [1, 10–12]. After this normalization, the system extracts an image representation from each frame. The algorithm takes a frame as input and returns a descriptor. The aim is to have a descriptor that maximize the differences between semantically different frames and minimize the differences between semantically similar ones. In [1] a definition of light conditions (and almost devices) independent representation is given. The method is based upon the observation that changes of light conditions or device directly change the RGB values of the frame, while order of sensors response remains the same. Finally, equalization of RGB channels, as described in [1], is performed. After the normalization of the color domain, the video streams are analysed in our approach in three phases (Fig. 1), detailed in the followings.

3.1 Intraflow Analysis

The intraflow analysis segments the sequence of frames of a single video stream (Fig. 1(a)). During intraflow analysis the frames of each video are processed comparing their visual contents. For each frame of the video flow, we extract keypoints using the SIFT detection algorithm [13]. The set of the extracted SIFT features represents a template for the acquired scene. In this way, the comparison between frames could be done as the comparison between SIFT templates. When the comparison between the current frame end the reference template generates a sensible variation of features (i.e., low matching score), then the algorithm refreshes the reference template and splits the video producing a new segment. To make the matching more reliable, we reject the matchings where the keypoints are too far in terms of spatial coordinates by assuming smooth transition between consecutive frames [14]. For major stability, a new template can be defined only

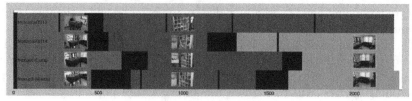

(a) Intraflow Analysis

(b) Interflow Analysis

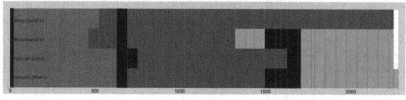

(c) Cluster Tracking

Fig. 1. RECfusion results applied on Foosball dataset. The chronograms show the results of the three main steps of RECfusion (intraflow analysis, interflow analysis and cluster tracking). Foosball dataset is composed by 4 video streams having a duration of ∼2300 frames (∼90 s). Each video stream is represented as a row in the chronograms. Vertical red lines mark the end of time-slots. (a) Intraframe analysis: red, blue and green frames are respectively the first, second and third scene of each video stream. Noisy frames are depicted in black. (b) Interframe analysis: yellow and green clusters are respectively the first and second cluster of each time-slot. (c) Cluster tracking: red, blue and green clusters are respectively the first, second and third cluster of the whole video set. Noisy clusters are depicted in black. (Color figure online)

if it has a duration greater than 2 s, otherwise it is considered as noise. In other words, a template is considered a stable template if the number of matching SIFTs do not change too much in time. A backward checking is required in order to understand if a new defined template regards a new scene or it is related to a previously observed one. The algorithm compares the new defined template with the past ones, starting from the last found template. Each reference template is labeled with a *SceneCounter* and all video frames achieving a robust match are labeled with the same *SceneCounter*. Note that all the frames required to decide if a template should be considered as a new or an updated one are labeled as a transition interval.

3.2   Interflow Analysis

The interflow analysis is computed for each time-slot. It segments video frames labeled by intraflow analysis and assigns a *ClusterCounter* with respect to all the video streams in that specific time-slot (Fig. 1(b)). We want to group together the devices that are looking at the same scene over time. The descriptor used in the interflow analysis is based on weighted color histograms [15]. In this context the device invariance should be granted as well as possible. For this reason we firstly apply an histogram equalization, as suggested in [1]. The equalization is followed by a quantization of the color space (8 colors for each channel). The weights are obtained by using a gradient map as suggested in [15]. The gradient map is useful to highlight the structures of the objects that appear in the scene, making more robust the descriptor.

The different scenes obtained with the intraflow analysis could be considered as nodes of a complete graph in which arcs are weighted with the interflow distances between the scenes acquired by the devices. The clustering procedure selects a frame among the unclustered frames and assigns it to the most similar cluster. We used an average linkage approach to compare a frame with a cluster: the distance between a frame and a cluster is given by the average distance between the frame and all the elements within the cluster [14].

3.3   Cluster Tracking

To understand the meaning of the Cluster Tracking module we have to step back to intraflow analysis. The intraflow analysis segments the sequence of frames of a single video stream, and assigns a *SceneCounter* to each segmented scene. However, frames taken by two different video streams but labeled with the same *SceneCounter* can represent different scenes, since *SceneCounters* are discriminative only within a single video stream. The interflow analysis segments video frames in a time-slot and assigns a *ClusterCounter* to the scenes of the video streams. Interflow analysis exploits the *SceneCounters* and the set of SIFT features templates from intraflow analysis. Similarly to *SceneCounters*, the *ClusterCounters* are to be considered only within a single time-slot. Therefore, we developed a cluster tracking procedure in order to track the clusters representing the same scene in every video stream and time-slot (Fig. 1(c)). In [16] a Graphical User Interface implementing the cluster tracking typical video player commands (like Start, Pause, Stop, . . .) is described (Fig. 2).

We propose a cluster tracking procedure based on a voting routine that combines the results of the intraflow and interflow analyses. Once interflow procedure has assigned a *ClusterCounter* to several *SceneCounters*, this set of scenes will characterize the same cluster also in further time-slots, so cluster tracking procedure an unique *LoggedCluster$_{ID}$* to this set of scenes. Differently from the *ClusterCounters*, the *LoggedCluster$_{IDs}$* are intended to be always discriminative. Cluster tracking procedure tracks the clusters in each time-slot assigning them *TrackedCluster$_{IDs}$* equals to the most similar *LoggedCluster$_{ID}$*. In order to define the most similar *LoggedCluster$_{ID}$*, cluster tracking procedure

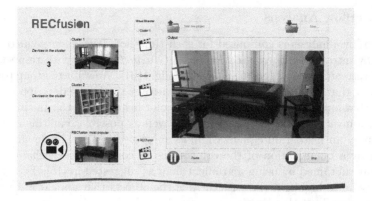

Fig. 2. RECfusion Graphical User Interface showing the Cluster Tracking framework. On the left, active clusters with respective amount of recording devices and automatically suggested video stream (called *RECfusion: most popular*) are shown. User can browse the Virtual Director panel to dinamically change the active video stream. On the right side, active video stream with classic video player commands is shown.

requires an initialization phase (at first time-slot). In this phase, the assigned $LoggedCluster_{IDs}$ are equals to the $ClusterCounters$. Then, from the second time-slot on, the clusters will be associated to an existent $LoggedCluster_{ID}$ or to a new one, depending on a voting routine. The same routine is also used to track the $LoggedCluster_{IDs}$ with proper $TrackedCluster_{IDs}$.

The voting routine can be divided into 2 phases: casting of vote and voting decision. In the former phase, for each time-slot, each scene votes with three different possible values: $TrackedCluster_{ID}$ at the previous time-slot, $LoggedCluster_{ID}$ or unlogged scene (V_N), if the scene is *Noise*, already logged or unlogged, respectively. Once all the votes are casted in a time-slot, then we look for a non ambiguous voting decision (i.e., a majority is found). Majority of unlogged scenes is not admitted, so in this case we simply remove these votes from the voting decision. Depending on the reached decision, new $LoggedCluster_{IDs}$ might be instantiated, while $TrackedCluster_{IDs}$ at current time-slot is eventually updated. We will compare the new proposed method with respect to a cluster tracking method based on a threshold T_{CT} [16]. This threshold was used as an hyperparameter to decide whenever to create a new $LoggedCluster_{ID}$ or not. The issue with this threshold employed in [16] is that its value should be fine tuned for each video set in order to achieve the best results in cluster tracking procedure.

4   Datasets

To perform experiments we have used the RECfusion dataset [14] which is publicly available at the following URL: http://iplab.dmi.unict.it/recfusionICIAP17. This dataset is made up of three video sets:

1. *Foosball*: indoor context, some people appear in the scene. The number of contributing devices for this video set is 4, with an average number of frames per video stream of 2250 (44 time-slots). There are three main subjects in this video set: a foosball, a couch and a bookcase.
2. *Meeting*: indoor context, two people appear in the scene. The number of contributing devices for this video set is 5, with an average number of frames per video stream of 2895 (60 time-slots). There are two main subjects in this video set (the two guys).
3. *S. Agata*: outdoor context, lots of people appear in the scene. The number of contributing devices for this video set is 7, with an average number of frames per video stream 1258 (34 time-slots). There are two main subjects in this video set: the reliquary of S. Agata and the facade of a church.

In the experiments we exploit also a video set from the dataset used in Ballan et al. [17]. This dataset is called *Magician*. It is related to an indoor context, where one person appear in the foreground. The number of contributing devices for this video set is 6, with a fixed number of 3800 frames per video stream (77 time-slots). There are two main points of view in this video set: one above and one in front of the magician. We have chosen *Magician* video set because it is slighty different from the videos currently in RECfusion dataset. In *Magician* all the video streams are focused on a single target and are acquired as a "casual multi-view video collection" [17]. This means that backgrounds in the video streams are very different from each other and that severe camera motion could often appear. The casually filmed events represent a challenging scenario for detector like SIFT (exploited in our intraflow analysis, see Sect. 3.1), so we add *Magician* video set to our tests in order to stress and evaluate scene analysis and cluster tracking performances. We have also compared the obtained results with the benchmark dataset proposed in Hoshen et al. [5]. This dataset has been acquired with wearable devices and, like *Magician* video set, it is challenging since every video is strongly affected by motion.

5   Experimental Settings and Results

We select the last instant of time for every time-slot as the representative of that interval. Validation are made exploiting the Ground Truth related to these representative frames. To evaluate the performances of the proposed method, we compute the two quality measures described in [14]. Specifically, for each clustering step we consider:

- P_r: ground truth popularity value (number of cameras looking at the most popular scene) obtained from manual labelling;
- P_a: popularity score computed by the system (number of the elements in the popular cluster);
- P_g: number of the correct videos in the popular cluster (i.e., the number of inliers in the popular cluster).

From the above defined scores, the weighted mean of the ratios P_a/P_r and P_g/P_r over all the clustering steps are computed. The ratio P_a/P_r provides a score for the popularity estimation, whereas the ratio P_g/P_r verifies the visual content of the videos in the popular cluster and provides a measure of the quality of the popular cluster. Note that P_a/P_r is a score: when is lower than 1 it means that system is under-estimating the popularity of the cluster, while, conversely, if it is higher than 1 it results in an over-estimation.

Table 1. Validation results of popularity estimation.

Scenario	Devices	Models	P_a/P_r	P_g/P_r
Foosball	4	2	1.02	1
Meeting	2	2	1.01	0.99
Meeting	4	4	0.99	0.95
Meeting	5	5	0.89	0.76
SAgata	7	6	1.05	1
Magician	6	6	0.73	0.73
Concert [5]	3	1	1.06	1
Lecture [5]	3	1	1.05	0.86
Seminar [5]	3	1	0.62	0.62

The results of the comparison between the tested video sets are shown in Table 1. The first five rows are related to RECfusion dataset, whereas the last three rows are related to the dataset proposed in [5]. Although the constantly head motion of the wearable recording devices in videos from [5], the framework reaches good results and seems to be promising room for improvement in the field of wearable devices. Conversely, we found a drop in the performances when there is a severe difference of scale between videos in a video set. Indeed, we exploited *Meeting* video set to evaluate the drawback in performances when there are high differences between resolution of devices. We compared three cases, with 2, 4 and all the 5 devices in *Meeting* video set, respectively. Other analysis outputs could be found at the following URL http://iplab.dmi.unict.it/recfusionICIAP17.

In the new proposed procedure we removed the threshold T_{CT}, used as an hyperparameter to decide whenever to create a new logged-cluster or not. In [16] the value of T_{CT} was empirically set equals to 0.15 founding the best overall value between True Positive Rate, True Negative Rate and Accuracy of clustering tracking procedure on RECfusion dataset. In Fig. 3 a comparison between the average values of *TPR (True Positive Rate, or Recall)*, *TNR (True Negative Rate, or Specificity)* and *ACC (Accuracy)* of RECfusion dataset and *Magician* video set whit several values of T_{CT} is shown. As can be seen, the value of T_{CT} equals to 0.15 is not the best value to be used by cluster tracking procedure, while $T_{CT} = 0.5$ should be used instead. For this reason we proposed the new

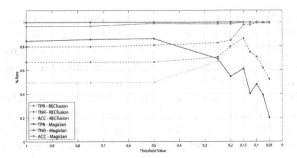

Fig. 3. A comparison of *TPR (True Positive Rate, or Recall), TNR (True Negative Rate, or Specificity)* and *ACC (Accuracy)* between RECfusion_dataset_2015 and *Magician* video set cluster tracking validations using the threshold-based procedure from [16]. As can be seen, *Magician* requires a fine tuned threshold to increase *TPR*, *TNR* and *ACC* values.

threshold independent cluster tracking procedure described in Sect. 3.3. We computed *TPR*, *TNR* and *ACC* values for each video set described in Sect. 4 and compared them with the results obtained in [16]. The comparative validation results are shown in Table 2.

Table 2. Validation results between cluster tracking procedure threshold-based and vote-based.

DS	Scene	TPR (Recall)		TNR (Specificity)		ACC (Accuracy)	
		[16]	Proposed	[16]	Proposed	[16]	Proposed
Foosball	1	0,91	**0,92**	0,70	**1,00**	0,69	**1,00**
	2	0,69	**0,97**	**0,98**	0,91	**0,99**	0,97
	3	0,41	**0,74**	**1,00**	**1,00**	0,50	**1,00**
	Mean	0,67	**0,87**	0,89	**0,97**	0,73	**0,99**
Meeting	1	0,99	**1,00**	**1,00**	**1,00**	**1,00**	**1,00**
	2	0,80	**1,00**	**0,95**	0,93	**0,83**	0,67
	3	0,43	**0,50**	**1,00**	**1,00**	0,70	**1,00**
	Mean	0,74	**0,83**	**0,98**	**0,98**	0,84	**0,89**
S.Agata	1	0,71	**1,00**	**1,00**	**1,00**	**1,00**	**1,00**
	2	0,87	**0,97**	**0,49**	0,14	**0,80**	0,68
	3	**0,48**	0,00	**1,00**	**1,00**	**0,60**	0,00
	Mean	**0,69**	0,66	**0,83**	0,71	**0,80**	0,56
Magician	1	0,73	**1,00**	**1,00**	**1,00**	**1,00**	**1,00**
	2	0,45	**0,56**	**1,00**	**1,00**	**0,98**	0,91
	Mean	0,59	**0,78**	**1,00**	**1,00**	**0,99**	0,96

These results show that the proposed vote-based cluster tracking procedure reaches *TPR* values much higher than the threshold-based procedure, while results on TNR and ACC are comparable between the two procedures. Just in the *Meeting* video set the proposed vote-based procedure is slighty outperformed: this is a limitation of the procedure. Indeed, cluster tracking procedure relies on intraflow analysis, so if the latter defines N scenes, then the former is able to distinguish at most N scenes. Hence, differently by threshold-based procedure used in [16], that can generate a bunch of small sparse clusters if T_{CT} is not fine tuned, in this case only a limited number of clusters is tracked. In *Meeting* video set two people are recorded and there are only two distinguished clusters focusing on each one of them. Sometimes interflow analysis generates a cluster containing both of the two people. This is treated by the cluster tracking vote-based procedure as *Noise*, since intraflow analysis has never labeled a scene in which the people are recorded together.

A final remark is about *Magician* video set. We added it to our dataset in order to evaluate scene analysis and cluster tracking performances in a video collection with a single scene, where all the user are focused on the same target and videos are affected by severe camera motion. Cluster tracking results with threshold-based procedure from [16] are really bad, indeed we got the worst average performance on this video set (Table 2). On the other hand, the proposed vote-based procedure reached good values of *TPR*, further assessing the soundness of this new cluster tracking approach. The output videos showing the result of cluster tracking vote-based procedure could be found at the following http:// iplab.dmi.unict.it/recfusionICIAP17.

6   Conclusion and Future Works

In this paper we described RECfusion, a framework designed for automatic video curation driven by the popularity of the scenes acquired by multiple devices. Given a set of video streams as input, the framework can group these video streams by means of similarity and popularity, then it automatically suggests a video stream to be used as output, acting like a "virtual director". We compared RECfusion intraflow and interflow analysis validations with Hoshen [5]. We have added a video set from Ballan et al. [17] to our RECfusion dataset showing that RECfusion is capable of recognize and track the scenes of a video collection even if there is a single scene, where all the user are focused on the same target and videos are affected by severe camera motion. We proposed a novel and alternative vote-based cluster tracking procedure and compared it with the one, threshold-based, described in [16]. From this comparison we found that vote-based procedure reaches very good results totally automatic and independently by a hyperparameter fine tuning phase, but with the tradeoff of be unable to create and track an unlimited number of clusters. As future works and possible applications, we are planning to augment the framework with features specifically focused on Assistive Technology or Security issues (i.e., highlight/track bad behaviour in the life style, log the visited places, search something or someone that appears in the scene).

Acknowledgments. This work has been performed in collaboration with Telecom Italia JOL WAVE in the project FIR2014-UNICT-DFA17D.

References

1. Finlayson, G., Hordley, S., Schaefer, G., Tian, G.Y.: Illuminant and device invariant colour using histogram equalisation. Pattern Recogn. **38**(2), 179–190 (2005)
2. Finlayson, G., Schaefer, G.: Colour indexing across devices and viewing conditions. In: International Workshop on Content-Based Multimedia Indexing (2001)
3. Arev, I., Park, H.S., Sheikh, Y., Hodgins, J., Shamir, A.: Automatic editing of footage from multiple social cameras. ACM Trans. Graph. **33**, 81 (2014)
4. Park, H.S., Jain, E., Sheikh, Y.: 3D social saliency from head-mounted cameras. In: Advances in Neural Information Processing Systems, pp. 431–439 (2012)
5. Hoshen, Y., Ben-Artzi, G., Peleg, S.: Wisdom of the crowd in egocentric video curation. In: IEEE Conference on Computer Vision and Pattern Recognition Workshops, pp. 587–593 (2014)
6. Saini, M.K., Gadde, R., Yan, S., Ooi, W.T.: Movimash: online mobile video mashup. In: ACM International Conference on Multimedia, pp. 139–148 (2012)
7. Bano, S., Cavallaro, A.: ViComp: composition of user-generated videos. Multimedia Tools Appl. **75**, 7187–7210 (2016)
8. Mittal, A., Moorthy, A.K., Bovik, A.C.: No-reference image quality assessment in the spatial domain. IEEE Trans. Image Process. **21**(12), 4695–4708 (2012)
9. Nagasaka, A., Miyatake, T.: Real-time video mosaics using luminance-projection correlation. Trans. Inst. Electron. Inf. Commun. Eng. **82**(10), 1572–1580 (1999). http://ci.nii.ac.jp/naid/110003183527/en/. ISSN 09151923
10. Farinella, G.M., Ravì, D., Tomaselli, V., Guarnera, M., Battiato, S.: Representing scenes for real-time context classification on mobile devices. Pattern Recogn. **48**(4), 1086–1100 (2015)
11. Farinella, G.M., Battiato, S.: Scene classification in compressed and constrained domain. Comput. Vis. **5**(5), 320–334 (2011)
12. Naccari, F., Battiato, S., Bruna, A., Capra, A., Castorina, A.: Natural scenes classification for color enhancement. IEEE Trans. Consum. Electron. **51**(1), 234–239 (2005)
13. Lowe, D.G.: Distinctive image features from scale-invariant keypoints. Int. J. Comput. Vis. **60**(2), 91–110 (2004)
14. Ortis, A., Farinella, G.M., D'Amico, V., Addesso, L., Torrisi, G., Battiato, S.: Recfusion: automatic video curation driven by visual content popularity. In: ACM Multimedia, MM 2015, pp. 1179–1182. ACM (2015)
15. Domke, J., Aloimonos, Y.: Deformation and viewpoint invariant color histograms. In: British Machine Vision Conference, pp. 509–518 (2006)
16. Milotta, F.L.M., Battiato, S., Stanco, F., D'Amico, V., Torrisi, G., Addesso, L.: RECfusion: automatic scene clustering and tracking in video from multiple sources. In: EI - Mobile Devices and Multimedia: Enabling Technologies, Algorithms, and Applications (2016)
17. Ballan, L., Brostow, G.J., Puwein, J., Pollefeys, M.: Unstructured video-based rendering: interactive exploration of casually captured videos. In: ACM Transactions on Graphics, pp. 1–11 (2010)

360° Tracking Using a Virtual PTZ Camera

Luca Greco and Marco La Cascia[✉]

DIID, Università degli Studi di Palermo, Viale delle Scienze, Palermo, Italy
{luca.greco,marco.lacascia}@unipa.it

Abstract. Object tracking using still or PTZ cameras is a hard task for large spaces and needs several devices to completely cover the area or to track multiple subjects. The introduction of 360° camera technology offers a complete view of the scene in a single image and can be useful to reduce the number of devices needed in the tracking problem. In this paper we present a framework using 360° cameras to simulate an unlimited number of PTZ cameras and to be used for tracking. The proposed method to track a single target process an equirectangular view of the scene and obtains a model of the moving object in the image plane. The target is tracked analyzing the next frame of the video sequence and estimating the P,T and Z shifts needed to keep the target in the center of the virtual camera view. The framework allows to use a single 360° device to obtain an equirectangular video sequence and to apply the proposed tracking strategy on each target simulating several virtual PTZ cameras.

Keywords: 360° cameras · Equirectangular projection · PTZ cameras · Object tracking

1 Introduction

The recent availability of 360° camera has provided a new type of images and videos and several visualization models. For still images, it is common to use the expanded version of the image (i.e. cylindrical or equirectangular projection) with the whole content directly visible to the user. Navigable players are mainly used for videos, allowing to change the direction of the view, different users can focus on different parts of the 360° video and have a totally different information and experience from the media.

In this paper we use 360° videos for tracking. Following the navigable players paradigm, we don't use the complete 360° information, but we simulate a traditional PTZ camera that can freely move in a 360° world. Using this approach it is possible to reduce the computational complexity of tracking analyzing only a small region of the video frames.

In real-time systems, a PTZ camera can miss the target during tracking. The re-acquisition of the target can be hard and time consuming due to mechanical limitations of PT motors. In 360° video, on the other hand, the target is still present (in a unknown position) and the re-acquisition of the target might be as simple as object detection in images.

© Springer International Publishing AG 2017
S. Battiato et al. (Eds.): ICIAP 2017, Part I, LNCS 10484, pp. 62–72, 2017.
https://doi.org/10.1007/978-3-319-68560-1_6

Finally, the use of this type of videos can provide practical advantages: reducing the number of cameras needed to cover a space (just one in a small place) and allowing multiple subjects tracking using only one video.

2   Related Works

Object tracking has a long history in computer vision. There are a lot of papers and surveys, [12,18] for example, explaining which are the main aspects to evaluate developing a new approach. Different methods use different information of the observed video, for example the appearance of the target, a model of the target, color information, keypoints detection and so on. 360° tracking is nowadays an open problem and the related works in this field is limited and in fast evolution. In this paper we use 360° video to simulate PTZ virtual cameras. In next paragraphs we report an overview on 360° video capturing and handling, and on classical PTZ tracking techniques.

2.1   360° Image Representation

Panoramic images are used to give a single representation of a wide scene, changing the representation coordinate system from the classic planar projection to a different one. Most commonly used coordinate systems are the cylindrical and the spherical ones. They were mainly used in 3D Graphics to reproduce a real environment or in panoramic stitching to create a synthetic wide-horizontal image from a collection of pictures, augmenting the real pan of the camera.

An overview on panoramic imaging, including both hardware and software specific tools, is presented in [8]. In particular, this work describes the capturing, processing, stitching [17] and reconstruction steps for single cameras or pairs of stereo cameras. Nowadays this function is often available on smartphones. 360° cameras can be considered off-the-shelf products. Image stitching [4] is now a less important problem, because these devices can directly provide a 360° image from a small number (two, in the principal commercial devices) of sensors. In our work we give only a brief description of the geometry of a specific panoramic image projection and we assume that the output of the 360° camera is calibrated and with a negligible distortion.

2.2   PTZ Tracking

The introduction of PTZ cameras offered the capability of focusing on the interesting part of the scene moving the camera and zooming in, having a higher resolution with respect to static cameras. The main drawback is the control of the PTZ parameters and the camera calibration. The typical surveillance system or simple tracking scenario usually uses a master-slave (static-PTZ) couple of camera [7,16], or a network of devices [14].

In this paper we want to focus only on tracking approaches that use PTZ or 360° cameras and, therefore, can observe a wide part of the scene. In [6] tracking is performed using a histogram based on the HSV values of the frame as target feature and the mean shift method to search the target in the next frame. The result of tracking is then used, separately, to estimate when the target is going out of the center of the field of view and which are the correct camera PTZ shifts to keep it inside.

In [3] authors work on the estimation of the PTZ parameters tracking a moving planar target. The dimensions of the object are known and the focus of the paper is in recovering from errors in moving the camera due to incorrect projection and to camera motors. This is done using an extended Kalman filter and estimating the camera state and the object position in a recursive Bayesian filter network.

An alternative method, proposed in [10], is based on adaptive background modelling and the detection of moving objects. Once moving blobs are detected, tracking is performed using color distribution by histogram intersection.

Authors of [13] proposed a virtual framework for PTZ tracking algorithm testing. They collect a large number of views of a specific background and generate a panoramic image. The moving target (foreground) is then inserted in the image and the PTZ camera can virtually move in it. In this case, possible camera motions is limited (left, right, up, down, zoom in and out) and with fixed steps.

An experimental framework to evaluate different tracking algorithms is presented in [15], consisting mainly in a specific hardware setup and a known ground-truth. Authors uses the following configuration: a PTZ camera is placed in front of a screen where a projector displays the ground-truth video, so the whole system is calibrated to allow the comparison of the tracking result (i.e. the velocities of the motors and the part of the screen seen by the camera) and the real position of the target on the screen. The paper provides the setting and calibration parameters estimated using a Cam-Shift based tracking approach and a Particle Filter one.

A system that combine the use of a face detector and a tracker, based on a single PTZ camera, is shown in [5]. Authors avoided the use of a static camera dividing the work of the PTZ one in two modes: zoom-out mode, that shows the whole scene, and the zoom-in mode, that points on a single face detection using the known faces in the scene and the calculated trajectories. So the camera passes from the wide angle mode to the zoomed mode on a single person using a scheduling algorithm. The authors focused mainly on the face to face and face to person association, on the camera mode handling and scheduling and on the real-time implementation of the system.

3   360° Camera Model

There are two principal ways to show the output of a 360° camera in a single video: the two-sphere version and the equirectangular projection. Examples of the former and the latter are shown in Fig. 1. In this case we are considering a 2-sensor camera (a common configuration), but it is possible to use cameras with a higher number of sensors obtaining several overlapping spheres. The equirectangular version, on the other hand, is always a single image so we decided to use it for the tracking problem.

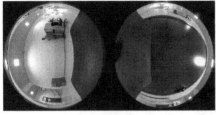

(a) two-sphere output   (b) equirectangular projection

Fig. 1. An example of output of a 360° device: (a) the two-sphere view, (b) the equirectangular projection.

3.1   Equirectangular Projection

Given a point $P = [x, y, z]$ in world coordinates, the equirectangular projection is defined by two angles: θ (polar angle), that is the angle defined by the projection of the P vector on the x-z plane and x axis; ϕ (azimuth angle) that is the one formed by the P vector and the x-z plane. Given P, the equirectangular coordinates θ and ϕ can be obtained using the following formulas:

$$\theta = acos(z/d)$$

$$\phi = atan(y/d)$$

where $d = sqrt(x^2 + z^2)$.

The equirectangular projection basically maps the θ and ϕ coordinates obtained by a spherical projection, as shown in Fig. 2(a), in a coordinate plane (the equirectangular projection) where θ varies from $-\pi$ to π and ϕ from $-\pi/2$ to $\pi/2$.

3.2   Equirectangular to Virtual Camera Plane Projection

The equirectangular projection is used to have in a single image a complete information of the scene. To simulate the PTZ camera it is necessary to retrieve the virtual output of the camera given the PTZ values. The part of the real world seen in the virtual camera plane according to the PTZ values, can be computed from a rectangular grid of points. The grid is projected in the equirectangular representation and the final camera plane image is computed interpolating the color value of the projected points.

One example of this transformation is shown in Fig. 3, where the virtual camera setting is: $P = -120°$, $T = 10°$ and $Z = 90°$. The projection of the virtual camera plane is the blue grid of Fig. 3(a), these values are then interpolated to obtain the planar surface in Fig. 3(b).

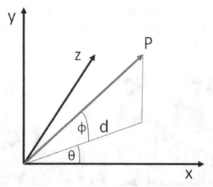

(a) World to spherical coordinates

(b) eqirectangular projection

Fig. 2. The world coordinates are transformed to spherical ones and then projected in the equirectangular version

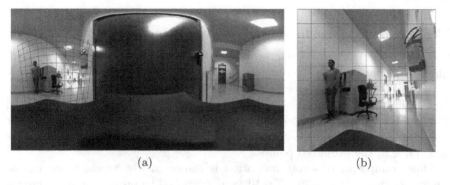

(a)   (b)

Fig. 3. The blue grid in (a) is the projection of the part of the scene seen by the virtual camera, (b) is the planar reconstruction. (Color figure online)

4   Tracking Model

The tracking model takes the equirectangular projection as input, but the actual tracking is performed in the virtual camera plane. The idea is to use the equirectangular image as a world representation and the virtual camera plane as if it was the output of a traditional PTZ camera.

Supposing that the virtual camera parameters Pan and Tilt coincide with the θ and ϕ of the equirectangular image and that Zoom coincides with the field of view of the projection, our tracking method estimates, for each frame, how to update the parameters to keep the target in the center of the virtual camera plane.

Our method initially computes the differences between the virtual camera plane with the initial values of PTZ set by hand or by any object detector and the views obtained shifting the virtual camera parameters, one at once, by a fixed value. These images, shown in Fig. 5 are assumed as representative of the moving object relatively to the virtual camera. In this case we move the camera to simulate the object motion, during tracking the object moves and the tracker has to adjust PTZ parameters to keep the object in the center of camera plane.

Our method is based on the assumption that the variation in virtual camera plane due to small object motion is approximately equivalent to the variation due to small virtual camera parameters shifts.

Figure 4 contains the original target image and the images obtained with virtual camera P parameter shifted by 0.5°, 1° and 2° in both directions. Figure 5 contains the difference images. It is possible to notice that small shifts in the virtual camera parameters leads to small values in the difference images (the brighter the value in the figure, the higher is the value of the difference) and vice versa. For this reason the difference associated to small values of parameter shift (0.5) are useful to describe slower movements and the ones associated to higher

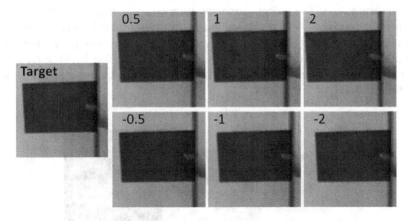

Fig. 4. The target and the virtual camera planes changing P value by 0.5, 1, 2, −0.5, −1 and −2.

values (2) are representative of faster movements. These difference images can then be considered as a sort of motion templates.

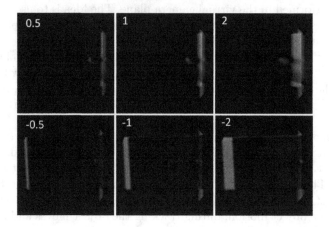

Fig. 5. Difference images changing P value by 0.5, 1, 2, −0.5, −1 and −2.

Our method tries to estimate the motion of the object from a weighted sum of basic motion templates. For each frame the motion is estimated selecting the most similar difference image calculated in the off-line step of the algorithm (Fig. 5), or selecting several difference images creating a mixture. If the target is not moving, no difference image is selected.

An example of this is present in Fig. 6: the left image is the target object, the central one is the new frame seen by the camera using the parameters at previous time instant and the right image is the difference. Looking at the latter, the position of the brighter parts (the horizontal and vertical segments in the figure) indicates that the object is moving upwards and slightly to the left.

It is easy to notice that a similar image can be obtained as the mixture of different difference images: a 0.5° in P and a 1°T, representing the motion in horizontal and vertical direction. The actual oblique motion is detected as the weighted mixture of the horizontal and vertical motion. A similar idea has been exploited in [9,11].

Fig. 6. Residual image obtained from the target and the current output of the camera

4.1   Tracking Algorithm

The proposed method implements this motion decomposition stating the problem as a least-square minimization one. We define:

- $[P_i, T_i, Z_i]$ as the camera parameters at time t_i;
- $I(P, T, Z, t)$ as the virtual camera plane with P,T,Z parameters and at time t;
- $s = [\Delta P, \Delta T, \Delta Z]$ as the parameter shifts needed to keep the target in the center of the virtual camera plane;
- $w = [w_P, w_T, w_Z]$ as the vector containing the difference steps (both positive and negative) used for P, T and Z parameter in computing the difference images;
- A as the matrix containing the difference images, represented as column vectors, repeated for different steps in the positive and negative direction.

The tracking algorithm can then be stated as follows:

1. Set $t_i = t_0$ (the first frame). Set the initial Pan, Tilt, Zoom parameters (P_0, T_0, Z_0) on the target obtaining $I_0 = I(P_0, T_0, Z_0, t_0)$ for the first frame;
2. Calculate A using the differences in w;
3. Calculate $I = I(P_i, T_i, Z_i, t_{i+1})$;
4. calculate $R = I - I_0$;
5. solve $x = (A^T A)^{-1} A^T R$;
6. $s = -xw^T$;
7. Set $[P_{i+1}, T_{i+1}, Z_{i+1}] = [Pi, Ti, Zi] + s$, $t_i = t_{i+1}$ and go to step 3.

5   Results

As discussed in Sect. 2, it is difficult to test the performances of 360° and PTZ tracking for different reasons: the lack of a common accepted benchmark for the former, the use of "real" scenes and actual camera movement for the latter. In particular for PTZ methods, the only way to compare different algorithms is testing them in real scenes and visually looking at the result, having a not repeatable experiment.

To test our method, we initially created a dataset of 59 360° videos using two cameras: a Ricoh Theta S (2 sensors with effective 12 MPixels [1], 26 videos) and a Samsung Gear 360° (2 sensors with effective 15 MPixels [2], 33 videos). The content of video is composed by outdoor (10) and indoor (49) scenes and the targets are in uniform or cluttered background.

The method shows good performance for well-shaped targets (i.e. object with corners) also in cluttered scenes, but the performances are worse when the target speed is too high (i.e. sudden changes of position).

We tested the method with our dataset using six difference images per parameter (three steps in positive and negative direction), so matrix A contains 6×3 difference images. A higher number of steps could lead to a more precise tracking, but with a higher computational complexity and the risk of numerical instability in matrix inversion.

Fig. 7. An example of two simulated PTZ cameras using a single equirectangular video

6   Conclusions

The large diffusion of the 360° technology will probably lead to a significant use of 360° videos in next years. This type of cameras could reduce the number of devices needed to cover a large area.

In this paper we showed how to simulate one or several PTZ cameras using a single 360° video and a tracking method based on this idea. The first task is important because 360° videos are very useful to compress in a single (equirectangular) video the informations of the whole 360° scene without significant loss of information. In fact the strong distortion, and consequent loss of information deriving from projection, is mainly in those part (top, bottom) that contains less important information (roof, sky, ground).

Moreover, simulating a PTZ or still cameras from a 360° video allows the use of well-known methods for tracking and video analytics for conventional video.

The contribution of the paper is two-fold: the realization of a simulated PTZ camera from a 360° video whit the explication of the geometric transformations needed to implement it and a simple object tracking method for 360° videos. The first one can be used as a starting point to apply known methods of the state of the art of PTZ or still cameras on 360° videos, the second one can be used to implement a low-complexity tracking method that can run in real time and on low power devices.

The entire proposed framework, in addition, uses a single equirectangular video as input and allows to have a dedicated virtual PTZ camera for each target to track. An example is shown in Fig. 7: on the left there is the equirectangular video, on the right the two simulated PTZ cameras tracking the book and the face simultaneously.

As future work, it is possible to fuse sophisticated tracking methods (i.e. using more informations of motion, keypoint features, recognition and so on) on the virtual camera plane.

This paper is (partially) funded on a research grant by the Italian Ministry of University and Research, namely project NEPTIS (Grant no. PON03PE 002143).

References

1. Ricoh theta s. https://theta360.com/en/about/theta/s.html
2. Samsung gear 360 specifications. http://www.samsung.com/global/galaxy/gear-360/#!/spec
3. Al Haj, M., Bagdanov, A.D., Gonzalez, J., Roca, F.X.: Reactive object tracking with a single PTZ camera. In: 2010 20th International Conference on Pattern Recognition (ICPR), pp. 1690–1693. IEEE (2010)
4. Brown, M., Lowe, D.G.: Automatic panoramic image stitching using invariant features. Int. J. Comput. Vis. **74**(1), 59–73 (2007)
5. Cai, Y., Medioni, G.: Demo: persistent people tracking and face capture using a PTZ camera. In: 2013 7th International Conference on Distributed Smart Cameras (ICDSC), pp. 1–3. IEEE (2013)
6. Chang, F., Zhang, G., Wang, X., Chen, Z.: PTZ camera target tracking in large complex scenes. In: 2010 8th World Congress on Intelligent Control and Automation (WCICA), pp. 2914–2918. IEEE (2010)
7. Funahasahi, T., Tominaga, M., Fujiwara, T., Koshimizu, H.: Hierarchical face tracking by using PTZ camera. In: 2004 Proceedings of 6th IEEE International Conference on Automatic Face and Gesture Recognition, pp. 427–432. IEEE (2004)
8. Gledhill, D., Tian, G.Y., Taylor, D., Clarke, D.: Panoramic imaging—a review. Comput. Graph. **27**(3), 435–445 (2003)
9. Gleicher, M.: Projective registration with difference decomposition. In: 1997 Proceedings of IEEE Computer Society Conference on Computer Vision and Pattern Recognition, pp. 331–337. IEEE (1997)
10. Kang, S., Paik, J.K., Koschan, A., Abidi, B.R., Abidi, M.A.: Real-time video tracking using PTZ cameras. In: Quality Control by Artificial Vision, pp. 103–111. International Society for Optics and Photonics (2003)
11. La Cascia, M., Sclaroff, S., Athitsos, V.: Fast, reliable head tracking under varying illumination: an approach based on registration of texture-mapped 3D models. IEEE Trans. Pattern Anal. Mach. Intell. **22**(4), 322–336 (2000)
12. Li, X., Hu, W., Shen, C., Zhang, Z., Dick, A., Hengel, A.V.D.: A survey of appearance models in visual object tracking. ACM Trans. Intell. Syst. Technol. (TIST) **4**(4), 58 (2013)
13. Liu, C., Cao, R., Jia, S., Zhang, Y., Wang, B., Zhao, Q.: The PTZ tracking algorithms evaluation virtual platform system. In: 2014 International Conference on Multisensor Fusion and Information Integration for Intelligent Systems (MFI), pp. 1–6. IEEE (2014)
14. Micheloni, C., Rinner, B., Foresti, G.L.: Video analysis in pan-tilt-zoom camera networks. IEEE Sig. Process. Mag. **27**(5), 78–90 (2010)
15. Salvagnini, P., Cristani, M., Del Bue, A., Murino, V.: An experimental framework for evaluating PTZ tracking algorithms. In: Crowley, J.L., Draper, B.A., Thonnat, M. (eds.) ICVS 2011. LNCS, vol. 6962, pp. 81–90. Springer, Heidelberg (2011). doi:10.1007/978-3-642-23968-7_9
16. Stillman, S.T., Tanawongsuwan, R., Essa, I.A.: A system for tracking and recognizing multiple people with multiple camera. Technical report, Georgia Institute of Technology (1998)

17. Szeliski, R., Shum, H.Y.: Creating full view panoramic image mosaics and environment maps. In: Proceedings of 24th Annual Conference on Computer Graphics and Interactive Techniques, pp. 251–258. ACM Press/Addison-Wesley Publishing Co. (1997)

18. Yilmaz, A., Javed, O., Shah, M.: Object tracking: a survey. ACM Comput. Surv. (CSUR) **38**(4), 13 (2006)

Benchmarking Two Algorithms for People Detection from Top-View Depth Cameras

Vincenzo Carletti, Luca Del Pizzo, Gennaro Percannella[✉], and Mario Vento

Department of Computer and Electrical Engineering and Applied Mathematics,
University of Salerno, Via Giovanni Paolo II, 132, 84084 Fisciano (SA), Italy
{vcarletti,ldelpizzo,pergen,mvento}@unisa.it

Abstract. Automatic people detection from videos is an important task
in many computer vision applications either for security and safety motivations or for business intelligence purposes. In order to achieve high person detection accuracy many authors propose the adoption of a depth
sensor mounted in a top-view position in order to mitigate the effects
of occlusions and illumination conditions on the performance. Unfortunately, most approaches presented so far in the scientific literature have
been tested on very small datasets which do not account for the typical
situations arising in real scenarios and consequently do not allow interested readers to figure out which method has to be used in the specific
scenario at hand. In this paper we benchmark two different approaches
available in the literature for people detection from a zenithal mounted
depth camera; the former is an unsupervised method aimed at finding
the head of persons defined as the local minimum regions in the depth
map, while the latter is based on the combination of the histograms of
oriented gradient description and the support vector machine classifier.
The benchmarking is performed on a public dataset of images captured
in two different lighting conditions and with varying number of persons;
this allows to assess the performance of the considered approaches under
different real world scenarios. A detailed analysis of the two methods is
reported in the experimental section of the paper allowing the reader
to comprehend the pros and cons of each approach on the considered
scenes.

1 Introduction

Computer vision algorithms for the automatic detection of the presence of persons within a scene captured by a camera represent the enabling technology
for several important real world applications. Some noticeable examples where
person detection is required as a preliminary step are: counting the number of
persons passing through a virtual line, determining the statistics regarding the
permanence times of persons in specific areas (as in front of shop windows),
detecting overcrowding conditions, etc. [2,6,9,14].

Unfortunately, accurate person detection is seriously hampered by a series
of problems that arise in real contexts. Among the most important issues we
highlight the occlusions, i.e. the situation when a subject is not detected since

© Springer International Publishing AG 2017
S. Battiato et al. (Eds.): ICIAP 2017, Part I, LNCS 10484, pp. 73–83, 2017.
https://doi.org/10.1007/978-3-319-68560-1_7

he/she is partly or completely obscured by the scene elements (typically other subjects) which are interposed between the camera and the subject. It is intuitive to understand that the probability of occurrence of this phenomenon is directly related to the density of people in the area. In order to mitigate such phenomenon several authors propose the installation of the camera in a zenithal position which allows to eliminate the occlusions in the area immediately below the camera, with only a gradual increase as far as the persons move apart from the optical axis of the device. A further phenomenon that typically has a significant impact on the reliability of the person detection methods in real contexts is represented by the variability of the lighting conditions of the scene due to light switching in indoor environments or to the slow variation of the solar illumination along the day in outdoor environments [15], so as the presence of shadows [12] and specular reflections [1]. In order to cope with such issues, in the recent years several authors have proposed to use the depth map image provided by the Microsoft Kinect device as it proves to be partially immune to the problems due to lighting. Furthermore, the availability of the depth information may ease the detection of the persons starting from the observation that in typical real world applications the head is the element of the person that is closest to the camera, if the latter is installed in a zenithal position. It is interesting to note that the adoption of top-view depth cameras has also an additional positive effect. In fact, it allows to easily overcome the stringent regulations on the privacy of the people applied in the vast majority of countries, making this preferable to other solutions based on the use of traditional optical cameras and mounted in such positions as to acquire the faces of persons.

All the above observations motivated several research groups in the recent years to propose solutions to the problem based on the use of top-view depth cameras. The recent literature on this topic can be divided in two main streams: on one side, there are papers proposing unsupervised approaches which find the persons by looking for their head, being the body part closest to the camera. Specifically, in [17], Zhang et al. propose an unsupervised approach to locate persons in the scene; the method simulates water filling with the aim of finding the local minimum regions in the input depth map, which should correspond to heads of people. Similarly, in [8], Galcík et al. propose a method that locates people in the scene by detecting maximal in the depth images followed by region growing. The found regions are considered heads if they satisfy criteria related to size, roundness, and there is evidence of being above shoulder-like structures. Lin and Jhuang in [10] assume that the shapes of the pedestrians in the scene are similar to ellipses and the area of projection of the upper portion of a person is normally smaller compared to the lower one, so they compare and stack the areas of every layer from low to high estimating the top portion of each object (head and shoulders). Nalepa et al. [11] determine local minima of pixel depth values and use a modified flood fill algorithm to append neighboring pixels to the found minima. Then, the method groups blobs representing various parts of a human body into a single connected component.

A second group of methods formulate the person localization in the scene as a detection problem. Rauter in [13] introduces simplified local ternary patterns, a new feature descriptor which is used for human tracking based on the head and shoulder part of the human body, then a support vector machine (SVM) for the classification stage. Also Vera et al. in [16] propose to use an SVM classifier to detect people, although in this case the description is based on the histograms of oriented gradients. In Zhu and Wong [18], the 3D map of a person is described by a data structure called the head and shoulder profile (HASP) based on Haar wavelet features. The classifier is based on Adaboost algorithm [7]. Unfortunately, most methods have been tested on private datasets, or in few cases on very small public datasets that do not allow to thoroughly assess performance of the approaches in realistic conditions and to have a detailed insight of the pros and cons of each method.

Contribution. This paper intends to face the latter issue by proposing the benchmarking of two alternative person detection methods selected from the recent literature which use depth based vision systems mounted in a top-view position. The methods have been selected to be representative of the two categories of approaches described before, i.e. unsupervised and supervised ones. To the best of our knowledge there is no paper providing a similar contribution in the literature. The benchmarking is carried out on a common and large dataset, publicly available, with the aim of providing the reader with a detailed view of the performance of each method subjected to the two main sources of errors, namely the lighting conditions and the people density.

The paper is organized as it follows: in Sect. 2 we provide basic information regarding the methods which have been considered for the benchmarking reported in this paper; then, in Sects. 3 and 4 we describe, respectively, the dataset adopted for the experimentations and the results achieved by the two methods focusing on their behaviors in two different lighting scenarios and under varying persons densities. Finally, in Sect. 5 we draw conclusions and delineate future directions of our research.

2   Methods Considered for the Benchmarking

In this Section we briefly describe the two methods [16,17] which have been considered for the benchmarking in this paper; for the details the interested reader may examine the original papers. Hereinfter, we will refer to the method proposed by Zhang et al. in [17] with the name WATERFILLING, and with the name HOG-SVM to the method by Vera et al. in [16]. The methods here considered for the benchmarking have been selected as representative of two complementary approaches: the WATERFILLING is an unsupervised method devised to locate the heads of the persons by searching for the local minimum regions within the depth map, while the HOG-SVM adopts a supervised approach based on support vector machine classifier.

2.1 WATERFILLING

The method moves from the idea that the head is the part of the human body that is closest to the top-view sensor, so the authors formulate person detection as the problem of searching the local minimum regions in the depth image; such regions should correspond to the head of the persons. Formally, the localization of a head into the depth image is done by finding a region A and its neighborhood N satisfying the following constraint:

$$E_A(f(x,y)) + \eta \leq E_{N \setminus A}(f(x,y)), A \in N \tag{1}$$

The operator $E(\cdot)$ allows to pool the depth information in the region to a real value that reflects the total depth information in the region. η is a predefined threshold to ensure that depth in A should be lower than $N \setminus A$ with a margin. The idea is that A and N represent the head and the shoulder, respectively. In order to find the local regions A, the authors employ a methodology based on the water filling process, which, starting from a representation of the depth map as a land with humps and hollows, simulates the falling of the raindrop over it. After the water fall simulation the hollow regions will gather the raindrops. The hollow regions sufficiently large and deep are considered as heads. For our test, the authors of the WATERFILLING method provided us the original code implemented in C++ and based on the OpenCV library.

2.2 HOG-SVM

The HOG-SVM is based on the method initially proposed by Dalal and Triggs in [3] for pedestrian detection and adapted by Vera et al. in [16] for people detection from top-view depth cameras. The HOG-SVM method describes a candidate in terms of the histograms of oriented gradients (HOG). The analysis is performed on patches of the image of fixed size (96×96 pixels); each patch is divided into blocks, which are divided in 2×2 cells each of 8×8 pixels. The blocks are partially overlapped; the amount of overlap corresponds to the size of the cell. The features are extracted by computing the gradient over the cells. The orientation of the gradient is clustered into nine-bin histograms. The frequency is weighted using the magnitude of the gradient blocks. At the end, a person is described by a feature vector of size 1089. Then classification is done using a support vector machine. The HOG-SVM person detector uses a sliding window which is moved around the image over a dense grid. At each position the HOG description is derived from the 96×96 pixels patch and used by the SVM to classify the patch as either person or not a person. In order to detect persons at different scales, the image is subsampled to multiple sizes and each of these subsampled images is searched for people. We provided our own implementation of the HOG-SVM method. Also in this case, the method has been implemented in C++ using the OpenCV library.

3 The Adopted Dataset

The experimental validation of the method has been carried out using the dataset presented in [4] and successively adopted in [5]. The dataset has been acquired by using two image sensors, namely a traditional RGB camera and the depth sensor of a Kinect device. Both acquisition devices are mounted in a zenithal position; video sequences were captured at 30 fps with a resolution of 640×480 pixels. Since in this paper we are interested only to the images provided by the depth sensor, the RGB images were not considered. The dataset includes scenes captured with either the prevalence of the solar illumination (OUTDOOR) or the artificial light (INDOOR). The dataset comprises sequences with a variable number of persons flowing within the area of interest in the same direction and/or in opposite directions. In particular, in the simplest case, there is a single person in the area framed by the camera, while in the most complex cases there are up to four persons moving within the area and proceeding either in the same direction, as in a queue, or walking in two opposite directions. As a consequence, the adoption of this dataset for our tests allows to characterize the accuracy of the analyzed methods under different illumination and crowding conditions. Example images from the INDOOR and the OUTDOOR environments are shown in Fig. 1, while in Table 1 we report the number of frames in the dataset containing the number of persons as specified in the leftmost columns.

Fig. 1. Examples of depth images acquired in the INDOOR (left image) and in the OUTDOOR (right image) scenarios. The OUTDOOR case is characterized by high noise due to the sunlight illumination and appearing in the form of numerous black spots.

The test dataset was originally devised to allow the test of the methods for counting people crossing a virtual line. Thus, in order to allow the benchmarking of the methods proposed in this paper, we augmented the ground truth of the dataset by providing information regarding the position of each person in each frame. In particular, for each person we added a smaller box containing the head of the person, and a second box including also the shoulders, as shown in Fig. 2.

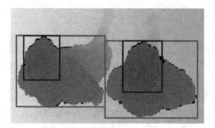

Fig. 2. Ground truth used for the people detection: the orange and red solid lines shows the head-shoulders ground truth used for the HOG-SVM method, the blue solid line shows the head ground truth used for the WATERFILLING method. (Color figure online)

Table 1. Dataset information: number of frames with 0 to 4 persons, under different illumination conditions (INDOOR/OUTDOOR).

Persons in the frame	INDOOR	OUTDOOR	TOTAL
0	23, 626	26, 010	49, 636
1	7, 788	8, 753	16, 541
2	1, 652	2, 501	4, 153
3	438	427	865
4	95	108	203

4   Experimental Analysis

In this section we report and analyze the results achieved by the two considered people detection approaches on the adopted dataset. Specifically, we first describe the performance indices used for comparing the methods, then we provide information regarding the configuration parameters of the methods, and finally, we report the performance and comment the pros and cons of both approaches.

4.1   Performance Indices

The figure of merit adopted for measuring the detection performance of the considered approaches is the *f-index* defined as the armonic mean of *Precision* and *Recall*. Following [16], we declare a person as correctly detected by a method if the following condition stands:

$$\frac{area(B_d \cap B_g)}{area(B_d \cup B_g)} > 0.5 \tag{2}$$

where B_d and B_g are the bounding boxes generated by the method and of the ground truth, respectively. It has to be noted that the outputs of the two methods considered for the benchmarking are not exactly the same. As a matter of fact the WATERFILLING method only provides the location of the head of the person, while the HOG-SVM method provides the head and shoulder area. Consequently in the evaluation of the performance the condition in Eq. (2) was checked using for each method the proper ground truth (head bounding box for WATERFILLING and shoulder bounding box for HOG-SVM). We also highlight that for our evaluation we did not consider the persons in the dataset with head and shoulder bounding box not completely contained into the capture area of the camera; coherently we did not care about the object detected by the methods laying across the borders of the frame.

4.2   Training Procedure

Both people detection methods required a training phase aimed at setting the optimal parameters to be used during the tests. To this aim, we extracted a total of 102 frames from the dataset described in the previous section. The frames, containing at least a person, were randomly selected within the whole dataset, preserving the original distribution of the number of persons present in each frame and equally distributed between the two scenarios. The frames used for the training stage were not used during the tests. Furthermore, the training dataset was also augmented using rotated and flipped version of the images; this was particular important for achieving higher generalization of the SVM stage of the HOG-SVM method from the given set of samples extracted from the original dataset. During the training phase we noticed that while the HOG-SVM method is able to cope with the high difference in the signal to noise ratio that characterizes the video sequences captured in the INDOOR and the OUTDOOR scenarios (see Fig. 1), in the case of the WATERFILLING approach we found that the optimal values of the parameters greatly change between the two scenarios. Consequently we used two different parameterizations for the latter method for the INDOOR or the OUTDOOR cases.

4.3   Analysis of the Experimental Results

Table 2 reports the overall performance achieved by the HOG-SVM and the WATERFILLING methods over the considered dataset expressed in terms of the indices defined before. We immediately notice the large difference between the two methods. The HOG-SVM largely outperforms the WATERFILLING approach with respect to all the three indices with a 28.9% relative improvement of the *f-index*.

Table 2. Overall performance of the HOG-SVM and the WATERFILLING methods over the considered dataset.

Method	Precision	Recall	f-index
HOG-SVM	0.978	0.992	0.985
WATERFILLING	0.687	0.859	0.764

Table 3. Performance of the HOG-SVM and the WATERFILLING methods over the considered dataset in the INDOOR e OUTDOOR cases.

Method	Scenario	Precision	Recall	f-index
HOG-SVM	INDOOR	0.969	1.000	0.984
	OUTDOOR	0.988	0.983	0.986
WATERFILLING	INDOOR	0.943	0.897	0.919
	OUTDOOR	0.561	0.830	0.670

In Table 3 we analyze performance of the two approaches with respect to the scenario, reporting the values of the indices separately for the INDOOR and the OUTDOOR cases. Focusing on the HOG-SVM method, we notice that its performance does not depend on the scenario; in fact, the variation of the f-$index$ between the two cases remains practically unchanged (0.984 vs 0.986), thus demonstrating to be highly robust to the image noise. Conversely, the WATERFILLING shows a very different behavior, being strongly affected by the noise, especially the one that characterizes the OUTDOOR scenario (see Fig. 1). This is demonstrated by the very high difference of the f-$index$ achieved in the INDOOR and in the OUTDOOR cases, 0.919 vs 0.670, respectively. Results in Table 3 shows that the strongest limitation of the WATERFILLING in the OUTDOOR scenario is the high incidence of false alarms and, to a lesser extent, the incidence of false negatives. The high alarm rate is motivated by the fact that the high noise level into the background often causes the fragmentation of the person's head in several connected components which generate spurious detections.

In Tables 4 and 5 we report the performance of the methods for the two scenarios and for the different number of persons simultaneously present into the scene. We notice that for both methods the number of persons in the scene does not have a significant influence over the performance. Specifically, in the case of HOG-SVM the value of the f-$index$ is bound to a narrow range, from

Table 4. Performance of the HOG-SVM method under different flow densities.

Scenario	Persons	Precision	Recall	f-index
INDOOR	1	0.975	1.000	0.987
	2	0.964	0.999	0.982
	3	0.946	0.999	0.972
	4	0.965	1.000	0.982
OUTDOOR	1	0.994	0.997	0.996
	2	0.987	0.974	0.980
	3	0.993	0.988	0.991
	4	0.988	0.968	0.978

Table 5. Performance of the WATERFILLING method under different flow densities.

Scenario	Persons	Precision	Recall	f-index
INDOOR	1	0.965	0.912	0.938
	2	0.901	0.863	0.882
	3	0.920	0.887	0.903
	4	0.944	0.904	0.923
OUTDOOR	1	0.551	0.804	0.654
	2	0.579	0.863	0.693
	3	0.551	0.870	0.675
	4	0.580	0.883	0.701

false negative examples		false positive examples	
HOG-SVM	WATERFILLING	HOG-SVM	WATERFILLING

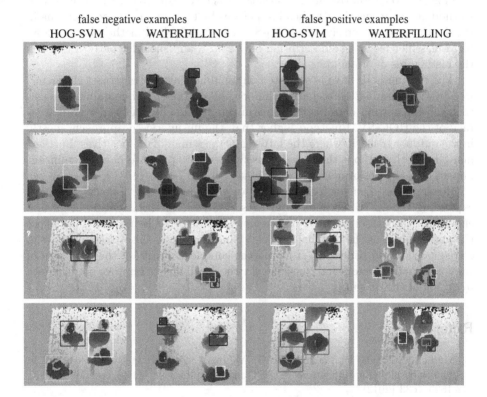

Fig. 3. The first two rows show the output in the INDOOR scenario, the last two rows show the output in the OUTDOOR scenario. In the first and third columns there are false negative and false positive events from HOG-SVM method, while in the second and fourth columns there are false negative and false positive events from WATERFILLING method.

0.972 to 0.996. Furthermore, the best values are obtained in case of a single person in the scene. This is motivated by the fact that in few cases when there are persons close to each other the method provides false detections in the region separating the two persons (see Fig. 3 for some examples of this situation). In the case of the WATERFILLING we notice that the values of the f-*index* varies in two relatively short ranges for the INDOOR scenario (from 0.882 to 0.983) and the OUTDOOR scenario (from 0.654 to 0.701) highlighting that the illumination source has an higher impact than the crowding level on the performance of this method.

5 Conclusions and Future Work

In this paper, we studied two methods available in the scientific literature for people detection from top-view depth cameras. The methods under consideration follow two alternative approaches: the WATERFILLING is an unsupervised method aimed at locating the head of persons by looking for the local minima in the depth map; conversely, the HOG-SVM is a supervised method based on an SVM classifier fed by the description of the head and shoulder pattern through the histograms of oriented gradients.

The two methods have been tested on a publicly available dataset characterized by two illumination scenarios (indoor and outdoor) and containing images with varying persons density. The experimental results highlight an overall accuracy of the HOG-SVM method higher than the unsupervised approach, mostly in the outdoor scenario where the latter generates many false positives. Furthermore, for both methods the crowd density does not appear to have a significant impact over the performance.

In our future benchmarking effort, we will consider the following aspects: expanding the set of methods from those available in the scientific literature, enlarging the dataset in order to account also for other issues that may affect the performance as the installation height and the depth sensor technology (e.g. stereo camera and Kinect 2), studying the complementarity of the responses of the considered detectors and consequently the possibility to improve performance by fusion of the outputs.

References

1. Conte, D., Foggia, P., Percannella, G., Vento, M.: Removing object reflections in videos by global optimization. IEEE Trans. Circuits Syst. Video Technol. **22**(11), 1623–1633 (2012)
2. Conte, D., Foggia, P., Percannella, G., Vento, M.: Counting moving persons in crowded scenes. Mach. Vis. Appl. **24**(5), 1029–1042 (2013)
3. Dalal, N., Triggs, B.: Histograms of oriented gradients for human detection. In: 2005 IEEE Computer Society Conference on Computer Vision and Pattern Recognition (CVPR 2005), vol. 1, pp. 886–893. IEEE (2005)
4. Del Pizzo, L., Foggia, P., Greco, A., Percannella, G., Vento, M.: A versatile and effective method for counting people on either RGB or depth overhead cameras. In: 2015 IEEE International Conference on Multimedia and Expo Workshops, ICMEW 2015 (2015)

5. Del Pizzo, L., Foggia, P., Greco, A., Percannella, G., Vento, M.: Counting people by RGB or depth overhead cameras. Pattern Recogn. Lett. **81**, 41–50 (2016)
6. Erickson, V.L., Lin, Y., Kamthe, A., Brahme, R., Surana, A., Cerpa, A.E., Sohn, M.D., Narayanan, S.: Energy efficient building environment control strategies using real-time occupancy measurements. In: Proceedings of 1st ACM Workshop on Embedded Sensing Systems for Energy-Efficiency in Buildings, BuildSys 2009, pp. 19–24. ACM, New York (2009)
7. Freund, Y., Schapire, R.E.: A desicion-theoretic generalization of on-line learning and an application to boosting. In: Vitányi, P. (ed.) EuroCOLT 1995. LNCS, vol. 904, pp. 23–37. Springer, Heidelberg (1995). doi:10.1007/3-540-59119-2_166
8. Galčík, F., Gargalík, R.: Real-time depth map based people counting. In: Blanc-Talon, J., Kasinski, A., Philips, W., Popescu, D., Scheunders, P. (eds.) ACIVS 2013. LNCS, vol. 8192, pp. 330–341. Springer, Cham (2013). doi:10.1007/978-3-319-02895-8_30
9. Karpagavalli, P., Ramprasad, A.: Estimating the density of the people and counting the number of people in a crowd environment for human safety. pp. 663–667 (2013)
10. Lin, D.-T., Jhuang, D.-H.: A novel layer-scanning method for improving real-time people counting. In: Stephanidis, C. (ed.) HCI 2013. CCIS, vol. 374, pp. 661–665. Springer, Heidelberg (2013). doi:10.1007/978-3-642-39476-8_133
11. Nalepa, J., Szymanek, J., Kawulok, M.: Real-time people counting from depth images. In: Kozielski, S., Mrozek, D., Kasprowski, P., Małysiak-Mrozek, B., Kostrzewa, D. (eds.) BDAS 2015. CCIS, vol. 521, pp. 387–397. Springer, Cham (2015). doi:10.1007/978-3-319-18422-7_34
12. Prati, A., Mikic, I., Trivedi, M.M., Cucchiara, R.: Detecting moving shadows: algorithms and evaluation. IEEE Trans. Pattern Anal. Mach. Intell. **25**(7), 918–923 (2003)
13. Rauter, M.: Reliable human detection and tracking in top-view depth images. In: Proceedings of IEEE Conference on Computer Vision and Pattern Recognition Workshops, pp. 529–534 (2013)
14. Saleh, S.A.M., Suandi, S.A., Ibrahim, H.: Recent survey on crowd density estimation and counting for visual surveillance. Eng. Appl. Artif. Intell. **41**, 103–114 (2015)
15. Toyama, K., Krumm, J., Brumitt, B., Meyers, B.: Wallflower: principles and practice of background maintenance. In: Proceedings of 7th IEEE International Conference on Computer Vision, vol. 1, pp. 255–261 (1999)
16. Vera, P., Zenteno, D., Salas, J.: Counting pedestrians in bidirectional scenarios using zenithal depth images. In: Carrasco-Ochoa, J.A., Martínez-Trinidad, J.F., Rodríguez, J.S., di Baja, G.S. (eds.) MCPR 2013. LNCS, vol. 7914, pp. 84–93. Springer, Heidelberg (2013). doi:10.1007/978-3-642-38989-4_9
17. Zhang, X., Yan, J., Feng, S., Lei, Z., Yi, D., Li, S.Z.: Water filling: unsupervised people counting via vertical KINECT sensor. In: 2012 IEEE Ninth International Conference on Advanced Video and Signal-Based Surveillance (AVSS), pp. 215–220. IEEE (2012)
18. Zhu, L., Wong, K.-H.: Human tracking and counting using the KINECT range sensor based on Adaboost and Kalman filter. In: Bebis, G., et al. (eds.) ISVC 2013. LNCS, vol. 8034, pp. 582–591. Springer, Heidelberg (2013). doi:10.1007/978-3-642-41939-3_57

Gesture Modelling and Recognition by Integrating Declarative Models and Pattern Recognition Algorithms

Alessandro Carcangiu[1]([✉]), Lucio Davide Spano[2], Giorgio Fumera[1],
and Fabio Roli[1]

[1] Department of Electrical and Electronic Engineering,
University of Cagliari, 09123 Cagliari, Italy
{alessandro.carcangiu,fumera,roli}@diee.unica.it
[2] Department of Mathematics and Computer Science,
University of Cagliari, 09124 Cagliari, Italy
davide.spano@unica.it
http://pralab.diee.unica.it, http://people.unica.it/davidespano/

Abstract. Gesture recognition approaches based on computer vision and machine learning mainly focus on recognition accuracy and robustness. Research on user interface development focuses instead on the orthogonal problem of providing guidance for performing and discovering interactive gestures, through compositional approaches that provide information on gesture sub-parts. We make a first step toward combining the advantages of both approaches. We introduce DEICTIC, a compositional and declarative gesture description model which uses basic Hidden Markov Models (HMMs) to recognize meaningful pre-defined primitives (gesture sub-parts), and uses a composition of basic HMMs to recognize complex gestures. Preliminary empirical results show that DEICTIC exhibits a similar recognition performance as "monolithic" HMMs used in state-of-the-art vision-based approaches, retaining at the same time the advantages of declarative approaches.

Keywords: Gesture recognition · Hidden Markov Models · Compositional · Declarative

1 Introduction

Gesture recognition is a long-standing research topic in the computer vision field, with many applications to Human-Computer Interaction (HCI) [16]. Vision-based approaches to gesture recognition can be categorized into appearance- and 3D model- (or tracking-) based [2,17]. In particular, the recognition of dynamic gestures (as opposed to static ones, that do not include a temporal dimension) has been addressed using techniques that explicitly consider the temporal dimension, like Hidden Markov Models (HMM), Dynamic Time Warping (DTW), Time-Delay Neural Networks (TDNN) and Finite-State Machines

© Springer International Publishing AG 2017
S. Battiato et al. (Eds.): ICIAP 2017, Part I, LNCS 10484, pp. 84–95, 2017.
https://doi.org/10.1007/978-3-319-68560-1_8

(FTM) [2,12,17], as well as traditional supervised classification algorithms like support vector machines (although they are more suited to static gestures [12]).

Vision-based gesture recognition poses a number of challenges, like coping with a large variety of gestures, and achieving invariance to lighting conditions, viewpoint changes, cluttered background and gesture speed; usually, a trade-off between accuracy, performance and usefulness has to be found, based on criteria like real-time processing capability and scalability [2,17]. Beside the above issues, under the viewpoint of user interface (UI) development it is very important to address the orthogonal problem of *usability*, which is related to the *meaning* of interactive gestures for users [15]: indeed, not all gestures that can be recognized by a machine have a meaning for the human counterpart. In particular, contrary to WIMP (Windows, Icons, Menus, Pointer device) interfaces, gestures are rarely self-revealing, and thus a guidance system for discovering what commands are available and how to trigger them can definitely improve their usability [3]. This implies that the underlying recognition system should be able to provide (through a graphical interface) *feedback* and *feedforward* information [22], i.e., information on which portion of a gesture has been completed, and on its potential completion, which may be more than one.

The two goals of an accurate/effective recognition and a usable gestural interface can be conflicting. Vision-based approaches usually provide a class label to a whole gesture pattern recognized in an input sequence, which is viewed as an atomic event even when the time dimension is internally taken into account (e.g., using HMMs). However, from the user point of view the performance of a gesture cannot be reduced to a single event, since it spans over a perceivable amount of time. On the other hand, compositional and/or declarative approaches have been proposed for modelling gestures, which explicitly take into account the subdivision of a gesture into meaningful sub-parts; however, to recognize sub-parts they rely on heuristic techniques that exhibit a lower effectiveness and robustness with respect to "monolithic" vision-based approaches. We survey the relevant literature on both approaches in Sect. 2.

In this paper we make a first step towards filling the gap between vision-based and compositional/declarative approaches. We start from the declarative and compositional gesture description model GestIT [20,21] (see Sect. 2.2), that solves the intermediate feedback problem and provides a superset of composition operators described in other approaches. We integrate GestIT with HMMs, using HMMs to recognize basic gesture segments ("primitives") instead of whole gestures (Sect. 3). We show that the resulting method, called DEICTIC (DEclarative and CompoSiTional Input Classifier), is capable of recognizing complex gestures made up of several primitives, rigorously defined according to GestIT operator semantics. Preliminary empirical results (Sect. 4) provide evidence that DEICTIC exhibits a recognition performance comparable to that of standard HMM classifiers, while retaining the advantages of declarative approaches.

2 Related Work

In this section we overview first the vision-based approaches most relevant to our work, i.e., the ones which subdivide gestures into sub-parts, and then the main approaches based on declarative models, including GestIT.

2.1 Vision-Based Gesture Recognition Approaches

Vision-based methods that identify a set of sub-parts (or primitives) common to different gestures have already been proposed, either for increasing the recognition rate or to reduce the training set size in learning-based approaches. Primitives can be broadly defined as a set of distinguishable patterns from which either a whole movement or a part of it can be reconstructed. Different, specific definitions of "primitive" have been considered in the literature: they may represent basic movements (e.g., raising a leg, moving an arm to the left), static poses, or characteristic patterns of low-level signals like the Fast Fourier Transform. In the following we give representative examples for each interpretation of the primitive concept.

In [23] primitives are identified using a bottom-up clustering approach aimed at reducing the training set size and at improving the organisation of unlabeled datasets for speeding up its processing. Gestures are then labelled with sequences of primitives, which is close to a representation useful also for building UIs. However, since primitives are identified automatically, they are difficult to understand for designers while creating feedback and feed-forward systems. In [1] primitives are defined in a context-grammar established in advance using a top down approach, which is more suitable to UI designers; however, grammars were not created taking into account the gesture meaning from the user perspective.

In [13] primitives are used together a three-level HMM classifier architecture for recognizing (i) the primitives, (ii) their composition and (iii) the pose or gesture. However, also in this case unsupervised learning was used for defining both primitives and their composition, which is not suitable for building UIs.

A set of primitives that better suits the understanding by designers includes 3D properties of the movement trajectory. For instance, in [14] primitives identified in a 2D video are used for classifying 3D movements. Here the primitives are functions on the 2D features that represent the user's state. A representation more linked to geometric features on the 3D space for identifying primitives was proposed in [5]; however, it requires the understanding of the underlying mathematical representation, which is not feasible for UI designers.

To our knowledge, the vision-based approach most similar to ours is the one of [9]. It decomposes gestures into application-specific "primitive strokes", and uses a distinct HMM for modelling each stroke; each gesture is then modelled by a composite HMM obtained by concatenating the corresponding stroke models. This technique is valid for describing stroke sequences. Our approach is able to define more complex composite gestures, including alternative (choice) and parallel definitions. In addition, we do not use a re-training step for avoiding a degradation in recognition performance.

2.2   Gesture Recognition: The Declarative Modelling Approach

Declarative approaches allow splitting a gesture into several sub-components. There are different compositional approaches based on heuristic gesture recognition. For instance, Kammer et al. [7] introduced GeForMT, a language for formalizing multitouch gesture for filling the gap between the high-level complex gestures and the low-level device events. GeForMT uses an Extended Backus-Naur form grammar, with five basic movements (move, point, hold, line, circle and semicircle), which are composed through parallel and sequence operators.

A rule-based approach for multitouch gestures has been introduced in Midas [18]. The rules work on different features, for example the 2D positions, the speed and the finger tracking state and consists of two components: a prerequisite part and an action part. The first defines the input fact pattern to be recognized while the second the UI behaviour. Mudra [6] is a follow-up research from the same group extending Midas for multimodal interfaces. It unifies the input stream coming from different devices, exploiting different modalities. Designers define both the low-level handling events and the high-levels rules, combining them into a single software architecture. Khandkar et al. proposed GDL [8] (Gesture Description Language), a domain-specific language designed to streamline the process of defining gestures. GDL separates the gesture recognition code from the definition of UI behaviour. This work defines three components: the gesture name, the code for the gesture validation and a return type. The last component represents the data notified with a callback to the application logic.

More structured and expressive declarative methods are Proton++ [11] and GestIT [20,21]. Proton++ separates the temporal sequencing of the event from the code which describes the UI behaviour. It also allows developers to declaratively describe custom gestures through regular expressions, using the operators of concatenation, alternation and Kleene's star. A regular expression if defined by a triplet: (i) the event type, (ii) the touch identifier, (iii) the interface item hit by the touch; An improved version of the framework [10] included means for calculating a set of attributes that may be associated to an expression literal.

In GestIT [20,21], gestures are modelled through expressions that define their temporal evolution, obtained by composing two main elements: ground and composite terms. A ground term is the smallest block for defining a gesture: it describes an atomic event which cannot be further decomposed. In general, it is associated to a value change of a *feature*, such as the pixel coordinates of a touch on the screen or the position and rotation of a skeleton joint. Composite terms are used for defining more complex gestures through a set of operators. We will use them in the rest of this work, since they are a superset of those included in Proton++ [21]. Considering two gestures g and h (either ground or composite terms): $g*$ is the iteration of g; $g \gg h$ is the sequence that connects g with h; $g \parallel h$ defines that g and h are performed in parallel; $g[\,]h$ is the choice between either g or h; $g[> h$ disables the iteration of g by performing h. Due to space limits, for further details about GestIT we refer the reader to [20,21]. Its main drawback is the heuristic recognition approach for ground terms, which do not guarantee a good recognition accuracy. We try to solve this problem in this work.

3   Combining the Two Approaches with DEICTIC

In defining DEICTIC, our main goal is to make a first step toward filling the gap between machine learning approaches and declarative description methods, combining the advantages for UI developers in providing gesture sub-parts notification (which is a feature of GestIT), together with the high recognition accuracy and the robustness to input noise offered by learning-based recognition approaches like HMMs. We focus in particular on stroke gestures, which may be segmented into sequences of basic components (e.g., points, lines or arcs).

To describe a gesture, DEICTIC uses the approach of GestIT [20,21]: ground terms are defined first, then they are combined through temporal operators for describing more complex ones. The key feature of DEICTIC is that each ground term is recognised using a distinct, "basic" HMM, trained on a set of examples; the same, basic HMMs are then used in different gesture recognizers, in the same way a line following a given direction may be used in more than one trajectory. In particular, more complex strokes are described through the combination of basic HMMs into a composite one, whose topology is defined according to the semantics of the GestIT composition operators (see below). This allows DEICTIC to provide information on the recognition of each single operand inside the composition. Moreover, contrary to a similar approach like the one of [9] (see Sect. 2.1), the composite HMM of DEICTIC does not require additional training with respect to the basic HMMs, providing additional temporal relationships between primitives besides the sequence. In the following we explain how to create basic HMMs for ground terms, and the proposed algorithms to define the topology of composite HMMs.

Ground Terms. For basic HMMs we use the left-to-right (or Bakis) topology, which is the most commonly used one for recognizing simple gestures like lines or arcs. It requires to specify the number of states, whereas the probability distributions of both transitions and observations may be learned from a dataset. We point out that in DEICTIC training data are needed only for ground terms.

Iterative Operator. Given a HMM trained to recognize a gesture g, the iterative operator allows recognizing the same gesture an indefinite number of times. Assuming that the starting state of g is s_0 and the ending state is g_f, the HMM for g^* is defined by adding a transition from all states in the backward star of s_f (represented in red in Fig. 1(a) to all states in the forward of s_0 (in green in Fig. 1(a). This creates a loop in the topology, while no changes are made to the probability distributions.

Sequence Operator. For recognizing a sequence of gestures in a specified order, we use the sequence operator. Given two HMMs trained to recognize respectively gesture g and h, an HMM that recognises the sequence $g \gg h$ is obtained by connecting the backward star of the ending state in g with the forward star of starting state in h. Such operation is depicted in Fig. 1(b). It guarantees that $g \gg h$ has only one starting and one ending state. Since the two HMMs may use a different set of features, the observations of the composed one are obtained by the union of all features considered by both g and h. In other words, the

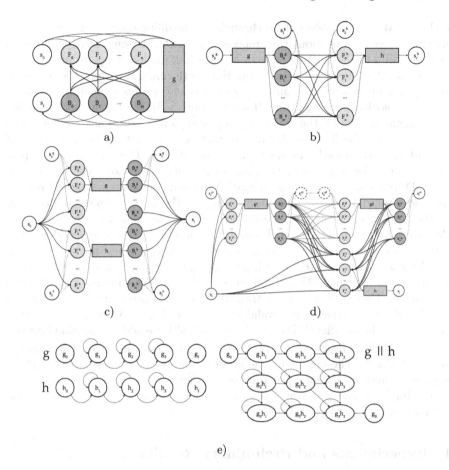

Fig. 1. Composite HMM topologies: (a) iterative, (b) sequence, (c) choice, (d) disabling, (e) parallel. We denote with g and h the gestures HMMs to be composed. (Color figure online)

composite model must specify an emission probability for each feature and for each state.

Choice Operator. A choice between two gestures, denoted by $g[]h$, allows recognizing either g or h. We obtain the corresponding HMM starting from HMMs trained to recognize g and h, and putting them in two separate recognition lines. No transition between the states of g and h is added. The composite HMM has one starting and one ending state. They are linked respectively with the forward star of the original starting state and the backward star of the original ending state in both g and h. All these transitions are equally likely. The choice topology is shown in Fig. 1(c).

Disabling Operator. In defines a gesture that stops the recognition of another one. In general, we use it for stopping an iteration loop. Given two gestures g and

h, the HMM that composes them through the disabling operator is obtained by inserting a set of transitions that represent a short-cut from each ground term contained in g to the starting state in h. Each link models the possibility for the user to stop the execution of g at any time performing h: before starting each one of the ground terms contained in g the model contains a possible transition to h that blocks its recognition. However, since a HMM has a single starting and a single final state, the real HMM topology is more complicated: one needs to consider also the forward star of the starting state and the backward star of the ending state in each ground term contained in g. The schema is shown in Fig. 1(d). In order to maintain the sum of the probabilities towards each state in the HMM equal to 1, we split the original transition likelihood among all involved arcs (both the old and the new ones). We apply the same completion procedure for the observation probability vector we used for the sequence operator.

Parallel Operator. The parallel operator defines the simultaneous performance of two or more independent gestures. Give two gestures g and h, the composite HMM for $g \parallel h$ has a state for each pair (s_g, s_h) where s_g is a state in g and s_h is a state in h. The new HMM represents all the possible combinations of states in g and h. Therefore, we add a transition between two states in the parallel HMM, only if the transition is valid both in g and in h. More precisely, given two states in the parallel HMM (s_i^g, s_j^h) and (s_x^g, s_y^h), we add a transition between them only if the transition from s_i^g to s_x^g exists in g and the transition from $s_i^h j$ to s_y^h exists in h. The observable values of $g \parallel h$ are the concatenation of those in g and h, and are independent from each other.

We finally point out again that, using the proposed approach, composite HMMs need not to be re-trained.

4   Experiments and Preliminary Results

We implemented the above composition algorithms using the Python Pomegranate HMM library [19]. For our first experiments on DEICTIC, we used a data set containing 60 repetitions of 10 stroke gestures, performed by 14 different people. The input sequences consist of the position of the tip of the user's dominant hand forefinger, tracked using a Leap Motion sensor. Our dateset contains the following gestures: swipe left ←, swipe right →, V, caret ∧, left square bracket [, right square bracket], X, delete ⌫, triangle △ and rectangle □.

We used *only one* ground term for describing such gestures, a left-to-right movement on the horizontal axis. In order to recognize such ground term, we created a Bakis HMM with six states, whose observation vector is composed by two normal distributions, one for the x and one for y coordinate of the finger/hand position. We then collected 14 training examples (separated from the gesture dataset) which, after a normalization and resampling step, were used for estimating the parameters of the ground term HMM. We then "cloned" such HMM and applied geometric transformations to the x and y distributions in the observation vector, such as scaling, translation and rotation, in order to represent different segments in a normalised 2D plane. In order to define the expressions for the considered gesture set, we used a cardinal direction notation and the x and y coordinates in brackets for representing the starting point (in the list below we always consider the origin, but in the gesture definitions they are positioned between 0 and 1 in both axes). They represent a geometrically transformed version of the same ground term HMM:

- $e(0,0)$, the original term, without any transformation
- $n(0,0)$, the original term rotated 90°
- $s(0,0)$, the original term rotated −90°
- $w(0,0)$, the original term rotated 180°
- $ne(0,0)$, the original term rotated 45°
- $se(0,0)$, the original term rotated −45°
- $nw(0,0)$, the original term rotated 135°
- $sw(0,0)$, the original term rotated −135°
- $nw60(0,0)$, the original term rotated 120°
- $sw60(0,0)$, the original term rotated −120°.

Table 1 shows the modelling expression for all gestures in our dataset. The third column shows the recognition rate for the DEICTIC HMMs, directly fed with samples in the dataset, without additional training besides the original ground term. For comparison, we also trained an ad-hoc HMM for each gesture type defined using the Bakis topology, and performed a 10-fold cross validation for each gesture sample of the same type. We consider these results as an upper limit for DEICTIC, since optimizing the HMM on real samples allows to better adapt the transition probabilities and emission distributions.

The recognition rates of DEICTIC and of the ad hoc HMM are reported in Table 1. The confusion matrix for DEICTIC is shown in Table 2. In summary, the recognition rates are similar, which provides a first evidence that DEICTIC does not significantly degrade the recognition performance with respect to the state-of-the-art HMM classification approach. We had only two errors on our dataset. In the first one, the classifier confused a delete gesture with a triangle. The sample had a small cross (see sample in Table 2), thus it was really similar to a triangle. In the second error (a rectangle confused again with a triangle), the sample had some initial noise that resembled a triangle.

In contrast, DEICTIC exhibits several advantages, in particular considering the development of gestural UIs. First, DEICTIC was able to recognize new gestures, significantly different from the samples included in the training set of each ground term. This is important for UI designers, who would be able to create gesture recognizers exploiting existing components, as they already do with UI widgets.

Second, DEICTIC allows the reconstruction of the most likely sequence of ground terms associated to a particular gestural input, using the Viterbi algorithm [4]. Indeed, by construction, each state in a composite HMM is associated to a single ground term, for each considered stroke (except for parallel gestures, that consider more than one stroke). Such information is not trivial when gestures are composed in choice or parallel, since the designer would have the

Table 1. Recognition rate comparison between HMM defined through DEICTIC and trained ad-hoc (HMM column).

Gesture	Model	DEICTIC	HMM
←	$w(1,0)$	100%	100%
→	$w(0,0)$	100%	100%
V	$se(0,1) \gg se(0.5,0)$	100%	100%
∧	$ne(0,0) \gg se(0.5,1)$	100%	100%
[	$w(1,1) \gg s(0,1) \gg e(0,0)$	100%	100%
]	$e(0,0) \gg n(1,0) \gg w(1,1)$	100%	100%
X	$ne(0,0) \gg s(1,1) \gg nw(1,0)$	100%	100%
⊠	$se(0,1) \gg w(1,0) \gg nw(0,0)$	98.34%	98.34%
△	$sw60(0.5,1) \gg e(0,0) \gg nw60(1,0)$	100%	100%
□	$s(0,1) \gg e(0,0) \gg n(1,0) \gg w(1,1)$	98.34%	100%

Table 2. Samples gestures included in the dataset and confusion matrix for the recognition using DEICTIC.

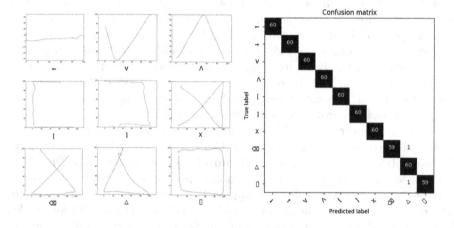

possibility to associate different feedback and feed-forward reactions to different ground terms. Such level of granularity is not supported by ad-hoc trained HMMs.

To further test all composition operators, we also considered a set of synthetic sequences produced as follows. First, we randomly grouped the gestures in a set of 5 pairs; then, for each pair we created a set of 14 sequences that should be recognized by the composition of two gestures, using the sequence, disabling and parallel operator. For creating the sequence samples, we simply concatenated those of the first gesture with those of the second one. For creating the disabling samples, we supposed to perform iteratively the first gesture, which should be blocked by the second one. Therefore, we randomly repeated the samples for the first gestures a random number of times between 3 and 5, concatenating the result with a sample of the second gesture. For the parallel operator, we juxtaposed the samples of both gestures, randomly shifting up or down the rows of the second gesture and filling the blanks with random values. The latter operation guarantees that the gestures may start at different times. We then built one composite HMM for the sequence with the disabling operator, and one for the parallel operator. Similarly to the previous experiment, beside using DEICTIC we also trained an ad-hoc HMM for each gesture category, and evaluated the recognition performance using the leave-one-out technique. Table 3 shows that also in this case our compositional approach did not introduce a sensible degradation of the recognition rate.

Table 3. Recognition rates for syntethic sequences.

| | DEICTIC | | | HMM | | |
Gesture	≫	[>	‖	≫	[>	‖
∧, ⌧	100%	100%	92.86%	92.86%	100%	92.86%
←, □	100%	100%	100%	100%	100%	92.86%
→, [	100%	100%	100%	100%	100%	100%
], △	84.62%	84.62%	100%	100%	100%	100%
V, X	100%	100%	100%	100%	100%	100%

5   Conclusions

We proposed DEICTIC, a declarative and compositional description model for interactive gestures, based on the composition of a set of basic gesture sub-parts (ground terms, or primitives) through a set of operators. We use HMMs, a state-of-the-art technique in vision-based approaches, to recognize ground terms; we then combine such "basic" HMMs into composite HMMs, according to the operators, to describe and recognize complex gestures, retaining at the same time the inspection capabilities on gesture sub-parts needed for providing feedback and feed-forward in user interfaces. The main contribution of our work is the definition of algorithms for defining the composite HMM topology according to the composition semantics of complex gestures, without requiring additional training for the resulting HMM with respect to the basic ones. Preliminary empirical evidence shows that our approach is a promising direction toward filling the gap between the higher recognition accuracy and robustness achieved by vision- and learning-based approaches, and the capability of providing information on meaningful gesture sub-parts exhibited by compositional approaches, which is very useful for user interface design. The main limitation of DEICTIC to be addressed in future work is that the number of states of the composite HMM grows linearly (for the sequence, disabling and choice operators) or quadratically (for the parallel operator) with respect to basic HMMs.

Acknowledgements. This work has been supported by the project D3P2, Sardinia Regional Government (CUP code F72F16002830002) with the funds of Regional Law 7/07, year 2015, "Capitale Umano ad Alta Qualificazione".

References

1. Chen, Q., Georganas, N.D., Petriu, E.M.: Real-time vision-based hand gesture recognition using haar-like features. In: Proceedings of IMTC 2007, pp. 1–6. IEEE (2007)
2. Cheng, H., Yang, L., Liu, Z.: Survey on 3D hand gesture recognition. IEEE Trans. Circuits Syst. Video Technol. **26**(9), 1659–1673 (2016)
3. Delamare, W., Coutrix, C., Nigay, L.: Designing guiding systems for gesture-based interaction. In: Proceedings of EICS 2015, pp. 44–53. ACM (2015)
4. Forney, G.D.: The Viterbi algorithm. Proc. IEEE **61**(3), 268–278 (1973)
5. Holte, M.B., Moeslund, T.B.: View invariant gesture recognition using 3D motion primitives. In: Proceedings of ICASSP 2008, pp. 797–800. IEEE (2008)
6. Hoste, L., Dumas, B., Signer, B.: Mudra: a unified multimodal interaction framework. In: Proceedings of ICMI 2011, pp. 97–104. ACM, New York (2011)
7. Kammer, D., Wojdziak, J., Keck, M., Groh, R., Taranko, S.: Towards a formalization of multi-touch gestures. In: Proceedings of ITS 2010, pp. 49–58. ACM, New York (2010)
8. Khandkar, S.H., Maurer, F.: A domain specific language to define gestures for multi-touch applications. In: Proceedings of DSM 2010, pp. 2:1–2:6. ACM, New York (2010)
9. Kim, I., Chien, S.: Analysis of 3D hand trajectory gestures using stroke-based composite hidden Markov models. Appl. Intell. **15**(2), 131–143 (2001)
10. Kin, K., Hartmann, B., DeRose, T., Agrawala, M.: Proton++: a customizable declarative multitouch framework. In: Proceedings of UIST 2012, pp. 477–486. ACM Press, Berkeley (2012)
11. Kin, K., Hartmann, B., DeRose, T., Agrawala, M.: Proton: multitouch gestures as regular expressions. In: Proceedings of CHI 2012, pp. 2885–2894. ACM Press, Austin (2012)
12. Mitra, S., Acharya, T.: Gesture recognition: a survey. IEEE Trans. Syst. Man Cybern. Part C **37**(3), 311–324 (2007)
13. Natarajan, P., Nevatia, R.: Online, real-time tracking and recognition of human actions. In: Proceedings of WMVC 2008, pp. 1–8. IEEE (2008)
14. Natarajan, P., Singh, V.K., Nevatia, R.: Learning 3D action models from a few 2D videos for view invariant action recognition. In: Proceedings of CVPR 2010, pp. 2006–2013. IEEE (2010)
15. Norman, D.A.: Natural user interfaces are not natural. Interactions **17**(3), 6–10 (2010)
16. Pavlovic, V., Sharma, R., Huang, T.S.: Visual interpretation of hand gestures for human-computer interaction: a review. IEEE Trans. Pattern Anal. Mach. Intell. **19**(7), 677–695 (1997)
17. Rautaray, S.S., Agrawal, A.: Vision based hand gesture recognition for human computer interaction: a survey. Artif. Intell. Rev. **43**(1), 1–54 (2015)
18. Scholliers, C., Hoste, L., Signer, B., De Meuter, W.: Midas: a declarative multi-touch interaction framework. In: Proceedings of TEI 2011, pp. 49–56. ACM, New York (2011)

19. Schreiber, J.: Pomegranate (2016). https://github.com/jmschrei/pomegranate
20. Spano, L.D., Cisternino, A., Paternò, F.: A compositional model for gesture definition. In: Winckler, M., Forbrig, P., Bernhaupt, R. (eds.) HCSE 2012. LNCS, vol. 7623, pp. 34–52. Springer, Heidelberg (2012). doi:10.1007/978-3-642-34347-6_3
21. Spano, L.D., Cisternino, A., Paternò, F., Fenu, G.: GestIT: a declarative and compositional framework for multiplatform gesture definition. In: Proceedings of EICS 2013, pp. 187–196. ACM (2013)
22. Vermeulen, J., Luyten, K., van den Hoven, E., Coninx, K.: Crossing the bridge over Norman's gulf of execution: revealing feedforward's true identity. In: Proceedings CHI 2013, pp. 1931–1940. ACM (2013)
23. Yang, Y., Saleemi, I., Shah, M.: Discovering motion primitives for unsupervised grouping and one-shot learning of human actions, gestures, and expressions. IEEE Trans. Pattern Anal. Mach. Intell. **35**(7), 1635–1648 (2013)

How Far Can You Get by Combining Change Detection Algorithms?

Simone Bianco, Gianluigi Ciocca$^{(\boxtimes)}$, and Raimondo Schettini

Department of Informatic Systems and Communications,
University of Milano-Bicocca, 20126 Milano, Italy
{bianco,ciocca,schettini}@disco.unimib.it

Abstract. Given the existence of many change detection algorithms, each with its own peculiarities and strengths, we propose a combination strategy, that we termed IUTIS (In Unity There Is Strength), based on a genetic Programming framework. This combination strategy is aimed at leveraging the strengths of the algorithms and compensate for their weakness. In this paper we show our findings in applying the proposed strategy in two different scenarios. The first scenario is purely performance-based. The second scenario performance and efficiency must be balanced. Results demonstrate that starting from simple algorithms we can achieve comparable results with respect to more complex state-of-the-art change detection algorithms, while keeping the computational complexity affordable for real-time applications.

Keywords: Video surveillance · Change detection · Algorithm combining and selection · Genetic Programming · CDNET

1 Introduction

Many computer vision applications require the detection of changes within video streams, e.g. video surveillance, smart environments, video indexing and retrieval [5,28]. For all these applications, a robust change detection algorithms with a low false alarm rate is required as a pre-processing step. Many algorithms have been proposed to solve the problem of video change detection. Most of them rely on background subtraction techniques to segment the scene into foreground and background components. The outputs of a change detection algorithm are usually binary images of the foreground areas corresponding to moving objects. These algorithms are designed to cope with the challenges that can be found in a real-world videos such as high variation in environmental conditions, illumination changes, shadows, camera movements and camera-induced distortions and so on. To this end, algorithms are becoming increasingly more complex and thus computationally expensive both in terms of time and memory space. Parallelization of background subtraction algorithms on GPU is a possible way to speed up the computation to make them usable in real-time applications (e.g. [24]).

© Springer International Publishing AG 2017
S. Battiato et al. (Eds.): ICIAP 2017, Part I, LNCS 10484, pp. 96–107, 2017.
https://doi.org/10.1007/978-3-319-68560-1_9

Notwithstanding the improvements, it is still difficult to design general-purpose background subtraction algorithms. These algorithms have been demonstrated to perform well on some types of videos but there is no single algorithm that is able to tackle all the challenges in a robust and computationally efficient way. This can be clearly seen in the CDNET 2014 competition [26] (ChangeDetection.net) where change detection algorithms are evaluated on a common dataset composed of different types of videos sequences and classified according to their performance. To date, more than 35 different algorithms have been evaluated, but many more exists in the literature.

Finally the output of a background subtraction algorithm is usually refined in order to reduce noisy patterns such as isolated pixels, holes, and jagged boundaries. To improve algorithm accuracy, post-processing of the foreground component, ranging from simple noise removal to more complex object-level techniques, has been investigated. Results indicate that significant improvements in performance are possible if a specific post-processing algorithm is designed and the corresponding parameters are set appropriately [18].

Given the existence of so many change detection algorithms, each which its own peculiarities and strengths, here we are interested in finding how far can we get, in terms of performances, by leveraging existing algorithms to create new ones. Instead of designing from scratch a new algorithm, we combine existing ones with the aim to build a better change detection algorithm. We are interested in testing this idea under two scenarios: a purely performance-based scenario and a performance/efficiency balanced scenario. In [3] we have investigated the first scenario by considering the best change detection algorithms in the CDNET 2014 competition, disregarding their computational complexity, and combining then under a Genetic Programming (GP) framework [11]. The resulting algorithms significantly outperform the algorithms used in the combination and even other, more recent, approaches. In this work, we present our findings with respect to the second scenario. We apply the same general approach used in [3] but considering state-of-the-art algorithms that are computationally efficient but not top-performing. We want to investigate if also in this scenario, we are able to create an effective algorithm and what kind of performances we can achieve.

2 The Proposed Approach

In this section, we summarize our GP-based combining approach. A detailed description of the method can be found in [3]. GP is a domain-independent evolutionary method that genetically breeds a population of functions, or more generally, computer programs to solve a given problem [11]. Evolution is driven by the best fit individuals according to an objective function (i.e. fitness function) that must be maximized or minimized. The solutions can be represented as trees, lines of code, expressions in prefix or postfix notations, strings of variable length, etc. GP has been widely used for finding suitable solutions for a wide range of problems needing optimization. For example, in image processing and computer vision applications GP has been used for: image segmentation, enhancement, layouting, classification, feature extraction, and object recognition [1,2,4,7,13]. For our purposes, we feed GP the set of the binary foreground images that correspond to the outputs of the single change detection algorithms, and a set operators represented by unary, binary, and n-ary functions that are used to combine the outputs (via logical AND, logical OR, etc...) as well as to perform post-processing (via filter operators).

More formally, given a set of n change detection algorithms $C = \{C_i\}_{i=1}^n$, the solutions evolved by GP are built using the set of functionals symbols $\mathcal{F}$ and the set of terminal symbols $\mathcal{T} = C$. We build the set of functionals symbols considering operators that work in the spatial neighborhood of the image pixel, or combine (stack) the information at the same pixel location but across different change detection algorithms. The list of functional symbols used is given below:

- ERO (Erosion): it requires one input, works in the spatial domain and performs morphological erosion with a 3×3 square structuring element;
- DIL (Dilation): it requires one input, works in the spatial domain and performs morphological dilation with a 3×3 square structuring element;
- MF (Median Filter): it requires one input, works in the spatial domain an performs median filtering with a 5×5 kernel;
- OR (Logical OR): it requires two inputs, works in the stack domain and performs the logical OR operation;
- AND (Logical AND): it requires two inputs, works in the stack domain and performs the logical AND operation;
- MV (Majority Vote): it requires two or more inputs, works in the stack domain and performs the majority vote operation;

We define the fitness function used in GP by taking inspiration from the CDNET website, where change detection algorithms are evaluated using different performance measures and ranked accordingly. Given a set of video sequences $\mathcal{V} = \{V_1, \ldots, V_S\}$, a set of performance measures $\mathcal{M} = \{m_1, \ldots, m_M\}$ the fitness function of a candidate solution C_0, $f(C_0)$ is defined as the average rank across video sequences and performance measures:

$$f(C_0) = \frac{1}{M} \sum_{j=1}^{M} \left(\text{rank}_{C_0}\left(C_0; \{m_j(C_k(\mathcal{V}))\}_{k=1}^n\right) + \sum_{i=1}^{2} w_i P_i(C_0) \right) \quad (1)$$

where $\text{rank}_{C_0}(\cdot)$ computes the rank of the candidate solution C_0 with respect to the set of algorithms C according to the measure m_j. $P_1(C_0)$ is defined as the signed distance between the candidate solution C_0 and the best algorithm in C according to the measure m_j:

$$P_1(C_0) = \begin{cases} -m_j(C_0(\mathcal{V})) + \max_{C_k \in C} m_j(C_k(\mathcal{V})) \\ \qquad \text{if the higher } m_j \text{ the better} \\ m_j(C_0(\mathcal{V})) - \min_{C_k \in C} m_j(C_k(\mathcal{V})) \\ \qquad \text{if the lower } m_j \text{ the better} \end{cases} \quad (2)$$

and $P_2(C_0)$ is a penalty term corresponding to the number of different algorithms selected for the candidate solution C_0:

$$P_2(C_0) = \frac{\# \text{ of algorithms selected in } C_0}{\# \text{ of algorithms in } \mathcal{C}} \tag{3}$$

The role of P_1 is to produce a fitness function $f(C_0) \in \mathbb{R}$, so that in case of candidate solutions having the same average rank, the one having better performance measures is considered a fitter individual in GP. The penalty term P_2 is used to force GP to select a small number of algorithms in $\mathcal{C}$ to build the candidate solutions. The relative importance of P_1 and P_2 is independently regulated by the weights w_1 and w_2 respectively.

3 Experimental Setup

Since we wanted to test computationally efficient and simple change detection algorithms, we chose the set of change detection algorithms $\mathcal{C}$ to be combined among those implemented in the BGSLibrary[1]. BGSLibrary is a free, open source and platform independent library which provides a C++ framework to perform background subtraction using code provided by different authors. We used the 1.9.1 version of the library which implements more than 30 different algorithms. We base our choice of the algorithms on the recent review paper of the authors of BGSLibrary [20] where the computational costs as well as the performances of the different algorithms have been assessed. The rationale is to use computationally efficient algorithms having above average performances, and possibly exploiting different background subtraction strategies. Based on the results in [20], and on some preliminary tests that we have performed, we selected the following algorithms: Static Frame Difference (SFD), Adaptive-Selective Background Learning (ASB), Adaptive Median (AM) [15], Gaussian Average (GA) [27], Gaussian Mixture Model (ZMM) [29], Gaussian Mixture Model (MoG) [6], Gaussian Mixture Model (GMM) [22], Eigenbackground/SL-PCA (EIG) [17], VuMeter (VM) [9], $\Sigma\Delta$ Background Estimation (SD) [12], Multiple Cues (MC) [16]. All the algorithms have been tested in [20] with the exception of SigmaDelta and SJNMultiCue algorithms. These have been added in recent versions of the BGSLibrary. We decide to include them since they show interesting performances although they are slightly more computationally intensive with respect to the simpler algorithms.

The performance measures $\mathcal{M}$ are computed using the framework of the CDNET 2014 challenge [26]. The framework implements the following seven different measures: recall, specificity, false positive ratio (FPR), false negative ratio (FNR), percentage of wrong classifications (PWC), precision, and F-measure. A ranking of the tested algorithms is also computed starting from the partial ranks on these measures. The CDNET 2014 dataset is composed of 11 video categories, with four to six videos sequences in each category. The categories exhibit different video contents and are chosen to test the background subtraction algorithms

---

[1] https://github.com/andrewssobral/bgslibrary.

under different operating conditions. The challenge rules impose that each algorithm should use only a single set of parameters for all the video sequences. For this reason we set the parameters of the algorithms to their default values, i.e. the values in the configuration files provided in the BGSLibrary.

We set the parameters of GP as in [3]. Also, the GP solutions are generated by considering the shortest video sequence in each of the 11 CDNET 2014 categories as training set. The images in this set are less than 10% of the total images in the whole dataset; this minimizes the over-fitting effect if more images were used. We name the best solution found by GP in this way as IUTIS-2 (the term IUTIS is derived by quoting the Greek fabulists Aesop 620BC-560BC: "In Unity There Is Strength"). We also created a different algorithm, IUTIS-1, by considering a smaller training set composed of all video sequences in the "Baseline" category. As the name suggests, this category contains basic video sequences. IUTIS-1 exhibits worse performances than IUTIS-2, and since its results are not directly comparable with the reported ones, it will not be further considered in the discussion.

4   Results

The tree structure for the IUTIS-2 solution is shown in Fig. 1. In the same figure an example of the output at each node on a sample frame is also reported. From the solution tree it is possible to notice that IUTIS-2 selected and combined a subset of four simple change detection algorithms out of the 11 available: it selected GA, ZMM, MC, and ASB. Concerning the tree structure, we can notice that IUTIS-2 presents a single long branch in its right-hand side. Starting from the functionals defined in Sect. 2, GP was able to create new ones. For instance, the solution tree uses a sequence of the operator MF, which can be seen as an approximation of what could be obtained using a larger kernel for the median filter. The detailed results of the IUTIS-2 algorithm, computed using the evaluation framework of the CDNET 2014 challenge on its 11 video categories, are reported in Table 1. The overall F-measure of IUTIS-2 and of

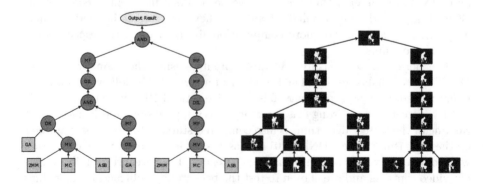

Fig. 1. IUTIS-2 solution tree and example masks.

Table 1. Detailed evaluation results of the IUTIS-2 algorithm for each category of the evaluation dataset.

Scenarios	Recall	Specificity	FPR	FNR	PWC	Precision	F-measure
Overall	0.6703	0.9846	0.0154	0.3297	2.9932	0.7191	0.6163
Bad weather	0.6388	0.9995	0.0005	0.3612	0.6290	0.9380	0.7525
Low framerate	0.7295	0.9947	0.0053	0.2705	1.2079	0.7160	0.6395
Night videos	0.6089	0.9861	0.0139	0.3911	2.2366	0.4706	0.5157
PTZ	0.8329	0.8939	0.1061	0.1671	10.6978	0.1884	0.2397
Turbulence	0.8444	0.9988	0.0012	0.1556	0.2349	0.7737	0.7967
Baseline	0.7452	0.9978	0.0022	0.2548	1.5115	0.9100	0.7913
Dynamic background	0.8027	0.9828	0.0172	0.1973	2.0051	0.5564	0.5741
Camera jitter	0.7209	0.9867	0.0133	0.2791	2.4236	0.7184	0.7165
Intermittent object motion	0.3735	0.9973	0.0027	0.6265	4.7669	0.8374	0.4836
Shadow	0.6636	0.9946	0.0054	0.3364	2.2199	0.8621	0.7393
Thermal	0.4125	0.9987	0.0013	0.5875	4.9923	0.9395	0.5306

Table 2. Overall F-measure all the change detection algorithms considered to build IUTIS-2 (left), and IUTIS-3, IUTIS-5 and IUTIS-7 (right). An empty circle means that the algorithm was in $\mathcal{C}$ but was not selected, a full circle otherwise.

Method	F-meas.	Used by IUTIS-2	Impr. by IUTIS-2	Method	F-meas.	Used by IUTIS-3/5/7	Impr. by IUTIS-3	Impr. by IUTIS-5
ASB	0.4501	●	0.1662	FTS [25]	0.7281	●/●/●	0.0413	0.0540
ZMM [29]	0.5175	●	0.0988	SBS [21]	0.7092	●/●/●	0.0602	0.0729
GA [27]	0.4535	●	0.1628	CWS [10]	0.7050	●/●/●	0.0644	0.0771
MC [16]	0.5444	●	0.0719	SPC [19]	0.6932	/●/●	0.0762	0.0889
SFD	0.2626	○	0.3537	AMB [24]	0.7058	/●/●	0.0636	0.0763
AM [15]	0.4029	○	0.2134	KNN [29]	0.5984	/ / ○	0.1710	0.1837
GMM [22]	0.4589	○	0.1574	SCS [14]	0.6572	/ / ○	0.1122	0.1249
EIG [17]	0.3215	○	0.2948	RMG [23]	0.6282		0.1412	0.1539
MoG [6]	0.4304	○	0.1859	KDE [8]	0.5689		0.2005	0.2132
VM [9]	0.3990	○	0.2173	MV-3	0.7496		0.0198	0.0325
SD [12]	0.3969	○	0.2194	MV-5	0.7569		0.0125	0.0252
MV-11	0.5098		0.1065	MV-7	0.7115		0.0579	0.0706
IUTIS-2	**0.6163**		-.—	IUTIS-3	0.7694		-.—	0.0127
				IUTIS-5	**0.7821**		-0.0127	-.—
				IUTIS-7	**0.7821**		-0.0127	0.0000

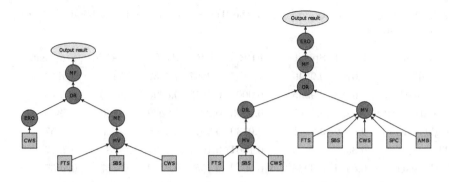

Fig. 2. IUTIS-3 solution tree (left) and IUTIS-5 solution tree (right).

all the change detection algorithms considered are reported in Table 2. From the values reported it is possible to see that our solution is better than the best algorithm fed to GP (i.e. MC), achieving a F-measure that is 7.2% higher.

For comparison, we also report here (see Fig. 2) the solution trees of IUTIS-3 and IUTIS-5, that we recall have been generated by the same method here described but in a purely performance-based scenario [3]. The set of algorithms $\mathcal{C}$ available for GP to build IUTIS-3 were the three top performing algorithms on CDNET 2014, i.e.: Flux Tensor with Split Gaussian models (FTS) [25], Self-Balanced SENsitivity SEgmenter (SBS) [21], and Change Detection with Weightless Neural Networks (CWS) [10]. The set $\mathcal{C}$ available for IUTIS-5 also included Change Detection based on Spectral Reflectaces (SPC) [19] and Extension of the Adapting Multi-resolution Background Extractor (AMB) [24].

From the comparison of the solution trees reported in Figs. 1 and 2, it is possible to notice that IUTIS-2 is more complex in terms of functionals used (see for example the right branch). This is due to the fact that very simple algorithms are used and more operations are needed on their output to achieve higher performance. The overall F-measure of IUTIS-3, IUTIS-5 and of all the change detection algorithms considered are reported in Table 2. From the results it is possible to see that IUTIS-3 is better than any other single algorithms, with a F-measure that is 4.13% higher than the best algorithm used by GP. In the case of IUTIS-5 this difference increases to 5.4%. It is worth noting that all our solutions are better than majority vote solutions (denoted with MV) applied to the corresponding sets $\mathcal{C}$. In [3] we also experimented with larger cardinalities of $\mathcal{C}$, i.e. $\#\mathcal{C} = 7$ and 9, but in both cases the corresponding solutions found by GP, i.e. IUTIS-7 and IUTIS-9, obtained identical performance with respect to IUTIS-5 and thus we only report them in Table 2.

Outputs of some of the tested algorithms on sample frames in the CDNET 2014 dataset, together with input images and ground truth masks, are shown in Fig. 3.

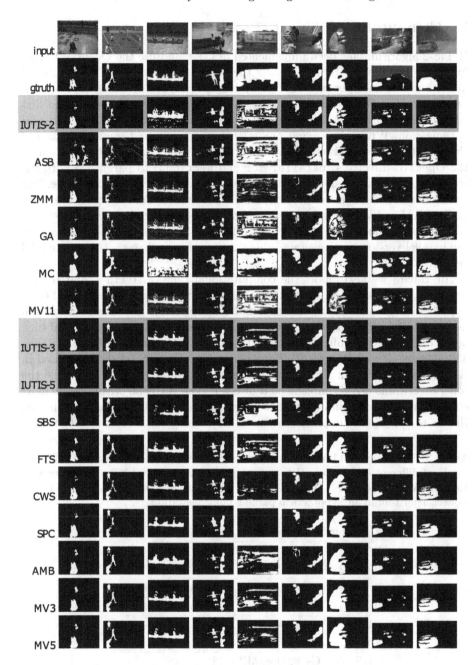

Fig. 3. Examples of binary masks created by the proposed algorithms and some of the algorithms used in the combination.

5   Computational Time

IUTIS-2 algorithm has been implemented in C++ and use the OpenCV library for image processing. Table 3 reports the computational time of the proposed algorithm in frames per seconds. For evaluation purpose, we have implemented two versions of IUTIS-2: a "Sequential" one, and a "Parallel" one. The first version refers to the implementation of the algorithms without any particular optimization. While the second one refers to an optimized implementation of the algorithms obtained by exploiting parallelism on a multicore CPU. We used the OpenMP directives (`parallel for` and `sections`) to parallelize both the computation of the masks, and the execution branches of the solution tree. The timing measurements are carried out on a 3.3 GHz Intel Core-i5 (quadcore) with 16 GB RAM and Windows 7 Professional operating system. As it can be seen, the IUTIS-2 algorithm can be efficiently parallelized. Specifically, the frame rates of the parallel version is, on average, about 2 times faster than that the sequential version.

In Table 3 we also report the computational time of IUTIS-3, and IUTIS-5 used in [3]. For these algorithms, the computational time is an estimated of an hypothetical parallel implementation and corresponds to the slowest algorithm used in the solution tree. For completeness we compare the computational time of the different IUTIS algorithms with the top five algorithms in Table 2 (right). The slowest algorithm, with 10 frame-per-seconds, is FTS (that is also used in IUTIS-3 and IUTIS-5). This algorithm is implemented in MATLAB while the other algorithms are all implemented in C++. The AMB algorithm is the most efficient one with an impressive 843 frame per second. This result is achieved thanks to the parallel implementation on GPU using the CUDA architecture.

Table 3. Computational time, in frames per seconds and at the resolution of 320×240 pixels, of different change detection algorithms on a i5-2500K@3.3Ghz computer with 16 GB RAM. For IUTIS-3 and IUTIS-5 we report an estimate corresponding to the slowest algorithm in an hypothetical parallel implementation.

Algorithm	Implementation	FPS@320x240
IUTIS-2	C++, Sequential	18
IUTIS-2	C++, Parallel	40
IUTIS-3	Misc	10 (Parallel Estimate)
IUTIS-5	Misc	10 (Parallel Estimate)
FTS	MATLAB	10
SBS	C++	31
CWS	C++, OpenMP	18
SPC	C++	12
AMB	C++, CUDA	843

6   Conclusion

In this paper we have presented an evolutionary approach, based on Genetic Programming, to combine simple change detection algorithms to create a more robust algorithm. The solutions provided by Genetic Programming allow us to select a subset of the simple algorithms. Moreover, we are able to automatically combine them in different ways, and perform post-processing on their outputs using suitable operators to produce the best results. Our combination strategy, is able to produce algorithms that are more effective in solving the change detection problem in different scenario. If we are interested in obtaining the maximum performance disregarding the computational complexity of the algorithms themselves, we can combining few top-performing algorithms and achieve the best overall performances (i.e. IUTIS-3 and IUTIS-5). On the contrary, if we want to improve the performances of existing algorithms while maintaining a limited computational complexity, we can effectively combine several simple algorithms and achieve comparable results of more complex state-of-the-art change detection algorithms (i.e. IUTIS-2). In particular, the parallelized version of IUTIS-2 exhibits remarkable performance while being computationally affordable for real-time applications.

References

1. Al-Sahaf, H., Al-Sahaf, A., Xue, B., Johnston, M., Zhang, M.: Automatically evolving rotation-invariant texture image descriptors by genetic programming. IEEE Trans. Evol. Comput. **21**(1), 83–101 (2017)
2. Amelio, A., Pizzuti, C.: An evolutionary approach for image segmentation. Evol. Comput. **22**(4), 525–557 (2014)
3. Bianco, S., Ciocca, G., Schettini, R.: Combination of video change detection algorithms by genetic programming. IEEE Trans. Evol. Comput. (2017). http://dx.doi.org/10.1109/TEVC.2017.2694160
4. Bianco, S., Ciocca, G.: User preferences modeling and learning for pleasing photo collage generation. Trans. Multimedia Comput. Commun. Appl. **12**(1), 1–23 (2015)
5. Bianco, S., Ciocca, G., Napoletano, P., Schettini, R., Margherita, R., Marini, G., Pantaleo, G.: Cooking action recognition with *i*VAT: an *interactive* video annotation tool. In: Petrosino, A. (ed.) ICIAP 2013. LNCS, vol. 8157, pp. 631–641. Springer, Heidelberg (2013). doi:10.1007/978-3-642-41184-7_64
6. Bouwmans, T., Baf, F.E., Vachon, B.: Background modeling using mixture of Gaussians for foreground detection a survey. Recent Pat. Comput. Sci. **1**, 219–237 (2008)
7. Corchs, S., Ciocca, G., Francesca, G.: A genetic programming approach to evaluate complexity of texture images. J. Electron. Imaging **25**(6), 061408 (2016)
8. Elgammal, A., Harwood, D., Davis, L.: Non-parametric model for background subtraction. In: Vernon, D. (ed.) ECCV 2000. LNCS, vol. 1843, pp. 751–767. Springer, Heidelberg (2000). doi:10.1007/3-540-45053-X_48
9. Goyat, Y., Chateau, T., Malaterre, L., Trassoudaine, L.: Vehicle trajectories evaluation by static video sensors. In: 2006 Intelligent Transportation Systems Conference, ITSC 2006, pp. 864–869. IEEE (2006)

10. Gregorio, M.D., Giordano, M.: Change detection with weightless neural networks. In: 2014 IEEE Conference on Computer Vision and Pattern Recognition Workshops (CVPRW), pp. 409–413. IEEE (2014)
11. Koza, J.R.: Genetic Programming: On the Programming Of Computers by Means of Natural Selection, vol. 1. MIT Press, Cambridge (1992)
12. Lacassagne, L., Manzanera, A., Dupret, A.: Motion detection: fast and robust algorithms for embedded systems. In: 2009 16th IEEE International Conference on Image Processing (ICIP), pp. 3265–3268 (2009)
13. Liu, L., Shao, L., Li, X., Lu, K.: Learning spatio-temporal representations for action recognition: a genetic programming approach. IEEE Trans. Cybern. **46**(1), 158–170 (2016)
14. Maddalena, L., Petrosino, A.: The SOBS algorithm: what are the limits? In: 2012 IEEE Computer Society Conference on Computer Vision and Pattern Recognition Workshops (CVPRW), pp. 21–26. IEEE (2012)
15. McFarlane, N., Schofield, C.: Segmentation and tracking of piglets in images. Mach. Vis. Appl. **8**(3), 187–193 (1995)
16. Noh, S.J., Jeon, M.: A new framework for background subtraction using multiple cues. In: Lee, K.M., Matsushita, Y., Rehg, J.M., Hu, Z. (eds.) ACCV 2012. LNCS, vol. 7726, pp. 493–506. Springer, Heidelberg (2013). doi:10.1007/978-3-642-37431-9_38
17. Oliver, N., Rosario, B., Pentland, A.: A Bayesian computer vision system for modeling human interactions. IEEE Trans. Pattern Anal. Mach. Intell. **22**(8), 831–843 (2000)
18. Parks, D.H., Fels, S.S.: Evaluation of background subtraction algorithms with post-processing. In: IEEE 5th International Conference on Advanced Video and Signal Based Surveillance, AVSS 2008, pp. 192–199. IEEE (2008)
19. Sedky, M., Moniri, M., Chibelushi, C.C.: Spectral-360: a physics-based technique for change detection. In: 2014 IEEE Conference on Computer Vision and Pattern Recognition Workshops (CVPRW), pp. 405–408. IEEE (2014)
20. Sobral, A., Vacavant, A.: A comprehensive review of background subtraction algorithms evaluated with synthetic and real videos. Comput. Vis. Image Underst. **122**, 4–21 (2014)
21. St-Charles, P.L., Bilodeau, G.A., Bergevin, R.: Subsense: a universal change detection method with local adaptive sensitivity. IEEE Trans. Image Process. **24**(1), 359–373 (2015)
22. Stauffer, C., Grimson, W.: Adaptive background mixture models for real-time tracking. In: 1999 IEEE Computer Society Conference on Computer Vision and Pattern Recognition, vol. 2, pp. 246–252 (1999)
23. Varadarajan, S., Miller, P., Zhou, H.: Spatial mixture of Gaussians for dynamic background modelling. In: 2013 10th IEEE International Conference on Advanced Video and Signal Based Surveillance (AVSS), pp. 63–68. IEEE (2013)
24. Wang, B., Dudek, P.: A fast self-tuning background subtraction algorithm. In: 2014 IEEE Conference on Computer Vision and Pattern Recognition Workshops (CVPRW), pp. 401–404. IEEE (2014)
25. Wang, R., Bunyak, F., Seetharaman, G., Palaniappan, K.: Static and moving object detection using flux tensor with split Gaussian models. In: 2014 IEEE Conference on Computer Vision and Pattern Recognition Workshops (CVPRW), pp. 420–424. IEEE (2014)

26. Wang, Y., Jodoin, P.M., Porikli, F., Konrad, J., Benezeth, Y., Ishwar, P.: CDnet 2014: an expanded change detection benchmark dataset. In: 2014 IEEE Conference on Computer Vision and Pattern Recognition Workshops (CVPRW), pp. 393–400. IEEE (2014)
27. Wren, C., Azarbayejani, A., Darrell, T., Pentland, A.: Pfinder: real-time tracking of the human body. IEEE Trans. Pattern Anal. Mach. Intell. **19**(7), 780–785 (1997)
28. Yu, M., Rhuma, A., Naqvi, S.M., Wang, L., Chambers, J.: A posture recognition-based fall detection system for monitoring an elderly person in a smart home environment. IEEE Trans. Inf Technol. Biomed. **16**(6), 1274–1286 (2012)
29. Zivkovic, Z., van der Heijden, F.: Efficient adaptive density estimation per image pixel for the task of background subtraction. Pattern Recogn. Lett. **27**(7), 773–780 (2006)

Robust Tracking of Walking Persons by Elite-Type Particle Filters and RGB-D Images

Akari Oshima[1(✉)], Shun'ichi Kaneko[1], and Masaya Itoh[2]

[1] Hokkaido University, Sapporo, Japan
oshima@hce.ist.hokudai.ac.jp, kaneko@ist.hokudai.ac.jp
[2] Hitachi Research Laboratory, Hitachi Ltd., Hitachi, Japan
masaya.itoh.pp@hitachi.com

Abstract. In this paper, we propose a robust real-time tracking system using RGB-D image sequence which are obtained through stereo camera. We apply 'Elite-type' particle filter, which is novel structure of particle filter, for tracking multiple persons. In Elite-type particle filter, to be robust to change of appearance and partial occlusion, likelihood is designed based on histogram and each particle possess their own model histogram. The system assign this particle filter to each person, and estimate state of the target person which vary from frame to frame. Furthermore, the system is able to measure the height of person's head, which is effective for analysis human behavior. Real-time tracking performance of multiple persons was confirmed by experiments which simulating a real shop.

Keywords: Tracking · Robustness · Particle filter · Color · Stereo sensing · Depth · Likelihood · Human behavior · Walking person · Shopper

1 Introduction

Recently, due to the growing awareness of safety and security or crime prevention requirement, surveillance cameras are introduced into many places increasingly. Accordingly, human behavior recognition and analysis technologies based on image sequences acquired from these have been studied [1–4]. Such technologies are used in a variety situations such as marketing design, security and healthcare management. We, for example, have tried to make a tracking system named ISZOT [5] by use of a calibrated single camera to measure rough 2D positions in the shop to analyze shopper behaviors. In the ISZOT system, shopper's zone trajectories could be effectively analyzed, which represent their purchasing and/or wondering behaviors in front of pre-specified zones.

In order to design any effective tracking algorithms, we have had to solve many ill-conditions in the real environment, such as illumination fluctuation, occlusion between walking persons, shadows, and so on. The image data acquired from monochrome or color cameras installed for many security-oriented monitoring, however, are not sufficient for making the tracking algorithms more robust

© Springer International Publishing AG 2017
S. Battiato et al. (Eds.): ICIAP 2017, Part I, LNCS 10484, pp. 108–118, 2017.
https://doi.org/10.1007/978-3-319-68560-1_10

against those ill-conditions due to their limitations of two dimensional (2D) observation. In recent decade, stereo sensing cameras become popular in the real world in performance and price in addition with the availability of rapid network environment to connect them from/to their central controllers. It is getting important to design any effective algorithms to introduce much more 3D real-time sensing functions into the above mentioned systems, and then to utilize the 3D data for robust capturing of the target continuous movements in the scene. In this paper, as our contribution to this field, based on a novel structure of particle filter, a robust real-time tracking system using RGB-D image sequence which are obtained through stereo camera sensing is proposed.

The rest of this paper is organized as follows: Sect. 2 describes algorithms of Elite-type particle filter, Sect. 3 describes how to manage these particle filters in order to track multiple free walking persons, Sect. 4 shows how to adjust the parameters based on target locations relative to the stereo sensors or cameras, Sect. 5 presents the experimental results in the laboratory, and then in Sect. 6 we discuss our conclusive remarks and future works.

2   Elite-Type Particle Filter

2.1   Overall Structure of Tracking Algorithm

In this research, we develop a tracking algorithm by applying particle filters which has been an approach to estimating the non-linear and transitional statistical distributions of object states by using a large number of particles distributed in the observation space. Many study have been reported in human tracking [6–8]. In general type of it one uses a simple likelihood because of its limited calculation cost for large number of particles. In this research, contrast to these conventional methods, we originally utilizes multiple filters, each of which consists of a smart few particles to follow simultaneously multiple persons. In this independent filter, each particle memorizes which part of the target or person it may be placed on at the previous frame or sampling time based on three likelihoods. The basic structure of the proposed tracking system is shown in Fig. 1. When the person detector finds a set of data probably representing a person in a subtracted depth image calculated from a RGB-D image, the system generates and places a particle filter around it. The updating process includes search and resampling of particles and the state estimation based on likelihoods are repeated in every sampled frame. Each process is described in detail later. In this system, 3D position information which xy plane represent floor is calculated from the RGB-D image, where through a calibrated adjustment by the stereo sensor. We call information of 3D position and color which associated with coordinate value in image space as a data point.

2.2   Basic Structure of Elite-Type Particle Filter

Figure 2 shows a stereo sensor installed on the ceiling to take images of walking persons on the floor and the basic concept and situation of the proposed particle

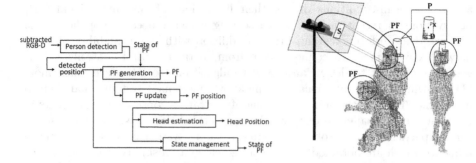

Fig. 1. Schematic of proposed algorithm   **Fig. 2.** Definition of particle filters

filter. We define a particle filter by coordinating in multiple and dispersive manner each particle of a cylindrical shape, which always lies in the vertical direction in the real space. Because of their frequent attitudes in standing and/or walking, we choose the cylinder of the designed size and shape for this purpose and then it is expected to strong against the changing direction of walking persons due to its invariance in shape in rotation about the vertical axis to the floor. The particle **P** is defined by following equation:

$$\mathbf{P} = \{^{p}\mathbf{x}, \mathbf{D}, \mathbf{S}\} \tag{1}$$

where $^{p}\mathbf{x}$, $\mathbf{S}$ and $\mathbf{D}$ define the center position, the image region where it is projected onto, and the set of data points included, respectively. Here, Smart Window Transform (SWT) [9] has been used to project particle to image, which enables to calculate $\mathbf{D}$ efficiently from $\mathbf{S}$. The particle filter (PF) is a set of the neighboring particles that can be coordinated not to belong to any other PF which is defined as $\mathbf{F}$ by the next equation.

$$\mathbf{F}_i = \{\mathbf{P}^i_j \mid j = 1, \dots N_p\} \tag{2}$$

where, N_p is the number of component particles. The position of $\mathbf{F}_i$ is defined as $^{f}\mathbf{x}_i$ which is the average of $^{p}\mathbf{x}^i$. In these Elite-type particles, through somewhat a taking care of each process, we aim to make not so many particles to govern themselves and follow each target autonomously.

A PF is typically generated when a person comes in the scene and the person detector, one of provided libraries, possibly detect him or her in the observation space. Some particles are defined and generated just around the portions of the head through the chest because the upper body of a walking person has less change in shape than the lower body has. In order to realize such arrangement, we use the highest data points in the detected area, where we call them 'head-top' of the tracked person. The particle layout is slightly controlled by introducing random factors in 3D space. The condition is represented in the next equation with respect to the number density of data points so that each particle can include enough amount of data points inside it.

$$|\mathbf{D}| > \delta_p \tag{3}$$

During this process, although the person detector find person-like area, if the condition shown in Eq. 3 cannot be satisfied for more than a certain period of time, it is regarded as an error of the detector and the generation process gets to be quitted partially only just in this area. Moreover, in order to prevent multiple placement of PF with severe overlap, we check whether the generated PF can follow the target and then in order to judge any successful PF generation relative arrangement of the existing PFs and the the new one $\mathbf{x}_d$ is used as]break follows:

$$\forall i, |\mathbf{x}_d - {}^f\mathbf{x}| > \lambda_s \tag{4}$$

where, the threshold λ_s is so important that it can control the relative arrangement of all of the existing PFs by keeping that their mutual distances should be larger than λ_s.

Since as the one of our applications of this algorithm we aim shopper behavior analysis for effective marketing, we have designed an estimator of head-top positions of persons in each sampled frame because the gaze orientation is one of the most important demands in such application, extending the possibility of our proposed algorithm. In order to do this, H-Mask is designed to cover a head of the average size of Japanese. Figure 3 show the procedure of estimating the head position ${}^h\mathbf{X}$ independently of any tracking process. We first calculate the head-top position of the target as in the same way as the PF generation. Secondly, the upper center of the H-Mask is defined to coincide it with the head-top. Finally, the position corresponding to the center part of the H-Mask is taken as ${}^h\mathbf{X}$.

2.3   Likelihoods

The particle memorizes its position in the previous sampled frame and then it searches its own possible location in the current frame for fixing itself in some range around the previous position. This position determination process is performed based on the likelihoods which evaluate three types of similarities with respect to color, height, and trajectory. The likelihood of color L_c uses color features in their 2D histogram $\mathbf{H}_c^{(t)} = \{h_c^{(t)}(i,j)\}$ is made from the data point set $\mathbf{D}^{(t)}$ at the frame t. We use color phase ab in the Lab color coordination for making bins of the histogram. Besides, L_c is calculated by the following equation which is the intersection evaluation [10] of $\mathbf{H}_c^{(t)}$ and $\mathbf{H}_c^{(t-1)}$.

$$L_c = \sum_{i=1}^{n_a} \sum_{j=1}^{n_b} \min(h_c^{(t)}(i,j), h_c^{(t-1)}(i,j)) \tag{5}$$

where n_a and n_b indicate the number of bins in the histogram, respectively. The likelihood of height L_h addresses the similarity based on the height histograms.

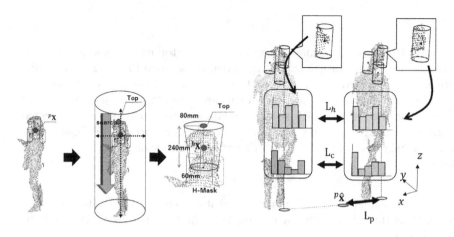

Fig. 3. Head-mask   **Fig. 4.** Three likelihood

Similarly to L_c, a height histogram $\mathbf{H}_h^{(t)} = \{h_h^{(t)}(i)\}$ is made from $\mathbf{D}^{(t)}$. It is calculated as

$$L_h = \sum_{i=1}^{n_h} \min(h_h^{(t)}(i), h_h^{(t-1)}(i)) \tag{6}$$

where n_h indicates the number of bin in the height histogram. The likelihood of trajectory L_p is calculated from the difference between the current position of the particle: $^p\mathbf{x}$ and the estimated position $^p\hat{\mathbf{x}}$ by use of the velocity and the position at the previous frame as follows:

$$L_p = \frac{1}{K_r|^p\hat{\mathbf{x}} - {}^p\mathbf{x}| + 1} \tag{7}$$

where K_r is a weight parameter. Here, we have $0 \leq L_p \leq 1$ and hence the likelihood decreases as the particle moves away from the estimated location. The total likelihood L is finally calculated by the following equation as the combination of the above three likelihoods.

$$L = \alpha_c L_c + \alpha_h L_h + \alpha_p L_p \tag{8}$$

where $\alpha_c, \alpha_h, \alpha_p, (\alpha_c + \alpha_h + \alpha_p = 1)$ are the weights for each likelihood, and in this paper we give those values empirically. Figure 4 shows the concept of three elemental likelihoods. Moreover, when multiple persons approach and then sometimes occlude each other, it is afraid that any PF tracking them may fall into ill-condition and then possibly lose their correct trajectories or follow another person vice versa as misrecognition. In order to deal with these cases, we utilize a prohibited area $\mathbf{C}_i$ for PF replacement as shown in the next expression.

$$\mathbf{C}_i = \{\mathbf{x}_s | |\mathbf{x}_s - {}^f\hat{\mathbf{x}}_j| \leq \lambda_c, j = 1, \ldots, N_f, j \neq i\} \tag{9}$$

where λ_c indicates the threshold and $^f\hat{\mathbf{x}}_j$ is estimated position of $\mathbf{F}_j$ which should be distinguished from the estimated position of particle in Eq. 7, for example. Since the probability of another target person's existence is high in the estimated position of their PF, it is regarded as prohibition area.

2.4 Resampling

In order to improve the tracking performance, the resampling process is performed after the searching as in the normal approaches. One can reproduce the particles by use of this process so that those placed at any 'wrong' location or at non-target ones could be moved to a possible location of the same target. The resampling process is very important to maintain Elite-type PF in better activity and in this paper we need the following three procedures: (1) Compatibility evaluation, (2) Grouping, and (3) Reliability check. First, we judge a compatibility of each particle in tracking by using the number density of data points included in the particle through the Eq. 3 which is same as PF generation. The condition seems as a simple one however we need introduce a novel scheme as shown in Sect. 4 to adjust the threshold values including the above one with respect to their distances from the stereo sensors. If all of the particles have disappeared at a frame, the PF is judged as it lost its target and then transited to 'standby' state, where the detail of this state transition mechanism will be described in the next Sect. 3.

Secondly, the particles belonging to the same PF are grouped according to their distances. Let $\mathbf{F}'$ be a set of all of the $\mathbf{P}$ that survive through the previous compatibility evaluation, $\mathbf{G}_i(i = 1, 2, \ldots, N_g)$ be the direct sum decomposition of $\mathbf{F}'$, and $\mathbf{I}_i$ be their subscript set. Here $\mathbf{G}$ means the group, each of which should satisfies the following condition.

$$\mathbf{G}_i = \{\mathbf{P}_n \mid n \in \mathbf{I}_i\}$$
$$n \in \mathbf{I}_i, m \in \mathbf{I}_j, i \neq j \Rightarrow |^p\mathbf{x}_n - {}^p\mathbf{x}_m| > \Gamma_p \tag{10}$$

The above condition means that any member particle to an arbitrary group and other non-member particles is separated to have larger distance than the threshold λ_g. By appropriately setting of this λ_g, it is possible to satisfactorily separate the particles placed in the target and non-target subjects.

Finally, we calculate a reliability γ_i for each group $\mathbf{G}_i$. This represents how firmly it is placed with fitting to just the target person as follows:

$$\gamma_i = \alpha_\gamma \gamma_i^p + (1 - \alpha_\gamma)\gamma_i^d \tag{11}$$

where α_γ, γ_i^p, and γ_i^d are a weighting coefficient, the position-based reliability, and the data-point-based reliability, respectively, and furthermore γ_i^p is calculated as follows:

$$\gamma_i^p = \frac{1}{K_r|^f\hat{\mathbf{x}} - {}^g\mathbf{x}_i| + 1} \tag{12}$$

where $^f\hat{\mathbf{x}}$ and $^g\mathbf{x}_i$ are the estimated position of PF and the average position of particles belonging to $\mathbf{G}_i$, respectively, and using the same K_r as used for PF generation. In addition, γ_i^d is calculated using the following equation.

$$\gamma_i^d = \frac{|^g\mathbf{D}_i|}{\sum_{j=1}^{n_g}|^g\mathbf{D}_j|} \tag{13}$$

where $|^g\mathbf{D}_i|$ is the total number of data points of particle belonging to the group. Only the particles belonging to $\mathbf{G}$ which have sufficiently high reliability to continue better tracking, and the other particles may be deleted. Finally, to supplement new particles is provided randomly around $^g\mathbf{x}$ of $\mathbf{G}$ so that the total number of particles in any PF is kept as N_p.

3   Transitional Management of PF States

When one imagines some indoor scenes having freely walking people, there may be some happenings in observation, such as appearing and disappearing in/from the scene, crossover between any two persons, and sudden stopping and standing. It is not so easy to deal with all of these cases, however, we try to attack some of the problems by recognizing transitional states of all of the PF under control of a management algorithm proposed in this paper. The states of PF can basically be divided into the following two: an active state S_a and a standby state S_r which are simply shown in Fig. 5. The former one has been described so far, however, the latter one needs to explain here. That is the state where any PF may lose its target person temporally.

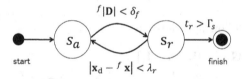

Fig. 5. State transition management of PF

As shown in Fig. 5, any PF must be in the active state in the beginning, and then during its 'active' life, it is expected to perform normal tracking. Since sometimes it gets to lose its target and to have only few data points, it must be transited to the standby state by checking the total number of data points belonging to it ($^f|\mathbf{D}|$) falls below the threshold δ_f. During the 'standby' life, the PF must be in exemption from any process for PF except for keeping in the position just before the transition to wait re-activation. Any standby PF can have two possibilities as follows: the transition again to the active state if its target person may be detected just in the neighboring range of the distance λ_r from its position. Or the disappearance from the scene if the elapsed time of its standby state t_r exceeds Γ_s, where we may find absence of the target person from the observation space. By managing the state of PF, it is possible to obtain a better adaptation as a real facility.

4 Parameter Adjustment

In any stereo sensing, the spatial resolution of measured coordinates basically decreases according to the increasing ego-centric distance from the camera to the objects, in addition, the measurement errors may have an opposite tendency to increase together with the distance. For not so short period to keep observation of moving persons by the stereo camera, we should have designed some scheme to adjust important parameters in our proposed mechanism to the change. From fundamental experiments, there must be some sensitive but important parameters as follows: the threshold value for compatibility check δ_p in Eq. 3, the specified distance between any two PF λ_s in Eq. 4, and then the threshold value to determine the standby state of PF in Fig. 5 δ_f. In the case of δ_p, the number of data points may increase as looming persons to the camera as their projected sizes on the camera plane increase. Thus, it is necessary to adjust their values smoothly within a predetermined range according to the distance from the camera. In order to realize this requirement we adopted the sigmoid function which has two representative values. For example, δ_p is calculated as follows:

$$\delta_p = \frac{\delta_{p2} - \delta_{p1}}{1 + e^{\alpha_s(|^p\mathbf{x}^c| - \lambda_d)}} + \delta_{p1} \tag{14}$$

where δ_{p1} and δ_{p2} are the lower and the upper limits, which can be used in the large and the small distance from the camera, respectively, according to the ego-centric distance $|^p\mathbf{x}^c|$ of $\mathbf{P}$ from the camera. In addition, λ_d gives the distance at which the controlled parameter has the middle value and α_s realizes an arbitrary rate of smooth change. Figure 6 shows how δ_p varies according to the distance from the camera. Even though two particles contain the same number of data points, the particle near to the camera is judged incompatible, while the other one far from the camera is judged compatible. The remaining two parameters

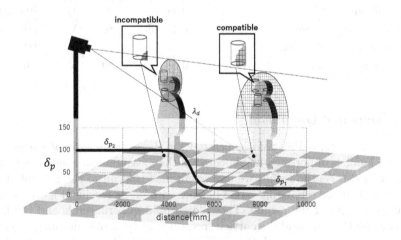

Fig. 6. Parameter adjustment

λ_s and δ_f could be adjusted by use of the similar mechanism above mentioned successfully in our experiments.

5 Experiment

5.1 Specifications

Since the developed prototype system is a total and somewhat complex one for the behavior analysis of shoppers, we could not find any other one developed for the same purpose so far. Therefore, for this reason, we could not include any simple comparison with the other methods in this paper. In the lab room, we have installed a stereo sensor near the ceiling for simulation of tracking multiple shoppers in the small shops. Eight zones were prepared by desks, pillars, and walls in the observation spaces together with different product-like items, such as stuffed toys and stationeries etc. Five walking persons have performed some types of shoppers, each of whom simply walks, searches around for their products, walks and stops often to be interested in their products, walks with accompanying persons, repeatedly stands and crouches, and then frequently picks up the products. In addition, some persons have entered twice into the observation space. Table 1 shows the specifications of the cameras installed in the stereo sensor. We have acquired the RGB-D image sequence of the scene in which the maximum five persons could walk at the same time as a typical complicated situation such as persons passing and occlusion is frequently occurred between them. We have tried analyzing the shopper's behaviors with using parameters shown in Table 2.

Table 1. Camera functions

Param	Value
Angle of dip	30°
Baseline length	150 mm
Size of image	640 × 480
Frame rate	10 fps

Table 2. Experimental specifications

Param	Value	Param	Value	Param	Value
α_c	0.48	δ_{p1}	10	λ_{s1}	700 mm
α_h	0.42	δ_{p2}	80	λ_{s2}	1800 mm
α_p	0.10	δ_{f1}	5	λ_g	180 mm
α_γ	0.38	δ_{f2}	20	K_r	0.0028

5.2 Results and Discussion

In the experiments, 6 PFs were generated for the same image data sequence, where the number of person who appeared in the measurement space was 6 (one appeared twice). As the result, we have confirmed that all of the generated PF could track their corresponding target persons without any losing. Figure 7 shows the sampled shots of the results. The colored bold quadrilaterals and the finer ones show the H-masks and the particle, respectively, both of which are projected onto the camera plane through the SWT. We could see that continuous tracking can be realized without losing, even if people pass each other. Table 3 shows

(a) t=617   (b) t=627   (c) t=637

Fig. 7. Experimental results (Color figure online)

Table 3. Distance error of the system

	xy	z
ID1	166	118
ID2	88	135
ID3	101	123
ID4	59	123
ID5	100	130
ID6	98	139
Average	102	128

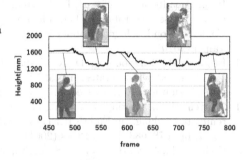

Fig. 8. Measurement trajectory (along z)

the distance errors in the xy plane as 102 mm and the z direction as 128 mm of each ID respectively, which were not so large in order to use in supermarkets and so on.

Due to the simple way of locating the H-mask, the error of z direction is larger than the ones in the xy plane. However, in Fig. 8, we have shown some measured profiles of variation of height of the head in each frame. We could see that the estimated values of the head could represent the person postures, such as standing, crouching, or being seated.

From the above results, one could find that by use of the proposed method described in this paper it may be possible to realize the simultaneous and robust tracking of multi-persons in the real environments.

6   Conclusions

A robust tracking approach of multiple walkers was proposed by using RGB-D image sequence obtained from a stereo sensor. An 'Elite-type' particle filter can be adopted in the proposed method, where three likelihoods based on color, height, and trajectory are effectively utilized and one can estimate positions of heads of the target persons by using some specialized mask operation. We designed our own state transition model for state management of PF for their application to the real facility where any target to track often changes its situation very frequently. In addition, we proposed a unique mechanism to adjust parameters according to the camera distance, which is one of the important process in using any stereo sensors. Experimental results simulating a real store showed the effectiveness of the proposed method.

References

1. Hampapur, A., Brown, L., Connell, J., Ekin, A., Haas, N., Lu, M., Merkl, H., Pankanti, S.: Smart video surveillance: exploring the concept of multiscale spatiotemporal tracking. IEEE Sig. Process. Mag. **22**, 38–51 (2005)
2. Jayawardena, C., Kuo, I.H., Unger, U., Igic, A., Wong, R., Watson, C.I., Stafford, R.Q., Broadbent, E., Tiwari, P., Warren, J., Sohn, J., MacDonald, B.A.: Deployment of a service robot to help older people. In: 2010 IEEE/RSJ International Conference on Intelligent Robots and Systems (IROS), pp. 5990–5995. IEEE Press, Taipei (2010)
3. Fieguth, P., Terzopoulos, D.: Color-based tracking of heads and other mobile objects at video frame rates. In: 1997 IEEE Computer Society Conference on Computer Vision and Pattern Recognition, pp. 21–27. IEEE Press (1997)
4. Nguyen, H.T., Worring, M., Van Den Boomgaard, R.: Occlusion robust adaptive template tracking. In: Eighth IEEE International Conference on Computer Vision, vol. 1, pp. 678–683. IEEE Press (2001)
5. Etchuya, T., Nara, H., Kaneko, S., Li, Y., Miyoshi, M., Fujiyoshi, H., Shishido, K.: Integration of image and ID-POS in ISZOT for behavior analysis of shoppers. In: Tutsch, R., Cho, Y.-J., Wang, W.-C., Cho, H. (eds.) Progress in Optomechatronic Technologies. LNEE, vol. 306, pp. 3–14. Springer, Cham (2014). doi:10.1007/978-3-319-05711-8_1
6. Nummiaro, K., Koller-Meier, E., Van Gool, L.: An adaptive color-based particle filter. Image Vis. Comput. **21**, 99–110 (2003)
7. Wang, J., Chen, X., Gao, W.: Online selecting discriminative tracking features using particle filter. In: IEEE Computer Society Conference on Computer Vision and Pattern Recognition, vol. 2, pp. 1037–1042. IEEE Press (2005)
8. Zhao, X., Satoh, Y., Takauji, H., Kaneko, S.: Hybrid feature and adaptive particle filter for robust object tracking. World Acad. Sci. Eng. Technol. **59**, 2486–2491 (2011)
9. Li, Y., Ito, M., Miyoshi, M., Fujiyoshi, H., Kaneko, S.: Human detection using smart window transform and edge-based classifier. In: Proceedings of JSPE Semestrial Meeting, vol. 2011A, pp. 920–921 (2011). (in Japanese)
10. Swain, M.J., Ballard, D.H.: Indexing via color histograms. In: Third International Conference on Computer Vision, pp. 390–393. IEEE Press (1990)

Video Saliency Detection Based on Boolean Map Theory

Rahma Kalboussi[(⊠)], Mehrez Abdellaoui, and Ali Douik

Networked Objects Control and Communication Systems Laboratory,
National Engineering School of Sousse, University of Sousse,
Pôle technologique de Sousse, Route de Ceinture Sahloul, 4054 Sousse, Tunisia
rahma.kalboussi@gmail.com

Abstract. In the last years, visual saliency has become a challenging research field, and a big number of computational models were developed. While detecting salient object in still images was well studied, video saliency detection is in the early stages. In this paper, we propose a novel video saliency detection method based on Boolean maps. Unlike still images, video frames are characterized by statistic and dynamic information. A set of Boolean maps are generated by thresholding feature channels (color and motion features). Using the gestalt principle for figure-ground segregation, saliency prediction is derived from the Boolean maps where connected regions are marked as salient. Our proposed method is evaluated over two video saliency benchmark datasets and compared to seven state-of-the-art methods. Results have shown that our method outperforms other methods on the two datasets.

1 Introduction

In the early 80's, Treisman and Gelade [23] proposed the feature integration theory for visual attention. Where they assume that the visual scene is initially coded along a number of separable dimensions such as color, orientation, spatial frequency, brightness, direction of movement, etc to provide the feature maps. Then, these maps are recombined to ensure the correct synthesis of features, and to provide final focal attention. Based on this theory, various visual saliency models are developed. Such models can be grouped into two categories: local and global approaches.

Local approaches measure the rarity of a region over its neighborhoods. Itti et al. [10] derived a bottom-up visual saliency model based on center surround difference through multi-scale image features. A bottom-up saliency model derived from a Bayesian framework is proposed in [26]. A saliency model that computes local descriptors from a given image in order to measure the similarity of a pixel to its neighborhoods was proposed in [22]. AWS method [5] is based on two biological mechanisms: the decorrelation and the distinctiveness of local responses. Harel et al. [6] propose GBVS which is a bottom-up saliency approach that consists of two steps: the activation maps generation over feature channels, and their normalization.

© Springer International Publishing AG 2017
S. Battiato et al. (Eds.): ICIAP 2017, Part I, LNCS 10484, pp. 119–128, 2017.
https://doi.org/10.1007/978-3-319-68560-1_11

In contrast, global approaches are based on the rarity and uniqueness of image regions with respect to the whole scene. Scharfenberger et al. [21] proposed a texture based saliency model, where an object is salient if it has a distinctive texture from the rest of the scene. Kim et al. [12] developed a method which separates the background from foreground to highlight the salient object. Cheng et al. [2] proposed a regional contrast based salient object detection model which assumes that human cortical cells preferentially respond to high contrast stimulus in their receptive fields. Hou and Zhang [7] introduced a spectral based method by analyzing the spectrum of the input image in order to extract the residual spectrum. Bruce and Tsotsos [1] proposed the AIM approach using Shanon's self-information measure to maximize the sampled information from a scene. Scharfenberger et al. [20] proposed a salient object detection model that uses the natural images structural and textural characteristics.

While there are many computational models that detect salient regions in still images, video saliency methods are in the early stages. In that context, Itti and Baldi in [9] assumed that salient object is related to surprising events and developed a model that detects objects influenced by surprising events. Rahtu et al. [19] developed a saliency model where they incorporated local saliency features into a conditionnal random field model. Mancas et al. [16] used the optical flow magnitude to highlight motion in a crowd. A video saliency model based on optical flow strength and static saliency features of an input video frame was proposed by Zhong et al. [27]. Beside motion, Fang et al. [3] have used color, luminance and texture to produce saliency model in compressed domain. Lee et al. [13] combine a set of spatial saliency features including rarity, compactness, and center prior with temporal features of motion intensity and motion contrast into an SVM regressor to detect each video frame's salient object. Kim et al. [11] developed a novel approach based on the random walk with restart to detect salient regions. First, a temporal saliency distribution is found using the motion distinctiveness, Then, that temporal saliency distribution is used as a restarting distribution of the random walker. The spatial features are used to design a transition probability matrix for the walker, to estimate the final spatiotemporal saliency distribution.

While the Feature Integration Theory for visual attention has led to the development of many saliency approaches (for videos and still images), the Boolean Map theory for visual attention [8] has attracted Researchers. Zhang and Sclaroff proposed the Boolean Map Saliency in their paper [25]. Also, Qi et al. [18] used Boolean maps to produce a multi-scale propagation method where a graph-inference is performed to produce final saliency maps.

Therefore, in this paper we propose a novel video saliency model based on Boolean maps. First, we compute the optical flow of each pair of frames, and use our proposed motion feature to remove noise caused by camera motion. The smoothed optical flow will serve to produce the motion Boolean map. Then, we generate the color Boolean maps by thresholding the input frame color channel map. Thereafter, we combine the color and motion Boolean maps into one global map. Finally, we use the Gestalt principle for figure-ground segregation to maintain the surrounded regions in each Boolean map and eliminate the unfenced regions. The saliency of each video frame is the mean of the processed global Boolean maps of each frame over the total number of the randomly generated Boolean maps. The main contribution of this paper is the evaluation of Boolean maps for video saliency. As an additional contribution we propose a new motion feature for saliency prediction. To evaluate the proposed approach we use two standard benchmark datasets for video saliency: SegTrack v2 [14] and Fukuchi [4].

The paper is organized as follow: First, we will present our method and explain how we proceeded to generate Boolean maps in Sect. 2. Then, we will discuss experimental results in Sect. 3. Finally, we provide conclusions in Sect. 4.

2   Boolean-Map Video Saliency

The Boolean Map Theory of visual attention was introduced by Huang and Pashler [8] where they assume that at a moment an observer's awareness of a scene can be represented by a Boolean map. From that assumption, we derive a video saliency model which highlights regions of interest in videos. We first, fix two saliency features which are motion and color, then, we build for each frame color Boolean map and motion Boolean map. These Boolean maps will define the saliency level of each region in the Boolean map according to its connectivity (connected regions belong to foreground).

2.1   Boolean Maps Generation

The novel Boolean map saliency method proposed by Zhang and Sclaroff in [25] used a color thresholding on the input image's feature maps on the Lab color space to produce a boolean-map. Lately, Qi et al. in [18] combined the RGB, Lab and HSV color spaces to generate moreprecise boolean-maps. Recent works on saliency detection using boolean map theory have been used on color cues for saliency computation. n this paper, we present a Boolean maps video saliency model, where we use motion and color cues to generate boolean maps. Recent video saliency detection works [11, 24], have proved that moving objects attract attention. Therefore, optical flow is used to estimate motion and determine moving object in the video frame. We use the ptical flow estimation method proposed by [15] to produce direction and velocity measures. In case of static camera, optical flow can be a perfect video saliecy indicator, but we work on benchmark datasets which include different scenarios where the camera is not static. And While the optical flow displays pixels that change position from one frame to another, the motion caused by the camera will also be displayed. To remove the noise caused by the optical flow, we propose a new motion feature.

If we consider O_t and M_t the orientation and the magnitude of the frame t, we define the motion strength as

$$S_t(x, y) = \sqrt{M_t(x, y)^2 - O_t(x, y)^2} \tag{1}$$

Since the optical flow noise caused by the camera motion will produce wrong measures, we use the motion feature proposed by Papazoglou and Ferrari [17] to extract the exact motion boundaries

$$M_t = 1 - exp(-\lambda_M * M_t(x,y)) \qquad (2)$$

λ_M is used to control the function's steepness which is set to 0.8 then our motion feature can be computed as

$$M_t = (M_t(x,y) * S_t(x,y))/max(M_t(x,y)) \qquad (3)$$

We define M_t as the computed motion at frame t, to determine the motion Boolean map we apply the function thresh$(,\theta)$ to assign 1 to the pixel if its value is greater than the threshold value θ and 0 otherwise see Eq. 4.

$$B_m(t) = thresh(M_t, \theta), \qquad (4)$$

In our experiments, we set θ to be between the maximum and the minimum of M_t. While motion is basic to predict saliency in videos, color cue is crucial in saliency prediction in still images. To strengthen the saliency prediction for our video frames, we opt to use a static saliency feature (color). We select the RGB and Lab color spaces to produce the color Boolean map.

We define a vector $F_c = \{[F_R, F_G, F_B], [F_L, F_a, F_b]\}$ where $c \in [1,6]$. Then we generate the feature map using a linear combination between $f_m(t)$ which is the feature channel of the frame t and the vector F_c

$$F_m(t) = f_m(t) * F_c, \qquad (5)$$

The color Boolean map can be computed as

$$B_c(t) = thresh(F_m(t), \beta), \qquad (6)$$

also the function thresh$(,\beta)$ assigns 1 to a pixel if its value is greater than β and 0 otherwise. The values of the feature map F_m are assumed to vary between 0 and 255 by a uniform distribution. The threshold β is set to be between the maximum and the minimum values of the feature map F_m. Given a color Boolean map and a motion Boolean map, the master Boolean map can be estimated as the union of both maps and can be defined as follow

$$B(t) = B_m(t) \cup B_c(t) \qquad (7)$$

2.2   Saliency Computation

The Gestalt principle for figure-ground segregation, assumes that connected regions belong to foreground and are more likely to be perceived as figures. While Boolean maps decomposes the input frame into selected or non selected regions, selected regions are defined as a connected region that has either a value of 0 or 1. Based on the Gestalt principle for figure-ground segregation saliency maps are computed. The first thing to do is to eliminate connected regions that touch the border and set them to be a part of the background. Then each pixel in the Boolean map, is marked by 1 if it belongs to a surrounded regions which means that it is salient and 0 to the rest of the map. For each Boolean map in each video frame a post-processed map R which highlights only important connected regions is deducted

$$R(x,y) = \begin{cases} 1 \ (x,y) \in SR \\ 0 \ \text{otherwise} \end{cases} \tag{8}$$

where SR is a surrounded region. The map R should be smoothed so that small areas get more accentuation. Thereby, we apply a dilatation over each post-processed Boolean map R, then we ensure a linear normalization so that small areas will get more accentuation.

The final saliency map can be defined as the mean of the post-processed Boolean map R over the whole number of generated Boolean maps and can be defined as follow

$$S(x,y) = \frac{1}{n} \sum_{i=1}^{n} R_i \tag{9}$$

where n is the number of Boolean maps.

3 Experimental Results

3.1 Experiments

Our method uses Boolean maps to predict saliency in videos. In this section we will evaluate the performance of our method by comparing the resultant saliency maps to seven state-of-the-art methods on two benchmark datasets in terms of Precion-Recall, ROC curves and Mean Absolute Error.

SegTrack v2 dataset [14] is a video segmentation and tracking dataset. It contains 14 videos with 976 frames. The videos are diversified, there is some videos with one dynamic object, others with more than one. Each video object has specific characteristics that can be Slow motion, Motion blur, change in Appearance, Complex deformation, Occlusion, and Interacting objects. In addition to video frames, a binarized ground truth for each frame is provided.

Fukuchi dataset [4] is a video saliency dataset which contains 10 video sequences with a total of 936 frames with a segmented ground truth.

PR-curve plots the Precision against the recall. To do so, each saliency map is binarized using a fixed set of thresholds variant from 0 to 255. The precision and the recall are then computed by comparing the binarized map S to the ground-truth G see Eqs. 10 and 11

$$\text{precision} = \frac{\sum\limits_{x,y} S(x,y)G(x,y)}{\sum\limits_{x,y} S(x,y)} \tag{10}$$

$$recall = \frac{\sum\limits_{x,y} S(x,y)G(x,y)}{\sum\limits_{x,y} G(x,y)} \tag{11}$$

Receiver operating characteristics (ROC) curve plots the false positive rate against the truth positive rate by varying a fixed threshold from 0 to 255. For a better estimate of the saliency map ground truth dissimilarity, we compute the Mean Absolute Error MAE which approximates the estimate level between the ground truth and the saliency map Eq. 12

$$MAE = \frac{|S - G|}{N} \tag{12}$$

where S and G are the saliency map and the ground truth, and N is the number of pixels in the video frame.

3.2   Results

Precision-Recall curves over the benchmark datasets are repotedin Fig. 1. They provide an efficient comparison of how the produced salient regions in the video frames are correctly predicted. These curves show that our proposed method outperforms other methods. When varying the fixed threshold from 0 to 255, the values of precision and recall are affected. When the value of the threshold is by 255, the recall values of [6, 9, 16] go down to 0 because their predicted salient objects do not highlight the exact or the right salient object. Our proposed method provides a minimum value of recall different from zero because our saliency maps point out the object of interest with a big response. Furthermore, our proposed method offers more precise saliency maps since we achieved the best precision rate (over 0.75).

Figure 2 presents our ROC curves against state-of-the-art methods curves over the two evaluation datasets. On SegTrack v2 dataset our curve has competitive shape with the BMS [25] curve. On Fukuchi dataset, our curve has similar shape with the BMS [25] in the begining and the end.

On **SegTrack v2** and **Fukuchi** datasets we outperform all other approaches with a big gap in terms of MAE values (see Fig. 3).

A visual comparison between our proposed method and the state-of-the-art methods where higher saliency predictions are indicated by bright pixels are reported in Fig. 4.

The GVS [24] used the spatial and temporal edges of each dynamic object in the video frame to compute saliency maps. In case of static camera and one moving object, this method converge to the exact salient object which explains the good results on Fukuchi dataset and the competitive results SegTrack v2 dataset which includes video frames with different conditions (as we explained in the last paragraph). The graph based method [6] is a saliency method which does not include motion cues in saliency map generation which leads to bad PR, ROC curves and high MAE values and bad saliency maps. While it can be used

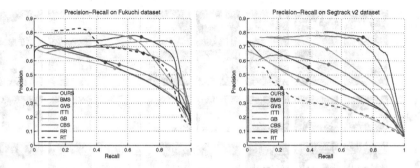

Fig. 1. Precision-Recall curves on Fukuchi and Segtrack v2 datasets

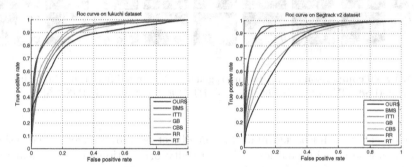

Fig. 2. ROC curves on Fukuchi and Segtrack v2 datasets

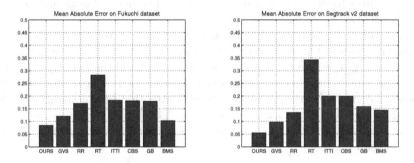

Fig. 3. Mean Absolute Error on Fukuchi and Segtrack v2 datasets

in saliency detection, this method is more suitable for still image saliency. [16] is a saliency method that defined region of interest as the region where moving object is focused. It uses the optical flow to detect moving objects. PR and ROC curves are not as good as our curves because optical flow with no smoothing can be useful only in case of videos with no moving camera.

In case of Fukuchi and Segtrack v2 datasets, a camera motion estimation or an optical flow smoothing should be added to improve saliency maps. The statistical framework proposed by [19] includes motion features to segment the

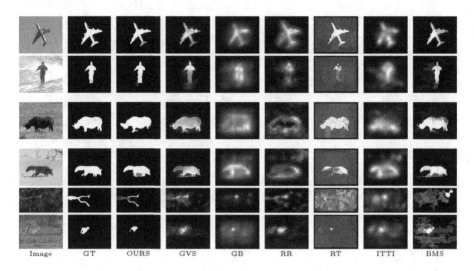

Fig. 4. Visual comparison of saliency maps generated from 6 different methods, including our method, GVS [24], GB [6], RR [16], RT [19], ITTI [9] and BMS [25]

salient object from its background. Salient object is characterized by suprising event in [9] where besides motion, color, intensity, orientation and flicker features are extracted to produce final saliency map. The saliency maps produced by [9,19] are not good indicative of salient object because they use spatial and motion features together, so a static pixel which belongs to background can be marked salient.

The BMS [25] which uses the Boolean maps theory to predict saliency in still images, does not provide good results in some video frames (e.g. the last two rows of Fig. 4) where the color of the background and the moving object are almost the same. Our Boolean map based method tried to heal this issue. We suppose that not only moving object attracts attention but, a change in color could also be important in saliency detection. The Eq. 7 introduces our global Boolean map which is the union of the motion based Boolean map and color based Boolean map.

4   Conclusion

In this paper, we presented a video saliency detection method using the Boolean map theory for visual attention. The proposed method combines color and motion cues to produce a set of Boolean maps for each video frame. Motion cue is generated from the optical flow and smoothed using our proposed motion feature. Then, each Boolean map is processed to highlight only the surrounded regions which are considered as salient. The final saliency map is a linear combination of all Boolean maps. We evaluate the performance of our method over two benchmark datasets against seven state-of-the-art methods. We revealed that the Boolean maps based video saliency can be effective using color and motion cues. As future work, we remain to test the influence of other features channels on saliency detection.

References

1. Bruce, N., Tsotsos, J.: Saliency based on information maximization. In: Advances in Neural Information Processing Systems, pp. 155–162 (2005)
2. Cheng, M.-M., Mitra, N.J., Huang, X., Torr, P.H., Hu, S.-M.: Global contrast based salient region detection. IEEE Trans. Pattern Anal. Mach. Intell. **37**, 569–582 (2015)
3. Fang, Y., Lin, W., Chen, Z., Tsai, C.-M., Lin, C.-W.: A video saliency detection model in compressed domain. IEEE Trans. Circuits Syst. Video Technol. **24**, 27–38 (2014)
4. Fukuchi, K., Miyazato, K., Kimura, A., Takagi, S., Yamato, J.: Saliency-based video segmentation with graph cuts and sequentially updated priors. In: 2009 IEEE International Conference on Multimedia and Expo, pp. 638–641. IEEE (2009)
5. Garcia-Diaz, A., Fdez-Vidal, X.R., Pardo, X.M., Dosil, R.: Decorrelation and distinctiveness provide with human-like saliency. In: Blanc-Talon, J., Philips, W., Popescu, D., Scheunders, P. (eds.) ACIVS 2009. LNCS, vol. 5807, pp. 343–354. Springer, Heidelberg (2009). doi:10.1007/978-3-642-04697-1_32
6. Harel, J., Koch, C., Perona, P.: Graph-based visual saliency. In: Advances in Neural Information Processing Systems, pp. 545–552 (2006)
7. Hou, X., Zhang, L.: Dynamic visual attention: searching for coding length increments. In: Advances in Neural Information Processing Systems, pp. 681–688 (2009)
8. Huang, L., Pashler, H.: A Boolean map theory of visual attention. Psychol. Rev. **114**, 599 (2007)
9. Itti, L., Baldi, P.: A principled approach to detecting surprising events in video. In: 2005 IEEE Computer Society Conference on Computer Vision and Pattern Recognition (CVPR 2005), vol. 1, pp. 631–637. IEEE (2005)
10. Itti, L., Koch, C., Niebur, E., et al.: A model of saliency-based visual attention for rapid scene analysis. IEEE Trans. Pattern Anal. Mach. Intell. **20**, 1254–1259 (1998)
11. Kim, H., Kim, Y., Sim, J.-Y., Kim, C.-S.: Spatiotemporal saliency detection for video sequences based on random walk with restart. IEEE Trans. Image Process. **24**, 2552–2564 (2015)
12. Kim, J., Han, D., Tai, Y.-W., Kim, J.: Salient region detection via high-dimensional color transform. In: Proceedings of the IEEE Conference on Computer Vision and Pattern Recognition, pp. 883–890 (2014)
13. Lee, S.-H., Kim, J.-H., Choi, K.P., Sim, J.-Y., Kim, C.-S.: Video saliency detection based on spatiotemporal feature learning. In: 2014 IEEE International Conference on Image Processing (ICIP), pp. 1120–1124. IEEE (2014)
14. Li, F., Kim, T., Humayun, A., Tsai, D., Rehg, J.M.: Video segmentation by tracking many figure-ground segments. In: Proceedings of the IEEE International Conference on Computer Vision, pp. 2192–2199 (2013)
15. Lucas, B.D., Kanade, T., et al.: An iterative image registration technique with an application to stereo vision. In: IJCAI, vol. 81, pp. 674–679 (1981)
16. Mancas, M., Riche, N., Leroy, J., Gosselin, B.: Abnormal motion selection in crowds using bottom-up saliency. In: 2011 18th IEEE International Conference on Image Processing, pp. 229–232. IEEE (2011)

17. Papazoglou, A., Ferrari, V.: Fast object segmentation in unconstrained video. In: Proceedings of the IEEE International Conference on Computer Vision, pp. 1777–1784 (2013)
18. Qi, W., Han, J., Zhang, Y., Bai, L.-F.: Graph-Boolean map for salient object detection. Signal Process.: Image Commun. **49**, 9–16 (2016)
19. Rahtu, E., Kannala, J., Salo, M., Heikkilä, J.: Segmenting salient objects from images and videos. In: Daniilidis, K., Maragos, P., Paragios, N. (eds.) ECCV 2010. LNCS, vol. 6315, pp. 366–379. Springer, Heidelberg (2010). doi:10.1007/978-3-642-15555-0_27
20. Scharfenberger, C., Wong, A., Clausi, D.A.: Structure-guided statistical textural distinctiveness for salient region detection in natural images. IEEE Trans. Image Process. **24**, 457–470 (2015)
21. Scharfenberger, C., Wong, A., Fergani, K., Zelek, J.S., Clausi, D.A.: Statistical textural distinctiveness for salient region detection in natural images. In: Proceedings of the IEEE Conference on Computer Vision and Pattern Recognition, pp. 979–986 (2013)
22. Seo, H.J., Milanfar, P.: Static and space-time visual saliency detection by self-resemblance. J. Vis. **9**, 15 (2009)
23. Treisman, A.M., Gelade, G.: A feature-integration theory of attention. Cogn. Psychol. **12**, 97–136 (1980)
24. Wang, W., Shen, J., Porikli, F.: Saliency-aware geodesic video object segmentation. In: Proceedings of the IEEE Conference on Computer Vision and Pattern Recognition, pp. 3395–3402 (2015)
25. Zhang, J., Sclaroff, S.: Saliency detection: a Boolean map approach. In: Proceedings of the IEEE International Conference on Computer Vision, pp. 153–160 (2013)
26. Zhang, L., Tong, M.H., Marks, T.K., Shan, H., Cottrell, G.W.: Sun: a Bayesian framework for saliency using natural statistics. J. Vis. **8**, 32 (2008)
27. Zhong, S.-H., Liu, Y., Ren, F., Zhang, J., Ren, T.: Video saliency detection via dynamic consistent spatio-temporal attention modelling. In: AAAI, pp. 1063–1069 (2013)

A System for Autonomous Landing of a UAV on a Moving Vehicle

Sebastiano Battiato[1], Luciano Cantelli[2], Fabio D'Urso[1],
Giovanni Maria Farinella[1], Luca Guarnera[1], Dario Guastella[2],
Carmelo Donato Melita[2], Giovanni Muscato[2], Alessandro Ortis[1(✉)],
Francesco Ragusa[1], and Corrado Santoro[1]

[1] Dipartimento di Matematica e Informatica, University of Catania,
Viale A. Doria 6, 95125 Catania, Italy
ortis@dmi.unict.it
[2] Dipartimento di Ingegneria Elettrica, Elettronica e Informatica,
University of Catania, Viale A. Doria 6, 95125 Catania, Italy

Abstract. This paper describes the approach employed to implement the autonomous landing of an Unmanned Aerial Vehicle (UAV) upon a moving ground vehicle. We consider an application scenario in which a target, made of a visual pattern, is mounted on the top of a ground vehicle which roams in an arena using a certain path and velocity; the UAV is asked to find the ground vehicle, by detecting the visual pattern, and then to track it in order to perform the approach and finalize the landing. To this aim, Computer Vision is adopted to perform both detection and tracking of the visual target; the algorithm used is based on the TLD (Tracking-Learning-Detection) approach, suitably integrated with an Hough Transform able to improve the precision of the identification of the 3D coordinates of the pattern. The output of the Computer Vision algorithm is then exploited by a Kalman filter which performs the estimation of the trajectory of the ground vehicle in order to let the UAV track, follow and approach it. The paper describes the software and hardware architecture of the overall application running on the UAV. The application described has been practically used with success in the context of the "Mohamed Bin Zayed" International Robotic Challenge (MBZIRC) which took place in March 2017 in Abu Dhabi.

1 Introduction

Autonomous landing on a moving vehicle is an important problem that has been investigated by different research groups worldwide [2–4]. Cooperation between UAVs and Unmanned Ground Vehicles (UGVs) to help humanitarian demining operations [5–7] and for aerial monitoring [8,9] are some of the main applications in this context. In this paper we describe the system we have designed and employed in the MBZIRC Challenge. The Mohamed Bin Zayed International Robotics Challenge 2017 (MBZIRC) is a robotic competition held in Abu Dhabi in March 2017. The team of the University of Catania has been selected to

© Springer International Publishing AG 2017
S. Battiato et al. (Eds.): ICIAP 2017, Part I, LNCS 10484, pp. 129–139, 2017.
https://doi.org/10.1007/978-3-319-68560-1_12

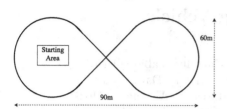
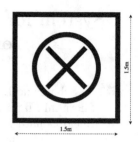

Fig. 1. Playing Area (left) and Visual Target (right)

participate to the Challenge 1 and this paper reports an overview and some details on the developed system. The Challenge 1 consists in the autonomous landing of an UAV on a moving platform [1]. According to MBZIRC rules, Challenge 1 requires a UAV to locate, track and land on a moving ground vehicle. The competition was performed in an open arena where a ground vehicle moves following an eight shaped trajectory, as shown in Fig. 1. On top of the vehicle, the landing area is a square of dimensions 1.5 m × 1.5 m indicated by a given target (see the Visual Target depicted in Fig. 1). The UAV takes off from a given position and autonomously lands on the target placed above the moving vehicle, in the shortest time possible.

In the following sections an overview of the developed system and its components will be presented. Focus will be given to the main modules related to the dynamical estimator and the vision system. In particular, a tracking module able to detect and track a known pattern is employed to select a region of interest within the whole image. Then, a circle Hough transform is used to detect the center of the target with high precision. This system resulted the best solution defined taking into account several constrains related to the considered task. Indeed, the addressed Challenge requires the definition of an hardware and software system able to detect the target and its coordinates with very high precision, and combine this information with other data coming from different sensors (e.g., UAV position, speed, altitude) in order to define the best trajectory for the UAV. Due to the nature of the Challenge, this software pipeline is performed in real-time, taking into account further limits caused by the need of a lightweight hardware. Therefore, we discarded approaches to visual object tracking existing in the literature that have been formulated making specific assumptions on the application domain. For the considered task, a method to update the target representation during the tracking is required, due to the pose and scale changes the target is subject to. Several solutions based on the state of the art in object tracking have been considered [12–15]; the final choice is an hardware and software vision system consisting of an Ocam camera (chosen due its wide Field of View), which rectified images are processed with a pre-trained TLD based detection and tracking of the target and the circle Hough transform. Results of simulations and of the on-field trials will be presented and commented.

2   System Architecture

Challenge 1 requires carefully taking into account of the control aspects, Computer Vision algorithms and the development and integration of suitable hardware needed to perform the autonomous task. The basic overall approach we followed consists in reaching the center of the path by using a precise RTK-DGPS at an altitude suitable for a global view of the environment, for a preliminary detection and localization of the target by means of a wide range camera. Then, a visual detection and tracking procedure is able to estimate the position of the target and generate a suitable trajectory for the UAV. A dynamic estimator merges the measurements of the vision algorithms with the inertial and positioning measurements of the UAV and the estimated trajectories of the UGV. Then, based on the UAV dynamic, the estimator generates the optimal trajectory to reach the target in real time. When the UAV is in proximity of the target, Computer Vision techniques are adopted for the accurate estimation of the 3D coordinates of the target center to be used for safe landing. Once landed, all motors are switched off. The emphasis has been put on the use of lightweight hardware platforms. To this aim, the Computer Vision and control algorithms are optimized to run effectively on a lightweight high performance embedded system.

2.1   Hardware Architecture

The multirotor frame chosen for the competition is the "Spreading Wings S900" by DJI, characterized by high payload and stability. The PixHawk is used as autopilot, it is a high-performance system able to deal with both the stabilization and the navigation of the UAV. This simple but powerful system can be connected to an on-board companion computer that, by running the high-level navigation algorithms, can easily drive the UAV. The "eyes" of the multirotor are represented by an Ocam camera, a fish-eye camera which allows the exploitation of a wide Field of View. The image processing algorithm is executed by a Jetson TX1, an embedded system developed by NVIDIA for visual computing which provides a high performance GPU computing. The computed target position is used by the high level control algorithms to give the proper commands to the Pixhawk autopilot by means of the Mavlink protocol. The accuracy in the localization of the multirotor is ensured by an on-board RTK-DGPS system, receiving the corrections from a base station. In Fig. 2 the whole hardware platforms selected are shown.

2.2   Software Architecture

The control software runs on the Jetson TX1. The software architecture is designed as a multi-thread C/C++ application and it is executed on a Linux environment. Furthermore, for simulation purposes, the software is able to run inside a SITL (Software In The Loop) environment, using Gazebo as physics engine.

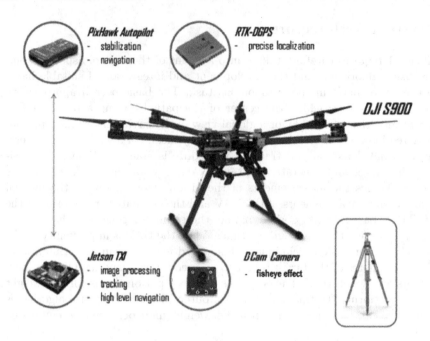

Fig. 2. Hardware platforms used.

The multi-thread process is composed by four threads, as shown in Fig. 3. MAVLINK, PLANNER and COMPUTER VISION are the threads that provide support to the STRATEGY one:

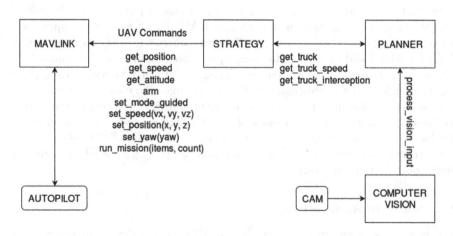

Fig. 3. Software architecture.

- The MAVLINK thread is used as an interface between the process and the autopilot. It allows translating messages from and to the autopilot through the Mavlink protocol.
- The COMPUTER VISION thread acquires and analyses images from the camera and provides the coordinates of the target to the PLANNER thread.
- The PLANNER thread is the interface between the main thread (STRATEGY) and a Finite State Machine (FSM). It receives coordinates from COMPUTER VISION in order to update the FSM and gives the position of the target over time as output to the main thread.
- The STRATEGY is the main thread and represents the decision-making module of the overall system. It has a continuous acknowledge of the state variables of both the system and the target. Its aim is to choose, in each condition, the best strategy to optimally achieve the result.

3   Dynamic Target Position Estimation

The output of the PLANNER thread consists on the estimated target position (in terms of latitude and longitude pairs) over time. This information is inferred by combining the data coming from the Computer Vision algorithms and the target trajectory estimation, which takes into account the known information about the path and the vehicle speed.

This thread is composed of the following basic software modules:

- *Target Detector* is the module handling the visual identification and tracking of the target;
- *Trajectory Predictor* is the estimator of the trajectory of the target that takes into account the (known) path and speed, and suitably adjust the position of the ground vehicle on the basis of the information given by the Target Detector.

The first module is described in depth in the following sections since it is the main objective of this paper, while the latter is briefly described here. The *Target Predictor* is a Kalman estimator that tries to determine the position of the target at each time instant. It basically implements the equation of the motion of the ground vehicle using a virtual point that drives on the path at the speed of 15 km/h. The output of the predictor is the expected Earth coordinates (latitude and longitude) of the target, information that is then used by the High-level Control to proper drive the UAV. These coordinates are continuously adjusted using data coming from the Target Detector: this module returns the center of the target, in local coordinates; a local-to-global transformation is then applied and the error between the detected and estimated coordinates is used to update the estimate. The Target Detector and the Target Predictor thus work in a tight cooperation according to the schema reported in Fig. 4.

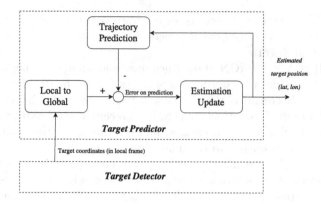

Fig. 4. Working schema of the target predictor

4   Target Detection and Tracking

The Target Detector module is aimed to detect and track the target from a live video stream. For each frame, this module provides to the system the coordinates of the target, according to the coordinate system of the camera (i.e., the target coordinates from the UAV point of view). These local coordinates are then transformed in global coordinates, referred to the global coordinate system.

4.1   Acquisition Hardware

The images processed by the Target Detector module are acquired by an Ocam camera (see Fig. 5). We selected this device due its large Field of View given by the fish-eye lens. The fish-eye lens produces a strong visual distortion in the acquired frames. Therefore, the first step of the vision module is a camera calibration aimed to perform a proper image rectification. Figure 6 an image frame acquired by the Ocam camera, in Fig. 7 the results of image rectification is shown.

Fig. 5. Exploited acquisition hardware consisting on an Ocam camera.

Fig. 6. Camera calibration: the chessboard pattern, with known squares dimensions, is exploited to perform the camera calibration (i.e., find the camera calibration parameters.

Fig. 7. Camera calibration: this figure shows the result obtained after the image rectification.

4.2 Video Analysis

The employed video analysis algorithm implements a combination of two different well known Computer Vision techniques for the detection and tracking of a known pattern. The aim of an object tracking algorithm is to estimate the trajectory of an object as it moves over time by identifying the object positions in different frames of an input video. Tracking objects can be complex depending on the application domain that can involve specific constrains. One of the main issue related to object tracking is to address with the appearance change of the target object. Generative tracking algorithms represents the target object in a specific feature space, and then perform a research of the best match within the image [17,18,21]. Discriminative tracking algorithms define a binary classification problem aimed to distinguish the target from the background [11,16,19,20]. In particular, the vision system exploits the Tracking Learning Detection (TLD) [11] algorithm to detect and continuously track the position of the target over time, considering both the vehicle and UAV movements. This algorithm implements a real-time detection and tracking of a given image pattern specified at the starting frame. In our system, the object of interest is provided by the initial detection of the target. It was possible because the TLD algorithm has been previously trained to detect the considered target. The TLD has been trained off-line, considering several target positions and distances. Furthermore, the TLD algorithm simultaneously tracks the object and learns the object appearances. As a result, the detection and tracking performance improve over time during the execution of the algorithm, allowing the system to learn from a large amount of target examples taken with huge acquisition variability. The TLD algorithm performs a fusion step, which combines the bounding box given by the tracker and the bounding box of the detector into a single output bounding box. When at least one of the two algorithms provide a bounding box, the fusion step outputs the maximally confident one, otherwise, if neither the tracker nor the detector provides a candidate bounding box, the object is declared as not visible by the system. The whole TLD pipeline is shown in Fig. 8.

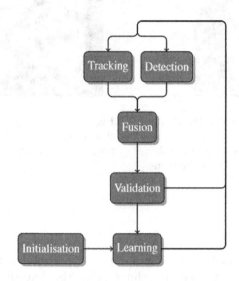

Fig. 8. Scheme of the TLD algorithm.

Once the TLD algorithm provides a bounding box containing the target, the system employs a circle Hough Transform [10] to detect a circular shaped pattern in the provided bounding box. This technique allows to find an image patch which contains an object with the shape of a circle taking into account imperfect shapes, low quality images and changes in the target pose. The aim of this step is to find the center of the target, corresponding to the center of the detected circle. The previous target detection and tracking provided by TLD gives robustness to the circle detection provided by the Hough transform. This, combined with the wide Field of View of the camera, allows to find the target and its position at almost any distance from the UAV with very high precision.

5   Basic Landing Strategy

The STRATEGY software module implements the high-level code to control the overall behavior of the UAV. The first state, that is achieved when the challenge is started, is *TAKE-OFF* and implies to drive the UAV to take-off and reach a certain starting altitude; immediately after this phase, the UAV is driven towards the center of the path[1] (i.e., the center of the eight shaped path) at an altitude of 10 m from the ground; then, the UAV waits for the passage of the ground vehicle. When the target is detected, the Target Predictor is initialized and the *intercept*

---

[1] The GPS coordinates of the area are known a priori.

position is computed (i.e., the position in which the vehicle can be intercepted again). When this event occurs, the UAV starts following the vehicle by tracking the target, also approaching the landing area by means of a descending path. During the approaching phase, the UAV trajectory is continuously modulated considering the output of the Target Predictor. When the landing gear touch sensors detect the successful touch-down event, it causes the turning-off of the propellers. For what concerns the Computer Vision module, when the target enter the visible area of the camera, the Detection Module performs the target detection exploiting the Hough Transform and provides the first target example to the tracker module, as well as the position of the target.

Starting from this first information, the tracker updates the position of the target over time. When the vehicle is in detected in the rectilinear part of the path, the UAV starts the landing phase. When the UAV touches the landing area, its motors are turned off.

6   Results

6.1   Simulations

Several simulations have been executed to test both the software architecture and the sub-blocks. PLANNER block has been extensively simulated in MAT-LAB/Simulink environment. The mission strategy has been improved by further simulations in both Gazebo (Fig. 9) and MATLAB (Fig. 10) environments by introducing the dynamical estimation of the target, to generate in real time the optimal trajectory to reach the target. The whole Software architecture has been initially simulated in Gazebo environment (Fig. 11).

6.2   On Field Trials

Several on field tests have been performed to acquire real images and data; moreover target tracking and landing on the mobile platform have been executed. Initially the videos have been acquired by using a Phantom 3 DJI UAV, and then the camera was mounted on an ASCTEC Firefly. The software architecture has been preliminary tested on a Raspberry PI board communicating to the Pixhawk autopilot and installed on two smaller UAVs (DJI F450 and DJI F550). Finally, the involved hardware and software solutions has been installed and tested on the selected DJI S900 platform. Several different trials have been also performed on the field arena concerning autonomous take-off, navigation and landing. The experiments highlighted the importance of the vision system during the target detection, tracking and the approaching of the landing area. The video of autonomous UVA in action during the MBZIRC competition is available at the following link: http://iplab.dmi.unict.it/MBZIRC/video.mp4.

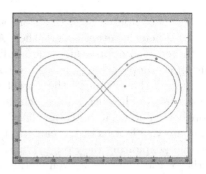

Fig. 9. GAZEBO simulations.

Fig. 10. MATLAB/SIMULINK simulations.

Fig. 11. S900 platform during the field trials.

7   Conclusions

The system described in this paper has been used during the above mentioned International Challenge in March 2017. The developed system reached the goal to land on the moving vehicle in 120" and the achieved result has been placed at the fourth position in the ranking of the International Challenge. The support of the visual module to the whole UAV driving system resulted a crucial factor for the achieved result during the attended competition. Indeed, all the teams that didn't exploit a vision system were unable to detect the target even when it was very close to the UAV, and often to land at all.

References

1. MBZIRC. www.mbzirc.com
2. Serra, P., Cunha, R., Hamel, T., Cabecinhas, D., Silvestre, C.: Landing of a quadrotor on a moving target using dynamic image-based visual servo control. IEEE Trans. Robot. **32**(6), 1524–1535 (2016)
3. Jin, S., Zhang, J., Shen, L., Li, T.: On-board vision autonomous landing techniques for quadrotor: a survey. In: IEEE 35th Chinese Control Conference (CCC), pp. 10284–10289, July 2016

4. Amidi, O., Kanade, T., Miller, R.: Vision-based autonomous helicopter research at Carnegie Mellon robotics institute 1991–1997. American Helicopter Society (1998)
5. Cantelli, L., Laudani, P., Melita, C.D., Muscato, G.: UAV/UGV cooperation to improve navigation capabilities of a mobile robot in unstructured environments. In: Proceedings of CLAWAR 2016, London, September 2016
6. Cantelli, L., Mangiameli, M., Melita, C.D., Muscato, G.: UAV/UGV cooperation for surveying operations in humanitarian demining. In: 11th IEEE International Symposium on Safety Security and Rescue Robotics, 21–26 October, Linkoping, Sweden (2013)
7. Cantelli, L., Lo Presti, M., Mangiameli, M., Melita, C.D., Muscato, G.: Autonomous cooperation between UAV and UGV to improve navigation and environmental monitoring in rough environments. In: 10th International Symposium on Humanitarian Demining coupled with the 11th IARP WS HUDEM2013, 23 April 2013, ibenik, Croatia (2013)
8. De Benedetti, M., D'Urso, F., Messina, F., Pappalardo, G., Santoro, C.: UAV-based aerial monitoring: a performance evaluation of a self-organising flocking algorithm. In: Proceedings of 2015 IEEE International Conference on P2P, Parallel, Grid, Cloud and Internet Computing (3PGCIC) (2015)
9. De Benedetti, M., D'Urso, F., Messina, F., Pappalardo, G., Santoro, C.: Self-Organising UAVs for wide area fault-tolerant aerial monitoring. In: WOA2015 CEUR Workshop Proceedings, Napoli, pp. 142–145, 17–19 Giugno 2015. ISSN 1613–0073
10. Hough, P.V.C.: Method and means for recognizing complex patterns. US Patent 3,069,654 (1962)
11. Kalal, Z., Mikolajczyk, K., Matas, J.: Tracking-learning-detection. IEEE Trans. Pattern Anal. Mach. Intell. 1409–1422
12. Battiato, S., et al.: An integrated system for vehicle tracking and classification. Expert Syst. Appl. $42(21)$, 7263–7275 (2015)
13. Smeulders, A.W.M., et al.: Visual tracking: an experimental survey. IEEE Trans. Pattern Anal. Mach. Intell. $36(7)$, 1442–1468 (2014)
14. Maggio, E., Cavallaro, A.: Video Tracking: Theory and Practice. Wiley, Hoboken (2011)
15. Yilmaz, A., Javed, O., Shah, M.: Object tracking: a survey. ACM Comput. Surv. (CSUR) $38(4)$, 13 (2006)
16. Babenko, B., Yang, M.-H., Belongie, S.: Robust object tracking with online multiple instance learning. IEEE Trans. Pattern Anal. Mach. Intell. $33(8)$, 1619–1632 (2011)
17. Bao, C., et al.: Real time robust l1 tracker using accelerated proximal gradient approach. In: 2012 IEEE Conference on Computer Vision and Pattern Recognition (CVPR). IEEE (2012)
18. Black, M.J., Jepson, A.D.: Eigentracking: robust matching and tracking of articulated objects using a view-based representation. Int. J. Comput. Vis. $26(1)$, 63–84 (1998)
19. Grabner, H., Leistner, C., Bischof, H.: Semi-supervised on-line boosting for robust tracking. In: Forsyth, D., Torr, P., Zisserman, A. (eds.) ECCV 2008. LNCS, vol. 5302, pp. 234–247. Springer, Heidelberg (2008). doi:10.1007/978-3-540-88682-2_19
20. Hare, S., et al.: Struck: structured output tracking with kernels. IEEE Trans. Pattern Anal. Mach. Intell. $38(10)$, 2096–2109 (2016)
21. Kwon, J., Lee, K.M.: Visual tracking decomposition. In: 2010 IEEE Conference on Computer Vision and Pattern Recognition (CVPR). IEEE (2010)

One-Step Time-Dependent Future Video Frame Prediction with a Convolutional Encoder-Decoder Neural Network

Vedran Vukotić[1,2,3($\boxtimes$)], Silvia-Laura Pintea[2], Christian Raymond[1,3], Guillaume Gravier[1,4], and Jan C. van Gemert[2]

[1] INRIA/IRISA Rennes, Rennes, France
{vedran.vukotic,christian.raymond,guillaume.gravier}@irisa.fr
[2] TU Delft, Delft, The Netherlands
[3] INSA Rennes, Rennes, France
[4] CNRS, Rennes, France
{s.l.pintea,j.c.vangemert}@tudelft.nl

Abstract. There is an inherent need for autonomous cars, drones, and other robots to have a notion of how their environment behaves and to anticipate changes in the near future. In this work, we focus on anticipating future appearance given the current frame of a video. Existing work focuses on either predicting the future appearance as the next frame of a video, or predicting future motion as optical flow or motion trajectories starting from a single video frame. This work stretches the ability of CNNs (Convolutional Neural Networks) to predict an anticipation of appearance at an arbitrarily given future time, not necessarily the next video frame. We condition our predicted future appearance on a continuous time variable that allows us to anticipate future frames at a given temporal distance, directly from the input video frame. We show that CNNs can learn an intrinsic representation of typical appearance changes over time and successfully generate realistic predictions at a deliberate time difference in the near future.

Keywords: Action forecasting · Future video frame prediction · Appearance prediction · Scene understanding · Generative models · CNNs

1 Introduction

For machines to successfully interact in the real world, anticipating actions and events and planning accordingly, is essential. This is a difficult task, despite the recent advances in deep and reinforcement learning, due to the demand of large annotated datasets. If we limit our task to anticipating future appearance, annotations are not needed anymore. Therefore, machines have a slight advantage, as they can employ the vast collection of unlabeled videos available, which is perfectly suited for unsupervised learning methods. To anticipate future appearance

© Springer International Publishing AG 2017
S. Battiato et al. (Eds.): ICIAP 2017, Part I, LNCS 10484, pp. 140–151, 2017.
https://doi.org/10.1007/978-3-319-68560-1_13

based on current visual information, a machine needs to successfully be able to recognize entities and their parts, as well as to develop an internal representation of how movement happens with respect to time.

We make the observation that time is continuous, and thus, video frame-rate is an arbitrary discretization that depends on the camera sensor only. Instead of predicting the next discrete frame from a given input video frame, we aim at predicting a future frame at a given continuous temporal distance Δt away from the current input frame. We achieve this by conditioning our video frame prediction on a time-related input variable.

In this work we explore one-step, long-term video frame prediction, from an input frame. This is beneficial both in terms of computational efficiency, as well as avoiding the propagation and accumulation of prediction errors, as in the case of sequential/iterative prediction of each subsequent frame from the previous predicted frame. Our work falls into the autoencoding category, where the current video frame is presented as input and an image resembling the anticipated future is provided as output. Our proposed method consists of: an encoding CNN (Convolutional Neural Network), a decoding CNN, and a separate branch, parallel to the encoder, which models time and allows us to generate predictions at a given time distance in future.

1.1 Related Work

Predicting Future Actions and Motion. In the context of action prediction, it has been shown that it is possible to use high-level embeddings to anticipate future actions up to one second before they begin [23]. Predicting the future event by retrieving similar videos and transferring this information, is proposed in [28]. In [8] a hierarchical representation is used for predicting future actions. Predicting a future activity based on analyzing object trajectories is proposed in [6]. In [3], the authors forecast human interaction by relying on body-pose trajectories. In the context of robotics, in [7] human activities are anticipated by considering the object affordances. While these methods focus on predicting high-level information—the action that will be taken next, we focus on predicting low-level information, a future video frame appearance at a given future temporal displacement from a given input video frame. This has the added value that it requires less supervision.

Anticipating future movement in the spatial domain, as close as possible to the real movement, has also been previously considered. Here, the methods start from an input image at the current time stamp and predict motion—optical flow or motion trajectories—at the next frame of a video. In [9] images are aligned to their nearest neighbour in a database and the motion prediction is obtained by transferring the motion from the nearest neighbor to the input image. In [12], structured random forests are used to predict optical flow vectors at the next time stamp. In [11], the use of LSTM (Long Short Term Memory Networks) is advised towards predicting Eulerian future motion. A custom deep convolutional neural network is proposed in [27] towards future optical flow prediction. Rather than predicting the motion at the next video frame through optical flow, in [25] the authors propose to predict motion trajectories using variational autoencoders. This is similar to predicting optical flow vectors, but given the temporal consistency of the trajectories, it offers greater accuracy. Dissimilar to these methods which predict future motion, we aim to predict the video appearance information at a given continuous future temporal displacement from an input video frame.

Predicting Future Appearance. One intuitive trend towards predicting future information is predicting future appearance. In [26], the authors propose to predict both appearance and motion for street scenes using top cameras. Predicting patch-based future video appearance, is proposed in [14], by relying on large visual dictionaries. In [29] future video appearance is predicted in a hirarchical manner, by first predicting the video structure, and subsequently the individual frames. Similar to these methods, we also aim at predicting the appearance of future video frames, however we condition our prediction on a time parameter than allows us to perform the prediction efficiently, in one step.

Rather than predicting future appearance from input appearance information, hallucinating possible images has been a recent focus. The novel work in [24] relies on the GAN (Generative Adversarial Network) model [13] to create not only the appearance of an image, but also the possible future motion. This is done using spatio-temporal convolutions that discriminate between foreground and background. Similarly, in [17] a temporal generative neural network is proposed towards generating more robust videos. These generative models can be conditioned on certain information, to generate feasible outputs given the specific conditioning input [15]. Dissimilar to them, we rely on an autoencoding model. Autoencoding methods encode the current image in a representation space that is suitable for learning appearance and motion, and decode such representations to retrieve the anticipated future. Here, we propose to use video frame appearance towards predicting future video frames. However, we condition it on a given time indicator which allows us to predict future appearance at given temporal distances in the future.

2   Time-Dependent Video Frame Prediction

To tackle the problem of anticipating future appearance at arbitrary temporal distances, we deploy an encoder-decoder architecture. The encoder has two separate branches: one to receive the input image, and one to receive the desired temporal displacement Δt of the prediction. The decoder takes the input from the encoder and generates a feasible prediction for the given input image and the desired temporal displacement. This is illustrated in Fig. 1. The network receives as inputs an image and a variable Δt, $\Delta t \in \mathbb{R}^+$, indicating the time difference from the time of the provided input image, t_0, to the time of the desired prediction. The network predicts an image at the anticipated future time $t_0 + \Delta t$. We use a similar architecture to the one proposed in [20]. However, while their architecture is made to encode RGB images and a continuous angle variable to produce RGBD as output, our architecture is designed to take as input a monochromatic image and a continuous time variable, Δt, and to produce a monochromatic image, resembling a future frame, as output.

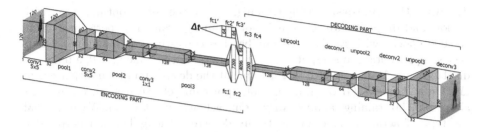

Fig. 1. Our proposed architecture consists of two parts: (i) an encoder part consisting of two branches: the first one taking the current image as input, and the second one taking as input an arbitrary time difference Δt to the desired prediction and (ii) a decoder part that generates an image, as anticipated, at the desired input time difference, Δt.

More specifically, the architecture consists of the following:

1. *an encoding part* composed of two branches:
   - *an image encoding branch* defined by 4 convolutional layers, 3 pooling layers and 2 fully-connected layers at the end;
   - *a time encoding branch* consisting of 3 fully-connected layers.

   The final layers of the two branches are concatenated together, forming one bigger layer that is then provided to the decoding part.
2. *a decoding part* composed of 2 fully-connected layers, 3 "unpooling" (upscaling) layers, and 3 "deconvolutional" (transpose convolutional) layers.

The input time-indicator variable is continuous and allows for appearance anticipations at arbitrary time differences. Training is performed by presenting to the network batches of $\{I_x, \Delta t, I_y\}$ tuples, where I_x represents an input image at current relative time t_0, and Δt represents a continuous variable indicating the time difference to the future video frame, and I_y represents the actual video frame at $t_0 + \Delta t$.

Predictions are obtained in one step. For every input image I_x and continuous time difference variable Δt, a $\{I, \Delta t\}$ pair is given to the network as input, and an image representing the appearance anticipation I_y after a time interval Δt is directly obtained as output. No iterative steps are performed.

3   Experiments

3.1   Experimental Setup

We evaluate our method by generating images of anticipated future appearances at multiple time distances, and comparing them both visually and through MSE

(Mean Squared Error) with the true future frames. We also compare to a CNN baseline that iteratively predicts the future video frame at $k\Delta t$ ($k = 1, 2, ...$) temporal displacements, from previous predictions.

Training Parameters. During training, we use the Adam optimizer [5], with L_2 loss and dropout rate set to 80% for training. Training is performed up to 500,000 epochs with randomized minibatches consisting of 16 samples, where each sample contains one input image at current relative time $t_0 = 0$, a temporal displacement Δt and the real target frame at the desired temporal displacement Δt. On a *Titan X* GPU, training took approximately 16 h with, on average, about 100,000 training samples (varying in each action category). We argue that the type of action can be automatically detected, and is better incorporated by training a network per action category. Thus, we opt to perform separate preliminary experiments for each action instead of training one heavy network to anticipate video frames corresponding to all the different possible actions.

Network Architecture. Given that the input, and thus also the output, image size is $120 \times 120 \times 1$ (120×120 grayscale images), in our encoder part, we stack convolutional and pooling layers that yield consecutive feature maps of the following decreasing sizes: 120×120, 60×60, 30×30 and 15×15, with an increasing number of feature maps per layer, namely 32, 64 and 128 respectively. Fully-connected layers of sizes 7,200 and 4,096 are added at the end. The separated branch of the encoder that models time consists of 4 fully connected layers of size 64, where the last layer is concatenated to the last fully-connected layer of the encoder convolutional neural network. This yields an embedding of size 4160 that is presented to the decoder. Kernel sizes used for the convolutional operations start at 5×5 in the first layers and decrease to 2×2 and 1×1 in the deeper layers of the encoder.

For the decoder, the kernel sizes are the same as for the encoder, but ordered in the opposite direction. The decoder consists of interchanging "unpooling" (upscaling) and "deconvolutiton" (transpose convolution) layers, yielding feature maps of the same sizes as the image-encoding branch of the encoder, only in the opposing direction. For simplicity, we implement pooling as a convolution with 2×2 strips and unpooling as a 2D transpose convolution.

3.2   Dataset

We use the KTH human action recognition dataset [18] for evaluating our proposed method. The dataset consists of 6 different human actions, namely: *walking*, *jogging*, *running*, *hand-clapping*, *hand-waving* and *boxing*. Each action is performed by 25 actors. There are 4 video recordings for each action performed by each actor. Inside every video recording, the action is performed multiple times and information about the time when each action starts and ends is provided with the dataset.

To evaluate our proposed method, we randomly split the dataset by actors, in a training set—with 80% of the actors, and a testing set—with 20% of the actors. By doing so, we ensure that no actor is present in both the training and the testing split and that the network can generalize well with different looking people and does not overfit to specific appearance characteristics of specific actors. The dataset provides video segments of each motion in two directions— e.g. walking from right to left, and from left to right. This ensures a good setup for checking if the network is able to understand human poses and locations, and correctly anticipate the direction of movement. The dataset was preprocessed as follows: frames of original size 160×120 px were cropped to 120×120 px, and the starting/ending time of each action were adjusted accordingly to match the new cropped area. Time was estimated based on the video frame-rate and the respective frame number.

3.3   Experimental Results

Our method is evaluated as follows: an image at a considered time, $t_0 = 0$ and a time difference Δt is given as input. The provided output represents the anticipated future frame at time $t_0 + \Delta t$, where Δt represents the number of milliseconds after the provided image.

The sequential encoder-decoder baseline method is evaluated by presenting solely an image, considered at time $t_0 = 0$ and expecting an image anticipating the future at $t_0 + \Delta t_b$ as output. This image is then fed back into the network in order to produce an anticipation of the future at time $t_0 + k\Delta t_b$, $k = 1, 2, 3, ...$

For simplicity, we consider $t_0 = 0$ ms and refer to Δt as simply t. It is important to note that our method models time as a continuous variable. This enables the model to predict future appearances at previously unseen time intervals, as in Fig. 3. The model is trained on temporal displacements defined by the framerate of the training videos. Due to the continuity of the temporal variable, it

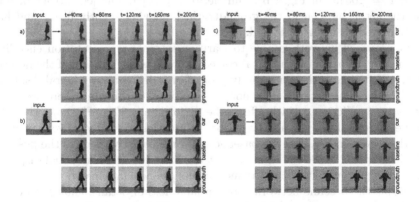

Fig. 2. Comparison of predictions for (a) a person walking to the left, (b) a person walking to the right, (c) a person waving their hands and (d) a person slowly clapping with their hands. The third set of images in each group represent the actual future frame—the groundtruth.

can successfully generate predictions for: (i) temporal displacements found in the videos (e.g. $t=\{40\,ms,\ 80\,ms,\ 120\,ms,\ 160\,ms,\ 200\,ms\}$), (ii) unseen temporal displacement within the values found in the training videos (e.g. $t=\{60\,ms,\ 100\,ms,\ 140\,ms,\ 180\,ms\}$) and (iii) unseen temporal displacement after the maximal value encountered during training (e.g. $t=220\,ms$).

Fig. 3. Prediction of seen and unseen temporal displacements.

Figure 2(a) illustrates a person moving from right to left, from the camera viewpoint, at walking speed. Despite the blurring, especially around the left leg when predicting for $t = 120$ ms, our network correctly estimates the location of the person and position of body parts. Figure 2(b) illustrates a person walking, from left to right. Our proposed network correctly localized the person and the body parts. The network is able to estimate the body pose, and thus the direction of movement and correctly predicts the displacement of the person to the right for any given time difference. The network captures the characteristics of the human gait, as it predicts correctly the alternation in the position of the legs. The anticipated future frame is realistic but not always perfect, as it is hard to perfectly estimate walking velocity solely from one static image. This can be seen at $t = 200$ ms in Fig. 2(b). Our network predicts one leg further behind while the actor, as seen in the groundtruth, is moving slightly faster and has already moved their leg past the knee of the other leg.

Our proposed network is able to learn an internal representation encoding the stance of the person such that it correctly predicts the location of the person, as well as anticipates their new body pose after a deliberate temporal displacement. The baseline network does not have a notion of time and therefore relies on iterative predictions, which affects the performance. Figure 2 shows that the baseline network loses the ability to correctly anticipate body movement after some time. Also in Fig. 2(a) the baseline network correctly predicts the position of the legs up to $t = 80$ ms, after that, it correctly predicts the global displacement of the person, but body part movements are not anticipated correctly. At $t > 160$ ms the baseline network shows a large loss of details, enough to cause its inability to correctly model body movement. Therefore, it displays fused legs where they should be separated, as part of the next step the actor is making. Our proposed architecture correctly models both global person displacement and body pose, even at $t = 200$ ms (Fig. 4).

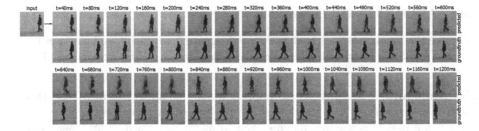

Fig. 4. Long distance predictions. For larger temporal displacements artifacting becomes visible. The anticipated location of the person begins to differ from the groundtruth towards the end of the total motion duration.

Figure 2(c) displays an actor *handwaving*. Our proposed network successfully predicts upward movement of the arms and generates images accordingly. Here however, more artifacts are noticeable due to the bidirectional motion of hands during *handwaving*, which is ambiguous. It is important to note that although every future anticipation is independent from the others, they are all consistent: i.e. it does not happen that the network predicts one movement for t_1 and a different movement for t_2 that is inconsistent with the t_1 prediction. This is a strong indicator that the network learns an embedding of appearance changes over time, the necessary filters relevant image areas and synthesizes correct future anticipations.

As expected, not every action is equally challenging for the proposed architecture. Table 1 illustrate MSE scores averaged over multiple time differences, t, and for different predictions from the KTH test set. MSE scores were computed on dilated edges of the groundtruth images to only analyze the part around the person and remove the influence of accumulated variations of the background. A Canny edge detector was used on the groundtruth images. The edges were dilated by 11 px and used as a mask for both the groundtruth image and the predicted image. MSE values were computed solely on the masked areas. We compare our proposed method with the baseline CNN architecture. The average MSE scores, given in Table 1, show that our proposed method outperforms the encoder-decoder CNN baseline by a margin of 13.41, on average, which is due to the iterative process of the baseline network.

Table 1. Average MSE over multiple time distances and multiple video predictions, on the different action categories of KTH. We compare our method with the iterative baseline CNN, and show that our method on average performs better than the baseline in terms of MSE (lower is better).

Method	Jogging	Running	Walking	Clapping	Waving	Boxing	Avg
Baseline	30.64	40.88	30.87	43.23	43.71	46.22	39.26
Our method	11.66	17.35	19.26	33.93	35.19	37.71	25.85

3.4   Ambiguities and Downsides

There are a few key factors that make prediction more difficult and cause either artifacts or loss of details in the predicted future frames. Here we analyze these factors.

(i) Ambiguities in body-pose happen when the subject is in a pose that does contain inherent information about the future. A typical example would be when a person is waving, moving their arms up and down. If an image with the arms at a near horizontal position is fed to the network as input, this can results in small artifacts, as visible in Fig. 2(c) where for larger time intervals t, there are visible artifacts that are part of a downward arm movement. A more extreme case is shown in Fig. 5(a) where not only does the network predict the movement wrong, but it also generates many artifacts with a significant loss of detail, which increases with the time difference, t.

(ii) Fast movement causes loss of details when the videos provided for training do not offer a high-enough framerate. Examples of this can be seen in Figs. 5(b) and (c) where the increased speed in jogging and an even higher speed in running generate significant loss of details. Although our proposed architecture can generate predictions at arbitrary time intervals t, the network is still trained on discretized time intervals derived from the video framerate. These may not be sufficient for the network to learn a good model. We believe this causes the loss of details and artifacts, and using higher framerate videos during training would alleviate this.

(iii) Decreased contrast between the subject and the background describes a case where the intensity values corresponding to the subject are similar to the ones of the background. This leads to an automatic decrease of MSE values, and a more difficult convergence of the network for such cases. Thus, this causes to

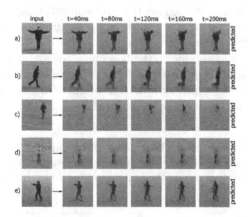

Fig. 5. Examples of poorly performing future anticipations: (a) loss of details in waving, (b) loss of details in jogging, (c) extreme loss of details in running, (d) loss of details with low contrast and (e) artifacts in boxing.

loss of details and artifacts. This can be seen in Fig. 5(d). Such effect would be less prominent in the case in which color images would be used during training. **(iv) Excessive localization of movements** happens when the movements of the subject are small and localized. A typical example is provided by the boxing action, as present in the KTH dataset. Since the hand movement is close to the face and just the hand gets sporadically extended, the network has more difficulties in tackling this. Despite the network predicting a feasible movement, often artifacts appear for bigger time intervals t, as visible in Fig. 5(e).

Despite the previously enumerated situations leading our proposed architecture to predictions that display loss of details and artifacts, most of these can be tackled and removed by either increasing the framerate, the resolution of the training videos, or using RGB information.

4   Conclusion

In this work, we present a convolutional encoder-decoder architecture with a separate input branch that models time in a continuous manner. The aim is to provide anticipations of future video frames for arbitrary positive temporal displacements Δt, given a single image at current time ($t_0 = 0$). We show that such an architecture can successfully learn time-dependant motion representations and synthesizes accurate anticipation of future appearance for arbitrary time differences $\Delta t > 0$. We compare our proposed architecture against a baseline consisting of an analogous convolutional encoder-decoder architecture that does not have a notion of time and relies on iterative predictions. We show that out method outperforms the baseline both in terms of visual similarity to the groundtruth future video frames, as well as in terms of mean squared error with respect to it. We additionally analyze the drawbacks of our architecture and present possible solutions to tackle them. This work shows that convolutional neural networks can inherently model time without having a clear time domain representation. This is a novel notion that can be extended further and that generates high quality anticipations of future video frames for arbitrary temporal displacements. This is achieved without explicitly modelling the time period between the provided input video frame and the requested anticipation.

References

1. Fouhey, D.F., Zitnick, C.L.: Predicting object dynamics in scenes. In: CVPR, pp. 2019–2026 (2014)
2. Gatys, L.A., Ecker, A.S., Bethge, M.: A neural algorithm of artistic style. CoRR (2015)
3. Huang, D.-A., Kitani, K.M.: Action-reaction: forecasting the dynamics of human interaction. In: Fleet, D., Pajdla, T., Schiele, B., Tuytelaars, T. (eds.) ECCV 2014. LNCS, vol. 8695, pp. 489–504. Springer, Cham (2014). doi:10.1007/978-3-319-10584-0_32

4. Johnson, J., Alahi, A., Fei-Fei, L.: Perceptual losses for real-time style transfer and super-resolution. CoRR (2016)
5. Kingma, D., Ba, J.: Adam: a method for stochastic optimization. CoRR (2014)
6. Kitani, K.M., Ziebart, B.D., Bagnell, J.A., Hebert, M.: Activity forecasting. In: Fitzgibbon, A., Lazebnik, S., Perona, P., Sato, Y., Schmid, C. (eds.) ECCV 2012. LNCS, vol. 7575, pp. 201–214. Springer, Heidelberg (2012). doi:10.1007/978-3-642-33765-9_15
7. Koppula, H.S., Saxena, A.: Anticipating human activities using object affordances for reactive robotic response. PAMI **38**(1), 14–29 (2016)
8. Lan, T., Chen, T.-C., Savarese, S.: A hierarchical representation for future action prediction. In: Fleet, D., Pajdla, T., Schiele, B., Tuytelaars, T. (eds.) ECCV 2014. LNCS, vol. 8691, pp. 689–704. Springer, Cham (2014). doi:10.1007/978-3-319-10578-9_45
9. Liu, C., Yuen, J., Torralba, A.: Sift flow: dense correspondence across scenes and its applications. PAMI **33**(5), 978–994 (2011)
10. Mottaghi, R., Bagherinezhad, H., Rastegari, M., Farhadi, A.: Newtonian image understanding: unfolding the dynamics of objects in static images. CoRR (2015)
11. Pintea, S.L., Gemert, J.C.: Making a case for learning motion representations with phase. In: Hua, G., Jégou, H. (eds.) ECCV 2016. LNCS, vol. 9915, pp. 55–64. Springer, Cham (2016). doi:10.1007/978-3-319-49409-8_8
12. Pintea, S.L., van Gemert, J.C., Smeulders, A.W.M.: Déjà Vu. In: Fleet, D., Pajdla, T., Schiele, B., Tuytelaars, T. (eds.) ECCV 2014. LNCS, vol. 8691, pp. 172–187. Springer, Cham (2014). doi:10.1007/978-3-319-10578-9_12
13. Radford, A., Metz, L., Chintala, S.: Unsupervised representation learning with deep convolutional generative adversarial networks. CoRR (2015)
14. Ranzato, M., Szlam, A., Bruna, J., Mathieu, M., Collobert, R., Chopra, S.: Video (language) modeling: a baseline for generative models of natural videos. CoRR (2014)
15. Reed, S., Akata, Z., Yan, X., Logeswaran, L., Schiele, B., Lee, H.: Generative adversarial text to image synthesis. CoRR (2016)
16. Ruder, M., Dosovitskiy, A., Brox, T.: Artistic style transfer for videos. CoRR (2016)
17. Saito, M., Matsumoto, E.: Temporal generative adversarial nets. CoRR (2016)
18. Schuldt, C., Laptev, I., Caputo, B.: Recognizing human actions: a local SVM approach. In: ICPR, vol. 3, pp. 32–36. IEEE (2004)
19. Springenberg, J.T., Dosovitskiy, A., Brox, T., Riedmiller, M.: Striving for simplicity: the all convolutional net. arXiv preprint (2014) arXiv:1412.6806
20. Tatarchenko, M., Dosovitskiy, A., Brox, T.: Multi-view 3D models from single images with a convolutional network. In: Leibe, B., Matas, J., Sebe, N., Welling, M. (eds.) ECCV 2016. LNCS, vol. 9911, pp. 322–337. Springer, Cham (2016). doi:10.1007/978-3-319-46478-7_20
21. van den Oord, A., Kalchbrenner, N., Kavukcuoglu, K.: Pixel recurrent neural networks. CoRR (2016)
22. van den Oord, A., Kalchbrenner, N., Vinyals, O., Espeholt, L., Graves, A., Kavukcuoglu, K.: Conditional image generation with pixelCNN decoders. CoRR (2016)
23. Vondrick, C., Pirsiavash, H., Torralba, A.: Anticipating the future by watching unlabeled video. CoRR (2015)
24. Vondrick, C., Pirsiavash, H., Torralba, A.: Generating videos with scene dynamics. In: NIPS, pp. 613–621 (2016)

25. Walker, J., Doersch, C., Gupta, A., Hebert, M.: An uncertain future: forecasting from static images using variational autoencoders. In: Leibe, B., Matas, J., Sebe, N., Welling, M. (eds.) ECCV 2016. LNCS, vol. 9911, pp. 835–851. Springer, Cham (2016). doi:10.1007/978-3-319-46478-7_51
26. Walker, J., Gupta, A., Hebert, M.: Patch to the future: unsupervised visual prediction. In: CVPR, pp. 3302–3309. IEEE (2014)
27. Walker, J., Gupta, A., Hebert, M.: Dense optical flow prediction from a static image. In: ICCV, pp. 2443–2451 (2015)
28. Yuen, J., Torralba, A.: A data-driven approach for event prediction. In: Daniilidis, K., Maragos, P., Paragios, N. (eds.) ECCV 2010. LNCS, vol. 6312, pp. 707–720. Springer, Heidelberg (2010). doi:10.1007/978-3-642-15552-9_51
29. Villegas, R., Yang, J., Zou, Y., Sohn, S., Lin, X., Lee, H.: Learning to generate long-term future via hierarchical prediction. In: ICML (2017)

Joint Orientations from Skeleton Data for Human Activity Recognition

Annalisa Franco[✉], Antonio Magnani, and Dario Maio

C.d.L. Ingegneria e Scienze Informatiche,
University of Bologna, Via Sacchi, 3, Cesena, FC, Italy
{annalisa.franco,antonio.magnani,dario.maio}@unibo.it

Abstract. The recognition of activities performed by humans, in a non-intrusive and non-cooperative way, is a very relevant task in the development of Ambient Intelligence applications aimed at improving the quality of life by realizing digital environments that are adaptive, sensitive and reactive to the presence (or absence) of the users and to their behavior. In this paper, we present an activity recognition approach where angle information is used to encode the human body posture, i.e. the relative position of its different parts; such information is extracted from skeleton data (joint orientations), acquired by a well known cost-effective depth sensor (Kinect). The system is evaluated on a well-known dataset (CAD-60 (Cornell Activity Dataset) for comparison with the state of the art; moreover, due to the lack of datasets including skeleton orientations, a new benchmark named OAD (Office Activity Dataset) has been internally acquired and will be released to the scientific community. The tests confirm the efficacy of the proposed model and its feasibility for scenarios of varying complexity.

1 Introduction

Automated high-level human activity analysis and recognition play a fundamental role in many relevant and heterogeneous application fields such as video-surveillance, ambient assisted living, automatic video annotation or human-computer interfaces. Of course different applications need specific approaches to be designed and implemented; general-purpose solutions, though highly desirable, are very difficult to implement due to the differences in the source of information, the requirements in terms of efficiency, the environmental factors which have a significant impact on performance, etc. This work focuses on human activity recognition in indoor environments which has typical applications in fall-detection of elderly people, abnormal human behavior detection or human computer interfaces. In our opinion unobtrusiveness is one of the most important and interesting features of ambient intelligence applications; to meet this requirement, the proposal of this paper is a vision-based technique where simple cameras are used as input devices and the users are not require to wear neither to actively interact with sensors of different nature.

© Springer International Publishing AG 2017
S. Battiato et al. (Eds.): ICIAP 2017, Part I, LNCS 10484, pp. 152–162, 2017.
https://doi.org/10.1007/978-3-319-68560-1_14

With respect to other application scenarios such as video-surveillance, indoor environments offer several advantages: the input data are somehow more "controlled" and easier to process (e.g. to segment the subjects in the scene), the number of possible users is generally limited and input devices, such as RGB-D cameras, can be successfully adopted for data acquisition. The problem of activity recognition is however still complex if we consider that the users are not cooperative and a real-time processing is needed to produce timely and useful information. This paper proposes an activity recognition technique based on the use of RGB-D cameras, and in particular the Kinect sensor, for data acquisition. To the best of our knowledge all the existing techniques based on skeleton data only exploit 3D joint position, while joint orientation is typically neglected. Aim of this work is to evaluate the reliability of the joint orientation estimates provided by Kinect and to verify their effectiveness for action recognition.

The paper is organized as follows: an overview of the state-of-the-art is provided in Sect. 2, Sect. 3 presents the proposed approach, the results of the experimental evaluation are given in Sect. 4 and finally Sect. 5 draws some conclusions an presents possible future research directions.

2 State of the Art

Vision-based activity recognition techniques do not require the use of special devices and the only source of information is represented by cameras placed in the environment which continuously acquire video sequences. Many works adopt common RGB cameras to acquire information from the environment, but undoubtedly the widespread diffusion of low-cost RGB-D sensors, such as the well-known Microsoft Kinect, greatly boosted the research on this topic. Even though a few hybrid approaches combining gray-scale and depth information have been proposed (e.g. [1]), RGB-D sensors alone have been widely used for activity analysis [2] and several benchmarks have been released to facilitate the comparative evaluation of recognition algorithms [3,4]. The most attractive feature of the Kinect sensor is the ability to capture depth images, coupled with the possibility of tracking rather accurately skeletons of individuals in the scene. The skeleton representation provided by Kinect which consists of a set of joints, each described in terms of position and orientation in the 3D space. Such information is extremely useful for human activity analysis as confirmed by many approaches in the literature. A few works exploit only the depth information (and not the skeleton), and typically perform an image segmentation to identify some relevant posture features from the human body [5]. Most of the approaches perform a skeleton analysis, adopting different representations of the set of joints such as the simple joint coordinates, normalized according to some body reference measure [6,7] or joint distances [8], EigenJoints in [9] where PCA is applied to static and dynamic posture features to create a motion model, histograms of 3D joints [10], kinematic features, obtained observing the angles between couples of joints [11], Gaussian Mixture Models representing the 3D positions of skeleton joints [12], Dynamic Bayesian Mixture Model of 3D skeleton features [13]

or spatio-temporal interest points and descriptors derived from the depth image
[14]. Another common approach is to adopt a hierarchical representation where
an activity is composed of a set of sub-activities, also called *actionlets* [15–18].
Finally a few works also analyze the interaction of humans with objects to obtain
a better scene understanding. The authors of [18] adopt a Markov random field
where the nodes represent objects and sub-activities, and the edges represent the
relationships between object affordances, their relations with sub-activities, and
their evolution over time is proposed; in [19] the authors propose a graph-based
representation.

3   Proposed Approach

The idea behind the proposed approach is to encode each frame of a video
sequence as a set of angles derived from the human skeleton, which summarize
the relative positions of the different body parts. This proposal presents some
advantages: the use of skeleton data ensures a higher level of privacy for the
user with respect to RGB sequences, and the angle information derived from
skeletons is intrinsically normalized and independent from the user's physical
build. The skeleton information extracted by the Kinect [20] consists of a set
of n joints $J = \{j_1, j_2, ..., j_n\}$ where the number n of joints depends on the
software used for the skeleton tracking (i.e. typical configurations include 15,
20 or 25 joints). Each joint $j_i = (\mathbf{p_i}, \overrightarrow{\mathbf{o_i}})$ is described by its 3D position $\mathbf{p_i}$
and its orientation $\overrightarrow{\mathbf{o_i}}$ with respect to "the world". Our approach exploits the
information given by joint orientations to compute relevant angles whose spatio-
temporal evolution characterizes an activity. We consider three different families
of angles (see Fig. 1a and b):

- θ_{ab}: angle between the orientations $\overrightarrow{\mathbf{o_a}}$ and $\overrightarrow{\mathbf{o_b}}$ of joints j_a and j_b. Angles θ_{ab}
  are computed for the following set of couples of joints:

$$A_\theta = \{(j_1, j_3), (j_1, j_5), (j_3, j_4), (j_5, j_6), (j_0, j_{11}), (j_0, j_{12}), (j_7, j_8), (j_9, j_{10})\}$$

- φ_{ab}: angle between the orientation $\overrightarrow{\mathbf{o_a}}$ of j_a and the segment $\overrightarrow{j_a j_b}$ connecting
  j_a to j_b (we can consider the segment as the bone that interconnects the two
  joints). Angles φ_{ab} are computed for the following set of couples of joints:

$$A_\varphi = \{(j_3, j_1), (j_3, j_4), (j_4, j_3), (j_4, j_{11}), (j_{11}, j_4), (j_5, j_1), (j_5, j_6), (j_6, j_5),$$

$$(j_6, j_{12}), (j_{12}, j_6), (j_2, j_7), (j_7, j_2), (j_7, j_8), (j_2, j_9), (j_9, j_2), (j_9, j_{10})\}$$

- α_{bac}: angle between the segment $\overrightarrow{j_a j_b}$ connecting j_a to j_b and $\overrightarrow{j_a j_c}$ that connects j_a to j_c. Angles α_{abc} are computed for the following triplets of joints:

$$A_\alpha = \{(j_2, j_7, j_8), (j_7, j_8, j_{13}), (j_2, j_9, j_{10}), (j_9, j_{10}, j_{14})\}$$

We consider only subset of the possible angles, mainly obtained from the
joints of the upper part of the body, because not all the angles are really informative: for example the angles between head and neck are almost constant over

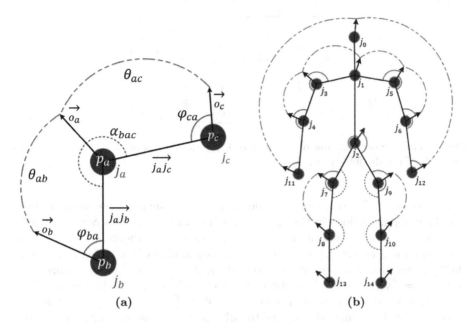

Fig. 1. (a) Representation of a subset of joints $j_a = (p_a, \overrightarrow{o_a})$, $j_b = (p_b, \overrightarrow{o_b})$ and $j_c = (p_c, \overrightarrow{o_c})$ and related angles θ, φ and α. (b) The 28 angles used in our experiments computed from a skeleton configuration with 15 joints.

time and does not provide useful information for activity discrimination. Different configurations of angles have been evaluated and compared in (see Sect. 5). Therefore, each frame f_i of the video sequence $S_i, i = 1, .., l$ is represented by a vector obtained as the ordered concatenation of the values of $\theta_i \mid i \in A_\theta$, $\varphi_j \mid j \in A_\varphi$, $\alpha_k \mid k \in A_\alpha$

$$\mathbf{v}_i = (\theta_1, ..., \theta_m, \varphi_1, ...\varphi_n, \alpha_1, ..., \alpha_s)$$

of size $(m + n + s)$.

It is worth noting that the number of frames for each video sequence can be extremely high and certainly not all the resulting feature vectors are significant: the variation of the angles between two subsequent frames is minimal and usually unnoticeable. We decided therefore to adopt a Bag of Word model [21] with a two-fold objective: minimizing the representation of each sequence keeping only the relevant information and producing fixed-length descriptor which can be used to train an activity classifier. The idea is to represent each activity as an histogram of occurrences of some reference postures (see Fig. 2 for a visual representation), derived from the analysis of the training set. A reference dictionary is first built by applying the K-means clustering algorithm [22] to the set of posture features extracted from the training sequences. Since some subjects could be left-handed, all the angle features are mirrored with respect to the x-axis. We denote with k the number of clusters determined (i.e. the

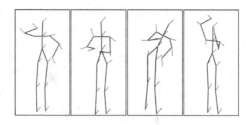

Fig. 2. Visual representation of a subset of key poses corresponding to some cluster centroids of the dictionary W.

dictionary size). The dictionary should encode the basic postures assumed during the different actions in the training set and will be used to represent each sequence as an histogram of occurrences of such basic elements. Given a set of training sequences $TS = \{S_i, i = 1, .., d\}$, representative of the different actions, the k-means clustering algorithm is applied to the associated set of feature vectors $FV = \{\mathbf{v_i}, i = 1, .., d\}$ to obtain a set of k clusters: the cluster centroids are used as words of the reference dictionary $W = \{w_i, i = 1, .., k\}$. The number of clusters k determines the size of the dictionary and is one of the most relevant parameters of the proposed approach. Each sequence is then encoded as a normalized histogram of occurrences of the words in W. Of course the angle features are continuous values and a precise correspondence between the words in the dictionary and the descriptors is very unlikely; therefore when computing the histogram each feature vector f_i is associated to the closest word w_j^* in the dictionary: $j^* = \mathrm{argmin}_j \|f_i - w_j\|$.

A Random Forest Classifier [23] is trained to discriminate the different activities represented in the training set; the classifier consists of an ensemble of decision trees, each trained on a subset of the patterns and a subset of the features and the final classification is obtained combining the decisions of the single sub-trees.

4   Experiments

Several experiments have been conducted to evaluate the sensitivity of the proposed approach to its main parameters (i.e. the set of angles selected and the dictionary size). Despite of the large number of existing benchmarks for activity recognition from skeleton information, joint orientations are generally not available. We used for testing the well-known CAD-60 [15,24], released by the Cornell University, and a newly acquired dataset. CAD-60 contains 60 RGB-D videos where 4 different subjects (two male and two female, one left-handed) perform 12 daily activities in 5 environments (office, kitchen, bedroom, bathroom and living room). The authors of the benchmark propose two settings named *new person*, where a leave-one-out cross-validation is adopted, and *have seen* where the training set includes data from all the subjects. We adopted the *new person* testing protocol, in accordance with all the related works in the literature, to allow for a comparison of the results. Moreover, analogously to other works, the recognition accuracy is measured separately for the different rooms.

4.1   Office Activity Dataset (OAD)

Due to the lack of datasets including information on joint orientations, we decided to acquire a new database of human activities to perform further tests. Data acquisition was carried out in a single environment (office) from several perspectives based on the action being performed. From this point of view the benchmark is more complex than CAD-60 because all the activities need to be compared for activity recognition and the higher number of subjects increases the variability of each action. It contains 14 different activities: *drinking, getting up, grabbing an object from the ground, pour a drink, scrolling book pages, sitting, stacking items, take objects from a shelf, talking on the phone, throwing something in the bin, waving hand, wearing coat, working on computer, writing on paper*. Data was collected from 10 different subjects (five males and five females) aged between 20 and 35, one subject left-handed. The volunteers received only basic information (e.g. *"pour yourself a drink"*) in order to be as natural as possible while performing actions. Each subject performs each activity twice, therefore we have collected overall 280 sequences.

The device used for data acquisition is the Microsoft Kinect V2 whose SDK allows to track 25 different joints (19 of which have their own orientation). For testing, we adopted the same *"new person"* setting of the CAD-60 dataset: a leave-one-out cross-validation with rotation of the test subject. The set of angles used for testing the proposed approach is however the same used for CAD-60. The dataset will be made available online in the Smart City Lab web site (http://smartcity.csr.unibo.it).

4.2   Results

Performance evaluation starts from the analysis of the confusion matrix M where a generic element $M(i,j)$ represents the percentage of patterns of class i classified by the system as belonging to class j. Further synthetic indicators can be derived from the confusion matrix; in particular, we computed precision P and recall R as follows:

$$P = \frac{TP}{TP + FP}, R = \frac{TP}{TP + FN}$$

where *TP, FP* and *FN* represent respectively the True Positives, False Positives and False Negatives which can be easily derived from the extra-diagonal elements of the confusion matrix. In analogy to the proposal in [8], each video sequence is partitioned into three subsequences which are used independently in the tests. The results obtained are summarized in Fig. 3 where the Precision (P) and Recall (R) values are reported for different experimental settings, i.e. variable dictionary size (k) and three subsets of angles considered for skeleton representation. In particular, the efficacy of the joint orientations is assessed by comparing the

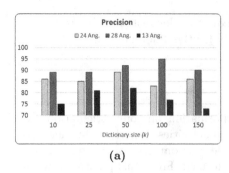

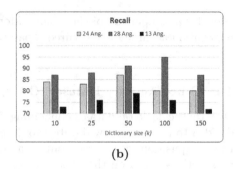

(a)   (b)

Fig. 3. Precision (a) and recall (b) values on CAD-60 with different configurations of angles, as a function of the dictionary size (k).

results of two different settings - 24 angles, (α angles omitted) and 28 angles - with those obtained using only A_α angles, computed between all the existing pairs of neighboring segments (13 angles, no joint orientation is used in this case). The results show that, overall, the accuracy of the proposed technique is good. As expected the dictionary size has a significant impact on the performance; it is worth noting that different actions have often very similar postures (e.g. drinking and talking on the phone) and a value of k excessively low probably determines the reference posture of such activities to collapse in a single word, thus making difficult to correctly distinguish them. On the other hand, a high value of k produces very sparse feature vectors, more sensitive to the presence of noise. The best results have been reached with a value of $k = 100$ which also allows to efficiently perform the classification task. Also the angle configuration is important; the use of 28 angles produces better results both in terms of precision and recall with respect to the version with 24 angles. The limited accuracy of the configuration with 13 angles, where the orientation is not exploited, confirm the effectiveness of joint orientation for accurate posture representation. These results also show that the significance of the angles varies greatly and a few strategical angles can greatly improve the recognition performance. As to the computational complexity, the proposed approach is very efficient, and all the angle configuration are suitable for a real time processing.

The confusion matrix, reported in Table 1, allows to analyze the main causes of errors. The mismatch occurred are all rather comprehensible since they are related to very similar activities (e.g. cooking-chopping, cooking-stirring). In these cases the skeleton information is probably too synthetic to discriminate the two actions which are very similar in terms of posture. A comparison with the state of the art is provided in Table 2 which summarizes the results published in the benchmark website. Despite of the very good accuracy reached by different approaches in recent years, the proposed approach outperforms existing methods, both in terms of precision and recall.

The results on the Office Activity Dataset are reported in Tables 3 and 4 for the standard configuration with 28 angles and $k = 100$. The overall results confirm that this benchmark is more difficult for several reasons: (i) the activities are

Table 1. Confusion matrix using $k = 100$ words and a configuration of 28 angles on CAD-60.

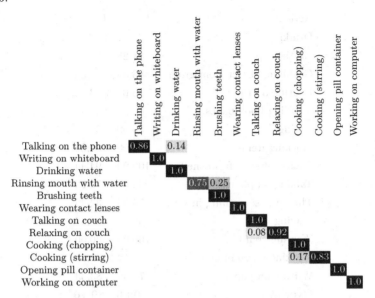

Table 2. Precision (P) and recall (R) of the proposed approach on CAD-60, compared to the results published in the benchmark website. *"*"* indicates that a different protocol was used.

Algorithm	P	R
Proposed approach	**95.0**	**95.0**
Sung et al. [15,24] - 2012	67.9	55.5
Koppula et al. [18] - 2012	80.8	71.4
Zhang and Tian [11] - 2012	86	84
Ni et al. [25] - 2012	Accur: 65.32	-
Yang and Tian [9] - 2013	71.9	66.6
Piyathilaka and Kodagoda [12] - 2013	70*	78*
Ni et al. [1] - 2013	75.9	69.5
Gupta et al. [5] - 2013	78.1	75.4
Wang et al. [17] - 2013	Accur: 74.70	-
Zhu et al. [14] - 2014	93.2	84.6
Faria et al. [13] - 2014	91.1	91.9
Shan and Akella [7] - 2014	93.8	94.5
Gaglio and Lo Re [6] Morana - 2014	77.3	76.7
Parisi et al. [26] - 2015	91.9	90.2
Cippitelli et al. [8] - 2016	93.9	93.5

Table 3. Precision (P) and Recall (R) values of the proposed approach for each activity on OAD.

Action	P	R
Drinking	60.87	77.78
Getting up	81.25	72.22
Grabbing object from ground	83.33	83.33
Pouring a drink	75.00	83.33
Scrolling book pages	80.95	94.44
Sitting	59.09	72.22
Stacking items	90.00	100.00
Taking objects from shelf	100.00	94.44
Talking on phone	86.67	72.22
Throwing something in bin	75.00	33.33
Waving	66.67	66.67
Wearing coat	100.00	100.00
Working on computer	94.12	88.89
Writing on paper	78.95	83.33
Overall	**80.85**	**80.16**

Table 4. Confusion matrix using $k = 100$ words and a configuration of 28 angles on OAD.

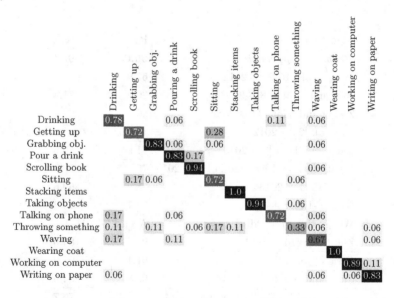

not partitioned according to the room where they are performed and the probability of misclassification increases; (ii) the number of subjects is higher and the variability in executing the actions increases proportionally. For instance the worst results have been measured for the activity "throwing something in bin" that the different subjects executed very differently. Other mismatches occur between the activities "sitting" and "getting up"; in principle the reference postures of the two actions are similar, but their temporal ordering in the execution is different and probably the BoW representation adopted is not able to capture this aspect. However in general the good performance of the proposed approach is confirmed on this dataset as well.

5   Conclusions

A human activity recognition technique based on skeleton information has been proposed in this work. In particular, the effectiveness of joint orientations, typically neglected by the works in the literature, has been evaluated on different benchmarks. The efficacy of the proposal have been confirmed; the results obtained overcome the state-of-the-art in the well-known CAD-60 benchmark and good accuracy levels can be reached also on the newly acquired OAD dataset. Future researches will be devoted to the study of techniques able to couple the human posture information (encoded according to the model proposed here) to the information from the surrounding environment (e.g. about interactions with objects or facial expressions) which would certainly increase the performance and enable a fine-grained classification of activities.

References

1. Ni, B., Pei, Y., Moulin, P., Yan, S.: Multilevel depth and image fusion for human activity detection. IEEE Trans. Cybern. **43**(5), 1383–1394 (2013)
2. Han, J., Shao, L., Xu, D., Shotton, J.: Enhanced computer vision with microsoft kinect sensor: a review. IEEE Trans. Cybern. **43**(5), 1318–1334 (2013)
3. Zhang, J., Li, W., Ogunbona, P.O., Wang, P., Tang, C.: RGB-D-based action recognition datasets: a survey. Pattern Recogn. **60**, 86–105 (2016)
4. Ni, B., Wang, G., Moulin, P.: RGBD-HuDaAct: a color-depth video database for human daily activity recognition. In: Fossati, A., Gall, J., Grabner, H., Ren, X., Konolige, K. (eds.) Consumer Depth Cameras for Computer Vision. ACVPR, pp. 193–208. Springer, London (2013). doi:10.1007/978-1-4471-4640-7_10
5. Gupta, R., Chia, A.Y.S., Rajan, D.: Human activities recognition using depth images. In: Proceedings of the 21st ACM International Conference on Multimedia, MM 2013, pp. 283–292. ACM, New York (2013)
6. Gaglio, S., Re, G.L., Morana, M.: Human activity recognition process using 3-D posture data. IEEE Trans. Hum.-Mach. Syst. **45**(5), 586–597 (2015)
7. Shan, J., Akella, S.: 3D human action segmentation and recognition using pose kinetic energy. In: 2014 IEEE International Workshop on Advanced Robotics and its Social Impacts, pp. 69–75 (2014)
8. Cippitelli, E., Gasparrini, S., Gambi, E., Spinsante, S.: A human activity recognition system using skeleton data from RGBD sensors. Intell. Neurosci. **2016**, 21 (2016)
9. Yang, X., Tian, Y.: Effective 3D action recognition using eigenjoints. J. Vis. Commun. Image Represent. **25**(1), 2–11 (2014). Visual Understanding and Applications with RGB-D Cameras

10. Xia, L., Chen, C.C., Aggarwal, J.K.: View invariant human action recognition using histograms of 3D joints. In: CVPR Workshops, pp. 20–27. IEEE Computer Society (2012)
11. Zhang, C., Tian, Y.: RGB-D camera-based daily living activity recognition. J. Comput. Vis. Image Process. **2**(4), 12 (2012)
12. Piyathilaka, L., Kodagoda, S.: Gaussian mixture based HMM for human daily activity recognition using 3D skeleton features. In: 2013 IEEE 8th Conference on Industrial Electronics and Applications (ICIEA), pp. 567–572 (2013)
13. Faria, D.R., Premebida, C., Nunes, U.: A probabilistic approach for human every-day activities recognition using body motion from RGB-D images. In: The 23rd IEEE International Symposium on Robot and Human Interactive Communication, pp. 732–737 (2014)
14. Zhu, Y., Chen, W., Guo, G.: Evaluating spatio-temporal interest point features for depth-based action recognition. Image Vis. Comput. **32**(8), 453–464 (2014)
15. Sung, J., Ponce, C., Selman, B., Saxena, A.: Unstructured human activity detection from RGBD images. In: 2012 IEEE International Conference on Robotics and Automation, pp. 842–849 (2012)
16. Wang, J., Liu, Z., Wu, Y., Yuan, J.: Mining actionlet ensemble for action recognition with depth cameras. In: 2012 IEEE Conference on Computer Vision and Pattern Recognition, pp. 1290–1297 (2012)
17. Wang, J., Liu, Z., Wu, Y.: Learning Actionlet Ensemble for 3D Human Action Recognition. In: Wang, J., Liu, Z., Wu, Y. (eds.) Human Action Recognition with Depth Cameras. SCS, pp. 11–40. Springer, Cham (2014). doi:10.1007/978-3-319-04561-0_2
18. Koppula, H.S., Gupta, R., Saxena, A.: Learning human activities and object affordances from RGB-D videos. CoRR abs/1210.1207 (2012)
19. Koppula, H., Saxena, A.: Learning spatio-temporal structure from RGB-D videos for human activity detection and anticipation. In: Proceedings of the 30th International Conference on Machine Learning (ICML 2013), JMLR Workshop and Conference Proceedings, vol. 28, pp. 792–800 (2013)
20. Shotton, J., Sharp, T., Kipman, A., Fitzgibbon, A., Finocchio, M., Blake, A., Cook, M., Moore, R.: Real-time human pose recognition in parts from single depth images. Commun. ACM **56**(1), 116–124 (2013)
21. Wang, H., Ullah, M.M., Klaser, A., Laptev, I., Schmid, C.: Evaluation of local spatio-temporal features for action recognition. In: BMVC 2009 - British Machine Vision Conference, London, United Kingdom, pp. 124.1–124.11. BMVA Press (2009)
22. Fukunaga, K.: Introduction to Statistical Pattern Recognition, 2nd edn. Academic Press Professional Inc., San Diego (1990)
23. Breiman, L.: Random forests. Mach. Learn. **45**(1), 5–32 (2001)
24. Sung, J., Ponce, C., Selman, B., Saxena, A.: Human activity detection from RGBD images. In: Proceedings of the 16th AAAI Conference on Plan, Activity, and Intent Recognition. AAAIWS 11-16, pp. 47–55. AAAI Press (2011)
25. Ni, B., Moulin, P., Yan, S.: Order-preserving sparse coding for sequence classification. In: Fitzgibbon, A., Lazebnik, S., Perona, P., Sato, Y., Schmid, C. (eds.) ECCV 2012. LNCS, pp. 173–187. Springer, Heidelberg (2012). doi:10.1007/978-3-642-33709-3_13
26. Parisi, G.I., Weber, C., Wermter, S.: Self-organizing neural integration of pose-motion features for human action recognition. Front. Neurorobot. **9**(3), 1–14 (2015)

A Tensor Framework for Data Stream Clustering and Compression

Bogusław Cyganek[1]([⊠]) and Michał Woźniak[2]

[1] AGH University of Science and Technology,
Al. Mickiewicza 30, 30-059 Kraków, Poland
cyganek@agh.edu.pl
[2] Wrocław University of Science and Technology,
Wybrzeże Wyspiańskiego 27, 50-370 Wrocław, Poland

Abstract. In the paper a tensor based method for video stream clustering and compression is presented. The method does video partitioning in temporal domain based on its content. Such coherent video partitions are amenable for better compression. The proposed method detects shot boundaries building a tensor model from a number of frames in the stream. To build the model, the best rank tensor decomposition is used. Each incoming tensor-frame is verified with the model based on the proposed concept drift detector – if it fits, then the model is updated with that frame. Otherwise, a model is rebuilt. This way obtained shots are then compressed also with the best rank tensor decomposition methods.

Keywords: Video shot detection · Signal compression · Tensor-frames · Best-rank tensor decomposition · Stream tensor analysis

1 Introduction

Enormous amounts of visual data streams put new challenges and requirements on their automatic analysis methods. In this context, one of the techniques is automatic video segmentation based on a measure of concise signal content. In video processing such methods are used for video shot detection which serves for automatic video summarization. In this scenario, a set of consecutive frames with sufficiently coherent contents is represented by a single representative frame, called a keyframe [1]. To build such video summaries, majority of the proposed methods utilize specific color and texture features [2, 9, 13, 15, 17, 20, 22]. On the other hand, large streams of data require data compression. In this work extend our previous work on shot detection [6] and propose to join the two activities, i.e. shot detection and data compression, under one framework of processing streams of multi-dimensional signals. In this framework we treat the frames holistically as 2D or 3D tensors – therefore we call them *tensor-frames*. Such a framework allows easy extensions to higher dimensions and any type of digital signals, though. In case of a video, seen as a 4D tensor, the method allows a uniform approach to its structure analysis, both in spatial and temporal dimensions. In our framework we rely on the best rank-$(R_1, R_2, ..., R_P)$ tensor decomposition as well as tensor stream analysis [14, 19].

© Springer International Publishing AG 2017
S. Battiato et al. (Eds.): ICIAP 2017, Part I, LNCS 10484, pp. 163–173, 2017.
https://doi.org/10.1007/978-3-319-68560-1_15

In respect to the related works, a description of the main tasks in video abstraction is provided in Truong and Venkatesh [20]. Valdes and Martinez discuss on efficient video summarization and retrieval tools [22]. A recent survey of video scene detection methods is provided in Fabro and Böszörmenyi [8]. Other works are by De Menthon *et al.* [16], video summarization by Mundur *et al.* [17], STIMO system proposed by Furini [9], as well as VSUMM proposed by de Avila *et al.* [2] and VSCAN proposed by Mahmoud *et al.* [15]. Tensors and their decompositions are presented in de Lathauwer *et al.* [14], Kolda *et al.* [12], as well as Cyganek [3, 4]. Finally, data streams are analyzed in the works by Gama [10].

2   Architecture of Tensor Stream Clustering and Compression

Figure 1 shows architecture of the proposed method. Each frame is represented as a 2D or 3D tensor-frame for monochrome and color versions, respectively.

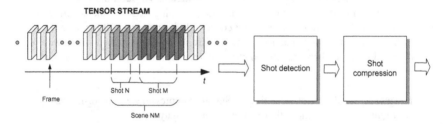

Fig. 1. Overview of the system architecture. Frames are represented as 2D or 3D tensor-frames for monochrome and color images, respectively. The frames are partitioned into shots based on the tensor model. Detected shots are then efficiently compressed due to their content coherency.

The frames are partitioned into shots based on the tensor model. Detected shots are then compressed. The method relies on tensor analysis which further details can be found in literature, such as [4, 12, 14]. The proposed system operates as follows. From the input stream of tensor data, a window of consecutive frame-tensors of size D is selected. All of them are used to build a tensor model from the best rank-$(R_1, R_2, ..., R_N)$ tensor decomposition method [14]. However, the main modification of this model, which we incorporate after the work by Sun *et al.* [19], consists of construction of covariance matrices from the flattened versions of all tensors from the window D. This way, for each of the tensor flattenings, a single covariance matrix is created from all the tensors in the input window. Thus, such covariance matrices convey statistical information on all of the input tensors. The next computational gain of this approach comes from the fact that the covariance matrices belong to the positive definite ones, for which a more effective eigenvalue decomposition method can be used, as will be discussed. However, tensor decompositions allow also for significant data compression, as proposed and analyzed in many works, for examples by Wang and Ahuja [23, 24] or by Cyganek [4]. In this paper we propose to join the video shot detection and then shot compression based on the tensor stream analysis and best rank-$(R_1, R_2, ..., R_P)$ tensor decomposition. Details will be discussed in consecutive sections.

3 Tensor Model Build and Update Scheme for Data Clustering

Algorithm 1 shows our proposed tensor model build and concept drift detection mechanisms for video shot detection. Its particular modules are further discussed.

Input:	Tensors $\mathcal{T}_n$ in a stream of the same valence P; \	
	Window size D	
	Maximal ranks for each dimension: $R_1, R_2, ..., R_P$;	
Output:	Tensor clusters (shots);	

    1. Build tensor model $\mathbf{S}_k$ in the window D (see Algorithm 2);

    2. for each tensor in the stream

        3. Compute model values in (16);

        4. if   model fit measure (15) does not hold   (1)
                 for G consecutive frames

            5. *Rebuild* model with Algorithm 2, starting at current position in the stream

      else

            6. *Update* model based on (13);

Algorithm 1. The algorithm for tensor stream temporal clusterization in a stream based on construction of tensor model and drift detection.

The above method relies on the best rank tensor decomposition, discussed in the consecutive parts of this paper. It is worth noticing that the initial rank values $R_1, R_2, ..., R_P$ in Algorithm 1 are the maximal possible ranks that are considered when building the model. However, real ranks are determined based on the automatic rank assessment mechanism and, in practice, are usually much smaller than the conservatively assumed initial ranks [6]. These depend on a type of the input signal. For instance, in our experiments with monochrome and color video, these were heuristically set to $R_1 = 0.2N_1$, $R_2 = 0.2N_2$, $R_3 = N_3$, where N_1 and N_2 denote the column and row dimensions, whereas N_3 corresponds to the color, respectively. To present our tensor decomposition method, let us define a P-dimensional tensor as a P-dimensional cube of data, with each of its k-th dimensions denoted by N_k, and for $1 \leq k \leq P$, as follows [4, 14]:

$$\mathcal{T} \in \Re^{N_1 \times N_2 \times ... N_P} . \tag{2}$$

An important role plays the so called tensor *flattening*, which for a P-th order tensor $\mathcal{T}$, is defined as the following matrix

$$\mathbf{T}_{(j)} \in \Re^{N_j \times \left(N_1 N_2 ... N_{j-1} N_{j+1} ... N_P \right)} . \tag{3}$$

In a similar way, a k-th modal product of a tensor and a matrix is denoted as $\mathcal{T} \times_k \mathbf{M}$ for a tensor $\mathcal{T} \in \Re^{N_1 \times N_2 \times ... N_P}$ and a matrix $\mathbf{M} \in \Re^{Q \times N_k}$ is defined as a tensor $\mathcal{S} \in \Re^{N_1 \times N_2 \times ... N_{k-1} \times Q \times N_{k+1} \times ... N_P}$, with the following elements

$$\mathcal{S}_{n_1 n_2 ... n_{k-1} q n_{k+1} ... n_P} = \left(\mathcal{T} \times_k \mathbf{M} \right)_{n_1 n_2 ... n_{k-1} q n_{k+1} ... n_P} = \sum_{n_k=1}^{N_k} t_{n_1 n_2 ... n_{k-1} n_k n_{k+1} ... n_P} m_{q n_k} . \tag{4}$$

Differently from the matrix analysis, in case of tensors there are at least three different concepts of their ranks. The r-th rank of a tensor $\mathcal{T}$ is a dimension of the vector space spanned by the columns of the r-th flattening $\mathbf{T}_{(r)}$ of this tensor. After these definitions of a tensor algebra, let us briefly focus upon the tensor decompositions.

For a given tensor $\mathcal{T} \in \Re^{N_1 \times N_2 \times ... \times N_P}$ its Tucker decomposition is its approximating tensor $\tilde{\mathcal{T}}$, given as follows [21]

$$\tilde{\mathcal{T}} = \mathcal{Z} \times_1 \mathbf{S}_1 \times_2 \mathbf{S}_2 ... \times_P \mathbf{S}_P , \tag{5}$$

where $\mathcal{Z}$ is a core tensor, $\mathbf{S}_i \in \Re^{N_i \times R_i}$ are so called mode matrices, and which minimizes the functional

$$\Theta(\mathcal{T}) = \left\| \mathcal{T} - \tilde{\mathcal{T}} \right\|_F^2 . \tag{6}$$

From (5) the following formula for the core tensor is easily obtained

$$\mathcal{Z} = \tilde{\mathcal{T}} \times_1 \mathbf{S}_1^T \times_2 \mathbf{S}_2^T ... \times_P \mathbf{S}_P^T . \tag{7}$$

Then, from (7) to (5) applied to (6), the following tensor fit measure is obtained

$$\Theta(\mathcal{T}) = \left\| \mathcal{T} - \mathcal{T} \prod_{k=1}^{P} \times_k \left(\mathbf{S}_k \mathbf{S}_k^T \right) \right\|_F^2 . \tag{8}$$

Nevertheless, in many applications it is important to request orthogonality or ask for specific rank of $\mathbf{S}_k$. If such a constraint is assumed, then the Tucker decomposition leads to its special version, called the best rank-$(R_1, R_2, ..., R_P)$ tensor decomposition [14]: A tensor $\tilde{\mathcal{T}}$ of ranks in each of its modes $rank_1\left(\tilde{\mathcal{T}} \right) = R_1$, $rank_2\left(\tilde{\mathcal{T}} \right) = R_2$, ..., $rank_P\left(\tilde{\mathcal{T}} \right) = R_P$, respectively, is the best rank-$(R_1, R_2, ..., R_P)$ approximation of a tensor $\mathcal{T} \in \Re^{N_1 \times N_2 \times ... \times N_P}$ if it minimizes the functional (8).

Implementation details of the best rank-$(R_1, R_2, ..., R_P)$ method are presented for instance in the publications [3, 12, 14]. However, this algorithm has been extended of computation of the covariance matrices from the input tensors, using all of their flattening modes [19]. Algorithm 2 presents the best rank-$(R_1, R_2, ..., R_P)$ tensor decomposition algorithm, with the mentioned modified for processing streams of tensor data. This is accomplished in steps (9) and (10) which compute the covariance matrices from each tensor in the input window D and at each of their P flattening modes. In the step (11) the dominating subspace is computed with help of the *fds* function. However, here we take benefit of processing of the covariance matrices which belong to the symmetrical positive definite group of matrices. Thus, instead of the standard SVD matrix decomposition, for the *fds* function a much faster fixed-point method is used, as described in our previous publication [5].

Input: Series of D tensors $\mathcal{T}_d$ of valence P, where $1 \leq d \leq D$

Assumed ranks of the mode matrices: $R_1, R_2, ..., R_P$.

Assumed reconstruction error: e_{thresh} ;

Output: Core tensor $\mathcal{Z}$ and P mode matrices $\mathbf{S}_k$

1. For each $1 \leq k \leq P$ randomly initialize the mode matrices $\mathbf{S}_k^{(0)} \in \mathfrak{R}^{N_k \times R_k}$;

Set t=0;

2. **do:**

   3. **for each** k, such that $1 \leq k \leq P$, **do:**

      4. **for each** n, such that $1 \leq d \leq D$ **do:**

$$_d\hat{\mathbf{S}}_k^{(t+1)} = {}_d\mathbf{T}_{(k)}\left[{}_d\mathbf{S}_{k-1}^{(t)} \otimes {}_d\mathbf{S}_{k-2}^{(t)} \otimes ... \otimes {}_d\mathbf{S}_1^{(t)} \otimes {}_d\mathbf{S}_P^{(t+1)} \otimes ... \otimes {}_d\mathbf{S}_{k+1}^{(t+1)} \right] \quad (9)$$

$$\mathbf{C}_k^{(t+1)} = \mathbf{C}_k^{(t+1)} + \left({}_n\hat{\mathbf{S}}_k^{(t+1)} \right)\left({}_n\hat{\mathbf{S}}_k^{(t+1)} \right)^T \quad (10)$$

$$\mathbf{S}_k^{(t+1)} = fds\left(\mathbf{C}_k^{(t+1)}, R_k, \mathbf{S}_k^{(t)} \right) \quad (11)$$

$$\mathcal{Z}_{t+1} = \mathcal{T} \times_1 \mathbf{S}_1^{(t+1)^T} \times_2 \mathbf{S}_2^{(t+1)^T} ... \times_P \mathbf{S}_P^{(t+1)^T} \quad (12)$$

while ($\left\| \mathcal{Z}_{t+1} \right\|^2 - \left\| \mathcal{Z}_t \right\|^2 > e_{thresh}$)

5. Output last values of $\mathbf{S}_k$.

Algorithm 2. The best rank-$(R_1, R_2, ..., R_p)$ tensor decomposition algorithm for processing streams of tensor data.

Based on Algorithm 2 a tensor model is built from a series of D tensors. If an incoming tensor does not fit to this model, the model needs to be rebuilt from scratch, starting at the position in the stream. On the other hand, if a new frame fits to the model, then the model needs only to be adjusted to account for this new tensor and to account for the changing data stream. However, in this case a faster model update procedure can be employed. Concretely, it simply relies on an update of the covariance matrices in (10) in Algorithm 2, in each of the flattening modes with the new tensor data, as follows [18].

$$\mathbf{C}_k^{(t+1)} = \alpha \mathbf{C}_k^{(t)} + \left(\mathbf{S}_k^{(t+1)}\right)\left(\mathbf{S}_k^{(t+1)}\right)^T. \tag{13}$$

In the above, α denotes a forgetting factor of the previous model, and $\mathbf{S}_k^{(t+1)}$ is a k-th mode matrix of a new (updating) tensor at a time stamp $t+1$.

Last part of the presented method is a concept drift detection function which role is to measure fitness of a test frame $\mathcal{X}$ to the tensor model. First, for all D tensors $\mathcal{T}_i$ used to build a model, their fit values (8) are computed. However, instead of the absolute error values Θ, the differences of $\Delta\Theta$ are taken into computations in the mean and standard deviation. Now the error function is defined as follows

$$\Delta\Theta_i \equiv \Theta_{i-1} - \Theta_i. \tag{14}$$

For right processing of the shots with slowly changing content, the following drift measure is proposed [6].

$$\left\|\Delta\Theta_{\mathcal{X}} - \bar{\Theta}_\Delta\right\| < a\,\sigma_\Delta + b. \tag{15}$$

where a is multiplicative factor (range 3.0–4.0) and b is an additive component (set to 0.2–2.5), $\bar{\Theta}_\Delta$ and σ_Δ are the mean and standard deviation computed from the differences of fit values in (14), which are defined as follows

$$\bar{\Theta}_\Delta = \frac{1}{D}\sum_{i=1}^{w} {}_i\Theta_\Delta, \text{ and } \sigma^2 = \frac{1}{D-1}\sum_{i=1}^{w}\left({}_i\Theta_\Delta - \bar{\Theta}_\Delta\right)^2. \tag{16}$$

4   Stream Compression on Shot Boundaries

The aforementioned best rank-$(R_1, R_2, ..., R_P)$ tensor decomposition can be also used for efficient data compression [4, 23, 24]. Comparing memory required to store the original tensor and its approximation (5), the following measures should be considered

$$Q_0 = N_1 N_2 ... N_P, \qquad Q_1 = R_1 R_2 ... R_P + \sum_{k=1}^{P} N_k R_k. \tag{17}$$

Assuming sufficiently small rank values R_k when compared to N_k – it holds that $Q_1 \ll Q_0$. This makes the best rank-$(R_1, R_2, ..., R_P)$ decomposition well suited also for data compression. Connection of the shot detection followed by the shot compression based on the tensor best rank-$(R_1, R_2, ..., R_P)$ decomposition is one of the novelties presented in this paper. Nevertheless, in practice, when considering storage of $\mathbf{S}_k$ and $\mathcal{Z}$ the dynamical range of their elements need to be considered. In other words, the following coefficient needs to be checked

$$C = B(Q_0)/B(Q_1), \tag{18}$$

where $B(Q)$ is a function that returns a number of bytes necessary to store Q elements. However, as shown in one of our previous publications, a significant memory savings are obtained after changing data format from the floating (*double* in C++) to the fixed-point representation (2 bytes in our implementation). These require the following tensor rescaling [4]

$$\tilde{\mathcal{T}} = \lambda \lfloor \mathcal{Z} \rfloor \times_1 \lfloor \mathbf{S}_1 \rfloor \times_2 \lfloor \mathbf{S}_2 \rfloor \times_3 \lfloor \mathbf{S}_3 \rfloor \tag{19}$$

where $\lfloor \mathbf{S}_i \rfloor$ denotes a scaled values of a matrix $\mathbf{S}_i$ to a certain range r, and λ denotes a scaling parameter. The scaling of the matrices can be achieved in the following steps:

1. Compute the maximum absolute value s_{max} of the matrix $\mathbf{S}_i$.
2. Multiply each element by r/s_{max}.

The same procedure is applied to the core tensor to find out a value of z_{max}. In result the additional scalar

$$\lambda = \frac{z_{max}}{r^{P+1}} \prod_{i=1}^{P} s_{max}^{(i)} \tag{20}$$

is obtained which also needs to be stored. However, its memory occupation is negligible. Thanks to this, each element of the video tensors can be stored on only *two bytes*. On the other hand, signal compression ratio is measured by means of the mean-square error (MSE) or the peak signal to noise ratio (PSNR) measures.

5 Experimental Results

The method was implemented in C++ with the *DeRecLib* [4, 7]. The experiments were performed on a computer with the Intel® Xeon® E-1545 processor with clock 2.9 GHz, memory 64 GB RAM, and OS 64-bit Windows 10. The experimental database contains videos from the Open Video Project [2, 11]. The videos are in the MPEG-1 format, 30 fps 352×240 pixels, of length 1 to 4 min.

Figure 2 depicts shots found by our method in the *The Great Web of Water* test video. In this case the color video was converted to monochrome for speed up. The model windows was set to $W = 13$, the model check parameters $(a, b) = (3.7, 0.2)$, and the $G = 3$. However, during the experiments we observed that it is possible to select hard shots from all of the detected shots by a simple thresholding since for these types of shots the fitness function is much larger. In Fig. 2 hard shot cuts detected with this method are shown in red, while the model fit values are in black. What we have also noticed is that the results depend on the chosen model window size W, but its value is not critical. That is, for many videos good results are obtained for W in a certain range, such as 7–15 in this case, rather than for a particular value. This happens because the model is continuously updated in accordance with the procedure described in Algorithm 1. Also, a choice of the parameter G, controlling the series of consecutive "not-fit" frames, is not a critical one. We set this value to 3 to avoid shot detection on some spurious but single frames, e.g. due to noise and scratches. On the other hand, the results depend much on the parameters a and b in (15). Especially, this choice determines how many soft shots will be detected. The shot detection was tested with the user annotated database of the [2], and compared with other methods in [9, 15, 17]. The obtained average F value is 0.73 and 0.70 for color and monochrome videos, respectively [6]. In this respect only VSCAN and for color video shows better parameters, but it relies on specific feature extraction while tensor methods take signal as it is.

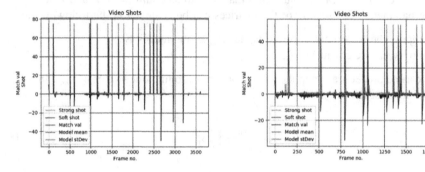

Fig. 2. Video shots found by our method for the *The Great Web of Water* (left) $W = 13$, $(a,b) =$ (3.7,1.3), sequence *Exotic Terrane* (*segment 04*) $W = 53$, $G = 3$, $(a,b) = (3.7,0.2)$, $\alpha = 0.95$ (right). Hard shot cuts denoted in red. Model fit value (14) shown in black.

Table 1 shows compression ratio and PSNR value for the exemplary shots shown in Figs. 3 and 4. Due to relatively small variation among the tensor-frames in a shot, the small rank values can be set which results in high compression ratio. At the same time, the MPEG compression for that file was 49. However, the latter provides better accuracy. This happens because the presented method is not optimized for particular type of signal, color images in this case. Thanks to this, the proposed method can be applied to any type of data, though. Figures 3 and 4 show exemplary frames from the original shot and after compression/decompression. Some artefacts are due to the used color space and very high compression exceeding 100.

Table 1. Compression for an exemplary shots.

Ranks	Compression ratio	PSNR
35-45-6-3	87.54	24.17
24-35-4-3	132.26	24.27

Further research is conducted to avoid these artefacts, as well as to test compression on different types of signals. However, in our opinion color video examples were the best to verify the idea behind the presented method.

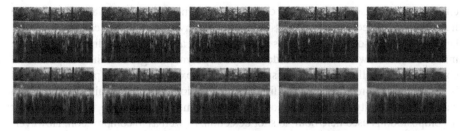

Fig. 3. Exemplary frames from a shot in the *The Great Web of Water* video from the *Open Video* database [11] – top row: original frames no. 2750-2754). Bottom row – the same frames after compression/decompression ratio $C = 87$ and PSNR = 24.17 with the tensor method.

Fig. 4. Exemplary frames from a shot in the *The Great Web of Water* video from the *Open Video* database [11] – top row: original frames. Bottom row – the same frames after compression/decompression with the tensor method, PSNR = 24.27, $C = 132$.

The average execution times for the shot detection are presented in Table 2. Unfortunately, execution of the compression module is an order of magnitude longer due to necessity of decomposing large tensor. Improvement to this is for future work.

Table 2. Average execution time for the monochrome 2D tensors, and color 3D tensors.

Video type	Mono (2D) 352 × 240	Color (3D) 352 × 240
Frames	1948	1948
Processing time [frames/s]	15.6	3.2

6 Conclusions

In this paper the tensor based framework for multidimensional stream *temporal clustering and compression* is presented. The main idea is to find content coherent shots in the input stream which then can be easier and better compressed due to its coherency. For this purpose we propose to build a tensor model for stream processing which then is either updated, or entirely rebuilt, based on the proposed tensor-frame fitness function. The found shots cluster the input stream into chunks which can be more efficiently compressed due to their coherence. The method was built and tested in the framework of video streams processing, although any type of signal can be used. The benefit of the tensor approach is its easy scaling to any number of dimensions of the input streams. However, this is burdened by the polynomial grow of the computational complexity, as well as memory consumption. The next feature of our proposed method is that it does not require any specific feature detection which is a main computational part of the majority of other methods of video summarization. The best rank tensor decomposition is used here both, to shot detection, as well as to compression. However, for specific signals, such as color video, a more accurate compression can be used instead. Regardless of this, the proposed method reaches high shot detection accuracy when compared to other methods specifically designed for processing of the color video streams. Further research will be conducted to improve accuracy and speed, possibly by developing more efficient model fitness functions, as well as faster methods for tensor decomposition. For the latter, the GPU acceleration is investigated.

Acknowledgement. This work was supported by the National Science Centre, Poland, under the grant no. 2016/21/B/ST6/01461.

References

1. Asghar, M.N., Hussain, F., Manton, R.: Video indexing: a survey. Int. J. Comput. Inf. Technol. **03**(01), 148–169 (2014)
2. de Avila, S.E.F., Lopes, A.P.B., da Luz Jr., A., Araújo, A.A.: VSUMM: a mechanism designed to produce static video summaries and a novel evaluation method. Pattern Recogn. Lett. **32**, 56–68 (2011)
3. Cyganek, B.: An analysis of the road signs classification based on the higher-order singular value decomposition of the deformable pattern tensors. In: Blanc-Talon, J., Bone, D., Philips, W., Popescu, D., Scheunders, P. (eds.) ACIVS 2010. LNCS, vol. 6475, pp. 191–202. Springer, Heidelberg (2010). doi:10.1007/978-3-642-17691-3_18
4. Cyganek, B.: Object Detection and Recognition in Digital Images. Theory and Practice. Wiley, Hoboken (2013)
5. Cyganek, B., Woźniak, M.: On robust computation of tensor classifiers based on the higher-order singular value decomposition. In: Silhavy, R., Senkerik, R., Oplatkova, Z., Silhavy, P., Prokopova, Z. (eds.) Software Engineering Perspectives and Application in Intelligent Systems. Advances in Intelligent Systems and Computing, vol. 465, pp. 193–201. Springer, Heidelberg (2016). doi:10.1007/978-3-319-33622-0_18
6. Cyganek, B., Woźniak, M.: Tensor Based Shot Boundary Detection in Video Streams. Springer, Submitted to the New Generation Computing (2017)
7. DeRecLib (2013). http://www.wiley.com/go/cyganekobject
8. Del Fabro, M., Böszörmenyi, L.: State-of-the-art and future challenges in video scene detection: a survey. Multimedia Syst. **19**(5), 427–454 (2013). Springer
9. Furini, M., Geraci, F., Montangero, M., Pellegrini, M.: STIMO: STIll and moving video storyboard for the web scenario. Multimedia Tools Appl. **46**(1), 47–69 (2010)
10. Gama, J.: Knowledge Discovery from Data Streams. CRC Press, Boca Raton (2010)
11. https://open-video.org/
12. Kolda, T.G., Bader, B.W.: Tensor decompositions and applications. SIAM Rev. **51**(3), 455–500 (2008)

13. Kuanar, S.K.: Video key frame extraction through dynamic Delaunay clustering with a structural constraint. J. Vis. Commun. Image Represent. **V24**(7), 1212–1227 (2013)
14. de Lathauwer, L., de Moor, B., Vandewalle, J.: On the best Rank-1 and Rank-(R1, R2, ..., RN) approximation of higher-order tensors. SIAM J. Matrix Anal. Appl. **21**(4), 1324–1342 (2000)
15. Mahmoud, K.M., Ismail, M.A., Ghanem, N.M.: VSCAN: an enhanced video summarization using density-based spatial clustering. In: Petrosino, A. (ed.) ICIAP 2013. LNCS, vol. 8156, pp. 733–742. Springer, Heidelberg (2013). doi:10.1007/978-3-642-41181-6_74
16. DeMenthon, D., Kobla, V., Doermann, D.: Video summarization by curve simplification. In: Proceedings of the Sixth ACM International Conference on Multimedia, pp. 211–218. ACM (1998)
17. Mundur, P., Rao, Y., Yesha, Y.: Keyframe-based video summarization using Delaunay clustering. Int. J. Dig. Libr. **6**(2), 219–232 (2006)
18. Sun, J., Tao, D., Faloutsos, C.: Beyond streams and graphs: dynamic tensor analysis. In: KDD 2006, Philadelphia, Pennsylvania, USA (2006)
19. Sun, J., Tao, D., Faloutsos, C.: Incremental tensor analysis: theory and applications. ACM Trans. Knowl. Discovery Data **2**(3), 11:1–11:37 (2008)
20. Truong, B.T., Venkatesh, S.: Video abstraction: a systematic review and classification. ACM Trans. Multimedia Comput. Comm. Appl. **3**(1) (2007)
21. Tucker, L.R.: Some mathematical notes on three-mode factor analysis. Psychometrika **31**, 279–311 (1966)
22. Valdes, V., Martinez, J.: Efficient video summarization and retrieval tools. In: International Workshop on Content-Based Multimedia Indexing, pp. 43–48 (2011)
23. Wang, H., Ahuja, N.: Compact representation of multidimensional data using tensor rank-one decomposition. In: Proceedings of the 17th International Conference on Pattern Recognition, vol. 1, pp. 44–47 (2004)
24. Wang, H., Ahuja, N.: A tensor approximation approach to dimensionality reduction. Int. J. Comput. Vis. **76**(3), 217–229 (2008)

Convex Polytope Ensembles for Spatio-Temporal Anomaly Detection

Francesco Turchini$^{(\boxtimes)}$, Lorenzo Seidenari, and Alberto Del Bimbo

University of Florence, Florence, Italy
{francesco.turchini,lorenzo.seidenari,alberto.delbimbo}@unifi.it

Abstract. Modern automated visual surveillance scenarios demand to process effectively a large set of visual stream with a limited amount of human resources. Actionable information is required in real-time, therefore abnormal pattern detection shall be performed in order to select the most useful streams for an operator to visually inspect. To tackle this challenging task we propose a novel method based on convex polytope ensembles to perform anomaly detection. Our method relies on local trajectory based features. We report State-of-the-Art results on pixel-level anomaly detection on the challenging publicly available UCSD Pedestrian dataset.

Keywords: Computer vision · Anomaly detection · Surveillance

1 Introduction and Related Work

Nowadays a huge effort is put in securing cities and public spaces. Apart from human engagement in security policy with police forces and other security personnel, a lot of spending is dedicated to surveillance system deployment. Unfortunately while growing the amount of operators may enhance the security, growing the amount of sensors alone is not obtaining much benefits. While cameras are often installed as a deterrent for crimes, the usual approach is to use footage as evidence in investigations. More actionable information could be gathered if real-time video analysis provided to surveillance operators a subset of frames to inspect. Dadashi *et al.* [6] conducted a study to understand the role of automatic and semi-automatic video analysis in security context. They have shown that when reliable automatically computed information is provided workload is greatly reduced. This kind of support to human operators is key since, as reported in [8] the attention of operators, viewing multiple streams, greatly degrades just after 20 min.

A very desirable feature in automatic visual surveillance system, is the ability to pick the right set of streams to watch. This can be casted as measuring the deviation of the most recent frames, from some nominal distribution of the imagery for the very same stream. More specifically an algorithm, selecting streams, should also provide localization of such anomalies. This is an important

© Springer International Publishing AG 2017
S. Battiato et al. (Eds.): ICIAP 2017, Part I, LNCS 10484, pp. 174–184, 2017.
https://doi.org/10.1007/978-3-319-68560-1_16

feature since it allows to use high resolution PTZ cameras able to directly frame, at a higher quality, the abnormal pattern.

Modeling complex patterns requires to learn the distribution characterizing a set of video sequences, taken from a certain view. It is usually assumed that the camera is fixed, this allows to make models which are simpler and can learn patterns which are scene specific. Anomaly detection is usually casted as a one-class learning problem over features extracted from video sequences.

Most of the recent works are based on motion or spatio-temporal features. The seminal work from Adam *et al.* [1], learned local optical flow statistics and compared them to the one computed on forthcoming frames. Optical flow has been used extensively as low-level feature on which contextual models are then built [9,13]. One of the main limitation of optical flow lies in the impossibility to model appearance abnormalities. Nonetheless, using just the appearance, is only suitable for low-frame rate scenarios [3], therefore many work resort to spatio-temporal representation, in order to jointly capture appearance and motion [2, 10–12,16].

Several models have been applied to solve one-class learning. Non-parametric approaches [2,3], model feature distribution implicitly, by looking at distance between features. Parametric models, have the advantage of a lower memory footprint, they typically fit a mixture of density functions on the extracted features. Li *et al.* [11] learn a mixture of dynamic textures, computing likelihood over unseen patterns to perform inference. Similarly, Kim and Grauman [9] learn a mixture of Principal Components Analyzers, which jointly learns the distribution and perform dimensionality reduction. Feature learning has been rarely used except for Xu *et al.* [16], which use autoencoders to directly learn the representation, obtaining high accuracy. In this work, we only consider methods not using anomaly labels in learning, in such cases, the problem becomes a binary classification task with much less challenge.

In the past, trajectories were the feature of choice to model patterns in visual surveillance scenarios [4]. Trajectory based anomaly detection unfortunately requires high quality object tracking and can not find appearance abnormal patterns. In action recognition, the use of short local trajectories, namely dense trajectories, to extract features has led to a sensible increase in performance [15]. Several approaches build on this features, showing interesting further improvements and localization capabilities [7,14]. Up to now we are not aware of such features being employed in unsupervised or semi-supervised tasks like anomaly detection.

Considering the relatively low computational requirement and high performance, we build on dense trajectories, which are known to be very well suited for a wide set of action recognition problems, since they are able to represent motion and appearance jointly. We propose to estimate the distribution of trajectory descriptors using convex polytopes [5]. Convex polytopes have been used in the past but never for computer vision problems. Our approach is inspired by [5], but is different since instead of modeling the distribution of data with a single polytope which is approximated using random projections, we consider explicitly an ensemble of low-dimensional models. This approach is more suited to model multi-modal distributions and it allows to merge multiple features in a single decision.

We report state of the art results on the UCSD dataset both at pixel and frame level anomaly detection. Interestingly we found that local trajectory shape can get very good detection rates, potentially reducing the computational cost for feature extraction.

2   Anomaly Detection with Convex Polytopes

We tackle anomaly detection and localization as a single-class classification problem in a fully unsupervised way. As we can only train our system on a single class of input points (the non-abnormal class), we choose to employ the polytope ensemble technique as modeling method. In particular, we make use of Polytope Ensemble technique [5]. Polytope Ensemble considers a set of convex polytopes representing an approximation of the space containing the input feature points. We want a representation which is shaped according to the distribution of the points we can observe; among the convex class of polytopes, the convex hull has the geometric structure which is best tailored to model this kind of data distribution.

2.1   Model Building

Given an input set of points $X = \{x_1, \ldots, x_m\}$, its convex hull is defined as

$$C(X) = \left\{ \sum_{i=1}^{|X|} \theta_i x_i | x_i \in X; \ \sum_i \theta_i = 1, \theta_i \geq 0 \forall i \right\} \tag{1}$$

By exploiting the convex hull properties, we can then identify an abnormal point simply checking whether it belongs to the convex hull or not.

Extended Convex Hull. To ensure robustness of the model, we follow the procedure of [5] and modify the structure of the convex hull, performing a shift of its vertices closer or farther from its centroid. This allows to avoid overfitting and tune our system to cope with different practical conditions. Considering the set of vertices $V \subset X$ and the centroid of the polytope c_i, we can calculate the expanded polytope setting an α parameter such that

$$V_\alpha = \{v + \alpha \frac{(v - c_i)}{||v - c_i||}, v \in V\} \tag{2}$$

The new polytope defined by vertices in V_α is a shrunken/enlarged version of the original convex hull. Negative values of α increase system sensivity, while positive values reduce it.

Ensemble Building. We rely on dense trajectory features [15]. We extract both motion and appearance descriptors using Improved Dense Trajectories algorithm. This allows us to jointly employ multiple features such as trajectory coordinates, HoG, HoF and MBH to achieve robust anomaly detection and localization. We set the ensemble size to T convex hulls. Then, for each feature and for each convex hull, we generate a random projection matrix P_i^f with norm 1 and size $d \times D_f$, where d is the size of the destination subspace, and D_f is the size of the feature f. We then apply this projections to the original data:

$$X_{P_i^f} = \{P_i^f x, \forall x \in X\} \tag{3}$$

The i-th convex hull is calculated on $X_{P_i^f}$. Each convex hull will be characterized by a unique shape, as we generate a different random projection matrix at every iteration of model learning. A set of different sensitivity ensembles can be obtained by the aforementioned shrinking/expansion procedure, based on different values of the α parameter. It is not required to have an α set for each polytope since, as can be seen in Eq. 2, shrinking factors are computed by scaling the distance of vertices from the centroid.

2.2 Anomaly Localization

At inference time, we test each extracted descriptor for inclusion in each convex hull of the ensemble, for each feature. We consider a local trajectory, with descriptors x_f as anomalous if the following condition is true:

$$x_f \notin C^f(X_{P_i^f}) \ \forall f, i \tag{4}$$

meaning that the descriptor is external to all the polytopes and that this happens for all the considered features (Trajectories, HoG, HoF, MBH).

These assumptions are rather strong, but they ensure that we reduce anomaly detection on unusual but yet ordinary patterns. When a descriptor is marked as abnormal, this detection lasts for the entire extent of the trajectory descriptor (15 frames by default). Detecting anomalies for individual trajectory descriptors allows to generate anomaly proposals in various areas of video frames, exploiting trajectory coordinates. We can then obtain an anomaly mask for each frame of each video by filtering these proposals. In Fig. 1 we represent the three main operations we perform to achieve anomaly detection and localization.

We take into consideration the set of trajectories $T_a = \{t_1, t_2, \ldots, t_N\}$ which have been marked as anomalous after testing their inclusion into the convex hulls of the ensemble. Each trajectory t_i is a sequence of M points, $t_i = \{p_{i1}, \ldots, p_{iM}\}$ lasting M video frames. At frame f, we consider the points of the active anomalous trajectories, that is to say the set of points

$$P_a = \{p_{in} \in t_i | n = f, \ t_i \in T_a\} \tag{5}$$

Points identified by active anomalous trajectories at frame f are clustered with K-means algorithm to locate potentially abnormal areas of the frame. K-Means yields a partition S_a of the anomalous points set P_a in K Voronoi cells:

$$S_a = \{S_1, \ldots, S_K | S_1 \cup \cdots \cup S_K = P_a, \ S_{k_1} \cap S_{k_2} = \emptyset \ \forall \ k_1, k_2\} \tag{6}$$

Polytope inclusion test

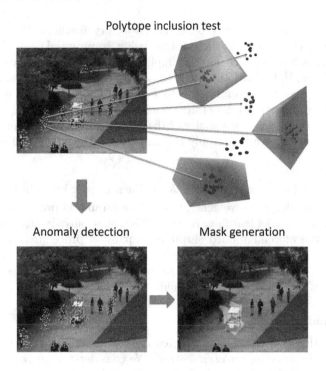

Anomaly detection   Mask generation

Fig. 1. Operating scheme of our anomaly detection and localization model

Each S_k represents an anomaly proposal for the considered frame. For each S_k, we verify if its cardinality is smaller than a fixed threshold, that is to say, if the anomaly proposal constitutes of a minimum number of points. We assume that small clusters are likely originated by spurious false positive detections, so we discard all the anomaly proposals S_k whose cardinality does not guarantee that the detection is reliable. Then, for each remaining S_k, we calculate the polygon described by its points. Each polygon represents an accepted anomaly proposal which contributes to the final anomaly mask creation for the frame.

3   Experimental Results

We conduct our experiments on the UCSD Pedestrian dataset. This dataset has been proposed by Mahadevan *et al.* [11], and it consist of two sets of videos, named Ped1 and Ped2, of pedestrian traffic. The dataset is not staged and features realistic scenarios. In the setting designed by the authors anomalous patterns are all the non-pedestrian entities appearing in the scene. We perform the evaluation on the Ped1 and Ped2 following the standard experimental protocol for this dataset which comprises two evaluation settings: frame-level and pixel-level [11].

In the frame-level criterion, detections are evaluated frame-wise, meaning that a frame is considered anomalous if at least an abnormal detection is predicted for that frame disregarding its location. In this setting it is possible to have "lucky guesses", predicting a frame correctly thanks to a detection which is spatially incorrect or with a too small overlap with the ground truth annotation.

Pixel-level evaluation is introduced to obtain a more detailed analysis of algorithm behavior. In this setting anomaly detections are compared with ground truth pixel masks. A frame is considered a true positive if there is at least 40% of pixel overlap between the ground truth and the predicted mask. A frame is considered a false positive in case anomalies are predicted in normal frames or if the overlap with ground truth masks is lower than 40%. We report the Receiver Operating Characteristic (ROC) curve of TPR and FPR varying system sensitivity, and the Rate of Detection (RD) of our system. We modify system sensitivity varying α in Eq. 2.

First we perform an analysis of the contribution of different features. For simplicity, we divide features in three groups: trajectories, motion and appearance. We test each kind of feature alone and in combination with the others on UCSDPed1. We report the results of feature evaluation in Table 1.

Table 1. Pixel level rate of detection for different descriptors on UCSDPed1

Trajectories	Motion (HoF, MBH)	Appearance (HoG)	RD
✓	-	-	57.9
-	✓	-	60.1
-	-	✓	48.9
✓	✓	✓	**62.2**

Interestingly, local trajectories show very good performance. Anyhow, it appears clearly that motion descriptors give the main contribution to anomaly localization; however, as expected, best results are obtained fusing the contributions of all descriptors. In the following, we will then perform other tests using all the descriptors extracted from the dense trajectory pipeline.

Regarding our model, there are two parameters that can affect the performance. In the following experiments we want to understand how projection size and ensemble cardinality influence the correct detection of anomalies.

All projection size tests were obtained fixing ensemble size to 10 convex hulls, while all ensemble size tests were obtained fixing projection size to 5. We report detection rate variation charts in Fig. 2. As we expected, increasing projection size leads to consistent gain in rate of detection results. On the contrary, bigger ensembles do not always guarantee performance improvements. This outcome may be caused by the unpredictable behavior of the random projections when we raise the number of random generated projection matrices. The best trade-off from a computational point of view is obtained keeping an ensemble of 10 convex hulls and a projection size of 5 dimensions. Increasing projection size over 7 causes convex hull generation and inclusion test to be nearly unfeasible due to very long computation time without bringing noticeable benefits.

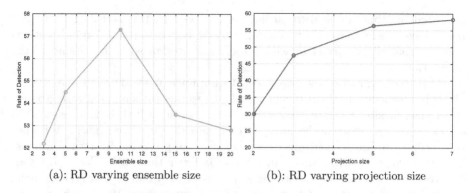

(a): RD varying ensemble size   (b): RD varying projection size

Fig. 2. Evaluation of ensemble size and projection size for our system on UCSDPed1.

With these settings fixed, we compare our results with the existing State-of-the-Art methods in fully unsupervised settings. First of all, it can be noted that with our method trajectory descriptors alone obtain very high Rate of Detection at the pixel level (57.9% as shown in Table 1), higher than most approaches on Ped1, excluding [12], and the deep learning based method by [16].

As we can see in Figs. 3 and 4, our method succeeds in limiting false positive detections, especially at low sensitivity, at the frame level. We detect and localize less than 20% of false positives facing more than 50% of true positives at lower sensitivity values on Ped1 setting. Our system behaves even better on Ped2 setting, where we correctly detect and localize more than 50% of true positive anomalies with less than 5% of mistakes. As we expect, false positive rate

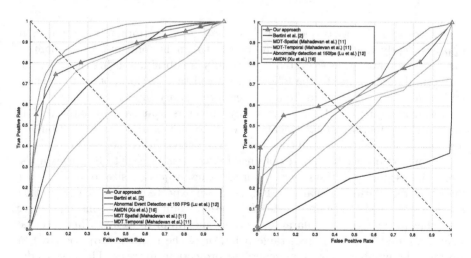

Fig. 3. TPR-FPR curves comparing our approach with various well-known methods on Ped1 setting. Left figure shows the Frame level criterion, right figure shows Pixel level criterion.

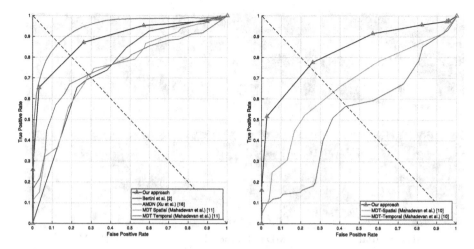

Fig. 4. TPR-FPR curves comparing our approach with various well-known methods on Ped2 setting. Left figure shows the Frame level criterion, right figure shows Pixel level criterion.

increases when our system becomes more sensitive to unseen patterns, however maintaining good robustness. Table 2 reports Rate of Detections for all considered methods for both datasets and both criteria, when reported by authors. Our method obtains a frame-level performance which is comparable to the State-of-the-Art and beat all existing methods on the more challenging pixel-level evaluation. Considering the evaluation protocol established in [11], frame level accuracy may not reflect the actual behavior of a method, because of lucky guesses, while the pixel-level criterion is stricter.

Table 2. RD comparison of our method versus various well-known State-of-the-Art techniques on Ped1 and Ped2 (where available) settings, frame-level and pixel-level criteria.

Method	Ped1		Ped2	
	Frame	Pixel	Frame	Pixel
Ours	78.1	**62.2**	80.7	**75.7**
Xu et al. [16]	78.0	59.9	**83.0**	-
MDT Spatial [11]	56.2	54.2	71.3	63.4
MDT Temporal [11]	77.1	48.2	72.1	56.8
150 fps [12]	**85.0**	59.1	-	-
Bertini et al. [2]	66.0	29.0	68.0	-
Mehran et al. [13]	63.5	40.9	65.0	27.6
Kim and Grauman [9]	64.4	23.2	64.2	22.4
Adam et al. [1]	61.1	32.6	54.2	22.4

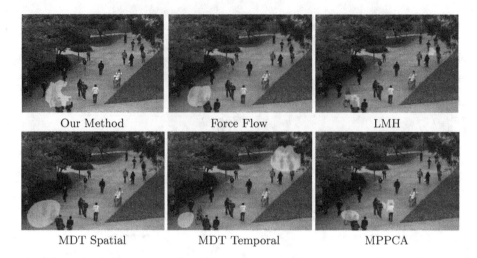

Fig. 5. Qualitative pixel level anomaly detection results on UCSD Ped1 comparing our method to previous approaches.

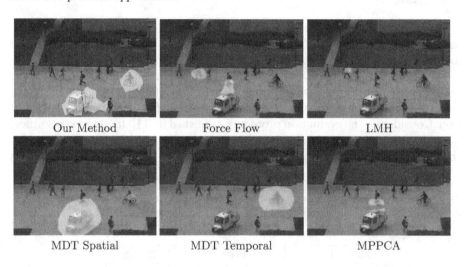

Fig. 6. Qualitative pixel level anomaly detection results on UCSD Ped2 comparing our method to previous approaches.

To show the high quality of our generated masks, we report a qualitative comparison on two frames. Notably our masks frame very tightly abnormal patterns, such as the bicycle rider and the truck in Figs. 5 and 6. With respect to [11] our masks are tighter. Methods such as MPPCA, Force Flow and LMH, are not able, especially in Ped2, to locate all anomalies. This is likely due to a lower quality of features employed.

4   Conclusion

In this paper we show a novel, low memory footprint method to exploit dense trajectory features in anomaly detection. Our method is able to model a complex multimodal distribution yielded by spatio-temporal descriptors using a simple convex polytope ensemble. Moreover, when multiple views of the same datum are available our approach seamlessly performs feature fusion. Indeed, our method is very flexible, as it allows to combine multiple features maintaining the same operating mechanisms, and is tunable by a simple geometric transformation of polytope hulls. Our system can thus be adapted to cope with various practical conditions without losing its benefits both for anomaly detection and localization tasks. We also propose a technique to obtain precise masks by clustering abnormal trajectories; this mask generation technique allows us to achieve good robustness against false positive detections and is shown to obtain State-of-the-Art results in term of pixel-wise detection rate.

References

1. Adam, A., Rivlin, E., Shimshoni, I., Reinitz, D.: Robust real-time unusual event detection using multiple fixed-location monitors. IEEE Trans. Pattern Anal. Mach. Intell. **30**(3), 555–560 (2008)
2. Bertini, M., Del Bimbo, A., Seidenari, L.: Multi-scale and real-time non-parametric approach for anomaly detection and localization. Comput. Vis. Image Underst. **116**(3), 320–329 (2012)
3. Breitenstein, M.D., Grabner, H., Van Gool, L.: Hunting nessie-real-time abnormality detection from webcams. In: Proceedings of ICCV Workshops, pp. 1243–1250. IEEE (2009)
4. Calderara, S., Prati, A., Cucchiara, R.: Mixtures of von mises distributions for people trajectory shape analysis. IEEE Trans. Circ. Syst. Video Technol. **21**(4), 457–471 (2011)
5. Casale, P., Pujol, O., Radeva, P.: Approximate polytope ensemble for one-class classification. Pattern Recogn. **47**(2), 854–864 (2014)
6. Dadashi, N., Stedmon, A.W., Pridmore, T.P.: Semi-automated CCTV surveillance: the effects of system confidence, system accuracy and task complexity on operator vigilance, reliance and workload. Appl. Ergon. **44**(5), 730–738 (2013)
7. Gaidon, A., Harchaoui, Z., Schmid, C.: Activity representation with motion hierarchies. Int. J. Comput. Vision **107**(3), 219–238 (2014)
8. Haering, N., Venetianer, P.L., Lipton, A.: The evolution of video surveillance: an overview. Mach. Vis. Appl. **19**(5), 279–290 (2008)
9. Kim, J., Grauman, K.: Observe locally, infer globally: a space-time MRF for detecting abnormal activities with incremental updates. In: Proceedings of CVPR, pp. 2921–2928. IEEE (2009)
10. Kratz, L., Nishino, K.: Anomaly detection in extremely crowded scenes using spatio-temporal motion pattern models. In: Proceedings of CVPR, pp. 1446–1453. IEEE (2009)
11. Li, W., Mahadevan, V., Vasconcelos, N.: Anomaly detection and localization in crowded scenes. IEEE Trans. Pattern Anal. Mach. Intell. **36**(1), 18–32 (2014)

12. Lu, C., Shi, J., Jia, J.: Abnormal event detection at 150 FPS in MATLAB. In: Proceedings of ICCV, pp. 2720–2727 (2013)
13. Mehran, R., Oyama, A., Shah, M.: Abnormal crowd behavior detection using social force model. In: 2009 IEEE Conference on Computer Vision and Pattern Recognition, CVPR 2009, pp. 935–942. IEEE (2009)
14. Turchini, F., Seidenari, L., Del Bimbo, A.: Understanding and localizing activities from correspondences of clustered trajectories. Comput. Vis. Image Underst. **159**, 128–142 (2017)
15. Wang, H., Oneata, D., Verbeek, J., Schmid, C.: A robust and efficient video representation for action recognition. Int. J. Comput. Vision **119**(3), 219–238 (2016)
16. Xu, D., Ricci, E., Yan, Y., Song, J., Sebe, N.: Learning deep representations of appearance and motion for anomalous event detection. Comput. Vis. Image Underst. **156**, 117–127 (2017)

Human Action Classification Using an Extended BoW Formalism

Raquel Almeida[1]([⊠]), Benjamin Bustos[2], Zenilton Kleber G. do Patrocínio Jr.[1],
and Silvio Jamil F. Guimarães[1]

[1] Audio-Visual Information Laboratory (VIPLAB),
Department of Computer Science, Pontifical Catholic University of Minas Gerais
(PUC Minas), Belo Horizonte, Brazil
`raquel.almeida.685026@sga.pucminas.br`, {`zenilton,sjamil`}`@pucminas.br`
[2] Department of Computer Science, University of Chile, Santiago, Chile
`bebustos@dcc.uchile.cl`

Abstract. In human action classification task, a video must be classified into a pre-determined class. To cope with this problem, we propose a mid-level representation which extends the Bag-of-Words formalism in order to better described the low-level features, exploring distance-to-codeword histograms. The main contribution of this article is the assembly of low-level features by a mid-level representation enriched with information about distances between descriptors and codewords. The proposed representation takes into account volumes of hyper-regions obtained from hyperspheres centered at codewords. Experimental results demonstrated that our strategy either has improved the classification rates more than 6% with respect to the compared mid-level representation for UCF Sports, or it is a competitive one, for KTH and UCF-11.

Keywords: Human action classification · Mid-level representation · Pooling strategy

1 Introduction

Human Action Classification is a pattern recognition task in which the main goal is to identify the action displayed in a media content. Regardless of the media source, such as images, sequential selection of video frames, raw video data or annotations, this task gained a lot of attention in the past few years. Here the focus is on video data, and one could define the main task as: given a video, one needs to classify the action displayed into one of a predetermined set of actions using only the content presented in the video.

To achieve this goal we explore ways to better represent the video content, transforming the original input information into a more suitable representation for the classifier. Usually researchers address this problem using two stages [9]:

The authors are grateful to FAPEMIG (PPM-00006-16), CNPq (Grant 421521/2016-3), PUC Minas and CAPES for the financial support to this work.

S. Battiato et al. (Eds.): ICIAP 2017, Part I, LNCS 10484, pp. 185–196, 2017.
https://doi.org/10.1007/978-3-319-68560-1_17

(i) feature extraction; and (ii) action classification. In a typical approach, feature extraction is performed directly on the raw data, here called low-level description, trying to avoid noise or irrelevant information. Action classification involves learning statistical models from the extracted features, and using those models to classify new feature observations.

The most discriminative low-level descriptors available in literature today rely on identifying regions of interest. Once these regions are identified, desirable features are then extracted around these regions. The output created by this process is a set of features, which are related to regions, representing the media. Facing this scenario, one popular approach is to map the set of local descriptors into one vector used as a global representation, so-called *mid-level representation*.

Among the methods for creating a mid-level representation, standout Bag-of-Words (BoW), Spatial Pyramids and Convolutional Networks for their notable results. As stated by [2], mid-level representations have three steps in common: (i) coding; (ii) pooling; and (iii) concatenation. Coding stands for the transformation locally applied into features vectors, extracting distribution characteristics. Pooling, in turn, explores the spatial relation between these characteristics; and concatenation constructs the final vector representation.

Here we explore a new strategy for the pooling step based on a volumetric partition of an hypersphere centered at codewords. The goal is to maintain the same probability of assignment to a given hyper-region. We argue that this kind of pooling could decrease the quantization error created during codification.

This paper is organized as follows. In Sect. 2, some related works involving mid-level representations and human action classification are described. While in Sect. 3, a formalization of traditional BoW is presented, in Sect. 4, the new mid-level representation is given. Experiments for human action recognition taking into account three well-known datasets are presented in Sect. 5, and finally, some conclusions are drawn in Sect. 6.

2   Related Work

Human action recognition is a popular topic in video processing, but it still an open problem due to the difficult in creating a representation able to capture and describe action motions in different scenarios. Among the most pronounced action descriptors, local spatio-temporal features, as proposed in [6,15], have been successfully used in several applications. In [6] the Space-Time Interest Point (STIP) descriptor are proposed. In STIP, interest points are detected in multiple scales and associated to a patch. Each path is described using Histogram of Oriented Gradients (HOG) and Histogram of Optical Flow (HOF). Regarding Dense Trajectories descriptor, proposed in [15], the trajectories are obtained by densely tracking sampled points obtained with optical flow fields. After tracking, feature point descriptors are extracted using HOG, HOF and Motion Boundary Histogram (MBH) around the trajectories.

These two description approaches represent action videos by a set of local features. Inspired by the success of mid-level representations of local features in image processing, they rely on vector quantization in a BoW scheme to create a global video representation. Although they achieve good results, the loss of information during codification still an open problem.

In order to deal with quantification errors, a wide range of methodologies for creating mid-level representation has been proposed. Most of these representations, when applied in action recognition task, follow the vector quantization based on BoW model, but try to preserve spatial temporal relations during the coding process taking into account multiple weighted representation.

In [18], it was proposed the combination of local histograms with body regions histograms in order to preserve spatial temporal relations between interest points. In [8], it was proposed a sparse coding with max pooling framework applied in multiple contexts, which are defined according to the spatial scale and nearest neighbors of local features. Moreover, it is constructed one vocabulary and one histogram for each defined context. At the end, these data are concatenated for the final representation.

In [17], it was proposed a method based on multiple hierarchical levels for creating histograms. The first level is constructed using the descriptors extracted from video cuboids. The other levels are created by applying a neighboring function regarding previous level description and by creating a new codebook and new histogram for the current level. This scheme is called hierarchical BoW. In [13] an hierarchical BoW is constructed by recursive computing partitions of depth maps sequence in temporal domain, called Temporal Bag-of-Words.

In order to explore a representation driven by the histogram information, in [20], it was defined a contextual domain surrounding a spatial temporal area. After that, a contextual distance is calculated by adding a penalty value proportional to the probability density function computed from the local descriptors and codewords of the contextual domain. The contextual information is also used in [19], however it is obtained by histogram intersections, using both spatial and temporal distances as weighted controlling factors. In [16], Term Frequency-Inverse Document Frequency (TF-IDF) of visual words is used to create histograms representing video segments. These histograms are applied in a continuous framework using a data stream algorithm to update the system knowledge based on the classification score obtained by the histogram. In [3] contextual information incorporate depth camera data and global frame descriptors in a BoW framework.

In image processing domain, enriched BoW representations with extra knowledge from the set of local descriptors have been explored on several approaches [11,21]. However, those works use parametric models leading to a very high-dimensional representation. On other hand, BossaNova model [1], which follows BoW formalism (coding/pooling), keeps more information than the traditional BoW during the pooling step. It estimates a probability density function by computing a histogram of distances between local descriptors and codewords. In addition to the pooling strategy, in [1], it also proposed a localized soft-assignment coding that considers only the k-nearest codewords for coding a local descriptor.

3   Traditional Bag-of-Words

In the traditional Bag-of-Words (BoW) model for mid-level representation, the input is a set of unordered local descriptors, representing the whole data. The BoW model first requires a dictionary learned from the feature points. The most common approach to create the dictionary is by an unsupervised clustering algorithm (*e.g.*, K-means algorithm). The dictionary is composed by a set of M codewords. More precisely, let $\mathbb{X} = \{\mathbf{x}_j \in \mathbb{R}^d\}_{j=1}^N$ be an unordered set of d-dimensional descriptors $\mathbf{x}_j$ extracted from the data and let $\mathbb{C} = \{\mathbf{c}_m \in \mathbb{R}^d\}_{m=1}^M$ and $\mathbb{Z} \in \mathbb{R}^M$ be the dictionary learned and the final vector representation, respectively. As formalized in [2], the mapping from $\mathbb{X}$ to $\mathbb{Z}$ can be decomposed into three sucessive steps: (i) coding; (ii) pooling; and (iii) concatenation, as follows:

$$\alpha_j = f(\mathbf{x}_j), j \in [1, N] \qquad \text{(coding)} \qquad (1)$$

$$h_m = g(\alpha_m = \{\alpha_{m,j}\}_{j=1}^N), m \in [1, M] \qquad \text{(pooling)} \qquad (2)$$

$$z = [h_1^T, \dots, h_M^T] \qquad \text{(concatenation)} \qquad (3)$$

In the traditional BoW framework [14], the coding function f minimizes the distance to a codebook, and the pooling function g computes the sum over the pooling region. As illustrated in Fig. 1, the coding and pooling functions can be visualized in terms of the matrix $\mathbf{H}$ with N column and M rows, in this example, the coding function f for a given descriptor $\mathbf{x}_j$ corresponds to information obtained from the j^{st} column. Moreover, the pooling function g for a given visual word $\mathbf{c}_m$ corresponds to the m^{st} row of the $\mathbf{H}$ matrix. Both functions could be, more precisely, defined as follows:

$$\alpha_j \in \{0,1\}^M = \alpha_{m,j} = 1 \text{ iff } j = \underset{m \leq M}{\arg\min} \parallel \mathbf{x}_j - \mathbf{c}_m \parallel_2^2 \qquad (4)$$

$$h_m = \frac{1}{N} \sum_{j=1}^N \alpha_{m,j} \qquad (5)$$

in which $\mathbf{c}_m$ denotes the m-th codeword.

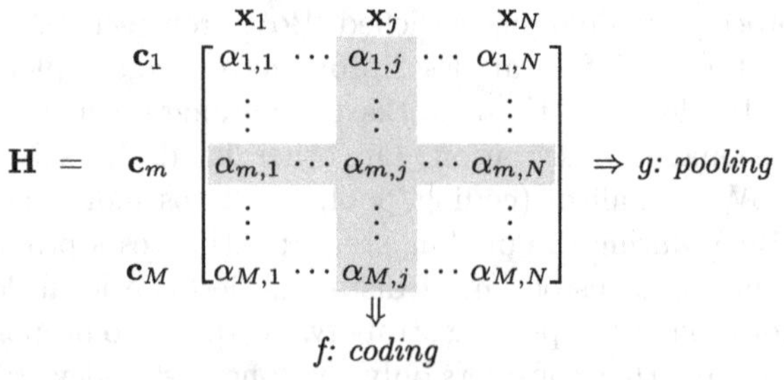

Fig. 1. Matrix $\mathbf{H}$ of BoW model in which the rows and columns are related to coding and pooling functions, respectively, as presented in [1].

In [4], some improvements were obtained by smoothing the distribution during the pooling function. This approach, called soft-assignment, models an ambiguity concept in the attribution, creating more expressive models for classification. They indicate that large visual vocabulary increases the probability of multiple relevant visual words to represent one feature point. This is called visual word uncertainty and can be formulated as follows:

$$\alpha_{m,j} = \frac{\exp(-\beta \parallel \mathbf{x}_j - \mathbf{c}_m \parallel_2)}{\sum_{k=1}^{M} \exp(-\beta \parallel \mathbf{x}_j - \mathbf{c}_k \parallel_2)} \tag{6}$$

where β is a parameter that controls the softness of the soft assignment (hard assignment is the limit when $\beta \to \infty$).

4   An Extended BoW Formalism

In the traditional BoW framework [14], the function g for pooling computes the number of descriptors over the pooling region, thus the mid-level representation could be defined by a concatenation of all values related to the codewords. Unfortunately, this pooling strategy is quite poor in terms of information inside each pooling region, mainly related to spatial distribution of the descriptors. To cope with this lack of information, we propose a new mid-level representation, so-called BOH (**B**ag **O**f local distribution of descriptors on concentric **H**yperspheres), which explores the descriptor position inside the largest hypersphere centred at each codeword for computing the pooling. For that, we propose to divide this hypersphere into equally probable hyper-regions in which the descriptors inside one hyper-region have similar distances to the codeword.

Let S_i and S_j be two hyperspheres centered at codeword $\mathbf{c}_m$ with radius r_i and r_j, respectively, in which $r_i < r_j$. We define the hyper-region $R_{i,j}$ between the hyperspheres S_i and S_j as the hyper-region computed by the difference of S_i and S_j. More precisely, a d-dimensional descriptor belongs to the $R_{i,j}$ if the distance to the codeword $\mathbf{c}_m$ is higher than r_i and smaller than or equal to r_j.

Two hyper-regions $R_{i,j}$ and $R_{i',j'}$ are considered equally probables if they have the same volume, i.e., $V(R_{i,j}) = V(R_{i',j'})$. Let E be the number of equally probable hyper-regions related to the codeword $\mathbf{c}_m$. Without loss of generality, let S_E and S_1 be two hyperspheres with radius r_E and r_1 centered at $\mathbf{c}_m$, $V(R_{E-1,E}) = V(R_{0,1})$ iff $V(S_E) = E \times V(S_1)$. From this definition, it is easy to show that $r_e = r_1 \times \sqrt[n]{e}, \forall\, e \in [1, E]$.

Considering these E equally probable hyper-regions, the proposed pooling strategy is the histogram of distances between the local descriptors and the codewords taking into account the radius of the largest n-dimensional hypersphere over the pooling region. Let $\mathbb{X} = \{\mathbf{x}_j\}$ be an unordered set of d-dimensional descriptors $\mathbf{x}_j$ extracted from a video, such that $j \in [1, N]$. The proposed strategy for pooling is defined by:

$$h_{m,e} = \mathrm{card}\left(\mathbf{x}_j \mid \alpha_{m,j} \in \left[r_E^{\mathbf{c}_m}\sqrt[N]{\frac{e}{E}}, r_E^{\mathbf{c}_m}\sqrt[N]{\frac{e+1}{E}}\right]\right), e \in [0, E-1] \tag{7}$$

in which $r_E^{\mathbf{c}_m}$ is the radius of largest n-dimensional hypersphere centered at codeword $\mathbf{c}_m$. The final representation $\mathbf{z}$ is given by:

$$\mathbf{z} = [h_{m,e}]^{\mathrm{T}}, \ (m, e) \in \{1, ..., M\} \times \{1, ..., E\} \tag{8}$$

where $\mathbf{z}$ is a vector of size $M \times E$.

When the number of equally probable hyper-regions is equal to 1, our pooling strategy is similar to the traditional BoW. As the number of hyper-regions and codewords increase, the vector $\mathbf{z}$ is more sparse but it approximates better the actual distribution of distances. Thus there is a trade-off between the sparsity and this size.

In order to exemplify the traditional BoW, the BossaNova and the proposed method pooling strategies, it is illustrated in Fig. 2 how the regions related to two codewords are divided. In this example, the coding is done by a hard-assignment in which the d-dimensional descriptor is associated with just one codeword. In the traditional BoW, as illustrated in Fig. 2(a), the codewords are represented by the number of descriptors which are assigned to them. For BossaNova and BOH each codeword is represented by a histogram of descriptors which are quantized according to their distance to the codeword. While the quantization used by BossaNova is based on linear function in terms of the distance-to-codeword, as shown in Fig. 2(b), the quantization used by BOH is based on the volumes of the hyper-regions obtained by hyperspheres centered at codewords, as illustrated in Fig. 2(c).

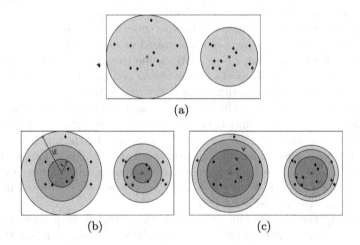

(a)

(b)   (c)

Fig. 2. Example of pooling strategy for d-dimensional descriptors taking into account BoW, BossaNova and BOH. For BossaNova the number of hyper-regions is equal to 3, and for BOH, the number of equally probable hyper-regions related to each codeword is also equal to 3.

(a) Outdoors   (b) Scaled   (c) Clothing   (d) Inside

Fig. 3. Example of hand waving with same subject in different scenarios

5 Experimental Analysis

In this section, we describe the three used datasets, the protocols for classification and the experimental setup. Moreover, a quantitative analysis, in terms of classification rates, comparing our method with the state-of-the-art approaches is given.

5.1 Datasets and Protocols

In order to validate the proposed method we tested our approach in three well-known action recognition datasets: (i) KTH [12]; (ii) UCF Sports [10]; and (iii) UCF 11 [7]. The datasets choices were made due their distinctive characteristics, such as video duration, intraclass variability and noise scene elements.

The KTH dataset [12] contains six types of human actions: walking, jogging, running, boxing, hand waving and hand clapping. These actions are performed by 25 different subjects in four scenarios: outdoors, outdoors with scale variation, outdoors with different clothes and inside. Some examples are illustrated in Fig. 3. There are totally 600 video clips with 160×120 pixels size and different video durations. We adopt the same experimental setup as in [12,15], so-called split, where the videos are divided into a training set (eight subjects), a validation set (eight subjects) and a test set (nine subjects).

The UCF sports dataset [10] contains ten different types of sports actions: swinging, diving, kicking, weight-lifting, horse-riding, running, skateboarding, swinging at the high bar, golf swinging and walking. The dataset consists of 150 real videos with a large intra-class variability. Each action class is performed in different ways, and the frequencies of various actions also differ considerably, as can be seen in Fig. 4. Contrary to what has been done in many works that apply their methods on this dataset, we do not extended the dataset with a flipped version of the videos, trying to prevent the classifier from learning the background instead of the actions. We adopt a split set dividing the dataset into 103 training and 47 test samples as in [5].

The UCF11 dataset [7] contains 11 action categories: biking/cycling, diving, golf swinging, soccer juggling, trampoline jumping, horse riding, basketball shooting, volleyball spiking, swinging, tennis swinging, and walking with a dog. This dataset is challenging due to large variations in camera motion, object appearance and pose, object scale, viewpoint, cluttered background, and illumination conditions. Some examples are illustrated in Fig. 5. The dataset contains

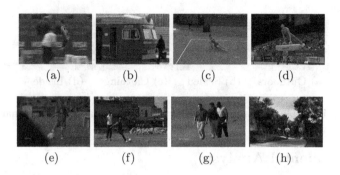

(a)   (b)   (c)   (d)

(e)   (f)   (g)   (h)

Fig. 4. Example of intra-class variability in UCF Sports dataset. (a) and (b) are both examples from running class, (c) and (d) from swinging, (e) and (f) from kicking; while (g) and (h) from walking.

(a)   (b)   (c)   (d)

(e)   (f)   (g)   (h)

Fig. 5. Example of UCF11 challenges, such as object appearence in (a) and (b), viewpoint in (c) and (d), cluttered background in (e) and (f), and illumination conditions in (g) and (h).

a total of 1646 videos. We adopt the original setup [7] using the leave-one-out cross-validation for a pre-defined set of 25 folds.

5.2   Experimental Setup

Regarding the feature descriptor, we have chosen to use an approach with a dense descriptor (dense trajectories [15]) because it is simple and achieved good results. After the feature extraction step, BoW, BossaNova and BOH are used to organize the low-level features to represent each video using the mid-level representation. Here, we used the following parameter values for computing the BossaNova: $\lambda_{min} = 0.4$, $\lambda_{max} = 2$, $knn = 10$ (semi-soft assignment), $B = \{2, 4\}$ and $M = \{512, 2048\}$ (number of visual codewords). For the proposed pooling strategy BOH, we used the following parameter values: $knn = 10$ (semi-soft assignment), $E = \{2, 4\}$ and $M = \{512, 2048\}$ (number of visual codewords).

For classification we used non-linear SVM with an RBF kernel which is a popular classifier that is used throughout different works for human action classification [15]. Since this classifier is vastly used in human action classification, it is interesting to use it to make fair comparisons between different approaches.

5.3 Comparison with the State-of-the-art

In order to compare the proposed method to some of the state-of-the-art approaches we adopted the classification rate (also called recognition rate). Usually, in literature, there is a bit confusion between the use of classification rate and accuracy. For the sake of clarification, in this work, the classification rate is the number of correct video classification by the number of videos. In Table 1, a comparison, in terms of classification rate, is presented. Except for BossaNova, the rates of the compared methods were obtained from the original paper. As one can note, ours give competitive rates for KTH and UCF 11, and much better results for UCF Sports. When compared to BossaNova, which uses a similar pooling strategy, our results are better in UCF Sports and UCF 11.

Table 1. The classification rates for the compared approaches.

Approach	Parameters	KTH	UCF sports	UCF 11
Dense trajectories [15]	-	94.2%	-	84.1%
BoW	$M = 2048$	85.7%	55.3%	53.9%
BossaNova [1]	$M = 512, B = 2$	94.9%	66.0%	78.4%
	$M = 512, B = 4$	96.3%	66.0%	81.0%
	$M = 2048, B = 2$	97.7%	70.2%	78.0%
	$M = 2048, B = 4$	97.7%	70.2%	75.7%
Ours	$M = 512, E = 2$	94.9%	72.3%	75.3%
	$M = 512, E = 4$	94.0%	72.3%	76.0%
	$M = 2048, E = 2$	96.8%	74.5%	79.3%
	$M = 2048, E = 4$	96.3%	76.6%	81.4%

The performances, in terms of classification rates, of the compared approaches applied to the KTH, UCF Sports and UCF 11 are illustrated in Fig. 6. As we can see, the rate for BOH increases when the number of codewords and hyper-regions increase. This behavior does not occurs for BossaNova since there is no a monotonic increasing neither for number of codewords nor for the number of hyper-regions. Furthermore, both methods are better than traditional BoW. In terms of time performance, there is no significant difference between BOH, BossaNova and BoW, once the main time consuming operation rely on calculating the distances between feature points and codewords during coding and are the same for all methods.

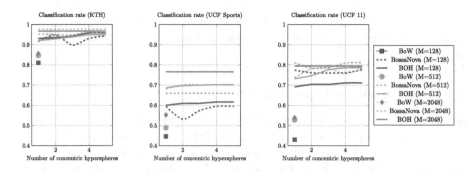

Fig. 6. A comparison between the proposed method and BossaNova concerning the classification rate according to the number of hyperpheres from 1 to 5. The classification rate for BoW is also illustrated (when the number of hypersphere is equal to 1).

6   Conclusions and Further Works

In this work, we addressed the task of human action classification using only the information present in the content of the video. Also, we focused on an intermediate stage between feature extraction and classification by using an extended BoW formalism, so-called BOH to generate a new mid-level video representation which is obtained directly from densely sampled features extracted around trajectories.

The idea is to increase the classification rate by careful use of a well disseminate motion descriptor. Here we explored a new strategy for the pooling step based on a volumetric partition of the hypersphere centered at codewords in order to maintain the same probability of assignment to a given hyper-region. The results indicates that this kind of pooling could decrease the quantization error of the descriptors.

Regarding classification protocols, we experimented the training and testing classification (here called split) for KTH and UCF Sports, and the leave-one-group-out cross-validation for UCF 11. Experimental results demonstrated that our strategy either has improved the recognition rates with respect to the BossaNova, expect for KTH.

For further endeavors, we will study different ways encoding quantization errors into video descriptors. Another interesting research path is to investigate the quality of video data used during (and filter it out before) training time for the classification step and its relationship with the support vectors needed to produce better accuracy results in human action classification.

References

1. Avila, S., Thome, N., Cord, M., Valle, E., Araújo, A.D.A.: Pooling in image representation: the visual codeword point of view. Comput. Vis. Image Underst. (CVIU) **117**(5), 453–465 (2013)

2. Boureau, Y.L., Bach, F., LeCun, Y., Ponce, J.: Learning mid-level features for recognition. In: 2010 IEEE Computer Society Conference on Computer Vision and Pattern Recognition, pp. 2559–2566, June 2010
3. Foggia, P., Percannella, G., Saggese, A., Vento, M.: Recognizing human actions by a bag of visual words. In: 2013 IEEE International Conference on Systems, Man, and Cybernetics (SMC), pp. 2910–2915. IEEE (2013)
4. van Gemert, J.C., Veenman, C.J., Smeulders, A.W.M., Geusebroek, J.M.: Visual word ambiguity. IEEE Trans. Pattern Anal. Mach. Intell. **32**(7), 1271–1283 (2010)
5. Lan, T., Wang, Y., Mori, G.: Discriminative figure-centric models for joint action localization and recognition. In: International Conference on Computer Vision (ICCV) (2011)
6. Laptev, I., Marszalek, M., Schmid, C., Rozenfeld, B.: Learning realistic human actions from movies. In: 2008 IEEE Conference on Computer Vision and Pattern Recognition, pp. 1–8 (2008)
7. Liu, J., Luo, J., Shah, M.: Recognizing realistic actions from videos. In: 2009 IEEE Computer Society Conference on Computer Vision and Pattern Recognition, CVPR 2009, 20–25 June 2009, Miami, Florida, USA, pp. 1996–2003 (2009)
8. Luo, H., Lu, H.: Multi-level sparse coding for human action recognition. In: 2016 8th International Conference on Intelligent Human-Machine Systems and Cybernetics (IHMSC), vol. 1, pp. 460–463. IEEE (2016)
9. Poppe, R.: A survey on vision-based human action recognition. Image Vis. Comput. **28**(6), 976–990 (2010)
10. Rodriguez, M.D., Ahmed, J., Shah, M.: Action mach a spatio-temporal maximum average correlation height filter for action recognition. In: 2008 IEEE Conference on Computer Vision and Pattern Recognition, CVPR 2008, pp. 1–8, June 2008
11. Sánchez, J., Perronnin, F., Mensink, T., Verbeek, J.: Image classification with the fisher vector: theory and practice. Int. J. Comput. Vis. (IJCV) **105**(3), 222–245 (2013)
12. Schuldt, C., Laptev, I., Caputo, B.: Recognizing human actions: a local SVM approach. In: 17th International Conference on Proceedings of the Pattern Recognition, ICPR 2004, Washington, DC, vol. 3, pp. 32–36. IEEE Computer Society (2004)
13. Shukla, P., Biswas, K.K., Kalra, P.K.: Action recognition using temporal bag-of-words from depth maps. In: MVA, pp. 41–44 (2013)
14. Sivic, J., Zisserman, A.: Video Google: a text retrieval approach to object matching in videos. In: Proceedings of the Ninth IEEE International Conference on Computer Vision, ICCV 2003, Washington, DC, vol. 2, p. 1470. IEEE Computer Society (2003)
15. Wang, H., Kläser, A., Schmid, C., Liu, C.L.: Action recognition by dense trajectories. In: 2011 IEEE Conference on Computer Vision and Pattern Recognition (CVPR), pp. 3169–3176. IEEE (2011)
16. Wiliem, A., Madasu, V., Boles, W., Yarlagadda, P.: Adaptive unsupervised learning of human actions. In: 3rd International Conference on Imaging for Crime Detection and Prevention, ICDP 2009, pp. 1–6, December 2009
17. Wu, J., Zhou, D., Xiao, G.: A hierarchical bag-of-words model based on local space-time features for human action recognition. In: 2013 International Conference on IT Convergence and Security (ICITCS), pp. 1–4. IEEE (2013)
18. Yan, X., Luo, Y.: Making full use of spatial-temporal interest points: an adaboost approach for action recognition. In: 2010 17th IEEE International Conference on Image Processing (ICIP), pp. 4677–4680. IEEE (2010)

19. Yi, T., Qiuqi, R.: Weight and context method for action recognition using histogram intersection. In: 5th IET International Conference on Wireless, Mobile and Multimedia Networks, ICWMMN 2013, pp. 229–233, November 2013
20. Zhang, Z., Wang, C., Xiao, B., Zhou, W., Liu, S.: Action recognition using context-constrained linear coding. IEEE Sig. Process. Lett. **19**(7), 439–442 (2012)
21. Zhou, X., Yu, K., Zhang, T., Huang, T.S.: Image classification using super-vector coding of local image descriptors. In: Daniilidis, K., Maragos, P., Paragios, N. (eds.) ECCV 2010. LNCS, vol. 6315, pp. 141–154. Springer, Heidelberg (2010). doi:10.1007/978-3-642-15555-0_11

Virtual EMG via Facial Video Analysis

Giuseppe Boccignone[1], Vittorio Cuculo[1,2], Giuliano Grossi[1],
Raffaella Lanzarotti[1(✉)], and Raffaella Migliaccio[1]

[1] PHuSe Lab - Dipartimento di Informatica, Università degli Studi di Milano,
Via Comelico 39/41, Milan, Italy
giuseppe.boccignone@unimi.it, {grossi,lanzarotti}@di.unimi.it,
raffaella.migliaccio@studenti.unimi.it
[2] Dipartimento di Matematica, Università degli Studi di Milano,
Via Cesare Saldini 50, Milan, Italy
vittorio.cuculo@unimi.it

Abstract. In this note, we address the problem of simulating electromyographic signals arising from muscles involved in facial expressions - markedly those conveying affective information -, by relying solely on facial landmarks detected on video sequences. We propose a method that uses the framework of Gaussian Process regression to predict the facial electromyographic signal from videos where people display nonposed affective expressions. To such end, experiments have been conducted on the OPEN EmoRec II multimodal corpus.

1 Introduction

The face is the locus of a great deal of emotional expressions and researchers in different fields crossing with affective science [9] have been keen on facial electromyographic measures of muscle activity, in particular those related to the *zygomaticus major* and the *corrugator supercilii* (see Fig. 1a). The motivation for such endeavour is straightforward: the *zygomaticus major* controls the corners of the mouth (e.g., by pulling them back and up into a smile), the *corrugator supercilii* hauls the brow down and together into a frown [18]. In brief, facial electromyography is a reliable detector of the affective state, either in the continuous dimension of valence (positive versus negative affective state) [18], or to reveal the discrete emotions [16].

Electromyography measures the electrical potentials arising from skeletal muscles [27]. Facial EMG (fEMG), is based on recording the difference in electrical potential pairs of electrodes that are placed close together on the target facial muscle (Fig. 1b). Main advantages of fEMG stem from (1) the capability of intercepting even very weak affective expression and (2) the very good time resolution that allows to reliably register sudden expression changes. On the other hand, the need of placing electrodes over the face limits the applicability of this sensor to laboratory acquisition only (see again Fig. 1b). Cogently, in this case and more generally, the option of monitoring physiological signals via non-contact means has promise for a variety of out-of-lab applications well beyond the affective computing realm [23].

© Springer International Publishing AG 2017
S. Battiato et al. (Eds.): ICIAP 2017, Part I, LNCS 10484, pp. 197–207, 2017.
https://doi.org/10.1007/978-3-319-68560-1_18

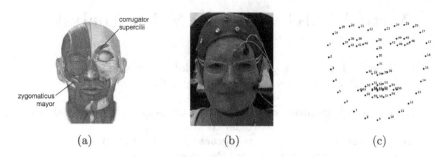

(a)   (b)   (c)

Fig. 1. (a) Anatomical location of facial muscles involved in this study. (b) Electrode placing to detect the activity of the *zygomaticus major* and the *corrugator supercilii* muscles. (c) Facial landmarks inferred by the method [8]

Whilst there is a number of works addressing noncontact, physiological measurements of heart rate, e.g. [23,26,30], to the best of our knowledge, this is the first attempt to estimate fEMG signals from video sequences.

We argue that, apart from the *per se* appealing issue of avoiding the obtrusiveness of fEMG, the idea of virtual fEMG derived from observing natural, non-posed facial expression, can be important for dealing with emotion understanding in a broader perspective (see Sect. 4, for a discussion). All things considered, this endeavour is at this stage affordable, given that in the last decade, the number of public repositories has grown larger, where behavioral data have been recorded by multiple modalities [7,29], hence providing adequate training sets and benchmarking, as will be detailed in Sect. 3.

In Sect. 2 the method we propose for the virtual fEMG generation is described; in Sect. 3 the experiments and the obtained results are shown and discussed. In Sect. 4 conclusive remarks on this preliminary study are given.

2   Method

Given a video stream $\mathbf{I}(t)$, fEMG signal generation is obtained by relying on perceived facial fiducial points, or landmarks. In a nutshell, landmarks are detected in a sparse coding framework and signal generation is obtained through Gaussian Process (GP) regression and prediction. More precisely, use the following random variables (RVs):

- $\mathbf{E}$: a set of fEMG data over time intervals, i.e. a set of signals $\mathbf{e}$;
- $\mathbf{L}$: a set of landmarks $\mathbf{l}$, over time intervals, each $\mathbf{l}^i$ being a landmark;
- $\mathbf{F}$: a set of feature responses $\mathbf{f}$, over time intervals, each $\mathbf{f}^i$ being a local feature response;
- $\mathbf{X} = [\mathbf{x}_1, \cdots, \mathbf{x}_N] \in \mathbb{R}^{D \times N}$: the matrix of observed training patches.
- $\mathbf{W} = [\mathbf{w}_1, \cdots, \mathbf{w}_L] \in \mathbb{R}^{D \times L}$: a dictionary; each column $\mathbf{w}_i$ is referred to as a basis vector or atom;
- $\mathbf{Z} = [\mathbf{z}_1, \cdots, \mathbf{z}_N] \in \mathbb{R}^{L \times N}$ the latent sparse code matrix associated to $\mathbf{W}$.

Then the proposed method can be summarised as the sampling of the virtual fEMG signal $\widetilde{\mathbf{e}} = [e(1), e(2), \cdots, e(T)]$ from the joint conditional distribution:

$$\widetilde{\mathbf{e}} \sim P(\mathbf{E}, \mathbf{L}, \mathbf{F}, \mathbf{W} \mid \mathbf{X}, \mathbf{I}). \tag{1}$$

The joint pdf can be factorised as follows:

$$P(\mathbf{E}, \mathbf{L}, \mathbf{F}, \mathbf{W} \mid \mathbf{X}, \mathbf{I}) = P(\mathbf{E} \mid \mathbf{L}) \times P(\mathbf{L} \mid \mathbf{F}) \times P(\mathbf{F} \mid \mathbf{W}, \mathbf{I}) \times P(\mathbf{W} \mid \mathbf{X}) \tag{2}$$

The method can be best explained by starting from the last factor on the r.h.s. of Eq. 2. In the sparse coding framework, such term supports dictionary inference given a set of training patches:

$$\mathbf{W}^* = \arg\max_{\mathbf{W}} P(\mathbf{W} \mid \mathbf{X}) \tag{3}$$

The problem of inferring dictionary $\mathbf{W}$ can be reduced to a maximum likelihood estimation $\mathbf{W}^* = \arg\max P(\mathbf{W} \mid \mathbf{X}) \approx \arg\max P(\mathbf{X} \mid \mathbf{W})$, where the observable patch vector $\mathbf{x}_i$ is approximated as a sparse combination of basis vectors $\mathbf{w}_i$, i.e. $\mathbf{x} = \mathbf{W}\mathbf{z} + \mathbf{v}$, $\mathbf{v}$ being a residual noise vector sampled from a zero mean Gaussian distribution $\mathcal{N}(0, \sigma^2 \mathbb{I})$. The dictionary can be derived under the Olshausen and Field approximation [21], $\log P(\mathbf{X}|\mathbf{W}) \approx \sum_{i=1}^{N} \max_{\mathbf{z}_i}[\log \mathcal{N}(\mathbf{x}_i|\mathbf{W}\mathbf{z}_i, \sigma^2 \mathbb{I}) + \log P(\mathbf{z}_i)]$, and turned in the minimization of the negative log-likelihood (NLL). This can be done efficiently by using either the K-SVD [3] or the R-SVD [15] algorithms as shown in [1,2,14].

The third factor represents the feature likelihood under the current observable video $\mathbf{I}$ and the inferred dictionary. The goal here is to compute feature responses

$$\mathbf{F}^* \sim P(\mathbf{F} \mid \mathbf{W}, \mathbf{I}) \tag{4}$$

at each frame in $\mathbf{I}$. Here, we adopt the Histograms of Sparse Codes (HSC) representation to sample the local response $\mathbf{f}^i$ [8].

The second factor accounts for the detection of landmarks given the observed $\mathbf{F}^*$. A part-based detection approach is adopted [8], where every facial landmark can be modeled as a part, and the locations $\mathbf{L}$ of parts of the face can be generated according to m views or poses by some similarity transformation τ, giving rise to the global model $\mathbf{L}_{k,\tau}$. The generation of $\mathbf{L}$ can be accomplished by marginalising over the set of m models, i.e., $P(\mathbf{L}|\mathbf{F}) = \sum_{k=1}^{m} \int_{\tau} P(\mathbf{L}|\mathbf{L}_{k,\tau}) P(\mathbf{L}_{k,\tau}|\mathbf{F}) d\tau$. The term $P(\mathbf{L}|\mathbf{L}_{k,\tau})$ accounts for dependence of $\mathbf{L}$ from the global configuration $\mathbf{L}_{k,\tau}$.

Assume that: (i) the locations of the parts $\{\mathbf{l}^i\}_{i=1}^l$ are conditionally independent of one another and the same holds for the detector responses $\mathbf{f}^i$; (ii) the relation between the transformed model landmark and the true landmark is translationally invariant, i.e., $P(\mathbf{l}_{k,\tau}^i|\mathbf{l}_{k,\tau})$ only depends on $\Delta\mathbf{l}_{k,\tau}^i = \mathbf{l}_{k,\tau}^i - \mathbf{l}^i$. Then, the following MAP solution can be derived,

$$\mathbf{L}^* = \arg\max_{\mathbf{L}} \sum_{k=1}^{m} \int_{\tau} \prod_{i=1}^{l} P(\Delta\mathbf{l}_{k,\tau}^i) P(\mathbf{l}^i|\mathbf{f}^i) d\tau, \tag{5}$$

where the prior $P(\Delta l^i_{k,\tau})$ accounts for the *shape* or global component of the model, and $P(l^i | f^i)$ for the *appearance* or local component. The latter relies on patches representing HSC responses to face landmarks.

Eventually, the first factor on the r.h.s. of Eq. 2, is the likelihood supporting the generation of the fEMG signal given the extracted landmarks. The generative model behind the conditional distribution $P(\mathbf{E} | \mathbf{L})$, under Gaussian assumption, assumes that a realisation of a target electromyographic signal $\mathbf{e}$ is generated by a latent function $\mathbf{g} = \{g(\mathbf{d}_n)\}$ of a suitable measurement $\mathbf{d}$ of the landmarks corrupted by additive Gaussian noise. Thus, at time (frame index) t:

$$e(t) = g(\mathbf{d}(\mathbf{l}_p(t))) + \nu(t), \quad \nu \sim \mathcal{N}(0, \sigma_e^2) \tag{6}$$

where, in our case, $\mathbf{d}(\mathbf{l}_p)$ is a vector of distances over the *pool* $\mathbf{l}_p$, a subset of the extracted landmarks $\mathbf{l}$, which is suitable to capture muscle activity. Note that the mapping function $g(\cdot)$ needs not to be linear. In other terms, the conditional distribution $P(\mathbf{E} | \mathbf{L})$ is defined as the marginal likelihood $P(\mathbf{E} | \mathbf{L}) = \int P(\mathbf{E} | \mathbf{g}, \mathbf{L}) P(\mathbf{g} | \mathbf{L}) d\mathbf{g}$, where the marginalisation over the function values g, can be performed by using a GP prior distribution over functions $P(\mathbf{g} | \mathbf{L}) = \mathcal{N}(\mu_g(\mathbf{L}), k(\mathbf{L}, \mathbf{L}))$, $k(\mathbf{L}, \mathbf{L})$ being the kernel function [24], i.e. in our case

$$g(\mathbf{d}(\mathbf{l}_p)) \sim \mathcal{GP}(\mu(\mathbf{d}(\mathbf{l}_p)), k(\mathbf{d}(\mathbf{l}_p), \mathbf{d}'(\mathbf{l}_p))), \tag{7}$$

and where the likelihood of the observed targets is $P(\mathbf{E} | \mathbf{g}, \mathbf{L}) = \mathcal{N}(\mathbf{g}, \sigma_e^2 \mathbb{I})$, from which Eq. 6 is obtained. Note that, due to analytical tractability of the Gaussian distribution, all the above computations are determined in closed form so that, prior to the prediction of the virtual fEMG signal $\tilde{\mathbf{e}}$, parameter learning can be efficiently performed on the given dataset $\{\mathbf{L}, \mathbf{E}\}$ (see Rasmussen and Williams [24] for details).

3   Experimental Work

(A) Experimental Setup. The experiments have been conducted on the multimodal corpus OPEN EmoRec II [25]. The dataset was designed to induce emotional responses in users involved in naturalistic-like human-computer interaction (HCI) according to two HCI-experimental settings. In the former, pictures taken from the IAPS set [17] were used to induce emotions. Stimulus sequences consisted of 10 pictures with similar ratings according to the 5 possible affective states: high valence and high arousal (HVHA), high valence and low arousal (HVLA), low valence and low arousal (LVLA), low valence and high arousal (LVHA) and neutral. In the second part of the experiment, the emotions were induced during a naturalistic-like HCI in a standardized environment. In both the experiments several data were recorded: video, audio, trigger information and physiological data, namely respiration, fEMG from *corrugator supercilii* activity, fEMG *zygomaticus major* activity, Blood Volume Pulse and Skin Conductance.

In this paper we refer to the data, videos and fEMG signals, acquired in the first experiment, that is the recording of 30 subjects, each one stimulated by 5 image sequences.

(B) Landmark Extraction. Given a video sequence of a facial expression, we account for Eqs. 3, 4 and 5 by applying the method described in [8] to infer the locations of facial landmarks (Fig. 1c). Such method extends in a sparse coding framework Zhu and Ramanan's technique [31], which jointly performs face and landmark detection. Once landmarks **L** have been detected, an adequate pool l_p of landmarks should be defined in order to provide related distance measures $d(l_p)$ as a "proxy" to muscle activity. Figures 2 and 3 below show the landmarks involved in measuring *corrugator supercilii* and *zygomaticus major* activities, respectively.

Fig. 2. Landmarks and distances accounting for the *corrugator supercilii* activity (Color figure online)

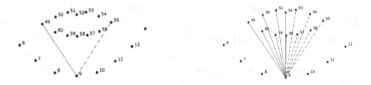

Fig. 3. Landmarks and distances accounting for the *zygomaticus major* activity (Color figure online)

The fEMG signal captures very local muscle movements and its simulation should derive from a small subset of facial landmarks with superposition to the muscle of interest. The most natural choice would be to consider the landmarks closest to the muscle as shown in Fig. 2, (blue dashed line, left panel) for the *corrugator supercilii*, and in Fig. 3, (blue dashed line, left panel) for the *zygomaticus major*. However, landmark displacements are noisy, due to the detection method and possible occlusions caused by the sensors. We thus investigate several pools of displacements aiming at pinpointing the most suitable ones for fEMG regression.

In the case of the *corrugator supercilii*, we thus consider the symmetric distance between the inner eye corners and the inner eyebrow landmarks (Fig. 2, left panel), the two distances coupled, and more global measures obtained considering the distances between the inner eye corners and the corresponding eyebrow landmarks, both separately and all together (Fig. 2, right panel). Similarly for the *zygomaticus major*, we take into account the symmetric distance as in Fig. 3, (red line, left panel), the two punctual distances coupled, and the distances between the chin and the two halves extern lip contour landmarks, both singularly and coupled (Fig. 3, right panel).

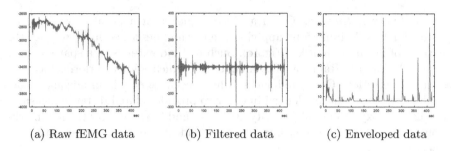

(a) Raw fEMG data   (b) Filtered data   (c) Enveloped data

Fig. 4. fEMG signal processing pipeline.

(C) fEMG Preprocessing. The raw data set of fEMG measurements derived from *corrugator supercilii* and *zygomaticus major* activities - which we denote $\mathbf{E}^c$ and $\mathbf{E}^z$, respectively - is a collection of 1-D signals captured at 512 Hz or more (Fig. 4a). The low frequencies are strongly influenced by artifacts such as motion potentials, eye movements, eye blinks, swallowing, and respiration, thus requiring a preliminary high-pass filtering to remove the strongest artifacts that would otherwise dominate the real facial EMG potentials. In the literature different cutoff frequencies are adopted for this purpose, ranging from 5 to 20 Hz [6,19,32], We use a 20 Hz cutoff frequency, guaranteeing artifact elimination. In addition, filtering has to be applied to remove the 50 Hz power line interference. To this aim, notch filtering is adopted (Fig. 4b). Further, when fEMG activation is addressed, the rectification and envelope are advised [5,20]. Eventually, to train the Gaussian process, the signals are down-sampled to 25 Hz so that the fEMG and the video frequencies are in correspondence (Fig. 4c).

(D) GP Model Learning and fEMG Prediction. Given a dataset of inputs and targets, $\{\mathbf{L}, \mathbf{E}\} = \{l_n, e_n \mid n = 1, \cdots, N\}$, we are interested in evaluating the mapping of S test sequences of landmarks $\mathbf{L}_{new} = \{l_{new,s} \mid s = 1, \cdots, S\}$ into fEMG sequences $\mathbf{E}_{new} = \{e_{new,s} \mid s = 1, \cdots, S\}$, where $\widetilde{e} = e_{new,s}$ is the desired virtual fEMG signal. Notice that here and in what follows, we thoroughly write $l_{p,new}$ in place of actual measurements $\mathbf{d}(l_{p,new})$ to simplify notation. Formally, we need to evaluate the predictive distribution $P(\mathbf{E}_{new}|\mathbf{L}, \mathbf{E}, \mathbf{L}_{new}) = \int P(\mathbf{E}_{new} \mid \mathbf{g}_{new})P(\mathbf{g}_{new} \mid \mathbf{L}, \mathbf{E}, \mathbf{L}_{new})d\mathbf{g}_{new}$, where $P(\mathbf{E}_{new} \mid \mathbf{g}_{new})$ is the likelihood given by Eq. 6. The posterior over functions $P(\mathbf{g}_{new} \mid \mathbf{L}, \mathbf{E}, \mathbf{L}_{new})$ is a Gaussian distribution $\mathcal{N}(\mu_{new}, k_{new})$, whose parameters can be written in closed form [24], namely, $\mu_{new} = k(\mathbf{L}_{new}, \mathbf{L})\left[k(\mathbf{L}, \mathbf{L}) + \sigma_e^2\mathbb{I}\right]^{-1}$ and $k_{new} = k(\mathbf{L}_{new}, \mathbf{L}_{new}) - k(\mathbf{L}_{new}, \mathbf{L})\left[k(\mathbf{L}, \mathbf{L}) + \sigma_e^2\mathbb{I}\right]^{-1}k(\mathbf{L}, \mathbf{L}_{new})$. Kernel functions and related hyperparameters are obtained from the training stage.

As to the latter, we train different models, varying the referred landmark pool, $p \in \{1, ..., 6\}$, associated with the related muscle, and exploring the GP behaviour by adopting the well-known Squared Exponential Kernel (k_{SE}), Rational Quadratic Kernel (k_{RQ}), and the Matern 3/2 kernel (k_{M32}) [24]. For each model, training and test sets are derived adopting the k-fold cross validation method, partitioning data into 10 subsets.

(E) Results. The quality of the virtual fEMG, $\tilde{e}$, with respect to the original fEMG filtered signal, e, is evaluated in terms of Mean Square Error (MSE), and by the Concordance Correlation Coefficient measures (CCC):

$$MSE(e, \tilde{e}) = \frac{1}{T} \sum_{t=1}^{T} (e(t) - \tilde{e}(t))^2 \qquad CCC(e, \tilde{e}) = \frac{2cov(e, \tilde{e})}{\sigma_e^2 + \sigma_{\tilde{e}}^2 + (\mu_e - \mu_{\tilde{e}})^2},$$

being μ_e and $\mu_{\tilde{e}}$ the signal means, σ_e^2 and $\sigma_{\tilde{e}}^2$ the variances, and $cov(e, \tilde{e})$ the covariance.

In Table 1 we report the performances obtained in simulating the *corrugator supercilii* fEMG, adopting the different learnt models. Those concerning the virtual generation of the *zygomaticus major* fEMG are shown in Table 2.

Analysing the behaviour of the models, we observe that the MSE and the CCC performances are always coherent. We can conclude that both in the simulations of the *corrugator supercilii* fEMG and of the *zygomaticus major*, best performances are achieved through the largest pool of landmark distances. This is likely to depend on the noise that characterizes landmarks localization, certainly attenuated by considering a pool of landmarks rather than punctual ones.

Table 1. Performances achieved in the virtual generation of the *corrugator supercilii* fEMG, referring to different pool of landmarks ($p \in \{1...6\}$), and different kernels (k_{SE}, k_{RQ}, k_{M32}). Performances are expressed as MSE and CCC.

Pool	MSE			CCC		
	SE	RQ	M32	SE	RQ	M32
1	42.0075	31.8062	36.2262	0.9855	0.9888	0.9873
2	4.9623	4.9285	4.9343	0.9983	0.9983	0.9983
3	3.8280	3.7601	3.7185	0.9987	0.9987	0.9987
4	2.0082	1.7696	1.7273	0.9993	0.9994	0.9994
5	1.9681	1.7991	1.7070	0.9993	0.9994	0.9994
6	1.0478	0.7167	**0.6522**	0.9996	0.9997	**0.9998**

Table 2. Performances achieved in the virtual generation of the *zygomaticus major* fEMG. Results are organized as in Table 1

Pool	MSE			CCC		
	SE	RQ	M32	SE	RQ	M32
1	11.5408	11.5193	11.5001	0.9969	0.9969	0.9969
2	13.3317	13.2269	13.1922	0.9965	0.9965	0.9965
3	8.1977	7.8251	7.7871	0.9978	0.9979	0.9979
4	2.2541	1.5721	1.4217	0.9994	0.9996	0.9996
5	2.5401	1.2763	1.3191	0.9993	0.9997	0.9997
6	1.0401	0.6291	**0.5617**	0.9997	0.9998	**0.9999**

In particular, we observe that the punctual distance $d = 1$ in the *corrugator supercilii* fEMG gives the worst performances, this because, in the considered dataset, the fEMG sensor often partial occludes the eyebrow. Also, it is worth noticing that system behaviour is robust to the use of different kernels.

Figures 5 and 6 illustrate typical fEMG signal reconstructions for both the *corrugator* and the *zygomaticus* muscles.

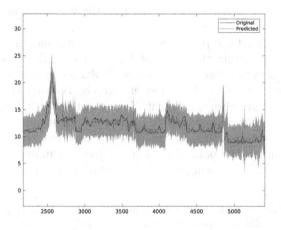

Fig. 5. Detail of fEMG reconstruction of the *corrugator supercilii* signal, using the Squared Exponential kernel and considering the 5-th landmark pool. The shaded area represents the pointwise mean plus and minus two times the standard deviation for each input value (corresponding to the 95% confidence region)

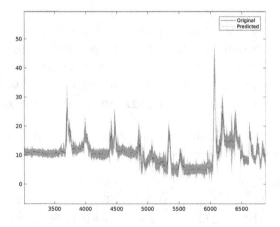

Fig. 6. Detail of fEMG reconstruction of the *zygomaticus* signal, using the Matern 3/2 kernel and considering the 6-th landmark pool

4   Discussion and Conclusions

We have presented a method for detecting the electromyographic signal arising from muscles involved in affective, non-posed, facial expressions, which only relies on the facial landmarks detected in videos. Preliminary experiments on the OPEN EmoRec II multimodal corpus [25] have given evidence of promising results.

Clearly, one should be aware that there are limitations in the detection capability of the method. It is known that real fEMG can intercept even very weak affective expressions, even below the visible display of the expression itself [18]; however, this limit is shared by all virtual methods that attempt at simulating *in vivo* measurements from visual input.

Apart from the appealing issue of avoiding the obtrusiveness of fEMG measurement, what is to be gained by such attempt in view of the affective computing problem? All things considered, as detailed in Sect. 2, the landmarks we rely upon for regressing the fEMG signal are nothing but a subset of the facial landmarks we collect, the latter, in principle, providing full information - at least that available from the video sequence - to further proceed with facial expression analysis for affective computing purposes. Under the circumstances, it is worth making clear the rationale behind this study. Affective computing aims at dealing with machines that might have the ability to (1) recognize emotions, (2) express emotions, (3) "have emotions", the latter being the "hardest stuff" [22]. So far, most current research focuses on (1) and (2), with image processing and pattern recognition-based affect detection playing a prominent role [7]. The research work fostering this study pursues a different approach, centred on simulation-based affect analysis [28]. According to embodied simulation theories, understanding emotions of others is supported by running the same emotional apparatus - possibly in reverse - that is already used to generate or experience the emotion, eventually causing a "reactivation" of the corresponding mental state [11–13]. Indeed, an emotion is a neural reaction to a certain stimulus, realised by a complex ensemble of neural activations in the brain. The latter often are preparations for (muscular, visceral) actions (facial expressions, heart rate increase, etc.), as a consequence, the body will be modified into an "observable" [10]. It is in such a broader perspective that it is particularly relevant to have available a variety of physiological signals, real or virtual, for building the latent continuous space of emotions [4]. fEMG, together with others that can be obtained by less obtrusive means (heart rate, skin conductance, respiratory rhythm, gaze scan path), is one such signal.

Acknowledgments. This research was carried out as part of the project "Interpreting emotions: a computational tool integrating facial expressions and biosignals based shape analysis and bayesian networks", supported by the Italian Government, managed by MIUR, financed by the *Future in Research* Fund.

References

1. Adamo, A., Grossi, G., Lanzarotti, R.: Local features and sparse representation for face recognition with partial occlusions. IEEE, September 2013
2. Adamo, A., Grossi, G., Lanzarotti, R., Lin, J.: Robust face recognition using sparse representation in LDA space. Mach. Vis. Appl. **26**(6), 837–847 (2015)
3. Aharon, M., Elad, M., Bruckstein, A.: K-SVD: an algorithm for designing overcomplete dictionaries for sparse representation. IEEE Trans. Sig. process. **54**(11), 4311–4322 (2006)
4. Anderson, D.J., Adolphs, R.: A framework for studying emotions across species. Cell **157**(1), 187–200 (2014)
5. Barzilay, O., Wolf, A.: A fast implementation for EMG signal linear envelope computation. J. Electromyogr. Kinesiol. **21**(4), 678–682 (2011)
6. van Boxtel, A.: Optimal signal bandwidth for the recording of surface EMG of facial, jaw, oral, and neck muscles. Psychophysiology **38**, 22–34 (2001)
7. Calvo, R., D'Mello, S.: Affect detection: an interdisciplinary review of models, methods, and their applications. IEEE Trans. Affect. Comput. **1**(1), 18–37 (2010)
8. Cuculo, V., Lanzarotti, R., Boccignone, G.: Using sparse coding for landmark localization in facial expressions. In: 5th European Workshop on Visual Information Processing (EUVIP), pp. 1–6, December 2014
9. Dalgleish, T., Dunn, B., Mobbs, D.: Affective neuroscience: past, present, and future. Emot. Rev. **1**(4), 355–368 (2009)
10. Damasio, A.R.: The Feeling of What Happens: Body and Emotion in the Making of Consciousness. Houghton Mifflin Harcourt, Boston (1999)
11. Gallese, V.: The 'shared manifold' hypothesis. From mirror neurons to empathy. J. Conscious. Stud. **8**(5–7), 33–50 (2001)
12. Gallese, V.: The manifold nature of interpersonal relations: the quest for a common mechanism. Philos. Trans. R. Soc. Lond. Ser. B: Biol. Sci. **358**(1431), 517–528 (2003)
13. Goldman, A.I., Sripada, C.S.: Simulationist models of face-based emotion recognition. Cognition **94**(3), 193–213 (2005)
14. Grossi, G., Lanzarotti, R., Lin, J.: Robust face recognition providing the identity and its reliability degree combining sparse representation and multiple features. Int. J. Pattern Recogn. Artif. Intell. **30**(10) (2016)
15. Grossi, G., Lanzarotti, R., Lin, J.: Orthogonal procrustes analysis for dictionary learning in sparse linear representation. PLoS One **12** (2017)
16. Hildebrandt, A., Recio, G., Sommer, W., Wilhelm, O., Ku, J.: Facial EMG responses to emotional expressions are related to emotion perception ability. PLoS One **9**(1) (2014)
17. Lang, P.J., Bradley, M.M., Cuthbert, B.N.: International affective picture system (IAPS): affective ratings of pictures and instruction manual. Technical Report A-8, The Center for Research in Psychophysiology, University of Florida, Gainesville, FL (2008)
18. Larsen, J., Norris, C., Cacioppo, J.: Effects of positive and negative affect on electromyography activity over zygomaticus major and corrugator supercilii. Psychophysiology **40**, 776–785 (2003)
19. Lu, G., Brittain, J.S., Holland, P., Yianni, J., Green, A.L., Stein, J.F., Aziz, T.Z., Wang, S.: Removing ECG noise from surface EMG signals using adaptive filtering. Neurosci. Lett. **462**, 14–19 (2009)

20. Myers, L., Lowery, M., O'Malley, M., Vaughan, C., Heneghan, C., Gibson, A.S.C., Harley, Y., Sreenivasan, R.: Rectification and non-linear pre-processing of EMG signals for cortico-muscular analysis. J. Neurosci. Methods **124**(2), 157–165 (2003)
21. Olshausen, B.A., Field, D.J.: Natural image statistics and efficient coding. Netw.: Comput. Neural Syst. **7**(2), 333–339 (1996)
22. Picard, R.W.: Affective Computing. MIT press, Cambridge (2000)
23. Poh, M.Z., McDuff, D.J., Picard, R.W.: Advancements in noncontact, multiparameter physiological measurements using a webcam. IEEE Trans. Biomed. Eng. **58**(1), 7–11 (2011)
24. Rasmussen, C.E., Williams, C.K.: Gaussian Processes for Machine Learning. The MIT Press, Cambridge (2006)
25. Rukavina, S., Gruss, S., Walter, S., Hoffmann, H., Traue, H.C.: OPEN EmoRec II - a multimodal corpus of human-computer interaction. Int. J. Comput. Electr. Autom. Control Inf. Eng. **9**(5), 1181–1187 (2015)
26. Sun, Y., Thakor, N.: Photoplethysmography revisited: from contact to noncontact, from point to imaging. IEEE Trans. Biomed. Eng. **63**(3), 463–477 (2016)
27. Tassinary, L.G., Cacioppo, J.T., Vanman, E.J.: The skeletomotor system: surface electromyography. In: Cacioppo, J.T., Tassinary, L.G., Berntson, G. (eds.) Handbook of Psychophysiology (Chap. 12), pp. 267–300. Cambridge University Press, Cambridge (2012)
28. Vitale, J., Williams, M.A., Johnston, B., Boccignone, G.: Affective facial expression processing via simulation: a probabilistic model. Biolog. Inspired Cogn. Archit. J. **10**, 30–41 (2014)
29. Wang, S., Ji, Q.: Video affective content analysis: a survey of state-of-the-art methods. IEEE Trans. Affect. Comput. **6**(4), 410–430 (2015)
30. Wu, H.Y., Rubinstein, M., Shih, E., Guttag, J., Durand, F., Freeman, W.: Eulerian video magnification for revealing subtle changes in the world. ACM Trans. Graph. (TOG) **31**(4), 65 (2012)
31. Zhu, X., Ramanan, D.: Face detection, pose estimation, and landmark localization in the wild. In: Proceedings of IEEE CVPR, pp. 2879–2886 (2012)
32. Zschorlich, V.R.: Digital filtering of EMG-signals. Electromyogr. Clin. Neurophysiol. **29**(April), 81–86 (1989)

Pattern Recognition and Machine Learning

A Compact Kernel Approximation for 3D Action Recognition

Jacopo Cavazza[1,2]([✉]), Pietro Morerio[1], and Vittorio Murino[1,3]

[1] Pattern Analysis and Computer Vision (PAVIS), Istituto Italiano di Tecnologia, Genova, Italy
{jacopo.cavazza,pietro.morerio,vittorio.murino}@iit.it
[2] Electrical, Electronics and Telecommunication Engineering and Naval Architecture Department (DITEN), Università degli Studi di Genova, Genova, Italy
[3] Computer Science Department, Università di Verona, Verona, Italy

Abstract. 3D action recognition was shown to benefit from a covariance representation of the input data (3D positions of the joints). A kernel machine fed with such feature is an effective paradigm for 3D action recognition, yielding state-of-the-art results. Yet, the whole framework is affected by the well-known scalability issue. In fact, in general, the kernel function has to be evaluated for all pairs of instances inducing a Gram matrix whose complexity is quadratic in the number of samples. In this work we reduce such complexity to be linear by proposing a novel and explicit feature map to approximate the kernel function. This allows to train a linear classifier with an explicit feature encoding, which implicitly implements a Log-Euclidean machine in a scalable fashion. Not only we prove that the proposed approximation is unbiased, but also we work out an explicit strong bound for its variance, attesting a theoretical superiority of our approach with respect to existing ones. Experimentally, we verify that our representation provides a compact encoding and outperforms other approximation schemes on a number of publicly available benchmark datasets for 3D action recognition.

Keywords: Action recognition · 3D · Kernel · Feature map

1 Introduction

Action recognition is a key research domain in video/image processing and computer vision, being nowadays ubiquitous in human-robot interaction, autonomous driving vehicles, elderly care and video-surveillance to name a few [21]. Yet, challenging difficulties arise due to visual ambiguities (illumination variations, texture of clothing, general background noise, view heterogeneity, occlusions). As an effective countermeasure, joint-based skeletal representations (extracted from depth images) are a viable solution.

Combined with a skeletal representation, the symmetric and positive definite (SPD) covariance operator scores a sound performance in 3D action recognition [5,9,22]. Indeed, while properly modeling the skeletal dynamics with a

© Springer International Publishing AG 2017
S. Battiato et al. (Eds.): ICIAP 2017, Part I, LNCS 10484, pp. 211–222, 2017.
https://doi.org/10.1007/978-3-319-68560-1_19

second order statistic, the covariance operator is also naturally able to handle different temporal duration of action instances. This avoids slow pre-processing stages such as time warping or interpolation [20]. In addition, the superiority of such representation can be attested by achieving state-of-the-art performance by means of a relative simple classification pipeline [5,22] where, basically[1], a non-linear Support Vector Machine (SVM) is trained using the Log-Euclidean kernel

$$K_{\ell E}(\mathbf{X}, \mathbf{Y}) = \exp\left(-\frac{1}{2\sigma^2}\|\log \mathbf{X} - \log \mathbf{Y}\|_F^2\right) \tag{1}$$

to compare covariance operators $\mathbf{X}$, $\mathbf{Y}$. In (1), for any SPD matrix $\mathbf{X}$, we define

$$\log \mathbf{X} = \mathbf{U} \mathrm{diag}(\log \lambda_1, \ldots, \log \lambda_d)\mathbf{U}^\top, \tag{2}$$

being $\mathbf{U}$ the matrix of eigenvectors which diagonalizes $\mathbf{X}$ in terms of the eigenvalues $\lambda_1 \geq \cdots \geq \lambda_d > 0$. Very intuitively, for any fixed bandwidth $\sigma > 0$, $K_{\ell E}(\mathbf{X}, \mathbf{Y})$ is actually computing a radial basis Gaussian function by comparing the covariance operators $\mathbf{X}$ and $\mathbf{Y}$ by means of the Frobenius norm $\|\cdot\|_F$ (after $\mathbf{X}, \mathbf{Y}$ have been log-projected). Computationally, the latter stage is not problematic (see Sect. 3) and can be performed for each covariance operator *before* computing the kernel. In addition to its formal properties in Riemannian geometry, this makes (1) widely used in practice [5,9,22].

However, the modern big data regime mines the applicability of such a kernel function. Indeed, since (1) has to be computed for *every pair* of instances in the dataset, the so produced Gram matrix has prohibitive size. So its storage becomes time- and memory-expensive and the related computations (required to train the model) are simply unfeasible.

The latter inconvenient can be solved as follows. According to the well established kernel theory [2], the Kernel (1) induces an infinite-dimension feature map φ, meaning that $K_{\ell E}(\mathbf{X}, \mathbf{Y}) = \langle \varphi(\mathbf{X}), \varphi(\mathbf{Y}) \rangle$. However, if we are able to obtain an explicit feature map Φ such that $K_{\ell E}(\mathbf{X}, \mathbf{Y}) \approx \langle \Phi(\mathbf{X}), \Phi(\mathbf{Y}) \rangle$, we can directly compute a finite-dimensional feature representation $\Phi(\mathbf{X})$ for each action instance separately. Then, with a compact Φ, we can train a linear SVM instead of its kernelized version. This is totally feasible and quite efficient even in the big data regime [7]. Therefore, the whole pipeline will actually provide a scalable implementation of a Log-Euclidean SVM, whose cost is reduced from quadratic to linear.

In our work we specifically tackle the aforementioned issue through the following main contributions.

1. We propose a novel compact and explicit feature map to approximate the Log-Euclidean kernel within a probabilistic framework.
2. We provide a rigorous mathematical formulation, proving that the proposed approximation has null bias and bounded variance.

---

[1] For the sake of precision, let us notice that [22] take advantage of multiple kernel learning in combining several low-level representations and [5] replaces the classical covariance operator with a kernelization.

3. We compare the proposed feature map approximation against alternative approximation schemes, showing the formal superiority of our framework.
4. We experimentally evaluate our method against the very same approximation schemes over six 3D action recognition datasets, confirming with practice our theoretical findings.

The rest of the paper is outlined as follows. In Sect. 2 we review the most relevant related literature. Section 3 proposes the novel approximation and discusses its foundation. We compare it with alternative paradigms in Sect. 4. Section 5 draws conclusions and the Appendix A reports all proofs of our theoretical results.

2   Related Work

In this Section, we summarize the most relevant works in both covariance-based 3D action recognition and kernels' approximations.

Originally envisaged for image classification and detection tasks, the covariance operator has experienced a growing interest for action recognition, experiencing many different research trends: [9] extends it to the infinite dimensional case, while [10] hierarchically combines it in a temporal pyramid; [12,22] investigate the conceptual analogy with trial-specific kernel matrices and [5] further proposes a new kernelization as to model arbitrary, non-linear relationships conveyed by the raw data. However, those kernel methods usually do not scale up easily to big datasets due to demanding storage and computational costs. As a solution, the exact kernel representation can be replaced by an approximated, more efficient version. In the literature, this is done according to the following mainstream approaches.

(i) The kernel Gram matrix is replaced with a surrogate low-rank version, in order to alleviate both memory and computational costs. Within these methods, [1] applied Cholesky decomposition and [24] exploited Nyström approximation.

(ii) Instead of the exact kernel function k, an explicit feature map Φ is computed, so that the induced linear kernel $\langle \Phi(\mathbf{x}), \Phi(\mathbf{y}) \rangle$ approximates $k(\mathbf{x}, \mathbf{y})$. Our work belong to this class of methods.

In this context, Rahimi and Recht [17] exploited the formalism of the Fourier Transform to approximate shift invariant kernels $k(\mathbf{x}, \mathbf{y}) = k(\mathbf{x} - \mathbf{y})$ through an expansion of trigonometric functions. Leveraging on a similar idea, Le et al. [13] sped up the computation by exploiting the Walsh-Hadamard transform, downgrading the running cost of [17] from linear to log-linear with respect to the data dimension. Recently, Kar and Karnick [11] proposed an approximated feature maps for dot product kernels $k(\mathbf{x}, \mathbf{y}) = k(\langle \mathbf{x}, \mathbf{y} \rangle)$ by directly exploiting the MacLaurin expansion of the kernel function.

Instead of considering a generic class of kernels, our work specifically focuses on the log-Euclidean one, approximating it through a novel unbiased estimator which ensures a explicit bound for variance (as only provided by [13]) and resulting in a superior classification performance with respect to [11,13,17].

3   The Proposed Approximated Feature Map

In this Section, we present the main theoretical contribution of this work, namely (i) a random, explicit feature map Φ such that $\langle \Phi(\mathbf{X}), \Phi(\mathbf{Y}) \rangle \approx K_{\ell E}(\mathbf{X}, \mathbf{Y})$, (ii) the proof of its unbiasedness and (iii) a strong theoretical bound on its variance.

Construction of the Approximated Feature Map. In order to construct a ν dimensional feature map $\mathbf{X} \mapsto \Phi(\mathbf{X}) = [\Phi_1(\mathbf{X}), \dots, \Phi_\nu(\mathbf{X})] \in \mathbb{R}^\nu$, for any $d \times d$ SPD matrix $\mathbf{X}$, fix a probability distribution ρ supported over $\mathbb{N}$. Precisely, each component $\Phi_1, \dots, \Phi_\nu$ of our ν-dimensional feature map Φ is independently computed according to the following algorithm.

    foreach $j = 1, \dots, \nu$ **do**

1   Sample n according to ρ

2   Sample the $d^n \times d^n$ matrix $\mathbf{W}$ of independent Gaussian distributed weights with zero mean and $\sigma^2 / \sqrt{n}$ variance

3   Compute $\log(\mathbf{X})^{\otimes n} = \log \mathbf{X} \otimes \cdots \otimes \log \mathbf{X}$, n times.

4   Assign

$$\Phi_j(\mathbf{X}) = \frac{1}{\sigma^{2n}} \sqrt{\frac{\exp(-\sigma^{-2})}{\nu \rho(n) n!}} \, \mathrm{tr}(\mathbf{W}^\top \log(\mathbf{X})^{\otimes n}). \tag{3}$$

    end

The genesis of (3) can be explained by inspecting the feature map φ associated to the kernel $K(x, y) = \exp\left(-\frac{1}{2\sigma^2}|x - y|^2\right)$, where $x, y \in \mathbb{R}$ for simplicity. It results $\varphi(x) \propto \left[1, \sqrt{\frac{1}{1!\sigma^2}}x, \sqrt{\frac{1}{2!\sigma^4}}x^2, \sqrt{\frac{1}{3!\sigma^6}}x^3, \dots\right]$. Intuitively, we can say that (3) approximates the infinite dimensional $\varphi(x)$ by randomly selecting one of its components: this is the role played by $n \sim \rho$. In addition, we introduce the log mapping and replace the exponentiation with a Kronecker product. As a consequence, the random weights $\mathbf{W}$ ensure that $\Phi(\mathbf{X})$ achieves a sound approximation of (1), in terms of unbiasedness and rapidly decreasing variance.

In the rest of the Section we discuss the theoretical foundation of our analysis, where all proofs have been moved to Appendix A for convenience.

Unbiased Estimation. In order for a statistical estimator to be reliable, we need it to be at least *unbiased*, *i.e.*, its expected value must be equal to the exact function it is approximating. The unbiasedness of the feature map Φ of Eq. (3) for the Log-Euclidean kernel (1) is proved by the following result.

Theorem 1. *Let ρ be a generic probability distribution over $\mathbb{N}$ and consider $\mathbf{X}$ and $\mathbf{Y}$, two generic SPD matrices such that $\|\log \mathbf{X}\|_F = \|\log \mathbf{Y}\|_F = 1$. Then, $\langle \Phi(\mathbf{X}), \Phi(\mathbf{Y}) \rangle$ is an unbiased estimator of (1). That is*

$$\mathbb{E}[\langle \Phi(\mathbf{X}), \Phi(\mathbf{Y}) \rangle] = K_{\ell E}(\mathbf{X}, \mathbf{Y}), \tag{4}$$

where the expectation is computed over n and $\mathbf{W}$ which define $\Phi_j(\mathbf{X})$ as in (3).

Once averaging upon all possible realizations of n sampled from ρ and the Gaussian weights $\mathbf{W}$, Theorem 1 guarantees that the linear kernel $\langle \Phi(\mathbf{X}), \Phi(\mathbf{Y}) \rangle$ induced by Φ is equal to $K_{\ell E}(\mathbf{X}, \mathbf{Y})$. This formalizes the unbiasedness of our approximation.

On the Assumption $\|\log \mathbf{X}\|_F = \|\log \mathbf{Y}\|_F = 1$. Under a practical point of view, this assumption may seem unfavorable, but this is not the case. The reason is provided by Eq. (2), which is very convenient to compute the logarithm of a SPD matrix. Since in (3), $\Phi(\mathbf{X})$ is explicitly dependent on $\log \mathbf{X}$, we can simply use (2) and then divide each entry of the obtained matrix by $\|\log \mathbf{X}\|_F$. This is a non-restrictive strategy to satisfy our assumption and actually analogous to require input vectors to have unitary norm, which is very common in machine learning [2].

Low-Variance. One can note that, in Theorem 1, even by choosing $\nu = 1$ (a scalar feature map), $\Phi(\mathbf{X}) = [\Phi_1(\mathbf{X})] \in \mathbb{R}$ is unbiased for (1). However, since Φ is an approximated finite version of the exact infinite feature map associated to (1), one would expect the quality of the approximation to be very bad in the scalar case, and to improve as ν grows larger. This is indeed true, as proved by the following statement.

Theorem 2. *The variance of* $\langle \Phi(\mathbf{X}), \Phi(\mathbf{Y}) \rangle$ *as estimator of* (1) *can be explicitly bounded. Precisely,*

$$\mathbb{V}_{n,\mathbf{W}}(K_\Phi(\mathbf{X}, \mathbf{Y})) \leq \frac{\mathcal{C}_\rho}{\nu^3} \exp\left(\frac{3 - 2\sigma^2}{\sigma^4} \right), \tag{5}$$

where $\|\log \mathbf{X}\|_F = \|\log \mathbf{Y}\|_F = 1$ *and the variance is computed over all possible realizations of* $n \sim \rho$ *and* $\mathbf{W}$, *the latter being element-wise sampled from a* $\mathcal{N}(0, \sigma^2/\sqrt{n})$ *distribution. Also,* $\mathcal{C}_\rho \stackrel{\text{def}}{=} \sum_{n=0}^{\infty} \frac{1}{\rho(n)n!}$, *the series being convergent.*

Let us discuss the bound on the variance provided by Theorem 2. Since the bandwidth σ of the kernel function (1) we want to approximate is fixed, the term $\exp\left(\frac{3-2\sigma^2}{\sigma^4}\right)$ can be left out from our analysis. The bound in (5) is linear in $\mathcal{C}_\rho$ and inversely cubic in ν. When ν grows, the increased dimensionality of our feature map Φ makes the variance rapidly vanishing, ensuring that the *approximated kernel* $K_\Phi(\mathbf{X}, \mathbf{Y}) = \langle \Phi(\mathbf{X}), \Phi(\mathbf{Y}) \rangle$ converges to the target one, that is $K_{\ell E}$. Such trend may be damaged by big values of $\mathcal{C}_\rho$. Since the latter depends on the distribution ρ, let us fix it to be the geometric distribution $\mathcal{G}(\theta)$ with parameter $0 \leq \theta < 1$. This yields

$$\mathcal{C}_\rho \propto \sum_{n=0}^{\infty} \frac{1}{(1-\theta)^n \cdot n!} = \exp\left(\frac{1}{1-\theta} \right). \tag{6}$$

There is a trade-off between a low variance (*i.e.*, $\mathcal{C}_\rho$ small) and a reduced computational cost for Φ (*i.e.*, n small). Indeed, choosing $\theta \approx 1$ makes $\mathcal{C}_\rho$ big in

(6). In this case, the integer n sampled from $\rho = \mathcal{G}(\theta)$ is small with great probability: this leads to a reduced number of Kronecker products to be computed in $\log(\mathbf{X})^{\otimes n}$. Conversely, when $\theta \approx 0$, despite n and the related computational cost of $\log(\mathbf{X})^{\otimes n}$ are likely to grow, $\mathcal{C}_\rho$ is small, ensuring a low variance for the estimator.

As a final theoretical result, Theorems 1 and 2 immediately yield that

$$\mathbb{P}[|K_\Phi(\mathbf{X}, \mathbf{Y}) - K_{\ell E}(\mathbf{X}, \mathbf{Y})| \geq \epsilon] \leq \frac{\mathcal{C}_\rho}{\nu^3 \epsilon^2} \exp\left(\frac{3 - 2\sigma^2}{\sigma^4}\right) \tag{7}$$

for every pairs of unitary Frobenius normed SPD matrices $\mathbf{X}, \mathbf{Y}$ and $\epsilon > 0$, as a straightforward implication of the Chebyshev inequality. This ensures that K_Φ differs in module from $K_{\ell E}$ by more than ϵ with a (low) probability $\mathbb{P}$, which is inversely cubic and quadratic in ν and ϵ, respectively.

Final Remarks. To sum up, we have presented a constructive algorithm to compute a ν-dimensional feature map Φ whose induced linear kernel is an unbiased estimator of the log-Euclidean one. Additionally, we ensure an explicit bound on the variance which rapidly vanishes as ν grows (inversely cubic decrease). This implies that $\langle \Phi(\mathbf{X}), \Phi(\mathbf{Y}) \rangle$ and $K_{\ell E}(\mathbf{X}, \mathbf{Y})$ are equal with very high probability, even at low ν values. This implements a Log-Euclidean kernel in a scalable manner, downgrading the quadratic cost of computing $K_{\ell E}(\mathbf{X}, \mathbf{Y})$ for every $\mathbf{X}, \mathbf{Y}$ into the linear cost of evaluating the feature map $\Phi(\mathbf{X})$ as in (3) for every $\mathbf{X}$. Practically, this achieve a linear implementation of the log-Euclidean SVM.

4   Results

In this Section, we compare our proposed approximated feature map versus the alternative ones by Rahimi and Recht [17], Kar and Karnick [11] and Le et al. [13] (see Sect. 2).

Theoretical Comparison. Let us notice that all approaches [11,13,17] are applicable also to the log-Euclidean kernel (1). Indeed, [13,17] includes our case of study since $K_{\ell E}(\mathbf{X}, \mathbf{Y}) = k(\log \mathbf{X} - \log \mathbf{Y})$ is logarithmic shift invariant. At the same time, thanks to the assumption $\|\log \mathbf{X}\|_F = \|\log \mathbf{Y}\|_F = 1$ as in Theorem 1, we obtain $K_{\ell E}(\mathbf{X}, \mathbf{Y}) = k(\langle \log \mathbf{X}, \log \mathbf{Y} \rangle)$ (see (13) in Appendix A), thus satisfying the hypothesis of Kar and Karnick [11].

As we proved in Theorem 1, all works [11,13,17] can also guarantee an unbiased estimation for the exact kernel function.

Actually, what makes our approach superior is the explicit bound on the variance (see Table 1). Indeed, [11,17] are totally lacking in this respect. Moreover, despite an analogous bound is provided in [13, Theorem 4], it only ensures a $O(1/\nu)$ decrease rate for the variance with respect to the feature dimensionality ν. Differently, we can guarantee a $O(1/\nu^3)$ trend. This implies that, *we achieve a better approximation of the kernel with a lower dimensional feature representation*, which ease the training of the linear SVM [7].

Table 1. Comparison of explicit bounds on variance between the proposed approximation and [11,13,17]: the quicker the decrease, the better the bound. Here, $\nu \geq 1$ denotes the dimensionality of the approximated feature vector.

Proposed	Rahimi and Recht [17]	Kar and Karninck [11]	Le et al. [13]
$O(1/\nu^3)$	missing	missing	$O(1/\nu)$

Experimental Comparison. We reported here the experimental comparison on 3D action recognition between our proposed approximation and the paradigms of [11,13,17].

Datasets. For the experiments, we considered UTKinect [23], Florence3D [19], MSR-Action-Pairs (MSR-*pairs*) [16], MSR-Action3D [14], [3], HDM-05 [15] and MSRC-Kinect12 [8] datasets.

For each, we follow the usual training and testing splits proposed in the literature. For Florence3D and UTKinect, we use the protocols of [20]. For MSR-Action3D, we adopt the splits originally proposed by [14]. On MSRC-Kinect12, once highly corrupted action instances are removed as in [10], training is performed on odd-index subject, while testing on the even-index ones. On HDM-05, the training exploited all instances of "bd" and "mm" subjects, being "bk", "dg" and "tr" left out for testing [22], using the 65 action classes protocol of [6].

Data Preprocessing. As a common pre-processing step, we normalize the data by computing the relative displacements of all joints $x - y - z$ coordinates and the ones of the hip (central) joint, for each timestamps.

Results. Figure 1 reports the quantitative performance while varying ν in the range 10, 20, 50, 100, 200, 500, 1000, 2000, 5000. When comparing with [13], since the data input size must be a multiple of a power of 2, we zero-padded our

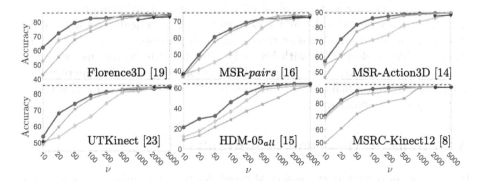

Fig. 1. Experimental comparison of our approximation (red curves) against the schemes ofr Rahimi and Recht [17] (pink curves), Kar and Karnick [11] (green curves) and Le et al. [13] (blue curves). Best viewed in colors.

vectorized representation to match 4096 and (whenever possible) 2048 and 1024 input dimensionality. These cases are then compared with the results related to $\nu = 5000, 2000, 1000$ for RGW and [11,17], respectively. Since all approaches have a random component, we performed ten repetitions for each method and dimensionality setup, averaging the scored classification performances obtained through a linear SVM with $C = 10$. We employ the publicly available codes for [11,13,17]. Finally, we also report the classification performance with the exact method obtained by feeding an SVM with the log-Euclidean kernel whose bandwidth σ is chosen via cross validation.

Discussion. For large ν values, all methods are able to reproduce the performance of the log-Euclidean kernel (black dotted line). Still, in almost all the cases, our approximation is able to outperform the competitors: for instance, we gapped Rahimi and Recht on both MSR-Pairs and MSR-Action3D, while Kar and Karnick scored a much lower performance on HDM-05 and Florence3D. If comparing to Le et al., the performance is actually closer, but this happens for all the methods which are able to cope the performance of the Log-Euclidean kernel with $\nu \geq 2000, 5000$. Precisely, the true superiority of our approach is evident in the case of a small ν value ($\nu = 10, 20, 50$). Indeed, our approximation always provides a much rapid growing accuracy (MSR-Action3D, Florence3D and UTKinect), with only a few cases where the gap is thinner (Kar and Karnick [11] on MSR-*pairs* and Rahimi and Recht [17] on MSRC-Kinect 12). Therefore, our approach ensures a more descriptive and compact representation, providing a superior classification performance.

5   Conclusions

In this work we propose a novel scalable implementation of a Log-Euclidean SVM to perform proficient classification of SPD (covariance) matrices. We achieve a linear complexity by providing an explicit random feature map whose induced linear kernel is an unbiased estimator of the exact kernel function.

   Our approach proved to be more effective than alternative approximations [11,13,17], both theoretically and experimentally. Theoretically, we achieve an explicit bound on the variance on the estimator (such result is totally absent in [11,17]), which is decreasing with inversely cubic pace versus the inverse linear of [13]. Experimentally, through a broad evaluation, we assess the superiority of our representation which is able to provide a superior classification performance at a lower dimensionality.

A   Proofs of All Theoretical Results

In this Appendix we report the formal proofs for both the unbiased approximation (Theorem 1) and the related rapidly decreasing variance (Theorem 2).

Proof of Theorem 1. Use the definition of (3) and the linearity of the expectation. We get that $\mathbb{E}_{n,\mathbf{W}}\left[\langle \Phi(\mathbf{X}), \Phi(\mathbf{Y})\rangle\right]$ equals to

$$\mathbb{E}_n\left[\frac{1}{\sigma^{4n}}\frac{\exp(-\sigma^{-2})}{\rho(n)n!}\mathbb{E}_{\mathbf{W}}\left[\mathrm{tr}\left(\mathbf{W}^\top \log(\mathbf{X})^{\otimes n}\right)\mathrm{tr}\left(\mathbf{W}^\top \log(\mathbf{Y})^{\otimes n}\right)\right]\right], \qquad (8)$$

by simply noticing that the dependence with respect to $\mathbf{W}$ involves the terms inside the trace operators only. Let us focus on the term $\mathrm{tr}\left(\mathbf{W}^\top \log(\mathbf{X})^{\otimes n}\right)$. We can expand

$$\mathrm{tr}\left(\mathbf{W}^\top \log(\mathbf{X})^{\otimes n}\right) = \sum_{i_1,\ldots,i_{2n}=1}^{d} w_{i_1,\ldots,i_{2n}} \log(\mathbf{X})_{i_1,i_2} \cdots \log(\mathbf{X})_{i_{2n-1},i_{2n}} \qquad (9)$$

by using the definition of $\log(\mathbf{X})^{\otimes n}$ and the properties of the trace operator. In Eq. (9), we replace the random coefficient $w_{i_1,\ldots,i_{2n}}$ with $u_{i_1,i_2}^{(1)},\ldots,u_{i_{2n-1},i_{2n}}^{(n)}$ independent and identically distributed (i.i.d.) according to a $\mathcal{N}(0,\sigma^2)$ distribution. We can notice that (9) can be rewritten as

$$\mathrm{tr}\left(\mathbf{W}^\top \log(\mathbf{X})^{\otimes n}\right) = \prod_{\alpha=1}^{n}\sum_{i,j=1}^{d} u_{i,j}^{(\alpha)} \log(\mathbf{X})_{ij}. \qquad (10)$$

Making use of (10) in (8), we can rewrite $\mathbb{E}_{n,\mathbf{W}}\left[K_\Phi(\mathbf{X},\mathbf{Y})\right]$ as

$$\mathbb{E}_n\left[\frac{1}{\sigma^{4n}}\frac{\exp(-\sigma^{-2})}{\rho(n)n!}\mathbb{E}_{\mathbf{W}}\left[\left(\sum_{i,j=1}^{d} u_{i,j}^{(1)} \log(\mathbf{X})_{ij}\right)\left(\sum_{h,k=1}^{d} u_{h,k}^{(1)} \log(\mathbf{Y})_{hk}\right)\right]^n\right] \qquad (11)$$

by also considering the independence of $u_{i,j}^{(\alpha)}$ are independent. By furthermore using the fact that $\mathbb{E}_{\mathbf{W}}\left[u_{i,j}^{(1)}u_{h,k}^{(1)}\right] = 0$ if $i \neq h$ and $j \neq k$ and the formula for the variance of a Gaussian distribution, we get

$$\mathbb{E}_{n,\mathbf{W}}\left[K_\Phi(\mathbf{X},\mathbf{Y})\right] = \mathbb{E}_n\left[\frac{1}{\sigma^{4n}}\frac{\exp(-\sigma^{-2})}{\rho(n)n!}\sigma^{2n}\left(\langle\log(\mathbf{X}),\log(\mathbf{Y})\rangle_F\right)^n\right], \qquad (12)$$

by introducing the Frobenius inner product $\langle \mathbf{A},\mathbf{B}\rangle_F = \sum_{i,j=1}^{d}\mathbf{A}_{ij}\mathbf{B}_{ij}$ between matrices $\mathbf{A}$ and $\mathbf{B}$. By expanding the expectation over ρ, (12) becomes

$$\mathbb{E}_{n,\mathbf{W}}\left[K_\Phi(\mathbf{X},\mathbf{Y})\right] = \sum_{n=0}^{\infty}\rho(n)\frac{1}{\sigma^{2n}}\frac{\exp(-\sigma^{-2})}{\rho(n)n!}\left(\langle\log(\mathbf{X}),\log(\mathbf{Y})\rangle_F\right)^n$$

$$= \exp\left(-\frac{1}{\sigma^2}\right)\sum_{n=0}^{\infty}\left(\frac{\langle\log(\mathbf{X}),\log(\mathbf{Y})\rangle_F}{\sigma^2}\right)^n\frac{1}{n!}. \qquad (13)$$

The thesis easily comes from (13) by using the Taylor expansion for the exponential function and the assumption $\|\log(\mathbf{X})\|_F = \|\log(\mathbf{Y})\|_F = 1$. $\qquad\square$

Proof of Theorem 2. Due to the independence of the components in Φ, by definition of inner product we get $\mathbb{V}_{n,\mathbf{W}}\left[\langle\Phi(\mathbf{X}),\Phi(\mathbf{Y})\rangle\right]=\nu\mathbb{V}_{n,\mathbf{W}}\left[\Phi_1(\mathbf{X})\Phi_1(\mathbf{Y})\right]$. But then $\mathbb{V}_{n,\mathbf{W}}\left[\langle\Phi(\mathbf{X}),\Phi(\mathbf{Y})\rangle\right]\leq\nu\mathbb{E}_{n,\mathbf{W}}\left[\Phi_1(\mathbf{X})^2\Phi_1(\mathbf{Y})^2\right]$ by definition of variance. Taking advantage of (3), yields to the equality between $\mathbb{V}_{n,\mathbf{W}}\left[K_\Phi(\mathbf{X},\mathbf{Y})\right]$ and

$$\frac{1}{\nu^3}\mathbb{E}_{n,\mathbf{U}}\left[\frac{1}{\sigma^{8n}}\frac{\exp(-2\sigma^{-2})}{(\rho(n)n!)^2}\prod_{\alpha=1}^{n}\left(\sum_{i,j=1}^{d}u_{i,j}^{(\alpha)}\log(\mathbf{X})_{ij}\right)^2\left(\sum_{h,k=1}^{d}u_{h,k}^{(\alpha)}\log(\mathbf{Y})_{hk}\right)^2\right],\quad(14)$$

where $u_{i_1,i_2}^{(1)},\ldots,u_{i_{2n-1},i_{2n}}^{(n)}$ are i.i.d. from $\mathcal{N}(0,\sigma^2)$ distribution used to re-parametrize the original weights $\mathbf{W}$. Exploit the independence of $u_{ij}^{(\alpha)}$ to rewrite (14) as

$$\frac{1}{\nu^3}\mathbb{E}_n\left[\frac{1}{\sigma^{8n}}\frac{\exp(-2\sigma^{-2})}{(\rho(n)n!)^2}\mathbb{E}_\mathbf{U}\left[\left(\sum_{i,j=1}^{d}u_{i,j}^{(1)}\log(\mathbf{X})_{ij}\right)^2\left(\sum_{h,k=1}^{d}u_{h,k}^{(1)}\log(\mathbf{Y})_{hk}\right)^2\right]^n\right].\quad(15)$$

By exploiting the zero correlation of the weights in $\mathbf{U}$ and the formula $\mathbb{E}[(\mathcal{N}(0,\sigma^2))^4]=3\sigma^4$ [4]. Thus,

$$\mathbb{V}_{n,\mathbf{W}}\left[K_\Phi(\mathbf{X},\mathbf{Y})\right]\leq\frac{1}{\nu^3}\mathbb{E}_n\left[\frac{1}{\sigma^{8n}}\frac{\exp(-2\sigma^{-2})}{(\rho(n)n!)^2}3^n\sigma^{4n}\left(\sum_{i,j=1}^{d}\log(\mathbf{X})_{ij}^2\log(\mathbf{Y})_{ij}^2\right)^n\right].\quad(16)$$

Since $\sum_{i,j=1}^{d}\log(\mathbf{X})_{ij}^2\log(\mathbf{Y})_{ij}^2\leq\left(\sum_{i,j=1}^{d}\log(\mathbf{X})_{ij}^2\right)\left(\sum_{i,j=1}^{d}\log(\mathbf{Y})_{ij}^2\right)=1$ due to the assumption of unitary Frobenius norm for both $\log\mathbf{X}$ and $\log\mathbf{Y}$, we get

$$\mathbb{V}_{n,\mathbf{W}}\left[K_\Phi(\mathbf{X},\mathbf{Y})\right]\leq\frac{1}{\nu^3}\mathbb{E}_n\left[\frac{1}{\sigma^{8n}}\frac{\exp(-2\sigma^{-2})}{(\rho(n)n!)^2}3^n\sigma^{4n}\right].\quad(17)$$

We can now expand the expectation over ρ in (17), achieving

$$\mathbb{V}_{n,\mathbf{W}}\left[K_\Phi(\mathbf{X},\mathbf{Y})\right]\leq\frac{\exp(-2\sigma^{-2})}{\nu^3}\sum_{n=0}^{\infty}\left(\frac{3}{\sigma^4}\right)^n\frac{1}{n!}\sum_{n=0}^{\infty}\frac{1}{\rho(n)n!},\quad(18)$$

since the series of the products is less than the product of the series, provided that both converge. This is actually true since, by exploiting the McLaurin expansion for the exponential function, we easily get $\sum_{n=0}^{\infty}\left(\frac{3}{\sigma^4}\right)^n\frac{1}{n!}=\exp\left(\frac{3}{\sigma^4}\right)$. On the other hand, since ρ is a probability distribution, it must be $\lim_{n\to\infty}\frac{\rho(n+1)}{\rho(n)}=L$ where $0<L\leq1$, being $\mathbb{N}$ the support of ρ and due to $\sum_{n=0}^{\infty}\rho(n)=1$. Then, since $\lim_{n\to\infty}\frac{\rho(n)}{\rho(n+1)}=\frac{1}{L}<\infty$ and $\lim_{n\to\infty}\frac{1}{n+1}=0$, by the ration criterion for positive-terms series [18], there must exist a constant $\mathcal{C}_\rho>0$ such that

$$\sum_{n=0}^{\infty}\frac{1}{\rho(n)n!}=\mathcal{C}_\rho.\quad(19)$$

Therefore, by combining (19) in (18), we obtain

$$\mathbb{V}_{n,\mathbf{W}}\left[K_\varPhi(\mathbf{X},\mathbf{Y})\right] \leq \frac{\exp(-2\sigma^{-2})}{\nu^3}\exp\left(\frac{3}{\sigma^4}\right)\mathcal{C}_\rho = \frac{\mathcal{C}_\rho}{\nu^3}\exp\left(\frac{3-2\sigma^2}{\sigma^4}\right),$$

which is the thesis. □

References

1. Bach, F.R., Jordan, M.I.: Predictive low-rank decomposition for kernel methods. In: ICML (2005)
2. Bishop, C.M.: Pattern Recognition and Machine Learning - Information Science and Statistics. Springer, New York (2006)
3. Bloom, V., Makris, D., Argyriou, V.: G3D: a gaming action dataset and real time action recognition evaluation framework. In: CVPR (2012)
4. Casella, G., Berger, R.: Statistical Inference. Duxbury Advanced Series in Statistics and Decision Sciences. Thomson Learning, Boston (2002)
5. Cavazza, J., Zunino, A., San Biagio, M., Murino, V.: Kernelized covariance for action recognition. In: ICPR (2016)
6. Cho, K., Chen, X.: Classifying and visualizing motion capture sequences using deep neural networks. CoRR 1306.3874 (2014)
7. Fan, R.E., Chang, K.W., Hsieh, C.J., Wang, X.R., Lin, C.J.: LIBLINEAR: a library for large linear classification. JMLR 9, 1871–1874 (2008)
8. Fothergill, S., Mentis, H.M., Kohli, P., Nowozin, S.: Instructing people for training gestural interactive systems. In: ACM-CHI (2012)
9. Harandi, M., Salzmann, M., Porikli, F.: Bregman divergences for infinite dimensional covariance matrices. In: CVPR (2014)
10. Hussein, M., Torki, M., Gowayyed, M., El-Saban., M.: Human action recognition using a temporal hierarchy of covariance descriptors on 3D joint locations. IJCAI (2013)
11. Kar, P., Karnick, H.: Random feature maps for dot product kernels. In: AISTATS (2012)
12. Koniusz, P., Cherian, A., Porikli, F.: Tensor representations via kernel linearization for action recognition from 3D skeletons. In: Leibe, B., Matas, J., Sebe, N., Welling, M. (eds.) ECCV 2016 Part IV. LNCS, vol. 9908, pp. 37–53. Springer, Cham (2016). doi:10.1007/978-3-319-46493-0_3
13. Le, Q., Sarlos, T., Smola, A.: Fastfood - approximating kernel expansion in loglinear time. In: ICML (2013)
14. Li, W., Zhang, Z., Liu, Z.: Action recognition based on a bag of 3D points. In: CVPR Workshop (2010)
15. Müller, M., Röder, T., Clausen, M., Eberhardt, B., Krüger, B., Weber, A.: HDM-05 doc. Technical report (2007)
16. Oreifej, O., Liu., Z.: HON4D: histogram of oriented 4D normals for activity recognition from depth sequences. In: CVPR (2013)
17. Rahimi, A., Recht, B.: Random features for large-scale kernel machines. In: NIPS (2007)
18. Rudin, W.: Real and Complex Analysis, 3rd edn. McGraw-Hill Inc., New York (1987)

19. Seidenari, L., Varano, V., Berretti, S., Bimbo, A.D., Pala, P.: Recognizing actions from depth cameras as weakly aligned multi-part bag-of-poses. In: CVPR Workshops (2013)
20. Vemulapalli, R., Arrate, F., Chellappa, R.: Human action recognition by representing 3D skeletons as points in a lie group. In: CVPR, June 2014
21. Vrigkas, M., Nikou, C., Kakadiaris, I.A.: A review of human activity recognition methods. Front. Robot. AI **2**, 28 (2015)
22. Wang, L., Zhang, J., Zhou, L., Tang, C., Li, W.: Beyond covariance: feature representation with nonlinear kernel matrices. In: ICCV (2015)
23. Xia, L., Chen, C.C., Aggarwal, J.: View invariant human action recognition using histograms of 3D joints. In: CVPR Workshops (2012)
24. Zhang, K., Tsang, I.W., Kwok, J.T.: Improved Nyström low-rank approximation. In: ICML (2008)

A Machine Learning Approach for the Online Separation of Handwriting from Freehand Drawing

Danilo Avola[1], Marco Bernardi[2], Luigi Cinque[2(✉)], Gian Luca Foresti[1],
Marco Raoul Marini[2], and Cristiano Massaroni[2]

[1] Department of Mathematics, Computer Science, and Physics, University of Udine,
Via Delle Scienze 206, 33100 Udine, Italy
{danilo.avola,gianluca.foresti}@uniud.it
[2] Department of Computer Science, Sapienza University,
Via Salaria 113, 00198 Rome, Italy
{bernardi,cinque,marini,massaroni}@di.uniroma1.it

Abstract. The automatic distinction (domain separation) between handwriting (textual domain) and freehand drawing (graphical domain) elements into the same layer is a topic of great interest that still requires further investigation. This paper describes a machine learning based approach for the online separation of domain elements. The proposed approach presents two main innovative contributions. First, a new set of discriminative features is presented. Second, the use of a Support Vector Machine (SVM) classifier to properly separate the different elements. Experimental results on a wide range of application domains show the robustness of the proposed method and prove the validity of the proposed approach.

Keywords: Domain separation · Handwriting · Textual domain · Freehand drawing · Graphical domain · SVM classifier

1 Introduction

Handwriting and freehand drawing are two modalities of communication that allow people to express concepts and ideas naturally. Each of them supports an ever-increasing number of popular desktop and mobile applications [8,9]. Actually, different fields of the technical design (e.g., mechanical engineering) together with an increasing number of professional applications (e.g., freehand annotation systems [15]) require that users are enabled to perform both handwriting and freehand drawing elements on the same interface with the aim to make their design experience as effective and efficient as possible. This paper describes an SVM classifier based approach for the online separation of handwriting from freehand drawing elements. In particular, the paper presents two main novelties with respect to the current literature. First, a new set of highly discriminative features is presented. Second, an SVM classifier to address this

© Springer International Publishing AG 2017
S. Battiato et al. (Eds.): ICIAP 2017, Part I, LNCS 10484, pp. 223–232, 2017.
https://doi.org/10.1007/978-3-319-68560-1_20

matter is adopted. Since all the measurements on the strokes that compose a scenario are computationally inexpensive, the system works in real-time without special hardware configuration. The experimental results were supported by 25 persons, i.e., 10 persons for the training set and 15 persons for the evaluation set. The experiments were performed on 6 scenarios: electronic circuits, mind maps, Venn diagrams, use cases, flowcharts, entity-relationship diagrams. The obtained results on the accuracy metric prove that this work is a concrete contribution to the current literature.

The rest of the paper is structured as follows. Section 2 provides an overview of the current state-of-the-art in domain separation. Section 3 describes the proposed method, including the set of features and the SVM classifier. Section 4 reports the experimental results obtained on the application domains. Finally, Sect. 5 concludes the paper.

2   Related Work

The online separation of domain elements is a topic that needs to be further investigated. The majority of the methods in the literature are focused on recognizing one or more domains with respect to a specific application context [14]. Examples of multi-domain sketch recognition are presented in [1,11]. In [11], a mixture of geometrical features and an extensible set of heuristics are used to identify a set of shapes by a fuzzy logic approach. The solution proposed in [1] can identify shapes through an innovative Bayesian network supported by structural descriptions. Unlike these works, the focus of the present paper regards the domain separation. In the current literature, few works are reported. A first approach for separating text and drawing patterns is presented in [17], where the textual domain is formed by Japanese characters. According to the nature of this vocabulary, each stroke is considered as a set of segments. Instead, the features are based on the relationships between the segment length, the number of segments, and the bounding-box size (i.e., the small rectangle that contains all segments). Following, the method proposed in [6] is based on the Multi-Layer Perceptron (MLP) and Hidden Markov Model (HMM). The MLP performs a text domain recognition on the feature vectors extracted from the strokes, instead the HMM discriminates each stroke of the digital ink into two classes: text and graphics. Another interesting work is proposed in [17], where the sum of the angles formed by two consecutive segments, the ratio between the stroke length and the bounding-box size, and the stroke direction on the x and y axis are considered as features. Differently, in [5], the authors perform a classification between shapes and text strokes, in the context of digital ink, by an entropy measure. The latter is obtained by the internal angles of the stroke, where a high value of entropy represents a text, while a low value is associated to a shape. The work proposed in [3] describes an online framework able to automatically distinguish freehand drawing from handwriting, where an interesting feature, called band-ratio, was introduced. This feature considers the distribution of the stroke points within three specific areas, i.e., top, middle, and bottom, of the bounding-box. The work in [7] uses the features proposed in [5] and introduces a new feature related to the acquisition by hardware mechanism, i.e., the pressure exerted by the user on the pen to create a stroke. This new set of features is used to perform the separation between text and freehand drawing. More specifically, the authors analysed a wide set of Machine Learning (ML) algorithms, including Bootstrap Aggregating, LADTree, LMT, LogitBoost, MLP, Random Forest, and Sequential Minimal Optimization (SMO), to check the discriminative power of the selected features. Finally, the framework presented in [2] shows different interesting stages to separate and recognize text and graphical symbols. In particular, the authors describe separation stage that uses two processes to detect how many and which objects are performed by users. Subsequently, the framework computes mathematical and statistical relationships on each candidate object to provide a reliable classification. Inspired by different works reported above [2,3], but unlike them, the presented work proposes the use of an SVM classifier to perform the separation task. SVM technique, respect other well-known techniques [5,13], can be considered an optimal solution for binary classification. In domain separation, the distinguishing between text strokes and graphical strokes can be seen as a binary classification problem where the features are considered as points of a hyperplane.

3   The Proposed Method

The definitions and terminologies used in this section are defined in [3]. As shown in Fig. 1, the proposed method is composed of four main stages: pre-processing, feature extraction, machine learning, and domain separation. The first deals with simplifying and aggregating each stroke. The second extracts the different features from a stoke and combines them into a single feature vector. The third adopts a learned SVM to classify each stroke in one of the two available classes: textual domain or graphical domain. Finally, the last provides a feedback to the user in real-time.

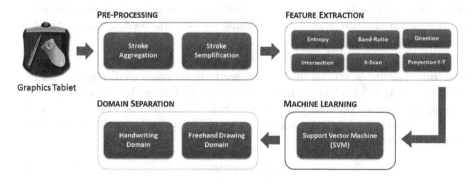

Fig. 1. Logical architecture of the proposed system composed of four main stages: pre-processing, feature-extraction, machine learning, and domain separation.

3.1 Pre-processing Stage

This stage is composed of two processes: *stroke aggregation* and *stroke simplification*. The first detects how many and which strokes must be aggregated. The second simplifies the lines of the stroke by deleting unnecessary points.

Multiple strokes are very frequent in both textual domain and graphical domain. The *stroke aggregation* process checks temporal and spatial relationships among strokes to provide one or more partitioned sets of strokes representing the candidate objects. More specifically, the process examines pairs of consecutive strokes and considers the time interval, linked to each bounding-box, elapsed between the end of the first stroke and the start of the second stroke. To evaluate if two strokes can be aggregated, one of these conditions must be respected:

- if part of a stroke crosses another one, and the areas of the two bounding-boxes have a difference of about 10%;
- if the time interval is less of 500 ms;
- if two strokes are overlapped of about 20%, and al least half of a bounding-box is contained within the other.

At the end of this process, a new set of strokes is created. Then, the latter is sent to the *stroke simplification* process. Often, the strokes are composed of a high number of unnecessary points that may affect the performance and precision of some features. The *stroke simplification* process (or *line fitting*) allows to delete these points thus simplifying curves and lines [12]. In the proposed work, two techniques are implemented [10]: Radial-Distance and Douglas-Peucker. The first (default option) provides an approximation of the elements less accurate but faster ($O(n)$). The second provides a more accurate approximation but with a high computational cost ($O(n^2)$).

3.2 Feature Extraction Stage

Feature extraction is a critical step that can influence the performance of the separation algorithm. In this work, two features, i.e., entropy [5] and band-ratio [2], are inherited by the current state-of-the-art due to their proven usefulness. The other four features have been ad-hoc created to provide a high discriminative feature vector. In this way, a new feature vector composed by six features is implemented (Fig. 2).

Entropy feature is defined in [5] as an accurate criterion to distinguish shapes and text strokes. This feature measures the angles formed by three consecutive points. For each of them, a letter based on its amplitude is assigned. So, each stroke is represented by a string of letters. For each representation of stroke, entropy is calculated as follows:

$$\sum_{x \in X} p_x log_2 p_x \tag{1}$$

where X is the set of letters, and p_x is the probability that a point is assigned to the letter x.

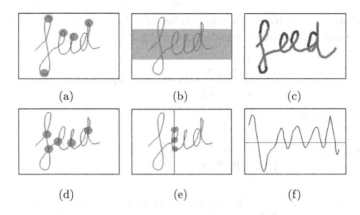

Fig. 2. Set of implemented features: (a) Entropy, (b) Band-ratio, (c) Direction, (d) Intersection, (e) X-Scan, and (f) Projection y-t.

Band-Ratio feature, is defined in [2]. It measures the distribution of the stroke style. This feature is computed from a vertical point, where the band is created. Subsequently, the band is increased until it covers 65% of the points of the whole stroke. The feature is calculated as follows:

$$f_{br} = \frac{h_{band}}{h_{bb}} \tag{2}$$

where h_{band} is the height of the band and h_{bb} is the height of bounding box of the stroke. Its value has a range between 0 and 1.

Direction feature measures represent the number of repeated forwards-backwards movements produced by the stroke. The number of these movements is constant and can be considered a very discriminant measure to distinguish text from drawing. This feature can be calculated as follows:

$$f_d = \frac{\sum_{i=2}^{|N|} |P| S(p_i, p_{i-1})}{l_{bb}} \tag{3}$$

where $|P|$ is the number of the points of the stroke, i is an integer within the interval $2 \leq i \leq |N|$, l_{bb} is the length of the bounding-box, and S is a function defined by the following expression:

$$S(u, v) = \begin{cases} d(u, v) & \text{if } u_x \leq v_x \\ -d(u, v) & \text{otherwise} \end{cases} \tag{4}$$

where u and v are two consecutive points, and $d(u, v)$ is the distance between them. In the case of text, the values are always positive and have a range between 0.1 and 0.5. For the drawing, they can be negative for irregular forms or they can have a range between 0.6 and 1.

Intersection feature measures the number of intersection points of a stroke. The feature calculation process is described as follows:

$$f_I = \sum_{i=2}^{|P|} \sum_{j=i+2}^{|P|} I(p_i, p_{i-1}, p_j, p_{j-1}) \tag{5}$$

where I is a function defined as:

$$I(u, v, w, z) = \begin{cases} 1 & \text{if the segment } uv \text{ intersect the segment } wz \\ 0 & \text{otherwise} \end{cases} \tag{6}$$

X-Scan feature. Given the imaginary vertical segments at regular intervals throughout the length of the bounding-box, the stroke will intersect them many times. The X-Scan feature measures the number of these segments. This features is calculated as follow:

$$f_{xs} = \frac{\sum_{v \in V} I_s(v)}{l_{bb}} \tag{7}$$

where v is the vertical segment considered, l_{bb} is the length of bounding-box, and I_s is a function defined as:

$$I_s(v) = \begin{cases} 1 & \text{if } v \text{ intersects the stroke more than once} \\ 0 & \text{otherwise} \end{cases} \tag{8}$$

Projection y-t feature measures analyses the horizontal movement of the stroke. To avoid the disturbance of the lateral movement, it operates a data transformation by replacing the x-axis with the acquisition time of the stroke. This transformation produces a sinusoidal-type curve for the text and more irregular patterns for the figures.

3.3   Machine Learning Stage

A good set of features is an optimum starting point, but it is necessary to create or adopt a suitable classifier to reach high level in accuracy and performance. In the proposed context, we have two main factors. The first regarding the natural amount of errors due to the handwriting and freehand drawing activities. The second concerning the binary nature of the matter. These reasons promoted the use of a SVM classifier to estimate, on one side, the values of the different features and their relationships and, on the other hand, to mitigate the propagation of the different errors by a robust hyperplane [4,16].

3.4   Domain Separation Stage

The domain separation stage manages the Graphics User Interface (GUI) and shows the processing results to the user. In Fig. 3 an example of domain separation is reported.

Fig. 3. Online separation GUIs: (a) scenario, (b) handwriting domain, (c) freehand drawing domain.

4 Experimental Results and Discussion

The main purposes of the experiments were the assessment of the set of salient features for the separation between textual and graphical domains, the overall robustness of the proposed approach, and its higher accuracy. The experiments were performed by using a challenging set of scenarios described in Sect. 4.1. A discussion of the results and a comparison of the proposed method with selected key works of the current state-of-the-art are reported in Sect. 4.2.

4.1 Dataset

Nowadays, there is not a common dataset in the field of domain separation. Consequently, in order to show the robustness of proposed approach, a new dataset was built. The dataset is based on the union of the six scenarios used by selected key works of the current literature [2,5–7,17]. The scenarios are shown in Fig. 4. From left to right are electronic circuits, mind maps, Venn diagrams, use cases, flowcharts, and entity-relationship diagrams, respectively. These scenarios were chosen for different reasons. First, they allow a comparison with the key works of the current state-of-the-art. Second, they are challenging in domain separation, for example, mind maps is a very difficult scenario because it is not a formalized diagram and each user can have a personal style in drawing the different shapes. In order to train the adopted SVM, a training set was created (in Fig. 5 some instances are shown). In particular, a set of 10 persons aged from 20 up to 30 years, 5 males, 5 females was selected. Each user had to perform, for 8 times, the whole set of graphical symbols represented by the 6 scenarios (an example is provided in Fig. 5a), and for 5 times, a set of summaries of about 1000 words in which the words presented different levels of grouping (an example is provided in Fig. 5b).

4.2 Results

In the evaluation step a set of 15 persons, different from the previous ones (i.e., training step) but with the same characteristics, was selected (9 males, 6 females). To evaluate the experiments, the accuracy metric was adopted [18].

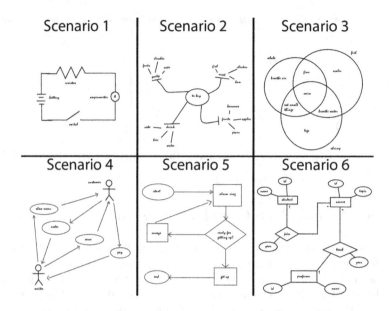

Fig. 4. Scenarios for separation between textual and graphical domains.

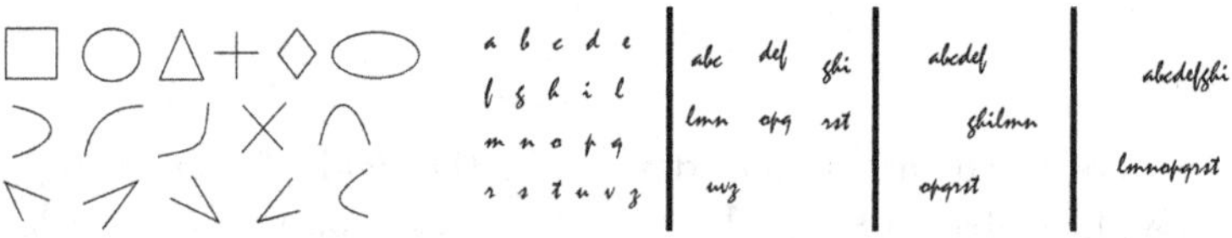

Fig. 5. Training set: (a) geometrical shapes, (b) four groups of words with different lengths.

As reported in Table 1, the method achieves an overall accuracy of the 97.3%. In particular, the use cases scenarios has achieved the best accuracy of 98.5%. Instead, electronic circuits and mind maps have obtained an accuracy of 96.5%.

Table 1. Comparison of the accuracy measures among state-of-the-art approaches.

Method	Accuracy
Proposed approach	97.3%
Bishop et al. [6]	97%
Bhat and Hammond [5]	92.1%
Blagojevic et al. [7]	90.5%
Machii et al. [17]	88%
Avola et al. [2]	85%

We have compared the proposed method with five key works of the current literature presented in [2, 5–7, 17]. The comparison is based on the benchmark of these works. In Table 1 the overall results are reported. They show that the proposed method is a concrete contribution to the current literature. As mentioned before, there is not a common dataset to have a direct comparison with these key works. In addition, some of these works, like the proposed one, are based on tests performed by a specific class of users (e.g., young people, computer science students). These factors can influence the experimental phases and often these details about the persons are not present in the other works. Furthermore, different data acquisition methods can be distinguished. Blagojevic et al. [7] use a system similar to that proposed. Instead, Machii et al. [17] and Bishop et al. [6] use an optical system for scanning of strokes. Another consideration regards the handwriting styles. Machii et al. [17] focused on Japanese writing. Instead, Bhat and Hammond [5] and Bishop et al. [6] focused on writing in block letters. All these aspects make the comparison a hard task. To obtain a comparative analysis, we built a dataset containing the contexts in which these works were tested. Blagojevic et al. [7] performed extensive evaluations on diagrams from 6 different domains (4 of these are used into the built dataset). Bishop et al. [6] used data collected among the employees at Microsoft Research in Cambridge, using a purpose-written piece of software and additional tests were also obtained from the Tablet PC Ink Parsing Team at Microsoft in Redmond. Machii et al. [17] used a dataset where they have chosen 18 patterns on which to perform the experiments. Finally, the presented results allow to give two considerations. First, the novel set of features is very discriminating. Second, the SVM is very suitable for this kind of binary separation domain.

5   Conclusions

This paper describes an SVM classifier based approach for the online separation of handwriting (textual domain) from freehand drawing (graphical domain) elements. The paper presents two main novelties with respect to the current literature. First, a new set of highly discriminative features. Second, the use of an SVM classifier. Despite the lack in literature of a dataset and the lack of a standard for the comparison of different approaches in this field, the authors of the present paper have produced wide efforts to provide a reasonable and reliable comparison between them. The experimental tests have provided a high accuracy of 97.3% which shown the concrete contribution to the current state-of-the-art.

References

1. Alvarado, C., Davis, R.: Dynamically constructed bayes nets for multi-domain sketch understanding. In: International Conference and Exhibition on Computer Graphics and Interactive Techniques (SIGGRAPH), pp. 1–6 (2006)

2. Avola, D., Cinque, L., Placidi, G.: SketchSPORE: a sketch based domain separation and recognition system for interactive interfaces. In: Petrosino, A. (ed.) ICIAP 2013. LNCS, vol. 8157, pp. 181–190. Springer, Heidelberg (2013). doi:10.1007/978-3-642-41184-7_19

3. Avola, D., Del Buono, A., Del Nostro, P., Wang, R.: A novel online textual/graphical domain separation approach for sketch-based interfaces. In: Damiani, E., Jeong, J., Howlett, R.J., Jain, L.C. (eds.) New Directions in Intelligent Interactive Multimedia Systems and Services - 2. SCI, vol. 226, pp. 167–176. Springer, Heidelberg (2009). doi:10.1007/978-3-642-02937-0_15

4. Bahlmann, C., Haasdonk, B., Burkhardt, H.: Online handwriting recognition with support vector machines - a kernel approach. In: Proceedings Eighth International Workshop on Frontiers in Handwriting Recognition, pp. 49–54 (2002)

5. Bhat, A., Hammond, T.: Using entropy to distinguish shape versus text in hand-drawn diagrams. In: International Joint Conference on Artificial Intelligence, pp. 1395–1400 (2009)

6. Bishop, C.M., Svensen, M., Hinton, G.E.: Distinguishing text from graphics in on-line handwritten ink. In: International Workshop on Frontiers in Handwriting Recognition, pp. 142–147 (2004)

7. Blagojevic, R., Plimmer, B., Grundy, J., Wang, Y.: Using data mining for digital ink recognition: dividing text and shapes in sketched diagrams. Comput. Graph. 35(5), 976–991 (2011)

8. Chen, Q., Shi, D., Feng, G., Zhao, X., Luo, B.: On-line handwritten flowchart recognition based on logical structure and graph grammar. In: International Conference on Information Science and Technology (ICIST), pp. 424–429 (2015)

9. Degtyarenko, I., Radyvonenko, O., Bokhan, K., Khomenko, V.: Text/shape classifier for mobile applications with handwriting input. Int. J. Doc. Anal. Recognit. 19(4), 369–379 (2016)

10. Douglas, D.H., Peucker, T.K.: Algorithms for the reduction of the number of points required to represent a digitized line or its caricature. In: Classics in Cartography, pp. 15–28 (2011)

11. Fonseca, M.J., Pimentel, C., Jorge, J.A.: Cali: an online scribble recognizer for calligraphic interfaces. In: Papers from the 2002 AAAI Spring Symposium on Sketch Understanding, pp. 51–58 (2002)

12. Foresti, G., Murino, V., Regazzoni, C., Vernazza, G.: Grouping of rectilinear segments by the labeled hough transform. CVGIP: Image Underst. 59(1), 22–42 (1994)

13. Foresti, G.L., Micheloni, C.: Generalized neural trees for pattern classification. IEEE Trans. Neural Netw. 13(6), 1540–1547 (2002)

14. Gabe, J., Mark, D.G., Jason, H., Ellen, Y.L.D.: Computational support for sketching in design: A review. In: Foundations and Trends in Human-Computer Interaction, pp. 1–93 (2009)

15. Grundel, M., Abulawi, J.: SkiPo - a sketch and flow based model to develop mechanical systems. INCOSE Int. Symp. 26(1), 399–414 (2016)

16. Ma, Y., Guo, G.: Support Vector Machines Applications. Springer, Cham (2014)

17. Machii, K., Fukushima, H., Nakagawa, M.: On-line text/drawings segmentation of handwritten patterns. In: International Conference on Document Analysis and Recognition, pp. 710–713 (1993)

18. Sokolova, M., Lapalme, G.: A systematic analysis of performance measures for classification tasks. Inf. Process. Manag. 45(4), 427–437 (2009)

Learning to Map Vehicles into Bird's Eye View

Andrea Palazzi$^{(\boxtimes)}$, Guido Borghi, Davide Abati,
Simone Calderara, and Rita Cucchiara

University of Modena and Reggio Emilia, Modena, Italy
{andrea.palazzi,guido.borghi,davide.abati,
simone.calderara,rita.cucchiara}@unimore.it

Abstract. Awareness of the road scene is an essential component for
both autonomous vehicles and Advances Driver Assistance Systems and
is gaining importance both for the academia and car companies. This
paper presents a way to learn a semantic-aware transformation which
maps detections from a dashboard camera view onto a broader bird's eye
occupancy map of the scene. To this end, a huge synthetic dataset featuring 1M couples of frames, taken from both car dashboard and bird's eye
view, has been collected and automatically annotated. A deep-network
is then trained to warp detections from the first to the second view. We
demonstrate the effectiveness of our model against several baselines and
observe that is able to generalize on real-world data despite having been
trained solely on synthetic ones.

1 Introduction

Vision-based algorithms and models have massively been adopted in current
generation ADAS solutions. Moreover, recent research achievements on scene
semantic segmentation [9,14], road obstacle detection [3,12] and driver's gaze,
pose and attention prediction [7,22] are likely to play a major role in the rise of
autonomous driving.

As suggested in [5], three major paradigms can be individuated for vision-
based autonomous driving systems: *mediated perception* approaches, based on
the total understanding of the scene around the car, *behavior reflex* methods,
in which driving action is regressed directly from the sensory input, and *direct
perception* techniques, that fuse elements of previous approaches and learn a
mapping between the input image and a set of interpretable indicators which
summarize the driving situation.

Following this last line of work, in this paper we develop a model for mapping vehicles across different views. In particular, our aim is to warp vehicles
detected from a dashboard camera view into a bird's eye occupancy map of
the surroundings, which is an easily interpretable proxy of the road state. Being
almost impossible to collect a dataset with this kind of information in real-world,
we exclusively rely on synthetic data for learning this projection. We aim to create a system close to surround vision monitoring ones, also called around view

© Springer International Publishing AG 2017
S. Battiato et al. (Eds.): ICIAP 2017, Part I, LNCS 10484, pp. 233–243, 2017.
https://doi.org/10.1007/978-3-319-68560-1_21

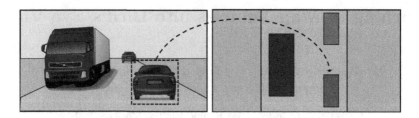

Fig. 1. Simple outline of our task. Vehicle detections in the frontal view (left) are mapped onto a bird-eye view (right), accounting for the positions and size.

cameras that can be useful tools for assisting drivers during maneuvers by, for example, performing trajectory analysis of vehicles out from own visual field. In this framework, our contribution is twofold:

- We make available a huge synthetic dataset (>1 million of examples) which consists of couple of frames corresponding to the same driving scene captured by two different views. Besides the vehicle location, auxiliary information such as the distance and yaw of each vehicle at each frame are also present.
- We propose a deep learning architecture for generating bird's eye occupancy maps of the surround in the context of autonomous and assisted driving. Our approach does not require a stereo camera, nor more sophisticated sensors like radar and lidar. Conversely, we learn how to project detections from the dashboard camera view onto a broader bird's eye view of the scene (see Fig. 1). To this aim we combine learned geometric transformation and visual cues that preserve objects size and orientation in the warping procedure.

Dataset, code and pre-trained model are publicly available and can be found at http://imagelab.ing.unimore.it/scene-awareness.

2   Related Work

Surround View. Few works in literature tackle the problem of the vehicle's surround view. Most of these approaches are vision and geometry based and are specifically tailored for helping drivers during parking manoeuvres. In particular, in [13] a perspective projection image is transformed into its corresponding bird's eye view, through a fitting parameters searching algorithm. In [16] exploited the calibration of six fish eye cameras to integrate six images into a single one, by a dynamic programming approach. In [17] were described algorithms for creating, storing and viewing surround images, thanks to synchronized and aligned different cameras. Sung *et al.* [20] proposed a camera model based algorithm to reconstruct and view multi-camera images. In [21], an homography matrix is used to perform a coordinate transformation: visible markers are required in input images during the camera calibration process. Recently, Zhang *et al.* [24] proposed a surround view camera solution designed for embedded systems, based on a geometric alignment, to correct lens distortions, a photometric alignment, to correct brightness and color mismatch and a composite view synthesis.

Videgames for Collecting Data. The use of synthetic data has recently gained considerable importance in the computer vision community for several reasons. First, modern open-world games exhibit constantly increasing realism - which does not only mean that they feature photorealistic lights/textures etc., but also show plausible game dynamics and lifelike autonomous entity AI [18,19]. Furthermore, most research fields in computer vision are now tackled by means of deep networks, which are notoriously data hungry in order to be properly trained. Particularly in the context of assisted and autonomous driving, the opportunity to exploit virtual yet realistic worlds for developing new techniques has been embraced widely: indeed, this makes possible to postpone the (very expensive) validation in real world to the moment in which a new algorithm already performs reasonably well in the simulated environment [8,23]. Building upon this tendency, [5] relies on TORCS simulator to learn an interpretable representation of the scene useful for the task of autonomous driving. However, while TORCS [23] is a powerful simulation tool, it's still severely limited by the fact that both its graphics and its game variety and dynamics are far from being realistic.

Many elements mark as original our approach. In principle, we want our surround view to include not only nearby elements, like commercial geometry-based systems, but also most of the elements detected into the acquired dashboard camera frame. Additionally, no specific initialization or alignment procedures are necessary: in particular, no camera calibration and no visible alignment points are required. Eventually, we aim to preserve the correct dimensions of detected objects, which shape is mapped onto the surround view consistently with their semantic class.

3   Proposed Dataset

In order to collect data, we exploit *Script Hook V* library [4], which allows to use Grand Theft Auto V (GTAV) video game native functions [1]. We develop a framework in which the game camera automatically toggle between frontal and bird-eye view at each game time step: in this way we are able to gather information about the spatial occupancy of the vehicles in the scene from both views (*i.e.* bounding boxes, distances, yaw rotations). We associate vehicles information across the two views by querying the game engine for entity IDs. More formally, for each frame t, we compute the set of entities which appear in both views as

$$E(t) = E_{frontal}(t) \cap E_{birdeye}(t) \qquad (1)$$

where $E_{frontal}(t)$ and $E_{birdeye}(t)$ are the sets of entities that appear at time t in frontal and bird's eye view, respectively. Entities $e(t) \in E(t)$ constitute the candidate set for frame t $C(t)$; other entities are discarded. Unfortunately, we found that raw data coming from the game engine are not always accurate (Fig. 2). To deal with this problem, we implement a post-processing pipeline in order to discard noisy data from the candidate set $C(t)$. We define a discriminator function

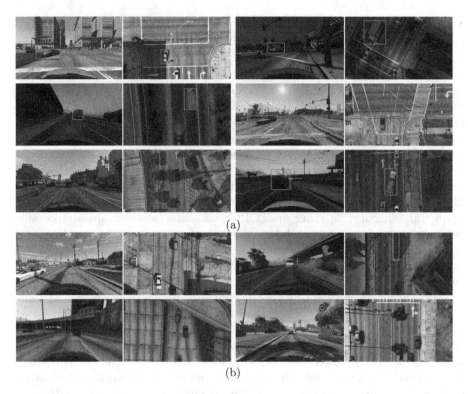

Fig. 2. (a) Randomly sampled couples from our GTAV dataset, which highlight the huge variety in terms of landscape, traffic condition, vehicle models etc. Each detection is treated as a separate training example (see Sect. 3 for details). (b) Random examples rejected during the post-processing phase.

$$f(e(t)) : C \mapsto \{0, 1\} \tag{2}$$

which is positive when information on dumped data $e(t)$ are reliable and zero otherwise. Thus we can define the final filtered dataset as

$$\bigcup_{t=0}^{T} D(t) \quad \text{where} \quad D(t) = \{c_i(i) \mid f(c_i(t)) > 0\} \tag{3}$$

being T the total number of frames recorded. From an implementation standpoint, we employ a rule-based ontology which leverage on entity information (*e.g.* vehicle model, distance etc.) to decide if the bounding box of that entity can be considered reasonable. This implementation has two main values: first it's lightweight and very fast in filtering massive amounts of data. Furthermore, rule parameters can be tuned to eventually generate different dataset distribution (*e.g.* removing all trucks, keeping only cars closer than 10 m, etc.).

Each entry of the dataset is a tuple containing:

- $frame_f$, $frame_b$: 1920 × 1080 frames from the frontal and bird's eye camera view, respectively;
- ID_e, $model_e$: identifiers of the entity (e) in the scene and of the vehicle's type;
- $frontal\_coords_e$, $birdeye\_coords_e$: the coordinates of the bounding box that encloses the entity;
- $distance_e$, yaw_e: distance and rotation of the entity w.r.t. the player.

Figure 3 shows the distributions of entity rotation and distance across the collected data.

Table 1. Overview of the statistics on the collected dataset. See text for details.

	Total
Number of runs	300
Number of bounding boxes	1125187
Unique entity IDs	56454
Unique entity models	198

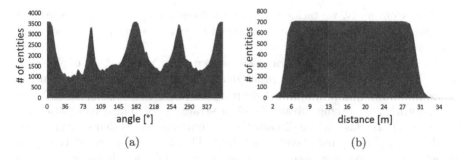

(a)   (b)

Fig. 3. Unnormalized distribution of vehicle orientation (a) and distances (b) present in the collected dataset. Distribution of angles conversely presents two prominent modes around 0°/360° and 180° respectively, due to the fact that the major part of vehicles encountered travel in parallel to the player's car, on the same (0/360°) or the opposite (180°) direction. Conversely, distance is almost uniformly distributed between 5 and 30 m.

4   Model

At a first glance, the problem we address could be mistaken with a bare geometric warping between different views. Indeed, this is not the case since targets are not completely visible from the dashboard camera view and their dimensions in the bird's eye map depend on both the object visual appearance and semantic category (*e.g.* a truck is longer than a car). Additionally, it cannot be cast as a correspondence problem, since no bird's eye view information are available at

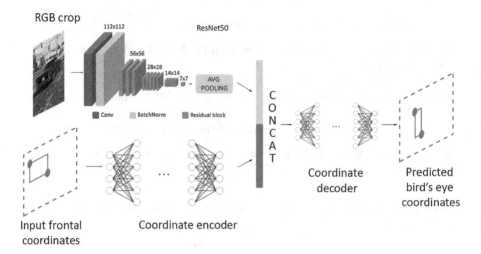

Fig. 4. A graphical representation of the proposed SDPN (see Sect. 4). All layers contain *ReLU* units, except for the top decoder layer which employs *tanh* activation. The number of fully connected units is $(256, 256, 256)$ and $(1024, 1024, 512, 256, 128, 4)$ for the coordinate encoder and decoder respectively.

test time. Conversely, we tackle the problem from a deep learning perspective: dashboard camera information are employed to learn a spatial occupancy map of the scene seen from above.

Our proposed architecture composes of two main branches, as depicted in Fig. 4. The first branch takes as input image crops of vehicles detected in the dashboard camera view. We extract deep representations by means of *ResNet50* deep network [10], taking advantage of pre-training for image recognition on ImageNet [6]. To this end we discard the top fully-connected dense layer which is tailored for the original classification task. This part of the model is able to extract semantic features from input images, even though it is unaware of the location of the bounding box in the scene.

Conversely, the second branch consists of a deep *Multi Layer Perceptron* (MLP), composed by 4 fully-connected layers, which is fed with bounding boxes coordinates (4 for each detection), learning to encode the input into a 256 dimensional feature space. Due to its input domain, this segment of the model is not aware of objects' semantic, and can only learn a spatial transformation between the two planes.

Both appearance features and encodings of bounding box coordinates are then merged through concatenation and undergo a further fully-connected decoder which predicts vehicles' locations in the bird's eye view. Since our model combines information about object's location with semantic hints on the content of the bounding box, we refer to it as *Semantic-aware Dense Projection Network* (SDPN in short).

Training Details: ImageNet [6] mean pixel value is subtracted from input crops, which are then resized to 224×224 before being fed to the network. During training, we freeze *ResNet50* parameters. Ground truth coordinates in the bird's eye view are normalized in range $[-1, 1]$. Dropout is applied after each fully-connected layer with drop probability 0.25. The whole model is trained end-to-end using *Mean Squared Error* as objective function and exploiting *Adam* [11] optimizer with the following parameters: $lr = 0.001, \beta_1 = 0.9, \beta_2 = 0.999$.

5   Experimental Results

We now assess our proposal comparing its performance against some baselines. Due to the peculiar nature of the task, the choice of competitor models is not trivial.

To validate the choice of a learning perspective against a geometrical one, we introduce a first baseline model that employs a projective transformation to estimate a mapping between corresponding points in the two views. Such correspondences are collected from bottom corners of both source and target boxes in the training set, then used to estimate an homography matrix in a least-squares fashion (*e.g.* minimizing reprojection error). Since correspondences mostly belong to the street, which is a planar region, the choice of the projective transformation seems reasonable. The height of the target box, however, cannot be recovered from the projection, thus it is cast as the average height among training examples. We refer to this model as *homography model*.

Additionally, we design second baseline by quantizing spatial locations in both views in a regular grid, and learn point mappings in a probabilistic fashion. For each cell G_i^f in the frontal view grid, a probability distribution is estimated over bird's eye grid cells G_j^b, encoding the probability of a pixel belonging to G_i^f to fall in the cell G_j^b. During training, top-left and bottom-right bounding box corners in both views are used to update such densities. At prediction stage, given a test point p_k which lies in cell G_i^f we predict destination point by sampling from the corresponding cell distribution. We fix grid resolution to 108×192, meaning a $10 \times$ quantization along both axes, and refer to this baseline as *grid model*. It could be questioned if the appearance of the bounding box content in the frontal view is needed at all in estimating the target coordinates, given sufficient training data and an enough powerful model. In order to determine the importance of the visual input in the process of estimating the bird's eye occupancy map, we also train an additional model with approximately the same number of trainable parameters of our proposed model SDPN, but fully connected from input to output coordinates. We refer to this last baseline as *MLP*.

For comparison, we rely on three metrics:

- *Intersection over Union* (IoU): measure of the quality of the predicted bounding box BB_p with respect to the target BB_t:

$$IoU(BB_p, BB_t) = \frac{A(BB_p \cap BB_t)}{A(BB_p \cup BB_t)}$$

where $A(R)$ refers to the area of the rectangle R;

- *Centroid Distance* (CD): distance in pixels between box centers, as an indicator of localization quality[1];
- *Height, Width Error* (hE,wE): average error on bounding box height and width respectively, expressed in percentage w.r.t. the ground truth BB_t size;
- *Aspect ratio mean Error* (arE): absolute difference in aspect ratio between BB_p and BB_t:

$$arE = \left| \frac{BB_p.w}{BB_p.h} - \frac{BB_t.w}{BB_t.h} \right| \tag{4}$$

The evaluation of baselines and proposed model is reported in Fig. 5(a). Results suggest that both *homography* and *grid* are too naive to capture the complexity of the task and fail in properly warping vehicles into the bird's eye view. In particular, *grid* baseline performs poorly as it only models a point-wise transformation between bounding box corners, disregarding information about the overall input bounding box size. On the contrary, MLP processes the bounding box in its whole and provides a reasonable estimation. However, it still misses the chance to properly recover the length of the bounding box in the bird's eye view, being unaware of entity's visual appearance. Instead, SDPN is able to capture the object's semantic, which is a primary cue for correctly inferring vehicle's location and shape in the target view.

A second experiment investigates how vehicle's distance affects the warping accuracy. Figure 5(b) highlights that all the models' performance degrades as the distance of target vehicles increases. Indeed, closer examples exhibit lower variance (*e.g.* are mostly related to the car ahead and the ones approaching from the opposite direction) and thus are easier to model. However, it can be noticed that moving forward along distance axis the gap between the SDPN and MLP gets wider. This suggests that the additional visual input adds robustness in these challenging situations. We refer the reader to Fig. 6 for a qualitative comparison.

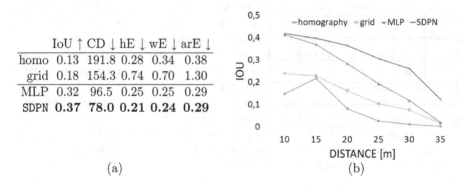

	IoU ↑	CD ↓	hE ↓	wE ↓	arE ↓
homo	0.13	191.8	0.28	0.34	0.38
grid	0.18	154.3	0.74	0.70	1.30
MLP	0.32	96.5	0.25	0.25	0.29
SDPN	**0.37**	**78.0**	**0.21**	**0.24**	**0.29**

(a)

(b)

Fig. 5. (a) Table summarizing results of proposed SDPN model against the baselines; (b) degradation of IoU performance as the distance to the detected vehicle increases.

---

[1] Please recall that images are 1920×1080 pixel size.

homography   grid   *MLP*   SDPN   *ground truth*

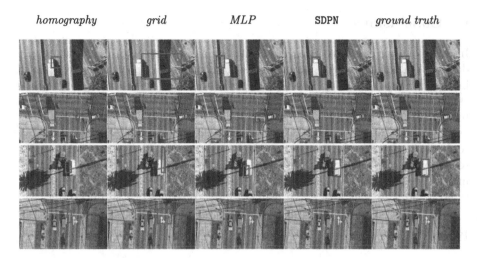

Fig. 6. Qualitative comparison between different models. Baselines often predict reasonable locations for the bounding boxes. SDPN is also able to learn the orientation and type of the vehicle (*e.g.* a truck is bigger than a car etc.).

A Real-World Case Study. In order to judge the capability of our model to generalize on real-world data, we test it using authentic driving videos taken from a roof-mounted camera [2]. We rely on state-of-the-art detector [15] to get the bounding boxes of vehicles in the frontal view. As the ground truth is not available for these sequences, performance is difficult to quantify precisely. Nonetheless, we show qualitative results in Fig. 7: it can be appreciated how the network is able to correctly localize other vehicles' positions, despite having been trained exclusively on synthetic data.

SDPN can perform inference at approximately 100 Hz on a NVIDIA TitanX GPU, which demonstrates the suitability of our model for being integrated in an actual assisted or autonomous driving pipeline.

Fig. 7. Qualitative results on real-world examples. Predictions look reasonable even if the whole training was conducted on synthetic data.

6 Conclusions

In this paper we presented two main contributions. A new high-quality synthetic dataset, featuring a huge amount of dashboard camera and bird's eye frames, in which the spatial occupancy of a variety of vehicles (i.e. bounding boxes, distance, yaw) is annotated. Furthermore, we presented a deep learning based model to tackle the problem of mapping detections onto a different view of the scene. We argue that these maps could be useful in an assisted driving context, in order to facilitate driver's decisions by making available in one place a concise representation of the road state. Furthermore, in an autonomous driving scenario, inferred vehicle positions could be integrated with other sensory data such as radar or lidar by means of *e.g.* a Kalman filter to reduce overall uncertainty.

References

1. DeepGTAV (2017). https://github.com/ai-tor/DeepGTAV
2. Alletto, S., Palazzi, A., Solera, F., Calderara, S., Cucchiara, R.: Dr (eye) ve: a dataset for attention-based tasks with applications to autonomous and assisted driving. In: Proceedings of the IEEE Conference on Computer Vision and Pattern Recognition Workshops, pp. 54–60 (2016)
3. Bernini, N., Bertozzi, M., Castangia, L., Patander, M., Sabbatelli, M.: Real-time obstacle detection using stereo vision for autonomous ground vehicles: a survey. In: 2014 IEEE 17th International Conference on Intelligent Transportation Systems (ITSC), pp. 873–878. IEEE (2014)
4. Blade, A.: Script Hook V (2017). http://www.dev-c.com/gtav/scripthookv/
5. Chen, C., Seff, A., Kornhauser, A., Xiao, J.: Deepdriving: learning affordance for direct perception in autonomous driving. In: Proceedings of the IEEE International Conference on Computer Vision, pp. 2722–2730 (2015)
6. Deng, J., Dong, W., Socher, R., Li, L.J., Li, K., Fei-Fei, L.: Imagenet: a large-scale hierarchical image database. In: IEEE Conference on Computer Vision and Pattern Recognition CVPR 2009, pp. 248–255. IEEE (2009)
7. Dong, Y., Hu, Z., Uchimura, K., Murayama, N.: Driver inattention monitoring system for intelligent vehicles: a review. IEEE Trans. Intell. Transp. Syst. **12**(2), 596–614 (2011)
8. Gaidon, A., Wang, Q., Cabon, Y., Vig, E.: Virtual worlds as proxy for multi-object tracking analysis. In: Proceedings of the IEEE Conference on Computer Vision and Pattern Recognition, pp. 4340–4349 (2016)
9. Ghiasi, G., Fowlkes, C.C.: Laplacian pyramid reconstruction and refinement for semantic segmentation. In: Leibe, B., Matas, J., Sebe, N., Welling, M. (eds.) ECCV 2016 Part III. LNCS, vol. 9907, pp. 519–534. Springer, Cham (2016). doi:10.1007/978-3-319-46487-9_32
10. He, K., Zhang, X., Ren, S., Sun, J.: Deep residual learning for image recognition. In: Proceedings of the IEEE Conference on Computer Vision and Pattern Recognition, pp. 770–778 (2016)
11. Kingma, D.P., Ba, J.: Adam: a method for stochastic optimization. CoRR abs/1412.6980 (2014). http://arxiv.org/abs/1412.6980
12. Levi, D., Garnett, N., Fetaya, E., Herzlyia, I.: StixelNet: a deep convolutional network for obstacle detection and road segmentation. In: BMVC, pp. 109.1–109.12 (2015)

13. Lin, C.C., Wang, M.S.: A vision based top-view transformation model for a vehicle parking assistant. Sensors **12**(4), 4431–4446 (2012)
14. Lin, G., Shen, C., van den Hengel, A., Reid, I.: Efficient piecewise training of deep structured models for semantic segmentation. In: Proceedings of the IEEE Conference on Computer Vision and Pattern Recognition, pp. 3194–3203 (2016)
15. Liu, W., Anguelov, D., Erhan, D., Szegedy, C., Reed, S., Fu, C.-Y., Berg, A.C.: SSD: single shot multibox detector. In: Leibe, B., Matas, J., Sebe, N., Welling, M. (eds.) ECCV 2016 Part I. LNCS, vol. 9905, pp. 21–37. Springer, Cham (2016). doi:10.1007/978-3-319-46448-0_2
16. Liu, Y.-C., Lin, K.-Y., Chen, Y.-S.: Bird's-eye view vision system for vehicle surrounding monitoring. In: Sommer, G., Klette, R. (eds.) RobVis 2008. LNCS, vol. 4931, pp. 207–218. Springer, Heidelberg (2008). doi:10.1007/978-3-540-78157-8_16
17. Nielsen, F.: Surround video: a multihead camera approach. Vis. Comput. **21**(1), 92–103 (2005)
18. Richter, S.R., Vineet, V., Roth, S., Koltun, V.: Playing for data: ground truth from computer games. In: Leibe, B., Matas, J., Sebe, N., Welling, M. (eds.) ECCV 2016 Part II. LNCS, vol. 9906, pp. 102–118. Springer, Cham (2016). doi:10.1007/978-3-319-46475-6_7
19. Ros, G., Sellart, L., Materzynska, J., Vazquez, D., Lopez, A.M.: The SYNTHIA dataset: a large collection of synthetic images for semantic segmentation of urban scenes. In: Proceedings of the IEEE Conference on Computer Vision and Pattern Recognition, pp. 3234–3243 (2016)
20. Sung, K., Lee, J., An, J., Chang, E.: Development of image synthesis algorithm with multi-camera. In: 2012 IEEE 75th Vehicular Technology Conference (VTC Spring), pp. 1–5. IEEE (2012)
21. Tseng, D.C., Chao, T.W., Chang, J.W.: Image-based parking guiding using Ackermann steering geometry. In: Applied Mechanics and Materials, vol. 437, pp. 823–826. Trans Tech Publ (2013)
22. Venturelli, M., Borghi, G., Vezzani, R., Cucchiara, R.: Deep head pose estimation from depth data for in-car automotive applications. In: Proceedings of the 2nd International Workshop on Understanding Human Activities through 3D Sensors (2016)
23. Wymann, B., Espié, E., Guionneau, C., Dimitrakakis, C., Coulom, R., Sumner, A.: Torcs, the open racing car simulator (2000). http://torcs.sourceforge.net
24. Zhang, B., Appia, V., Pekkucuksen, I., Liu, Y., Umit Batur, A., Shastry, P., Liu, S., Sivasankaran, S., Chitnis, K.: A surround view camera solution for embedded systems. In: Proceedings of the IEEE Conference on Computer Vision and Pattern Recognition Workshops, pp. 662–667 (2014)

Linear Regularized Compression of Deep Convolutional Neural Networks

Claudio Ceruti[✉], Paola Campadelli, and Elena Casiraghi

Dipartimento di Informatica, Università degli Studi di Milano, Milan, Italy
{claudio.ceruti,paola.campadelli,elena.casiraghi}@unimi.it

Abstract. In the last years, deep neural networks have revolutionized machine learning tasks. However, the design of deep neural network architectures is still based on try-and-error procedures, and they are usually complex models with high computational cost. This is the reason behind the efforts that are made in the deep learning community to create small and compact models with comparable accuracy to the current deep neural networks. In literature, different methods to reach this goal are presented; among them, techniques based on low rank factorization are used in order to compress pre trained models with the aim to provide a more compact version of them without losing their effectiveness. Despite their promising results, these techniques produce auxiliary structures between network layers; this work shows that is possible to overcome the need for such elements by using simple regularization techniques. We tested our approach on the VGG16 model obtaining a four times faster reduction without loss in accuracy and avoiding supplementary structures between the network layers.

1 Introduction

Following the breakthrough achieved by Alexnet [16] on the Imagenet challenge [25], during the last years deep learning has revolutionized computer vision, and has taken the lead on an extensive number of machine learning tasks, such as speech recognition, natural language processing, image caption generation, domain adaptation and many others [18].

Despite this widespread use, we are still facing a severe lack of insight on the design of neural network architectures, having to rely on trial-and-error procedures in order to achieve a specific aim; the training of a network is generally done by selecting a very large model with several layers, and then by solving a difficult non-convex optimization problem, treating the architecture's parameters as hyperparameters to be learned. Nevertheless, recent advances foster the current trend to train and model very deep neural networks, that is architectures composed by a great number of layers: in comparison with the 5 convolutional layers architecture of the pioneering Alexnet, two of the most recent, well-known and effective models, GoogleNet [31] and Resnet [10], are composed respectively of 22 layers and up to 152 layers.

© Springer International Publishing AG 2017
S. Battiato et al. (Eds.): ICIAP 2017, Part I, LNCS 10484, pp. 244–253, 2017.
https://doi.org/10.1007/978-3-319-68560-1_22

The price to be paid for these effective but complex models is the required computational cost; these networks are indeed composed by millions, and even billions of learned parameters, which cause an issue in terms of required space for storing them and in terms of execution time, which is dominated by the evaluation of convolutional layers.

This poses serious drawbacks on the use of these powerful models on mobile and embedded devices, such as smartphone operated through speech, robots and self-driving cars that perform real time object recognition, and medical devices that collect and analyze patient data. To overcome these limitations, several research works are investigating ways to design compact and efficient deep neural networks or to compress existing models without loss in effectiveness.

In this paper we present a novel technique for the compression of deep convolutional neural networks with little or none loss in accuracy. The proposed method is based on a linear reduction technique, relying on the low-rank characteristic of the weights of the convolutional layers [14], followed by regularizations techniques which ease the fine-tuning of the parameters in order to regain the accuracy of the original model. This paper is composed as follows: in Sect. 2, we are going to revise the state of the art in the compression of deep neural networks, focusing on the techniques that inspired our work; in Sect. 3, a detailed description of our technique will be provided; in Sect. 4, experimental results showing the quality of our proposal will be reported; Sect. 5 will be devoted to conclusions and future works that could follow from the presented technique.

2   Related Works

As stated before, a lot of efforts are devoted to the development of compact and efficient deep neural network models. These attempts are based on the high redundancy of the parameters of a learned model as shown in several works [3,5]. So far, the proposed approaches can be roughly divided in two classes. Methods in the first class aim at a design strategy to produce a compact deep neural network from scratch, using specific layer structures or by reducing the network as it learns, while methods in the second class are based on model compression, that is the reduction of a pretrained model guided by the amount of network accuracy that has to be retained. Our proposal belongs to the latter class, since it reduces weight matrixes of a pretrained deep convolutional network.

There are many methods that try to generate compact models from scratch, and generally they starts from oversized architectures using regularization functions in the objective function that guides training. Such regularizers reduce redundancy [2], enforce sparsity of the parameters [32], generate linear separable convolutional filters [23] or low-rank filters [13], or constrain the network to have a predetermined number of parameters [4,30]. Similar to the latter methods, evolutionary algorithms are used to select a compact network from a set of models in which the network parameters are treated as genetic traits of a population of network architectures [27,28]. Other approaches exploit particular network structures such in [12], or the use of multilayer perceptrons between layers [21].

In the model compression domain, one of the first work to be reported is Optimal Brain Damage [19] in which the authors prune away neural connections whose weights are labeled as unimportant in terms of network capabilities, alongside with the more recent Optimal Brain Surgeon [9], which uses information conveyed by second order derivatives as a measure of the influence of the connection. The pruning of irrelevant weights is a simple and effective method, still analyzed today [1,8,17,20], but that depends on the definition of what is an essential connection, a concept which is difficult to generalize due to the fact that is based on thresholds relative to the specific task of the network to be reduced. A novel technique, called Knowledge Distillation [11], trains a smaller model in order to mimic the behavior of a bigger model, i.e. forcing a more compact architecture to reproduce the output of the original model using the same training set. Recently, this technique was improved in [24], where the authors showed that is possible to train a new model layer-wise, using the output produced by hidden layers instead of using only the output of the last one.

The work proposed by [6] shows that the weight matrix of a fully connected layer can be approximated using Singular Value Decomposition, without affecting the network effectiveness. Following this work, several attempts were made in order to reduce the weight matrix of a convolutional layer using low-rank decomposition as in [14,15,34]. In [34] the authors use Singular Value Decomposition on the response of a convolutional layer to find its low-rank approximation. More specifically, they approximate a convolutional layer of dimension $D \times D$ by substituting it with two convolutional layers $D \times 1$ and $1 \times D$ that, after retraining each reduced layer one at a time, are shown to obtain the same prediction accuracy as the original model. In [15] the authors perform a Tucker Decomposition both on the fully connected and convolutional layers, being able to further reduce the computational complexity of the whole network at a cost of using three matrixes for each layer, one in the input, one in the output, and one for the reduced layer.

As in the aforementioned works, our method for the reduction of the weight matrix of both convolutional and fully connected layers uses a low-rank factorization. However, we introduce simple techniques to overcome the need for auxiliary matrices between layers after their reduction, used in the preceding techniques to maintain the original layer dimensionality. By doing so we are able to lower the computational complexity and the required space for the learned parameters of the reduced model, with small or even without loss in the prediction accuracy of the original model, being able to cope with both convolutional and fully connected layers.

3   Our Approach

A deep convolutional network is composed of L layers ($l = 1, \cdots, L$, where 1 indicates the input layer and L the output one) which can be convolutional or fully connected. A convolutional layer is composed by D sets of convolutional filters (called filter banks) of dimension $k \times k$ which operate on a C dimensional input; the layer weights are stored in a 4-dimensional matrix $k \times k \times C \times D$. A fully connected layer has a more simpler structure in which each neural unit is connected to all the output units of the preceding layer, having a 2-dimensional weight matrix $D \times D'$ where D is the input dimension and D' is the output dimension.

Our approach is based on linear dimensionality reduction of each layer in a sequential way, starting from the input layer till the last one, reducing one layer and transforming the subsequent one accordingly. We use the same dimensionality reduction procedure with both convolutional and fully connected layers: in order to do so, we reshape the 4-dimensional weight matrix of a convolutional layer in a 2-dimensional matrix $N \times D$ with $N = k^2 C$.

At each step of the procedure we take the weight matrix for the layer l, namely W_l and center it on its mean obtaining $\overline{W}_l$; we note that in our case centering has a negligible effect due to the weight matrices having mean value around zero. Then we estimate the covariance matrix $Cov = \frac{1}{N}\overline{W}_l^T\overline{W}_l$ and we perform a Singular Value Decomposition on it, thus obtaining $Cov = U\Sigma U^T$. The diagonal values of Σ are the eigenvalues for the eigenvectors stored in U; to perform dimensionality reduction we take only the top d eigenvectors corresponding to the top d eigenvalues, scaled by their respective eigenvalues. We obtain a projection matrix $P = \hat{U}\hat{\Sigma}$ of dimension $D \times d$, which is used to reduce the dimensionality of the original weight matrix, obtaining $w_l = W_l P$.

It is not possible to simply substitute the original weight matrix W_l with the reduced one w_l, due to the fact that the subsequent layer $l+1$ is expecting a D-dimensional input; several works overcame this problem adding a $d \times D$ matrix after the reduced layer in order to restore the original input dimension, bringing the complexity of the layer from $O(k^2cD)$ to $O(k^2cd) + O(dD)$. We choose not to insert additional elements between layers, but we use the projection matrix to transform the subsequent layer, obtaining a complexity of $O(k^2cd)$.

More precisely, if the subsequent layer $l+1$ is a convolutional layer having a $k \times k \times D \times D'$ weight matrix, we take the D' filter banks and perform on each of them the same dimensionality reduction projection of the previous layer; more in details, in the layer $l+1$ we multiply the D' filter banks F_i $(i = 1, \cdots, D')$, which have dimension $k \times k \times D$, with the projection matrix P; thus we obtain what we call the channel-reduced layer composed by D' filter banks f_i of dimension $k \times k \times d$. If the subsequent layer is a fully connected layer with weight matrix $D \times D'$ we simply perform the projection on the input dimension, obtaining a new weight matrix of dimension $d \times D$.

After this step we obtain the reduced layer l and the channel-reduced layer $l+1$, without the need for further structures as in the aforementioned works, and we can proceed on reducing the dimensionality of the channel-reduced layer $l+1$ and to transform the layer $l+2$ accordingly, until we reach the end of the network.

A crucial parameter is the dimension d for the layer reduction: choosing d too small results in a dimensionality reduction that doesn't preserve the information conveyed by the original layer, therefore putting at risk the capability of the reduced network to regain the original accuracy. In order to avoid this issue, we measure the amount of information conveyed by increasing sets of eigenvectors by calculating the ratio between the sum of their eigenvalues and the overall sum of all the eigenvalues. More formally, given a set of D eigenvectores e_i, and their respective eigenvalues λ_i, we choose d such that:

$$min_d \frac{\Sigma_{i=1}^d \lambda_i}{\Sigma_{i=1}^D \lambda_i} \geq \epsilon \tag{1}$$

where ϵ is the amount of information we want to preserve for the specific layer l; its value is empirically set for each layer, based on the desired accuracy to be retained after the model reduction. It is worth noting that, due to the high redundancy of the layers' weights, only a small set of eigenvectors retains most part of information, and that this number of relevant eigenvector is not changed by the channel-reduction. In the next section we'll cover these statements in details with the aid of experimental results.

After the reduction of the network's layers, we need to perform a retraining of the reduced network in order to regain the accuracy of the original model. It is possible to perform a retraining after each layer substitution or a retraining of all the reduced network at once. In both cases we first need to perform a regularization of each layer to ease the training of the reduced model by lowering the number of needed iterations to regain the performance of the original model. We do so by scaling the weight matrix of each layer in order to have the same standard deviation of the weight matrix of the original layers. More formally

$$w_l = w_l * \sigma(W_l)/\sigma(w_l) \tag{2}$$

where $\sigma(W)$ is the standard deviation for the matrix W. This simple scaling has proven to be very effective in reducing the number of necessary iterations. In some cases, we weren't even able to train the reduced network without this scaling procedure. This regularization is inspired by the well-known Xavier initialization [7], which can be seen as a way to keep the relation between input and output through a network layer.

4   Experimental Results

In our experiments we used the VGG16 network [29], due to its widespread use in numerous tasks. The VGG16 model is composed by 13 convolutional layers and 3 fully connected layers. Every filter has dimension 3×3, in Table 1 the layers' size are reported.

We tested the network and the reduced model with the Imagenet2012 train set [25] for retraining, and we used the validation set for assessing the model's accuracy. We also tested the transfer learning capability of the reduced model in comparision with the original one. More precisely, we trained the original and reduced networks on a different task than Imagenet, namely the 102 Category Flower Dataset [22]. The reduced model is able to obtain the same accuracy

Table 1. VGG16 layers' size (first two columns) and reduced layers' size (last two columns).

Layer	Channels	Filter banks	Channels	Filter banks
$conv1_1$	3	64	3	11
$conv1_2$	64	64	11	22
$conv2_1$	64	128	22	39
$conv2_2$	128	128	39	58
$conv3_1$	128	256	58	138
$conv3_2$	256	256	138	132
$conv3_3$	256	256	132	148
$conv4_1$	256	512	148	212
$conv4_2$	512	512	212	207
$conv4_3$	512	512	207	185
$conv5_1$	512	512	185	172
$conv5_2$	512	512	172	170
$conv5_3$	512	512	170	120
Layer	Input	Output	Input	Output
$fc6$	4096	4096	–	1986
$fc7$	4096	4096	1986	1653
$fc8$	4096	1000	1653	1000

of the original model on this new task, as shown in Fig. 1b. In Table 1 the size of channels and filter banks are reported before and after reduction, whilst in Table 2 we show the comparison between the original and the reduced model. The model retraining is accomplished in two ways. In the first setting, we retrained the reduced layer and the subsequent channel-reduced layer keeping the other layers fixed, until the original accuracy is regained; we proceed reducing and retraining in this layer-wise manner until we reach the output layer. In Fig. 1a we report the top 5 test accuracy of the reduced model during the retraining of the convolutional layer $conv5_3$, after having reduced the preceding layers. In the second framework, we perform what we call "one-shot" retraining; indeed, we skip the layer retraining step in favor of reducing the whole network and then carry out a single retraining pass for the entire reduced model. With the first

Table 2. Comparison of VGG16 and the reduced model in terms of required space of the model parameters and average execution time using a NVIDIA Tesla P100 GPU.

	VGG16	Reduced model
Required space	528 MB	72 MB
Execution time	2174.22 ms	569.764 ms

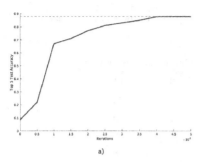
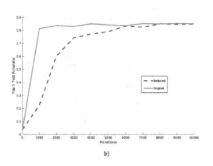

a)   b)

Fig. 1. (a) Top 5 test accuracy during retraining after the reduction of layer $conv5_3$, with the preceding layers already reduced and retrained. Dashed line is original accuracy. (b) Top 5 test accuracy of original and reduced model on the 102 Category Flower Dataset. Reduced model has a slower learning than original, although it reaches the same accuracy level of the original model.

strategy we were able to achieve the original accuracy after less than 15 epochs in total. Conversely, we didn't manage to replicate the VGG16 results after the same number of iterations using "one-shot" retraining, resulting in a loss of 5% in the top-5 and 8% in the top-1 classification error, probably due to the training being stuck in a local minima.

The choice of the number of filter banks in the reduced model is related to the ratio in Eq. 1; we show the amount of the ratio for the selected value of each layer in Table 3. In order to restore the original accuracy, we have to maintain an higher ratio on the first convolutionl layers, namely from the input layer to $conv4_1$, than

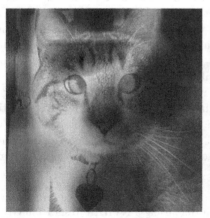
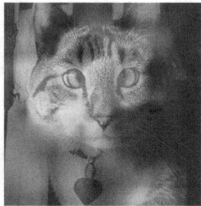

Fig. 2. Input area selected as most "important" for prediction of class "tabby cat" from Imagenet dataset, with red (maximum) and blue (minimum). On the left, visualization of original model, on the right, visualization of reduced model. Areas related to the reduced model are more fine grained than the ones of the original. Image best seen in color.

Table 3. Ratio of the selected number of filter banks for each layer.

Layer	Dimension	Ratio
$conv1_1$	11	99.96%
$conv1_2$	22	96.85%
$conv2_1$	39	95.13%
$conv2_2$	58	92.21%
$conv3_1$	128	97.04%
$conv3_2$	132	94.75%
$conv3_3$	148	93.15%
$conv4_1$	212	91.80%
$conv4_2$	207	84.25%
$conv4_3$	185	75.09%
$conv5_1$	172	70.29%
$conv5_2$	170	70.84%
$conv5_3$	120	59.96%
fc_6	1986	87.58%
fc_7	1653	92.40%

on the last ones, as shown in Table 3. A possible hypothesis for this could be found in [33], in which the authors state that the first layers of a convolutional neural network learn general purpose filters, whilst the last layers show a task-dependent behavior. The selected number of filter banks were empirically chosen following the aforementioned hypothesis of performing conservative reductions in the first layers. [33].

In Fig. 2 we show the input regions of "importance" for prediction either of the reduced and the original model on the same input image. We followed the technique described in [35] and its extension [26], observing more fine grained areas with the reduced model than the ones used by the original.

5   Conclusions and Future Works

In this work we propose a novel method for linear compression of deep convolutional neural networks, based on linear dimensionality reduction and simple regularization techniques to avoid the need for auxiliary structures. We show that our proposal doesn't affect the prediction accuracy of the original network. We also show that our reduction technique preserves the transfer learning capability of the original network.

Further studies will be devoted to a more accurate method for the selection of the number of reduced filter banks without relying on empirical procedures. In order to perform even further reductions we are going to investigate the use of non linear method for dimensionality reduction.

References

1. Babaeizadeh, M., Smaragdis, P., Campbell, R.H.: NoiseOut: a simple way to prune neural networks. arXiv preprint arXiv:1611.06211 (2016)
2. Chen, W., Wilson, J.T., Tyree, S., Weinberger, K.Q., Chen, Y.: Compressing convolutional neural networks. arXiv preprint arXiv:1506.04449 (2015)
3. Cheng, Y., Yu, F.X., Feris, R.S., Kumar, S., Choudhary, A., Chang, S.F.: An exploration of parameter redundancy in deep networks with circulant projections. In: Proceedings of the IEEE International Conference on Computer Vision, pp. 2857–2865 (2015)
4. Cortes, C., Gonzalvo, X., Kuznetsov, V., Mohri, M., Yang, S.: Adanet: adaptive structural learning of artificial neural networks. arXiv preprint arXiv:1607.01097 (2016)
5. Denil, M., Shakibi, B., Dinh, L., de Freitas, N., et al.: Predicting parameters in deep learning. In: Advances in Neural Information Processing Systems, pp. 2148–2156 (2013)
6. Denton, E.L., Zaremba, W., Bruna, J., LeCun, Y., Fergus, R.: Exploiting linear structure within convolutional networks for efficient evaluation. In: Advances in Neural Information Processing Systems, pp. 1269–1277 (2014)
7. Glorot, X., Bengio, Y.: Understanding the difficulty of training deep feedforward neural networks. In: AISTATS, 9, 249–256 (2010)
8. Han, S., Pool, J., Tran, J., Dally, W.: Learning both weights and connections for efficient neural network. In: Advances in Neural Information Processing Systems, pp. 1135–1143 (2015)
9. Hassibi, B., Stork, D.G., et al.: Second order derivatives for network pruning: optimal brain surgeon. In: Advances in Neural Information Processing Systems, pp. 164–164 (1993)
10. He, K., Zhang, X., Ren, S., Sun, J.: Deep residual learning for image recognition. In: Proceedings of the IEEE Conference on Computer Vision and Pattern Recognition, pp. 770–778 (2016)
11. Hinton, G., Vinyals, O., Dean, J.: Distilling the knowledge in a neural network. arXiv preprint arXiv:1503.02531 (2015)
12. Iandola, F.N., Han, S., Moskewicz, M.W., Ashraf, K., Dally, W.J., Keutzer, K.: Squeezenet: alexnet-level accuracy with 50x fewer parameters and <0.5 mb model size. arXiv preprint arXiv:1602.07360 (2016)
13. Ioannou, Y., Robertson, D., Shotton, J., Cipolla, R., Criminisi, A.: Training CNNs with low-rank filters for efficient image classification. arXiv preprint arXiv:1511.06744 (2015)
14. Jaderberg, M., Vedaldi, A., Zisserman, A.: Speeding up convolutional neural networks with low rank expansions. arXiv preprint arXiv:1405.3866 (2014)
15. Kim, Y.-D., Park, E., Yoo, S., Choi, T., Yang, L., Shin, D.: Compression of deep convolutional neural networks for fast and low power mobile applications. arXiv preprint arXiv:1511.06530 (2015)
16. Krizhevsky, A., Sutskever, I., Hinton, G.E.: Imagenet classification with deep convolutional neural networks. In: Advances in neural information processing systems, pp. 1097–1105 (2012)
17. Lebedev, V., Lempitsky, V.: Fast convnets using group-wise brain damage. In: Proceedings of the IEEE Conference on Computer Vision and Pattern Recognition, pp. 2554–2564 (2016)

18. LeCun, Y., Bengio, Y., Hinton, G.: Deep learning. Nature **521**(7553), 436–444 (2015)
19. LeCun, Y., Denker, J.S., Solla, S.A., Howard, R.E., Jackel, L.D.: Optimal brain damage. In: NIPS, vol. 2, pp. 598–605 (1989)
20. Li, H., Kadav, A., Durdanovic, I., Samet, H., Graf, H.P.: Pruning filters for efficient convnets. arXiv preprint arXiv:1608.08710 (2016)
21. Lin, M., Chen, Q., Yan, S.: Network in network. arXiv preprint arXiv:1312.4400 (2013)
22. Nilsback, M.E., Zisserman, A.: Automated flower classification over a large number of classes. In: Sixth Indian Conference on Computer Vision, Graphics & Image Processing, ICVGIP 2008, pp. 722–729. IEEE (2008)
23. Rigamonti, R., Sironi, A., Lepetit, V., Fua, P.: Learning separable filters. In: Proceedings of the IEEE Conference on Computer Vision and Pattern Recognition, pp. 2754–2761 (2013)
24. Romero, A., Ballas, N., Kahou, S.E., Chassang, A., Gatta, C., Bengio, Y.: Fitnets: Hints for thin deep nets. arXiv preprint arXiv:1412.6550 (2014)
25. Russakovsky, O., Deng, J., Hao, S., Krause, J., Satheesh, S., Ma, S., Huang, Z., Karpathy, A., Khosla, A., Bernstein, M., et al.: Imagenet large scale visual recognition challenge. Int. J. Comput. Vis. **115**(3), 211–252 (2015)
26. Selvaraju, R.R., Das, A., Vedantam, R., Cogswell, M., Parikh, D., Batra, D.: Grad-CAM: why did you say that? arXiv preprint arXiv:1611.07450 (2016)
27. Shafiee, M.J., Mishra, A., Wong, A.: Deep learning with darwin: evolutionary synthesis of deep neural networks. arXiv preprint arXiv:1606.04393 (2016)
28. Shafiee, M.J., Wong, A.: Evolutionary synthesis of deep neural networks via synaptic cluster-driven genetic encoding. arXiv preprint arXiv:1609.01360 (2016)
29. Simonyan, K., Zisserman, A.: Very deep convolutional networks for large-scale image recognition. arXiv preprint arXiv:1409.1556 (2014)
30. Soulié, G., Gripon, V., Robert, M.: Compression of deep neural networks on the fly. In: Villa, A.E.P., Masulli, P., Pons Rivero, A.J. (eds.) ICANN 2016. LNCS, vol. 9887, pp. 153–160. Springer, Cham (2016). doi:10.1007/978-3-319-44781-0_19
31. Szegedy, C., Liu, W., Jia, Y., Sermanet, P., Reed, S., Anguelov, D., Erhan, D., Vanhoucke, V., Rabinovich, A.: Going deeper with convolutions. In: Proceedings of the IEEE Conference on Computer Vision and Pattern Recognition, pp. 1–9 (2015)
32. Wen, W., Wu, C., Wang, Y., Chen, Y., Li, H.: Learning structured sparsity in deep neural networks. In: Advances in Neural Information Processing Systems, pp. 2074–2082 (2016)
33. Yosinski, J., Clune, J., Bengio, Y., Lipson, H.: How transferable are features in deep neural networks? In: Advances in Neural Information Processing Systems, pp. 3320–3328 (2014)
34. Zhang, X., Zou, J., Ming, X., He, K., Sun, J.: Efficient and accurate approximations of nonlinear convolutional networks. In: Proceedings of the IEEE Conference on Computer Vision and Pattern Recognition, pp. 1984–1992 (2015)
35. Zhou, B., Khosla, A., Lapedriza, A., Oliva, A., Torralba, A.: Learning deep features for discriminative localization. In: Proceedings of the IEEE Conference on Computer Vision and Pattern Recognition, pp. 2921–2929 (2016)

Network Edge Entropy from Maxwell-Boltzmann Statistics

Jianjia Wang[✉], Richard C. Wilson, and Edwin R. Hancock

Department of Computer Science, University of York, York YO10 5DD, UK
jw1157@york.ac.uk

Abstract. In prior work, we have shown how to compute global network entropy using a heat bath analogy and Maxwell-Boltzmann statistics. In this work, we show how to project out edge-entropy components so that the detailed distribution of entropy across the edges of a network can be computed. This is particularly useful if the analysis of non-homogeneous networks with a strong community as hub structure is being attempted. To commence, we view the normalized Laplacian matrix as the network Hamiltonian operator which specifies a set of energy states with the Laplacian eigenvalues. The network is assumed to be in thermodynamic equilibrium with a heat bath. According to this heat bath analogy, particles can populate the energy levels according to the classical Maxwell-Boltzmann distribution, and this distribution together with the energy states determines thermodynamic variables of the network such as entropy and average energy. We show how the entropy can be decomposed into components arising from individual edges using the eigenvectors of the normalized Laplacian. Compared to previous work based on the von Neumann entropy, this thermodynamic analysis is more effective in characterizing changes of network structure since it better represents the edge entropy variance associated with edges connecting nodes of large degree. Numerical experiments on real-world datasets are presented to evaluate the qualitative and quantitative differences in performance.

Keywords: Network edge entropy · Maxwell-Boltzmann statistics

1 Introduction

There has been a considerable recent interest in computing the entropy associated with different types of network structure [2,3,5]. Network entropy has been extensively used to characterize the salient features of the structure of static and dynamic of network systems arising in biology, physics, and the social sciences [1–3]. For example, the von Neumann entropy can be used as an effective characterization of network structure, commencing from a quantum analog in which the Laplacian matrix on graphs [1] plays the role of the density matrix. Further development of this idea has shown the link between the von Neumann entropy and the degree statistics of pairs of nodes forming edges in a network [2], which can be efficiently computed for both directed and undirected graphs [3]. Since

© Springer International Publishing AG 2017
S. Battiato et al. (Eds.): ICIAP 2017, Part I, LNCS 10484, pp. 254–264, 2017.
https://doi.org/10.1007/978-3-319-68560-1_23

the eigenvalues of the density matrix reflect the energy states of a network, this approach is closely related to the heat bath analogy in statistical mechanics. This provides a convenient route to network characterization [3,5]. By populating the energy states with particles which are in thermal equilibrium with a heat bath, this thermalization, of the occupation statistics for the energy states can be computed using the Maxwell-Boltzmann distribution [4,5]. The properties of this physical heat bath system are described by a partition function with the energy microstates of the network represented by a suitably chosen Hamiltonian. Usually, the Hamiltonian is computed from the adjacency or Laplacian matrix of the network, but recently, Ye et al. [4], have shown how the partition function can be computed from a characteristic matrix polynomial instead.

Although entropic analysis of the heat bath analogy provides a useful global characterization of network structure, it does not allow the entropy of edge or subnetwork structure to be easily computed. In this paper, we explore a novel edge entropy projection which can be applied to the global network entropy computed from Maxwell-Boltzmann statistics. We use this technique to analyze the distribution of edge entropy within a network and explore how this distribution encodes the intrinsic structural properties of different types of network.

The remainder of the paper is organized as follows. In Sect. 2, we briefly introduce the von Neumann entropy with its approximate degrees of nodes connected by an edge. In Sect. 3, we develop an entropic network characterization from the heat bath analogy and Maxwell-Boltzmann statistics, and then describe our edge entropy projection. In Sect. 4, we undertake experiments to demonstrate the usefulness of this novel method. Finally, in Sect. 5 we conclude our paper with a summary of our contribution and suggestions for future work.

2 Preliminaries

2.1 Von Neumann Entropy

Let $G(V, E)$ be an undirected graph with node set V and edge set $E \subseteq V \times V$, and let $|V|$ represent the total number of nodes on graph $G(V, E)$. The $|V| \times |V|$ adjacency matrix A of a graph is defined as

$$A = \begin{cases} 0 & \text{if} (u, v) \in E \\ 1 & \text{otherwise.} \end{cases} \qquad (1)$$

Then the degree of node u is $d_u = \sum_{v \in V} A_{uv}$.

The normalized Laplacian matrix $\tilde{L}$ of the graph G is defined as

$$\tilde{L} = D^{-\frac{1}{2}} L D^{\frac{1}{2}} = \Phi \tilde{\Lambda} \Phi^T \qquad (2)$$

where $L = D - A$ is the Laplacian matrix and D denotes the degree diagonal matrix whose elements are given by $D(u, u) = d_u$ and zeros elsewhere. $\tilde{\Lambda} = diag(\lambda_1, \lambda_2, \ldots \lambda_{|V|})$ is the diagonal matrix with the ordered eigenvalues as elements and $\Phi = (\varphi_1, \varphi_2, \ldots, \varphi_{|V|})$ is the matrix with the ordered eigenvectors as columns.

In quantum mechanics, the density matrix is used to describe a system with the probability of pure quantum states. Passerini and Severini [1] have extended this idea to the graph domain. Specifically, they show that a density matrix for a graph or network can be obtained by scaling the combinatorial Laplacian matrix by the reciprocal of the number of nodes in the graph.

With this notation, the specified density matrix is obtained by scaling the normalized Laplacian matrix by the number of nodes, i.e.

$$\rho = \frac{\tilde{L}}{|V|} \tag{3}$$

When defined in this way the density matrix is Hermitian i.e. $\rho = \rho\dagger$ and $\rho \geq 0, \mathrm{Tr}\rho = 1$. It plays an important role in the quantum observation process, which can be used to calculate the expectation value of measurable quantity.

The interpretation of the scaled normalized Laplacian as a density operator, opens up the possibility of characterizing a graph using the von Neumann entropy from quantum information theory. The von Neumann entropy is defined as the entropy of the density matrix associated with the state vector of a system. As noted above, Passerini and Severini [1] suggest how the von Neumann entropy can be computed by scaling the normalized discrete Laplacian matrix for a network. As a result the von Neumann entropy is given in terms of the eigenvalues $\lambda_1, \ldots, \lambda_{|V|}$ of the density matrix ρ,

$$S^{VN} = -\mathrm{Tr}(\rho \log \rho) = -\sum_{i=1}^{|V|} \frac{\hat{\lambda}_i}{|V|} \log \frac{\hat{\lambda}_i}{|V|} \tag{4}$$

The von Neumann entropy [1] computed from the normalized Laplacian spectrum has been shown to be effective for network characterization. In fact, Han et al. [2] have shown how to approximate the calculation of von Neumann entropy in terms of simple degree statistics. Their approximation allows the cubic complexity of computing the von Neumann entropy from the Laplacian spectrum, to be reduced to one of quadratic complexity using simple edge degree statistics, i.e.

$$S^{VN} = 1 - \frac{1}{|V|} - \frac{1}{|V|^2} \sum_{(u,v) \in E} \frac{1}{d_u d_v} \tag{5}$$

This expression for the von Neumann entropy allows the approximate entropy of the network to be efficiently computed and has been shown to be an effective tool for characterizing structural property of networks, with extremal values for the cycle and fully connected graphs.

Thus, the edge entropy decomposition is given as

$$S^{VN}_{edge}(u, v) = \frac{1}{|E|} - \frac{1}{|V||E|} - \frac{1}{|E||V|^2} \frac{1}{d_u d_v} \tag{6}$$

where $S^{VN} = \sum_{(u,v)\in E} S^{VN}_{edge}(u,v)$. This expression decomposes the global parameter of von Neumann entropy on each edge with the relation to the degrees from the connection of two vertexes.

3   Network Entropy in Maxwell-Boltzmann Statistics

3.1   Thermodynamic Representation

Thermodynamic analogies provide powerful tools for analyzing complex networks. The underpinning idea is that statistical thermodynamics can be combined with network theory to characterize both static and time-evolving networks [6].

Here we consider the thermodynamic system specified by a system of N particles with energy states given by the network Hamiltonian and immersed in a heat bath with temperature T. The ensemble is represented by a partition function $Z(\beta, N)$, where $\beta = 1/k_B T$ is an inverse of temperature parameter [5].

When specified in this way, the various thermodynamic characterizations of the network can be computed. For instance, the average energy of the network can be expressed in terms of the density matrix and the Hamiltonian operator,

$$\langle U \rangle = \langle H \rangle = \text{Tr}\,(\rho H) = \left[-\frac{\partial}{\partial \beta} \log Z \right]_N \tag{7}$$

and the thermodynamic entropy by

$$S = k_B \left[\log Z + \beta \langle U \rangle \right] \tag{8}$$

Both the energy and the entropy can be regarded as weighted functions of the Laplacian eigenvalues which characterize the network structure in different ways. In the following sections, we set the Boltzmann constant to the unity, i.e., $k_B = 1$, and explore the thermodynamic entropy in more detail to represent the intrinsic structure of networks.

3.2   Maxwell-Boltzmann Statistics

The Maxwell-Boltzmann distribution relates the microscopic properties of particles to the macroscopic thermodynamic properties of matter [4]. It applies to systems consisting of a fixed number of weakly interacting distinguishable particles. These particles occupy the energy levels associated with a Hamiltonian and in our case the Hamiltonian of the network, which is in contact with a thermal bath [7].

Taking the Hamiltonian to be the normalized Laplacian of the network, the canonical partition function for Maxwell-Boltzmann occupation statistics of the energy levels is

$$Z_{MB} = \text{Tr}\left[\exp(-\beta \tilde{L})^N \right] = \left(\sum_{i=1}^{|V|} e^{-\beta \lambda_i} \right)^N \tag{9}$$

where $\beta = 1/k_B T$ is the reciprocal of the temperature T with k_B as the Boltzmann constant; N is the total number of particles and λ_i denotes the microscopic energy of system at each microstate i with energy λ_i. Derived from Eq. (8), the entropy of the system with N particles is

$$S_{MB} = \log Z - \beta \frac{\partial \log Z}{\partial \beta} = -N \mathrm{Tr}\left\{ \frac{\exp(-\beta \tilde{L})}{\mathrm{Tr}[\exp(-\beta \tilde{L})]} \log \frac{\exp(-\beta \tilde{L})}{\mathrm{Tr}[\exp(-\beta \tilde{L})]} \right\}$$

$$= -N \sum_{i=1}^{|V|} \frac{e^{-\beta \lambda_i}}{\sum_{i=1}^{|V|} e^{-\beta \lambda_i}} \log \frac{e^{-\beta \lambda_i}}{\sum_{i=1}^{|V|} e^{-\beta \lambda_i}} \tag{10}$$

For a single particle, the density matrix is

$$\boldsymbol{\rho}_{MB} = \frac{\exp(-\beta \tilde{L})}{\mathrm{Tr}[\exp(-\beta \tilde{L})]} \tag{11}$$

Since the density matrix commutes with the Hamiltonian operator, we have $\partial \rho / \partial t = 0$ and the system can be viewed as in equilibrium. So the entropy in the Maxwell-Boltzmann system is simply N times the von Neumann entropy of a single particle, as we might expect.

3.3   Edge Entropy Analysis

Our goal is to project the global network entropy onto the edges of the network. In matrix form for Maxwell-Boltzmann statistics in Eq. (10), the entropy can be written as,

$$S^{MB} = -\mathrm{Tr}\left[\boldsymbol{\rho}_{MB} \log \boldsymbol{\rho}_{MB}\right] = -\mathrm{Tr}[\boldsymbol{\Sigma}_{MB}] \tag{12}$$

Since the spectral decomposition of the normalized Laplacian matrix is

$$\tilde{L} = \Phi \tilde{\Lambda} \Phi^T \tag{13}$$

We can decompose the matrix $\boldsymbol{\Sigma}_{MB}$ as follows

$$\boldsymbol{\Sigma}_{MB} = \Phi \sigma_{MB}(\tilde{\Lambda}) \Phi^T \tag{14}$$

where

$$\sigma_{MB}(\lambda_i) = -N \frac{e^{-\beta \lambda_i}}{\sum_{i=1}^{|V|} e^{-\beta \lambda_i}} \log \frac{e^{-\beta \lambda_i}}{\sum_{i=1}^{|V|} e^{-\beta \lambda_i}}$$

for Maxwell-Boltzmann statistics. As a result, we can perform edge entropy projection of the Maxwell-Boltzmann statistical model using the Laplacian eigenvectors, with the result that the entropy of edge (uv) is given as,

$$S_{edge}^{MB}(u, v) = \sum_{i=1}^{|V|} \sigma_{MB}(\lambda_i) \varphi_i \varphi_i^T \tag{15}$$

Thus, the global entropy can be projected on the edges of the network system. This provides useful measures for local entropic characterization of network structure in a relatively straightforward manner.

4 Experiments and Evaluations

4.1 Data Sets

Data-Set 1: The PPIs dataset extracted from STRING–8.2 [8] consisting of networks which describe the interaction relationships between histidine kinase and other proteins. There are 173 PPIs in this dataset and they are collected from 4 different kinds of bacteria with the following evolution order (from older to more recent). Aquifex and Thermotoga-8 PPIs from Aquifex aelicus and Thermotoga maritima, Gram-Positive-52 PPIs from Staphylococcus aureus, Cyanobacteria-73 PPIs from Anabaena variabilis and Proteobacteria-40 PPIs from Acidovorax avenae [9].

Data-Set 2: The New York Stock Exchange dataset consists of the daily prices of 3,799 stocks traded continuously on the New York Stock Exchange over 6000 trading days. The stock prices were obtained from the Yahoo! financial database (http://finance.yahoo.com) [10]. A total of 347 stock were selected from this set, for which historical stock prices from January 1986 to February 2011 are available. In our network representation, the nodes correspond to stock and the edges indicate that there is a statistical similarity between the time series associated with the stock closing prices [10]. To determine the edge structure of the network, we use a time window of 20 days to compute the cross-correlation coefficients between the time-series for each pair of stock. Connections are created between a pair of stock if the cross-correlation exceeds an empirically determined threshold. In our experiments, we set the correlation coefficient threshold to the value to $\xi = 0.85$. This yields a time-varying stock market network with a fixed number of 347 nodes and varying edge structure for each of 6,000 trading days. The edges of the network, therefore, represent how the closing prices of the stock follow each other.

4.2 Experimental Results

We first investigate the temperature dependence of edge entropy for the PPI networks. We select three different types of edges with different values of degrees at the vertices and explore how the entropy changes with temperature.

Figure 1(a) plots three selected edge entropies versus temperature with Maxwell-Boltzmann occupation statistics. The three edges show a similar dependence of entropy on the temperature. As the inverse of temperature (β) increases, the edge entropy reaches a maximum value. The edge entropy for vertices with the high degree increases faster than that for the low-degree in the high-temperature region. In the low-temperature limit, entropy approaches zero. This is because when the temperature decreases the configuration of particle occupation becomes identical as the particles always state at the low energy levels since the thermalization effects vanish.

Figure 1(b) shows the relationship between the edge entropies in the Maxwell-Boltzmann and von Neumann cases. There is a transition in the relationship between two entropies with temperature. At high temperature (i.e., $\beta = 0.1$),

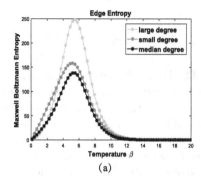 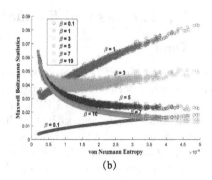

(a)   (b)

Fig. 1. (Color online) (a) Edge entropy with a different degree on both nodes for Maxwell-Boltzmann statistics. The red line represents the high-degree edge; the blue line is the low-degree edge and the black line is the median value of degree on the edge ends; (b) Scatter plot of edge entropies from Maxwell-Boltzmann vs. von Neumann entropy with different value of temperatures.

the Maxwell-Boltzmann entropy is roughly in linear proportion to von Neumann entropy. However, as the temperature reduces, it takes on an approximately exponential dependence. The Maxwell-Boltzmann edge entropy decreases monotonically with the von Neumann edge entropy in the low-temperature region ($\beta = 10$).

Further exploration of the relationship between Maxwell-Boltzmann edge entropy and von Neumann entropy is shown in Fig. 2, which shows the 3D plots of edge entropy with the vertex degree. The figure compares the edge entropy between Maxwell-Boltzmann statistics and von Neumann entropy with node degree connection for each edge in the network. The observation is that both entropies have a similar tendency with the degrees at the end. The Maxwell-Boltzmann edge entropy is more sensitive to the degree variance than the von Neumann entropy in the high degree region. The reason for this is the constant term in the von Neumann entropy formula dominates the value of edge entropy

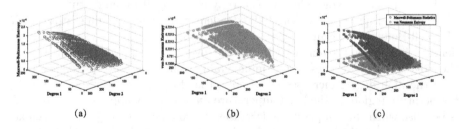

(a)   (b)   (c)

Fig. 2. (Color online) 3D scatter plot of edge entropy from Maxwell-Boltzmann statistics and von Neumann entropy. (a) Edge entropy in Maxwell-Boltzmann statistics. (b) Edge entropy from von Neumann formula. (c) The comparison of edge entropy between Maxwell-Boltzmann statistics and von Neumann entropy.

when the degrees are large. Thus, the Maxwell-Boltzmann edge entropy is better suited to represent the differences in graph structure associated with large degree nodes.

When compared to the von Neumann edge entropy, the Maxwell-Boltzmann edge entropy is distributed rather differently. Figure 3 shows two examples of PPI networks, namely Anabaena variabilis and Aquifex aelicus together with their associated edge entropy histograms. The Maxwell-Boltzmann edge entropies are more sensitive to the presence of edges associated with high degree nodes, which provides better edge discrimination. This effect is manifest in the differences of edge entropy histograms. In the Maxwell-Boltzmann case, the histogram shows two peaks in the edge entropy distribution, while the von Neumann edge entropy is concentrated at low values and has just a single peak. In other words, the von Neumann edge entropy offers less salient structure.

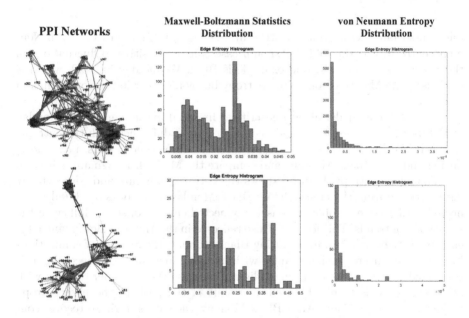

Fig. 3. (Color online) Examples of protein-protein interaction networks with the edge entropy distribution from von Neumann entropy and Maxwell-Boltzmann statistics.

Next, we turn our attention to the time evolution of networks. We take the NYSE network as an example to explore the entropic characterization in the network structure. Figure 4 plots the total network for the Maxwell-Boltzmann and von Neumann cases. Both entropies reflect the positions of significant global financial events such as Black Monday, Friday 13th mini-crash, Early 1990s Recession, 1997 Asian Crisis, 9.11 Attacks, Downturn of 2002–2003, 2007 Financial Crisis, the Bankruptcy of Lehman Brothers and the European Debt Crisis. In each case, the entropy undergoes significant fluctuations during the financial

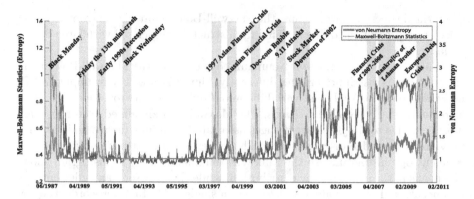

Fig. 4. (Color online) Entropy from Maxwell-Boltzmann statistics and von Neumann entropy for NYSE (1987–2011). Number of particle is $N = 1$ and temperature is $\beta = 10$.

crises, associated with dramatic structural changes. Compared to the von Neumann entropy, the Maxwell-Boltzmann case is more sensitive to fluctuations in the network structure. A good example is Black Wednesday in 1992, which is obvious in the Maxwell-Boltzmann entropy but is not clear in the von Neumann case.

We now focus in detail on one critical financial event, i.e., Black Monday in October 1987, to explore the dynamic structural difference with the entropic variance. We visualize the network structure at three-time epochs, i.e., before, during and after Black Monday, and compare the Maxwell-Boltzmann with von Neumann edge entropy. Figure 5 shows the network structure and edge entropy distribution during the crisis. Before Black Monday, the stocks are highly connected with a large number of densely connected clusters of stocks following the same trading trends. This feature is also reflected in the Maxwell-Boltzmann edge entropy distribution. However, during Black Monday, the number of connections between stock decrease significantly with large numbers of nodes becoming disconnected. Some stocks do though slightly increase their number of links with other stocks. This manifests itself as a shift of the peak to the high entropy region of the distribution. After Black Monday, the stocks begin to recover connections with another. The node degree distribution also returns to its previous shape. In contrast, the von Neuman edge entropy distribution does not completely reflect the details of these critical structural changes. Compared to the Maxwell-Boltzmann edge entropy, the distribution of von Neumann edge entropy does not change significantly during Black Monday and hence does not effectively characterize the dynamic structure on the network.

In conclusion, both the Maxwell-Boltzmann and von Neumann edge entropies can be used to represent changes in network structure. Compared to the von Neumann edge entropy, the Maxwell-Boltzmann edge entropy is more sensitive to variance associated with the degree distribution. In the low-temperature region, the Maxwell-Boltzmann edge-entropy has similar degree sensitivity to the von Neumann edge entropy. However, it is more sensitive to high degree variations.

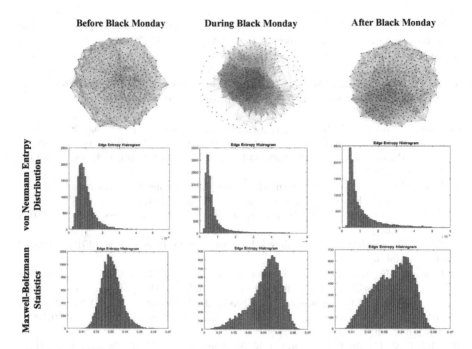

Fig. 5. (Color online) Visualization of network structure before, during and after Black Monday. The edge entropy distribution is computed from von Neumann entropy and Maxwell-Boltzmann statistics. The statistical model such as Maxwell-Boltzmann case is more sensitive to represent the dynamic structure in the networks.

5   Conclusion

This paper has explored the thermodynamic characterizations of networks resulting from Maxwell-Boltzmann statistics, and specifically those associated with the thermalization effects of the heat bath on the occupation of the normalized Laplacian energy states. We view the normalized Laplacian matrix as the Hamiltonian operator of the network with associated energy states which can be occupied by classical distinguishable particles. This extends the use of entropy as a tool to characterize network structures in both static and time series data. To compare with the extensively studied von Neuman entropy, we conduct the experiments which demonstrate that the thermodynamic edge entropy is better suited to represent the intrinsic structural properties associated to long-tailed degree distributions. Future work will focus on exploring non-classical alternatives to the Maxwell-Boltzmann occupation statistics and the detailed distribution of the entropic characterization for different types of complex networks.

References

1. Passerini, F., Severini, S.: The von Neumann entropy of networks. Int. J. Agent Technol. Syst. **1**, 1–6 (2008)

2. Han, L., Hancock, E., Wilson, R.: Characterizing graphs using approximate von Neumann entropy. Pattern Recogn. Lett. **33**, 1958–1967 (2012)
3. Ye, C., Wilson, R.C., Comin, C.H., Costa, L.D.F., Hancock, E.R.: Approximate von Neumann entropy for directed graphs. Phys. Rev. E **89**(5), 052804 (2014)
4. Ye, C., Wilson, R.C., Comin, C.H., Costa, L.F., Hancock, E.R.: Thermodynamic characterization of networks using graph polynomials. Phys. Rev. E **92**, 032810 (2015)
5. Wang, J., Wilson, R.: Hancock, E.R.: Network entropy analysis using the maxwell-boltzmann partition function. In: Davis, L., Bimbo, A.D., Lovell, B. (eds.) Proceedings 23rd International Conference on Pattern Recognition, pp. 1–6, Los Alamitos, CA, USA (2016)
6. Mikulecky, D.C.: Network thermodynamics and complexity: a transition to relational systems theory. Comput. Chem. **25**, 369–391 (2001)
7. Park, J., Newman, M.: Statistical mechanics of networks. Phys. Rev. E **70**(6), 066117 (2004)
8. STRING - Known and Predicted Protein-Protein Interactions (2010). http://string.embl.de
9. Escolano, F., Lozano, M.A., Hancock, E.R., Giorgi, D.: What is the complexity of a network? The heat flow-thermodynamic depth approach. In: Hancock, E.R., Wilson, R.C., Windeatt, T., Ulusoy, I., Escolano, F. (eds.) SSPR /SPR 2010. LNCS, vol. 6218, pp. 286–295. Springer, Heidelberg (2010). doi:10.1007/978-3-642-14980-1_27
10. Silva, F.N., Comin, C.H., Peron, T.K., Rodrigues, F.A., Ye, C., Wilson, R.C., Hancock, E., Costa, L.D.F.: Modular dynamics of financial market networks, physics and society (2015)

Learning from Enhanced Contextual Similarity in Brain Imaging Data for Classification of Schizophrenia

Tewodros Mulugeta Dagnew[1]([✉]), Letizia Squarcina[2], Massimo W. Rivolta[1],
Paolo Brambilla[3], and Roberto Sassi[1]

[1] Computer Science Department, Università degli Studi di Milano, Crema, Italy
{tewodros.dagnew, massimo.rivolta, roberto.sassi}@unimi.it
[2] Scientific Institute IRCCS "E. Medea", Bosisio Parini, Italy
letizia.squarcina@gmail.com
[3] Fondazione IRCCS Ca' Granda Ospedale Maggiore Policlinico,
Università degli Studi di Milano, Milan, Italy
paolo.brambilla1@unimi.it

Abstract. In certain severe mental diseases, like schizophrenia, structural alterations of the brain are detectable by magnetic resonance imaging (MRI). In this work, we try to automatically distinguish, by using anatomical features obtained from MRI images, schizophrenia patients from healthy controls. We do so by exploiting contextual similarity of imaging data, enhanced with a distance metric learning strategy (DML - by providing "must-be-in-the-same-class" and "must-not-be-in-the-same-class" pairs of subjects). To learn from contextual similarity of the subjects brain anatomy, we use a graph-based semi-supervised label propagation algorithm (graph transduction, GT) and compare it to standard supervised techniques (SVM and K-nearest neighbor, KNN). We performed out tests on a population of 20 schizophrenia patients and 20 healthy controls. DML+GT achieved a statistically significant advantage in classification performance (Accuracy: 0.74, Sensitivity: 0.79, Specificity: 0.69, Ck: 0.48). Enhanced contextual similarity improved performance of GT, SVM and KNN offering promising perspectives for MRI images analysis.

1 Introduction

Schizophrenia (SCZ) is a severe, chronic and debilitating mental illness affecting around 0.4% of the population [1]. Magnetic resonance imaging (MRI) studies consistently observed alterations in cortical and subcortical brain areas, especially frontal [21] and temporal [14] regions. The capability of detecting these pathological alterations in brain images would be of high relevance in accelerating the diagnostic process, with clear benefits for both patients and psychiatrists. Given the complexity and multidimensionality of the problem, machine learning (ML) analysis of magnetic resonance (MR) images is recently becoming popular in the understanding of this domain.

© Springer International Publishing AG 2017
S. Battiato et al. (Eds.): ICIAP 2017, Part I, LNCS 10484, pp. 265–275, 2017.
https://doi.org/10.1007/978-3-319-68560-1_24

ML algorithms have been used in SCZ studies [18] with the aim of detecting sets of features which could be discriminative in the diagnosis. In the literature, the majority of ML applications to psychiatric data are purely supervised methods that learn only from labeled data, with promising and interesting results [11,13,16]. However, while these findings have been received with great optimism within the neuropsychiatric community, a major criticism has been that these algorithms are ordinarily "trained" to categorize patients based on a symptom-based diagnosis. As such, there are inevitable uncertainty in the "gold standard": learning from the unlabeled data seems a possibility to mitigate the problem. In these situations, classification performances might improve when the learning process incorporates unlabeled data. Moreover, semi-supervised and unsupervised schemes could provide a better phenotype identification and classification of diseases [20].

In this paper, we propose to exploit learning from both labeled and unlabeled MR images. The addition of learning from unlabeled data will decrease the risk of circular analysis, by exploiting similarities between data without prior information on the class. To do so, we applied graph transduction (GT), *i.e.*, a data-driven graph-based semi-supervised label propagation algorithm [4], which can learn from the contextual similarity (CS) of the imaging data. However, the problem with label propagation methods is that their performance heavily depends on the pre-existing CS of the input data. To deal with this problem, we applied a *distance metric learning* (DML) strategy, to enhance CS information of features obtained from MRI images, by providing "must-be-in-the-same-class" and "must-not-be-in-the-same-class" pairs of subjects (*i.e.*, healthy controls and SCZ patients), thus increasing the intra-cluster similarity and decreasing the inter-cluster similarity. The formalization of GT is inspired from game theoretic notions [4], where the final labeling corresponds to the Nash equilibrium of a non cooperative game. The players of the game correspond to data features (or nodes of the graph) and the class labels correspond to available strategies. In our case, we map the problem of classifying MRI images, where the brain imaging data of each subject correspond to a player who can choose a strategy to maximize its pay-off (the pair-wise similarity of the image features between subjects).

Authors of [3] showed a similar concept of what we present here, to solve a problem in object recognition and scene classification (a general computer vision problem), confirming the importance of enhancing CS to improve the performance of a label propagation algorithm. In our study, we implemented one of the latest and most robust metric learning [19] and label propagation algorithm [4], to be then applied to MRI data.

To the best of our knowledge, this is the first study to address classification of SCZ patients and healthy subjects applying a metric learning and graph-based semi-supervised learning strategy to structural MRI data.

2 Learning from Enhanced Contextual Similarity

In this section, we present a scheme of classification that exploits contextual brain anatomical similarities of subjects from MR images, so as to differentiate healthy controls from SCZ patients. A set of features, characterizing the anatomy of the brain, was obtained from the MR images of every single subject. Then, we used a DML technique, specifically the one proposed in [19], to enhance the CS of the input MRI data and apply the GT algorithm [4] on top of this new metric space to learn from the newly enhanced context. The overall scheme is depicted in Fig. 1 and described step-by-step in the next sections.

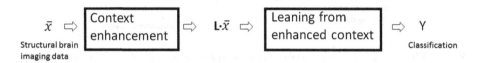

Fig. 1. The proposed schizophrenia classification scheme using structural brain imaging data.

2.1 CS Enhancement Using DML

DML represents a useful technique widely exploited in pattern recognition, which aims to find a metric that maximizes the distance between features belonging to different classes (and viceversa, minimizes the distance between features belonging to the same class). With this aim, linear and non-linear metrics had been investigated. On one hand, linear metrics can be computationally less expensive, but often provide lower performances. On the other hand, non-linear algorithm might perform better but they are computationally expensive and application-dependent.

In the linear domain, DML remaps features using a linear combination carried out by the transformation matrix L, as follows:

$$\bar{x}' = L\bar{x},$$

where $\bar{x}$ is the input feature vector and $\bar{x}'$ is the transformed feature vector. If the matrix L is full rank, it is possible to show that the Euclidean distance between two elements in the transformed space,

$$D(\bar{x}_i, \bar{x}_j) = ||L(\bar{x}_i - \bar{x}_j)||_2,$$

represents a valid metric. Furthermore, the Euclidean distance can be rewritten using a matrix notation which becomes the so-called *Mahalanobis* distance. Such distance is defined as

$$D_M(x_i, x_j) = \sqrt{(\bar{x}_i - \bar{x}_j)^\top M (\bar{x}_i - \bar{x}_j)}$$

being $M = L'L$ the Mahalanobis positive semidefinite matrix. The effect of such transformation is shown in Fig. 2. When L is the identity matrix, the Mahalanobis distance becomes the standard Euclidean distance.

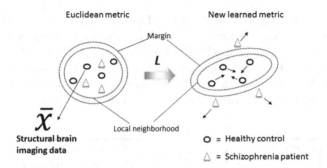

Fig. 2. Illustration of feature context enhancement by means of large margin nearest neighbor (LMNN) distance metric learning. Before training (left) and after training (right).

In this study, we used a linear DML to modify the pre-existing neighbouring structure of MRI data before feeding it to GT, aiming to achieve classification improvements. In order to determine the transformation matrix L, we used the *Large Margin Nearest Neighbor* DML method described in [19]. The algorithm makes use of the following equations

$$pullpush(L) = (1 - \mu)\, pull(L) + \mu\, push(L) \tag{1}$$

with

$$
\begin{aligned}
pull(L) &= \sum_{i,j \to i} \|L(\bar{x}_i - \bar{x}_j)\|^2 \\
push(L) &= \sum_{i,j \to i} \sum_{k} (1 - \delta_{ik})[1 + \|L(\bar{x}_i - \bar{x}_j)\|^2 \\
&\quad - \|L(\bar{x}_i - \bar{x}_k)\|^2]_+
\end{aligned}
\tag{2}
$$

where y_i is the class to which $\bar{x}_i$ belongs and $\delta_{ik} = 1$ if $y_i = y_k$ or $\delta_{ik} = 0$ otherwise. $[f]_+$ implies a hinge-loss such that $[f]_+ = \max(0, f)$. The term $j \to i$ in Eq. (2) implies that j belongs to the same class where i belongs too. Finally, the parameter μ sets the trade-off between the pulling and pushing objectives and was set to 0.5 as suggested in [19].

The process of getting the transformation metric L involves minimizing the overall objective function in Eq. (1). The first term pulls subjects with the same class label closer in terms of the Mahalanobis distance. The second term pushes away differently-labeled instances by a large margin, so that they are located further apart in the transformed space (Fig. 2).

As stated in [19], it is worth noting that Eq. (1) does not define a convex optimization problem in terms of L. However, it can be rephrased in a convex fashion using a semi-definitive programming approach by determining M instead of L. Then, L can be computed using matrix factorization of M.

2.2   Learning from Enhanced CS of MR Images Using GT

The aim of GT is to address the problem of consistent labeling, with the aim of predicting or propagating class membership to unlabeled data by exploiting learning both from the labeled and unlabeled samples. Such methodology involves three different areas: (i) graph theory; (ii) evolutionary game theory; and (iii) dynamical systems and optimization.

The main idea behind GT is to consider the samples of the dataset as nodes of a graph, and to propagate class labels to unlabeled nodes, by considering the CS among the samples. In particular, it exploits CS among data features to perform label propagation in a consistent way, relying on a common *a priori* assumption known as the "cluster assumption" (a reminiscent of the homophily principle used in social network analysis): nodes that are close to each other, in the same cluster or on the same manifold are expected to have the same label. Each node is then a feature vector $\in \mathbb{R}^d$ (with d being the number of features). Moreover, each node can select a strategy, *i.e.*, class membership, that maximizes its CS. Finally, the output labeling corresponds to the Nash equilibrium of the game.

Input features are represented with graph nodes $\mathcal{G} = (\mathcal{V})$, where the vertex set $\mathcal{V}$ is composed of $n = l + u$ elements $\in \mathbb{R}^d$ and consists of a first labeled set $\{(x_1, y_1), ..., (x_l, y_l)\}$ of l elements and a second unlabeled set $\{(x_{l+1}, ..., (x_{l+u})\}$ of u elements. Then, the similarity matrix $\mathcal{E}$ between pairs of nodes is computed, after having selected a similarity metric. A simple and effective optimization algorithm to propagate the labels through the graph is given by the so-called replicator dynamics, developed and studied in evolutionary game theory, which has proven to be effective in many applications [7,23].

In practice, as explained in Sect. 2.1, labeled examples in the form of "must-be-in-the-same-class" and "must-not-be-in-the-same-class" pairs of subjects are provided to the DML framework, to learn the best feature space transformation matrix L using Eq. (1). Afterwards, the class label propagation occurs on such transformed feature space (*i.e.*, $L\bar{x}$) by constructing the fully connected graph $\mathcal{G} = (\mathcal{V})$, where $\mathcal{V}$ is now the set of graph nodes representing the transformed feature vectors, and $\mathcal{E}$ encodes the brain anatomy similarity between subjects by means of the edge weights (similarity matrix) as depicted in Fig. 3b. $\mathcal{E}$ is constructed in the following manner (for simplicity we show how an edge is constructed between two transformed feature vectors):

$$\mathcal{E}_{ij} = \exp\left[-\frac{d(L\bar{x}_i, L\bar{x}_j)^2}{2\sigma^2}\right] \tag{3}$$

where $d(L\bar{x}_i, L\bar{x}_j)$ is the Euclidean distance. For estimating σ, which is a critical parameter of the graph's ability in representing the CS between data points, we adopted an automatic self tuning method as proposed in [22].

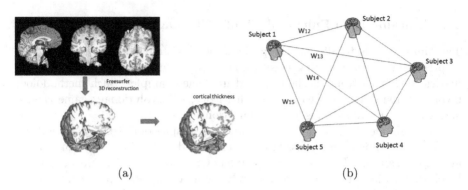

(a)   (b)

Fig. 3. (a) ROI and cortical thickness feature extraction from brain images. (b) Representation of brain anatomy similarity between subjects.

3   Experiments

3.1   Dataset and Representation

The dataset consisted in T1-weigthed MR images of 20 healthy control subjects (35.8 ± 13, 8 males) and 20 SCZ patients (37.9 ± 11, 13 males). The size of this dataset is in line with the dimensionality of datasets used in academic works aimed at medical applications [9,10,17]: in particular, it is not straightforward to obtain consistent MRI data of psychiatric patients, due to difficulties in recruitment and feasibility of MRI acquisitions in this kind of patients. The data were collected at the Psychiatric department of Ospedale di Verona (Verona, Italy). All involved subjects signed an informed consent, following the principle of the Helsinki's declaration.

The T1-weigthed images were preprocessed using the software FreeSurfer[1] as depicted in Fig. 3a. Based on prior knowledge on schizophrenia [14,21], we considered the average cortical grey matter thickness of frontal and temporal regions (namely: *caudal middle frontal, inferior temporal, middle temporal, rostral middle frontal* and *superior frontal* of the left hemisphere) as features in the classification task. The ROI thickness measurement of the subjects is reported in Table 1. Also, in order to take into account the effect of age on the cortical thickness, we corrected all the data for age differences using a generalized linear model [8].

3.2   Experimental Analysis

We performed two series of comparisons to assess the performances of the proposed classification scheme in differentiating healthy controls from SCZ subjects. First, we verified whether learning from CS (from both labeled and unlabeled data) might provide better classification results than just learning from labeled

---

[1] http://surfer.nmr.mgh.harvard.edu/.

Table 1. Grey matter cortical thickness of ROIs (in mm) of healthy controls and schizophrenia patients.

ROI	HC (mean ± std)	SCZ (mean ± std)
Caudal middle frontal	2.56 ± 0.21	2.44 ± 0.18
Inferior temporal	2.85 ± 0.16	2.64 ± 0.17
Middle temporal	2.89 ± 0.17	2.73 ± 0.18
Rostral middle frontal	2.47 ± 0.19	2.34 ± 0.17
Superior frontal	2.82 ± 0.18	2.65 ± 0.20

data. Second, we tested if the enhancement of CS by DML might provide further improvements. To do so, we compared the proposed classification scheme (DML+GT) with both GT [4] and KNN, with and without metric learning (KNN [6], DML+KNN [19]), linear SVM and DML+SVM.

We evaluated the classification performances by using accuracy (Acc), sensitivity (Se), specificity (Sp) and Cohen's kappa (Ck) coefficients. Sensitivity refers to the true recognition of SCZ patients.

We considered first 70%, then 80% of the data from each class for training and input labeling of GT, while the rest of data was left to be predicted. In fact, GT was found to perform sufficiently well when the labeled data were just a small fraction of the dataset [4]. However, given the small size of our dataset, we considered labeling 70% and 80% of the data at disposal. We repeated this procedure by randomly sampling the dataset 100 times and computed the average performance. In all the experiments we avoided the risk of circular analysis [5]. For KNN, we chose $K = 3$, for limiting the possible overfitting due to the relatively small sample size.

3.3 Experimental Results

The average and standard error of the classification performance for DML+GT (proposed scheme), DML+SVM, DML+KNN, GT, SVM and KNN (used for comparison), when 70% and 80% of the samples in each class are labeled are reported in Table 2 and Fig. 4.

As expected, performance got better when using higher percentage of labeled data on small datasets for DML+GT and GT. Moreover, in our proposed scheme, sensitivity was always lower than specificity (Fig. 4c and d), meaning that some subjects with schizophrenia were classified as healthy, regardless the labeled sample size. In addition, the increase of training data provided different relative improvements between sensitivity and specificity (Fig. 4c and d). This means that these methodologies, under the settings we considered, are capable of recognizing the healthy subjects more easily than the schizophrenia patients.

GT was more affected by the training set's size (fourth bar in each plot of Fig. 4) than the other methods. However, when DML was applied before GT,

Table 2. Average test-set classification performance ($\pm$ standard deviation across subjects) on brain sMRI data features using 70% and 80% of the data for training.

Methods		70%	80%
DML+GT	Acc	0.70 ± 0.01	0.74 ± 0.01
	Se	0.66 ± 0.02	0.69 ± 0.02
	Sp	0.73 ± 0.02	0.79 ± 0.02
	Ck	0.39 ± 0.02	0.48 ± 0.03
DML+SVM	Acc	0.71 ± 0.01	0.71 ± 0.02
	Se	0.70 ± 0.02	0.74 ± 0.02
	Sp	0.73 ± 0.02	0.68 ± 0.02
	Ck	0.42 ± 0.02	0.42 ± 0.03
DML+KNN	Acc	0.68 ± 0.01	0.70 ± 0.02
	Se	0.64 ± 0.02	0.66 ± 0.02
	Sp	0.72 ± 0.02	0.75 ± 0.02
	Ck	0.36 ± 0.02	0.41 ± 0.03
GT	Acc	0.61 ± 0.01	0.67 ± 0.01
	Se	0.57 ± 0.04	0.64 ± 0.03
	Sp	0.65 ± 0.03	0.71 ± 0.03
	Ck	0.22 ± 0.02	0.35 ± 0.03
SVM	Acc	0.69 ± 0.01	0.69 ± 0.01
	Se	0.70 ± 0.02	0.73 ± 0.02
	Sp	0.69 ± 0.02	0.66 ± 0.02
	Ck	0.38 ± 0.02	0.38 ± 0.03
KNN	Acc	0.65 ± 0.01	0.65 ± 0.01
	Se	0.64 ± 0.02	0.62 ± 0.02
	Sp	0.65 ± 0.02	0.69 ± 0.02
	Ck	0.29 ± 0.03	0.30 ± 0.03

we obtained a drastic classification improvement of all measures except the sensitivity, even with a smaller training set. Furthermore, the use of DML resulted in a higher performance in all the cases except sensitivity (Fig. 4).

Finally, when 80% of the data is used as training, CS learning, *i.e.*, learning from unlabeled data as well, enhanced with DML outperformed both SVM and KNN with DML (first bar vs second and third bar).

3.4   Discussion of the Experimental Results

This work supports the finding that DML+KNN is better than KNN (*i.e.*, with respect to every evaluation metric considered), as found by other authors [19]. In particular, we showed that this finding holds true when applied to thickness features extracted from MRI data.

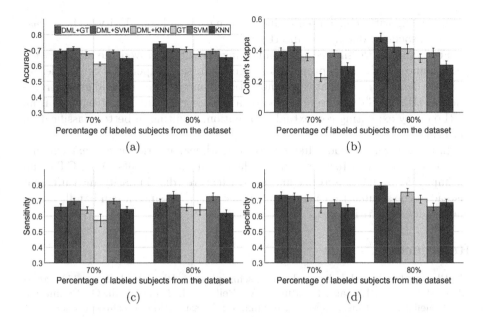

Fig. 4. Classification results for healthy controls vs. schizophrenia patients. Average performances and standard errors of the mean are reported.

Moreover, GT is consistently improved by the proposed scheme (DML+GT), which suggests that CS enhancement of MRI data coupled with learning from unlabeled samples, can result in a better performance of classifying schizophrenia. This result is also supported in [3], within the computer vision domain (object recognition and scene classification).

Finally, DML+GT performances are higher likely due to the additional information obtained from the unlabeled MRI data features. This confirms that DML and CS has the potential to improve schizophrenia classification.

The results obtained are comparable to the state-of-the-art in classification of schizophrenia. For example, in [12] using functional MRI (fMRI), they obtained an average classification accuracy of 0.59 and 0.84 using both static and dynamic resting-state functional network connectivity approach respectively and linear SVM. In [15] they obtained up to 0.75 accuracy (combining ROI thickness features) using 1.5 T sMRI and covariate multiple kernel learning approach using SVM. In [2], they achieved 0.75 accuracy considering the left hemisphere.

4   Conclusion

In this study, we designed a classification scheme to discriminate healthy controls from schizophrenia patients using MR images-derived data as features. We believe that learning from contextual anatomical similarity of subjects (*i.e.*, learning both from labeled and unlabeled MRI data features) has a great potential in dealing with schizophrenia, due to the nature and complexity of the disease and its associated diagnostic uncertainty.

Furthermore, we showed that enhancing the CS improved the classification performances of the label propagation algorithm (semi-supervised context learning). We demonstrated that the combination of metric learning and graph transduction (DML+GT) is useful to learn a meaningful underlying pattern from MRI data by exploiting contextual information, resulting in better classification performances.

In the future, we would like to test a non-linear metric for context enhancing to assess if it can further improve the classification results. Also, GT could be improved by using another anatomical feature (dis)similarity measurement instead of the symmetric Euclidean distance of Eq. (3) since it can handle asymmetric (dis)similarities also.

References

1. Bhugra, D.: The global prevalence of schizophrenia. PLoS Med. **2**(5), e151 (2005)
2. Castellani, U., Rossato, E., Murino, V., Bellani, M., Rambaldelli, G., Perlini, C., Tomelleri, L., Tansella, M., Brambilla, P.: Classification of schizophrenia using feature-based morphometry. J. Neural Transm. (Vienna) **119**(3), 395–404 (2012)
3. Ebert, S., Fritz, M., Schiele, B.: Pick your neighborhood-improving labels and neighborhood structure for label propagation. In: ICPR, pp. 152–162 (2011)
4. Erdem, A., Pelillo, M.: Graph transduction as a noncooperative game. Neural Comput. **24**(3), 700–723 (2012)
5. Kriegeskorte, N., Simmons, W.K., Bellgowan, P.S., Baker, C.I.: Circular analysis in systems neuroscience: the dangers of double dipping. Nat. Neurosci. **12**(5), 535–540 (2009)
6. Larose, D.T.: K-nearest neighbor algorithm. In: Discovering Knowledge in Data: An Introduction to Data Mining, pp. 90–106. Wiley (2005)
7. Zemene, E., Bulò, S.R., Pelillo, M.: Dominant-set clustering using multiple affinity matrices. In: Feragen, A., Pelillo, M., Loog, M. (eds.) SIMBAD 2015. LNCS, vol. 9370, pp. 186–198. Springer, Cham (2015). doi:10.1007/978-3-319-24261-3_15
8. Montgomery, D.C., Peck, E.A., Vining, G.G.: Introduction to Linear Regression Analysis. Wiley, Hoboken (2015)
9. Mwangi, B., Spiker, D., Zunta-Soares, G.B., Soares, J.C.: Prediction of pediatric bipolar disorder using neuroanatomical signatures of the amygdala. Bipolar Disord. **16**(7), 713–721 (2014)
10. Mwangi, B., Wu, M.J., Bauer, I.E., Modi, H., Zeni, C.P., Zunta-Soares, G.B., Hasan, K.M., Soares, J.C.: Predictive classification of pediatric bipolar disorder using atlas-based diffusion weighted imaging and support vector machines. Psychiatry Res. **234**(2), 265–271 (2015)
11. Peruzzo, D., Castellani, U., Perlini, C., Bellani, M., Marinelli, V., Rambaldelli, G., Lasalvia, A., Tosato, S., De Santi, K., Murino, V., et al.: Classification of first-episode psychosis: a multi-modal multi-feature approach integrating structural and diffusion imaging. J. Neural Transm. **122**(6), 897–905 (2015)
12. Rashid, B., Arbabshirani, M.R., Damaraju, E., Cetin, M.S., Miller, R., Pearlson, G.D., Calhoun, V.D.: Classification of schizophrenia and bipolar patients using static and dynamic resting-state fmri brain connectivity. Neuroimage **134**, 645–657 (2016)

13. Schnack, H.G., Van Haren, N.E., Nieuwenhuis, M., Hulshoff Pol, H.E., Cahn, W., Kahn, R.S.: Accelerated brain aging in schizophrenia: a longitudinal pattern recognition study. Am. J. Psychiatry **173**(6), 607–616 (2016)
14. Shenton, M.E., Kikinis, R., Jolesz, F.A., Pollak, S.D., LeMay, M., Wible, C.G., Hokama, H., Martin, J., Metcalf, D., Coleman, M., et al.: Abnormalities of the left temporal lobe and thought disorder in schizophrenia: a quantitative magnetic resonance imaging study. N. Engl. J. Med. **327**(9), 604–612 (1992)
15. Squarcina, L., Castellani, U., Bellani, M., Perlini, C., Lasalvia, A., Dusi, N., Bonetto, C., Cristofalo, D., Tosato, S., Rambaldelli, G., et al.: Classification of first-episode psychosis in a large cohort of patients using support vector machine and multiple kernel learning techniques. Neuroimage **145**(Part B), 238–245 (2015)
16. Squarcina, L., Perlini, C., Bellani, M., Lasalvia, A., Ruggeri, M., Brambilla, P., Castellani, U.: Learning with heterogeneous data for longitudinal studies. In: Navab, N., Hornegger, J., Wells, W.M., Frangi, A.F. (eds.) MICCAI 2015. LNCS, vol. 9351, pp. 535–542. Springer, Cham (2015). doi:10.1007/978-3-319-24574-4_64
17. Valli, I., Marquand, A.F., Mechelli, A., Raffin, M., Allen, P., Seal, M.L., McGuire, P.: Identifying individuals at high risk of psychosis: predictive utility of support vector machine using structural and functional MRI data. Front. Psychiatry **7**, 52 (2016)
18. Veronese, E., Castellani, U., Peruzzo, D., Bellani, M., Brambilla, P.: Machine learning approaches: from theory to application in schizophrenia. Comput. Math. Methods Med. **2013**, 867924 (2013)
19. Weinberger, K., Saul, L.: Distance metric learning for large margin nearest neighbor classification. J. Mach. Learn. Res. **10**, 207–244 (2009)
20. Wu, M.J., Mwangi, B., Bauer, I.E., Passos, I.C., Sanches, M., Zunta-Soares, G.B., Meyer, T.D., Hasan, K.M., Soares, J.C.: Identification and individualized prediction of clinical phenotypes in bipolar disorders using neurocognitive data, neuroimaging scans and machine learning. Neuroimage **145**, 254–264 (2017)
21. Xiao, Y., Lui, S., Deng, W., Yao, L., Zhang, W., Li, S., Wu, M., Xie, T., He, Y., Huang, X., et al.: Altered cortical thickness related to clinical severity but not the untreated disease duration in schizophrenia. Schizophr. Bull. **41**(1), 201–210 (2015)
22. Zelnik-Manor, L., Perona, P.: Self-tuning spectral clustering. In: NIPS, pp. 1601–1608 (2004)
23. Zemene, E., Tariku, Y., Idrees, H., Prati, A., Pelillo, M., Shah, M.: Large-scale image geo-localization using dominant sets. CoRR abs/1702.01238 (2017)

3D Object Detection Method Using LiDAR Information in Multiple Frames

Jung-Un Kim[1], Jihong Min[2], and Hang-Bong Kang[1(✉)]

[1] The Catholic University of Korea, Bucheon, Gyeonggi-do, Republic of Korea
{amysh,hbkang}@catholic.ac.kr
[2] Agency for Defense Development, Daejeon, Republic of Korea
happymin77@gmail.com

Abstract. For a safe autonomous navigation, it is important to understand the configuration of the environment and quickly, accurately grasp the information regarding the location, direction, and size of each constituent object. Recent studies on autonomous navigation were performed to not only detect and classify objects, but also to segment and evaluate their properties. However, in these studies, pre-processing was required, which incurred a considerable amount of computational cost. Moreover, the 3D shape model was further analyzed. In other words, more computation cost and computing power are required. In this study, we propose a new method for detecting and estimating the pose of a 3D object using LiDAR information via charge-coupled-device (CCD) in real-time environment. We classified objects into classes (e.g., car, pedestrian, and cyclist), and the 3D pose of an object is quickly estimated without requiring a separate 3D-shape model. From the multiple frames obtained using the LiDAR and CCD, we design a method to robustly reconstruct the 3D environment in real time by aligning the object information of the previously obtained frames with the current frame through an optical-flow method. Our method helps in complementing the limitations of CCD-based classifiers and correcting the defects by increasing the density of the 3D-LiDAR point cloud. We compared the results obtained using our method with the state-of-the-art results of the KITTI data set; which were in good agreement in terms of speed and accuracy. This comparison shows that the 3D pose of a box can be generated with better speed and accuracy using the reconstructed 3D-point-cloud clusters proposed in our method.

Keywords: Object detection · Deep learning · Optical flow · Sensor fusion

1 Introduction

Detecting 3D objects is particularly important in the field of robotics wherein real-time interaction with nearby objects is required in an autonomous environment. In early studies, the focus was on detecting and classifying objects in 2D

© Springer International Publishing AG 2017
S. Battiato et al. (Eds.): ICIAP 2017, Part I, LNCS 10484, pp. 276–286, 2017.
https://doi.org/10.1007/978-3-319-68560-1_25

images. In recent studies, 3D objects were classified in terms of their 3D pose and size. In 2D-based detection algorithms, view-point estimation is considered. Hence, there is a scope for development in detecting and estimating the pose of a 3D object. One of the important approaches [14] involves dividing the 3D pose into sub-categories with advanced two-dimensional object classifiers. In this method, the samples of many hypotheses are grouped and classified through sub-categories. So, it is necessary to have a projected 3D-shape model of different types and various viewing angles. However, combining a single CCD image with a shape model requires significant amount of manual work to generate learning data; moreover, subclasses with new shapes cannot be added easily. Another approach involves employing stereo images. For example, in 3DOP [1], a stereo image is used to create a depth map and each pixel of the RGB image is projected onto a 3D space. They use several properties to define the relationship between the pixels as an energy function of the Markov random field (MRF) [3] and classify the objects using the linear support vector machine (SVM) [2]. This is because the sophisticated stereo-based depth-map generation used in the 3DOP requires considerable amount of computational resources. In another example like 3DVP [13], a 3D CAD model is superimposed on a CCD image. A CAD model can be expressed in a voxel form to identify specific areas such as occlusion or truncation. This decreases the speed of overall process, and a considerable amount of manual work is required for learning the data.

To handle these problems, we propose a method to classify the objects and reconstruct a 3D pose in a driving environment using LiDAR and CCD information. In Fig. 1(a), our approach involves detecting objects by matching the LiDAR point cloud to the CCD images. The LiDAR point clouds are clustered in the form of an edge shape, which is used to express the shape of an object. After that, an object proposal is generated and is classified using the CCD-based Convolutional Neural Networks (CNN) classifier [8]. The universal size of the object is estimated through the class label of the classified object, and it is reconstructed using the 3D-space bounding box in the 3D space to overlay the partially captured LiDAR point. Through this process, the 3D objects around a vehicle can be restored. We develop a model to match the LiDAR point cloud with the CCD image and collect information from multiple frames. Using a local optical flow [7] in the image space to obtain any lost LiDAR point information due to long distance, diffuse reflection, and occlusion. The computational cost incurred in our proposed method is much lower than that in existing 3D-geometry models wherein subcategories are employed.

The contributions of our work are as follows:

- We match the uneven LiDAR point cloud with the CCD image by mapping the points of two coordinate systems into an ordered edge form and easily cluster the object to offer a proposal for the 2D classifier. Subsequently, we estimate the pose of a 3D object using the classification results and 3D-edge orientations.
- Our method helps in clearly representing the objects by improving the point cloud of the present time by matching objects from multiple frames. This

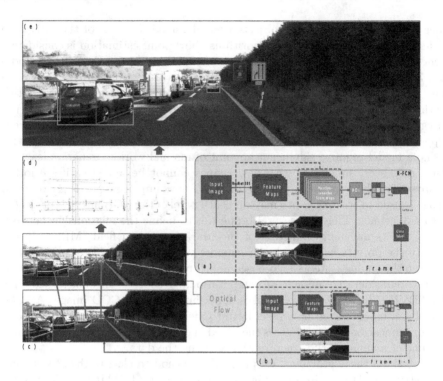

Fig. 1. System overview, (a) a two-dimensional object classification process in a t frame, (b) object classification process in a $t-1$ frame, (c) object matching by optical flow, (d) the alignment of the points that make up the matched object and the cumulative result at t frame (e) classification result image drawn through 3D plane matched to LiDAR Edge. Blue box : side of vehicle facing corner of 2D bounding box. Orange box : front or rear of vehicle.

helps in more accurately estimating the orientation in 3D space and supplementing the information of the obscured area compared to using single-frame information.

This paper is organized as follows. In Sect. 2, we describe the proposed method restore a 3D pose of the object by generating proposals, classifying the objects, and connecting them to adjacent frames. The experimental results are presented in Sect. 3.

2   Proposed Method

2.1   2D-Object Detection and 3D-Environment Reconstruction Using LiDAR

In this section, we explain the method of classifying objects in 2D and 3D spaces through a multi-frame LiDAR point cloud and a CCD image. We generate a proposal for classifying two-dimensional objects using a widely distributed LiDAR point cloud, and subsequently, track the objects in multiple frames with optical flow using convolutional feature maps in R-FCN [9]. Thereafter, we reconstruct the surrounding 3D objects more accurately by improving the unstable 3D-LiDAR-shape model.

Development of 2D LiDAR Edges. This section describes the process of generating proposals for classifying 2D objects with sparse and widely distributed LiDAR point cloud set P. A LiDAR point of an object with a particular height appears in the top view as a complex shape depending on the shape of the object, but the boundary shape of the object is clear. If the noise around the points that make up the boundary shape of an object can be effectively removed, this area can be projected onto a 2D space and use it as a robust 2D-object proposal. So, we unified the height of LiDAR point p_{xy}^z to 40 cm to eliminate the ground without missing the objects found in the driving environment. The process can be expressed as follows.

$$p_{xy}^z = \begin{cases} 40 & \text{if } p_{xy}^z >= 40 \\ empty & \text{if } p_{xy}^z < 40 \end{cases}, \tag{1}$$

where p_{xy}^z represents the coordinates of each point on the three-dimensional space comprising the x, y, and z axes. This compressed point cloud helps in removing the ground and increasing the density of the sparse shape. Any noise point other than the desired shape can be removed.

To remove any noise at CCD space, the point cloud should be projected onto a CCD image through a calibration matrix. Observing the projected point cloud, the shape of the object viewed from the viewpoint of the vehicle on which the sensor is mounted becomes the lowermost edge of the projected LiDAR points. When selected only the lowest point for all the x-axis pixels of the CCD image, it is possible to obtain an edge close to the peripheral shape of the object as shown in Fig. 2(a). Because of the difference in height based on the position of the LiDAR sensor and various factors that interfere with sensing, many noise points other than the shapes around the object are occurred as shown in Fig. 2(b). This problem can be solved by using a median filter along the x-axis of the CCD. This process changes the position of the LiDAR point on the CCD image. So, we merge the point where the coordinate movement occurs in the CCD image to the adjacent point having the median value in the 3D space.

Proposal Generation with LiDAR Edges. Because of the characteristics of the CCD space represented by the pixel unit, a change in one pixel varies significantly depending on the depth. This means that there is a limit in determining the continuity in the CCD space. Each point constituting the LiDAR edge in the CCD space is labeled by number along the x-axis and mapped to the same point in the 3D space. In the order of number, we form a merged edge until the sum of the differences in orientation between points is above the threshold $(\pi/2)$. Small groups are merged with adjacent groups. This method is similar to

Fig. 2. (a) Image after removing the noise; the shape of the object can be seen as an edge. (b) Image captured after 2D projection. Noise is observed above the bottom edge.

initial edge grouping on an edgebox [15]. Next, we divide the area by evaluating the affinity of the two neighboring edges. The following Eq. 2 is used to compare the connections of the two edges by employing the difference between the two average angles θ_i and θ_j and the two center coordinates x_i and x_j. The affinity score is computed as follows:

$$a(s_i, s_j) = \frac{|\cos(\theta_i - \theta_{ij})\cos(\theta_j - \theta_{ij})|^\gamma}{\lambda m_p}, \tag{2}$$

where θ_{ij} represents the angle between x_i and x_j. p_i and p_j are the adjacent points of the two edges s_i and s_j, respectively. m_p represents the distance between two edge boundaries. Using affinity score, a proposal is generated based on the boundary line where the score changes significantly. Based on an image with a resolution of 376×1241 in the KITTI dataset [5], the pixel height is determined as follows :

$$C^h_{pixel^i} = \frac{C^h_{real^i} \times 750}{d_i} \tag{3}$$

where $C^h_{real^i}$ represents the actual height of the class C_i to be classified, and $C^h_{pixel^i}$ is the pixel size to be projected on the CCD image.

Object Classification and Point Labeling. The generated proposal is used as an input to the region-based CNN classifier. Although several high-performance classifiers exist, we use R-FCN implemented using residual net [6]. R-FCN is a classifier based on the R-CNN, and it uses vote mechanism whether an object belongs to the ROI or not. When classifying the objects, the R-FCN helps in determining the class by considering the translation variance of the object in the ROI area. This shows a filtering effect among proposals consisting of clutterred backgrounds due to noise. As shown in Fig. 1(a), the ROI of the R-FCN architecture is divided into n grids, and the convolutional features are trained through each grid cell. In the test phase, if the class is classified through the proposal, the class label is mapped to the LiDAR edge used in generating the proposal. From the label-mapped LiDAR edges, we create 3D bounding boxes.

3D Pose Restoration. The most accurate way to perceive the information around a vehicle in the driving situation is to accurately restore the 3D pose of the object. We obtained the required information to restore the 3D pose around the vehicle in the previous section. We created LiDAR Edge s_i using Eq. 2 and obtained the average direction θ_i. Based on this, a 2D bounding box type proposal was generated to classify the class. The class labels were also mapped to the LiDAR edges for generating the proposal. Here, if a rigid body model and the aspect ratio and actual size information are given for each class, a 3D bounding box can be created based on the orientation of the edge. The origin of the 3D bounding box is determined by two edge points located adjacent to the boundary in the 2D space. Considering p_i^h, p_j^h be the height of the two points facing the boundary, the base point P_{xy}^o and angle of the 3D bounding box θ_i^o are calculated as follows:

$$p_{xy}^o = \begin{cases} p_i & \text{if } p_i^h < p_j^h \\ p_j & \text{if } p_i^h > p_j^h \end{cases}, \tag{4}$$

$$\theta_{xy}^o = \theta_i \tag{5}$$

where P_{xy}^o is the point corresponding to the foreground edge of the two-dimensional boundary line, and θ_i is the average orientation of the edge i including p_{xy}^o, which is the average angle made by the straight line touching the two-dimensional boundary line.

2.2   Supplementing Information Through Object Matching in Multi Frame

Because of the LiDAR sensor characteristics, the information of object pose is incomplete and the edge is not formed or the orientation is inaccurate. To compensate for this imperfection, we propose a method of matching the information of the point cloud of the adjacent frames and accumulating them. This method helps in identifying the same object with a simple optical flow trace and assigns an ID. Figure 3 shows this process.

Update 3D Bounding Box Using Optical Flow. We used a simple method to connect objects between the two frames F^t and F^{t-1}. In the previous process, F^t and F^{t-1} already have label-mapped bounding box of the object $b_i^t \in B^t$. The bounding box of the classified objects in the image and internal feature points

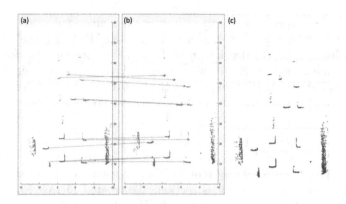

Fig. 3. Process of aligning LiDAR edges of previous frame and current frame in 3D space by using optical flow.

are matched using the optical flow in the two frames F^t and F^{t-1}, and an ID is assigned to the most matched box. To track the points through the optical flow, a feature map is needed to compare the similarity between the two frames. In the proposed system, a convolutional feature map for learning is generated from the R-FCN architecture. This feature map contains many characteristics that can represent a 2D class. The convolutional feature map of the R-FCN can be shared without any separate process as shown in Fig. 1(c). We select 10 feature points with the highest response in the bounding box. The feature points are traced through the optical flow to obtain the results in the adjacent frame. For this reason, optical flow has a simple configuration because all processes except the searching process are omitted. This can be summarized as follows.

$$b_i^t = max(\sum(OF(b_i^{t-1}))) \qquad (6)$$

where b_i^t is the ID of the bounding box at time t. This explores 10 points in each bounding box b_i^{t-1} at time $t-1$ through the optical flow using the Lucas Kanade method [10] at time t. The box with the largest number of points is given the same ID as the bounding box at time $t-1$. If the ID is matched, we accumulate the points by warping s_i^{t-1} contained in the bounding box to the reference coordinates of s_i^t at time t. This process makes the density of the LiDAR point edges uniform. After that, s_i^{t-1}, which is the edge in the bounding box of F^{t-1}, is matched with s_i of F^t, and the average orientation of the merged edges is newly calculated to update the 3D bounding box. Finally, the 3D environment is reconstructed.

3 Experimental Results

3.1 Implementation Details

We implemented our proposed method using the caffe framework with the NVIDIA TITAN X GPU. The Faster R-CNN [11] or R-FCN was originally trained on the PASCAL VOC dataset [4]. In our method, we trained and tested R-FCN on KITTI Dataset to utilize the 3D LiDAR data in the driving environment and test the orientation of the vehicle, which is a rigid body model. The performance is quantitatively evaluated by comparing the results of other studies on the KITTI benchmark with mean Average Precision (mAP) and Average Orientation similarity (AOS).

3.2 KITTI Detection Accuracy

We first trained the R-FCN using the KITTI object data set and tracking sequence, which is a common data set wherein the CCD and LiDAR information are employed. In the training phase, it was trained in four classes: car, pedestrian, cyclist, and background. In the KITTI object dataset, 6,000 of the 7,481 target training sets were used for training and the remaining 1481 were used for the testing. We also used 8,008 training sets for the tracking sequence. As the proposed method does not require additional learning for LiDAR, there is no need to change the structure of the R-FCN to learn the CCD images. Because the valid range of the LiDAR sensor data given in the KITTI data set is approximately 50m, we cannot detect the objects outside the LiDAR sensing range, even though they exist in the CCD image. Table 1 compares the object detection rate with that obtained using the state-of-the-art method by measuring the mean average precision (mAP) of the three classes. The accuracy of the box with the Intersection-of-Union (IOU) threshold of 50% or above with respect to the ground truth area was measured, and the results were good for car and pedestrian. However, in the case of a cyclist, the mAP score is relatively low because a box proposal is not often generated in a sufficient size.

Table 1. Results from the KITTI database of this study are compared with those obtained using state-of-the-art methods. As the effective measurement distance of the LIDAR SENSOR is approximately 50 m, hard difficulty is not tested

	Car		Pedestrian		Cyclist	
Method	E	M	E	M	E	M
Regionlet	84.75	76.45	73.14	61.15	70.41	58.72
3DVP	87.46	75.77				
SubCat	84.14	75.46				
SDP	90.33	83.53	77.74	64.19	**74.08**	**61.31**
Ours	**95.44**	**88.56**	**81.76**	**65.70**	72.09	60.91

Table 1 compares the object detection rate with the state-of-the-art methods by measuring the mean average precision (mAP) of the three classes. The accuracy of the box with the IOU threshold of 50% or above with respect to the ground truth area was measured, and the results were good for car and pedestrian. However, in the case of a cyclist, the LiDAR points are not uniformly distributed depending on the spoke shape of the bicycle wheels, and the mAP score is relatively low because a box proposal is not often generated in a sufficient size.

Table 2 presents the comparison of the accuracy and computation time of this study with those of other proposal generators. All classification networks used R-FCN. Table 3 presents the mAP changes with the variation in the IOU ratio compared to the ground truth data. As listed in Table 3, the overlap ratio is higher than that of the conventional region proposal network (RPN) [11] or selective search (SS) [12]. These methods are used to generate proposals using the internal convolutional feature map of the CNN and conventional image features. This is because the bounding box near the original class is created and fitted using the sub information of each class. In addition, as shown in Table 2, our method is faster than existing methods with use separate proposal generators.

Table 2. Comparison of mAPs in KITTI dataset by proposal-generation method used for R-FCN. For comparing with other studies, learning of R-FCN was conducted at PASCAL VOC 2007 and 2012.

	Training data	Test data	mAP (%)	Test time (sec.)
RPN+Faster R-CNN	07+12	KITTI	75.7	0.37
RPN+R-FCN	07+12	KITTI	77.4	0.20
SS+R-FCN	07+12	KITTI	80.4	2.21
Ours+R-FCN	07+12	KITTI	**82.4**	**0.17**

Table 3. mAP change based on IOU of CAR category. The proposed method shows a higher overlap ratio than the image-based proposal generator (e.g., SS) in the overlap ratio with ground truth.

	Training data	Test data	AP@ 0.5	AP@ 0.7	AP@ 0.9
RPN+R-FCN	07+12	KITTI (car)	84.8	77.4	55.2
SS+R-FCN	07+12	KITTI (car)	86.3	80.4	58.4
Ours+R-FCN	07+12	KITTI (car)	**89.7**	**82.4**	**77.2**

3.3 KITTI Orientation Accuracy

The official 3D measure of the KITTI data set is the average orientation similarity (AOS), which combines average cosine distance similarity with the 2D-object detection performance. In other words, as the AOS is performed after object detection and classification, the upper limit of the AOS is limited to the AP. The Orientation Score (OS) can be calculated by $(1 + \cos(\Delta\theta)/2$, and the error for the angle can be calculated by $(2 \times OS - 1)$. In the KITTI dataset, the error tolerance is 3/6 for easy/moderate case, respectively. Table 4 shows the results. It can be confirmed that the performance of our method is better than the latest approaches in evaluating a car.

Table 4. AOS comparison in car and cyclist classes. The performance of the car class near the straight line of the shape is very good, and the result obtained for the cyclist with the shape change shows a relatively bad result.

Method	Car		Cyclist	
	E	M	E	M
3DOP	91.44	86.10	70.13	58.68
Mono3D	91.01	86.62	65.56	54.97
SubCNN	90.67	88.62	**72.00**	**63.65**
Deep3Dbox	92.90	88.75	69.16	59.87
Ours	**94.22**	**88.95**	70.11	54.34

4 Conclusion

In this study, we have described a simple yet effective way to detect and estimate the pose of a 3D object for autonomous driving. We created LiDAR point clouds in 2D and 3D spaces to generate proposals based on similarities between the 2D corners made up of the LiDAR points. We used the proposal as an input to the R-FCN to classify the objects. Thereafter, we added sub-category information corresponding to the class label and edge orientation to create a 3D-bounding box. We also made the 3D LiDAR information more robust by matching the LiDAR edges of the previous and current frames. As the LiDAR edge is updated, the orientation of the LiDAR edge becomes more sophisticated to estimate the 3D pose. As a result, our research has achieved better results in terms of accuracy and speed compared with the latest research. The 3D bounding boxes helps in restoring the 3D object to near-ground truth at the correct position even though they are only partially represented in the CCD camera view because of the occlusion or truncation. In addition, our method is efficient because the proposals are generated without duplication in the area where the object could exist. Hence, the computation time is significantly reduced. However, an alternative method is needed to find objects outside the sensing range of the LiDAR. Moreover, additional studies should be conducted to discriminate the front and back sides of the 3D bounding box.

Acknowledgment. This research was supported by a grant from Agency for Defense Development, under contract UD150016ID.

References

1. Chen, X., Kundu, K., Zhu, Y., Berneshawi, A.G., Ma, H., Fidler, S., Urtasun, R.: 3d object proposals for accurate object class detection. In: Advances in Neural Information Processing Systems, pp. 424–432 (2015)
2. Cortes, C., Vapnik, V.: Support-vector networks. Mach. Learn. **20**(3), 273–297 (1995)

3. Cross, G.R., Jain, A.K.: Markov random field texture models. IEEE Trans. Pattern Anal. Mach. Intell. **1**, 25–39 (1983)
4. Everingham, M., Van Gool, L., Williams, C.K., Winn, J., Zisserman, A.: The pascal visual object classes (voc) challenge. Int. J. Comput. Vis. **88**(2), 303–338 (2010)
5. Geiger, A., Lenz, P., Urtasun, R.: Are we ready for autonomous driving? the kitti vision benchmark suite. In: 2012 IEEE Conference on Computer Vision and Pattern Recognition (CVPR), pp. 3354–3361. IEEE (2012)
6. He, K., Zhang, X., Ren, S., Sun, J.: Deep residual learning for image recognition. In: Proceedings of IEEE Conference on Computer Vision and Pattern Recognition, pp. 770–778 (2016)
7. Horn, B.K., Schunck, B.G.: Determining optical flow. Artif. Intell. **17**(1–3), 185–203 (1981)
8. Krizhevsky, A., Sutskever, I., Hinton, G.E.: Imagenet classification with deep convolutional neural networks. In: Advances in neural information processing systems, pp. 1097–1105 (2012)
9. Li, Y., He, K., Sun, J., et al.: R-FCN: Object detection via region-based fully convolutional networks. In: Advances in Neural Information Processing Systems, pp. 379–387 (2016)
10. Lucas, B.D., Kanade, T., et al.: An iterative image registration technique with an application to stereo vision (1981)
11. Ren, S., He, K., Girshick, R., Sun, J.: Faster R-CNN: towards real-time object detection with region proposal networks. In: Advances in neural information processing systems, pp. 91–99 (2015)
12. Uijlings, J.R., Van De Sande, K.E., Gevers, T., Smeulders, A.W.: Selective search for object recognition. Int. J. Comput. Vis. **104**(2), 154–171 (2013)
13. Xiang, Y., Choi, W., Lin, Y., Savarese, S.: Data-driven 3d voxel patterns for object category recognition. In: Proceedings of IEEE Conference on Computer Vision and Pattern Recognition, pp. 1903–1911 (2015)
14. Xiang, Y., Choi, W., Lin, Y., Savarese, S.: Subcategory-aware convolutional neural networks for object proposals and detection. arXiv preprint (2016). arXiv:1604.04693
15. Zitnick, C.L., Dollár, P.: Edge boxes: locating object proposals from edges. In: Fleet, D., Pajdla, T., Schiele, B., Tuytelaars, T. (eds.) ECCV 2014. LNCS, vol. 8693, pp. 391–405. Springer, Cham (2014). doi:10.1007/978-3-319-10602-1_26

Colorizing Infrared Images Through a Triplet Conditional DCGAN Architecture

Patricia L. Suárez[1], Angel D. Sappa[1,2(⊠)], and Boris X. Vintimilla[1]

[1] Escuela Superior Politécnica del Litoral, ESPOL, Facultad de
Ingeniería en Electricidad y Computación, CIDIS, Campus
Gustavo Galindo, 09-01-5863 Guayaquil, Ecuador
{plsuarez,asappa,boris.vintimilla}@espol.edu.ec
[2] Computer Vision Center,
Edifici O, Campus UAB, 08193 Bellaterra, Barcelona, Spain

Abstract. This paper focuses on near infrared (NIR) image colorization
by using a Conditional Deep Convolutional Generative Adversarial Network (CDCGAN) architecture model. The proposed architecture is based
on the usage of a conditional probabilistic generative model. Firstly, it
learns to colorize the given input image, by using a triplet model architecture that tackle every channel in an independent way. In the proposed model, the final layer of red channel consider the infrared image
to enhance the details, resulting in a sharp RGB image. Then, in the
second stage, a discriminative model is used to estimate the probability
that the generated image came from the training dataset, rather than the
image automatically generated. Experimental results with a large set of
real images are provided showing the validity of the proposed approach.
Additionally, the proposed approach is compared with a state of the art
approach showing better results.

Keywords: CNN in multispectral imaging · Image colorization

1 Introduction

Image acquisition devices have largely expanded in recent years, mainly due to
the decrease in price of electronics together with the increase in computational
power. This increase in sensor technology has resulted in a large family of images,
able to capture different information (from different spectral bands) or complementary information (2D, 3D, 4D); hence, we can have: HD 2D images; video
sequences at a high frame rate; panoramic 3D images; multispectral images;
just to mention a few. In spite of the large amount of possibilities, when the
information needs to be provided to a final user, the classical RGB representation is preferred. This preference is supported by the fact that human visual
perception system is sensitive to (400–700 nm); hence, representing the information in that range helps user understanding. In this context, the current paper
tackles the near infrared (NIR) image colorization, trying to generate realistic RGB representations. Different applications could take advantage of this

© Springer International Publishing AG 2017
S. Battiato et al. (Eds.): ICIAP 2017, Part I, LNCS 10484, pp. 287–297, 2017.
https://doi.org/10.1007/978-3-319-68560-1_26

contribution—infrared sensors can be incorporated for instance in driving assistance applications by providing realistic colored representations to the driver, while the image processing can be automatically performed by the system in the infrared domain (e.g., semantic segmentation at the material level avoiding classical problems related with the color of the object surface).

The NIR spectral band is the closest in wavelength to the radiation detectable by the human eye; hence, NIR images share several properties with visible images. The interest of using NIR images is related with their capability to segment images according to the object's material. Surface reflection in the NIR spectral band is material dependent, for instance, most pigments used for material colorization are somewhat transparent to NIR. This means that the difference in the NIR intensities is not only due to the particular color of the material, but also to the absorption and reflectance of dyes.

The absorption/reflectance properties mentioned above are used for instance in remote sensing applications for crop stress and weed/pest infestations. NIR images are also widely used in video surveillance applications since it is easier to detect different objects from a given scene. In these two contexts (i.e., remote sensing and video surveillance), it is quite difficult for users to orientate when NIR images are provided, since the lack of color discrimination or wrong color deploy. In this work a neural network based approach for NIR image colorization is proposed. Although the problem shares some particularities with image colorization (e.g., [1,2]) and color correction/transfer (e.g., [3,4]) there are some notable differences. First, in the image colorization domain—gray scale image to RGB—there are some clues, such as the fact that luminance is given by grayscale input, so only the chrominance needs to be estimated. Secondly, in the case of color correction/transfer techniques, in general three channels are given as input to obtain the new representation in the new three dimensional space. In the particular problem tackled in this work (NIR to visible spectrum representation) a single channel is mapped into a three dimensional space, making it a difficult and challenging problem. The manuscript is organized as follows. Related works are presented in Sect. 2. Then, the proposed approach is detailed in Sect. 3. Experimental results with a large set of images are presented in Sect. 4. Finally, conclusions are given in Sect. 5.

2   Related Work

The problem tackled in this paper is related with infrared image colorization, as mentioned before, somehow it shares some common problems with monocromatic image colorization that has been largely studied during last decades. Colorization technique algorithms mostly differ in the ways they obtain and treat the data for modeling the correspondences between grayscale and color. There have been a lot of techniques, like spatial and frequency based variational methods, in which obtain perceptually inspired color and contrast enhancement of digital images, and the color logarithmic image processing (CoLIP) and antagonist space, Gavet et al. [5] design a framework that defines a vectorial space for color images.

It illustrates the representation of the chromaticity diagram with color modification application, namely white balance correction and color transfer. Another technique is the grayscale image matting and colorization, Chen et al. [6] present a variation of a matting algorithm with the introduction of alpha's distribution and gradient into the Bayesian framework and an efficient optimization scheme. It can effectively handle objects with intricate and vision sensitive boundaries, such as hair strands or facial organs, plus they combine this algorithm with the color transferring techniques to obtain his colorization scheme. Welsh et al. [7] describe a semi-automatic technique for colorizing a grayscale image by transferring color from a reference color image. They examine the luminance values in the neighborhood of each pixel in the target image and transfer the color from pixels with matching neighborhoods in the reference image. This technique works well on images where differently colored regions give rise to distinct luminance clusters, or possess distinct textures. In other cases, the user must direct the search for matching pixels by specifying swatches indicating corresponding regions in the two images. It is also difficult to fine-tune the outcome selectively in problematic areas. There are other approaches like colorization by example; in [8] an algorithm that colorizes one or more input grayscale images is presented. It is based on a partially segmented reference color image. By partial segmentation they assume that one or more mutually disjoint regions in the image have been established, and each region has been assigned to a unique label.

The approaches presented above have been implemented using classical image processing techniques. However, recently Convolutional Neural Network (CNN) based approaches are becoming the dominant paradigm in almost every computer vision task. CNNs have shown outstanding results in various and diverse computer vision tasks such as stereo vision [9], image classification [10] or even difficult problems related with cross-spectral domains [11] outperforming conventional hand-made approaches. Hence, we can find some recent image colorization approaches based on deep learning, exploiting to the maximum the capacities of this type of convolutional neural networks. As an example, we can mention the work presented in [2]. The authors propose a fully automatic approach that produces brilliant and sharpen image color. They model the unknown uncertainty of the desaturated colorization levels designing it as a classification task and use class-rebalancing at training time to augment the diversity of colors in the result. On the contrary, [12] presents a technique that combines both global priors and local image features. Based on a CNN a fusion layer merges local information, dependent on small image patches, with global priors computed using the entire image. The model is trained in an end-to-end fashion, so this architecture can process images of any resolution. They leverage an existing large-scale scene classification database to train the model, exploiting the class labels of the dataset to more efficiently and discriminatively learn the global priors. In [13], a recent research on colorization, addressing images from the infrared spectrum, has been presented. It uses convolutional neural networks to perform an automatic integrated colorization from a single channel NIR image to RGB images. The approach is based on a deep multi-scale convolutional neural network to perform a direct estimation of the low RGB frequency values. The main problem with this approach lies on the blur results generated by the multi-scale approach. For that reason it requires a final step that filters the raw output of obtained image from the CNN and transfers the details of the input image to the final output image. Finally, also based on the usage of the CNN framework, [14] proposes a NIR image colorization using a Deep Convolutional Generative Adversarial Network (DCGAN). In that work, a colorization model is obtained based on a flat GAN architecture where all the colors are learned at once from the given input NIR image. This architecture has limitations since all the colors are considered together.

Generative Adversarial Networks (GANs) are a class of neural networks which have gained popularity in recent years. They allow a network to learn to generate data with the same internal structure as other data. GANs are powerful and flexible tools, one of its more common applications is image generation. It is a framework presented on [15] for estimating generative models via an adversarial process, in which simultaneously two models are trained: a generative model G that captures the data distribution, and a discriminative model D that estimates the probability that a sample came from the training data rather than G. The training procedure for G is to maximize the probability of D making a mistake. This framework corresponds to a minimax two-player game. In the space of arbitrary functions G and D, a unique solution exists, with G recovering the training data distribution and D equal to 1/2 everywhere. According to [16], to learn the generator's distribution p_g over data x, the generator builds a mapping function from a prior noise distribution $_pz(z)$ to a data space $G(z; \theta_g)$. And the discriminator, $D(x; \theta_d)$, outputs a single scalar representing the probability that x came from training data rather than p_g. G and D are both trained simultaneously, the parameters for G are adjusted to minimize $log(1 - D(G(z)))$ and for D to minimize $logD(X)$ with a value function $V(G, D)$:

$$\frac{min}{G} \frac{max}{D} V(D, G) = \mathbb{E}_x \sim_{p\ \text{data}(_x)}[logD(x)] + \quad (1)$$
$$\mathbb{E}_z \sim_{p\ \text{data}(_z)}[log(1 - D(G(z)))].$$

Generative adversarial nets can be extended to a conditional model if both the generator and discriminator are conditioned on some extra information y. This information could be any kind of auxiliary information, such as class labels or data from other modalities. We can perform the conditioning by feeding y into both discriminator and generator as additional input layer. The objective function of a two-player minimax game would be as :

$$\frac{min}{G} \frac{max}{D} V(D, G) = \mathbb{E}_x \sim_{p\ \text{data}(_x)}[logD(x|y)] + \quad (2)$$
$$\mathbb{E}_z \sim_{p\ \text{data}(_z)}[log(1 - D(G(z|y)))].$$

In order to improve the efficiency of the generative adversarial networks, [17] proposes some techniques, one of them named the virtual batch normalization;

it allows to significantly improve the network optimization using the statistics of each set of training batches. The main disadvantage is that this process is computationally expensive. Our proposal is based on designing a generative adversarial deep learning architecture that allows the colorization of images of the near infrared spectrum, so that they can be represented in the visible spectrum. The following section will explain in detail the proposed network model.

3   Proposed Approach

This section presents the approach proposed for NIR image colorization. As mentioned above, a recent work on colorization [14] has proposed the usage of a deep convolutional adversarial generative learning network. It is based on a traditional scheme of layers in a deep network. In the current work we also propose the usage of a conditional DCGAN but in a triplet learning layers architecture scheme. These models have been used to solve other types of problems such as learning local characteristics, feature extraction, similarity learning, face recognition, etc. Based on the results that have been obtained on this type of solutions, where improvements in accuracy and performance have been obtained, we propose the usage of a learning model that allows the multiple representation of each of the channels of an image of the visible spectrum (R, G, B). Therefore, the model will receive as input a near infrared patch (NIR), with a Gaussian noise added in each channel of the image patch to generate the necessary variability to generate more diversity of colors, to be able to generalize the learning of the colorization process. A $l1$ regularization term has been added on a single layer in order to prevent the coefficients to fit so perfectly to overfit, which can improve the generalization capability of the model.

A Conditional DCGAN network based architecture is selected due to several reasons: (i) the learning is conditioned on NIR images from the source domain; (ii) its fast convergence capability; (iii) the capacity of the generator model to easily serve as a density model of the training data; and (iv) sampling is simple and efficient. The network is intended to learn to generate new samples from an unknown probability distribution. In our case, the generator network has been modified to use a triplet to represent the learning of each image channel independently; at the output of the generator network, the three resulting image channels are recombined to generate the RGB image. This will be validated by the discriminative network, which will evaluate the probability that the colorized image (RGB), is similar to the real one that is used as ground truth. Additionally, in the generator model, in order to obtain a true color, the DCGAN framework is reformulated for a conditional generative image modeling tuple. In other words, the generative model $G(z; \theta_g)$ is trained from a near infrared image plus Gaussian noise, in order to produce a colored RGB image; additionally, a discriminative model $D(z; \theta_d)$ is trained to assign the correct label to the generated colored image, according to the provided real color image, which is used as a ground truth. Variables (θ_g) and (θ_d) represent the weighting values for the generative and discriminative networks.

The CDCGAN network has been trained using Stochastic AdamOptimazer since it prevents overfitting and leads to convergence faster. Furthermore, it is computationally efficient, has little memory requirements, is invariant to diagonal rescaling of the gradients, and is well suited for problems that are large in terms of data and/or parameters. Our image dataset was normalized in a $(-1,1)$ range and an additive Gaussian Distribution noise with a standard deviation of 0.00011, 0.00012, 0.00013 added to each image channel of the proposed triplet model. The following hyper-parameters were used during the learning process: learning rate 0.0002 for the generator and the discriminator networks respectively; epsilon = 1e-08; exponential decay rate for the 1st moment momentum 0.5 for discriminator and 0.4 for the generator; weight initializer with a standard deviation of 0.00282; $l1$ weight regularizer; weight decay 1e-5; leak relu 0.2 and patch's size of 64×64.

The Triplet architecture of the baseline model is conformed by convolutional, de-convolutional, relu, leak-relu, fully connected and activation function tanh and sigmoid for generator and discriminator networks respectively. Additionally, every layer of the model uses batch normalization for training any type of mapping that consists of multiple composition of affine transformation with element-wise nonlinearity and do not stuck on saturation mode. It is very important to maintain the spatial information in the generator model, there is not pooling and drop-out layers and only the stride of 1 is used to avoid downsize the image shape. To prevent overfitting we have added a l1 regularization term (λ) in the generator model, this regularization has the particularity that the weights matrix end up using only a small subset of their most important inputs and become quite resistant to noise in the inputs; this characteristics is very useful when the network try to learn which features are contributing to the learning process. Park and Kang [18], present a color restoration method that estimates the spectral intensity of the NIR band in each RGB color channel to effectively restores natural colors. According to the spectral sensitivity of conventional cameras with the IR cut-off filter, the contribution of the NIR spectral energy in each RGB color channel is greater in the red channel, hence our architecture add the NIR band at the final red channel layer, this improve the details of generated images, color and hue saturation. Figure 1 presents an illustration of the proposed Triplet GAN architecture.

The generator (G) and discriminator (D) are both feedforward deep neural networks that play a min-max game between one another. The generator takes as an input a near infrared image blurred with a Gaussian noise patch of 64×64 pixels, and transforms it into the form of the data we are interested in imitating, in our case a RGB image. The discriminator takes as an input a set of data, either real image (z) or generated image $(G(z))$, and produces a probability of that data being real $(P(z))$. The discriminator is optimized in order to increase the likelihood of giving a high probability to the real data (the ground truth given image) and a low probability to the fake generated data (wrongly colored NIR image), as introduced in [16]; thus, the conditional discriminator network it is updated as follow:

Conditional Deep Convolutional Generative Adversarial Network Architecture:

(G) Generator Network with Triplet Model

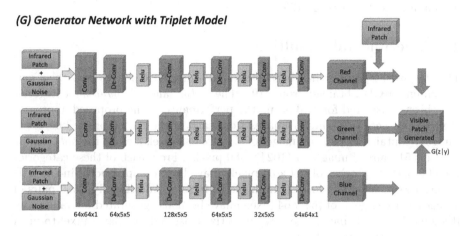

(D) Discriminator Network

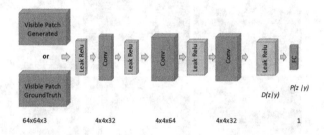

Fig. 1. Illustration of the network architecture used for NIR image colorization.

$$\nabla_{\theta_d} \frac{1}{m} \sum_{i=1}^{m} [\log D(x^{(i)}) + \log(1 - D(G(y^{(i)}, z^{(i)})))], \qquad (3)$$

where m is the number of patches in each batch, x is the ground truth image and y is the colored NIR image generated by the network and z is the randomly Gaussian sampled noise. The weights of the discriminator network (D) are updated by ascending its stochastic gradient. On the other hand, the generator is then optimized in order to increase the probability of the generated data being highly rated, it is updated as follow:

$$\nabla_{\theta_g} \frac{1}{m} \sum_{i=1}^{m} \log(1 - D(G(y^{(i)}, z^{(i)}))), \qquad (4)$$

where m is the number of samples in each batch and y is the colored NIR image generated by the network and z is the randomly Gaussian sampled noise. Like in the previous case, the weights of the generator network (G) are updated by descending its stochastic gradient.

4   Experimental Results

The proposed approach has been evaluated using NIR images and their corresponding RGB obtained from [19]. The *urban* and *old-building* categories have been considered for evaluating the performance of the proposed approach. Figure 2 presents two pairs of images from each of these categories. The *urban* category contains 58 pairs of images of (1024×680 pixels), while the *old-building* contains 51 pairs of images of (1024×680 pixels). From each of these categories 250.000 pairs of patches of (64×64 pixels) have been cropped both, in the NIR images as well as in the corresponding RGB images. Additionally, 2500 pairs of patches, per category, of (64×64 pixels) have been also generated for validation. It should be noted that images are correctly registered, so that a pixel-to-pixel correspondence is guaranteed.

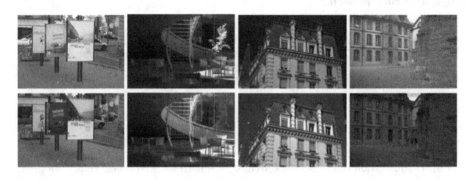

Fig. 2. Pair of images (1024×680 pixels) from [19], *urban* category (the two images in the left side) and *old-building* category (the two images in the right side): (*top*) NIR images to colorize; (*bottom*) RGB images used as ground truth. (Color figure online)

The CDCGAN network proposed in the current work for NIR image colorization has been trained using a 3.2 eight core processor with 16 GB of memory with a NVIDIA GeForce GTX970 GPU. On average every training process took about 28 h. Results from the proposed architecture have been compared with those obtained with the GAN model presented in [14].

Colored images are referred to as (RGB_{NIR}) while the corresponding RGB images, provided in the given data sets, are referred to as (RGB_{GT}) and used as ground truth. The quantitative evaluation consists of measuring at every pixel the angular error (AE) between the obtained result (RGB_{NIR}) and the corresponding ground truth value (RGB_{GT}):

$$Angular\,Error = \cos^{-1}\left(\frac{\text{dot}(RGB_{NIR}, RGB_{GT})}{\text{norm}(RGB_{NIR}) * \text{norm}(RGB_{GT})}\right). \tag{5}$$

This angular error is computed over every single pixel of the whole set of images used for validation. Table 1 presents the average angular errors (AE) obtained with the proposed approach for the two categories together with the results obtained with [14] for the same categories. It can be appreciated that in all the cases the results with the proposed CDCGAN are better that those obtained with [14].

Table 1. Average angular errors obtained with the approach presented in [14] (flat DCGAN) and with the proposed Triplet based CDCGAN architecture.

Category	[14]	Prop. Approach (CDCGAN)
Urban	6.15	5.94
Old-building	6.95	5.71

Qualitative results are presented in Figs. 3 and 4. Figure 3 shows NIR images from the *urban* category colorized with the proposed CDCGAN network and with the approach presented in [14]; ground truth images (last column) are depicted to appreciate the similarity reached with the proposed approach. Similar results have been obtained when images from the *old-building* category are colorized

Fig. 3. (*1st.Col*) NIR patches from the **Urban category**. (*2nd.Col*) Results from the approach presented in [14] (flat DCGAN). (*3rd.Col*) Colorization obtained with the proposed approach (CDCGAN network). (*4th.Col*) Ground truth images. (Color figure online)

Fig. 4. (*1st.Col*) NIR patches from the ***Old-Building category***. (*2nd.Col*) Results from the approach presented in [14] (flat DCGAN). (*3rd.Col*) Colorization obtained with the proposed approach (CDCGAN network). (*4th.Col*) Ground truth images. (Color figure online)

with the proposed CDCGAN network (see Fig. 4). As mentioned above, the usage of a conditional triplet model allows to improve results with respect to the flat model [14]. This improvement can be particularly appreciated in both the color and the edges of the colorized images.

5   Conclusions

This paper tackles the challenging problem of NIR image colorization by using a novel Conditional Generative Adversarial Network model. Results have shown that in most of the cases the network is able to obtain a reliable RGB representation of the given NIR image. Comparisons with a previous approach shows considerable improvements. Future work will be focused on evaluating others network architectures, like auto-encoders, cycle-consistent adversarial networks, which have shown appealing results in recent works. Additionally, increasing the number of images to train, in particular the color variability, will be considered. Finally, the proposed approach will be tested in other image categories.

Acknowledgments. This work has been partially supported: by the ESPOL projects PRAIM and KISHWAR; by the Spanish Government under Project TIN2014-56919-C3-2-R; and by the "CERCA Programme / Generalitat de Catalunya".

References

1. Cheng, Z., Yang, Q., Sheng, B.: Deep colorization. In: IEEE International Conference on Computer Vision (ICCV), pp. 415–423 (2015)
2. Zhang, R., Isola, P., Efros, A.A.: Colorful image colorization. In: Leibe, B., Matas, J., Sebe, N., Welling, M. (eds.) ECCV 2016. LNCS, vol. 9907, pp. 649–666. Springer, Cham (2016). doi:10.1007/978-3-319-46487-9_40
3. Oliveira, M., Sappa, A.D., Santos, V.: Unsupervised local color correction for coarsely registered images. In: Computer Vision and Pattern Recognition (CVPR), pp. 201–208. IEEE (2011)
4. Oliveira, M., Sappa, A.D., Santos, V.: A probabilistic approach for color correction in image mosaicking applications. IEEE Trans. Image Process. **24**, 508–523 (2015)
5. Gavet, Y., Debayle, J., Pinoli, J.-C.: The color logarithmic image processing (CoLIP) antagonist space. In: Celebi, M.E., Lecca, M., Smolka, B. (eds.) Color Image and Video Enhancement, pp. 155–182. Springer, Cham (2015). doi:10.1007/978-3-319-09363-5_6
6. Chen, T., Wang, Y., Schillings, V., Meinel, C.: Grayscale image matting and colorization. In: Asian Conference on Computer Vision (2004)
7. Welsh, T., Ashikhmin, M., Mueller, K.: Transferring color to greyscale images. ACM Trans. Graph. (TOG) **21**, 277–280 (2002)
8. Ironi, R., Cohen-Or, D., Lischinski, D.: Colorization by example. In: Rendering Techniques, pp. 201–210 (2005)
9. Zbontar, J., LeCun, Y.: Stereo matching by training a convolutional neural network to compare image patches. J. Mach. Learn. Res. **17**, 2 (2016)
10. Szegedy, C., Liu, W., Jia, Y., Sermanet, P., Reed, S., Anguelov, D., Erhan, D., Vanhoucke, V., Rabinovich, A.: Going deeper with convolutions. CoRR abs/1409.4842 (2014)
11. Aguilera, C.A., Aguilera, F.J., Sappa, A.D., Aguilera, C., Toledo, R.: Learning cross-spectral similarity measures with deep convolutional neural networks. In: Conference on Computer Vision and Pattern Recognition Workshops
12. Iizuka, S., Simo-Serra, E., Ishikawa, H.: Let there be color!: joint end-to-end learning of global and local image priors for automatic image colorization with simultaneous classification. ACM Trans. Graph. **35**, 110 (2016). Proceedings of SIGGRAPH 2016
13. Limmer, M., Lensch, H.: Infrared colorization using deep convolutional neural networks. arXiv preprint (2016). arXiv:1604.02245
14. Suarez, P.L., Sappa, A.D., Vintimilla, B.X.: Learning to colorize infrared images. In: 15th International Conference on Practical Applications of Agents and Multi-Agent Systems (2017)
15. Goodfellow, I., Pouget-Abadie, J., Mirza, M., Xu, B., Warde-Farley, D., Ozair, S., Courville, A., Bengio, Y.: Generative adversarial nets. In: Advances in Neural Information Processing Systems, pp. 2672–2680 (2014)
16. Mirza, M., Osindero, S.: Conditional generative adversarial nets. arXiv preprint (2014). arXiv:1411.1784
17. Salimans, T., Goodfellow, I., Zaremba, W., Cheung, V., Radford, A., Chen, X.: Improved techniques for training GANs. In: Advances in Neural Information Processing Systems, pp. 2226–2234 (2016)
18. Park, C., Kang, M.G.: Color restoration of RGBN multispectral filter array sensor images based on spectral decomposition. Sensors **16**, 719 (2016)
19. Brown, M., Süsstrunk, S.: Multi-spectral SIFT for scene category recognition. In: Computer Vision and Pattern Recognition (CVPR), pp. 177–184. IEEE (2011)

Complexity and Accuracy of Hand-Crafted Detection Methods Compared to Convolutional Neural Networks

Valeria Tomaselli[1]([✉]), Emanuele Plebani[2], Mauro Strano[1], and Danilo Pau[2]

[1] STMicroelectronics, Catania, Italy
valeria.tomaselli@st.com
[2] STMicroelectronics, Agrate Brianza, Italy
http://www.st.com

Abstract. Even though Convolutional Neural Networks have had the best accuracy in the last few years, they have a price in term of computational complexity and memory footprint, due to a large number of multiply-accumulate operations and model parameters. For embedded systems, this complexity severely limits the opportunities to reduce power consumption, which is dominated by memory read and write operations. Anticipating the oncoming integration into intelligent sensor devices, we compare hand-crafted features for the detection of a limited number of objects against some typical convolutional neural network architectures. Experiments on some state-of-the-art datasets, addressing detection tasks, show that for some problems the increased complexity of neural networks is not reflected by a large increase in accuracy. Moreover, our analysis suggests that for embedded devices hand-crafted features are still competitive in terms of accuracy/complexity trade-offs.

Keywords: Aggregated channel features · Convolutional neural networks · Detection

1 Introduction

The accuracy of object detection algorithms has improved over the years, on one hand thanks to enriched feature representations (multi-channel, multi-resolution, multi-orientation, etc.) and on the other hand due to the adoption of Convolutional neural networks (CNN), at the price of an increased computational cost, especially in the case of neural-based approaches. The complexity and execution time of detection algorithms have a great impact on many visual recognition applications, such as robotics, automotive safety, and human-computer interaction. In these contexts, real-time execution is crucial.

In this work, we perform an analysis and comparison of feature-based versus CNN-based approaches for object detection both in terms of accuracy and execution time. We focus on the automotive use case, where the task consists in the localization and recognition of three main categories: pedestrians, cars and traffic signs. For the hand-crafted approaches we rely on the well performing

© Springer International Publishing AG 2017
S. Battiato et al. (Eds.): ICIAP 2017, Part I, LNCS 10484, pp. 298–308, 2017.
https://doi.org/10.1007/978-3-319-68560-1_27

Aggregated Channel Features (ACF) detector [1], further optimizing it in terms of memory and speed, and we train three different optimized ACF detectors, one for each class. Then, following the approach implemented by Tomé et al. [2], the three detectors generate region proposals for fine-tuned AlexNet [3] networks, incrementally trained to classify pedestrians, cars and traffic signs against the background. Finally, we test an approach entirely based on neural networks, You Only Look Once (YOLO) [4] retrained on the same three categories.

The remainder of the paper is structured as follows: Sect. 2, after a brief excursus on state-of-the-art object detection methods, describes in depth the hand-crafted and CNN-based approaches exploited in this work; Sect. 3 reports the experimental evaluation of the detectors, in terms of accuracy and complexity, on a choice of publicly available datasets; finally, Sect. 4 concludes the paper with some remarks and hints on future work.

2   Object Detection

Given the importance of detecting pedestrians, cars, and traffic signs in automotive, a large number of approaches have been tried over the years. Among the three categories of objects analyzed in this paper, the most important and challenging, because of its large intra-class variability, is "pedestrian". For this reason, most of the efforts in developing new approaches have been focused on it.

2.1   Traditional Approaches

Traditional approaches for object detection usually employ a region proposal algorithm which selects regions from the input image at multiple scales. A high-level feature representation is extracted from the region, which is finally sent to a classifier to establish if that region contains the object or not.

Different region proposal algorithms exist: some of them are class-agnostic and hence they can be quickly adapted to any object detection tasks, but they have the drawback of giving to the subsequent stages an excessive number of negative regions that have to be successively rejected. The problem is lessened by designing a region proposal method tailored on the specific object detection task, to reject most of the negative regions early in the pipeline but preserving as many positive regions as possible.

The region proposal stage is followed by features extraction, which has greatly improved over the years thanks to enriched feature representations. The box-shaped filters, proposed by Viola and Jones [5] in 2003, have been superseded by more complex and powerful features, such as Histogram of Oriented Gradients (HOG) [6]. HOG features in turn have been the starting point of even richer and more complex approaches. For example, Felzenszwalb et al. [7] have improved accuracy by combining HOG with a Deformable Part Model and Dollár et al. [1] have proposed the Aggregated Channel Features (ACF) descriptor, which combines HOG with normalized gradient magnitude and LUV color channels.

A classifier, such as AdaBoost, Support Vector Machines (SVM), etc. finally decides if the object of interest is in the current region.

2.2 Neural Networks

In the last five years, Convolutional Neural Networks have shown their superiority, in terms of accuracy over hand-crafted features, in a variety of computer vision tasks such as image classification, object detection and semantic segmentation.

In the field of object detection, Regions with CNN (R-CNN) [8] has been widely used due to its generality and fairly good performances. Similarly to traditional approaches, a region proposal algorithm (typically Selective Search [9]) extracts candidate regions, on which CNN features are extracted. Either a SVM classifier is trained on the candidate region features to separate between object classes and background, or the CNN can be directly fine-tuned to discriminate the classes of interest from the background. After classification a non-maxima suppression stage is usually applied to refine the selected bounding boxes.

This complex pipeline is quite slow, especially if the number of proposed regions is high. One possible solution is to drastically reduce the number of regions, by e.g. applying a task-specific region proposal method to reject most of the negative examples. Tomé *et al.* [2] analyze different region proposal methods followed by a CNN-based representation, comparing them in terms of accuracy and efficiency for real-time applications and demonstrating that tailored region proposal algorithms (such as Local Decorrelated Channel Features, LDCF [10], or ACF) consistently outperform general purpose approaches (e.g. sliding window or Selective Search) and they achieve much lower miss rates after the CNN stage. Moreover, LDCF and ACF optimizations further speed up the execution of region proposal.

The running time of R-CNN can be reduced by sharing convolutions across proposals, as done in Spatial Pyramid Pooling [11] and Fast R-CNN [12]. To reduce the execution time of the region proposal stage itself, the Faster R-CNN [13] approach introduces the Region Proposal Network (RPN), which share the same convolutional layers of the classification stage. An even more integrated approach is YOLO [4], where the object detection problem is reformulated as a regression problem matching spatially separated bounding boxes and class probabilities to the ground truth. In this way, a single network is optimized end-to-end directly on detection performances.

2.3 Optimized Aggregated Channel Feature Detection

The Aggregated Channel Features (ACF) detector has been optimized in memory by compressing the classifier parameters, which represent a large part of ACF's memory requirements, with a non-linear scalar quantization.

ACF extracts a set of features from non-overlapped blocks on a multi-resolution pyramid constructed from the input image. The ACF classifier is a boosted cascade of small decision trees: each tree is a set of nodes defined as $\{(i, v), s\}$ where the tuple (i, v) represents an intermediate node as a feature lookup index i and a comparison threshold v and leaf nodes are represented by a final score s. The index i assumes values between 0 and whc, where w and h are the width and height of the detection window measured in feature blocks and c is the number of features per block. See Table 1 for a summary of the tested models and their parameters, including the minimum amount of bits required to encode the feature index values.

Table 1. Parameters and size of the ACF models.

Model	Trees	Block size	Scores	Thresholds	Index bits	Size (KB)
INRIA	4	2048	8192	6144	13	65.8
Caltech	2	4096	113252	109156	13	1042.0
Compcars	4	2048	8192	6144	12	65.0
Traffic signs	4	400	3200	2800	10	26.9

A trained tree cascade can have a large number of parameters, ranging from 10 s of KB to a few MBs, but the model size can be reduced by employing *scalar quantization* on the parameters of the trees. Thresholds and scores are quantized separately, because they have different ranges and statistics; for the same reason, different centroids for thresholds of different types of features (color channels, gradient magnitude and HOG) are used. If b is the original element size in bits and N_c is the number of centroids in the scalar quantization, the theoretical compression ratio is:

$$r = \frac{\lceil \log_2 N_c \rceil}{b} \tag{1}$$

However, the real compression ratio will be lower, as the index bits cannot be compressed.

There is a notable relation between the centroids of the tree thresholds and the quantization of the features; more precisely, the *centroids* of the quantized thresholds are the *quantization thresholds* for a feature quantization scheme. To see this, consider that already in the uncompressed case the set of all the thresholds in a tree cascade is partitioning the space of the feature values in a number of discrete intervals. If N distinct thresholds are present in the cascade and they are sorted in ascending order, the intervals are $[-\infty, t_1[, [t_1, t_2[, \ldots [t_N, \infty[$ and they implicitly quantize the features in $N + 1$ levels. However, if the thresholds are quantized e.g. in $2^k - 1 \ll N$ levels, there will be only 2^k distinct intervals, which is equivalent to quantize features with 2^k bits. In this case, both features and threshold will be represented by k bits.

The difference between feature values inside an interval has no impact on the result of the classifier, and thus after threshold quantization, the additional feature quantization has no further penalty on accuracy. Moreover, the comparison may be performed directly in the compressed domain if threshold and feature centroids are coded in ascending order. Let the feature centroid be n_f ($n \in \{0, \ldots, 2^n - 1\}$) and the code of the threshold in the current node be n_t: the left branch in the tree will be selected if $n_f \leq n_t$ and the right branch otherwise. The only operation required is an integer magnitude comparator, which can be efficiently implemented on specialized hardware.

2.4   Convolutional Neural Network Detection

We decided to compare two different CNN-based approaches, a region-based one (R-CNN) and one completely based on neural networks (YOLO), trying to address low complexity target platforms and real time applications.

In R-CNN, we decided to exploit the already trained ACF detectors in the proposal stage to discard most false positives in the early stages of the pipeline. Since learning the parameters of a CNN from scratch requires large annotated datasets, we start from the general-purpose AlexNet neural network, trained on the Imagenet dataset [17] and we fine-tuned it for a few epochs on the target dataset using a small learning rate to adapt the network parameters to the new task. Moreover, we trained three models to incrementally classify pedestrians, pedestrians and cars and finally all the three classes against the background. For training we used the well-known Caffe framework [23].

The CNN has been trained on windows cropped from the images in the dataset; the windows have been generated by the ACF detectors for pedestrians, cars and traffic signs ran with a low classification threshold. The ground truth annotations (described in Sect. 3.1) have been used to assign the windows to the right category, using the "background" class for false positives. By doing so, the CNN classifier learns to reject most of the false positives generated by the region proposal algorithm and it increases accuracy. As done by Tomé et al. in [2], the regions identified by the detectors have been enlarged with padding pixels to mitigate the issue of imprecise localization. In addition to the false positives generated by the ACF detectors, the background category has been populated with random negative regions and the final dataset is further refined by a quick visual inspection.

To assess the performances of a fully CNN-based detection approach we selected YOLO, because its low complexity is well suited for a real-time application. In particular, we decided to exploit a low-complexity version, tiny-YOLO, which is much faster than the original YOLO model but less accurate. This model achieves in classification mode the same top-1 and top-5 accuracy as AlexNet but with 1/10th of the parameters, since it lacks the large fully connected layers at the end. Starting from the pre-trained model on Imagenet, we fine-tune the network on the *Cityscapes* dataset (see Sect. 3.1) using the Darknet framework [24].

3   Experimental Evaluation

3.1   Datasets

We have have exploited different state of the art datasets for object detection and in particular we have trained three ACF detectors on object-specific dataset.

The pedestrian detector has been trained on the *INRIA* dataset and the more challenging *Caltech Pedestrians* dataset. In Caltech, we used the "Reasonable" setting for the train and test sets, which subsamples the original sequences by 30×. The car detector has been trained on the front/rear views in the Comprehensive Car (*CompCars*) dataset. The restriction on the viewpoint is justified by the fact that a single ACF detector does not recognize well both front and lateral views due to their different aspect ratio, and thus an additional detector must be specifically trained on lateral views. Of the original 136,726 images, we selected around 1500 rear views for training and test.

Following Mathias *et al.* [18], the traffic sign detector has been trained on the German Traffic Signs Detection (GTSD) [19] and the Belgian Traffic Signs Detection (BTSD) [20] datasets, two large image datasets captured in different German and Belgian cities and containing a variety of light conditions. We have merged the training and test sets of the two datasets to obtain a more robust detector. We use the three annotated super-classes: *mandatory* (M), *danger* (D) and *prohibitory* (P) and we have disregarded traffic signs which do not belong to those super-classes.

Training and test splits for all datasets are shown in Table 2.

Table 2. Training and test splits for INRIA, Caltech, Compcars, the merged GTSD and BTSD and Cityscapes datasets.

	Training		Testing	
	Pos.	*Neg.*	*Pos.*	*Neg.*
INRIA	614	1218	288	453
Caltech	4250		4024	
Compcars	968		484	
GTSD+BTSD	2915	3594	1804	648
Cityscapes	5000		500	

To evaluate the R-CNN detector, we trained and tested the pedestrian networks separately on the INRIA and Caltech dataset; we trained the car network on the CompCars dataset restricted to rear views and the traffic sign detector on *Cityscapes* [21]. The *Cityscapes* dataset contains urban street scenes exhibiting a high variability, in terms of places (50 cities), weather conditions, seasons and daytime light, hence it is suitable to mimic the behavior of the trained model in real scenarios. Only segmentation annotations are currently available and thus we generated object bounding boxes by extracting the rectangles enclosing the segmentation polygons, which are annotated in Javascript Object Notation (JSON) format.

The *Cityscapes* dataset with generated object annotations has also been exploited to fine-tune the tiny-YOLO CNN, by including only the three chosen classes (pedestrians, cars and traffic signs) and with the addition of the "car-group" and "person-group" categories. We choose this dataset for YOLO as it is the only one to include annotations for all the three classes.

3.2 Metrics for Complexity and Accuracy

To assess computational complexity we compared the analyzed approaches on an NVIDIA Jetson TK1, a development platform equipped with a 192-core NVIDIA Kepler GPU, an NVIDIA quad-core ARM Cortex-A15 CPU and 2 GB of memory. We ported the detectors on this platform and we measured the average time to process all the frames of a reference VGA video containing objects from the three categories; then, we estimated the average frame rate.

More precisely, the ACF detector has been ported on ARM using the NEON Single Instruction, Multiple Data (SIMD) instructions and multi-threading. For the neural network approaches, the Caffe framework [23] has been compiled on the platform with CUDA support and used to test both R-CNN and tiny-YOLO. As the tiny-YOLO model is implemented in the Darknet framework [24], which is not optimized for ARM platforms, we converted the trained model in the Caffe format using our own version of the *Caffe-yolo* project [25] to support a wider range of network architectures.

To assess accuracy, we chose the Log Average Miss Rate (LAMR) evaluation metric proposed by Dollár *et al.* [22]. This metric summarizes detector performance by averaging in the range 10^{-2} to 1 the miss rate at nine points in the False Positives Per Image (FPPI) axis, evenly spaced in log-space. If the curves are approximately linear in this range, the LAMR metric is a smoothed estimate of the miss rate at 10^{-1} FPPI.

3.3 Accuracy Results

Table 3 reports the results in terms of LAMR and frames per second (*fps*) for the ACF, R-CNN and YOLO detectors trained to recognize different objects (pedestrians, cars and traffic signs) on different datasets. Performances are heavily dependent on the specific dataset, as shown e.g. by the fact that the LAMR of the ACF pedestrian detector is lower for the simpler *INRIA* dataset than it is for the more challenging *Caltech* dataset. Moreover, the ACF traffic sign detector is sensitive to the choice of training set: initially, we trained it only on the GTSD dataset, achieving 9.21% on its test set; however, when the same model was tested on the BTSD dataset to assess its generalization capability, performances dropped to 16.52%. This large difference in performances can be ascribed to the big discrepancy between the two datasets. To obtain a more robust detector which can localize traffic signs even in adverse conditions (e.g. back-light), we merged the two training sets, increasing data variability. By doing so, the performances, assessed on the merged test sets, improved back to 9.21% LAMR, now on a much challenging dataset.

As explained in Sect. 2.3, the ACF models have been optimized in memory, by compressing thresholds and scores. Figure 1(a) shows the change in accuracy of ACF models with increasing compression, measured in bits per parameter (both

Table 3. Detection accuracy an speed on different datasets. The frames per second (*fps*) measure is cumulative, that is, for traffic signs is the speed of running a detector recognizing also cars and pedestrians (ACF trained on INRIA dataset).

Object	Dataset	ACF		R-CNN		YOLO	
		LAMR	fps	LAMR	fps	LAMR	fps
Pedestrians	*INRIA*	16.82	11	30.92	5.2	62.02	1.4
	Caltech	29.2	2.6	28.3	1.98	84.7	
Cars	*CompCars*	2.93	10	2.06	4.11	12.55	
Traffic signs	*GTSD+BTSD*	9.21	8.5	10.97	3.99	27.35	

for thresholds and scores). Up to moderate compression levels (4 bits/element) the impact on accuracy is small, and higher compression rates affect mostly models trained on difficult datasets (Caltech). Moreover, for compression rates of 3 bits/element or higher, the relative reduction in size is smaller, as the indexing bits (term i in Sect. 2.3) starts to dominate the total size, as shown by Fig. 1(b) and (c). The optimal compression level is thus 4 bits/element, which allows a model reduction of around 4× with a loss in LAMR of less than 2% over different datasets and object classes.

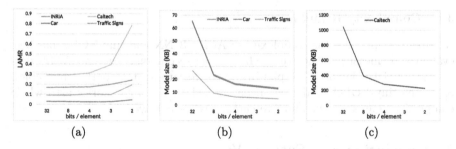

(a)   (b)   (c)

Fig. 1. Results of compression: classification error (LAMR) and model size. In (a) classification error (LAMR) vs bits per element; in (b), model size for the INRIA pedestrian, car and traffic sign models; in (c), model size for the Caltech pedestrian model.

Table 3 also shows LAMR results for R-CNN. R-CNN LAMR is lower than the ACF one both for pedestrians and cars categories, with the notable exception of the INRIA dataset. The annotations in INRIA are incomplete [15] and a large number of pedestrians in the background or partially occluded are not labeled. A quick inspection of the negative examples selected from ACF proposals shows significant overlap with the image of a person in a large fraction of them; as CNN performance is sensitive to label noise, the mislabeled examples end up decreasing the R-CNN accuracy.

For traffic signs the LAMR of the R-CNN approach (10.97%) is slighlty greater than the corresponding ACF detector (9.21%), because the CNN was fine-tuned on the *Cityscapes* and has been evaluated on the merged German and Belgian traffic signs test sets, by running the network on the regions proposed by the ACF traffic signs detector. The network only discards false positives and it cannot increase recall beyond the ACF one.

The tiny-YOLO detector has been fine-tuned on the challenging *Cityscapes* dataset and its accuracy performances have been evaluated on the available test set, obtaining high LAMR values for pedestrians, cars and traffic signs categories (74.92%, 32.66% and 72.48%, respectively). These results mostly depend on the high complexity of the dataset, which contains cars, pedestrians and traffic signs in a variety of views (e.g. front/back/lateral), and the traffic signs category includes many different kinds of street signs, not limited to the mandatory/prohibitory/danger sub-classes. In order to have a fairer comparison between ACF/R-CNN on one side and tiny-YOLO on the other, we have tested the latter on the *INRIA, Caltech, CompCars* and *GTSD+BTSD* datasets, obtaining the results reported in Table 3. Despite having been trained on a completely different dataset, tiny-YOLO shows acceptable performances in detecting cars, but performances drop in detecting traffic signs and especially pedestrians. This is line with the well-known structural limit of YOLO and tiny-YOLO networks in recognizing small objects.

3.4 Complexity Results

As already explained, the complexity of ACF, RCNN and tiny-YOLO has been evaluated by measuring the average frames per second on a reference VGA video on the NVIDIA Jetson TK1 platform. Results are again reported in Table 3. Since the tiny-YOLO model has been trained on the 3 categories as a whole, complexity figures for one and two categories are not available.

4 Conclusion and Future Work

Our analysis on object detection based on convolutional neural networks reveals that, when compared to an approach based on aggregation of hand crafted features followed by a cascade-of-trees classifier, the latter can provide an accurate, low memory and computational light detector in the automotive applications modelized by the adopted datasets.

Not surprisingly, YOLO shows the worst performances in term of LAMR and fps across all datasets and we had to go through multiple iterations to achieve satisfying results, as the performances we obtained initially were worse than the ones reported. Excluding YOLO, R-CNN on pedestrians decremented LAMR by only 3% compared to ACF, while on traffic signs LAMR incremented by 19% and by 84% on pedestrians/Inria. These are considered poor performances from an implementation point of view, since R-CNN frame rate ranges between 41% on cars and 76% on Caltech Pedestrians when compared to the ACF frame rate achieved on the NVIDIA Jetson TK1. This result is further exacerbated by the fact that R-CNN is also exploiting the computational power of the embedded GPU on top of the optimized ACF detector we developed, increasing the costs in power consumption and silicon area required to implement the detector.

Our experiments confirmed the initial hypothesis that the increased complexity of neural networks as implemented on the embedded systems under consideration is not justified by a remarkable increase in accuracy, and in some cases, such as traffic sign detection, neural networks even increased the miss ratio. We are also aware that new neural network accelerators are designed and implemented on non-GPU architectures for future smart sensors in order to overcome the aforementioned issues: these architectures will certainly exploit the massive parallelism of multiply and accumulate operations which dominates CNNs. In this direction, further investigation of low-power and low-precision implementations may be promising, such as the binary neural networks approaches [26] aimed at dramatically reducing memory and complexity costs while maintaining adequate accuracy and robustness to noise.

References

1. Dollár, P., Appel, R., Belongie, S., Perona, P.: Fast feature pyramids for object detection. IEEE Trans. Pattern Anal. Mach. Intell. **36**(8), 1532–1545 (2014)
2. Tomé, D., Monti, F., Baroffio, L., Bondi, L., Tagliasacchi, M., Tubaro, S.: Deep convolutional neural networks for pedestrian detection. Technical report, Politecnico di Milano (2015)
3. Krizhevsky, A., Sutskever, I., Hinton, G.: Imagenet classification with deep convolutional neural networks. In: NIPS, pp. 1–9 (2012)
4. Redmon, J., Divvala, S., Girshick, R., Farhadi, A.: You only look once: unified, real-time object detection. In: Proceedings of the IEEE Conference on Computer Vision and Pattern Recognition, pp. 779–788 (2016)
5. Viola, P., Jones, M., Snow, D.: Detecting pedestrians using patterns of motion and appearance. Int. J. Comput. Vis. **2**, 734–741 (2003)
6. Dalal, N., Triggs, B.: Histograms of oriented gradients for human detection. IEEE Comput. Soc. Conf. Comput. Vis. Pattern Recognit. **1**, 886–893 (2005)
7. Felzenszwalb, P., Girshick, R., McAllester, D., Ramanan, D.: Object detection with discriminatively trained part-based models. IEEE Trans. Pattern Anal. Mach. Intell. **32**(9), 1627–1645 (2010)
8. Girshick, R., Donahue, J., Darrell, T., Malik, J.: Rich feature hierarchies for accurate object detection and semantic segmentation. In: IEEE Conference on Computer Vision and Pattern Recognition (2014)
9. Uijlings, J., van de Sande, K., Gevers, T., Smeulders, A.: Selective search for object recognition. Int. J. Comput. Vis. **104**, 154–171 (2013)
10. Nam, W., Dollár, P., Han, J.H.: Local decorrelation for improved pedestrian detection. In: 28th Annual Conference on Neural Information Processing Systems (2014)
11. He, K., Zhang, X., Ren, S., Sun, J.: Spatial pyramid pooling in deep convolutional networks for visual recognition. arXiv:1406.4729v4 (2014)
12. Girshick, R.: Fast R-CNN. arXiv:1504.08083 (2015)
13. Ren, S., Ross, K.H., Sun, G.J.: Faster R-CNN: towards real-time object detection with region proposal networks. arXiv:1506.01497v2 (2015)

14. Dalal, L., Triggs, B.: Histograms of oriented gradients for human detection. In: IEEE Conference on Computer Vision and Pattern Recognition (2005)
15. Taiana, M., Nascimento, J.C., Bernardino, A.: An improved labelling for the INRIA person data set for pedestrian detection. In: Sanches, J.M., Micó, L., Cardoso, J.S. (eds.) IbPRIA 2013. LNCS, vol. 7887, pp. 286–295. Springer, Heidelberg (2013). doi:10.1007/978-3-642-38628-2_34
16. Yang, L., Luo, P., Loy, C.C., Tang, X.: A large-scale car dataset for fine-grained categorization and verification. In: IEEE Conference on Computer Vision and Pattern Recognition (2015)
17. Deng, J., Dong, W., Socher, R., Li, L.-J., Li, K., Fei-Fei, L.: ImageNet: a large-scale hierarchical image database. IEEE Conference on Computer Vision and Pattern Recognition (2009)
18. Mathias, M., Timofte, R., Benenson, R., Van Gool, L.: Traffic sign recognition - how far are we from the solution? In: International Joint Conference on Neural Networks (IJCNN), Dallas, USA, (2013)
19. Houben, S., Stallkamp, J., Salmen, J., Schlipsing, M., Igel, C.: Detection of traffic signs in real-world images: the German traffic sign detection benchmark. In: International Joint Conference on Neural Networks (2013)
20. Timofte, R., Zimmermann, K., Van Gool, L.: Multi-view traffic sign detection, recognition, and 3D localisation. Mach. Vis. Appl. **25**, 633–647 (2011)
21. Cordts, M., Omran, M., Ramos, S., Rehfeld, T., Enzweiler, M., Benenson, R., Franke, U., Roth, S., Schiele, B.: The cityscapes dataset for semantic urban scene understanding. In: IEEE Conference on Computer Vision and Pattern Recognition (2016)
22. Dollár, P., Wojek, C., Schiele, B., Perona, P.: Pedestrian detection: an evaluation of the state of the art. Pattern Anal. Mach. Intell. **34**(4), 743–761 (2012)
23. Jia, Y., Shelhamer, E., Donahue, J., Karayev, S., Long, J., Girshick, R., Guadarrama, S., Darrell, T.: Caffe: convolutional architecture for fast feature embedding. In: Proceedings of the 22nd ACM International Conference on Multimedia, pp. 675–678 (2014)
24. Redmon, J.: Darknet: Open Source Neural Networks in C (2013–2016). https://pjreddie.com/darknet/
25. Xing, W., Plebani, E.: YOLO (Real-Time Object Detection) in caffe. https://github.com/Banus/caffe-yolo
26. Courbariaux, M., Hubara, I., Soudry, D., El-Yaniv, R., Bengio, Y.: Binarized Neural Networks: Training Deep Neural Networks with Weights and Activations Constrained to +1 or −1 (2016). arXiv preprint arXiv:1602.02830

Emotion Recognition Based on Occluded Facial Expressions

Jadisha Yarif Ramírez Cornejo and Helio Pedrini[✉]

Institute of Computing, University of Campinas, Campinas, SP 13083-852, Brazil
helio@ic.unicamp.br

Abstract. Facial expressions provide important indications about human emotions. The development of an automatic emotion recognition method is a challenging task and has applications in several domains of knowledge, such as behavior prediction, pattern recognition, entertainment, interpersonal relations and human-computer interactions. An automatic approach to emotion recognition based on facial expressions robust to occlusions is proposed and evaluated in this work. Robust Principal Component Analysis is employed to reconstruct the occluded facial expressions. Facial expressions are extracted through different features (Gabor Filters, Local Binary Patterns and Histogram of Oriented Gradients), which are used to recognize the expressions by Support Vector Machine (SVM) and K-Nearest Neighbor (KNN) classifiers. Experiments conducted on three public datasets demonstrate the effectiveness of the proposed methodology.

Keywords: Emotion recognition · Facial expression · Occluded images · Facial features · Pattern recognition · Image classification

1 Introduction

Emotions can be expressed by means of different forms, for instance, body gestures, speech, cardiac rhythm, respiration, and facial expressions [16]. Facial expressions allow humans to express emotions in an effective and natural nonverbal communication.

Automatic recognition of human emotion plays an important role for research on affective computing and has been recently investigated in several applications, such as entertainment, human-computer interactions, behavior prediction, security, among others. The universality hypothesis considers that there are seven basic human facial expressions of emotions (anger, disgust, contempt, fear, happiness, sadness and surprise) expressed through similar facial movements independent on culture, age and gender.

The recognition of facial expressions can be classified into two main categories: sequence-based and frame-based. Frame-based approaches identify facial expressions from a single image, whereas sequence-based recognition employs temporal information over several images [16], such as head movement, skin color variation, facial muscle movement, among other factors.

© Springer International Publishing AG 2017
S. Battiato et al. (Eds.): ICIAP 2017, Part I, LNCS 10484, pp. 309–319, 2017.
https://doi.org/10.1007/978-3-319-68560-1_28

Automatic recognition systems of facial expression commonly involve three major stages: (i) facial detection, (ii) facial expression feature extraction and representation, (iii) and expression recognition [16]. Most of the existing systems do not deal with faces occluded, for instance, by sunglasses, hat, scarf, hands and beard during the training process, which could affect the facial expression recognition accuracy.

As main contribution of this work, a facial expression recognition approach robust to occlusions composed of five main stages is proposed in this work. The first step aims to perform the reconstruction of the facial expression under occlusion based on the Dual Algorithm using Robust Principal Component Analysis (RPCA) principles. The second one involves the automatic detection of facial fiducial points. The third stage extracts three types of features: Gabor Filters, Local Binary Patterns and Histogram of Oriented Gradients. The fourth step performs a dimensionality reduction through Principal Component Analysis (PCA) and Linear Discriminant Analysis (LDA). The latter stage is focused on occluded and non-occluded facial expression recognition, using Support Vector Machine (SVM) and K-Nearest Neighbor (KNN) classifiers. The proposed methodology was evaluated on three facial expression databases: Cohn-Kanade (CK+) [10], Japanese Female Facial Expression (JAFFE) [11] and MUG Facial Expression [1] datasets.

Facial occlusions can deteriorate significantly the performance of a facial expression recognition system. Despite being a challenging problem, our methodology was able to achieve high recognition accuracy rates for both occluded and non-occluded images on the evaluated datasets. The results obtained with our method were compared against other approaches available in the literature.

The remainder of the paper is structured as follows. Section 2 describes relevant work related to the topic under investigation. Section 3 presents the methodology proposed in this work, including details on preprocessing, facial expression reconstruction, facial feature extraction, feature reduction and facial expression classification. Experiments conducted on three public datasets are described and discussed in Sect. 4. Finally, conclusions and directions for future research are presented in Sect. 5.

2   Related Work

Some approaches of the literature have explored the problem of emotion recognition under the presence of partial obstruction (sunglasses, shadows, scarves, facial hair, lights), since occlusion is frequent in real-world scenarios.

Bourel et al. [3] proposed a method for facial expression recognition with occlusions of mouth, upper face and left/right half of the face from video frames, based on a localized representation of facial expression features and on data fusion. For tracking and recovering facial fiducial points, an enhanced Kanade-Lucas tracker was used. Independent local spatio-temporal vectors were created from geometrical relations between facial fiducial points. Local rank-weighted KNN classifiers were employed in the classification step. Bourel et al. [4] also presented a technique for facial expression recognition robust to partial facial occlusions and noisy data from image sequences, based on a state-based feature model of spatially-localized facial dynamics, that consists in a scalar quantization of the temporal evolution of geometric facial features.

Towner and Slater [14] described three techniques based on PCA to recover the positions of the upper and lower facial fiducial points. The results showed that more facial expression information is contained in the lower half of the face, being less accurately the reconstruction of that part of the face. Zhang et al. [15] proposed a method robust to occlusions using a Monte Carlo algorithm to extract a set of Gabor based templates from image datasets, then template matching is applied to find the most similar features located within a space around the extracted templates, generating features robust to occlusion. This approach conducted experiments on the Cohn-Kanade (CK) [7] and the Japanese Female Facial Expression (JAFFE) [11] datasets by considering different occluded facial regions, for instance, eyes, mouth, randomized patches of different sizes, and transparent and solid glasses. Results showed that the method is robust to eyes or mouth occlusions, achieving accuracy rates of 95.1% (eye occlusion) and 90.80% (mouth occlusion) for CK dataset; and 80.30% (eye occlusion) and 78.40% (mouth occlusion) for JAFFE dataset. However, by randomly applying occluded patches over faces in both training and testing phases (matched strategy), this approach obtained 75.00% and 48.80% recognition rates for CK and JAFFE databases, respectively.

There are other techniques that focus on reconstructing texture appearance features. Mao et al. [12] proposed an approach to robust facial expression recognition. Initially, occlusions were detected using RPCA algorithm and saliency detection. Occluded regions were filled by RPCA projection and a reweighted AdaBoost algorithm was used for classification. The method was trained and tested on both the Beihang University Facial Expression (BHUFE) and JAFFE databases, performing experiments with hand, hair and sunglasses occlusions separately, achieving accuracy rates of 59.30%, 84.80% and 68.80% respectively.

Jiang and Jia [6] performed several experiments considering eye and mouth occlusions separately, where occluded facial regions were reconstructed through PCA, Probabilistic PCA, RPCA, Dual and Augmented Lagrange Multiplier (ALM) algorithms. Eigenfaces and Fisherfaces algorithms were then used for feature extraction, whereas KNN and SVM classifiers were employed in the classification stage. The accuracy rates for eye and mouth occlusions were not superior to 76.57% and 72.73%, respectively.

Kotsia et al. [8] presented an analysis of partial occlusion effect on facial expression recognition. It was concluded that occlusions on the left/right side of the face did not affect recognition rates, i.e., that both regions contained less discriminant information for facial expression recognition. Furthermore, mouth occlusion caused a higher decrease in facial expression recognition performance than eye occlusion, because mouth occlusion affected more the emotions of anger, fear, happiness and sadness, whereas eye occlusion affected disgust and surprise. Experiments were conducted on Cohn-Kanade [7] and JAFFE [11] databases, using Gabor wavelets and Discriminant Non-negative Matrix Factorization (DNMF) algorithm for feature extraction and SVM classifier.

Zhang et al. [15] also performed an analysis on the effects of occlusions for both matched and mis-matched train and test strategies. Their method did not learn very well the sample patterns to reduce the effect of randomized patch occlusion, which followed the mis-matched strategy, i.e., using non-occluded images for training and partial occluded images for testing. Thus, recognition rates were worse than following the matched strategy. Furthermore, it was concluded that occluded facial expression recognition performance depends on the occluded region size. It was recommend to use the same type of occlusions during training phase as that expected to be present in tested samples.

Moore and Bowden [13] presented an analysis on the effects of head poses and multi-view on facial expression recognition through variations of Local Binary Patterns (LBP) and Local Gabor Binary Patterns (LGBP) for feature extraction. Experiments conducted on the BU3DFE database showed that frontal view was optimal for facial expression recognition. However, some emotions, such as sadness and anger, performed better at non-frontal views.

3   Methodology

The proposed facial expression recognition methodology with occlusions is described in this section. The main steps of the method are illustrated in Fig. 1 and detailed as follows.

3.1   Preprocessing

The image preprocessing step is fundamental to the expression recognition task, whose main objective is to generate randomized occluded facial expression images with aligned faces, as well as uniform shape and size. This stage is primordial toward the success of facial expression recognition.

Initially, we perform an automatic fiducial point detection over all facial expression image sets with Chehra Face and Eyes Tracking Software [2], which is a fully automatic system that tracks 49 facial landmark points and 10 eye fiducial points. Each facial expression image is aligned according to the left eye and right eye coordinates.

For each image dataset, we scale all images proportionally to the minimum distance between eye coordinates. Facial expression regions are cropped through a proper rectangle. Color images are converted into grayscale images. Finally, randomized black rectangles are applied over different facial expression regions, including bottom left side of the face, bottom right side of the face or bottom side of the face, left eye, right eye and two eyes, to simulate occlusions.

3.2   Facial Expression Reconstruction

The PCA technique is commonly used to reduce high-dimensional feature spaces into more compact descriptors. However, PCA does not operate well under corrupted observations, for instance, variations of facial expressions, occluded faces,

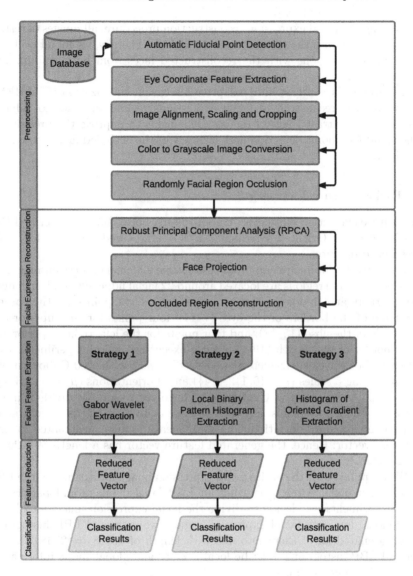

Fig. 1. Diagram of the facial expression recognition methodology.

image noise, illumination problems, among others. On the other hand, RPCA, an extension of the PCA technique, has demonstrated to be robust to outliers and missing data.

We applied the RPCA algorithm using 150 iterations and λ selected as follows

$$\lambda = \frac{1}{\sqrt{\max{(m, n)}}} \tag{1}$$

where m and n are the dimensions of a matrix D.

Following the facial expression reconstruction procedure, all images of testing set are projected onto the space created by RPCA. Thereby, we fill all occluded facial expression regions with the reconstructed faces from both training and testing sets.

Furthermore, a contrast-limit adaptive histogram equalization (CLAHE) is applied over the reconstructed facial expressions regions in order to enhance sharpness and contrast levels of images. This helps to improve the precision of the facial fiducial point detection and accuracy of the occluded facial expression recognition.

3.3   Facial Feature Extraction

Three feature extraction strategies - Gabor Filters, Local Binary Patterns (LBP) and Histogram of Oriented Gradients (HOG) - are used for occluded facial expression recognition.

Gabor wavelet filters are employed to convolve 22 facial expression regions of 15×15 pixels. These regions are located around 22 facial fiducial points: six points for the corners and middle of the eyebrows (1–6); eight points for the corners and middle of the borders of the eyes (7–14); four points for the superior and inferior side of the nose (15–18); and four points for the left, right, superior and inferior border of the mouth (19–22). After executing several experiments with different Gabor wavelet parameters, we select to work with a 20 Gabor wavelet kernel set, using 5 scales ($v = \{0, 1, 2, 3, 4\}$) and 4 orientations ($\mu = \{1, 2, 3, 4\}$), with $\sigma = k_{\max} = \pi$, and $f = \sqrt{2}$. For each convolved region, we divide it into 9 ($= 3 \times 3$) blocks of 5×5 pixels. For each of these equivalent blocks, we extracted the mean and standard deviation. These two measures are concatenated to form the feature vector. Hence, the generated feature vector has a length of 7920 ($= 2 \times 9 \times 20 \times 22$).

LBP is applied over the entire facial expression image for extracting the LBP code for each pixel. After generating an LBP labeled image and performing several experiments, we decide to divide the facial expression image into 63 ($= 7 \times 9$) regions. For each facial expression region, we extracted LBP histograms and concatenated all of them into one with length of 16128 ($= 256 \times 63$). The generated LBP histograms describe local texture and global shape information of the facial expression image.

We extract HOG features using the following parameter set: block size (bs) $= 2 \times 2$, cell size (cs) $= 8 \times 8$, block overlap (bo) $= 1 \times 1$, bin number (bn) $= 9$, and block normalization (bn) $= $ L2. The HOG feature vector encodes local shape information from regions within an image. The length N of the feature vector for an image I is expressed as

$$N = bpi * bs * bn \tag{2}$$

$$bpi = ((size(I)/cs) - bs)/((bs - bo) + 1) \tag{3}$$

where $size(.)$ is the matrix dimension.

3.4 Feature Reduction

Once the feature vector is obtained, it is simplified by applying feature dimensionality reduction techniques. This process modifies the data representation, such that the new set of features presents a smaller number of dimensions compared to the original representation, while maintaining the most representative features.

Two approaches were considered to perform feature reduction: PCA and PCA+LDA. First, PCA is applied to each feature vector set - Gabor, LBP and HOG - independently, obtaining principal (feature) vectors. Similarly, PCA is applied over the combination of feature vectors. Additionally, LDA is employed over the PCA reduced feature vectors, generating new reduced feature spaces.

3.5 Classification

SVM and KNN classifiers were employed to compare the occluded and non-occluded facial expression recognition rates. This process requires stages for training and testing, such that we selected 80% of image data for training and the remaining 20% for testing.

We established estimation models based on SVM and KNN, which are trained from the reduced training feature vectors. Thus, using reduced testing feature vectors, we performed SVM and KNN multiclass classification based on the trained SVM and KNN models. Afterwards, we obtain the recognition results to assess the accuracy.

Along this process, we used different feature combination sets, considering Gabor, LBP and HOG features, whose dimensionality was reduced before the training and testing stages.

4 Results

Experiments were conducted on three datasets to evaluate the proposed methodology: the Cohn-Kanade (CK+) [10] dataset, the Japanese Female Facial Expression (JAFFE) [11] dataset and the MUG Facial Expression dataset [1].

The CK [10] dataset is available in two versions, such that we used the second one (CK+). The difference between these two versions is that the second one contains posed and non-posed (spontaneous) expressions and different metadata types. The CK+ dataset consists of 593 sequences of labeled face images from 123 subjects, categorized into one of seven facial expressions: anger, disgust, contempt, happy, fear, surprise and sadness. Each image sequence incorporates the neutral expression to generate a facial expression. The CK+ is a comprehensive set that also includes some metadata, such as 68 facial fiducial points [10].

The JAFFE dataset is composed of 213 images performed by 10 Japanese female models, labeled as one of seven facial expressions: anger, disgust, fear, happiness, neutral, sadness and surprise [11].

The MUG dataset is an image sequence collection of 86 subjects performing seven facial expressions as the JAFFE database, without occlusions. The MUG database also offers 80 facial landmarks [1].

For each of the three datasets, we randomly select 80% of samples of each class for the training set, whereas the remaining 20% for the testing set. Moreover, 50% of the training set samples of each class are occluded and a similar procedure is performed to the testing set. We set 20 different randomized images of occluded and non-occluded data to conduct experiments on each dataset.

For each occluded and non-occluded image collection, we perform experiments using each strategy shown in Fig. 1, that is, Gabor Filters, Local Binary Patterns (LBP) and Histogram of Oriented Gradients (HOG) through four proposed classification schemes: PCA+KNN, PCA+LDA+KNN, PCA+SVM and PCA+LDA+SVM. The results are presented in Tables 1 and 2, whose values correspond to the average facial expression recognition accuracy, after executing 20 experiments on both randomized training and testing collections.

Table 1. Average accuracy (in percentage) for non-occluded images using Gabor filters, LBP and HOG for each dataset.

Recognition method	CK+			JAFFE			MUG		
	Gabor	LBP	HOG	Gabor	LBP	HOG	Gabor	LBP	HOG
PCA+KNN	59.71	43.74	**64.63**	**86.20**	64.41	83.10	79.06	79.69	**80.89**
PCA+LDA+KNN	**92.76**	92.62	90.45	95.36	93.00	**96.43**	**91.84**	91.40	91.02
PCA+SVM	**86.12**	77.17	83.36	**93.21**	84.18	92.74	**85.95**	85.70	85.26
PCA+LDA+SVM	**94.03**	92.84	91.20	95.12	92.50	**95.60**	**91.33**	90.07	89.12

Table 2. Average accuracy (in percentage) for occluded images using Gabor wavelets, LBP and HOG for each dataset.

Recognition method	CK+			JAFFE			MUG		
	Gabor	LBP	HOG	Gabor	LBP	HOG	Gabor	LBP	HOG
PCA+KNN	50.17	42.06	**61.57**	48.58	45.84	**60.01**	55.76	58.93	**63.04**
PCA+LDA+KNN	84.63	88.06	**88.29**	82.51	83.10	**89.05**	81.21	**85.39**	85.15
PCA+SVM	76.87	75.01	**80.38**	73.48	70.60	**79.30**	67.76	**77.98**	77.66
PCA+LDA+SVM	85.68	88.44	**88.74**	82.86	81.44	**88.46**	81.02	84.00	**84.18**

From our experiments with occlusions, it is important to state that RPCA was always applied to facial reconstruction independently of the evaluated feature reduction and classification methods. From Table 1, we can observe that the recognition accuracy rate for non-occluded images using Gabor wavelets is generally slightly better than other features, except for the JAFFE database, where HOG is slightly superior. On the other hand, we can see from Table 2 that a recognition rate of occluded collections using LBP and HOG features, independently, is much better than using Gabor filters.

From the experiments on both occluded and non-occluded collections, we can observe that the PCA+LDA approach achieves higher accuracy rate than just using PCA. Furthermore, it is possible to see that, in some cases, there is no significant difference between the non-occluded and occluded facial expression recognition rates using LBP and HOG features. However, when using Gabor filters, there is a difference of approximately 10% between the recognition rates of non-occluded and occluded collections.

High non-occluded facial expression recognition accuracy rates using Gabor filters were achieved due to an accurate fiducial point detection. On the other hand, the results achieved through HOG features for non-occluded and occluded collections were competitive since there was not much image background suppression. This allowed to encode local information, such as shape. Moreover, facial reconstruction for occluded sets influenced positively in the occluded accuracy rate.

It is also important to mention that the results achieved with LBP features are due to the use of PCA approach that allows to select the most relevant features, instead of assigning different weights to each LBP sub-region. We also conducted experiments by combining Gabor, LBP and HOG features, however,

Table 3. Accuracy rates (in percentage) for non-occluded images and for comparable methods that work with random partial occlusions of the faces in both training and testing phases.

Dataset	Approach	Strategy	Non-occlusion	Occlusion
CK+	Proposed method	HOG+PCA+LDA+SVM	91.20	88.74
	Proposed method	LBP+PCA+LDA+SVM	92.62	88.44
	Proposed method	Gabor+PCA+LDA+SVM	94.03	85.68
	Liu et al. [9]	Maximum likelihood estimation sparse representation	94.29	85.24
	Cornejo et al. [5]	Local gradient coding of horizontal and diagonal gradient priority	87.17	80.18
JAFFE	Proposed method	HOG+PCA+LDA+KNN	96.43	89.05
	Cornejo et al. [5]	Local gradient coding of horizontal and diagonal gradient priority	88.34	88.57
	Liu et al. [9]	Maximum likelihood estimation sparse representation	93.42	86.73
	Proposed method	LBP+PCA+LDA+KNN	93.00	83.10
	Proposed method	Gabor+PCA+LDA+SVM	95.12	82.86
	Zhang et al. [15]	Gabor template and SVM	81.20	48.80
MUG	Proposed method	LBP+PCA+LDA+KNN	91.40	85.39
	Proposed method	HOG+PCA+LDA+KNN	91.02	85.15
	Proposed method	Gabor+PCA+LDA+KNN	91.84	81.21

the obtained results did not contribute to a significant improvement in terms of recognition accuracy rate.

We compared our method to others available in the literature that apply random partial occlusions to faces in both training and testing phases. Table 3 shows a comparision of the results. There are only few similar works that consider occlusions in the training stage. It can be observed that our method achieves the best results for CK+ and JAFFE datasets, not only for occluded images, but also for non-occluded images.

Table 3 is sorted in descending order by accuracy rate for recognition with occlusion. Some approaches adopt different protocols on the same data and employ specific preprocessing stages to the data, such as alignment or cropping of the images, feature normalization and illumination adjustments. Besides being used to reconstruct occluded facial expressions, it is possible to observe that our method achieves good results for non-occluded images.

We also conducted experiments with the combination of LBP, HOG and Gabor descriptors. For CK+ dataset, the accuracy rate had an improvement in terms of recognition accuracy rate to 90.00% for the occluded images with PCA+LDA+KNN. For JAFFE and MUG datasets, the combination of the descriptors produced results equivalent to the application of each descriptor individually.

5   Conclusions and Future Work

This work described and evaluated an emotion recognition method using facial expressions robust to occlusions. Facial reconstruction was performed by Robust Principal Component Analysis (RPCA). Different features were applied over the reconstructed facial expression images and the resulting feature vector was reduced through a number of techniques, allowing high accuracy rates of facial expression recognition. Experiments were conducted on three public datasets to evaluate the effectiveness of the proposed methodology.

As directions for future work, we intend to investigate new facial fiducial point sets, the use of different features, as well as better facial reconstruction parameters. Additionally, we plan to conduct experiments using dynamic features for facial expression recognition in video scenes.

Acknowledgements. The authors are thankful to FAPESP (grants #2014/12236-1 and #2016/19947-6) and CNPq (grant #305169/2015-7) for their financial support.

References

1. Aifanti, N., Papachristou, C., Delopoulos, A.: The MUG facial expression database. In: 11th International Workshop on Image and Audio Analysis for Multimedia Interactive Services, pp. 1–4, Desenzano del Garda, Italy, April 2010
2. Asthana, A., Zafeiriou, S., Cheng, S., Pantic, M.: Incremental face alignment in the wild. In: IEEE International Conference on Computer Vision and Pattern Recognition, pp. 1859–1866, June 2014

3. Bourel, F., Chibelushi, C.C., Low, A.A.: Recognition of facial expressions in the presence of occlusion. In: British Machine Vision Conference, pp. 1–10 (2001)
4. Bourel, F., Chibelushi, C.C., Low, A.A.: Robust facial expression recognition using a state-based model of spatially-localized facial dynamics. In: 5th IEEE International Conference on Automatic Face and Gesture Recognition, pp. 106–111 (2002)
5. Cornejo, J.Y.R., Pedrini, H.: Recognition of occluded facial expressions based on centrist features. In: 2016 IEEE International Conference on Acoustics, Speech and Signal Processing (ICASSP), pp. 1298–1302. IEEE (2016)
6. Jiang, B., Jia, K.: Research of robust facial expression recognition under facial occlusion condition. In: Zhong, N., Callaghan, V., Ghorbani, A.A., Hu, B. (eds.) AMT 2011. LNCS, vol. 6890, pp. 92–100. Springer, Heidelberg (2011). doi:10.1007/978-3-642-23620-4_13
7. Kanade, T., Cohn, J., Tian, Y.: Comprehensive database for facial expression analysis. In: Fourth IEEE International Conference on Automatic Face and Gesture Recognition, pp. 46–53, Grenoble, France (2000)
8. Kotsia, I., Buciu, I., Pitas, I.: An analysis of facial expression recognition under partial facial image occlusion. Image Vis. Comput. 26(7), 1052–1067 (2008)
9. Liu, S., Zhang, Y., Liu, K.: Facial expression recognition under random block occlusion based on maximum likelihood estimation sparse representation. In: International Joint Conference on Neural Networks, pp. 1285–1290, July 2014
10. Lucey, P., Cohn, J., Kanade, T., Saragih, J., Ambadar, Z., Matthews, I.: The extended cohn-kande dataset (CK+): a complete facial expression dataset for action unit and emotion-specified expression. In: Third IEEE Workshop on Computer Vision and Pattern Recognition for Human Communicative Behavior Analysis, San Francisco, CA, USA, June 2010
11. Lyons, M., Kamachi, M., Gyoba, J.: Japanese female facial expressions (JAFFE). Database of Digital Images (1997). http://www.kasrl.org/jaffe.html
12. Mao, X., Xue, Y., Li, Z., Huang, K., Lv, S.: Robust facial expression recognition based on RPCA and AdaBoost. In: 10th Workshop on Image Analysis for Multimedia Interactive Services, pp. 113–116, May 2009
13. Moore, S., Bowden, R.: Local binary patterns for multi-view facial expression recognition. Comput. Vis. Image Underst. 115(4), 541–558 (2011)
14. Towner, H., Slater, M.: Reconstruction and recognition of occluded facial expressions using PCA. In: Paiva, A.C.R., Prada, R., Picard, R.W. (eds.) ACII 2007. LNCS, vol. 4738, pp. 36–47. Springer, Heidelberg (2007). doi:10.1007/978-3-540-74889-2_4
15. Zhang, L., Tjondronegoro, D., Chandran, V.: Random Gabor based templates for facial expression recognition in images with facial occlusion. Neurocomputing 145, 451–464 (2014)
16. Zhang, S., Zhao, X., Lei, B.: Facial expression recognition using sparse representation. WSEAS Trans. Syst. Control 11(8), 440–441 (2012)

Exploiting Context Information for Image Description

Andrea Apicella, Anna Corazza, Francesco Isgrò[⊠], and Giuseppe Vettigli

Università di Napoli Federico II, Napoli, Italy
francesco.isgro@unina.it

Abstract. Integrating ontological knowledge is a promising research direction to improve automatic image description. In particular, when probabilistic ontologies are available, the corresponding probabilities could be combined with the probabilities produced by a multi-class classifier applied to different parts in an image. This combination not only provides the relations existing between the different segments, but can also improve the classification accuracy. In fact, the context often gives cues suggesting the correct class of the segment. This paper discusses a possible implementation of this integration, and the first experimental results shows its effectiveness when the classifier accuracy is relatively low. For the assessment of the performance we constructed a simulated classifier which allows the a priori decision of its performance with a sufficient precision.

1 Introduction

This paper tackles the problem of recognising the content of a digital image, and being able to produce a schematic textual description. Because of the large number of images available on-line, this is a very hot research topic at the moment, as shown by the references in Sect. 2, and well performing systems using deep learning producing description in natural language have been proposed. In this work we start considering a new way to exploit context information to improve performance of classification based approaches.

When the aim is to design and implement a framework for the recognition of some of the components of a natural image, simply applying classification is not a solution as natural images classifiers only based on information extracted from the images, can be, in the most general case, error prone. The framework presented in this work aims at integrating the output of standard classifiers on different image parts with some domain knowledge, encoded in a probabilistic ontology. In fact, while standard ontologies are quite widespread as a means to manage a-priori information, they fail in the important task of dealing with real world uncertainty. Probabilistic ontologies aim at filling this gap by associating probabilities to the coded information, and provide then an adequate solution to the issue of coding the context information necessary to correctly understand the content of an image. Such information is then combined with the classifier output in order to correct possible classification errors on the basis of surrounding objects.

© Springer International Publishing AG 2017
S. Battiato et al. (Eds.): ICIAP 2017, Part I, LNCS 10484, pp. 320–331, 2017.
https://doi.org/10.1007/978-3-319-68560-1_29

In conclusion, our aim is to improve the performance of a natural images classifier introducing in the loop knowledge coming from the real world, expressed in terms of probability of a set of spatial relations between the objects in the images. Not only the probabilistic ontology can be made available for the considered domain: it could also be built or enriched by using entities and relations extracted from a document related to the image. For example, the picture could have been extracted from a technical report or a book, where the text gives information which are related to the considered images. We wish to stress the fact that we are not thinking of a text directly commenting or describing the image, but of a text which is completed and illustrated by the image. In this case, both the classes of objects which can appear in the image and the relations connecting them could be mentioned in the text and could therefore be automatically extracted [2]. A probability can then be associated to them on the basis of the reliability of the extraction or the frequency of the item in the text.

The system we are considering, the logical scheme of which is depicted in Fig. 1, and better detailed in Sect. 3, aims at determining a set of keywords describing the content of an image and the relations existing among them. The idea is to design a system that, starting from an image, will first hypothesize the presence of some objects in the scene through a battery of image based classifiers. Considering for example the image of a building close to a water poll with some boats, it is likely that a classifier might label the reflection of the building on the water beneath the boats as a building, that is a wrong classification. We advocate that such a mis-classification can be corrected introducing the spatial relation between the boat-segment and reflected building, and the external knowledge that an image segment beneath a boat and surrounded by water is more likely to be water than a building. This world knowledge, that we plan to formalise in a probabilistic ontology [9], together with the output of the classifier, will be fed to a probabilistic model [4], in order to improve the performance of the single classifiers.

The classes associated to each segment combined by the spatial relations which can be directly extracted from an analysis of the image are eventually organized in a schematic description of its content. Relations could be further specialized by better specifing the reciprocal position of the segments. For example, the fact that a segment is in the middle, or in the upper right part of the picture, and so on.

The framework presents two main aspects of novelty. First, the use of a probabilistic ontology for a computer vision problem has, at the best of our knowledge, never been proposed before. A second element of novelty is the integration of a probabilistic model with a probabilistic ontology. A preliminary description of the general idea of the approach has been sketched in [1] in a very concise way. Here, we discuss all details and a first preliminary experimental assessment.

In the following section, we discuss related work. Section 3 is devoted to the description of the different modules of the system, with a few details about the probabilistic ontology (Sect. 3.1), and to the model adopted to combine classification and ontology probabilities (Sect. 3.2). Experimental assessment is considered in Sect. 4. Some conclusions and proposals for extensions of the presented work conclude the paper.

2   Related Work

Human beings express their knowledge and communicate using natural language, and in fact they find usually easy to describe the content of images with simple and concise sentences. Because of this human skill it is not difficult for a human user, when using an image search engine, to formulate a query by means of natural language.

Due to the large amount of images available on the web, for answering to textual image queries, it will be very helpful being able to automatically describe the content of an image. However such a task is not easy at all for a machine, as it requires a visual understanding of the scene, that is almost each object in the image must be recognised, how the objects relate to each other in the scene, and in what they are involved must be understood [27]. This task is tackled in two different ways. The most classical one [10,12,13,17] tries to solve the single sub-problems separately and combines the solutions to obtain a description of an image. A different approach [6,15,27] proposes a framework that incorporates all the sub-problems in a single joint model. A method trying to merge the two main approaches has been proposed recently in [30] using a semantic attention model. The problem is, however, very far from being solved.

In the context of textual image queries, it can be enough to extract from the images a less complex description (image annotation [28]), such as a list of entities represented in the image, and information about their position and mutual spatial relation in the image. The work proposed in this document addresses this task, that is also, as mentioned above, a necessary sub-task of the more general problem of generating a description in natural language.

The use of ontologies in the context of image recognition is not new [25]. For instance, in [20] it is proposed a framework for an ontology based image retrieval for natural images, where a domain ontology was developed to model qualitative semantic image descriptions. An ontology of spatial relations, in order to guide image interpretation and the recognition of the structures it contains was proposed in [14]. In [18], low-level features describing the color, position, size and shape of segmented regions are extracted and automatically mapped to descriptors forming a simple vocabulary termed object ontology. At the best of our knowledge, a probabilistic ontology has never been used for the task of image recognition and annotation.

Contextual information have been used in image recognition for long time [19,26], and it has been already shown [3] that the use of spatial relations can decrease the response time and error rate, and that the presence of objects that have a unique interpretation improves the identification of ambiguous objects in the scene. Just to mention a few application domains, contextual information has been used for face recognition [24], medical image analysis [5], analysis of group activity [7].

In the same way the use of probabilistic models is not new in computer vision, in particular a probabilistic model combining the statistics of local appearance and position of objects was proposed already in [22] for the task of face recognition, and in [21] in an image retrieval task, showing that adding a probabilistic model in the loop would improve the recognition rate. In [32] it is proposed a probabilistic semantic model in which the visual features and the textual words are connected via a hidden layer. More recently in the context of 3D object recognition, a system that builds a probabilistic model for each object based on the distribution of its views was proposed in [29]. In [31] a weakly supervised segmentation model learning the semantic associations between sets of spatially neighbouring pixels, that is the probability of these sets to share the same semantic label. Finally [11], in the context of action recognition, presents a generative model that allows for characterising joint distributions of regions of interest, local image features, and human actions.

3   System Architecture

The proposed framework, depicted in Fig. 1, is a chain of several logical modules, each corresponding to an element of a computational pipeline. The first step is a classifier, or a set of classifiers, detecting a predefined set of interesting objects in the image, identifying then a set of segments of interest in the image.

The hypotheses formulated for each segment in the image by a statistical classifier are then fed to a probabilistic model, that has been trained off-line. The task of this module is to validate, or correct, the hypothesis formulated in the previous step, integrating the output of the classifier with the world knowledge given by a probabilistic ontology, and expressed in terms of probability of a spatial relationship between instances of two classes of image objects. The class associated to each segment, together with the relations existing between segment pairs, constitute the image description output by the system.

3.1   Probabilistic Ontology

This section discusses the construction of a fragment of Probabilistic Ontology (PO) providing the information needed by our system. We need such fragment for the experimental assessment.

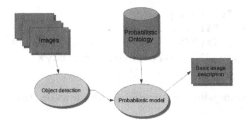

Fig. 1. Scheme of the proposed framework.

Table 1. Data set statistics.

Class	# of items
Sink	371
Chair	3,604
Table	558
Computer/monitor	256+417=673
Bed	407
Flower	1,822
Total	7,435

The main drawback of ontologies when facing real world problems is related to their inability to cope with uncertain information. Due to this, in the last years much work has been devoted to the design of effective tools to attach probabilities to the information contained in ontologies, among whose the most important is probably PrOWL [8]. From the so obtained POs, it is therefore possible to obtain a priori knowledge for applications effective also in complex contexts.

As a consequence, the research area concerning POs is very active and we expect that a number of POs in different domains will be available soon. However, we need a PO in the domain of the image data set we will adopt to assess the system performance, before we can start experimentation. We therefore design and implement an ontology to use in the experiments. In particular, the schema of the ontology will contain the classes to associate to the segments and the spatial relations among them considered in our analysis. On the other hand, probabilities are estimated from the training set after segments are automatically classified and spatial relations are constructed between segment pairs. In particular, we estimate the probability that two classes are in a given relation by the frequency of such event in the data set. No smoothing have been applied. More precisely denoting with D a set of segments used to compute the probabilities, with $R = \{r_1, \ldots, r_i\}$ the set of types of relation, with C the set of segments classes, we compute the probability that $c_1 \in C$ is in relation $r \in R$ with $c_2 \in C$ as:

$$\Pr(r, c_1, c_2) = \frac{D_r(c_1, c_2)}{\sum\limits_{c_x \in C, c_y \in C} D_r(c_x, c_y)} \tag{1}$$

where $D_r(c_x, c_y)$ is the number of times that pairs of segments in D of classes respectively c_1 and c_2 satisfy the relation r. In general, as the relations are not necessarily symmetric, we have $\Pr(r, c_1, c_2) \neq \Pr(r, c_2, c_1)$.

Since there are no tools for directly constructing a PO, we use Protégé [1] for the construction of the schema of the ontology, while we use Pronto [16] as a reasoner for POs, as it adopts the standard OWL 1.1. The import of the schema developed by Protégé into Pronto is performed by editing the corresponding XML files and adding the probabilities. An example is given in Fig. 2, where the element tagged `pronto:certainty` is added to the axiom prepared by Protégé.

```
<owl11:Axiom>
    <rdf:subject rdf:resource="URI#x"/>
    <rdf:predicate rdf:resource="&rdfs;subClassOf"/>
    <rdf:object rdf:resource="URI#y"/>
    <pronto:certainty>0.070990;0.070990</pronto:certainty>
</owl11:Axiom>
```

Fig. 2. Piece of the XML of the PO corresponding to an axiom with an associated probability.

---

[1] Freely available from http://protege.stanford.edu/.

Although Pronto accepts probability ranges, as we use simple values, the two extremes of the interval coincides $(0.070990; 0.070990$ in the example).

3.2 Combination Models

This section investigates which model to use to integrate the classifiers and the ontological knowledge.

In the task we are considering the role of POs requires providing probabilities describing the domain of interest, to be integrated with the ones associated by the classifier to each class for each input segment. The main goal of our system is the classification of the segments in the input image. We aim to exploit the relations between pairs of segments to improve this classification. More formally, every image contains a set of segments S and there are a number of possible relations R connecting segment pairs.

For each segment in the image, the classifier associates a probability distribution to the set of all possible classes C. When we consider only the classification step, we classify the segment with the most probable class: this represents our baseline, as it only considers the classifier output, without any information coming from the PO. However, we can see the output of the classifier for each segment s in the image as a random variable $c(s)$ with values in C. In the following we discuss how such random variable is integrated with the ontological probabilities.

In fact, the ontology produces, for every pair of classes $c_1, c_2 \in C$ and every possible relation $r \in R$, the probability $\Pr(r, c_1, c_2)$ that in the real world two segments of classes c_1 and c_2 respectively are in relation r: its expression is given in Eq. 1. By integrating this information with the probabilities computed by the classifier, the classification performance could improve. Moreover, the solution output by this integration is likely to be consistent with the ontological knowledge, which can be an important feature in systems where the post-processing requires a set of properties on the considered candidates. In fact, whenever a relation can not hold between two classes, the corresponding ontological probability is null, and this also lowers the probability of the corresponding couple of classes.

We associate the following log-linear probability to the two classes associated to each context $x = (s_1, s_2, r : r(s_1, s_2))$ built around the relation type r connecting segments s_1 and s_2:

$$\Pr(c_1, c_2 | x) = \frac{e^{v_{c_1} f_C(s_1, c_1) + v_{c_2} f_C(s_2, c_2) + v_{r, c_1, c_2} f_{PO}(r(s_1, s_2), c_1, c_2)}}{Z_{x, c_1, c_2}} \quad (2)$$

where $f_C(s, c) = \Pr(c(s) = c)$ and $f_{PO}(r, c_1, c_2) = \Pr(r(c_1, c_2))$, while Z_{x, c_1, c_2} is a normalisation factor depending on x and on the classes assigned to the two segments. Note that the features $f_C(\cdot)$ are produced by the classifier, while $f_{PO}(\cdot)$ depends on the probabilistic ontology. In conclusion, we consider two families of parameters: *class parameters* v_c for each class c and *relation parameters* v_{r, c_1, c_2} for each type of relation r and pair of classes (c_1, c_2). All in all, there are $|C|$ class parameters and $|R||C|^2$ *relation parameters*.

The parameters are estimated during the training, which maximises the like-lihood of the training set. For this optimisation, we use the *Toolkit for Advanced Optimisation* (TAO) library, which implements a variety of optimisation algo-rithms for several classes of problems (unconstrained, bound-constrained, and PDE-constrained minimisation, nonlinear least-squares, and complementarity). In our work we focus on unconstrained minimisation methods which are used to minimise a function of many variables without any constraints on the vari-ables. The method that we have used is *Limited Memory Variable Metric*, it is a *quasi-Newton* optimisation solver and it solves the Newton step using an approximation factor which is composed using the *BFGS* update formula.

Once we have estimated all the parameters $V = \{v_c, v_{r,c_i,c_j}\}$ with $c, c_i, c_j \in C$ and $r \in R$, we aim to assign the correct class to each segment in the input image. To do so, we consider two different models: in the former, to which we refer as M1, we assign to the classes in a given context a score which is equal to the $\Pr(c_1, c_2|x)$ as given by Eq. 2, while in the latter, M2, the score is given by its logarithm. In fact, when adopting, as in our case, a log-linear expression, only considering exponents is much more efficient than directly summing probabil-ities. We therefore obtain the following expressions for the scores sc_1 and sc_2 respectively corresponding to $M1$ and $M2$.

$$sc_1(c_1, c_2|x) = \Pr(c_1, c_2|x) = \frac{e^{v_{c_1} f_C(s_1,c_1) + v_{c_2} f_C(s_2,c_2) + v_{r,c_1,c_2} f_{PO}(r(s_1,s_2),c_1,c_2)}}{Z_{x,c_1,c_2}}$$

$$sc_2(c_1, c_2|x) = \log \Pr(c_1, c_2|x) = v_{c_1} f_C(s_1, c_1) + v_{c_2} f_C(s_2, c_2)$$
$$+ v_{r,c_1,c_2} f_{PO}(r(s_1,s_2), c_1, c_2) - \log Z_{x,c_1,c_2} \quad (3)$$

For each context x, we then compute the score that a given class c is associated to one segment, by summing the scores that every class is associated to each segment and that the relation assumes any of all possible relation types. We then associate to the first segment the class which maximises such a score in all segment pairs including it:

$$SC(c|s) = \max_{s_2: \exists r, r(s_1,s_2)} \sum_{c_2 \in C} \sum_{r \in R} sc(c, c_2|(s_1, s_2, r : r(s_1, s_2)). \quad (4)$$

In this expression, sc stays for sc_1 or sc_2 depending on the adopted model. Note that since all relation types we consider are symmetrical, for every context $x = (s_1, s_2, r : r(s_1, s_2))$ also the symmetrical one $x' = (s_2, s_1, r(s_2, s_1))$ is defined, and therefore we can express the score as considering only the first of the two cases. However, when asymmetrical relations are also considered, the expressions can be easily generalised.

Finally, we assign to each segment the class which maximises the score of the class given the segment:

$$c^*(s) = \arg \max_{c \in C} SC(c|s) \quad (5)$$

To complete the textual description, the relations existing between segment pairs and used for determining the contexts defined above are added.

4   Experimental Assessment

This section describes and discusses the quantitative assessment of the performance of the proposed approach.

For this first experimental assessment of the combination model proposed we chose a data-set where interesting objects have been manually segmented and labelled, so to have a reliable ground-truth for estimating the performance of our model. The data set chosen is the *MIT-Indoor* including 1,700 manually segmented images These pictures are taken in indoor surroundings, including kitchens, bedrooms, libraries, gymns and so on. Whenever an actual system based on the proposed approach is implemented, the best available solution for the segmentation will be included. We randomly divided the data in three parts: two of them, containing each the 30% of the data, are used to train the PO and the combination model respectively, while the remaining 40% of the data are used to assess the system performance. Note that in our view it is important that the data used to train the PO and the combination models are different, as in actual domains they usually have different origins.

The system performance is evaluated in terms of classification accuracy, i.e. the rate of segments which have been correctly classified. In particular, we considered six classes obtained by clustering the data set ones and then taking the six with a larger number of items: the adopted classes and the number of times they occur in the data set are reported in Table 1. Furthermore, we considered three relation types corresponding to the relative position of two segments in an image: *near*, *very near* and *intersecting*. Clearly, all three the relations are symmetrical.

The role of the classifier in our system is to produce a probability distribution on the set of classes for every input segment. The literature on object recognition is very rich [23]. The risk in choosing one approach or the other is that the final results would depend on this choice and its influence can not be distinguished by the one of the combination model. We therefore decided to substitute the actual classification with a random simulation able to produce any given performance. In this way, it is possible to describe the dependence of the system performance on the classification accuracy. All in all, we therefore need a method to simulate the behaviour of a multi-class classifier with an assigned accuracy a.

For this goal, we use the strategy described by the pseudo-code in Fig. 3. Given a segment, we randomly choose a score in $[0, 1]$ by the function $U(0, 1)$ for each class in the class set C. We then assign, with a probability given by the desired accuracy a, the maximum score to the gold class, while the other scores are randomly assigned to the remaining classes. The scores are finally normalised to obtain a probability distribution. As the classifier assigns to each segment the maximum probability class, we have that this corresponds to the right choice in the a percentage of cases, resulting in the desired accuracy. The use of a simulated classifier is not novel (see, for instance, [33]).

As we aim to assess the improvement we can obtain by introducing the ontological knowledge, we compare the system performance with a baseline consisting

```
maxClassProb←0.0;
BestClass← ∅;
for CurrentClass ∈ ClassSet do
    NewClassProb ∼ U(0, 1);
    ClassProb[CurrentClass]← newClassProb;
    if ClassProb[CurrentClass] > MaxClassProb then
        MaxClassProbValue← ClassProb[CurrentClass];
        BestClass ← CurrentClass;
    else
        if TossingACoin == Head then
            RandomClass ← CurrentClass;
        end if
    end if
end for
Accuracy ∼ U(0, 1);
Gold ← GoldClass(Segment);
if Accuracy < DesiredAccuracy then
    Swap(ClassProb[BestClass],ClassProb[Gold]);
else
    swap(ClassProb[RandomClass],ClassProb[Gold]);
end if
normalize(ClassProb);
```

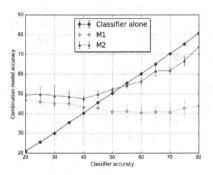

Fig. 3. Pseudo-code of the simulated classifier.

Fig. 4. Performance of the two systems compared with the baseline. Error bars give the 95% confidence intervals.

in the (simulated) classifier alone. The two approaches discussed in Sect. 3.2 are applied to combine the PO into the system: M1 and M2.

4.1   Results and Discussion

The system accuracy of the approaches proposed in this paper are depicted in Fig. 4 and compared with the accuracy of the statistical classifier applied alone.

For the sake of completeness, we considered a very wide range of accuracies for the simulated classifier: from 20% up to 80%, even if in actual conditions, the values of classifiers accuracy is more likely under $50-60\%$. However, in any case, we see that the M2 outperforms the M1, whose performance even deteriorates when the classifier accuracy improves. A possible explication for this behaviour could be that too much confidence is given to the a priori PO score with respect to the actual input data evidence.

On the other hand, the M2 improves on the simple classifier when the latter performance are inferior than about 55%, that is in realistic experimental conditions. We can observe how performance of this model are much better than the classifier alone when the latter performance are worse than 30%, and this can be the case when the task is not too easy. Even for classifiers obtaining an accuracy between 30% and 55%, the adoption of an approach integrating PO knowledge is advantageous.

Last, but not least, we observe that even when M2 performs worse than the classifier alone, its accuracy improves with the classifier accuracy, so that the two curves are approximately parallel. This could suggest that a better ontology design, resulting in a better PO, could help the system to overcome the performance obtained by the classifier alone.

5   Conclusions and Future Work

In this paper, we proposed and experimentally evaluated two different probabilistic models to integrate the probabilities derived from a probabilistic ontology with the ones produced by a statistical classifier. One of the two proved to perform in an acceptable way and could be used in an actual system.

For the sake of obtaining a clear view of the integration module performance, we tried to minimise the effect of the other modules. Therefore, we started from images which had been manually segmented and simulated a classifier in such a way that its accuracy could be controlled. As a future work, we plan to assess the performance of the proposed approach when coupled with state-of-the-art modules.

A fragment of a probabilistic ontology has been built by using three relations which could be automatically recognised in the input images, while the corresponding probabilities have been estimated from their frequencies. When more sophisticated ontologies will be available, containing information from large data sets, we expect the integration to give even better results.

Acknowledgments. We are grateful to M. Benerecetti e P. A. Bonatti for useful discussions about the most effective ways to represent knowledge. The research presented in this paper was partially supported by the national project CHIS - Cultural Heritage Information System and Perception, the national project Perception, Performativity and Cognitive Sciences (PRIN Bando 2015, 2015TM24JS).

References

1. Apicella, A., Corazza, A., Isgrò, F., Vettigli, G.: Integrating a priori probabilistic knowledge into classification for image description. In: Proceedings of the 26th IEEE WETICE Conference (2017)
2. Bach, N., Badaskar, S.: A review of relation extraction. Carnegie Mellon University, Language Technologies Institute (2007)
3. Bar, M., Ullman, S.: Spatial context in recognition. Perception **25**(3), 343–352 (1996)
4. Bishop, C.M.: Pattern Recognition and Machine Learning. Springer, New York (2006)
5. Bloch, I., Colliot, O., Camara, O., Graud, T.: Fusion of spatial relationships for guiding recognition, example of brain structure recognition in 3D MRI. Pattern Recognit. Lett. **26**(4), 449–457 (2005)
6. Chen, X., Zitnick, C.L.: Mind's eye: a recurrent visual representation for image caption generation. In: IEEE Conference on Computer Vision and Pattern Recognition (CVPR), pp. 2422–2431, June 2015
7. Choi, W., Shahid, K., Savarese, S.: Learning context for collective activity recognition. CVPR **2011**, 3273–3280 (2011)
8. Da Costa, P.C.G.: Bayesian semantics for the semantic web. Ph.D. thesis, George Mason University, Fairfax, VA, USA, aAI3179141 (2005)
9. Ding, Z., Peng, Y.: A probabilistic extension to ontology language owl. In: Proceedings of the 37th Annual Hawaii International Conference on System Sciences (HICSS 2004), Track 4, vol. 4, p. 40111.1 (2004)

10. Elliott, D., Keller, F.: Image description using visual dependency representations. EMNLP **13**, 1292–1302 (2013)
11. Eweiwi, A., Cheema, M.S., Bauckhage, C.: Action recognition in still images by learning spatial interest regions from videos. Pattern Recognit. Lett. **51**, 8–15 (2015)
12. Fang, H., Gupta, S., Iandola, F., Srivastava, R.K., Deng, L., Dollar, P., Gao, J., He, X., Mitchell, M., Platt, J.C., Lawrence Zitnick, C., Zweig, G.: From captions to visual concepts and back. In: The IEEE Conference on Computer Vision and Pattern Recognition (CVPR), pp. 1473–1482, June 2015
13. Farhadi, A., Hejrati, M., Sadeghi, M.A., Young, P., Rashtchian, C., Hockenmaier, J., Forsyth, D.: Every picture tells a story: generating sentences from images. In: Daniilidis, K., Maragos, P., Paragios, N. (eds.) ECCV 2010. LNCS, vol. 6314, pp. 15–29. Springer, Heidelberg (2010). doi:10.1007/978-3-642-15561-1_2
14. Hudelot, C., Atif, J., Bloch, I.: Fuzzy spatial relation ontology for image interpretation. Fuzzy Sets Syst. **159**(15), 1929–1951 (2008)
15. Karpathy, A., Fei-Fei, L.: Deep visual-semantic alignments for generating image descriptions. IEEE Trans. Pattern Anal. Mach. Intell. **39**(4), 664–676 (2017)
16. Bobillo, F., Costa, P.C.G., d'Amato, C., Fanizzi, N., Laskey, K.B., Laskey, K.J., Lukasiewicz, T., Nickles, M., Pool, M. (eds.): Uncertainty Reasoning for the Semantic Web II: International Workshops URSW 2008-2010 Held at ISWC and UniDL 2010 Held at Floc, Revised Selected Papers. Springer, Heidelberg (2013)
17. Kulkarni, G., Premraj, V., Dhar, S., Li, S., Choi, Y., Berg, A.C., Berg, T.L.: Baby talk: understanding and generating simple image descriptions. In: Proceedings of the 2011 IEEE Conference on Computer Vision and Pattern Recognition, CVPR 2011, pp. 1601–1608 (2011)
18. Mezaris, V., Kompatsiaris, I., Strintzis, M.G.: An ontology approach to object-based image retrieval. In: Proceedings of 2003 International Conference on Image Processing, ICIP 2003, vol. 2, pp. 511–514, September 2003
19. Oliva, A., Torralba, A.: The role of context in object recognition. Trends Cogn. Sci. **11**(12), 520–527 (2007)
20. Sarwar, S., Qayyum, Z.U., Majeed, S.: Ontology based image retrieval framework using qualitative semantic image descriptions. Proced. Comput. Sci. **22**, 285–294 (2013)
21. Schmid, C.: A structured probabilistic model for recognition. In: IEEE Computer Society Conference on Computer Vision and Pattern Recognition, vol. 2, p. 490 (1999)
22. Schneiderman, H., Kanade, T.: Probabilistic modeling of local appearance and spatial relationships for object recognition. In: Proceedings of the IEEE Computer Society Conference on Computer Vision and Pattern Recognition, p. 45. IEEE Computer Society (1998)
23. Szeliski, R.: Computer Vision: Algorithms and Applications, 1st edn. Springer, New York Inc. (2010)
24. Tanaka, J.W., Sengco, J.A.: Features and their configuration in face recognition. Mem. Cogn. **25**(5), 583–592 (1997)
25. Tousch, A.M., Herbin, S., Audibert, J.Y.: Semantic hierarchies for image annotation: a survey. Pattern Recogn. **45**(1), 333–345 (2012)
26. Toussaint, G.: The use of context in pattern recognition. Pattern Recogn. **10**(3), 189–204 (1978)
27. Vinyals, O., Toshev, A., Bengio, S., Erhan, D.: Show and tell: a neural image caption generator. In: IEEE Conference on Computer Vision and Pattern Recognition, CVPR, pp. 3156–3164 (2015)

28. Wang, C., Blei, D.M., Fei-Fei, L.: Simultaneous image classification and annotation. In: CVPR. pp. 1903–1910. IEEE Computer Society (2009)
29. Wang, M., Gao, Y., Lu, K., Rui, Y.: View-based discriminative probabilistic modeling for 3D object retrieval and recognition. Trans. Image Proc. **22**(4), 1395–1407 (2013)
30. You, Q., Jin, H., Wang, Z., Fang, C., Luo, J.: Image captioning with semantic attention. In: The IEEE Conference on Computer Vision and Pattern Recognition (CVPR), pp. 4651–4659, June 2016
31. Zhang, L., Yang, Y., Gao, Y., Yu, Y., Wang, C., Li, X.: A probabilistic associative model for segmenting weakly supervised images. IEEE Trans. Image Proc. **23**(9), 4150–4159 (2014)
32. Zhang, R., Zhang, Z., Li, M., Ma, W.Y., Zhang, H.J.: A probabilistic semantic model for image annotation and multimodal image retrieval. In: Tenth IEEE International Conference on Computer Vision, ICCV 2005, vol. 1, pp. 846–851, October 2005
33. Zouari, H., Heutte, L., Lecourtier, Y.: Simulating classifier ensembles of fixed diversity for studying plurality voting performance. In: Proceedings of the 17th International Conference on Pattern Recognition, ICPR 2004, vol. 1, pp. 232–235, August 2004

Generating Knowledge-Enriched Image Annotations for Fine-Grained Visual Classification

Francesca Murabito[✉], Simone Palazzo,
Concetto Spampinato, and Daniela Giordano

Pattern Recognition and Computer Vision (PeRCeiVe Lab),
Department of Electric Electronic and Computer Engineering,
University of Catania, Catania, Italy
{fmurabito,palazzosim,cspampin,dgiordan}@dieei.unict.it

Abstract. Exploiting high-level visual knowledge is the key for a great leap in image classification, in particular, and computer vision, in general. In this paper, we present a tool for generating knowledge-enriched visual annotations and use it to build a benchmarking dataset for a complex classification problem that cannot be solved by learning low and middle-level visual descriptor distributions only. The resulting *VegImage* dataset contains 3,872 images of 24 fruit varieties, over than 60,000 bounding boxes (portraying the different varieties of fruits as well as context objects such as leaves, etc.) and a large knowledge base (over 1,000,000 OWL triples) containing a-priori knowledge about object visual appearance. We also tested existing fine-grained and CNN-based classification methods on this dataset, showing the difficulty of purely visual-based methods in tackling it.

1 Introduction

Object recognition and image classification have been hot research topics in the last two decades. Recently, deep-learning methods have been able to achieve impressive performance on thousands of object classes from the ImageNet dataset. In spite of such progress, classification approaches are still predominantly based on visual features, leveraging the power of statistical machine learning to learn distributions of low and middle-level features. While this has proved to be an effective strategy even for fine-grained classification problems [13,17,31], there are cases where relying on visual appearance only might fail, especially in specialized application domains (such as fruit variety identification). For example, Fig. 1 (left image) shows four different varieties of cherry (namely, bing, black tartarian, burlat and lapin) that cannot be easily identified by only exploiting statistical distribution of visual descriptors. However, objects in the "real-world" do not appear as isolated items, but come in a rich context (the right-hand image in Fig. 1 shows the same cherry varieties in their natural context), which is largely exploited by humans for visual categorization.

© Springer International Publishing AG 2017
S. Battiato et al. (Eds.): ICIAP 2017, Part I, LNCS 10484, pp. 332–344, 2017.
https://doi.org/10.1007/978-3-319-68560-1_30

4 cherry varieties   **Cherry varieties in context**

Fig. 1. Example of fine-grained problem tackled in this paper. Left: Four different cheery varieties, namely (left to right, top to bottom), bing, black tartarian, burlat and lapin. Right: The same varieties in their natural context. Information about leaf shape, distance between fruit and tree branches, peduncle length may support the disambiguation between the four object classes.

Our main assumption is that, for a real breakthrough in computer vision, computers need to emulate human visual process by combining perceptive elements (visual descriptors) and cognitive factors (structured knowledge). Such combined perceptive-cognitive knowledge can be then exploited to solve complex visual recognition tasks when low-level visual description fails to express the differences among classes. However, while it is relatively easy to describe visually images, e.g., by identifying variations in shapes, colors, etc., it is more challenging and complex to annotate images according to specific knowledge as the ones depicted in Fig. 1, which only experts, making use of domain knowledge, would be able to do. Nevertheless, domain experts often do not wish to spend time to provide image annotations, so *how can we generate knowledge-enriched visual annotations necessary to train machine learning techniques?*. This paper aims at addressing the above question, specifically through:

– An annotation tool which guides the visual annotation process according to specialized domain knowledge model defined as a formal ontology, and which allows non-experts to generate large-scale domain-specific annotations.
– A knowledge-enriched fine-grained image dataset for fruit variety classification, which is hard to solve with typical visual-oriented approaches (e.g., GoogLeNet, Overfeat, VLFeat PHOW, KDES) without the use of domain knowledge.

2   Related Work

The goal of this paper is three-fold: *(a)* proposing a new semantic annotation tool driven by *(b)* domain knowledge through a formal ontology for *(c)* generating a fine-grained image dataset enriched with a large knowledge base. The importance of visual world semantics (and of context especially) in automated visual recognition has been long acknowledged [6,14]. Recently, there have been significant advances in modeling rich semantics using contextual relationships [18,25] such as object-object [6,20] and object-attribute [9,16] applied to scene classification [12] or object recognition [27]. In [27], the authors proved that context information is more relevant for object recognition than the intrinsic object representation, especially in cases of poor image quality (e.g., blurred images due to large distances, occlusions, illumination, shadows). However, visual scenes provide richer semantics than object-object or object-attribute relationships, which most of the existing methods do not take into account or do not exploit effectively as they try to solve the recognition problem by brute force. Nevertheless, one of the limitations to a larger use of high-level knowledge in computer vision is the lack of structured resources modeling exhaustively the semantics of our world. Indeed, so far, the largest resource of structured visual knowledge is the ImageNet dataset that, however, captures only limited semantic relations, ignoring, for instance, co-occurrence, dependency, mutual exclusion. The need for exhaustive knowledge is also highlighted by the recent sprout of methods employing high-level knowledge (mainly unstructured) for computer vision tasks: knowledge transfer methods [10,15] and semantic relation graphs [4,21] have been adopted to deal with the limits of traditional multi-class or binary models, which suffer from being overly restrictive or overly relaxed, respectively. Compared to scene graphs, computational ontologies are able to describe deeper scene semantics by defining high-level attributes and imposing constraints about real-world object appearance and their contextual and semantic relations, in an interoperable and generalizable way.

However, including high-level knowledge in the learning loop needs large semantically-annotated visual datasets, whose generation is an expensive process: beside annotating objects in images, other semantic information, such as color, shape, related-objects and their visual properties, etc., needs to be collected. This, especially, holds in specialized application domains (e.g., fruit variety, bird, medical images, etc.) where high precision is necessary to avoid affecting the learning process. In such cases, annotations should be provided by domain experts, who do not have enough time to spend into the process. To tackle this problem, one possible solution is to extract and use domain-knowledge to guide/constrain annotations done by non-expert users. So far, only few ontology-based image annotation tools have been proposed [3,7,19], which are, however, mainly thought for information retrieval rather than for computer vision.

Our proposed tool differs from the above ones and traditional tools [8,22] in that it constrains and guides the annotation process according to specific domain knowledge (codified as a formal ontology) where the visual attributes are inferred through ontology reasoning, thus reducing greatly the knowledge required to carry out the task.

We used our tool to generate knowledge-enriched visual annotations on fruit variety images, thus providing a complex benchmark for fine-grained recognition.

There are several benchmarking datasets for fine-grained classification of birds, stonefly larvae, etc. [5,17,31] but they mainly contain per-instance segmentations and do to provide any semantic visual descriptions of objects and their context. The datasets most similar to ours are the ones for semantic scene labeling [2,24], which, despite including context information, no exhaustive semantic relations are defined.

3   Generating Knowledge-Enriched Visual Annotations

In this section we present a formal OWL ontology encoding specialized knowledge for fruit variety categorization. The combination of such ontology with a tool able to guide and constrain the annotation process allows to minimize expert user intervention, thus providing the chance to create large-scale fine-grained annotations by involving mainly non-expert users.

3.1   The Fruit Ontology

An ontology is a formalism providing, for a specific domain, a common machine-processable vocabulary and a formal agreement about the meaning of the used terms, which include important concepts, their properties, mutual relations and constraints. Basic concepts of a domain correspond to *owl:Class*, whose expressiveness can be enhanced by adding attributes (as *owl:DataProperty*) and relations to other *owl:Class* (as *owl:ObjectProperty*). The vocabulary is designed and validated by human users through axioms expressed in a logic language and the concepts and properties can be enriched using natural language descriptions[1] and links (e.g. *rdfs:seeAlso* property).

We developed a new ontology describing visually the fruit application domain by involving three expert agronomists, who also supported the generation of correct instances for the considered fruit varieties. Figure 2 shows the ontology's VOWL (Visual OWL) representation and some statistics generated using Protégé[2]. We embedded this ontology in an annotation tool to speed up the labeling process, making annotation of domain-specific images accessible for non-expert users (see Sect. 3.2). Before describing the *Fruit Ontology*, let us introduce some terminology to avoid ambiguities. We refer to an *owl:Class* as an *ontology class*, and to an image class (i.e., a fruit variety) as a *dataset class*. Furthermore, we use *target class* to indicate the main object class we want to recognize (in our case *Fruit*), and *context class* for all the object classes that can be considered as part of the context (in our case, *Peduncle, Leaf, Petiole*) of the *target class*. Typically, *target classes* are objects which are spatially well-defined, easily-recognizable and possibly not a constituent part of a larger object (e.g., a dog rather than its tail, a fruit rather than its peduncle). *Context classes* are, instead, those that either are not classification targets or are more easily identified in relation to a target class. The Fruit Ontology contains two main class

---

[1] http://www.w3.org/2005/Incubator/mmsem/XGR-image-annotation/.

[2] http://protege.stanford.edu/.

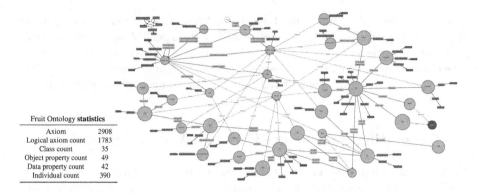

Fruit Ontology **statistics**	
Axiom	2908
Logical axiom count	1783
Class count	35
Object property count	49
Data property count	42
Individual count	390

Fig. 2. The fruit ontology VOWL representation. High resolution image: zoom in, too see classes and properties.

categories: the ones for visually describing target and context objects, and the ones needed for the annotation process.

Application Domain Classes and Properties. Domain classes and properties encode *a-priori* and expert knowledge on fruit varieties in terms of both visual appearance (colors, shape, edges, etc.) and their context relations (with *Leaf, Peduncle, Petiole*, etc.). Three expert agronomists supported us in the ontology design process by identifying for each target class (i.e., *Fruit*), the set of related context classes and the visual features describing their appearance. Both the target and context classes were mapped to ontology classes (*Fruit, Leaf, Peduncle, Petiole* are defined as subclasses of a domain-agnostic *PhysicalObject* class) and were enriched with as many *owl:DataProperty* (e.g., *fruitHasStripes, fruitHasColourDescription, fruitHasOvercolourDescription*, etc.) as needed to represent their visual appearance. Most of the physical features are defined as classes themselves (e.g. *Shape, Edge, FruitRusset*, etc., similarly defined as subclasses of *PhysicalProperty*) for defining more articulated visual characteristics (e.g., *fruitRussetHasDistribution, fruitRussetHasType*, etc.).

Ontology classes mapping target or context objects only differ in that target classes include the relations *fruitHasSpecies* and *fruitHasVariety* (easily generalizable to other domains) to *Species* and *Variety* ontology classes, which in turn are defined as skos[3] concepts in order to include a taxonomy of varieties (each taxonomy term corresponds to a dataset image class). The *physicalObjectHasPart* and its inverse *physicalObjectIsPartOf* and their specialized sub-properties (e.g., *fruitIsInTree, fruitHasPeduncle*) are used by the ontology reasoner to infer, starting from the target class and exploiting property transitivity, all ontology classes (e.g. Leaf, Peduncle, Petiole) related to its context.

---

[3] http://www.w3.org/2004/02/skos/.

Annotation Tool Specific Classes and Properties. The link between user annotations and entities in the ontology is represented by the *AnnotationSample* class, whose properties *hasBB* (for "bounding box") and *hasImage* specify the location of an annotated object in an image. The *AnnotationSample* class is specifically designed to speed up the annotation phase. For each new annotation, an instance of *AnnotationSample* is created and associated with the corresponding *PhysicalObject* subclass instance; this allows the tool to infer all corresponding *PhysicalObject* subclass instance properties encoded into the ontology without the need to specify manually all its properties. Annotator intervention is needed only in cases a property may assume multiple values (e.g., *Russet* for *Canadian Reinette* apple), from which, however, the tool displays samples (also encoded in the ontology) to simplify the labeling work for non-experts (see right-hand side in the bottom part of Fig. 3).

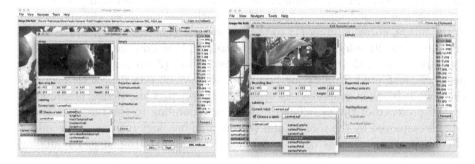

Fig. 3. User interface of the ontology-driven annotation tool. (Left image) Bounding box annotation of a target object (ideally performed by an expert user). (Right image). Annotation of context class objects (e.g., a leaf), with automatically-suggested labels inferred from the one assigned to the bounding box associated to an object of the target class. The right-hand side part is for disambiguating all those properties that can assume multiple values (as per Instance definition) through visual comparison with sample images (also encoded in the ontology instance).

Although the whole annotation schema and representation may seem overly complex (especially if compared to the current "flat-structure" annotations made public by dataset providers), they enable encoding annotations as ontology instances, with one great advantage: the annotation correctness and meaningfulness is implicitly validated, as they have to match the ontology schema.

3.2   The Annotation Tool

The presented ontology-driven annotation tool aims at guiding and constraining users the labeling process within the concepts enforced by the ontology. It basically provides means to draw and assign labels (most of them are automatically inferred through ontology reasoning) to bounding boxes for target and context classes and to specify attributes for them. Similarly to other annotation tools [8,22], the interface presents the user with an image to work on, together with several tools for browsing through images, zooming in and out, adding, editing and removing annotations. However, unlike those other tools, part of the label assignment responsibility is moved from the user to the tool itself, through a two-phase annotation process. The two phases of the annotation process differ by the degree of expert knowledge required and the amount of annotation work to be carried out. The first annotation phase consists in assigning a dataset image class (e.g., a fruit variety) to each image. This initial task requires expert knowledge necessary to distinguish between dataset classes differing only by subtle details. However, the amount of data to annotate is relatively small, since the user is only asked to draw one bounding box per image and select the corresponding dataset class, thus limiting the expert employment only to a fraction of the whole labeling process. Once annotations have been "bootstrapped" by specifying labels for the bounding boxes containing objects belonging to the target class, the second phase consists of annotating all the other objects present in the image, corresponding (1) to the target class (i.e., the fruit), whose labels are automatically inferred by ontology reasoning, based on the assumption that they are equal to the one provided by experts; and (2) to context classes (i.e., peduncle, leaf, etc.). Annotating bounding boxes of objects related to a context class, while being in general a task which requires expert knowledge, is simplified by the presence of the associated object corresponding to the target class: its label is employed by the tool to infer (through an ontology reasoner[4]) the subset of context class instances which can be used to annotate the current bounding box.

Figure 3 shows how the interface implements the above procedure. Firstly, the (expert) user annotates (left part in Fig. 3) one object related to a target class with the corresponding fine-grained class, by simply drawing a bounding box around the object and selecting its label from a list (dynamically built from the provided domain ontology), e.g. "cameoFruit". Then, the (not necessarily expert) user can continue the process by adding annotations for the other objects in the image, whose labels are inferred based on the target class instance assigned by the expert and on the ontology (right part in Fig. 3). In the example, the inferred labels are "cameoLeaf", "cameoPeduncle", "cameoPetal", since the target class instance was labeled by the expert as "CameoFruit'. As a final consideration, it should be noted that the above process transcends the specific application domain for which the tool is employed, and the concepts to be annotated can be simply configured at setup time by providing a custom ontology and specifying the set of target classes (namely, those for which properties *physicalObjectHasSpecies* or *physicalObjectHasVariety* are defined) and context classes (related to the target classes throughout a series of subproperties of *physicalObjectHasPart* and *physicalObjectIsPartOf*.

---

[4] http://owlapi.sourceforge.net/.

3.3   The Fruit Image Dataset

The *VegImage* dataset is a collection of 3,872 images of three common fruit species, namely, *malus domestica* (apple), *prunus avium* (cherry) and *pyrus communis* (pear). For each fruit species several fruit varieties were included, 10 for *malus domestica*, 7 for *prunus avium* and 7 for *pyrus communis*. Together with fruit images, we also provide over than 60,000 bounding boxes (depicting the different varieties of fruits, leaves, peduncles, etc.) and a large a knowledge base (over 1,000,000 OWL triples) containing a-priori knowledge about colors, shapes as well as context objects for the considered fruit varieties. A detailed list of fruit varieties is shown in Fig. 4.

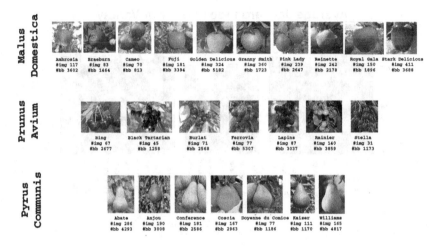

Fig. 4. Example images from the fruit image dataset. Numbers in red are number of images per class while in green the number of bounding boxes. (Color figure online)

Dataset Collection. The fruit variety images were mainly downloaded from Google Images, Flickr, ImageNet, Yahoo Images. To increase appearance variability, we also downloaded YouTube documentary videos, from which we manually selected key frames to avoid near duplicates in the dataset. For each of the 27 fruit varieties, about 1,000 images were manually selected to be included in the dataset. Low-quality images or images depicting multiple fruit varieties or people as main subjects were filtered out. After this screening, we asked three expert agronomists to manually check all the resulting images. Thus, we collected up to 500 images for each fruit variety.

Dataset Annotation. We performed a two-stage annotation phase using the tool described in the previous section: (a) **Image labeling**: in this step, the three agronomists annotated each image with a label decided through consensus among them; (b) **Bounding box annotation**: ten non-expert users were asked to draw bounding boxes (a distribution over fruit varieties is given in Fig. 4) for objects of both target (*Fruit*) and context classes (*Peduncle, Leaf* and *Petiole*), and to disambiguate multi-valued attributes defined in the Fruit Ontology (e.g., russet for *Canadian Reinette* apple), which were finally double-checked by the experts, being the only kind of annotations which could be subject to errors.

To test automatically the quality of the generated bounding boxes, we compared them with the ones provided by Selective Search [28]. In detail, for each image we ran selective search (SS) object localization (2,000 object proposals per image) and we computed the maximum Intersection over Union (IoU) index between each annotated bounding box and the ones provided by SS. The average IoU for each fruit class is given in Fig. 5 showing the high-quality of our annotations taking into account also SS failures.

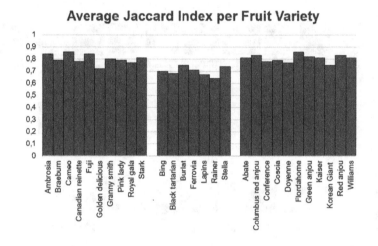

Fig. 5. Average IoU between generated bounding boxes and SS'ones.

Annotation Effort and Times. To test the performance of our annotation tool, we used as evaluation criteria: (1) shifting working time from experts to non-experts, while keeping annotation accuracy high and (2) reducing non-expert annotation time.

Domain experts manually annotated 3,872 fruit images, while over 60,000 bounding boxes were provided by ten non-expert users. Bounding box attributes were inferred automatically by the reasoner (through deductive inference) after the corresponding bounding box class (e.g. *Leaf*) and variety (e.g. *Cameo*) were specified. The annotations of 3,872 fruit images by the three experts took about 13 days (average of 1.3 h per day per expert) for a total of 51 worker hours, while the annotation of 105,284 bounding boxes took about 20 days (average of 4 h per day per annotator). In total, annotating the whole image dataset took 861 worker hours: 810 (about 94% of the total) hours provided by non-experts and the remaining 51 h by experts.

The average annotation time per bounding box for non-experts was 27.7 s, which is impressive given that the Fruit Image Dataset deals with a specific and complex application domain, and considering that in COCO [11] the annotators spent, on average, about 80 s per bounding box.

As a final note, our tool allows to tackle the issue recently reported in [29], i.e., high-quality annotations on domain-specific applications should be performed, if not by experts, at least by citizen scientists, since unskilled workers perform extremely bad. While this may hold for "traditional" annotation approaches, encoding and incorporating domain knowledge in a tool able to constrain the labeling process is a valid alternative, which allows non-expert annotators to provide high-quality annotations, thus saving significantly expensive resources.

4   Comparison to Existing Datasets for Fine-Grained Recognition

Table 1 compares the proposed Fruit Image dataset with three popular benchmarking datasets for fine-grained image classification: Oxford-IIIT Pet [17], Oxford Flower 102 [13] and Caltech-UCSD Birds [31]. Although the three datasets all have a comparable number of images, the Fruit Image dataset is more complete in the type of annotations it includes, as it contains several examples of images with multiple objects and all objects have associated parts (as context objects) and attributes, beside being enriched with a large knowledge base. Furthermore, although the number of images in the Fruit Image dataset is much smaller than popular image classification (not fine-grained) datasets, e.g., COCO (see Table 1), the number of annotations are comparable, especially since our dataset provides several object annotations per image, completed with bounding box locations, class labels and class-specific attributes. Such achievements would not have been practical if only few experts were asked to perform all annotations; the approach described in Sect. 3.2 allowed us to involve non-experts in a fine-grained annotation process, thus greatly speeding up the whole task.

Table 1. Comparison between popular fine-grained (and not) datasets and our dataset. Key: $\#C$: number of classes; $\#I$: number of images; I/C: average number of images per class; O/I: average number of objects per image; P/O: average number of parts per object; A/O: average number of attributes per object. For our dataset, the O/I value refers to the number of target objects (i.e., fruits), whereas the P/O value counts context objects as object parts; object attributes are the OWL triples, mostly inferred by ontology reasoning, and only a tiny part manually annotated.

	$\#C$	$\#I$	I/C	O/I	P/O	A/O
PET	37	7, 349	198.6	1.0	0.0	0.0
Flower_102	102	8, 189	80.3	1.0	0.0	0.0
Birds	200	11, 788	58.9	1.0	12.0	31.5
COCO	80	123, 287	1, 541.1	7.3	–	–
Fruit Image	24	3, 872	161.3	8.0	1.14	11.0

Table 2. Results obtained by VLFeat PHOW, KDES, OverFeat and GoogleNet on the proposed dataset and on three other fine-grained datasets.

Dataset	Method			
	VLFEAT	KDES	OverFeat	GoogleNet
Oxford-IIIT Pet	39.25%	45.47%	70.48%	86.14%
Oxford Flower 102	56.68%	24.63%	79.02%	90.04%
Caltech-UCSD Birds	14.62%	7.11%	59.2%	70.2%
Fruit Dataset	4.21%	24.4%	26.6%	36.1%

In order to test the complexity of the proposed dataset, we evaluated four state-of-the-art classification methods on these four datasets: VLFeat PHOW [30], KDES [1], OverFeat [23] and GoogleNet [26]. The comparison, in terms of mean classification accuracy (see Table 2) shows that the tested algorithms fail to tackle the proposed dataset. We believe that a cause for this failure is that, unlike current fine-grained datasets, the proposed fruit dataset describes an application domain where class discrimination is strongly based on a context dependency between objects, which needs to be encoded and integrated into the classification methods as *a priori* information.

5 Conclusions

In this paper we present a knowledge-driven annotation tool which exploits specialized domain knowledge to generate semantic fine-grained annotations, greatly reducing the efforts of domain experts, for classification problems that cannot be solved by using only low and middle-level features. The tool was used by three expert agronomists to provide high-level and coarse annotations and by ten non-expert users who provided fine-grained annotations without any knowledge on the application domain. The resulting *VegImage* dataset contains 3,872 images, over than 60,000 bounding boxes, and over than 1,000,000 OWL triples, representing, to the best of our knowledge, one of the most comprehensive resources for fine-grained classification and one the most exhaustive knowledge bases in computer vision. As future work, we are working on building semantic machine learning classifiers integrating classic learning methods with reasoning approaches able to convert a set of detections into an ontology instance describing the application domain to be matched against correct instances as provided by domain experts. The annotation tool, the image dataset, the knowledge base, and the Fruit ontology will be made publicly available.

References

1. Bo, L., Ren, X., Fox, D.: Kernel descriptors for visual recognition. In: NIPS (2010)
2. Brostow, G.J., Fauqueur, J., Cipolla, R.: Semantic object classes in video: a high-definition ground truth database. Pattern Recogn. Lett. **30**(2), 88–97 (2009)

3. Dasiopoulou, S., Giannakidou, E., Litos, G., Malasioti, P., Kompatsiaris, Y.: A survey of semantic image and video annotation tools. In: Paliouras, G., Spyropoulos, C.D., Tsatsaronis, G. (eds.) Knowledge-Driven Multimedia Information Extraction and Ontology Evolution. LNCS, vol. 6050, pp. 196–239. Springer, Heidelberg (2011). doi:10.1007/978-3-642-20795-2_8
4. Deng, J., et al.: Large-scale object classification using label relation graphs. In: Fleet, D., Pajdla, T., Schiele, B., Tuytelaars, T. (eds.) ECCV 2014. LNCS, vol. 8689, pp. 48–64. Springer, Cham (2014). doi:10.1007/978-3-319-10590-1_4
5. Deng, J., Dong, W., Socher, R., Li, L.J., Li, K., Fei-Fei, L.: Imagenet: A large-scale hierarchical image database. In: 2009 CVPR, pp. 248–255 (2009)
6. Galleguillos, C., Belongie, S.: Context based object categorization: a critical survey. Comput. Vis. Image Underst. **114**(6), 712–722 (2010)
7. Halaschek-Wiener, C., Golbeck, J., Schain, A., Grove, M., Parsia, B., Hendler, J.A.: PhotoStuff-an image annotation tool for the semantic web. In: 2005 International Semantic Web Conference (2005)
8. Kavasidis, I., Palazzo, S., Salvo, R., Giordano, D., Spampinato, C.: An innovative web-based collaborative platform for video annotation. MTAP **70**(1), 413–432 (2014)
9. Lampert, C., Nickisch, H., Harmeling, S.: Learning to detect unseen object classes by between-class attribute transfer. In: CVPR 2009, pp. 951–958 (2009)
10. Lampert, C., Nickisch, H., Harmeling, S.: Attribute-based classification for zero-shot visual object categorization. IEEE PAMI **36**(3), 453–465 (2014)
11. Lin, T.-Y., Maire, M., Belongie, S., Hays, J., Perona, P., Ramanan, D., Dollár, P., Zitnick, C.L.: Microsoft COCO: common objects in context. In: Fleet, D., Pajdla, T., Schiele, B., Tuytelaars, T. (eds.) ECCV 2014. LNCS, vol. 8693, pp. 740–755. Springer, Cham (2014). doi:10.1007/978-3-319-10602-1_48
12. Martin, D., Fowlkes, C., Malik, J.: Learning to detect natural image boundaries using local brightness, color, and texture cues. IEEE PAMI **26**(5), 530–549 (2004)
13. Nilsback, M.E., Zisserman, A.: Automated flower classification over a large number of classes. In: 2008 Indian Conference on Computer Vision, Graphics and Image Processing (2008)
14. Oliva, A., Torralba, A.: The role of context in object recognition. Trends Cogn. Sci. **11**(12), 520–527 (2007). (Regul. Ed.)
15. Oquab, M., Bottou, L., Laptev, I., Sivic, J.: Learning and transferring mid-level image representations using convolutional neural networks. In: 2014 CVPR (2014)
16. Parikh, D., Grauman, K.: Relative attributes. In: ICCV 2011. vol. 0, pp. 503–510. IEEE Computer Society, Los Alamitos, CA, USA (2011)
17. Parkhi, O.M., Vedaldi, A., Zisserman, A., Jawahar, C.V.: Cats and Dogs. In: CVPR (2012)
18. Patterson, G., Hays, J.: Sun attribute database: discovering, annotating, and recognizing scene attributes. In: 2012 CVPR (2012)
19. Petridis, K., Anastasopoulos, D., Saathoff, C., Timmermann, N., Kompatsiaris, Y., Staab, S.: M-OntoMat-Annotizer: image annotation linking ontologies and multimedia low-level features. In: Gabrys, B., Howlett, R.J., Jain, L.C. (eds.) KES 2006. LNCS, vol. 4253, pp. 633–640. Springer, Heidelberg (2006). doi:10.1007/11893011_80
20. Rabinovich, A., Vedaldi, A., Galleguillos, C., Wiewiora, E., Belongie, S.: Objects in context. In: 2007 ICCV (2007)
21. Ramanathan, V., Li, C., Deng, J., Han, W., Li, Z., Gu, K., Song, Y., Bengio, S., Rossenberg, C., Fei-Fei, L.: Learning semantic relationships for better action retrieval in images. In: 2015 CVPR, June 2015

22. Russell, B.C., Torralba, A., Murphy, K.P., Freeman, W.T.: LabelMe: a database and web-based tool for image annotation. In: IJCV 2008. vol. 77, pp. 157–173 (2008)
23. Sermanet, P., Eigen, D., Zhang, X., Mathieu, M., Fergus, R., LeCun, Y.: Overfeat: integrated recognition, localization and detection using convolutional networks. CoRR abs/1312.6229 (2013)
24. Shotton, J., Winn, J., Rother, C., Criminisi, A.: TextonBoost for image understanding: multi-class object recognition and segmentation by jointly modeling texture, layout, and context. In: IJCV 2009, January 2009
25. Sudderth, E., Torralba, A., Freeman, W., Willsky, A.: Learning hierarchical models of scenes, objects, and parts. In: ICCV 2005, vol. 2, pp. 1331–1338, October 2005
26. Szegedy, C., Liu, W., Jia, Y., Sermanet, P., Reed, S., Anguelov, D., Erhan, D., Vanhoucke, V., Rabinovich, A.: Going deeper with convolutions. In: CVPR 2015, pp. 1–9 (2015)
27. Torralba, A.: Contextual priming for object detection. In: IJCV 2003, vol. 53, pp. 169–191 (2003)
28. Uijlings, J., van de Sande, K., Gevers, T., Smeulders, A.: Selective Search for Object Recognition. In: 2013 IJCV (2013). http://www.huppelen.nl/publications/selectiveSearchDraft.pdf
29. Horn, V., et al.: Building a bird recognition app and large scale dataset with citizen scientists: the fine print in fine-grained dataset collection. In: CVPR (2015)
30. Vedaldi, A., Fulkerson, B.: VLFeat: an open and portable library of computer vision algorithms. In: ACM International Conference on Multimedia (2010)
31. Wah, C., Branson, S., Welinder, P., Perona, P., Belongie, S.: The Caltech-UCSD birds-200-2011 dataset. Technical report CNS-TR-2011-001 (2011)

Histological Image Analysis by Invariant Descriptors

Cecilia Di Ruberto, Andrea Loddo, and Lorenzo Putzu$^{(\boxtimes)}$

Department of Mathematics and Computer Science,
University of Cagliari, Via Ospedale 72, Cagliari, Italy
{dirubert,andrea.loddo,lorenzo.putzu}@unica.it

Abstract. In this work we propose a comparative study between different descriptors in analysing histological images. In particular, our study is focused on measuring the accuracy of moments (Hu, Legendre, Zernike), Local Binary Patterns and co-occurrence matrices in classifying histological images. The experimentation has been conducted on well known public datasets: HistologyDS, Pap-smear, Lymphoma, Liver Aging Female, Liver Aging Male, Liver Gender AL and Liver Gender CR. The comparison results show that when combined with co-occurrence matrices and extracted from the RGB images, the orthogonal moments improve the classification performance considerably, imposing themselves as very powerful descriptors for histological image analysis.

Keywords: Medical image analysis · Texture descriptors · Moments · Local binary pattern · Co-occurence matrix · Classification

1 Introduction

Histological image analysis is a process that allows to evaluate if microscopic structures at the sub-cellular, cellular, tissue and organs level are affected by diseases, through various computer assisted methods. Tissue image analysis could be used to measure the cancer cells in a biopsy of a cancerous tumour taken from a patient and it can significantly reduce uncertainty in characterizing tumours compared to evaluations done by histologists, or improve the prediction recurrence rate of some cancers. Image analysis involves complex algorithms which identify and characterize cellular colour, shape and quantity of the tissue sample using image pattern recognition technology. In [1] global features are used to automatically discriminate lymphoma, in [2] wavelet features are used for the detection of tumours in endoscopic images and in [3] image texture informations are used to automatically discriminate polyps in colonoscopy images. Over the past few years moment functions have been used in medical image analysis with promising performance. They are statistical measures used to obtain the relevant information of an object. Since the introduction of invariant moments in image analysis [4], moment functions have been widely used in image processing and pattern classification applications, as discriminative descriptors, such as

© Springer International Publishing AG 2017
S. Battiato et al. (Eds.): ICIAP 2017, Part I, LNCS 10484, pp. 345–356, 2017.
https://doi.org/10.1007/978-3-319-68560-1_31

the geometric moments [5] for texture classification, or the complex moments for texture segmentation [6]. However, both geometric and complex moments contain redundant information and are sensitive to noise, due to the fact that the kernel polynomials are not orthogonal. For these reasons many different moments have been proposed, such as the discrete Tchebichef moments [7], the discrete moments known as Krawtchouk moments [8] or orthogonal moments like Legendre and Zernike moments [9].

Orthogonal moments are shown to be less sensitive to noise and have an efficient capability in feature representation with minimum redundancy. Zernike moments have been widely used in different types of applications, in shape-based image retrieval [10,11] and in pattern recognition [12] tasks. In medical image analysis the orthogonal moments have been used to reconstruct noisy CT, MRI, X-ray medical images [13], to describe the texture of a CT liver image [14] or prostate ultrasound [15], to detect tumours in brain images [16] or in mammography images [17], to recognize parasites [18] and spermatogonium [19].

In this work we propose a comparative study between different descriptors based on texture information for histological image classification. In particular, our study is focused on measuring the accuracy of moments (Hu, Legendre, Zernike), Local Binary Patterns (LPBs) and co-occurrence matrices in classifying histopathological images. The experimental results show that the combination of orthogonal moments with co-occurrence matrices reaches a very high level of accuracy on all the tested datasets, overcoming the most common and used descriptors. The rest of paper is organized as follows. Next section presents the texture descriptors definition used throughout this work. Section 3 presents the experimental evaluation, describing the utilized datasets, showing the experimental measures, the implementation details of each descriptor, and the results achieved for the specific collections. Finally, in Sect. 4 we give the conclusions.

2   Texture Descriptors

In this section we describe three important classes of texture descriptors: image geometric and orthogonal moments, cooccurence matrices and local binary patterns.

2.1   Image Moments

The moments are widely used in many applications for features extraction due to their invariance to scale, rotation and reflection change. The use of moments for image analysis and pattern recognition was inspired by Hu [4]. Hu's, Legendre's and Zernike's are the most common moments.

Hu moments. They are derived and calculated from geometric moments. The two-dimensional geometric moments of order $(p+q)$ of an image of $M \times N$ pixels with intensity function $f(x, y)$ are defined as:

$$m_{pq} = \sum_{x=0}^{M-1} \sum_{y=0}^{N-1} x^p y^q f(x, y), \tag{1}$$

where $p, q = 0, 1, 2, \ldots$. A set of n moments consists of all m_{pq} for $p + q \leq n$. The corresponding central moments are defined as:

$$\mu_{pq} = \sum_{x=0}^{M-1} \sum_{y=0}^{N-1} (x - \overline{x})^p (y - \overline{y})^q f(x, y), \qquad (2)$$

where $\overline{x} = m_{10}/m_{00}$ and $\overline{y} = m_{01}/m_{00}$ are the coordinates of the centre of mass of the image. The central moments μ_{pq} defined in Eq. 2 are invariant under the translation of coordinates. They can be normalized to preserve the invariance by scaling. For $p + q = 2, 3, \ldots$ the normalized central moments of an image are given by:

$$\eta_{pq} = \frac{\mu_{pq}}{\mu_{00}^{\gamma}} \quad \text{with} \quad \gamma = \frac{p + q}{2} + 1. \qquad (3)$$

Hu defined seven functions that are invariant to scale, translation and rotation changes [4], from the normalized central moments through the order three.

Legendre Moments. Legendre moments are orthogonal moments first introduced by Teague [9]. They were used in several patterns recognition [4] tasks. The Legendre moment of order $(p + q)$ of an image of $M \times N$ pixels with intensity function $f(x, y)$ is defined on the square $[-1, +1] \times [-1, +1]$, by:

$$L_{pq} = \frac{(2p + 1)(2q + 1)}{M \times N} \sum_{i=0}^{M-1} \sum_{j=0}^{N-1} P_p(x_i) P_q(y_j) f(x_i, y_j) \qquad (4)$$

where x_i and y_j denote the normalized pixel coordinates in the range of $[-1, +1]$, which are given by:

$$x_i = \frac{2i - (M - 1)}{M - 1}, \qquad y_j = \frac{2j - (N - 1)}{N - 1} \qquad (5)$$

and

$$P_p(x_i) = \sum_{k=0}^{p} \frac{(-1)^{\frac{p-k}{2}} x^k (p + k)!}{2^p k! \left(\frac{p-k}{2}\right)! \left(\frac{p+k}{2}\right)!} \qquad (6)$$

with $p - k$ even.

Zernike Moments. Zernike moments are the mapping of an image onto a set of complex Zernike polynomials. As these Zernike polynomials are orthogonal to each other, Zernike moments can represent the properties of an image with no redundancy or overlapping of information between the moments [9]. Due to these characteristics, Zernike moments have been used as features set in many applications. The computation of Zernike moments from an input image consists of three steps: computation of radial polynomials, computation of Zernike polynomials and computation of Zernike moments by projecting the image onto the Zernike polynomials [20]. The real-valued radial polynomial is defined as:

$$R_{p,q}(r) = \sum_{s=0}^{(p-|q|)/2} \frac{(-1)^s (p-s)! r^{p-2s}}{s!\left(\frac{p+|q|}{2} - s\right)!\left(\frac{p-|q|}{2} - s\right)!} \tag{7}$$

with $R_{p,q}(r) = R_{p,-q}(r)$, and p, q generally called order and repetition, respectively. The order p is a non-negative integer, and the repetition q is an integer satisfying $p - |q|$ even and $|q| \leq p$. The discrete form of the Zernike moments of an image of size $M \times N$ is expressed as follows:

$$Z_{pq} = \frac{p+1}{\lambda} \sum_{x=0}^{M-1} \sum_{y=0}^{N-1} R_{pq}(r_{xy}) e^{-jq\theta_{xy}} f(x,y) \tag{8}$$

where $0 \leq r_{xy} \leq 1$ and λ is a normalization factor. In the discrete implementation of Zernike moments, the normalization factor λ must be the number of pixels located in the unit circle by the mapping transformation and corresponds to the area of a unit circle π in the continuous domain. The transformed θ_{xy} phase and the distance r_{xy} at the pixel coordinates (x, y) are given by:

$$\theta_{xy} = tan^{-1}\left(\frac{(2y - (N-1))/(N-1)}{(2x - (M-1))/(M-1)}\right) \tag{9}$$

$$r_{xy} = \sqrt{\left(\frac{2x - (M-1)}{M-1}\right)^2 + \left(\frac{2y - (N-1)}{N-1}\right)^2}. \tag{10}$$

2.2 Co-occurrence Matrices

One of the earliest method for texture descriptors extraction was proposed by Haralick et al. [21]. His method is based on the creation of the grey level co-occurrence matrices, GLCMs, from which features representing some image aspects can be calculated. A GLCM represents the probability of finding two pixels i, j with distance d and orientation θ. Obviously, the d and θ values can assume different values, but the most used are $d = 1$ and $\theta = [0°, 45°, 90°, 135°]$. A GLCM for an image of size $M \times N$ with N_g grey levels is a 2D array of size $Ng \times Ng$. Haralick proposed thirteen descriptors that can be extracted from these matrices. Interesting methods have already been presented in order to extend the original implementation of GLCM. In [22] different values for the distance parameter influencing the matrices computation are evaluated, in [23] the GLCM descriptors are extracted by calculating the weighted sum of GLCM elements, in [24] the GLCM features are calculated by using the local gradient of the matrix. Furthermore, the GLCM has been extracted using the colour information from single channels [25] or by combining them in pairs [26,27]. Considering that invariant descriptors have our main focus in this work, we compute the GLCM only using the grey level intensities, and convert the rotation-dependent descriptors in rotationally invariant by the following approach. We start considering all the possible circular shifts of a feature vector $f_k = [f_1, \ldots, f_m]$. Then, we construct a matrix of size $m \times m$ in which all the circular shifts of the vector f_k are present and disposed regularly, generating a symmetric matrix as follows:

$$
\begin{pmatrix}
f_1 & f_2 & \cdots & f_{m-1} & f_m \\
f_2 & \cdots & \cdots & f_m & f_1 \\
\cdots & \cdots & \cdots & \cdots & \cdots \\
f_{m-1} & f_m & \cdots & \cdots & \cdots \\
f_m & f_1 & \cdots & \cdots & f_{m-1}
\end{pmatrix}
$$

Hence, the eigenvalues of such matrix are the new invariant descriptors, GLCMri, as they preserve dimension and direction of the original feature vector.

2.3 LBP Descriptors

The LBPs, instead, are a quite recent tool for texture analysis, originally proposed in [28] and widely used for grey level texture classification, due to its simplicity and robustness. This operator transforms the image by thresholding the neighbourhood of each pixel and by coding the result as a binary number. The resulting image histogram can be used as a feature vector for texture classification. Moreover, radius and number of neighbourhood pixels are two main parameters needed for the LBP operator. Although the LBP have been extended in many different ways, the most useful version, proposed by the same authors [29], realizes a rotation invariant descriptor, called LBPri. The LBPri is easily obtained through an iterative rotation of the binary digits, until the smallest value has been reached.

3 Datasets

The experimentation has been carried out on seven of the most famous colour histology image databases: HistologyDS, Pap-smear, Lymphoma, Liver Aging Female, Liver Aging Male, Liver Gender AL and Liver Gender CR that represent a set of really different computer vision problems.

HystologyDS (HIS) database [30] is a collection of 20,000 histology images for the study of fundamental tissues. It is provided in a subset of 2828 images annotated by four fundamental tissues: connective, epithelial, muscular and nervous. Each tissue is captured in a 24-bit RGB image of size 720×480. Some sample tissue images from HIS database are showed in Fig. 1.

Pap-smear (PAP) database [31] is a collection of pap-smear images acquired from healthy and cancerous smears coming from the Herlev University Hospital. It is composed of 917 images containing cells, annotated into seven classes: four

| Connective | Epithelial | Muscular | Nervous |

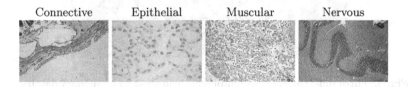

Fig. 1. Four different tissues from HistologyDS database.

Light Dysplastic Moderate Dysplastic Severe Dysplastic Carcinoma

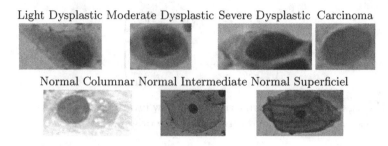

Normal Columnar Normal Intermediate Normal Superficiel

Fig. 2. The seven classes of cells belonging to Pap-smear database: first four abnormal and last three normal.

represent abnormal cells and three represent normal cases. Nevertheless, from the medical diagnosis viewpoint the most important requirement corresponds to the general two-class problem of correct separation between normal from abnormal cells. For this reason we have considered only the binary case. Each cell was captured in a 24-bit RGB image without a fixed size, that ranges from about 50×50 to about 300×300. Some examples are showed in Fig. 2.

Lymphoma (LYM) database [1] is a collection of tissues affected by malignant lymphoma, a cancer affecting lymph nodes. Three types of malignant lymphoma are represented in the set: Chronic Lymphocytic Leukemia (CLL), Follicular Lymphoma (FL) and Mantle Cell Lymphoma (MCL). This dataset presents a collection of samples from biopsies sectioned and stained with (H&E), realized in different laboratories by several pathologists. Only the most expert pathologists specialised in these types of lymphomas are able to consistently and accurately classify these three lymphoma types from H&E-stained biopsies. This slide collection contains significant variation in sectioning and staining and for this reason it is more representative of slides commonly encountered in a clinical setting. This database contains a collection of 374 slides captured in a 24-bit RGB image of size 1380×1040. In Fig. 3 a randomly selected image from each class is showed.

AGEMAP Atlas of Gene Expression in Mouse Aging Project [32] is a study by the National Institute on Aging, involving 48 male and female mice, of four ages (1, 6, 16, and 24 months), on ad-libitum or caloric restriction diets. Fifty

CLL   FL   MCL

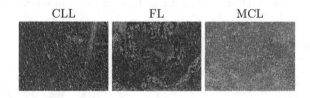

Fig. 3. Three different kinds of lymphoma belonging to lymphoma database.

colour images from 30 livers were manually acquired using a Carl Zeiss Axiovert 200 microscope and 40x objective, for a total of 1500 images. Each image is of size 1388 × 1040 in TIFF format with a 36-bit RGB colour depth. As the acquisition was done using 12 bits of quantization per colour channel, the histograms have been compressed so as to cover the 8 bits encoding. All the slides were prepared by the same person, thus staining variability in this dataset is very limited. AGEMAP images can be analysed across multiple axis of differentiation: age, gender, diet, or individual mice to construct a variety of classification problems. For these reasons the datasets' authors proposed three different experiments using three different subsets of the original images:

– **Liver Aging Female** (LAF) experiment consists on a 4-way classification problem using the four classes (1, 6, 16 and 24 months) of images of female mice on ad libitum diet. This set is composed of 529 images.
– **Liver Gender AL** (LGAL) experiment consists on a 2-way classifier which classifies the gender of the mouse based on the images of 6-month old male and female mice on ad-libitum diet. This set is composed of 265 images.
– **Liver Gender CR** (LGCR) experiment consists on a 2-way classifier which classifies the gender of the mouse based on the images of 6-month old male and female mice on caloric restriction diet. This set is composed of 303 images.

One more experiment has been added, **Liver Aging Male** (LAM), to the previously mentioned. It consists on a 4-way classification problem, like the first one, even though four classes (1, 6, 16 and 24 months) of images of male mice on an ad libitum diet have been used. This set is composed of 499 images.

4   Experimental Evaluation

The performance of the described descriptors has been evaluated following two strategies. The first one has been performed converting each image in grayscale and extracting each descriptor from the converted image. The second one has been performed over the RGB images applying the computation scheme for the three R, G, B channels and linking the descriptors into a single vector in order to take into account the colour information as we proposed in [27]. Classification performances have been evaluated by calculating the accuracy, which offers a good indication of the performance since it considers each class of equal importance. Thus the classification accuracy have been estimated through a k-Nearest Neighbour (k-NN) classifier, with $k = 1$ and using the euclidean distance. k-NN strategy has been preferred over more complex classifier in order to produce the results more representative of the effectiveness of the descriptors than of the classifiers themselves. Both analysing the grayscale images and the colour images we first tested the Hu, Zernike (up to order 10) and Legendre (up to order 8) moments and GLCMri and LBPri texture descriptors individually to assess the performances of the state-of-the-art methods. Then, we evaluated if the previous descriptors could benefit from a combination of them. In particular, we evaluated if the invariant moments could be more discriminative if extracted starting from a different representation instead of directly from the original images. Thus, we computed Hu, Zernike and Legendre moments starting from the LBP images and from the GLCM computed with angles $0°, 45°, 90°$ and $135°$. In order to better understand the behaviour of the single descriptors and their combination we report a plot in Fig. 5 where the average accuracy calculated on every descriptor applied to each of the datasets from the grayscale images is showed. As it can be observed, all the invariant moments, and in particular Zernike and Legendre moments, are more discriminant if extracted from a different representation. However, in order to further improve the classification performances, a second experiment has been conducted. We extracted the features considering the R, G, B, channels colour information, by computing every descriptor for each of the colour channels and then concatenating the results of the three channels in the same feature vector as we proposed in [27]. In that work we demonstrated that the performance of a descriptor depends on the used color model. So, in order to make a fair comparison of our descriptors, in this work we have chosen the RGB color space. A plot that sums up this experiment is presented in Fig. 6. The performance of all the descriptors improves considerably by using colour information.

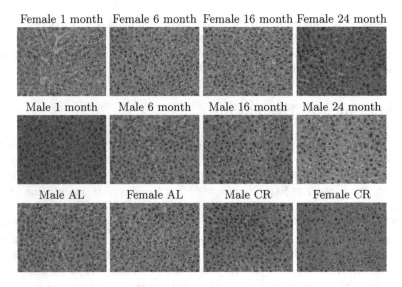

Fig. 4. Four liver images representing: female mice of the different ages (top), male mice of the different ages (center), male and female mice on Ad-libitum diet and on caloric restriction diet.

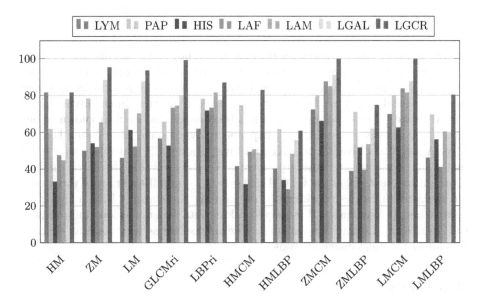

Fig. 5. Average performances of the descriptors extracted from the grayscale converted images. (Color figure online)

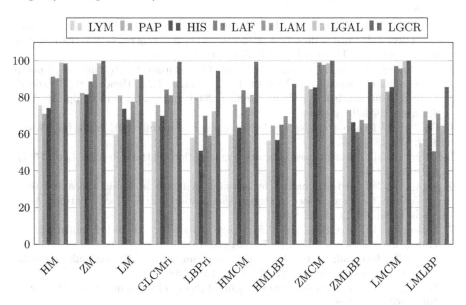

Fig. 6. Average performances of the descriptors extracted from the RGB images. (Color figure online)

5   Conclusions

In this work we have proposed a comparative study between different descriptors in analysing histological images. We focused the comparison on descriptors invariant to image rotations and, in particular, we measured the accuracy of moments, local binary patterns and co-occurrence matrices in classifying histological images. The experimentation has been conducted on well known public biomedical datasets: HistologyDS, Pap-smear, Lymphoma, Liver Aging Female, Liver Aging Male, Liver Gender AL and Liver Gender CR that represent a set of really different computer vision problems. We observed that, by extracting the invariant moments from the GLCM matrices, the overall accuracy of the invariant moments increases considerably, overcoming the classical LBP ang GLCM approaches. In particular, if extracted taking into account colour information, the Zernike and Legendre moments impose themselves as very powerful descriptors for histological image analysis.

References

1. Shamir, L., Orlov, N., Eckley, D.M., Macura, T., Goldberg, I.G.: A proposed benchmark suite for biological image analysis. Med. Biol. Eng. Comput. **46**(9), 943–947 (2008)
2. Karkanis, S.A., Iakovidis, D.K., Maroulis, D.E., Karras, D.A., Tzivras, M.: Computer-aided tumor detection in endoscopic video using colour wavelet features. IEEE Trans. Inf. Technol. BioMed. **7**(3), 141–152 (2003)
3. Ameling, S., Wirth, S., Paulus, D., Lacey, G., Vilarino, F.: Texture-based polyp detection in colonoscopy. In: Meinzer, H.P., Deserno, T.M., Handels, H., Tolxdorff, T. (eds.) Bildverarbeitung für die Medizin 2009. Informatik aktuell, pp. 346–350. Springer, Heidelberg (2009)
4. Hu, M.K.: Visual pattern recognition by moment invariants. IRE Trans. Inf. Theory **8**(2), 179–187 (1962)
5. Tuceryan, M.: Moment based texture segmentation. Pattern Recogn. Lett. **15**(7), 659–668 (1994)
6. Bigun, J.: N-folded symmetrics by complex moments in Gabor space and their application to unsupervised texture segmentation. IEEE Trans. Pattern Anal. Mach. Intell. **16**(1), 80–87 (1994)
7. Mukundan, R., Ong, S.H., Lee, P.A.: Image analysis by Tchebichef moment. IEEE Trans. Image Process. **10**(9), 1357–1364 (2001)
8. Yap, P.T., Raveendran, P., Ong, S.H.: Image analysis by Krawtchouk moments. IEEE Trans. Image Process. **12**(11), 1367–1377 (2003)
9. Teague, M.R.: Image analysis via the general theory of moments. J. Opt. Soc. Am. **70**(8), 920–930 (1980)
10. Di Ruberto, C., Morgera, A.: A comparison of 2-D moment-based description techniques. In: Roli, F., Vitulano, S. (eds.) ICIAP 2005. LNCS, vol. 3617, pp. 212–219. Springer, Heidelberg (2005). doi:10.1007/11553595_26
11. Di Ruberto, C., Morgera, A.: Moment-based techniques for image retrieval. In: 19th International Workshop on Database and Expert Systems Applications, pp. 155–159 (2008)

12. Haddadnia, J., Ahmadi, M., Faez, K.: An efficient feature extraction method with pseudo-Zernike moment in RBF neural network-based human face recognition system. J. Appl. Signal Process. **9**, 890–901 (2003)
13. Hosny, K.M., Papakostas, G.A., Koulouriotis, D.E.: Accurate reconstruction of noisy medical images using orthogonal moments. In: 18th International Conference on Digital Signal Processing (DSP), pp. 1–6, 1–3 July 2013
14. Vijayalakshmi, B., Bharathi, V.S.: Classification of CT liver images using local binary pattern with Legendre moments. Curr. Sci. **110**(4), 687 (2016)
15. Wu, K., Garnier, C., Coatrieux, J.L.: A preliminary study of moment-based texture analysis for medical images. In: 32nd Annual International Conference of the IEEE EMBS, pp. 5581–5584 (2010)
16. Iscan, Z., Dokur, Z., Olmez, T.: Tumor detection by using Zernike moments on segmented magnetic resonance brain images. Expert Syst. Appl. **37**(3), 2540–2549 (2010)
17. Tahmasbi, A., Saki, F., Shokouhi, S.B.: Classification of benign and malignant masses based on Zernike moments. Comput. Biol. Med. **41**, 726–735 (2001)
18. Dogantekin, E., Yilmaz, M., Dogantekin, A., Avci, E., Sengur, A.: A robust technique based on invariant moments - ANFIS for recognition of human parasite eggs in microscopic images. Expert Syst. Appl. **35**(3), 728–738 (2008)
19. Liyun, W., Hefei, L., Fuhao, Z., Zhengding, L., Zhendi, W.: Spermatogonium image recognition using Zernike moments. Comput. Methods Progra. Biomed. **95**(1), 10–22 (2009)
20. Oujaoura, M., Minaoui, B., Fakir, M.: Image annotation by moments. In: Papakostas, G.A. (ed.) Moments and Moment Invariants - Theory and Applications, vol. 1, no. 10, pp. 227–252. Science Gate Publishing (2014)
21. Haralick, R.M., Shanmugam, K., Dinstein, I.: Textural features for image classification. IEEE Trans. Syst. Man Cybern. **3**(6), 610–621 (1973)
22. Gelzinis, A., Verikas, A., Bacauskiene, M.: Increasing the discrimination power of the co-occurrence matrix-based features. Pattern Recogn. **40**(9), 2367–2372 (2007)
23. Walker, R., Jackway, P., Longstaff, D.: Genetic algorithm optimization of adaptive multi-scale GLCM features. Int. J. Pattern Recogn. Artif. Intell. **17**(1), 17–39 (2003)
24. Chen, S., Chengdong, W., Chen, D., Tan, W.: Scene classification based on gray level-gradient co-occurrence matrix in the neighborhood of interest points. In: IEEE International Conference on Intelligent Computing and Intelligent Systems, pp. 482–485 (2009)
25. Benco, M., Hudec, R.: Novel method for colour textures features extraction based on GLCM. Radioengineering **4**(16), 64–67 (2007)
26. Di Ruberto, C., Fodde, G., Putzu, L.: Comparison of statistical features for medical colour image classification. In: Nalpantidis, L., Krüger, V., Eklundh, J.-O., Gasteratos, A. (eds.) ICVS 2015. LNCS, vol. 9163, pp. 3–13. Springer, Cham (2015). doi:10.1007/978-3-319-20904-3_1
27. Di Ruberto, C., Fodde, G., Putzu, L.: On different colour spaces for medical colour image classification. In: Azzopardi, G., Petkov, N. (eds.) CAIP 2015. LNCS, vol. 9256, pp. 477–488. Springer, Cham (2015). doi:10.1007/978-3-319-23192-1_40
28. Ojala, T., Pietikäinen, M., Harwood, D.: A comparative study of texture measures with classification based on feature distributions. Pattern Recogn. **29**, 51–59 (1996)
29. Ojala, T., Pietikäinen, M., Mäenpää, T.: Multiresolution gray-scale and rotation invariant texture classification with local binary patterns. IEEE Trans. Pattern Anal. Mach. Intell. **24**(7), 971–987 (2002)

30. Cruz-Roa, A., Caicedo, J.C., González, F.A.: Visual pattern mining in histology image collections using bag of features. J. Artif. Intell. Med. **52**, 91–106 (2011)
31. Jantzen, J., Dounias, G.: Analysis of pap-smear data. In: NISIS 2006, Puerto de la Cruz, Tenerife, Spain (2006)
32. Zahn, J.M., Poosala, S., Owen, A.B., Ingram, D.K., Lustig, A., Carter, A., Becker, K.G., et al.: AGEMAP: a gene expression database for aging in mice. PLoS Genet. **3**(11), e201 (2007)

Just DIAL: DomaIn Alignment Layers for Unsupervised Domain Adaptation

Fabio Maria Carlucci[1](✉), Lorenzo Porzi[2], Barbara Caputo[1],
Elisa Ricci[3,4], and Samuel Rota Bulò[3]

[1] Sapienza University, Rome, Italy
`fabiom.carlucci@dis.uniroma1.it`
[2] Institut de Robotica I Informatica Industrial CSIC-UPC, Barcelona, Spain
[3] Fondazione Bruno Kessler, Trento, Italy
[4] University of Perugia, Perugia, Italy

Abstract. The empirical fact that classifiers, trained on given data collections, perform poorly when tested on data acquired in different settings is theoretically explained in domain adaptation through a shift among distributions of the source and target domains. Alleviating the domain shift problem, especially in the challenging setting where no labeled data are available for the target domain, is paramount for having visual recognition systems working in the wild. As the problem stems from a shift among distributions, intuitively one should try to align them. In the literature, this has resulted in a stream of works attempting to align the feature representations learned from the source and target domains by introducing appropriate regularization terms in the objective function. In this work we propose a different strategy and we act directly at the distribution level by introducing *DomaIn Alignment Layers* (DIAL) which reduce the domain shift by matching the source and target feature distributions to a canonical one. Our experimental evaluation, conducted on a widely used public benchmark, demonstrates the advantages of the proposed domain adaptation strategy.

Keywords: Unsupervised domain adaptation · Deep models · Feature normalization · Entropy loss

1 Introduction

Many scientists today believe we are witnessing the golden age of computer vision. The massive adoption of machine learning and, in particular, of deep learning techniques as well as the availability of large fully annotated datasets have enabled amazing progresses in the field. A natural question is if the novel generation of computer vision technologies is robust enough to operate in real world scenarios. One of the fundamental requirements for developing systems working in the wild is devising computational models which are immune to the domain shift problem, *i.e.* which are accurate when test data are drawn from a (slightly) different data distribution than training samples. Unfortunately, recent

© Springer International Publishing AG 2017
S. Battiato et al. (Eds.): ICIAP 2017, Part I, LNCS 10484, pp. 357–369, 2017.
https://doi.org/10.1007/978-3-319-68560-1_32

studies in the literature have shown that, even with powerful deep architectures, the domain shift problem can only be alleviated but not entirely solved [1] and several methods for deep domain adaptation have been developed.

Domain adaptation focuses on learning classification or regression models on some target data by exploiting additional knowledge derived from a related source task. In particular, unsupervised domain adaptation focuses on the challenging scenario where no labeled data are available in the target domain. Several approaches have been proposed for unsupervised domain adaptation in the past, the most successful of which are based on deep architectures [2–5]. Previous unsupervised domain adaptation methods can be roughly divided in two categories. The first category includes methods which attempt to reduce the discrepancy between source and target distributions by minimizing the distance between the mean embeddings of the learned representations, *i.e.* the so-called Maximum Mean Discrepancy (MMD) [2,5]. A second class of methods learns domain invariant features by maximizing a domain-confusion objective function, modelling the loss of an auxiliary classifier which should discriminate if a sample belongs to the source or to the target domain [3,4].

Following these recent approaches, in this paper we present a domain adaptation method which simultaneously learns discriminative deep representations while coping with domain shift in the unsupervised setting. Differently from previous works, we do not focus on learning domain-invariant features by *explicitly* optimizing additional loss terms (*e.g.* MMD, domain-confusion). We argue instead that domain adaptation can be achieved by embedding in the network some *Domain Alignment* layers (DA-layers) which operate by aligning both source and target distributions to a canonical one. We also show that several different transformations can be employed in our DA-layers to match source and target data distributions to the reference, thus highlighting the generality of our approach. We call our algorithm DIAL – DomaIn Alignment Layers. Our experimental evaluation, conducted on the most widely used domain adaptation benchmark, *i.e.* the **Office-31** [6] dataset, demonstrates that DIAL greatly alleviates the domain discrepancy and outperforms most state of the art techniques.

2   Related Work

In the last decade unsupervised domain adaptation have received considerable interest in the computer vision community as in many applications labeled data are not available in the target domain [2–4,7–13].

Traditional methods for unsupervised domain adaptation attempt to reduce the domain shift by adopting two main approaches. A first strategy, the so-called instance re-weighting [7–11], is based on building models for the target domain by adopting appropriately re-weighted source samples. The idea is to assign different importance to source samples such as to reflect their similarity with the target data. This approach has been proposed in [7] where a nonparametric method called Kernel Mean Matching is used to set weights without explicitly estimating the data distributions. Similarly, Gong *et al.* [10] introduced the notion of landmark datapoints, a subset of source samples which are similar to target data, and proposed a landmark-based domain adaptation method. Chu *et al.* [8] presented a framework for joint source sample selection and classifier learning. While these works considered hand-crafted features, similar ideas can be also exploited in the case of deep architectures. An example is the work in [11] where deep autoencoders are used to build source sample weights.

The large majority of previous unsupervised domain adaptation methods are based on feature alignment, *i.e.* domain shift is reduced by projecting source and target data in a common subspace. Several feature alignment methods have been proposed in the past, both considering shallow models [14–16] and deep architectures [2–4]. Focusing on works adopting deep architectures, most methods align source and target feature representations by adding in the objective function a regularization term attempting to (i) reduce Maximum Mean Discrepancy [2,5,17] or (ii) maximize a domain confusion loss [3,4]. Recent studies have also investigated alternative methodologies, such as building specific encoder-decoder networks to jointly learn source labels and reconstruct unsupervised target images [18,19]. Our approach significantly departs from previous works by reducing the discrepancy between source and target distributions through the introduction of our DA-layers. The most similar work to ours is [20] where Li *et al.* proposed to revisit batch normalization for deep domain adaptation: BN layers are used to independently align source and target distributions to a standard normal distribution, by matching the first- and second-order moments. While our approach develops from a similar intuition, our method can be regarded as a generalization of [20], as we consider several transformation in our DA layers and we introduce a prior over the network parameters in order to benefit from the target samples during training. Experiments presented in Sect. 4 show the significant added value of our idea.

3   DIAL: DomaIn Alignment Layers

Let $\mathcal{X}$ and $\mathcal{Y}$ denote the input space (*e.g.* images) and the output space (*e.g.* image categories) of our learning task, respectively. We consider an unsupervised domain adaptation setting, where we have a *source domain* described in terms of a probability distribution p^s_{xy} over $\mathcal{X} \times \mathcal{Y}$ and a *target domain* following p^t_{xy}. The source and target distributions differ in general and are unknown, but we are provided with n labeled observations $\mathcal{S} = \{(x^s_1, y^s_1), \ldots, (x^s_n, y^s_n)\}$ from the source domain, *i.e.* they are sampled from p^s_{xy}, and m unlabeled observations $\mathcal{T} = \{x^t_1, \ldots, x^t_m\}$ sampled from the marginal distribution p^t_x. The goal of the learning task is to estimate a predictor for the target domain, using the observations in $\mathcal{S}$ and $\mathcal{T}$. This task is particularly challenging because we lack observed labels from the target domain and the discrepancy between the source and target domains, which in general exists, prevents predictors trained on the source domain to be readily applicable to samples from the target domain.

One key element for the success of an unsupervised domain adaptation algorithm is its ability of reducing the discrepancy between source and target domains. There are different approaches to achieve this goal, but we focus on aligning the domains at the feature level. Within this family of methods the most successful ones *couple* the training process and the domain adaptation step within *deep* neural architectures [2,4,5], yielding alignments at different level of abstractions. Our method is close in spirit to this line of works but we distinguish from them by *(a)* not relying on the covariate shift assumption, *i.e.* we in general assume $p_{y|x}^s \neq p_{y|x}^t$, and by *(b)* hard-coding the domain-invariance properties directly into our deep neural network. The rationale behind the former choice is the impossibility theorem for domain adaptation given in [21], which intuitively states that no domain adaptation algorithm can succeed (in terms of the notion of learnability) if it relies on the covariate shift assumption and achieves a low discrepancy between the source and target unlabeled distributions, *i.e.* p_x^s and p_x^t, respectively. Since the latter assumption is what one implicitly pursues by performing domain alignments at the feature level, we drop the former assumption. The other distinguishing aspect of our method is an architectural solution to achieve domain-invariance, which contrasts with the majority of approaches that rely on additional loss terms (*e.g.* MMD-type losses [2] or adversarial losses [3,4]) that induce an *external* pressure on the networks' parameters at training time to fulfill the domain-invariance requirement. Works exists that do not rely on the covariate-shift assumption and take a loss-based approach to feature alignment, but those typically implement the source and target predictors using different sets of parameters (not necessarily disjoint) [5,22]. Instead, the method we propose is able to avoid the covariate shift assumption and at the same time have the set of learnable network parameters, denoted by θ in this work, that is totally shared. The key element of our method is the domain-alignment layer that we describe below.

3.1   Source and Target Predictors

We implement source and target predictors as two deep neural networks that share the same structure and the same parameters given by θ. However, the two networks differ by having a number of layers that perform a domain-specific operation. Those layers are called *Domain-Alignment Layers* (DA-layers) and their role is to apply a data transformation that aligns the input distribution to a pre-fixed reference distribution. In Fig. 1, we provide an illustration of the basic principle. In general, the input distributions to DA-layers in the source and target predictors differ, but the reference distribution remains fixed. As a result, the data transformations that are applied in the DA-layers of the source and target predictors differ. Consequently, source and target predictors implement different functions, thus violating the aforementioned covariate shift assumption, while still sharing the same set of learnable parameters. More details about the neural network architectures will be provided in the experimental section.

   To better understand how the domain-alignment transformation works, we consider a single DA-layer in isolation. The desired output distribution, namely the reference distribution, is decided a priori and thus known. The input distribution instead is unknown, but we can rely on a sample $\mathcal{D}$ thereof. Now given a transformation g from a family of transformations $\mathcal{G}$ we can push the reference distribution into the pre-image under g via a variable change. This yields

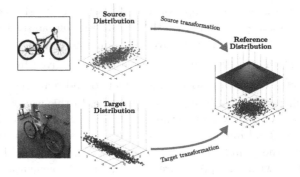

Fig. 1. DIAL learn a pair of transformations that shift the observed source and target distribution to match a desired reference distribution.

a family of distributions among which we can select the one, say $\hat{g}$, that most likely represents sample $\mathcal{D}$. In other words, if v is a random variable following the reference distribution and we assume that the input observations in $\mathcal{D}$ are realizations of random variable $u = g^{-1}(v)$, then we can determine the transformation $\hat{g} \in \mathcal{G}$ as the one that maximizes the likelihood $p_u(\mathcal{D}|g)$. We can alternatively encode some prior knowledge about the transformation by taking a Maximum-A-Posteriori (MAP) approach and thus maximize $p_u(g|\mathcal{D}, \psi)$, where ψ encodes hyper-parameters governing the prior over g.

This idea paves the way to a number of transformations that could be obtained by playing with different reference distributions and families of transformations. In this work, we restrict our focus to some families of DA-layers. In all the cases we consider in this work we assume that $\mathcal{G}$ consists of channel-wise linear transformations of the form $\mathcal{G} = \{u \mapsto \operatorname{diag}(a)^{-\frac{1}{2}}(u-b) : a, b \in \mathbb{R}^d, a > 0\}$ where $\operatorname{diag}(a)$ is a diagonal matrix with diagonal elements given by a. A first family of approaches is obtained by imposing the standard normal distribution as reference distribution and depending on the prior knowledge we inject we obtain the following variations of DA-layers:

Batch normalization. By pushing the standard normal distribution, *i.e.* the reference distribution of v, into the pre-image under $g \in \mathcal{G}$ we obtain a distribution for random variable $u = g^{-1}(v)$ that is normal with mean b and covariance $\operatorname{diag}(a)$. The maximum likelihood estimates of a and b given sample $\mathcal{D}$, consisting of i.i.d. realizations of u, are given by $\hat{a} = \sigma^2(\mathcal{D})$ and $\hat{b} = \mu(\mathcal{D})$ respectively, where $\mu(\mathcal{D})$ and $\sigma^2(\mathcal{D})$ represent the sample mean and the diagonal of the sample covariance of $\mathcal{D}$, respectively. The resulting domain-alignment transformation is $\hat{g}(u) = \operatorname{diag}(\sigma^2(\mathcal{D}))^{-\frac{1}{2}}[u - \mu(\mathcal{D})]$. This transformation corresponds to the well-known *batch-normalization* layer [23], when $\mathcal{D}$ is the mini-batch of a training iteration.

Batch normalization with prior on variance. This setting is similar to the previous one, but instead of considering a maximum likelihood estimate of the transformation parameter a we opt for a MAP estimate. To this end we introduce an Inverse-Gamma(α,β) prior on the transformation parameter a, yielding a posterior distribution for a that is Inverse-Gamma($\bar{\alpha},\bar{\beta}$) with $\bar{\alpha} = \alpha + \frac{|\mathcal{D}|}{2}$ and $\bar{\beta} = \beta + \frac{|\mathcal{D}|}{2}\sigma^2(\mathcal{D})$. The corresponding MAP estimate is given by $\hat{a} = \frac{\bar{\beta}}{\bar{\alpha}+1}$. The hyperparameters of the prior distribution, namely α and β are set to $\alpha = \frac{|\mathcal{D}|}{2} - 1$ and $\beta = \epsilon\frac{\mathcal{D}}{2}$, where ϵ is intuitively a prior variance. In this way we have that $\hat{\beta}$ gives approximately equal weight to the sample variance and the prior variance, yielding $\hat{\beta} = \epsilon + \sigma^2(\mathcal{D})$. Finally, the estimate of b remains the maximum likelihood estimate, namely the sample mean, *i.e.* $\hat{b} = \mu(\mathcal{D})$. Note that the data transformation we obtain with this procedure is the actual implementation of batch normalization that we find in most deep learning frameworks, for ϵ typically appears as a small additive constant for the variance that prevents numerical issues. In our case, however ϵ is not necessarily set to a small constant as we will see in the experimental section.

A second family of approaches is obtained by imposing the Laplace distribution as reference distribution. In this case we do not explore variations involving prior knowledge, although it would be possible.

Laplace Batch normalization. If we assume that the reference distribution follows a standard Laplace distribution, then the maximum likelihood estimate $\hat{b}$ corresponds to the sample median, while the maximum likelihood estimate of a is given by the mean absolute value deviation from the sample median, *i.e.* $\hat{a} = \frac{1}{|\mathcal{D}|}\sum_{x\in\mathcal{D}}|x - \hat{b}|$.

3.2   Training and Inference

Training. During the training phase we consider the datasets $\mathcal{S}$ and $\mathcal{T}$ and we estimate the neural network weights θ. Note that these parameters are shared by the source and the target predictors. To compute θ we define a posterior distribution of θ given the observations $\mathcal{S}$ and $\mathcal{T}$, $\pi(\theta|\mathcal{S},\mathcal{T})$, and maximize it over Θ to obtain a MAP estimate $\hat{\theta}$:

$$\hat{\theta} \in \arg\max_{\theta\in\Theta} \pi(\theta|\mathcal{S},\mathcal{T}). \tag{1}$$

The posterior distribution is defined as $\pi(\theta|\mathcal{S},\mathcal{T}) \propto \pi(y_\mathcal{S}|x_\mathcal{S},\theta)\pi(\theta|\mathcal{T})$, where $y_\mathcal{S} = \{y_1,\dots,y_n\}$ and $x_\mathcal{S} = \{x_1,\dots,x_n\}$ indicate the set of labels and data points in $\mathcal{S}$, respectively. The term $\pi(y_\mathcal{S}|x_\mathcal{S},\theta)$ is the likelihood of θ with respect to the source dataset, while $\pi(\theta|\mathcal{T})$ is a prior term depending on the unlabeled target samples. Assuming the data samples to be *i.i.d.*, the likelihood term is given by

$$\pi(y_\mathcal{S}|x_\mathcal{S},\theta) = \prod_{i=1}^{n} f_{y_i^s}^\theta(x_i^s;x_\mathcal{S}), \tag{2}$$

where $f_{y_i^s}^\theta(x_i^s;x_\mathcal{S})$ is the probability that sample point x_i^s takes label y_i^s according to the source predictor.

In analogy to previous works on semi-supervised learning [24] and unsupervised domain adaptation [5], the prior distribution $\pi(\theta|\mathcal{T})$ is defined in order to promote models that exhibit well separated classes. This is achieved by defining $\pi(\theta|\mathcal{T}) \propto \exp(-\lambda h(\theta|\mathcal{T}))$, where λ is a user-defined parameter and $h(\theta|\mathcal{T})$ is the empirical entropy of $y|\theta$ conditioned on x, $i.e.$:

$$h(\theta|\mathcal{T}) = -\frac{1}{m} \sum_{i=1}^{m} \sum_{y \in \mathcal{Y}} f_y^{\theta}(x_i^t; x_{\mathcal{T}}) \log f_y^{\theta}(x_i^t; x_{\mathcal{T}}), \tag{3}$$

where $f_y(x_i^t; \mathcal{T})$ represents the probability that sample point x_i^t takes label y according to the target predictor.

Inference. Once the optimal network parameters $\hat{\theta}$ are estimated by solving (1), the dependence of the target predictor $f_y^{\theta}(x; x_{\mathcal{T}})$ from $x_{\mathcal{T}}$ can be removed. In fact, after fixing $\hat{\theta}$, the input distribution to each DA-layer also becomes fixed, and we can thus compute and store the required transformation once and for all. *E.g.* , for the special case of *Batch normalization* discussed in Sect. 3.1, this means simply to store the values of $\mu(\mathcal{D})$ and $\sigma(\mathcal{D})$.

4 Experiments

In this section we extensively evaluate our approach and compare it with state-of-the-art unsupervised domain adaptation methods. We also provide a detailed analysis of the proposed framework, performing a sensitivity study and demonstrating empirically the effect of our domain alignment strategy.

4.1 Experimental Setup

To evaluate the proposed approach, we consider the **Office-31** [6] dataset. Office-31 is a standard benchmark for testing domain adaptation methods. It contains 4652 images organized in 31 classes from three different domains: Amazon (A), DSRL (D) and Webcam (W). Amazon images are collected from amazon.com, Webcam and DSLR images were manually gathered in an office environment. In our experiments we consider all possible source/target combinations of these domains and adopt the *full protocol* setting [10], *i.e.* we train on the entire labeled source and unlabeled target data and test on annotated target samples.

Networks and Training. We apply the proposed method to two state-of-the-art CNNs, *i.e.* AlexNet [25] and Inception-BN [23]. We train our networks using mini-batch stochastic gradient descent with momentum, as implemented in the Caffe library, using the following meta-parameters: weight decay 5×10^{-4}, momentum 0.9, initial learning rate 10^{-3}. We augment the input data by scaling all images to 256×256 pixels, randomly cropping 227×227 pixels (for AlexNet) or 224×224 pixels (Inception-BN) patches and performing random flips. In all experiments we choose the parameter λ, which is fixed for tests of a given setting, by cross-validation.

AlexNet [25] is a well-know architecture with five convolutional and three fully-connected layers, denoted as fc6, fc7 and fc8. The outputs of fc6 and fc7 are commonly used in the domain-adaptation literature as pre-trained feature representations [1,26] for traditional machine learning approaches. In our experiments we modify AlexNet by appending a DA-layer to each fully-connected layer. Differently from the original AlexNet, we *do not* perform dropout on the outputs of fc6 and fc7. We initialize the network parameters from a publicly-available model trained on the ILSVRC-2012 data, we finetune all layers, and learn from scratch the last fc layer (we increase its learning rate by a factor of 10). During training, each mini-batch contains a number of source and target samples proportional to the size of the corresponding dataset, while the batch size remains fixed at 256. We train for a total of 60 epochs (where "epoch" refers to a complete pass over the source set), reducing the learning rate by a factor 10 after 54 epochs.

Inception-BN [23] is a very deep architecture obtained by concatenating "inception" blocks. Each block is composed of several parallel convolutions with batch normalization and pooling layers. To apply the proposed method to Inception-BN, we replace each batch-normalization layer with a DA-layer. Similarly to AlexNet, we initialize the network's parameters from a publicly-available model trained on the ILSVRC-2012 data and freeze the first three inception blocks. Each batch is composed of 32 source images and 16 target images. In the Office-31 experiments we train for 20 epochs, reducing the learning rate by a factor 10 every 33% of the total number of iterations.

DIAL Variations. To evaluate the robustness of our framework, we tested the 3 DIAL variations we discussed in Sect. 3.1: classical Batch Normalization, reported as *BN*, Batch Normalization with prior on variance, reported as *Epsilon*[1], Laplacian Batch Normalization, reported as *Laplacian BN*.

Furthermore, we also tested a new sparse regularizer that has been recently proposed in [27], which operates at level of the centered features in the batch-normalization layer (before normalization by the variance). This is beneficial in terms of decorrelating the features and can be integrated readily in our framework. We consider the new regularizer for our DA-layers that are based on batch-normalization and regard them as Batch Normalization with sparsity, reported as *sparse* and Batch Normalization with prior on variance and sparsity, reported as *Epsilon sparse*.

4.2   Results

Comparison with State-of-the Art Methods. In our first series of experiments, summarized in Table 1, we compare our approach, applied to both

---

[1] The ϵ parameter is set to 1 for all experiments.

Table 1. Results on the Office-31 dataset using the full protocol.

Method	Source Target	Amazon Webcam	Amazon DSLR	Webcam Amazon	Webcam DSLR	DSLR Amazon	DSLR Webcam	Average
AlexNet – source [25]		61.6	63.8	49.8	99.0	51.1	95.4	70.1
DDC [28]		61.8	64.4	52.2	98.5	52.1	95.0	70.6
DAN [2]		68.5	67.0	53.1	99.0	54.0	96.0	72.9
ReverseGrad [4]		73.0	–	–	99.2	–	96.4	–
DIAL – AlexNet *sparse*		76.5	72.4	55.9	99.4	58.6	97.0	76.5
Inception-BN – source [23]		70.3	70.5	57.9	**100.0**	60.1	94.3	75.5
AdaBN [20]		74.2	73.1	57.4	99.8	59.8	95.7	76.7
AdaBN + CORAL [20]		75.4	72.7	60.5	99.6	59.0	96.2	77.2
DIAL – Inception-BN *BN*		**82.9**	**87.3**	**62.6**	99.9	**63.1**	**98.2**	**82.4**

AlexNet and Inception-BN, with several state-of-the-art methods on the Office-31 dataset. In particular, we consider: several deep methods based on AlexNet-like architectures, *i.e.* Deep Adaptation Networks (DAN) [2], Deep Domain Confusion (DDC) [28], the ReverseGrad network [4]; a recent deep method based on the Inception-BN architecture, *i.e.* AdaBN [20] with and without CORAL feature alignment [26]. We compare these baselines to the AlexNet and Inception-BN networks modified with our approach as explained in Sect. 4.1, reporting the best results among the DA-layer variations we experimented with (see Table 2). In the table our approach is denoted as DIAL – AlexNet and DIAL – Inception-BN. As a reference, we further report the results obtained considering standard AlexNet and Inception-BN networks trained only on source data.

Among the deep methods based on the AlexNet architecture, DIAL – AlexNet shows the best average performance. Among the methods based on Inception-BN, our approach considerably outperforms the others, with an average accuracy of five points higher than the second best, and improvements on the single experiments as high as ten points. It is interesting to note that the relative increase in accuracy from the source-only Inception-BN to DIAL – Inception-BN is higher than that from the source only AlexNet to DIAL – AlexNet. The considerable success of our method in conjunction with Inception-BN can be attributed to the fact that, differently from AlexNet, this network is pre-trained with batch normalization, and thus initialized with weights that are already calibrated for normalized features.

In-Depth Analysis of DA-Layers. In our second series of experiments we aim to characterize the effects of different variations of the proposed DA-layers. To do this, we perform an ablation study considering all possible combinations of the following network variations: (i) with and without the entropy term on the target samples in the loss function; (ii) with and without DA-layers; (iii) with the DA-layer variations (Sect. 4.1).

The results are reported in Table 2, and further synthesized in Fig. 2. As anticipated in the previous section, the *DIAL – AlexNet sparse* variant achieves the best accuracy. Overall, independently from the particular DA-layer variant, the networks utilizing our proposal in its full extent (*i.e.* those in the "With entropy

Table 2. Analysis of the different variants of the proposed DA layers on the Office-31 dataset using the full protocol.

| Method | Source | Amazon | Amazon | Webcam | Webcam | DSLR | DSLR | Average |
	Target	Webcam	DSLR	Amazon	DSLR	Amazon	Webcam	
Baselines								
AlexNet – source [25]		61.6	63.8	49.8	99.0	51.1	95.4	70.1
AlexNet – Entropy loss		63.7	65.6	35.5	96.6	42.9	99.6	67.3
With entropy loss								
DIAL – AlexNet *BN*		73.2	71.7	56.2	99.3	59.6	95.9	76.0
DIAL – AlexNet *Epsilon*		71.6	71.7	56.7	99.3	59.4	99.2	76.3
DIAL – AlexNet *sparse*		76.5	72.4	55.9	99.4	58.6	97.0	76.5
DIAL – AlexNet *Epsilon sparse*		72.1	72.3	57.0	99.7	59.0	97.2	76.2
DIAL – AlexNet *Laplacian BN*		73.0	72.0	55.1	98.7	56.7	96.6	75.4
Without entropy loss								
DIAL – AlexNet *BN*		62.2	65.5	47.1	99.2	47.6	95.2	69.5
DIAL – AlexNet *Epsilon*		65.3	64.5	47.3	99.5	48.4	95.0	70.0
DIAL – AlexNet *sparse*		60.6	64.0	47.0	99.3	48.1	95.6	69.1
DIAL – AlexNet *Epsilon sparse*		64.6	65.3	46.9	99.7	48.4	95.7	70.1
DIAL – AlexNet *Laplacian BN*		61.8	65.3	46.8	98.4	46.8	94.8	69.0

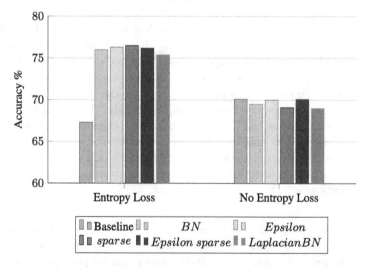

Fig. 2. Comparison of the different variants of the proposed method on the Office-31 dataset (average accuracy across different transfer tasks)

loss" section of Table 2) consistently outperform the others, further confirming the validity of our domain adaptation approach.

From the results in Table 2, we see that the use of an entropy loss term by itself does not provide any advantage over the baseline approach. On the contrary, an average drop in accuracy of about 3% is observed when comparing AlexNet – Entropy loss to AlexNet – source, with partial results greatly varying depending on the particular source/target pair. Interestingly, AlexNet – Entropy loss shows better accuracy compared to AlexNet – source in all the settings in which the target dataset is smaller than the source dataset, i.e. A→W, A→D and D→W. This may be explained by the fact that the entropy term is more effective when there are sufficient source samples to appropriately bias the decision boundary. As shown in Fig. 2, the best performance between the proposed variations of our domain alignment layers are obtained when considering BN with sparse activations. Adding a sparse regularizer on the activations helps to decorrelate the filter responses [27] and our results demonstrate that it has a positive effect on domain adaptation tasks.

5   Conclusions

In this paper we presented DIAL, a general framework for unsupervised, deep domain adaptation. Our main contribution is the introduction of novel, domain-alignment layers, which reduce domain shift by matching source and target distributions to a freely definable reference distribution. We also show that improved performance can be obtained by exploiting unlabeled target data introducing an entropy loss in the objective function. We evaluated the proposed approach devising a simple implementation of our DA-layers based on multiple batch normalization transformations. The results of our experiments demonstrate that DIAL outperforms state-of-the-art domain adaptation methods. Future works will investigate how to extend the proposed approach to a multi-source/multi-target setting. We also plan to consider other reference distributions for domain alignment in order to further improve performance.

References

1. Donahue, J., Jia, Y., Vinyals, O., Hoffman, J., Zhang, N., Tzeng, E., Darrell, T.: Decaf: a deep convolutional activation feature for generic visual recognition. In: ICML (2014)
2. Long, M., Wang, J.: Learning transferable features with deep adaptation networks. In: ICML (2015)
3. Tzeng, E., Hoffman, J., Darrell, T., Saenko, K.: Simultaneous deep transfer across domains and tasks. In: ICCV (2015)
4. Ganin, Y., Lempitsky, V.: Unsupervised domain adaptation by backpropagation. In: ICML (2015)
5. Long, M., Wang, J., Jordan, M.I.: Unsupervised domain adaptation with residual transfer networks. In: NIPS (2016)

6. Saenko, K., Kulis, B., Fritz, M., Darrell, T.: Adapting visual category models to new domains. In: Daniilidis, K., Maragos, P., Paragios, N. (eds.) ECCV 2010. LNCS, vol. 6314, pp. 213–226. Springer, Heidelberg (2010). doi:10.1007/978-3-642-15561-1_16

7. Huang, J., Gretton, A., Borgwardt, K.M., Schölkopf, B., Smola, A.J.: Correcting sample selection bias by unlabeled data. In: NIPS (2006)

8. Chu, W.S., De la Torre, F., Cohn, J.F.: Selective transfer machine for personalized facial action unit detection. In: CVPR (2013)

9. Yamada, M., Sigal, L., Raptis, M.: No bias left behind: covariate shift adaptation for discriminative 3D pose estimation. In: Fitzgibbon, A., Lazebnik, S., Perona, P., Sato, Y., Schmid, C. (eds.) ECCV 2012. LNCS, vol. 7575, pp. 674–687. Springer, Heidelberg (2012). doi:10.1007/978-3-642-33765-9_48

10. Gong, B., Grauman, K., Sha, F.: Connecting the dots with landmarks: discriminatively learning domain-invariant features for unsupervised domain adaptation. In: ICML (2013)

11. Zeng, X., Ouyang, W., Wang, M., Wang, X.: Deep learning of scene-specific classifier for pedestrian detection. In: Fleet, D., Pajdla, T., Schiele, B., Tuytelaars, T. (eds.) ECCV 2014. LNCS, vol. 8691, pp. 472–487. Springer, Cham (2014). doi:10.1007/978-3-319-10578-9_31

12. Zen, G., Sangineto, E., Ricci, E., Sebe, N.: Unsupervised domain adaptation for personalized facial emotion recognition. In: ICMI (2014)

13. Costante, G., Ciarfuglia, T.A., Valigi, P., Ricci, E.: A transfer learning approach for multi-cue semantic place recognition. In: IROS (2013)

14. Gong, B., Shi, Y., Sha, F., Grauman, K.: Geodesic flow kernel for unsupervised domain adaptation. In: CVPR (2012)

15. Long, M., Ding, G., Wang, J., Sun, J., Guo, Y., Yu, P.S.: Transfer sparse coding for robust image representation. In: CVPR (2013)

16. Fernando, B., Habrard, A., Sebban, M., Tuytelaars, T.: Unsupervised visual domain adaptation using subspace alignment. In: ICCV (2013)

17. Sun, B., Saenko, K.: Deep coral: correlation alignment for deep domain adaptation. arXiv preprint (2016). arXiv:1607.01719

18. Ghifary, M., Kleijn, W.B., Zhang, M., Balduzzi, D., Li, W.: Deep reconstruction-classification networks for unsupervised domain adaptation. In: Leibe, B., Matas, J., Sebe, N., Welling, M. (eds.) ECCV 2016. LNCS, vol. 9908, pp. 597–613. Springer, Cham (2016). doi:10.1007/978-3-319-46493-0_36

19. Bousmalis, K., Trigeorgis, G., Silberman, N., Krishnan, D., Erhan, D.: Domain separation networks. In: NIPS (2016)

20. Li, Y., Wang, N., Shi, J., Liu, J., Hou, X.: Revisiting batch normalization for practical domain adaptation. arXiv preprint (2016). arXiv:1603.04779

21. Ben-David, S., Lu, T., Luu, T., Pál, D.: Impossibility theorems for domain adaptation. In: AISTATS (2010)

22. Rozantsev, A., Salzmann, M., Fua, P.: Beyond sharing weights for deep domain adaptation. arXiv preprint (2016). arXiv:1603.06432

23. Ioffe, S., Szegedy, C.: Batch normalization: accelerating deep network training by reducing internal covariate shift. arXiv preprint (2015). arXiv:1502.03167

24. Grandvalet, Y., Bengio, Y.: Semi-supervised learning by entropy minimization. In: NIPS (2004)

25. Krizhevsky, A., Sutskever, I., Hinton, G.E.: Imagenet classification with deep convolutional neural networks. In: NIPS (2012)

26. Sun, B., Feng, J., Saenko, K.: Return of frustratingly easy domain adaptation. In: AAAI (2016)

27. Ren, M., Liao, R., Urtasun, R., Sinz, F.H., Zemel, R.S.: Normalizing the normalizers: comparing and extending network normalization schemes. arXiv preprint (2016). arXiv:1611.04520

28. Tzeng, E., Hoffman, J., Zhang, N., Saenko, K., Darrell, T.: Deep domain confusion: maximizing for domain invariance. arXiv preprint (2014). arXiv:1412.3474

Multi-stage Neural Networks with Single-Sided Classifiers for False Positive Reduction and Its Evaluation Using Lung X-Ray CT Images

Masaharu Sakamoto[1]([✉]), Hiroki Nakano[1], Kun Zhao[2], and Taro Sekiyama[2]

[1] IBM Tokyo Laboratory, Watson Health, Tokyo 103-8510, Japan
sakamoto@jp.ibm.com
[2] IBM Research - Tokyo, Tokyo 103-8510, Japan

Abstract. Lung nodule classification is a class imbalanced problem because nodules are found in much lower frequency than non-nodules. In the class imbalanced problem, conventional classifiers tend to be overwhelmed by the majority class and ignore the minority class. We therefore propose cascaded convolutional neural networks to cope with the class imbalanced problem. In the proposed approach, multi-stage convolutional neural networks perform as single-sided classifiers to filter out obvious non-nodules. Successively, a convolutional neural network trained with a balanced data set calculates nodule probabilities. The proposed method achieved the sensitivity of 92.4% and 94.5% at 4 and 8 false positives per scan in Free Receiver Operating Characteristics (FROC) curve analysis, respectively.

Keywords: Multi-stage training · Convolutional neural network · False positive reduction · Computer-aided diagnosis · Lung nodule

1 Introduction

Lung cancer occupies a high percentage in the mortality rates of cancer on a worldwide basis [1]. Early detection is one of the most promising strategies to reduce lung cancer mortality [2]. In recent years, along with performance improvements of CT equipment, increasingly large numbers of tomographic images are being taken (e.g., at slice intervals of 1 mm), resulting in improvements in the ability of radiologists to distinguish nodules. However, there is a limitation on interpreting a large number of images (e.g., 300–500 slices/scan) by relying on humans. Computer-aided diagnosis (CAD) systems show promise for the urgent task of time-efficient interpretation of CT scans. In one study [2], six computer-aided detection algorithms of lung nodules in computed tomography scans were compared. These methods extract features in lung nodule images with a signal processing technique and classify nodule candidates by using pattern matching based on statistics or a machine learning method such as the k-nearest neighbor algorithm (k-NN) and neural networks. By combining six computer-aided diagnosis algorithms, they obtained detection sensitivities of 81.6% and

© Springer International Publishing AG 2017
S. Battiato et al. (Eds.): ICIAP 2017, Part I, LNCS 10484, pp. 370–379, 2017.
https://doi.org/10.1007/978-3-319-68560-1_33

87.0% at 4 and 8 false positives per scan in Free Receiver Operating Characteristics (FROC) curve [2], respectively. In recent years, Convolutional Neural Network (CNN) has become available thanks to high speed and large capacity computing resources and it is showing superior performance to conventional technology in computer vision applications [3]. This is because CNN can be trained end-to-end in a supervised fashion while learning highly discriminative features, thus removing the need for handcrafting nodule feature descriptors. Setio et al. [4] used a CNN specifically trained for lung nodule classification. On 888 scans of a publicly available data set (the data set is the same as we used in this study.), their method reached high sensitivities of 90.1% and 91.5% at 4 and 8 false positives per scan in FROC curve, respectively. Dou et al. [19] proposed a method employing 3D CNNs for false positive reduction in automated pulmonary nodule detection from volumetric CT scans. Their detection sensitivity reached 90.7% and 92.2% at 4 and 8 false positives per scan in FROC curve, respectively.

In this paper, we focus on the task of lung nodule classification. The candidate locations are computed using three existing candidate detection algorithms [4]. Lung nodule classification is a class imbalanced problem, as nodules are found in much lower frequency than non-nodules among the candidate images. In other words, many irregular lesions that are visible in CT images are non-nodules, such as blood vessels or ribs. In the class imbalanced problem, conventional classifiers tend to be overwhelmed by the majority class and ignore the minority class. Several approaches have been proposed to deal with such problems in rare medical diagnosis [8], detection of oil spills in satellite radar images [9] and the detection of fraudulent calls [10]. Japkowicz [6] showed that oversampling the minority class and subsampling the majority class are both very effective methods of coping with the problem. Chawla et al. [11] proposed SMOTE (Synthetic Minority Over-sampling Technique) algorithm that is artificially creating minor class and randomly sub-sample majority class. Kubat and Matwin [7] proposed a one-sided selection method that keeps all minor class samples and subsamples the majority class samples. Sun et al. [12] comprehensively reviewed the class imbalanced problems.

As one method to cope with the class imbalanced problem in lung nodule classification, we propose a method to aim to filter out obvious non-nodules. Our method is completely different from previous methods. It positively utilizes deterioration of classification performance caused by learning of class imbalanced. We call such classifier as single-sided classifier because it filters out majority (nonnodules) class samples only. The single-sided classifier consists of CNN that outputs nodule probabilities and filter that filters out the majority class samples by using a threshold in nodules probability. It has two kinds of outputs: the obvious non-nodules and suspicious nodule candidates which consist of nodules and non-nodules. To implement such single-sided classifiers, the CNN is trained with an inversely imbalanced data set consisting of many nodule images and a few non-nodule images. By "inverse" we mean that the ratio of the number of nodules and non-nodules is reversed against the original data set. As the results, the single-sided classifiers work well for nodule samples, but do not work well for non-nodule samples. By using a threshold operation in nodule probability, the non-nodule samples are classified into obvious non-nodules and suspicious nodule candidates. The obvious non-nodules are dismissed and assigned zero probabilities, the suspicious nodule candidates are passed to the down-stream classifiers. This filtering mechanism contribute to the false positive reduction. In addition, the single-sided classifiers are concatenated in cascade arrangement. Figure 1 shows an illustration of our method. The obvious non-nodules (white circles) are filtered out in each stage, finally suspicious nodule candidates (gray circles) remain. The aim of our method is not to balance the number of samples in majority and minority (nodule) class, rather we want to filter out obvious non-nodules. In the final stage, the CNN trained by a balanced data set extracted from the suspicious nodule candidates calculates the nodule probabilities. By "balanced" we mean that the number of nodules is almost equal to the number of non-nodules. We rely on our newly designed CNNs which have excellent classification ability to calculate nodule probabilities of suspicious nodule candidates. As the result, our method can achieve low false positives while maintaining high sensitivity. It helps decreasing the burden of image interpretation on radiologists.

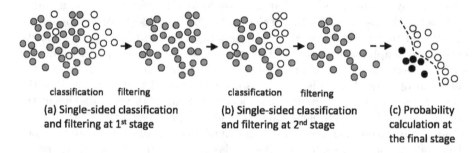

| classification   filtering | classification   filtering | |
| **(a) Single-sided classification and filtering at 1ˢᵗ stage** | **(b) Single-sided classification and filtering at 2ⁿᵈ stage** | **(c) Probability calculation at the final stage** |

Fig. 1. Multi-stage processing with single-sided classifiers. It classifies the samples as suspicious nodule candidates (gray circles) and obvious non-nodules (white circles). The obvious non-nodules are filtered out and the suspicious nodules are passed to the next stage. At the final stage, the suspicious nodules are classified.

2   Multi-stage Neural Networks with Single-Sided Classifiers

Figure 2 shows the schematic diagram of our method. Stage 1, Stage 2 and Stage n are CNNs that perform as single-sided classifiers and gates to filter out low nodule probability samples and pass the other samples to down stage. The final stage is the CNN that calculates nodule probabilities. At Stage 1, by using the CNN that performs as single-sided classifier, the test data set is classified, and then, the samples whose probabilities fall below a threshold are removed from the test data set as the obvious non-nodules. The nodule probabilities of the removed samples are assigned zero. At Stage 2, the same procedures are applied again to remove further the obvious non-nodules from the test data set. In the final stage, the CNN trained by a balanced data set calculates the probabilities of the remaining suspicious nodule candidates. The lower part of Fig. 2 shows the structure of the CNN. There are three main operations in the CNN: (1) Convolution with rectified linear unit, (2) Pooling or sub-sampling (3) classification by fully connected layer. The input to the CNN is extracted 2-D patches from three consecutive slices of X-ray CT scan images. The convolution layer will compute the output of neurons that are connected to local regions in the input, each computing a dot product between their weights and a small region they are connected to in the input volume. The purpose of convolution is to extract features from the input image. The pooling layer performs a sub-sampling operation along the spatial dimensions (width, height), resulting in size of single channel becoming half of input. The last fully-connected layers will compute the nodule probabilities. The same CNN structure with different network is used as the single-sided classifier for the probability calculation at the final stage.

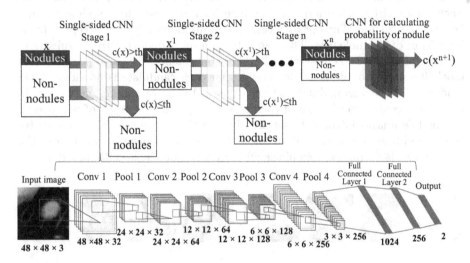

Fig. 2. Schematic diagram of cascaded multi-stage CNNs. Stage 1, Stage 2 and Stage n are CNNs that perform as single-sided classifiers to filter out non-nodule lesions. The final stage is a CNN to calculate nodule probabilities. $c(x)$ is nodule probability of nodule candidate x. th is a threshold value to filter out obvious non-nodules. The lower part shows the structure of the CNN. The numbers at lowest part show number of neurons in three dimensions (width, height and channel).

The unique points of our method are that it uses cascaded multi-stage CNNs that perform as single-sided classifiers and uses the inversely imbalanced data as the training data. In contrast, there are some works (e.g. Viola-Jones [13] and Wu et al. [14]) using the weak classifiers to construct boosted cascade layer with simple features. Compared with the weak classifiers with simple features, convolutional neural network can automatically capture features from the CT images, which can provide higher accuracy for the detection results. As for the cascaded CNN structure, Li et al. [15] have proposed a cascaded CNN structure for face detection. They use 6 CNNs in the cascade including 3 CNNs to detect the face and 3 CNNs to calibrate the bounding box separately. The bounding box calibration is not needed in our proposed method. The application of cascaded CNN for face detection [16] [17] and other kind of image feature detection [18] can also be found. However, the class imbalanced problem is not addressed in those works.

3   Experiments

3.1   Lung CT Image Data Set

We used the lung CT scan data set obtained from Lung Nodule Analysis 2016 [5]. This set includes 888 CT scan images along with annotations that were collected during a two-phase annotation process overseen by four experienced radiologists. Each radiologist marked lesions, they identified as non-nodule, nodule <3 mm, and nodule ≥3 mm. The data set consists of all nodules ≥3 mm accepted by at least 3 out of 4 radiologists. The complete data set is divided into ten subsets to be used for the 10-fold cross-validation. For convenience, the corresponding class label (0 for non-nodule and 1 for nodule) for each candidate is provided. 1,348 lesions are labeled as nodules and the other 551,062 are non-nodule lesions. In this study, center coordinates of each lesion are given. Examples of non-nodule and nodule images in the data set are given in Fig. 3. We use three consecutive slices to obtain volumetric information. Size of each image cropped from CT scan images is 48 pixels × 48 pixels with a central on the nodule candidate.

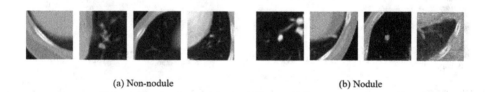

(a) Non-nodule   (b) Nodule

Fig. 3. Example of lesion images in data set of Lung Nodule Analysis 2016 [5].

3.2   Proposed Multi-stage Classifiers

The model of single-sided classifiers and the final stage classifiers are trained and validated by 10-fold cross-validation. In the cross-validation, eight subsets were used for training, and one subset is used for network regularization. The remaining subset is used for calculating nodules probabilities of each image sample. 10 CNN models are made by using the holdout procedures. To prepare the training data set for the single-sided classifiers, non-nodule samples in the subset are subsampled to 50 samples, and nodule samples are oversampled nine times by randomly rotating and scaling original images. As the result, the number of nodules is about twenty-four times the number of non-nodules in the training data set. if the probability value of a nodule candidate falls below a specific threshold value, it is classified as an obvious non-nodule, and removed from the subset and assigned zero probability. The threshold value is determined from a standard deviation σ of the nodule probability distribution of non-nodule samples at each stage. One-tenth of the standard deviation is set as the threshold value. In subsequent stage, the same procedures are repeated on the filtered data set of the previous stage. In the learning loops of the single-sided classifiers, CNN models having the best nodule classification accuracy over all learning epochs are stored. There are 20 epochs in each training in the test phase of single-sided classifiers. At the final stage, the CNN is trained by a balanced data set, extracted from the filtered data set at the previous stage. The CNN models having the best classification accuracy (nodules and non-nodules) over all learning epochs are stored for calculating the probabilities of the nodule candidates of the filtered data set at the previous stage.

3.3   Baseline Classifiers

For performance comparison, the same CNNs are trained and tested by using the same data set in manner of the 10-fold cross-validation. We call this conventional method as baseline. The CNNs are trained using a balanced data set with subsampled non-nodules and oversampled nodules. All of the nodule samples are oversampled nine times by randomly rotating and scaling original images and non-nodules are subsampled to balance the number of oversampled nodules. In the training phase, the CNN models having the best classification accuracy over all learning epochs are recorded.

4   Experimental Results

Figure 4 shows histograms of nodules (class 1) and non-nodules (class 0) probabilities calculated by the single-sided classifiers at the first stage. The probabilities of non-nodule class (class 0) are separated around 0.0 and 1.0. We assume the samples around probability 0.0 can be accepted as obvious non-nodules. At the same time, nodule samples (class 1) around probability 0.0 are erroneously classified as obvious non-nodules. This is what causes the false negatives in our method. Figure 5 shows histograms of the nodule probabilities calculated by the baseline classifiers. Although most of the non-nodule samples are concentrated around probability 0.0, a little concentration is also seen around 1.0 as shown in Fig. 5(b). This is what causes the false positives in the baseline classifiers. The nodules samples around probability 0.0 are more than that of single sided classifiers as shown in Fig. 5(a). This is what cause low sensitivity in the baseline classifiers.

We investigated the performance of proposed method with different ratios of number of class samples in training of single-sided classifiers. The ratios of

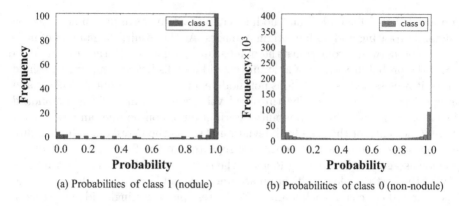

(a) Probabilities of class 1 (nodule)   (b) Probabilities of class 0 (non-nodule)

Fig. 4. Histogram of nodule probabilities calculated by single-sided classifiers.

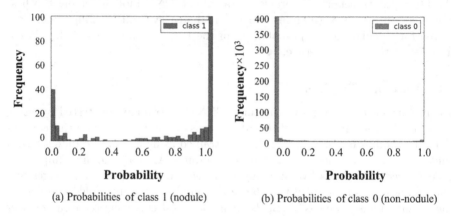

(a) Probabilities of class 1 (nodule)   (b) Probabilities of class 0 (non-nodule)

Fig. 5. Histogram of nodule probabilities calculated by baseline classifiers.

number of nodules to non-nodules are 6 to 1, 12 to 1 and 24 to 1. Figure 6 shows histograms of nodule probabilities calculated by the single-sided classifiers at the first stage. The nodule samples with small probability decrease as the class sample ratio becomes larger as shown in Fig. 6(a). At the same time, the non-nodules samples with small probability decrease as the ratio becomes larger as shown in Fig. 6(b). As the result, we obtained the reduction rate of number of class 0 (non-nodule) samples at each stage with different class sample ratios as shown in Fig. 7(a). The number of class 0 samples are decreased less than half at the first stage. And, by cascading the single-sided classifiers and filtering obvious non-nodule samples in class 0, the number of class 0 samples decreases further. The sample reduction rate of nodule samples reaches under 0.25 at 3-stage. At the same time, as shown in Fig. 7(b), the number of class 1 (nodules) samples decreases with the number of stages. This is a side effect of our method. However, the side effect decreases as the ratio increases.

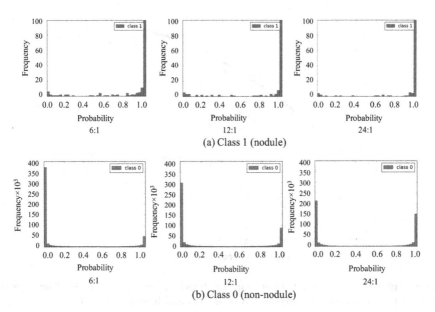

Fig. 6. Histograms of nodule probabilities of: (a) class 1 (nodule) and (b) class 0 (non-nodule), calculated by the single-sided classifier at the first stage. Three kind of class sample ratios, 6 to 1, 12 to 1 and 24 to 1, are compared.

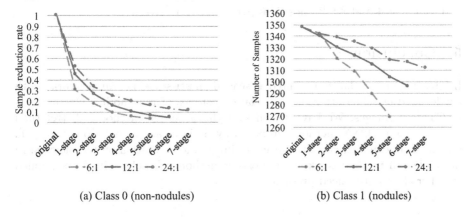

Fig. 7. Sample reduction rate of class 0 (non-nodules) and number of samples of class 1 (nodule) at each stage with deferent class sample ratios.

Figure 8 shows FROC curves at 4-stage with 6:1 ratio, 6-stage with ratio 12:1 and 3-stage with ratio 24:1. Each stage number showed the highest average sensitivity among each ratio. As shown in Fig. 7, as the ratio increases the number of nodule samples (class 1) that are erroneously removed decreases, while the number of obvious non-nodules (class 0) that are filtered decreases. As the results, the 3-stage single sided classifiers with 24:1 ratio reaches the best perfor-

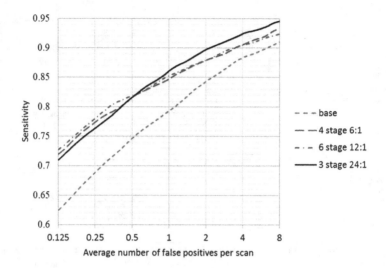

Fig. 8. FROC curves of multi-layer neural networks with different ratio of number of class samples in training of single-sided classifiers and baseline method.

mance, the sensitivities are 92.4% and 94.5% at 4 and 8 false positives per scan, respectively. On the other hand, the baseline achieves the sensitivity of 88.4% and 91.1% at 4 and 8 false positives per scan, respectively.

5   Conclusion

In this paper, we have presented cascaded multi-stage neural networks with single-sided classifiers to reduce the false positives of lung nodule classification in CT scan images. We have shown that the proposed method is better than state of the art CNN methods proposed by Setio et al. [4] and Dou et al. [19]. Our method can decrease the burden of image interpretation on radiologists. In principle, our method is a kind of boosting method. We believe it can be applied to other class imbalanced problems.

References

1. Lung cancer mortality statistics. http://www.cancerresearchuk.org/health-professio nal/cancer-statistics/statistics-by-cancer-type/lung-cancer/mortality. Accessed 3 Feb 2017
2. Ginneken, B.V., et al.: Comparing and combining algorithms for computer-aided detection of pulmonary nodules in computed tomography scans: the ANODE09 study. Med. Image Anal. **14**, 707–722 (2010)
3. Krizhevsky, A., Sutskeveret, I., Hinton, G.E.: ImageNet classification with deep convolutional neural networks. Advances in Neural Information Processing Systems, pp. 1097–1105. Curran Associates (2012)

4. Setio, A.A.A., et al.: Pulmonary nodule detection in CT images: false positive reduction using multi-view convolutional networks. IEEE Trans. Med. Imaging **35**(5), 1160–1169 (2016)
5. Lung nodule analysis (2016). http://luna16.grand-challenge.org/. Accessed 16 June 2017
6. Japkowicz, N.: Learning from imbalanced datasets: a comparison of various strategies. In: AAAI Workshop on Learning from Imbalanced Data Sets, Menlo Park, CA, vol. 68, pp. 10–15 (2000)
7. Kubat, M., Matwin, S.: Addressing the curse of imbalanced training sets: one-sided sampling. In: Proceedings of Fourteenth International Conference on Machine Learning, pp. 179–186. Morgan Kaufmann (1997)
8. Rahman, M., Davis, D.N.: Addressing the class imbalance problem in medical datasets. Int. J. Mach. Learn. Comput. **3**(2), 224–228 (2013)
9. Kubat, M., Holte, R.C., Matwin, S.: Machine learning for the detection of oil spills in satellite radar images. Mach. Learn. **30**, 195–215 (1998)
10. Fawcett, T.E., Provost, F.: Adaptive fraud detection. Data Min. Knowl. Disc. **1**(3), 291–316 (1997). Springer
11. Chawla, N., Bowyer, K., Hall, L., Kegelmeyer, W.P.: SMOTE: synthetic minority over-sampling technique. J. Artif. Intell. Res. **16**, 321–357 (2002)
12. Sun, Y., Wong, A.K.C., Kamel, M.S.: Classification of imbalanced data: a review. Int. J. Pattern Recognit. Artif. Intell. **23**(4), 687–719 (2009)
13. Viola, P., Jones, M.: Rapid object detection using a boosted cascade of simple
14. Wu, J., Rehg, J.M., Mullin, M.D.: Learning a rare event detection cascade by direct feature selection. In: Proceedings of Advances in Neural Information Processing Systems (NIPS), vol. 4, pp. 855–861 (2003)
15. Li, H., Linz, Z., Shen, X., Brandt, J., Hua, G.: A convolutional neural network cascade for face detection. In: 2015 IEEE Conference on Computer Vision and Pattern Recognition, pp. 5325–5334 (2015)
16. Qin, H., Yan, J., Li, X., Hu, X.: Joint training of cascaded CNN for face detection. In: IEEE Conference on Computer Vision and Pattern Recognition (CVPR), pp. 3456–3465 (2016)
17. Kalinovskii, I.A., Spitsyn, V.G.: Compact convolutional neural network cascade for face detection. arXiv preprint arXiv:1508.01292 (2015)
18. Chen, H., Dou, Q., Wang, X., Qin, J., Heng, P.A.: Mitosis detection in breast cancer histology images via deep cascaded networks. In: Proceedings of Thirtieth AAAI Conference on Artificial Intelligence, pp. 1160–1166. AAAI Press (2016)
19. Dou, Q., Chen, H., Yu, L., Qin, J., Heng, P.A.: Multi-level contextual 3D CNNs for false positive reduction in pulmonary nodule detection. IEEE Trans. Biomed. Eng. **64**, 1558–1567 (2016)

On the Importance of Domain Adaptation in Texture Classification

Barbara Caputo[1], Claudio Cusano[2], Martina Lanzi[1], Paolo Napoletano[3(✉)], and Raimondo Schettini[3]

[1] Department of Computer, Control, and Management Engineering,
Sapienza Rome University, via Ariosto 25, 00185 Rome, Italy
caputo@dis.uniroma1.it

[2] Department of Electrical, Computer and Biomedical Engineering,
University of Pavia, Via Ferrata 1, 2700 Pavia, Italy
claudio.cusano@unipv.it

[3] Department of Informatics, Systems and Communication,
University of Milano–Bicocca, Viale Sarca 336, 20125 Milan, Italy
{napoletano,schettini}@disco.unimib.it

Abstract. Texture classification algorithms require generalization abilities in order to be reliably used in real world applications. This paper casts this problem in the domain adaptation setting and presents the first study investigating (a) up to which extent this visual recognition problem suffers from this issue, and (b) the effectiveness of existing domain adaptation algorithms in mitigating it. We focus on domain adaptation methods based on shallow classifiers, and test their performance on deep and non deep features. Results obtained on a newly created domain adaptation texture setup show the superiority of deep features compared to other well known approaches, and highlights the importance of factoring in the domain shift when dealing with textures in the wild.

Keywords: Color texture classification · Domain adaptation · Generalization

1 Introduction

The ability to recognize materials and their texture based on their visual appearance is crucial in several applications, from robot manipulation to industrial production, to food recognition and so on. While the topic has historically been widely researched in computer vision, the generalization abilities obtained so far are still not up to what would be desirable for moving from research labs to commercial applications at large [18,32].

The generalization problem, i.e. the experimental fact that classifiers trained on a given dataset do not perform very well when tested on a new database, received a renewed attention in the visual learning community since 2012, when it has been casted into the domain adaptation framework [14,31]. Here, the key assumption is that images depicting the same visual classes, but acquired in

© Springer International Publishing AG 2017
S. Battiato et al. (Eds.): ICIAP 2017, Part I, LNCS 10484, pp. 380–390, 2017.
https://doi.org/10.1007/978-3-319-68560-1_34

different settings, at different times and with different devices, are generated by two related but different probability distributions. Hence, domain adaptation approaches attempt to close the shift among the two distributions. Although domain adaptation by its very nature is pervasive in visual recognition, to the best of our knowledge the problem has not been investigated so far in the texture classification scenario.

This paper aims at filling this gap, presenting a domain adaptation setting for material recognition, and studying how different state of the art non-deep domain adaptation algorithms perform in this scenario. We test all methods using shallow as well as deep features, and we compare our results with off-the-shelf classifiers not explicitly addressing the domain shift between training and test data. Our results clearly show that domain adaptation is a very real problem for classification of textures in the wild, and that the use of domain adaptive classifiers lead to an increase in performance of up to 6.87%.

The rest of the paper is organized as follows: Sect. 2 describes the data, features and classifiers used in our benchmark evaluation. Section 3 reports our experimental findings, clearly demonstrating the presence of a domain shift in this setting and the ability of existing domain adaptation algorithm to alleviate it. We conclude the paper with an overall discussion and proposing possible future research directions.

2 Materials and Methods

2.1 Databases

The goodness of a domain adaptation technique is evaluated by measuring the classification accuracy when trained on a given database and tested on another one that contains the same texture classes. To this end, we have analyzed most of the existing texture databases in order to identify those that share the highest number of texture classes. As a result of this process we found 23 classes in common between the ALOT [3] and RawFooT [11] databases and about ten classes in common between CureT [12] and ALOT, CureT [12] and KTH-TIPS2b [4], STex [16] and CureT. For the evaluation presented in this paper we considered the 23 classes in common between ALOT and RawFooT. Examples of these 23 texture classes of both databases are displayed in Fig. 1.

The Raw Food Texture database (RawFooT), has been specially designed to investigate the robustness of descriptors and classification methods with respect to variations in the lighting conditions [8–11]. Classes correspond to 68 samples of raw food, including various kind of meat, fish, cereals, fruit etc. Samples taken under D65 at light direction $\theta = 24°$ are showed in Fig. 2. The database includes images of 68 samples of textures, acquired under 46 lighting conditions which may differ in:

1. the light direction: 24, 30, 36, 42, 48, 54, 60, 66, and 90°;
2. illuminant color: 9 outdoor illuminants: D40, D45, ..., D95; 6 indoor illuminants: 2700 K, 3000 K, 4000 K, 5000 K, 5700 K and 6500 K, we will refer to these as L27, L30, ..., L65;

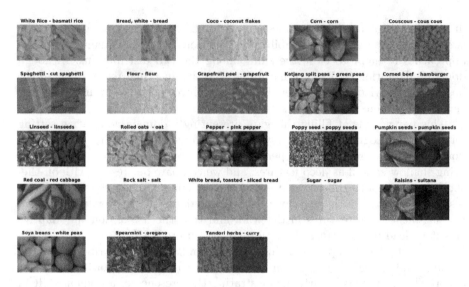

Fig. 1. Examples of the 23 classes in common between ALOT (left) and RawFooT (right)

Fig. 2. Overview of the 68 classes included in the Raw Food Texture database. For each class it is shown the image taken under D65 at direction $\theta = 24°$.

3. intensity: 100%, 75%, 50% and 25% of the maximum achievable level;
4. combination of these factors.

For each of the 23 classes in common with ALOT we considered 16 patches obtained by dividing the original texture image, that is of size 800×800 pixels, in 16 non-overlapping squares of size 200×200 pixels. We selected images taken under half of the imaging conditions for training (indicated as set1, a total of 3496 images) and the remaining for testing (set2, a total of 3496 images). For each class we selected eight patches for training and eight for testing by following a chessboard pattern (white positions are indicated as W, black positions as B).

The Amsterdam Library of Textures (ALOT) is a color image collection of 250 rough textures. In order to capture the sensory variation in object recordings, the authors systematically varied viewing angle, illumination angle, and illumination color for each material. This collection is similar in spirit as the CURET collection [3]. Examples from the 250 classes is displayed in Fig. 3.

The textures were placed on a turn table, and recordings were made for aspects of 0, 60, 120, and 180°. Four cameras were used, three perpendicular to the light bow at 0° azimuth and 80, 60, 40° altitude. Furthermore, one is mounted

Fig. 3. The 250 classes of the ALOT database. (Color figure online)

at 60° azimuth and 60° altitude. Combined with five illumination directions and one semi-hemispherical illumination, a sparse sampling of the BTF is obtained.

Each object was recorded with only one out of five lights turned on, yielding five different illumination angles. Furthermore, turning on all lights yields a sort of hemispherical illumination, although restricted to a more narrow illumination sector than true hemisphere. Each texture was recorded with 3075 K illumination color temperature, at which the cameras were white balanced. One image for each camera is recorded with all lights turned on, at a reddish spectrum of 2175 K color temperature.

For each of the 23 classes shared with RawFooT, we considered 6 patches obtained by dividing the original texture image, in 6 non-overlapping squares of size 200×200 pixels. For each class we have 100 textures acquired under different imaging conditions. For each texture we selected three patches for training and three for testing by following a chessboard pattern (white positions are indicated as W, black positions as B). We obtained a training set made of 6900 images (W positions) and a test set made of 6900 images (B positions).

The evaluation is performed on each single pair DB1 → DB2:

1. R → A: RawFooT used for training and ALOT used for test;
2. A → R: ALOT used for training and RawFooT used for test;

For each pair DB1 → DB2 we have 4 subsets:

1. *training using DB1:* set1 at positions W; *test using DB2:* set2 at positions B;
2. *training using DB1:* set1 at positions B; *test using DB2:* set2 at positions W;
3. *training using DB1:* set2 at positions W; *test using DB2:* set1 at positions B;
4. *training using DB1:* set2 at positions B; *test using DB2:* set1 at positions W;

this setup, even though is not required for this work, makes it possible to design unbiased inter-dataset experiments by excluding the possibility that the same portion of the texture samples or the same acquisition condition are included in both the training and the test set.

2.2   Features

The majority of texture analysis methods entails the computation of numerical representations, called features, that capture the distinctive properties of texture images. Many features have been proposed in the literature. These were traditionally divided into statistical, spectral, structural and hybrid [23]. Among traditional features the most widely known are probably those based on histograms, Gabor filters [2], co-occurrence matrices [15], and Local Binary Patterns [24].

More recent works approached the problem of texture classification by using features originally designed for scene and object recognition. For instance, Sharan et al. [27] used SIFT and HOG descriptors for material classification, while Sharma et al. [29] used a variation of the Fisher Vector approach for texture and face classification. Cimpoi et al. [5] shown how SIFT descriptors aggregated with the improved Fisher vector method greatly outperform previous descriptors in the state of the art on a variety of texture classification tasks. This direction of research further progressed with the replacement of image features explicitly designed with features automatically learned from data with methods based on deep learning [17]. Cimpoi et al., for instance, used Fisher Vectors to pool features computed by a convolutional neural network (CNN) originally trained for object recognition [6]. Lin and Maji used the same underlying CNN features and summarized them as Gram matrices [19]. In this work we considered three different images features: (i) Local Binary Patterns, (ii) Bag of SIFT descriptors, (iii) features computed by a CNN.

Local Binary Patterns (LBP) represent one of the most widely used method for the representation of textures [22]. LBPs are computed by thresholding the gray values in a circular neighborhood of pixels with the gray value of the central pixel. The resulting bits are arranged to form a binary representation that can be interpreted as a numeric code. The final descriptor is a histogram of the numeric codes. More in detail, we considered a neighborhood of 16 pixels at a distance of two pixels from the central one. Moreover, in forming the final histogram we considered only the "uniform" patterns that are those that include only at most two 0/1 transitions between adjacent bits and that, therefore, correspond to simple patterns. With this configuration, the feature vector is a histogram of 243 bins.

One of the most successful approach for image recognition is the use of the bag of visual words model [7]. Within this approach local descriptors extracted from an image are aggregated to form a histogram representing their distribution. More precisely, a codebook of visual words is formed by clustering the descriptors extracted on a set of training images. Then, given a new image, its descriptors are assigned to the closest visual word in the codebook, and the counts of descriptors assigned to each word form the final descriptor. In this work, we built a codebook of 1024 visual words by clustering the SIFT descriptors [21] extracted from a set of 20000 images from Flickr containing various content, such as sunset, countryside, etc. Therefore, the final feature vector is represented by the 1024 bins of the normalized histogram.

For the third feature vector, we followed the approach explored by Sharif Razavian et al. [28] that consists in using the intermediate representation computed by a convolutional neural network trained for image recognition. We used the VGG-16 network model [30] trained to identify the 1000 categories of the ILSVRC image recognition challenge [26]. As a feature vector we used the activations of the 4096 units forming the last layer before the computation of the final probability estimates.

2.3 Domain Adaptation Classifiers

We considered several domain adaptation methods:

Geodesic Flow Kernel (GFK): this method consists of embedding the source and target datasets in a Grassman manifold and model data with linear subspaces, and then constructing a geodesic flow between the two points, integrating an infinite number of subspaces along the flow. The geodesic flow represents incremental changes in geometric and statistical properties between the two domains. Then, the features are projected into this subspaces to form an infinite-dimensional feature vectors, and the inner product between these feature vectors define a kernel function that can be computed over the original feature space [14]. GFK is one of the most widely used domain adaptation methods in the literature; recent work showed that, when used over deep features, it is competitive with several deep domain adaptation approaches.

Subspace Alignment (SA): here, by using PCA we select, for each domain, the d eigenvectors corresponding to the d largest eigenvalues. These eigenvectors are used as bases of the source and target subspaces. Each source and target data are projected to its respective subspace. It is then learned a transformation matrix to map the source subspace to the target one. This allows to compare the source domain data directly to the target domain data, and to build classifiers on source data and apply them on the target domain. The advantages of the Subspaces Alignment are the robustness of the classifier which is not affected by local perturbations and the absence of regularization parameters [13].

Landmark-based Kernelized Subspace Alignment (LSSA): both methods described above have also some limitations. In the GFK algorithm, the search for the subspaces that lie on the geodesic flow is computationally costly and subject to a local perturbations. The SA algorithm assumes that the shift between the two distributions can be corrected by a linear transformation and in most of the cases only a subset of source data are distributed similarly to the target domain. So, the LSSA algorithm proposes: (i) selection of landmarks extracted from both domains so as to reduce the discrepancy between the source and target distributions, (ii) projecting the source and the target data onto a shared space using a Gaussian Kernel respect to the selected landmarks, (iii) learning a linear mapping function to align the source and target subspaces. This is done by simply computing inner products between source and target eigenvectors [1].

Transfer Component Analysis (TCA): this method tries to learn some transfer components across domains in a Reproducing Kernel Hilbert Space (RKHS) using Maximum Mean Discrepancy (MMD). TCA is a dimensionality reduction method for domain adaptation such that in the latent space spanned by these learned components, the variance of the data can be preserved as much as possible and the distance between different distributions across domain can be reduced [25].

Transfer Joint Matching (TJM): it aims at reducing the domain difference using jointly two learning strategies for domain adaptation: feature matching and instance re-weighting. Feature matching discovers a shared feature representation by jointly reducing the distribution difference and preserving the important properties of input data. Matching the feature distributions based on MMD minimization is not enough for domain adaptation, since it can only match the first-and high-order statistics, and the distribution matching is far from perfect. An instance re-weighted procedure should be cooperated to minimize the distribution difference by re-weighting the source data and then training a classifier on the re-weighted source data [20].

To fully assess the effect of each of the domain adaptation methods described above, we also used a linear SVM trained on the source data, and we tested it on the target data. We refer in the following to these experiments as "NA results". The C parameter of SVM was set by doing cross-validation on the source domain with following values ϵ {0.0001 0.001 0.01 0.1 1.0 10 100 1000 10000}, using the LIBSVM library.

2.4   Experimental Setup

As described before, we evaluated the different DA methods by comparing their performance with that of the linear classifier SVM for the no adaptation results, where we use the original input space without learning a new representation. The z-normalization is the first important step for the all domain adaptation algorithms and PCA is the method used for the dimensionality reduction. For each type of feature we set different parameters for each domain adaptation algorithm:

- In the GFK the dimensionality of the subspaces was set to 120 for the LBP features, 300 for the SIFT features and 200 for the CNN features. We evaluate the accuracy of this method on the target domain over 5 random trials for each type of features.
- In the SA we set the dimensionality of the subpaces to 150 for the LBP and CNN features and 300 for the SIFT features. Also for the SA algorithm the evaluation was performed over 5 random trials for each type of features.

- In the LSSA algorithm an important parameter is the threshold for measuring the quality of a candidate landmark. We set it to 0,5. If the quality measure of the candidate is above this threshold , it is kept as a landmark. The dimensionality of the subpaces was set to the number of matched landmarks.
- In the TCA we used the linear kernel on inputs and fixed $\mu = 0, 3$ (tradeoff parameter) to construct the transformation matrix. The dimensionalities of the latent spaces are fixed to 150 for LBP and SIFT features, 200 for CNN features.
- The TJM approach involves two model parameter: subspaces bases k and λ regularization parameter . We set λ by searching $\lambda \in \{0.01, 0.1, 1, 10, 100\}$. The k parameter was set to 100 for each type of feature. The evaluation was performed over 5 random trials.

3   Results

The results obtained, for each domain adaptation method, are illustrated in Table 1 when using the LBP features, in Table 2 when using the SIFT features and in Table 3 when using the CNN features. We see that the GFK method outperforms on average the other approaches with all type of features when the ALOT database is the source domain and the Rawfoot database is the target domain. In the opposite case we get the best result with the JTM algorithm for the CNN features. In fact we can note that the type of feature has an important role in the evaluation of the DA methods: when using the CNN features, we achieve the greatest improvement for all methods.

Figure 4 shows the confusion matrices, where an element of a matrix with position (i,j) is a count of observations known to be in group i (true label) but predicted to be in group j (predicted label), for NA and GFK (top row) and NA and JTM (bottom row), using deep features. These are the cases where we see the greater advantage in using Da approaches for the ALOT -Rawfoot

Table 1. Domain adaptation results with the LBP features. A: Alot database, R: Rawfoot Database

Dataset	NA	GFK	SA	TCA	LSSA	JTM
A → R	41,27%	**41,76%**	41,45%	37,24%	40,43%	31,34%
R → A	22,29%	22,26%	**22,38%**	18,15%	21,30%	18,59%

Table 2. Domain adaptation results with the SIFT features. A: Alot database, R: Rawfoot Database

Dataset	NA	GFK	SA	TCA	LSSA	JTM
A → R	55,36%	**60,79%**	60,66%	52,94%	58,09%	55,64%
R → A	46,82%	51,66%	51,83%	44,85%	**52,03%**	44,59%

Table 3. Domain adaptation results with the CNN features. A: Alot database, R: Raw-foot Database

Dataset	NA	GFK	SA	TCA	LSSA	JTM
A → R	67,53%	**74,40%**	74,18%	71,98%	73,67%	72,90%
R → A	71,73%	76,31%	76,83%	74,24%	78,06%	**78,17%**

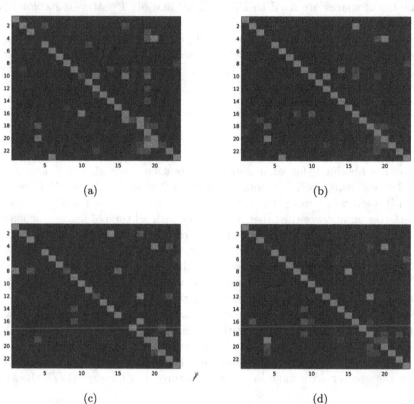

(a)

(b)

(c)

(d)

Fig. 4. Confusion matrices for No Adapt (a) and GFK (b) evaluation with the CNN features when ALOT database is the source domain and for No Adapt (c) and JTM (d) evaluation when Rawfoot database is the source domain.

and Rawfoot-ALOT settings, respectively. We see that both domain adaptation algorithms significantly reduce the domain shift, alleviating the misclassification compared to the case where the domain shift is not taken into consideration.

4   Conclusion

This paper addressed the issue of generalization in texture classification, in the context of domain adaptation. We presented a new benchmark setting that permits to study the problem in this domain, and a benchmark evaluation of shallow algorithms using handcrafted as well as deep features. Our results confirm the existence of the domain shift, as well as the superior generalization abilities of deep features and the effectiveness of domain adaptation algorithms in increasing the generalization across datasets. Future work will extend this study adding deep domain adaptation approaches, as well as designing larger experimental setups.

References

1. Aljundi, R., Emonet, R., Muselet, D., Sebban, M.: Landmarks-based Kernelized Subspace Alignment for Unsupervised Domain Adaptation. In: Computer Vision and Pattern Recognition (CVPR 2015) (2015)
2. Bianconi, F., Fernández, A.: Evaluation of the effects of Gabor filter parameters on texture classification. Pattern Recognit. **40**(12), 3325–3335 (2007)
3. Burghouts, G.J., Geusebroek, J.M.: Material-specific adaptation of color invariant features. Pattern Recognit. Lett. **30**(3), 306–313 (2009)
4. Caputo, B., Hayman, E., Mallikarjuna, P.: Class-specific material categorisation. In: 2005 Tenth IEEE International Conference on Computer Vision, ICCV 2005, vol. 2, pp. 1597–1604. IEEE (2005)
5. Cimpoi, M., Maji, S., Kokkinos, I., Mohamed, S., Vedaldi, A.: Describing textures in the wild. In: The IEEE Conference on Computer Vision and Pattern Recognition (CVPR) (2014)
6. Cimpoi, M., Maji, S., Kokkinos, I., Vedaldi, A.: Deep filter banks for texture recognition, description, and segmentation. Int. J. Comput. Vision **118**(1), 65–94 (2016)
7. Csurka, G., Dance, C., Fan, L., Willamowski, J., Bray, C.: Visual categorization with bags of keypoints. Workshop on statistical learning in computer vision, ECCV, vol. 22, pp. 1–2 (2004)
8. Cusano, C., Napoletano, P., Schettini, R.: Illuminant invariant descriptors for color texture classification. In: Tominaga, S., Schettini, R., Trémeau, A. (eds.) CCIW 2013. LNCS, vol. 7786, pp. 239–249. Springer, Heidelberg (2013). doi:10.1007/978-3-642-36700-7_19
9. Cusano, C., Napoletano, P., Schettini, R.: Intensity and color descriptors for texture classification. In: IS&T/SPIE Electronic Imaging, p. 866113. International Society for Optics and Photonics (2013)
10. Cusano, C., Napoletano, P., Schettini, R.: Combining local binary patterns and local color contrast for texture classification under varying illumination. JOSA A **31**(7), 1453–1461 (2014)
11. Cusano, C., Napoletano, P., Schettini, R.: Evaluating color texture descriptors under large variations of controlled lighting conditions. J. Opt. Soc. Am. A **33**(1), 17–30 (2016)
12. Dana, K.J., Van Ginneken, B., Nayar, S.K., Koenderink, J.J.: Reflectance and texture of real-world surfaces. ACM Trans. Graph. (TOG) **18**(1), 1–34 (1999)
13. Fernando, B., Habrard, A., Sebban, M., Tuytelaars, T.: Unsupervised visual domain adaptation using subspace alignment. In: ICCV (2013)
14. Gong, B., Shi, Y., Sha, F., Grauman, K.: Geodesic flow kernel for unsupervised domain adaptation. In: CVPR, pp. 2066–2073 (2012)
15. Haralick, R.M., Shanmugam, K., Dinstein, I.: Textural features for image classification. IEEE Trans. Syst. Man Cybern. **3**(6), 610–621 (1973)

16. Kwitt, R., Meerwald, P.: Salzburg texture image database (2012)
17. LeCun, Y., Bengio, Y., Hinton, G.: Deep learning. Nature **521**(7553), 436–444 (2015)
18. Li, W., Fritz, M.: Recognizing materials from virtual examples. In: Fitzgibbon, A., Lazebnik, S., Perona, P., Sato, Y., Schmid, C. (eds.) ECCV 2012. LNCS, vol. 7575, pp. 345–358. Springer, Heidelberg (2012). doi:10.1007/978-3-642-33765-9_25
19. Lin, T.Y., Maji, S.: Visualizing and understanding deep texture representations. In: Proceedings of the CVPR, pp. 2791–2799 (2016)
20. Long, M., Wang, J., Ding, G., Sun, J., Yu, P.S.: Transfer joint matching for unsupervised domain adaptation. In: 2014 IEEE Conference on Computer Vision and Pattern Recognition, CVPR 2014, 23–28 June 2014, Columbus, OH, USA, pp. 1410–1417 (2014)
21. Lowe, D.G.: Object recognition from local scale-invariant features. In: The Proceedings of the Seventh IEEE International Conference on Computer Vision, 1999, vol. 2, pp. 1150–1157 (1999)
22. Mäenpää, T., Pietikäinen, M.: Texture analysis with local binary patterns. In: Chen, C.H., Wang, P.S.P. (eds.) Handbook of Pattern Recognition and Computer Vision, 3rd edn, pp. 197–216. World Scientific Publishing, Singapore (2005)
23. Mirmehdi, M., Xie, X., Suri, J.: Handbook of Texture Analysis. Imperial College Press, London (2009)
24. Ojala, T., Pietikäinen, M., Mäenpää, T.: Multiresolution gray-scale and rotation invariant texture classification with local binary patterns. IEEE Trans. Pattern Anal. Mach. Intell. **24**(7), 971–987 (2002)
25. Pan, S.J., Tsang, I.W., Kwok, J.T., Yang, Q.: Domain adaptation via transfer component analysis. In: Proceedings of the 21st International Jont Conference on Artifical Intelligence, IJCAI 2009, pp. 1187–1192 (2009)
26. Russakovsky, O., Deng, J., Su, H., Krause, J., Satheesh, S., Ma, S., Huang, Z., Karpathy, A., Khosla, A., Bernstein, M., Berg, A.C., Fei-Fei, L.: Imagenet large scale visual recognition challenge. Int. J. Comput. Vision (IJCV) **115**(3), 211–252 (2015)
27. Sharan, L., Liu, C., Rosenholtz, R., Adelson, E.H.: Recognizing materials using perceptually inspired features. Int. J. Comput. Vision **103**(3), 348–371 (2013)
28. Sharif Razavian, A., Azizpour, H., Sullivan, J., Carlsson, S.: CNN features off-the-shelf: an astounding baseline for recognition. In: The IEEE Conference on Computer Vision and Pattern Recognition (CVPR) Workshops (2014)
29. Sharma, G., ul Hussain, S., Jurie, F.: Local higher-order statistics (LHS) for texture categorization and facial analysis. In: Fitzgibbon, A., Lazebnik, S., Perona, P., Sato, Y., Schmid, C. (eds.) ECCV 2012. LNCS, vol. 7578, pp. 1–12. Springer, Heidelberg (2012). doi:10.1007/978-3-642-33786-4_1
30. Simonyan, K., Zisserman, A.: Very deep convolutional networks for large-scale image recognition. arXiv preprint arXiv:1409.1556 (2014)
31. Tommasi, T., Caputo, B.: Frustratingly easy nbnn domain adaptation. In: ICCV (2013)
32. Weinmann, M., Gall, J., Klein, R.: Material classification based on training data synthesized using a BTF database. In: Fleet, D., Pajdla, T., Schiele, B., Tuytelaars, T. (eds.) ECCV 2014. LNCS, vol. 8691, pp. 156–171. Springer, Cham (2014). doi:10.1007/978-3-319-10578-9_11

Rotation Invariant Co-occurrence Matrix Features

Lorenzo Putzu[✉] and Cecilia Di Ruberto

Department of Mathematics and Computer Science,
University of Cagliari, via Ospedale 72, 09124 Cagliari, Italy
{lorenzo.putzu,dirubert}@unica.it

Abstract. Grey level co-occurrence matrix (GLCM) has been one of the most used texture descriptor. GLCMs continue to be very common and extended in various directions, in order to find the best displacement for co-occurrence extraction and a way to describe this co-occurrence that takes into account variation in orientation. In this paper we present a method to improve accuracy for image classification. Rotation dependent features have been combined using various approaches in order to obtain rotation invariant ones. Then we evaluated different ways for co-occurrence extraction using displacements that try to simulate as much as possible the shape of a real circle. We tested our method on six different datasets of images. Experimental results show that our approach for features combination is more robust against rotation than the standard co-occurrence matrix features outperforming also the state-of-the-art. Moreover the proposed procedure for co-occurrence extraction performs better than the previous approaches present in literature, able to give a good approximation of real circles for different distance values.

Keywords: Co-occurrence matrix · Feature extraction · Rotation invariance · Texture classification

1 Introduction

Texture is a feature that helps to analyse an image and although there isn't a specific definition of texture accepted by all, it can be viewed as a global descriptor belonging from the repetition of local patterns. Texture is an any and repetitive geometric arrangement of the grey levels of an image. It provides important information about the spatial disposition of the grey levels and the relationship with their neighbourhood. Human visual system determines and recognizes easily different types of textures but although for a human observer it is very simple to associate a surface with a texture, to give a rigorous definition for this is very complex. Typically a qualitative definition is used to describe textures. As it can be easily guessed a quantitative analysis of texture passes through statistical and structural relations among the basic elements of what we call just texture. Texture analysis is an important and useful area of image processing that leads the classification of images through the identification of

© Springer International Publishing AG 2017
S. Battiato et al. (Eds.): ICIAP 2017, Part I, LNCS 10484, pp. 391–401, 2017.
https://doi.org/10.1007/978-3-319-68560-1_35

their properties. The most important aspect of texture analysis is classification that concerns the search for particular regions of texture among different predefined classes of texture. Classification is carried out using statistical methods that define the descriptors of the texture. Many different methods for managing texture have been developed that are based on the various ways texture can be characterized. Although there are many powerful methods reported in the literature for texture analysis, including the scale-invariant feature transform (SIFT) [13], speeded up robust feature (SURF) [14], histogram of oriented gradients (HOG) [15], local binary patterns (LBP) [16], Gabor filters [19] and others, in this work we focus on improving one of the earliest method used for the analysis of grey level texture based on statistical approaches, that is the Grey-Level Co-Occurrence Matrix (GLCM). Motivated by the wide diffusion of this method and by the increasing numbers of rotation invariant descriptors, able to achieve good performances in various situations (eg. rotation invariant LBP [17]), we wished to investigate how it was possible to improve accuracy and robustness against rotation of co-occurrence matrix. Some interesting methods have been presented in order to extend the original implementation of GLCM, such as the method proposed in [7] where the authors evaluated different values for the distance parameter that influence the matrices computation, in [12] the GLCM descriptors are extracted by calculating the weighted sum of GLCM elements, in [6] the GLCM features are calculated by using the local gradient of the matrix. In [10] to calculate the features, the grey levels and the edge orientation of the image are considered. In [9] the authors propose to use a variable window size by multiple scales to extract descriptors by GLCM. The method in [8] uses the colour gradient to extract statistical features from GLCM. In [11] various types of GLCM descriptors (classical Haralick features and features from 3D co-occurrence matrix) and grey-level run-length features are extracted. Furthermore the GLCM has been extracted using the colour information from single channels [5] or by combining the channels in pairs [21,22]. Since in this work we are more interested on invariant descriptors, we compute the GLCM using just the gray level intensity. Thus, starting from the rotation dependent GLCM features we investigate different approaches to compute more efficient rotationally invariant features, as proposed in [18], and finally we propose our new approach to compute rotationally invariant features. This approach contains also a new formulation for displacement computation able to simulate as much as possible real circles. To validate our method we have used six different databases of images, Brodatz, Mondial Marmi, Outex, Vectorial, Kylberg Sintorn and ALOT that present different materials and so they represent different classification problems. The rest of the paper is organized as follows. In Sect. 2 we report some background information necessary to introduce the existing methods used. Section 3 shows the proposed approaches to extend the original features to rotation invariant. Section 4 provides the experiments realised to asses the classification performances. Finally, in Sect. 5 we present our conclusions and some possible future works.

2   Background

A feature is defined as a function of one or more measurements, specifying some quantifiable property of an object or an image. Features can be classified into low-level features and high-level features. Low-level features can be extracted directly from the original images, whereas high-level feature extraction must be based on low-level features. There are various methods for features extraction and texture classification and the most important are based on statistical approach. When we talk about texture analysis we cannot forget Haralick et al. [4], who has crafted in 1973 a first very simple and very powerful mathematical model with which all have faced, revising it, correcting it and making it more efficient. In fact, even after 40 years Haralick's method is the most powerful model for texture analysis finding many areas of application from biomedicine to remote sensing, to industrial or materials inspection. Haralick defined a type of matrix called Spatial Gray-Level Dependence Matrix (SGLDM), while the current definition of Gray Level Co-occurrence Matrix (GLCM) is attributed to Gonzalez et al. [1]. This method involves two steps for features extraction: in the first one the GLCMs are calculated and in the second one features are computed using the matrices calculated in the first phase. A GLCM represent the probability of finding two pixels i and j with distance d and orientation θ and is denoted with $p_{d,\theta}(i,j)$. A GLCM for an image of size $N \times M$ with N_g gray levels is a 2D array of size $Ng \times Ng$.

2.1   Displacement Type

The GLCM can be defined in eight directions (0, 45, 90, 135, 180, 225, 270 and 315), but in the original formulation Haralick proposed to use only four directions spaced at angular intervals of $45°$ considering the other four directions obtainable in a symmetrical way. Also the distance could present a wide range of value but the most used are $d = 1, 2, 3$. The displacement obtained with this formulation are shown in Fig. 1. The main drawback is that only four displacements for each distance value can be computed. Thus, all the others co-occurrences are not considered, loosing important texture information.

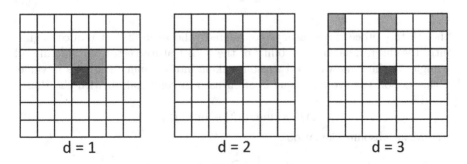

$d = 1$ $d = 2$ $d = 3$

Fig. 1. Types of displacements using the original Haralick's formulation.

For this reason Petrou [20] proposed the use of digital circles in which the displacements are calculated in a totally different way. In fact the displacements describe a circle around the central pixel. Since the definition of circle from the continuous space cannot be represented immediately into the digital domain there could be many different definitions of digital circles. In this formulation Petrou considered as valid displacement for a circle of radius d all those pixels included in the range $[d - 1/2, d + 1/2)$. Thus, the number of displacement increases as the d value increases, as it can be seen from Fig. 2.

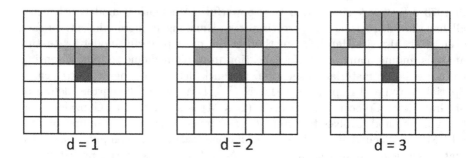

Fig. 2. Types of displacements using the Petrou's formulation.

2.2 Features Extraction

Once the m GLCMs have been computed with the chosen m angles and distance, it is possible to extract the features from each GLCM. So considering that from each GLCM n features could be extracted the amount of features is $m \times n$. In the original formulation this amount of features could be reduced, by combining them through a simple average operation, obtaining as a result a feature vector of size n which is rotationally invariant. So considering $f^k = [f_1^k, ..., f_m^k]$ as the feature vector obtained computing only the k-th descriptor from all the m GLCM, the final $\bar{f}^k$ can be computed as follows

$$\bar{f}^k = \frac{1}{m} \sum_{i=0}^{m} f_i^k \qquad (1)$$

Although this procedure is always considered valid as in [2,3] it reduces significantly the discrimination capability of the final feature vector, considering all the angles in the same way and without taking into account variations in value. For this reason the average value is often used in combination with the Range value [18]

$$\Delta f^k = max(f_i^k) - min(f_i^k) \qquad (2)$$

or the Mean absolute deviation

$$\delta f^k = \frac{1}{m} \sum_{i=0}^{m} |f_i^k - \bar{f}^k|. \qquad (3)$$

Another means to obtain rotation invariant features is the absolute values of the Discrete Fourier Transform (DFT) [18]

$$\hat{f}^k = \sum_{i=0}^{m} f_i^k e^{-\sqrt{1}\frac{2\pi(m-1)(i-1)}{m}} \tag{4}$$

for which the coefficient $\hat{f}^k$ should be invariant to any circular shift of the input vector f^k.

3 Our Approach

With both of the approaches previously presented the number of displacements heavily depends on the used method.

3.1 Proposed Displacement

The approach proposed in this work for displacement calculation comes from the idea that the digital circle is not enough to classify correctly fine texture. In fact, as it can be seen in Figs. 1 and 2, the displacements with distance value $d = 1$ are the same. Thus, we considered the approach proposed by Ojala et al. in [17] for the computation of invariant LBP suitable for this purpose, by using the circular symmetric neighbour set as a new way for representing our displacement. The displacement can be rewritten as the p grey values equally spaced on a circle with radius equal to the distance value d. Since the symmetrical property of GLCM is still valid we considered only the semicircles. Some examples of displacements can be seen in Fig. 3. Obviously the diagonal grey values cannot be computed directly but they are determined by interpolation as in [17]. The main advantage of this approach is that the number of displacement is not longer related to the distance value, as it was with the digital circle, but it can be decided specifying the p value, that can assume also higher values in order to better describe fine textures.

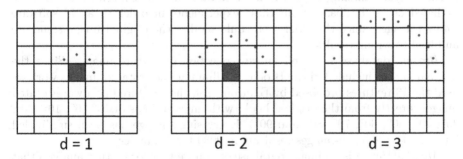

Fig. 3. Types of displacements using our formulation.

3.2   Proposed Feature Combination

Our features combination approach starts considering all the possible circular shifts of the input vector f^k. So, if this vector is of length m, we construct a matrix of size $m \times m$ denoted as f^{kk} in which all the circular shifts of the input vector f^k are present. The construction of the f^{kk} is not random as it disposes each circular shift regularly, in order to compose a symmetric matrix. Thus, for $f^k = [f_1, ..., f_m]$ we should have

$$\begin{pmatrix} f_1 & f_2 & \cdots & f_{m-1} & f_m \\ f_2 & \cdots & \cdots & f_m & f_1 \\ \cdots & \cdots \cdots & \cdots & \cdots \\ f_{m-1} & f_m & \cdots & \cdots & \cdots \\ f_m & f_1 & \cdots & \cdots & f_{m-1} \end{pmatrix}$$

This kind of matrix presents many important properties being square and symmetric. A first idea was to compute the absolute value of the determinant, denoted as f^D, in order to preserve the orientation of each possible transformation. But being a single value, the determinant could not be very discriminant for a classification task. For this reason we computed also the eigenvalues of the previous matrix, denoted as f^E, obtaining always a number of features equal to the original number of features and preserving the original direction of the feature vector. It must be noted that both the determinant and the eigenvalues are also invariant to mirroring, bringing to the new feature vectors an higher generalization capability.

4   Datasets

To validate our method we have used six different databases of images, Brodatz, Mondial Marmi, OuteX, Vectorial, Kylberg Sintorn and ALOT. They present different materials and textures and so they represent different classification problems. These databases contain hardware-rotated images taken at nine different rotation, making them the most suitable for our experiments. We have excluded other databases from these experiments in order to avoid software rotated images, since this operation could modify the original image structure and lead to wrong results.

Brodatz database is a well known collection of texture images. Since the original images included in the Brodatz's album are not rotated, in this work we used the 13 textures proposed by Bianconi [18], that acquired hardware-rotated images directly from the original book, with angular steps of 10° (0°, 10°, 20°, 30°, 40°, 50°, 60°, 70°, 80° and 90°). Every single image has been subdivided into 16 205 × 205 sub-images resulting in 2080 total samples.

MondialMarmi is an image database of granite tiles for texture analysis that includes 12 classes. Every class is represented by 4 textures, that have been acquired in a 24-bit RGB image of size 544 × 544 using nine rotation angles (0°, 5°, 10°, 15°, 30°, 45°, 60°, 75° and 90°). To create the dataset each image has been converted into grayscale, and divided into four non-overlapping sub-images of size 272 × 272, for a total image count of 1728.

Outex database contains a collection of 320 textures acquired with well defined variations in terms of illumination, rotation and spatial resolution. Each texture is captured in a 24-bit RGB image of size 538 × 746 using three different simulated illuminations, six spatial resolutions and nine rotation angles $(0°, 5°, 10°, 15°, 30°, 45°, 60°, 75°$ and $90°)$, for a total of 51840 images. Since we focused only on invariance to rotations and given the considerable size of this database, we used a test suites proposed by the same Outex authors called OUTEX00045. It contains 45 texture classes, the original images have been converted into grey levels and divided in 20 non overlapping sub-images of size 128 × 128, for a total count of 8100 images.

Vectorial database is a collection of 20 artificial texture classes proposed, here again, by Bianconi [18]. Considering that it is not composed of raster images, they have been software rotated with angular steps of $10°$ $(0°, 10°, 20°, 30°, 40°, 50°, 60°, 70°, 80°$ and $90°)$. The rotated images have been converted to raster with a resolution of 300 dpi, and subdivided into 16 255 × 255 sub-images, resulting in 16 samples per class and 3200 total images.

Kylberg Sintorn database is a collection of 25 textural classes of materials such as fabric, grains, sugar, rice, etc. The images are provided with nine rotation angles, but in this case the images have been rotated with angular steps of $40°$ $(0°, 40°, 80°, 120°, 160°, 200°, 240°, 280°, 320°$ and $90°)$. The original images (one for each class) are 24-bit RGB with a resolution of 5184 × 3456 pixels, but they have been provided also in small subsets for texture classification, presenting 400 images for each angle and thus 16 samples per class. The final dataset contains, therefore, 3600 images.

ALOT contains 250 textures, each one with 100 images obtained under different illumination conditions. For our experiments we considered a subset of the original dataset that contains only 80 textures provided with four rotation angles, that have been rotated with angular steps of $60°$ $(0°, 60°, 120°$ and $180°)$. The original images have been converted into grayscale and divided into 16 181 × 181 sub-images, for a total count of 5120 images.

5   Experimental Evaluation

To make a comparison we tested the proposed approach against the other proposed formulations and various combinations. We performed a set of image classification experiments to evaluate accuracy and robustness against rotation of the presented descriptors. The features used in our experiments are only the five most frequently used in literature, the *Angular Second Moment* that is the squares sum of the matrix values, the *Contrast* that is the weighted average of all diagonals parallel to the main one which emphasizes the correlation between the different tones, the *Correlation* that measures how a pixel is in correlation with its neighbor across the image, the *Inverse Difference Moment* that measures the proximity of the distribution from GLCM elements to the GLCM diagonal and the *Entropy* that measures the entropy of the entire matrix. Thus, we computed the GLCM both with original methods proposed by Haralick (HAR) and both with the digital circles (DC) and the proposed approaches for circular displacement extraction (CIR), using a distance value ranging from 1 to 3. The rotation dependent descriptors have been extracted from all the GLCMs and converted in rotationally invariant descriptors using the various approaches described previously and our approach. To highlight the performances of the proposed approach, for the state-of-the-art approaches we reported only the most powerful combinations, $\bar{f}||\Delta f$ reported as $\bar{f}\Delta$ and $\bar{f}||\Delta f||\delta f$ reported as $\bar{f}\Delta\delta$. For each dataset we performed 100 experiments and for each one training and test sets are represented respectively by the half of the original samples. The dataset has been divided with a stratified sampling which guarantees that each class is properly represented both in the training set and in test set. To better study the effects of image rotation, the classifier is always trained with features extracted from images acquired at orientation 0° and then tested with feature extracted from images acquired at all orientations. The classification performances have been evaluated by calculating the accuracy, which gives us a good indication of the performance since it considers each class of equal importance. Thus the classification accuracy has been estimated through a k-Nearest Neighbour (k-NN) classifier, with $k = 1$ and using the euclidean distance. Th k-NN has been preferred to a more complex classifier in order to make the results more representative of the effectiveness of the proposed approach than of the classifier itself. The results of this first experiment are reported in Table 1, that shows, for each descriptor, dataset, displacement and distance, the mean and standard deviation of the classification accuracy. As it can be observed the proposed approach to convert the rotation dependent descriptors into rotation invariant f^E outperforms the other approaches in most of the experiments. In particular it can be observed that it performs better than the other approaches when applied with HAR and DC methods for displacement extraction. Although the f^E does not produce the best results in all the datasets when applied with the CIR methods, it can be observed that in many cases the other approaches benefits from this kind of displacement. To further highlight the performances obtained in the previous experiment we compared our results against one of the newest and most used invariant texture descriptor that is the Local Binary Pattern LBP^{ri} (LBP). To make a direct comparison with our results, we take into account the rotation invariant LBP with a neighbourhood of 8 and distance 1 and 2 denoted as $LBP^{ri}_{8,1}$ and $LBP^{ri}_{8,2}$ respectively, and the rotation invariant LBP with a neighbourhood of 16 and distance 1 and 2 denoted as $LBP^{ri}_{16,1}$ and $LBP^{ri}_{16,2}$ respectively. The average accuracy values reported in Tables 1 and 2 shows an already known trend for co-occurrence features, that are fairly accurate in classification problems involving relatively few classes, but less accurate in classification problems involving more classes. Indeed, as it can be observed, the rotation invariant GLCM outperforms LBP in just 4 datasets out of 6, since the other 2 datasets presents an higher number of classes. It must be noted that in this work we used just five co-occurrence descriptors, because our main goal was to establish a good approach for the extraction of rotation invariant

Table 1. Comparison of displacement extraction and features combination.

		Brodatz	Mondial	Outex	Vectorial	Kylberg	ALOT
HAR $d=1$	$\bar{f}$	76.8 ± 2.6	80.7 ± 2.0	56.7 ± 1.2	76.9 ± 3.1	88.7 ± 4.3	52.2 ± 2.9
	$\bar{f}\Delta$	79.0 ± 12.8	87.0 ± 2.3	60.4 ± 4.5	80.2 ± 3.7	90.1 ± 4.4	58.2 ± 4.4
	$\bar{f}\Delta\delta$	78.7 ± 13.0	87.2 ± 2.5	60.9 ± 4.5	80.3 ± 3.7	90.6 ± 4.2	59.2 ± 4.4
	$\hat{f}$	82.6 ± 8.7	85.2 ± 2.0	60.5 ± 3.3	79.7 ± 3.2	90.7 ± 4.1	59.1 ± 3.5
	f^E	**88.8 ± 6.7**	**90.1 ± 2.5**	**62.0 ± 7.5**	**81.4 ± 3.4**	**91.1 ± 5.1**	**60.8 ± 5.2**
DC $d=1$	$\bar{f}$	76.8 ± 2.6	80.7 ± 2.0	56.7 ± 1.2	76.9 ± 3.1	88.7 ± 4.3	52.2 ± 2.9
	$\bar{f}\Delta$	79.0 ± 12.8	87.0 ± 2.3	60.4 ± 4.5	80.2 ± 3.7	90.1 ± 4.4	58.2 ± 4.4
	$\bar{f}\Delta\delta$	78.7 ± 13.0	87.2 ± 2.5	60.9 ± 4.5	80.3 ± 3.7	90.6 ± 4.2	59.2 ± 4.4
	$\hat{f}$	82.6 ± 8.7	85.2 ± 2.0	60.5 ± 3.3	79.7 ± 3.2	90.7 ± 4.1	59.1 ± 3.5
	f^E	**88.8 ± 6.7**	**90.1 ± 2.5**	**62.0 ± 7.5**	**81.4 ± 3.4**	**91.1 ± 5.1**	**60.8 ± 5.2**
CIR $d=1$	$\bar{f}$	78.3 ± 3.1	81.8 ± 1.9	60.0 ± 1.5	79.2 ± 4.1	87.4 ± 4.5	53.2 ± 3.4
	$\bar{f}\Delta$	87.7 ± 5.8	86.6 ± 2.3	60.2 ± 5.2	82.8 ± 3.5	88.1 ± 5.1	59.5 ± 4.7
	$\bar{f}\Delta\delta$	87.9 ± 5.5	87.2 ± 2.5	61.2 ± 5.8	83.2 ± 3.8	88.6 ± 5.2	**60.4 ± 4.7**
	$\hat{f}$	88.0 ± 6.0	87.7 ± 3.6	61.3 ± 5.7	81.0 ± 3.8	**89.5 ± 4.4**	59.4 ± 4.6
	f^E	**91.2 ± 5.5**	**87.9 ± 3.4**	**61.4 ± 8.2**	**83.9 ± 4.2**	88.8 ± 6.5	60.1 ± 4.9
HAR $d=2$	$\bar{f}$	81.4 ± 2.9	82.0 ± 2.8	56.8 ± 1.0	80.2 ± 2.4	92.0 ± 3.7	55.5 ± 2.3
	$\bar{f}\Delta$	84.5 ± 8.7	88.9 ± 2.1	60.2 ± 3.7	82.8 ± 3.4	95.2 ± 3.2	64.3 ± 3.2
	$\bar{f}\Delta\delta$	84.1 ± 9.1	89.5 ± 2.2	60.7 ± 3.6	82.8 ± 3.5	**95.2 ± 3.1**	65.8 ± 3.1
	$\hat{f}$	87.5 ± 6.4	88.5 ± 2.7	60.7 ± 2.2	82.9 ± 2.9	94.8 ± 3.3	64.6 ± 2.8
	f^E	**87.7 ± 7.2**	**90.2 ± 2.3**	**61.5 ± 6.5**	**83.1 ± 3.1**	95.0 ± 4.2	**66.2 ± 4.6**
DC $d=2$	$\bar{f}$	81.6 ± 1.6	82.6 ± 2.4	57.0 ± 0.9	79.8 ± 2.6	91.9 ± 3.6	55.5 ± 2.4
	$\bar{f}\Delta$	91.2 ± 5.8	88.3 ± 2.7	57.8 ± 6.4	82.5 ± 3.0	93.0 ± 4.4	61.0 ± 3.5
	$\bar{f}\Delta\delta$	91.3 ± 5.7	88.6 ± 2.7	57.9 ± 6.5	82.5 ± 3.1	93.1 ± 4.4	61.2 ± 3.6
	$\hat{f}$	93.4 ± 2.9	88.0 ± 2.3	59.9 ± 2.8	82.8 ± 2.5	**94.1 ± 3.4**	61.3 ± 3.1
	f^E	**95.3 ± 4.5**	**91.1 ± 3.5**	**61.7 ± 4.7**	**83.9 ± 4.6**	93.5 ± 4.2	**66.1 ± 4.1**
CIR $d=2$	$\bar{f}$	85.2 ± 1.6	84.4 ± 3.1	61.4 ± 1.5	82.4 ± 3.6	92.5 ± 3.4	58.8 ± 2.8
	$\bar{f}\Delta$	97.4 ± 1.6	89.0 ± 3.0	60.5 ± 7.0	85.1 ± 3.8	93.4 ± 4.4	63.8 ± 4.6
	$\bar{f}\Delta\delta$	97.6 ± 1.3	89.0 ± 3.1	60.9 ± 6.9	**85.3 ± 4.0**	93.6 ± 4.3	63.7 ± 5.2
	$\hat{f}$	97.3 ± 1.3	89.8 ± 2.9	**63.5 ± 3.7**	84.9 ± 3.2	**94.8 ± 3.3**	**65.4 ± 4.1**
	f^E	**98.1 ± 1.0**	**92.4 ± 2.7**	62.6 ± 7.8	85.1 ± 4.2	92.2 ± 4.6	64.6 ± 3.7
HAR $d=3$	$\bar{f}$	81.4 ± 2.3	79.7 ± 1.7	52.7 ± 1.1	81.6 ± 2.2	92.4 ± 3.9	55.3 ± 1.4
	$\bar{f}\Delta$	86.1 ± 7.3	88.6 ± 1.7	57.0 ± 1.4	84.1 ± 3.3	94.4 ± 3.2	65.9 ± 2.1
	$\bar{f}\Delta\delta$	85.8 ± 7.5	89.2 ± 1.6	58.1 ± 1.5	84.1 ± 3.4	94.2 ± 3.4	67.2 ± 2.1
	$\hat{f}$	87.6 ± 6.8	89.1 ± 1.8	58.7 ± 1.2	83.9 ± 2.7	94.6 ± 3.5	66.1 ± 2.0
	f^E	**87.6 ± 6.7**	**91.2 ± 2.7**	**59.2 ± 2.9**	**88.1 ± 3.4**	**95.8 ± 3.5**	**69.0 ± 3.0**
DC $d=3$	$\bar{f}$	82.6 ± 1.4	80.9 ± 2.5	55.0 ± 1.2	81.4 ± 2.1	91.4 ± 4.1	55.7 ± 1.9
	$\bar{f}\Delta$	95.0 ± 2.5	89.5 ± 2.1	59.5 ± 1.6	84.5 ± 2.1	**94.4 ± 4.0**	63.0 ± 2.7
	$\bar{f}\Delta\delta$	95.1 ± 2.5	89.6 ± 2.1	59.5 ± 1.8	84.5 ± 2.2	94.3 ± 4.0	63.0 ± 2.9
	$\hat{f}$	95.3 ± 1.1	88.6 ± 2.2	59.0 ± 1.3	84.4 ± 1.8	93.4 ± 3.7	61.9 ± 2.2
	f^E	**96.9 ± 2.4**	**92.2 ± 2.6**	**59.8 ± 4.1**	**86.8 ± 3.5**	93.8 ± 4.6	**63.2 ± 3.6**
CIR $d=3$	$\bar{f}$	87.9 ± 1.9	82.8 ± 3.3	59.0 ± 1.6	84.5 ± 2.9	92.1 ± 3.9	58.7 ± 1.7
	$\bar{f}\Delta$	97.4 ± 1.6	92.1 ± 2.3	62.5 ± 3.3	**86.8 ± 3.9**	95.0 ± 3.4	65.4 ± 3.1
	$\bar{f}\Delta\delta$	97.5 ± 1.6	92.2 ± 2.2	62.3 ± 3.5	86.6 ± 3.9	**95.1 ± 3.5**	65.4 ± 3.5
	$\hat{f}$	97.2 ± 1.2	91.2 ± 2.1	**63.1 ± 1.9**	86.6 ± 3.1	93.9 ± 3.7	**65.8 ± 2.7**
	f^E	**98.0 ± 0.9**	**92.6 ± 2.2**	62.0 ± 2.7	85.4 ± 4.7	94.0 ± 5.0	64.7 ± 3.6

Table 2. Rotation invariant LBP with different displacements.

	Brodatz	Mondial	Outex	Vectorial	Kylberg	ALOT
$LBP_{8,1}^{ri}$	85.1 ± 7.5	87.5 ± 2.9	78.4 ± 3.3	76.3 ± 6.7	92.1 ± 3.1	75.4 ± 3.9
$LBP_{16,1}^{ri}$	90.4 ± 5.0	85.5 ± 2.1	79.1 ± 3.1	76.1 ± 7.2	$\mathbf{94.5 \pm 3.4}$	78.7 ± 3.4
$LBP_{8,2}^{ri}$	87.1 ± 6.6	90.9 ± 1.7	81.9 ± 4.5	$\mathbf{81.0 \pm 5.8}$	86.1 ± 6.0	76.4 ± 2.5
$LBP_{16,2}^{ri}$	$\mathbf{96.8 \pm 1.5}$	$\mathbf{91.6 \pm 1.9}$	$\mathbf{84.1 \pm 3.6}$	81.0 ± 6.4	87.3 ± 5.9	$\mathbf{80.7 \pm 1.9}$

co-occurrence features, rather than an excellent approach for texture classification. In fact in a recent work [21] we demonstrated that an increased number of co-occurrence features could also obtain excellent performances for classification.

6   Conclusion

In this work we focused on GLCM, that is one of the oldest and still on of the most used texture descriptor. It continues to be very common and extended in various directions. In this work we investigated some approaches to improve accuracy of GLCM for texture classification, in particular with the presence of rotated images. Thus, starting from the rotation dependent GLCM features we investigated some approaches to compute more efficient rotationally invariant features, already present in literature, and finally we proposed our new approach to compute rotationally invariant features. We introduced also an approach to find the best displacement for co-occurrence extraction that try to simulate as much as possible the shape of a real circle. We tested our method on six different datasets of images. The results obtained are really encouraging, since the proposed approach for features combination is more robust against rotation than the standard co-occurrence matrix features outperforming also the state-of-the-art. Moreover, the new procedure for co-occurrence extraction is able to preserve the real position of co-occurrence, minimizing the influence of rotation for the co-occurrence extraction. Furthermore, this procedure allows to study textures more accurately. Indeed, increasing the value of the parameter p it is possible to extract an higher number of co-occurrences even with small distance values, which is very useful to characterise fine textures.

References

1. Gonzalez, R.C., Woods, R.E., Eddins, S.L.: Digital Image Processing Using MAT-LAB. Pearson Prentice Hall Pearson Education Inc., New Jersey (2004)
2. Alam, F.I., Faruqui, R.U.: Optimized calculations of haralick texture features. Eur. J. Sci. Res. **50**(4), 543–553 (2011)
3. Bino Sebastian, V., Unnikrishnan, A., Balakrishnan, K.: Gray level co-occurrence matrices: generalisation and some new features. J. CoRR abs/1205.4831 (2012)
4. Haralick, R.M., Shanmugam, K., Dinstein, I.: Textural features for image classification. IEEE Trans. Syst. Man Cybern. **3**(6), 610–621 (1973)

5. Benco, M., Hudec, R.: Novel method for color textures features extraction based on GLCM. Radioengineering **4**, 64–67 (2007)
6. Chen, S., Chengdong, W., Chen, D., Tan, W.: Scene classification based on gray level-gradient co-occurrence matrix in the neighborhood of interest points. In: IEEE International Conference on Intelligent Computing and Intelligent Systems, pp. 482–485 (2009)
7. Gelzinis, A., Verikas, A., Bacauskiene, M.: Increasing the discrimination power of the co-occurrence matrix-based features. Pattern Recogn. **40**, 2367–2372 (2007)
8. Gong, R., Wang, H.: Steganalysis for GIF images based on colors-gradient co-occurrence matrix. Optix Commun. **285**, 4961–4965 (2012)
9. Hu, Y.: Unsupervised texture classification by combining multi-scale features and k-means classifier. In: Chinese Conference on Pattern Recognition, pp. 1–5 (2009)
10. Mitrea, D., Mitrea, P., Nedevschi, S., Badea, R., Lupsor, M.: Abdominal tumor characterization and recognition using superior-order cooccurrence matrices, based on ultrasound images. In: Computational and Mathematical Methods in Medecine (2012)
11. Nanni, L., Brahnam, S., Ghidoni, S., Menegatti, E., Barrier, T.: Different approaches for extracting information from the co-occurrence matrix. PLoS ONE **8**(12), e83554 (2013)
12. Walker, R., Jackway, P., Longstaff, D.: Genetic algorithm optimization of adaptive multi-scale GLCM features. Int. J. Pattern Recognit. Artif. Intell. **17**, 17–39 (2003)
13. Lowe, D.G.: Distinctive image features from scale-invariant keypoints. Int. J. Comput. Vis. **60**(2), 91–110 (2004)
14. Bay, H., Tuytelaars, T., Van Gool, L.: SURF: speeded up robust features. In: Leonardis, A., Bischof, H., Pinz, A. (eds.) ECCV 2006. LNCS, vol. 3951, pp. 404–417. Springer, Heidelberg (2006). doi:10.1007/11744023_32
15. Dalal, N., Triggs, B.: Histograms of oriented gradients for human detection. In: Conference on Computer Vision and Pattern Recognition, vol. 1, pp. 886–893 (2005)
16. Ojala, T., Pietikäinen, M., Harwood, D.: A comparative study of texture measures with classification based on featured distributions. Pattern Recogn. **20**(1), 51–59 (1996)
17. Ojala, T., Pietikäinen, M., Maenpaa, T.: Multiresolution gray-scale and rotation invariant texture classification with local binary patterns. IEEE Trans. Pattern Anal. Mach. Intell. **24**(7), 971–987 (2002)
18. Bianconi, F., Fernández, A.: Rotation invariant co-occurrence features based on digital circles and discrete Fourier transform. Pattern Recogn. Lett. **48**, 34–41 (2014)
19. Jain, A.K., Farrokhnia, F.: Unsupervised texture segmentation using Gabor filters. In: IEEE International Conference on Systems, Man and Cybernetics, pp. 14–19 (1990)
20. Petrou, M., Sevilla, P.G.: Image Processing: Dealing with Texture. Wiley Interscience, Hoboken (2006)
21. Di Ruberto, C., Fodde, G., Putzu, L.: Comparison of statistical features for medical colour image classification. In: Nalpantidis, L., Krüger, V., Eklundh, J.-O., Gasteratos, A. (eds.) ICVS 2015. LNCS, vol. 9163, pp. 3–13. Springer, Cham (2015). doi:10.1007/978-3-319-20904-3_1
22. Di Ruberto, C., Fodde, G., Putzu, L.: On different colour spaces for medical colour image classification. In: Azzopardi, G., Petkov, N. (eds.) CAIP 2015. LNCS, vol. 9256, pp. 477–488. Springer, Cham (2015). doi:10.1007/978-3-319-23192-1_40

Visual and Textual Sentiment Analysis of Brand-Related Social Media Pictures Using Deep Convolutional Neural Networks

Marina Paolanti[1]([✉]), Carolin Kaiser[2], René Schallner[2], Emanuele Frontoni[1], and Primo Zingaretti[1]

[1] Department of Information Engineering, Università Politecnica delle Marche,
Via Brecce Bianche 12, 60131 Ancona, Italy
m.paolanti@pm.univpm.it, {e.frontoni, p.zingaretti}@univpm.it
[2] GfK Verein, Schnieglinger Str. 57, 90419 Nürnberg, Germany
{carolin.kaiser,rene.schallner}@gfk-verein.org

Abstract. Social media pictures represent a rich source of knowledge for companies to understand consumers' opinions, as they are available in real time and at low costs and represent an active feedback which is of importance not only for companies developing products, but also to their rivals and potential consumers. In order to estimate the overall sentiment of a picture, it is essential to not only judge the sentiment of the visual elements but also to understand the meaning of the included text. This paper introduces an approach to estimate the overall sentiment of brand-related pictures from social media based on both visual and textual clues. In contrast to existing papers, we do not consider text accompanying a picture, but text embedded in a picture, which is more challenging since the text has to be detected and recognized first, before its sentiment can be identified. Based on visual and textual features extracted from two trained Deep Convolutional Neural Networks (DCNNs), the sentiment of a picture is identified by a machine learning classifier. The approach was applied and tested on a newly collected dataset, "GfK Verein Dataset" and several machine learning algorithms are compared. The experiments yield high accuracy, demonstrating the effectiveness and suitability of the proposed approach.

1 Introduction

The advent of Social Media has enabled everyone with a smartphone, tablet or computer to easily create and share their ideas, opinions and contents with millions of other people around the world. Recent years have witnessed the explosive popularity of image-sharing services such as Instagram[1] and Flickr[2]. These images do not only reflect people social lives, but also express their opinions about products and brands. Social media pictures represent a rich source of

---

[1] www.instagram.com.

[2] www.flickr.com.

© Springer International Publishing AG 2017
S. Battiato et al. (Eds.): ICIAP 2017, Part I, LNCS 10484, pp. 402–413, 2017.
https://doi.org/10.1007/978-3-319-68560-1_36

knowledge for companies to understand consumers' opinions [1]. The multitude of pictures makes a manual approach infeasible and increases the attractiveness of automated sentiment analysis [2,3].

In the past, companies have conducted consumer surveys for this purpose. Although well-designed surveys can provide high quality estimations, they can be time-consuming and costly, especially if a large volume of survey data is gathered [4]. In contrast, social media pictures are available in real time and at low costs and represent an active feedback, which is of importance not only to companies developing products, but also to their rivals and potential consumers [5]. Algorithms to identify sentiment are crucial for understanding consumer behaviour and are widely applicable to many domains, such as retail [6], behaviour targeting [7], and viral marketing [8].

Sentiment analysis is the task of evaluating this goldmine of information. It retrieves opinions about certain products and classifies them as positive, negative, or neutral. Existing research papers [9,10], have focused on sentiment analysis of textual postings such as reviews in shopping platforms and comments in discussion boards. However, with the increasing popularity of social networks and image sharing platforms [11,12] more and more opinions are expressed by pictures. Several researchers have now started to propose solutions for the sentiment analysis of visual content. However, a multitude of consumers' pictures does not only include visual elements, but also textual elements. For example, people take pictures of advertisement posters or insert text into photos with the aid of photo editing software. In order to estimate the overall sentiment of a picture, it is essential to not only judge the sentiment of the visual elements but also to understand the meaning of the included text. While a picture showing a cosmetic product next to a cute rabbit might be positive, the same picture containing the words "animal testing" might be negative.

This paper introduces an approach to estimate the overall sentiment of a picture based on both visual and textual information. While many studies have performed sentiment analysis, most existing methods focus on either only textual content or only visual content. To the best of our knowledge, this is the first approach to consider visual and textual information in pictures at the same time. The sentiment of a picture is identified by a machine learning classifier based on visual and textual features extracted from two specially trained Deep Convolutional Neural Networks (DCNNs). The visual feature extractor is based on the VGG16 network architecture [13] and it is trained by fine-tuning a model pretrained on the ImageNet dataset [14]. While the visual feature extractor is applied to the whole image, the textual feature extractor detects and recognizes texts before extracting features. The textual feature extractor is based on the DCNN architecture proposed by [15] and is created by fine-tuning a model which has been previously trained on synthesized social media images. Based on these features, six state-of-the-art classifiers, namely kNearest Neighbors (kNN) [16,17], Support Vector Machine (SVM) [18], Decision Tree (DT) [19], Random Forest (RF) [20], Naïve Bayes (NB) [21] and Artificial Neural Network (ANN) [22,23], are compared to recognize the overall sentiment of the images.

The approach has been applied to a newly collected dataset "GfK Verein Dataset" of consumer-generated pictures from Instagram which show commercial products. This dataset comprises 4200 images containing visual and textual elements. In contrast to many existing datasets, the true sentiment is not automatically judged by the accompanying texts or hash-tags but has been manually estimated by human annotators, thus providing a more precise dataset. The application of our approach to this dataset yields good results in terms of precision, recall and F1-score and demonstrates the effectiveness of the proposed approach.

The paper is organized as follows: Sect. 2 is an overview of the research status of textual and visual sentiment analysis; Sect. 3 introduces our approach consisting of a visual model (Subsect. 3.1), a textual model (Subsect. 3.2) and a fusion model (Subsect. 3.3) and gives details on the "GfK Verein Dataset" (Subsect. 3.4); final sections present results (Sect. 4) and conclusions (Sect. 5) with future works.

2 Related Work

Sentiment analysis aims at the detection of polarity and can be achieved in many different ways. Approaches for sentiment analysis can be differentiated with respect to the used methods and data sources. From a methodological perspective, we can distinguish between knowledge-based techniques and statistical methods [24]. Knowledge-based techniques, such as WordNet Affect [25] and SentiWordNet [26], rely on semantic knowledge resources to determine the sentiment. For example, in textual sentiment analysis, the sentiment of text is classified based on the presence of affective words from a lexicon. These methods are popular because of their easy application and accessibility, but their validity depends on a comprehensive knowledge base and rich knowledge representation. Statistical methods are trained with the aid of annotated corpora to identify the sentiment. These powerful methods are widely applied in research, but their performance depends on a sufficiently large training corpus [27]. While in former times shallow feature representations such as bag-of-words combined with support vector machines have been the mainstream in textual sentiment analysis, deep learning methods are becoming increasingly popular in recent years. In [28], the authors use a Convolutional Neural Network (CNN) to extract sentence features and perform sentiment analysis of Twitter messages. An ensemble system to detect the sentiment of a text document from a dataset of IMDB movie reviews is built in [29]. CNNs have also been applied to visual sentiment analysis. A deep CNN model called DeepSentiBank is trained to classify visual sentiment concepts by Chen [30]. A visual sentiment prediction framework is introduced in [8]. It performs transfer learning from a pre-trained CNN with millions of parameters.

With respect to the underlying data sources, sentiment analysis approaches can be divided into unimodal and multimodal [31]. While unimodal approaches consider only one data source, mulitmodal models take several types of data sources into account when determining the sentiment. In [32] the authors employ both images and text to predict sentiment by fine-tuning a CNN for image sentiment analysis and by training a paragraph vector model for textual sentiment analysis. In [33], the authors employ deep learning to analyze the sentiment of Chinese microblogs from both textual and visual content.

In this work, we focus on sentiment analysis for both visual and textual information of brand-related pictures from social media. In contrast to [32,33], however, we do not consider text accompanying a picture, but text included in a picture, which is more challenging since the text has to be detected and recognized first, before its sentiment can be identified.

3 Methods

In this section, we introduce the joint visual and textual sentiment analysis framework as well as the dataset used for evaluation. The framework is depicted in Fig. 1 and comprises three main components: the visual feature extractor, the textual feature extractor, and the overall sentiment classifier. We use especially trained DCNNs for visual and textual feature extraction. The visual and textual features are fused and fed into the overall sentiment classifier. We compare common machine learning algorithms for the overall sentiment classification. Further details are given in the following subsections.

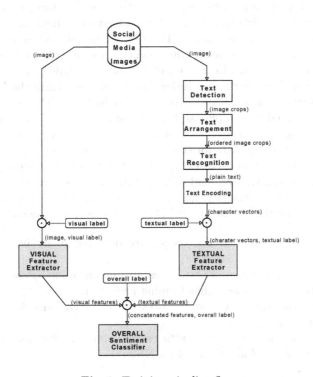

Fig. 1. Training pipeline flow

The framework is comprehensively evaluated on the "GfK Verein Dataset", a proprietary dataset collected for this work. The details of the data collection and ground truth labeling are discussed in Subsect. 3.4.

3.1   Visual Feature Extractor

The visual feature extractor aims at providing information about the visual sentiment of a picture and is therefore trained with image labels indicating the visual sentiment of the images. The training is performed by fine-tuning a VGG16 net [13] that has been pre-trained on the ImageNet dataset [14] to classify images into 1000 categories. We fine-tune by cutting off the final classification layer (fc8) and replacing it by a fully connected layer with 3 outputs (one for each sentiment class). In addition, the learning rate multipliers are increased for that layer so that it learns more aggressively than all the other layers. Finally, loss and accuracy layers are adapted to take input from the new fc8 layer. Since the image classifier serves as feature extractor, the output of the next to last fc7 layer is passed to the overall sentiment classifier. The image feature extractor is implemented using standard Caffe[3] tools.

3.2   Textual Feature Extractor

The goal of the textual feature extractor is to provide information about the textual sentiment of a picture. It is therefore trained with image labels indicating the textual sentiment of the images. The textual feature extractor consists of multiple components. The central component is a character-level DCNN with an architecture as described in [15], which has been extended by one additional convolution layer. The extra convolution layer, inserted before the last pooling layer, has a kernel size of 3 and produces 256 features. The textual feature extractor was trained in two phases: first training a base model on synthesized social media images and then fine-tuning that base model on our dataset. In order to generate training data for the base model, accompanying captions from brand-related social media pictures were inserted into social media pictures in varying fonts, font-sizes, colors and slight rotations. Since the text is embedded in the picture as pixels, the text has to be transformed to characters before it can be processed by the character-level DCNN. We perform the following steps:

1. *Text Detection*: individual text boxes are detected in an image with the TextBoxes Caffe model [34].
2. *Text Arrangement*: detected text boxes are put in order based on a left-to-right, top-to-bottom policy, thus forming logical lines.
3. *Text Recognition*: each text box is processed by the OCR model [35] to transcribe the text of the box.
4. *Text Encoding*: the recognized text is encoded into one-hot vectors based on the alphabet of the character-level DCNN.

---

[3] http://caffe.berkeleyvision.org/.

The textual features of the next to last layer of the character-level DCNN are passed to the final sentiment classifier.

3.3 Overall Sentiment Classifier

On the basis of the visual and textual features, the overall sentiment classifier aims at estimating the overall sentiment of an image. For this purpose, it is trained with labels indicating the overall sentiment of the images. The number of visual and textual features is illustrated in Table 1.

Table 1. Number of features

Model	Layer	Number of features
Image	fc7	4096
Text	ip4	1024

Based on the fused features, six state-of-the art classifiers, namely kNN, SVM, DT, RF, NB and ANN are used to recognize the overall sentiment of the images and compared with respect to precision, recall and F1-score.

3.4 GfK Verein Dataset

In this work, we provide, to the best of our knowledge, the first study on sentiment analysis of brand-related pictures on Instagram. As discussed in Sect. 1, Instagram provides a rich repository of images and captions that are associated with users' sentiments. We construct a visual and textual sentiment dataset from the pictures on Instagram. We utilize the captions of the Instagram posts to pre-select images that have detectable sentiment content about well-known brands from the industry of fast moving consumer goods. Typically, the image captions indicate the users' sentiment for the uploaded images. The "GfK Verein Dataset" is composed of brand related social media images as follows:

- 1400 images with positive sentiment;
- 1400 images with neutral sentiment;
- 1400 images with negative sentiment.

To obtain the ground truth of the collected pictures, the true sentiment has been manually estimated by human annotators, thus providing a more precise and less noisy dataset compared to automatically generated labels from image captions or hashtags. All pictures are annotated with respect to their visual, textual and overall sentiment.

Figure 2 shows three examples of brand related social media pictures of "GfK Verein Dataset". As can be seen, the overall sentiment towards a brand or product does not only depend on the visual content of a picture but also on its textual content.

(a)   (b)   (c)

Fig. 2. Brand Related Social Media Pictures of "GfK Verein Dataset". Figure 2a is an example of a picture with overall negative sentiment, Fig. 2b represents an image with overall neutral sentiment, and Fig. 2c is a picture with overall positive sentiment

Since sentiment estimation is a subjective task where different persons may assign different sentiments to images, we asked two persons to judge the sentiment of the images and measured their agreement. The inter-annotator-agreement is a common approach to determine the reliability of a dataset and the difficulty of the classification task [36]. We calculate Cohen's Kappa Coefficient k which measures the agreement between two annotators beyond chance [37]. The values of Kappa range from -1 to 1, with 1 indicating perfect agreement, 0 indicating agreement expected by chance, and negative values indicating systematic disagreement. The inter-annotator-agreement for the visual ($k = 0.82$), textual ($k = 0.82$) and overall ($k = 0.84$) sentiment assignment is high, assuring good quality of the dataset and feasibility of the machine learning task.

4   Results and Discussion

In this section, the results of the experiments conducted on "GfK Verein Dataset" are reported. In addition to the performance of the overall sentiment classifier, we also present the performance of the visual and textual sentiment classifiers which form the basis of the visual and textual feature extractors and are key to the overall sentiment classification.

The experiments are based only on these images of the dataset, where both annotators have agreed on the overall, visual and textual sentiment. By removing pictures with ambiguous sentiment, we increase the quality of the dataset and ensure the validity of the experiments. The final dataset is comprised of a total amount of 3452 pictures, including 1149 pictures with overall positive sentiment, 1225 pictures with overall neutral sentiment and 1078 pictures with overall negative sentiment.

We perform the experiments by splitting the labeled dataset into a training set and a test set. Each classifier will only be trained based on the training set. Likewise, the test set is also fixed in the beginning and used for all test purposes. The dataset is split into 80% training and 20% test images, taking into account all permutations of overall, visual, and textual annotations.

In order to create the visual feature extractor we trained a DCNN to classify the visual sentiment of a picture. The performance of the visual sentiment classification is reported in Table 2. As can be seen, high values of precision and recall can be achieved, especially for pictures with positive and neutral visual sentiment. The recognition of visually negative pictures is more difficult due to the smaller amount of available training data and the higher variation in motives. Consumers tend to express their overall negative sentiment towards brands by adding negative text to neutral or positive motives. As people avoid posting pictures with negative facial expressions on social media, the most frequent form of visual negative sentiment is graphics with many different motives.

Table 2. Performance of the visual DCNN model, predicting visual sentiment based only on visual features

Category	Precision	Recall	F1-Score
Positive	0.83	0.82	0.82
Neutral	0.86	0.89	0.88
Negative	0.72	0.67	0.69
MEAN()	**0.81**	**0.79**	**0.80**

For creating a textual feature extractor, we trained a DCNN to estimate the sentiment of the text in the pictures. Table 3 depicts precision and recall of the textual sentiment classification. The performance of the textual sentiment classification is good, but lower than the performance of the visual sentiment classification. While the judgment of visual and textual sentiment is equally difficult for humans, the classification of text in pictures is much more challenging for machines as the text has to be detected and recognized first before it can be classified, thus being more error-prone. Comparing the different classes reveals that negative and neutral texts can be recognized better than positive texts. This fact is also reflected by the characteristics of the dataset. As consumers prefer visual clues such as happy people or smileys to textual clues for showing their overall positive sentiment towards brands, positive texts are less expressive.

Table 3. Performance of the textual DCNN model, predicting textual sentiment based only on textual features

Category	Precision	Recall	F1-Score
Positive	0.71	0.68	0.70
Neutral	0.84	0.61	0.71
Negative	0.67	0.89	0.76
MEAN()	**0.74**	**0.73**	**0.74**

Based on the visual and textual features, a machine learning classifier is trained to identify the overall sentiment of a picture. We train several classifiers, namely SVM, DT, NB, RF, and ANN and compare their performance for different parameter settings. Table 4 reports the results of the best parameter setting for each classifier. As can be seen, the performance of all classifiers is good, with F1-Scores ranging from 0.72 for NB to 0.79 for ANN, thus demonstrating the effectiveness and the suitability of the proposed approach. The performance of the overall sentiment classification is much higher than the performance of the textual sentiment classification but slightly lower than the performance of the visual sentiment classification. This comparison shows that recognizing the overall sentiment is more challenging than only the visual sentiment. Estimating the overall sentiment, however, is crucial for understanding consumers' attitudes towards brands. Relying on the visual sentiment only can be misleasing in many cases since consumers often embed text in their pictures to verbalize their sentiment. Especially, overall negative sentiments are often expressed by adding negative text to neutral or positive visual motives.

Table 4. Performance of the overall classifier, predicting overall sentiment based on both visual and textual features

Classifier	Precision	Recall	F1-Score
NB	0.72	0.72	0.72
DT	0.72	0.72	0.72
RF	0.74	0.74	0.74
SVM	0.77	0.77	0.77
kNN	0.78	0.78	0.78
ANN	**0.79**	**0.79**	**0.79**

5   Conclusions

Multimodal sentiment analysis of social media content represents a challenging but rewarding task enabling companies to gain deeper insights into consumer behavior. In this paper, we introduce a deep learning approach for recognizing the sentiment of brand-related social media pictures by taking visual as well as textual information into account. The sentiment of a picture is identified by a machine learning classifier based on visual and textual features extracted from two trained DCNNs. By combining DCNNs with machine learning algorithms such as kNN, SVM, DT, RF, NB, and ANN, the approach is able to learn a high level representation of both visual and textual content and to achieve high precision and recall for sentiment classification. The experiments on the "GfK Verein Dataset" yield high accuracies and demonstrate the effectiveness and suitability of our approach. Further investigation will be devoted to improve our approach by employing a larger dataset and extracting additional informative features such as peoples' emotions as well as positive and negative symbols. Moreover, we will extend the evaluation by comparing our visual and textual classifiers with other existing systems for visual and textual sentiment analysis.

Acknowledgement. This work was funded by GfK Verein (www.gfk-verein.org). The authors would like to thank Lara Enzingmüller and Regina Schreder for their help with data preparation.

References

1. Carolin Kaiser, R.W.: Gaining marketing-relevant knowledge from social media photos - a picture is worth a thousand words. In: Proceedings of the 2016 ESOMAR Congress, New Orleans (2016)
2. Yang, Y., Jia, J., Zhang, S., Wu, B., Chen, Q., Li, J., Xing, C., Tang, J.: How do your friends on social media disclose your emotions? In: AAAI, vol. 14, pp. 1–7 (2014)
3. You, Q., Luo, J., Jin, H., Yang, J.: Robust image sentiment analysis using progressively trained and domain transferred deep networks. arXiv preprint arXiv:1509.06041 (2015)
4. Yi, J., Nasukawa, T., Bunescu, R., Niblack, W.: Sentiment analyzer: extracting sentiments about a given topic using natural language processing techniques. In: Third IEEE International Conference on Data Mining, ICDM 2003, pp. 427–434. IEEE (2003)
5. Mukherjee, S., Bhattacharyya, P.: Feature specific sentiment analysis for product reviews. In: Gelbukh, A. (ed.) CICLing 2012. LNCS, vol. 7181, pp. 475–487. Springer, Heidelberg (2012). doi:10.1007/978-3-642-28604-9_39
6. Liciotti, D., Paolanti, M., Frontoni, E., Mancini, A., Zingaretti, P.: Person re-identification dataset with RGB-D camera in a top-view configuration. In: Nasrollahi, K., Distante, C., Hua, G., Cavallaro, A., Moeslund, T.B., Battiato, S., Ji, Q. (eds.) FFER/VAAM -2016. LNCS, vol. 10165, pp. 1–11. Springer, Cham (2017). doi:10.1007/978-3-319-56687-0_1
7. Naspetti, S., Pierdicca, R., Mandolesi, S., Paolanti, M., Frontoni, E., Zanoli, R.: Automatic analysis of eye-tracking data for augmented reality applications: a prospective outlook. In: De Paolis, L.T., Mongelli, A. (eds.) AVR 2016. LNCS, vol. 9769, pp. 217–230. Springer, Cham (2016). doi:10.1007/978-3-319-40651-0_17
8. Xu, C., Cetintas, S., Lee, K.C., Li, L.J.: Visual sentiment prediction with deep convolutional neural networks. arXiv preprint arXiv:1411.5731 (2014)
9. Pang, B., Lee, L., et al.: Opinion mining and sentiment analysis. Found. Trends® Inf. Retrieval **2**(1–2), 1–135 (2008)
10. Thelwall, M., Buckley, K., Paltoglou, G., Cai, D., Kappas, A.: Sentiment strength detection in short informal text. J. Am. Soc. Inform. Sci. Technol. **61**(12), 2544–2558 (2010)
11. Yuan, J., Mcdonough, S., You, Q., Luo, J.: Sentribute: image sentiment analysis from a mid-level perspective. In: Proceedings of the Second International Workshop on Issues of Sentiment Discovery and Opinion Mining, p. 10. ACM (2013)
12. Chang, Y., Tang, L., Inagaki, Y., Liu, Y.: What is tumblr: a statistical overview and comparison. ACM SIGKDD Explor. Newsl. **16**(1), 21–29 (2014)
13. Simonyan, K., Zisserman, A.: Very deep convolutional networks for large-scale image recognition. arXiv preprint arXiv:1409.1556 (2014)

14. Krizhevsky, A., Sutskever, I., Hinton, G.E.: Imagenet classification with deep convolutional neural networks. In: Advances in Neural Information Processing Systems, pp. 1097–1105 (2012)
15. Zhang, X., Zhao, J., LeCun, Y.: Character-level convolutional networks for text classification. In: Advances in Neural Information Processing Systems, pp. 649–657 (2015)
16. Bø, T.H., Dysvik, B., Jonassen, I.: LSimpute: accurate estimation of missing values in microarray data with least squares methods. Nucleic Acids Res. **32**(3), e34 (2004)
17. Troyanskaya, O., Cantor, M., Sherlock, G., Brown, P., Hastie, T., Tibshirani, R., Botstein, D., Altman, R.B.: Missing value estimation methods for dna microarrays. Bioinformatics **17**(6), 520–525 (2001)
18. Cortes, C., Vapnik, V.: Support-vector networks. Mach. Learn. **20**(3), 273–297 (1995)
19. Quinlan, J.R.: Induction of decision trees. Mach. Learn. **1**(1), 81–106 (1986)
20. Breiman, L.: Random forests. Mach. Learn. **45**(1), 5–32 (2001)
21. Rish, I.: An empirical study of the naive Bayes classifier. In: IJCAI 2001 Workshop on Empirical Methods in Artificial Intelligence, vol. 3, pp. 41–46. IBM, New York (2001)
22. Lippmann, R.: An introduction to computing with neural nets. IEEE Assp Mag. **4**(2), 4–22 (1987)
23. Paolanti, M., Frontoni, E., Mancini, A., Pierdicca, R., Zingaretti, P.: Automatic classification for anti mixup events in advanced manufacturing system. In: ASME 2015 International Design Engineering Technical Conferences and Computers and Information in Engineering Conference, p. V009T07A061. American Society of Mechanical Engineers (2015)
24. Cambria, E.: Affective computing and sentiment analysis. IEEE Intell. Syst. **31**(2), 102–107 (2016)
25. Strapparava, C., Valitutti, A., et al.: Wordnet affect: an affective extension of wordnet. In: LREC, vol. 4, pp. 1083–1086. Citeseer (2004)
26. Esuli, A.: Sentiwordnet: a publicly available lexical resource for opinion mining. In: Proceedings of Language Resources And Evaluation (LREC), Genoa, Italy, pp. 24–26 (2006)
27. Cambria, E., White, B.: Jumping NLP curves: a review of natural language processing research [review article]. IEEE Comput. Intell. Mag. **9**(2), 48–57 (2014)
28. Kim, Y.: Convolutional neural networks for sentence classification. arXiv preprint arXiv:1408.5882 (2014)
29. Mesnil, G., Mikolov, T., Ranzato, M., Bengio, Y.: Ensemble of generative and discriminative techniques for sentiment analysis of movie reviews. arXiv preprint arXiv:1412.5335 (2014)
30. Chen, T., Borth, D., Darrell, T., Chang, S.F.: Deepsentibank: visual sentiment concept classification with deep convolutional neural networks. arXiv preprint arXiv:1410.8586 (2014)
31. Cambria, E., Poria, S., Bisio, F., Bajpai, R., Chaturvedi, I.: The CLSA model: a novel framework for concept-level sentiment analysis. In: Gelbukh, A. (ed.) CICLing 2015. LNCS, vol. 9042, pp. 3–22. Springer, Cham (2015). doi:10.1007/978-3-319-18117-2_1
32. You, Q., Luo, J., Jin, H., Yang, J.: Joint visual-textual sentiment analysis with deep neural networks. In: Proceedings of the 23rd ACM International Conference on Multimedia, pp. 1071–1074. ACM (2015)

33. Yu, Y., Lin, H., Meng, J., Zhao, Z.: Visual and textual sentiment analysis of a microblog using deep convolutional neural networks. Algorithms **9**(2), 41 (2016)
34. Liao, M., Shi, B., Bai, X., Wang, X., Liu, W.: Textboxes: a fast text detector with a single deep neural network. arXiv preprint arXiv:1611.06779 (2016)
35. Jaderberg, M., Simonyan, K., Vedaldi, A., Zisserman, A.: Reading text in the wild with convolutional neural networks. Int. J. Comput. Vis. **116**(1), 1–20 (2016)
36. Bhowmick, P.K., Mitra, P., Basu, A.: An agreement measure for determining inter-annotator reliability of human judgements on affective text. In: Proceedings of the Workshop on Human Judgements in Computational Linguistics, pp. 58–65. Association for Computational Linguistics (2008)
37. Cohen, J.: A coefficient of agreement for nominal scales. Educ. Psychol. Measur. **20**(1), 37–46 (1960)

Deep Multibranch Neural Network for Painting Categorization

Simone Bianco$^{(\boxtimes)}$, Davide Mazzini, and Raimondo Schettini

Dipartimento di Informatica, Sistemistica e Comunicazione,
Università degli Studi di Milano-Bicocca, viale Sarca 336, 20126 Milan, Italy
{simone.bianco,davide.mazzini,raimondo.schettini}@disco.unimib.it

Abstract. Coarse features, such as scene composition and subject together with fine details, such as strokes and line styles, are useful clues for painter and style categorization. In this work, to automatically predict painting's artist and style, we propose a novel deep multibranch neural network, where the different branches process the input image at different scales to jointly model the fine and coarse features of the painting. Experiments for both artist and style classification tasks are performed on the challenging Painting-91 dataset, that includes 91 different painters and 13 diverse painting styles. Our method outperforms the best method in the state of the art by 14.0% and 9.6% on artist and style classification respectively.

Keywords: Painting categorization · Painting style classification · Painter recognition · Deep convolutional neural network · Multiresolution

1 Introduction

Research on digital analysis of paintings is gaining increasing attention due to the large quantities of visual artistic data [4,10,12], made available from art museums digitizing their collection for cultural heritage, and the need of automatic tools to organize and manage them. In this work, we approach the problem of categorizing a painting by automatically predicting its artist and style given solely the digital version of the painting itself [1]. Both these tasks are very challenging due to the large amount both inter- and intra-class variations, e.g. the different personal styles in the same art movement, or the same artist adhering to different schools in different periods in his/her production. Artist classification consists in automatically associate the painting to its painter. In this task factors such as stroke patterns, the color palette used, the scene composition, and the subject must be taken into account. Style classification consists in automatically categorize a painting into the school or art movement it belongs to. Art theorists define an artistic style as the combination of iconographic, technical and compositional features that give to a work its character [20]. Style categorization is complicated by the fact that styles may not remain pure but could be influenced by others.

© Springer International Publishing AG 2017
S. Battiato et al. (Eds.): ICIAP 2017, Part I, LNCS 10484, pp. 414–423, 2017.
https://doi.org/10.1007/978-3-319-68560-1_37

1.1   Contribution

We propose a multiresolution approach to solve the tasks of artist and style categorization. A particular random-crop strategy permits to gather clues from low-level texture details and, at the same time, exploit the coarse layout of the painting. The classification process is carried on by a specifically-taylored multibranch neural network.

Experiments are performed on the challenging painting-91 dataset [10]. On both artist and style classification tasks our approach improves the mean classification accuracy by 14.3% and 10.2% respectively, compared to the previous state-of-the-art models.

1.2   Prior Works

The problem of painter or style categorization has been faced using different tecniques. Some existing approaches make use of traditional handcrafted features [4,10] whereas more recent works relay on the use of deep networks [1,14,15,18, 19]. Zhao et al. [21] used a pretrained neural network in a two-step bootstrap approach to categorize ancient illustration from the British Library. Peng and Chen [15] use a multiresolution approach to exploit both small details and the overall image structure. A more sophisticated technique is used by [1] where the use of a deformable part model is adopted in order to combine low-level details and an holistic representation of the whole painting. Deep CNNs have been widely used as features extractors to solve different tasks [3,16], Peng and Chen [15] and Anwer et al. [1] relay on pretrained deep CNNs to deal with the small quantity of images of the Painting-91 dataset. Tan et al. [18] made different experiments by training a network from scratch or finetuning an existing network for the task of style and painter recognition. They adopted a network structure similar to the one used by Krizhevsky et al. [11]. Hentschel et al. [8] performed interesting experiments about the quantity of data needed to fine-tune the network by Krizhevsky et al. [11] for the task of style classification.

2   Our Approach

The scene composition and the subject depicted are important clues to recognize a particular author or a painting style. These elements need to be extracted from the whole painting. At the same time finer details, such as stroke patterns or the line styles, are also very good clues. Obviously a powerful discriminative model should consider both the coarse level and fine details. On the basis of these considerations we decided to adopt a multiresolution approach: first, a predefined number of squared "small" crops are extracted from the high-resolution image. Then, the image is downsampled and another "large" crop is extracted from the low-resolution image (see Sect. 2.1). All the crops are then fed to the branches of a deep neural network that extracts the corresponding features. The outputs of the branches are collected by a join layer and fed to a deep neural network that carries on the categorization process.

2.1 Input Preprocessing

The first preprocessing step consists in normalizing the input image by subtracting the mean and dividing by the standard deviation of the pixel distribution of whole training set. This contrast normalization preprocessing is known to improve CNNs accuracy in different domains [2] by limiting the variability of the input range. The second step consists in a particular cropping strategy. Crops are taken at multiple resolutions to capture both fine details and coarse structures. Since paintings exhibit high variability in terms of aspect-ratios, the input image is resized such as the minimum side is 512 pixels and the aspect ratio is preserved. From the resulting image we extract two squared random crops of 227 pixels side. Then the image is further downsampled, using an average pooling layer, such as the minimum side is 256 pixels and another squared crop of 227 by 227 pixels is extracted. All the crops are squared, independently from the original aspect ratio of the input image. This is done to improve the computational efficiency allocating GPU memory blocks only once. Images and crops sizes has been choosen as a tradeoff to exploit fine details and to limit the computational burden accordingly to the size and quality of the original images. The coordinates of the crops inside the input image are randomly chosen with the only constraint that crops coming from the same scale do not overlap. The rational behind this choice is that the salient details can be anywhere inside the painting, and the extraction of crops at random locations permits the implementation of a consensus strategy by simply processing the same input image several times. The consensus strategy consists in averaging the output of the last fully-connected layer for the multiple passes of the same image trough the network, resulting in a feature vector that is then fed to the softmax layer to get the final prediction.

2.2 Deep Network Structure

We propose a novel network whose structure is shown in Fig. 1. It is composed of five modules: three branches to extract the low level structures of the painting crops, a join module to gather the output of the three branches and a classification module to make the prediction. Each branch is trained with crops from a specific scale, thus becoming specialized in processing texture patterns at that specific resolution. We decided to use only two scales since, in our preliminary experiments, the use of higher scales brought a slight improvement compared to the exponential increase of computational burden.

In the three branches and in the classification model our deep network makes use of Residual Blocks which have been shown to be an effective architectural choice to build very deep networks [7] and tackle the problem of vanishing gradients by using shortcut connections. In particular, we used "bottleneck" Residual Blocks, which allow the network architecture to be even deeper [7]. Each skip connection has four times the number of channels with respect to the internal elements of the block. This permits a large troughput of information among layers while mantaining a low computational complexity and low memory use

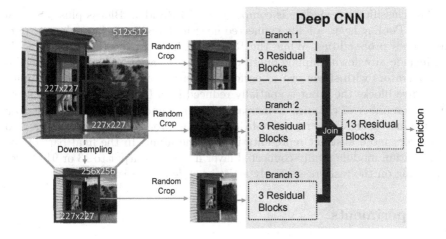

Fig. 1. Scheme of our deep multibranch neural network

insde each block. Our Residual Block structure is different from the one used by He et al. [7]: we moved the Batch Normalization layer [9] after the sum with the skip connection because, in our experiments, the resulting configuration has shown better performances.

The Residual Block we used is shown in Fig. 2. In our network (see Fig. 1) each of the three branches is composed by three Residual Blocks plus four layers near the input which perform the first processing (Convolution + BatchNorm [9] + ReLU [13]) and an initial downsampling (Max Pooling). The join module is a particular Residual Block which gathers the output of the three branches. It stacks the output features and then converts them to a smaller-dimensional feature space by compressing information along the channel dimension. The reason behind this operation is to make the computations feasible in the following layers by reducing the channel dimension of the output by a factor of three.

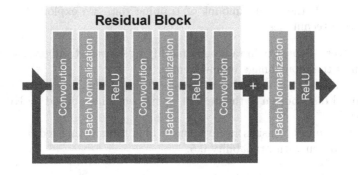

Fig. 2. The type of residual block used in our deep neural network

The classification module is composed by 13 Residual Blocks plus a Spatial Average Pooling layer, a Fully-connected layer and a Softmax layer that outputs the classes probabilities. While the Residual Blocks in the three branches do not include any downsampling operator, the classification module uses convolution operators with stride two to perform a spatial downsampling of the input. Every five blocks the input is spatially reduced by a factor of two. At the same time the number of channels is increased by the same amount. This leads to a gradual increasing of the receptive-fields of the network in the deeper layers and also favors more abstract representations of the input. In the final part of the classification module a fully-connected layer maps the output to 13 or 91 classes depending on the task, respectively artist or style categorization.

3   Experiments

3.1   Dataset

We evaluate our recognition pipeline on the challenging Painting91 dataset [10] for both artist and style classification tasks. The dataset consists of 4266 paintings of 91 painters. As train and test split we used those provided by the authors which are in both cases nearly 50%. For the task of artist recognition, the whole dataset is used whereas for the task of style recognition only 2338 groundtruth are provided.

3.2   Training

Our training procedure was carried on in two phases. We first pretrained our deep network on the Kaggle dataset Painterbynumbers.[1] This dataset is intended for a similar task, i.e. painter verification, but it is much bigger. It contains more than 1500 authors and a training set of 79433 images. Then we finetuned it two times (one for each of the two tasks) on the Painting91 dataset, substituting the last fully connected layer with a new one that matched the number of classes needed for each task.

To cope with the small amount of training data we exploited some data augmentation techniques:

– Color Jitter. It consists in randomly modifying constrast, brightness and saturation of the input image indipendently.
– Lighting noise. It is a pixelwise transform based on the eigenvalues of the RGB pixel distribution of the dataset. It has been introduced by Krizhevsky et al. [11].
– Gaussian Blur. It consists in applying a blur filter with fixed σ to random images choosen with probability 0.5.

---

[1] https://www.kaggle.com/. We took part to the Painterbynumbers competition and ended among the top positions. Our method, that is disclosed here, achieves an accuracy of 53.8% on validation set for the task of artist classification.

- Geometric transforms. It includes small changes in scale and aspect-ratio of the input image.

As explained in the Subsect. 2.1 our network exploits random crops. Therefore if the same input is processed several times by the same network, the final prediction vectors can be averaged before being fed to the last softmax layer. In Table 1 we report the performance in terms of accuracy at different number of passes. Results are averaged over ten independent runs. The biggest improvement is obtained by exploiting two passes with respect to the single one. The best performance are obtained using four passes.

Table 1. Accuracy vs number of passes trough the network. Each value represents the average of 10 runs.

Passes	1	2	4	8
Artist	77.5	78.1	**78.5**	78.3
Style	83.6	84.1	**84.4**	84.3

3.3 Results

In Table 2 we report the performances of our method with respect to the state-of-the-art on the Paintings-91 dataset. Concerning our method, we report the average accuracy over ten independent runs together with the minimum and maximum values. Considering our average performance, our method outperforms the best method in literature by 14.0% and 9.6% on the task of artist and style categorization respectively.

Table 2. Comparison with the state of the art. Average classification rates on the Paintings-91 dataset for the tasks of Artist and Style recognition. Our values are obtained as the maximum of 10 runs.

Method	Artist	Style
VGG-16 FC [17]	51.7	67.2
MF [10]	53.1	62.2
CL-CNN [14]	56.4	69.2
MS-MCNN [15]	58.1	71.0
MOP [6]	59.7	68.8
Holistic [5]	61.8	70.1
Holistic + Part Based [1]	64.5	74.8
Ours (worst performance among 10 runs)	77.9	83.8
Ours (average performance among 10 runs)	78.5	84.4
Ours (best performance among 10 runs)	**78.8**	**85.0**

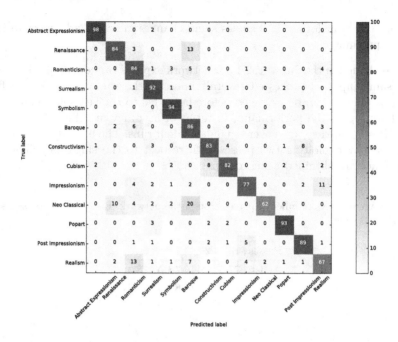

Fig. 3. Confusion matrix for the task of style recognition. The highest error rates are between Neo-Classical paintings, Baroque and Renaissance.

Figure 3 shows the confusion matrix for the style recognition task. The highest classification errors are between the Neo-Classical, Baroque and Renaissance classes. This seems to agree with styles' contaminations and influences as studied by art historians. For example Caravaggio paintings are classified as Baroque in Paintings-91 groundtruth. Actually he lived at the end of the Renaissance era, having a great influence on future Baroque painters.

Figure 4 shows the confusion matrix for the task of artist recognition. The highest error rates are between Memling and Van Eyck (27%), and Zurbaran and Vermeer (30%). Memling and Van Eyck are contemporaneous and both belonging to the Dutch and Flemish Renaissance, while Zurbaran and Vermeer are coeval painters, both belonging to the Baroque movement. To be able to actually discriminate between the last two painters, the network should be aware that Vermeer paintings are usually about indoor every-day life scenes whereas Zurbaran mostly painted religious subjects.

Figure 5 shows in the top row the highest scored errors. To better denote the complexity of the task, we also reported the highest scored and correctly classified example for the corresponding painter. Most confusions are between coeval painters. Even for an untrained human it could be difficult to predict the correct artist for a new unseen painting.

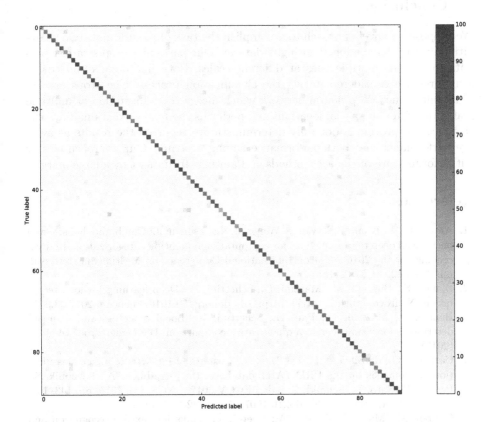

Fig. 4. Confusion matrix for the task of artist recognition. The highest error rates are between Zurbaran and Vermeer, Memling and Van Eyck. These painters are coeval and belongs to the same artistic movement.

Fig. 5. Top row: highest scored errors for the task of painters classification. Bottom row: for each of the predicted painters, we report the correctly classified example with the highest score.

4   Conclusions

We proposed a novel approach to accomplish the task of painter and style recognition on the challenging Painting91 dataset. Our particular crop strategy permits to exploit multiple cues at different scales. Both fine details and coarse structures are considered during the classification process. The crops are fed to a multibranch deep neural network which merge the information at multiple scales and different spatial locations and performs the final prediction. Since the classification process is not fully deterministic we reported the results as average performance and best performance among ten runs. Our approach clearly outperforms state-of-the-art methods on Paintings-91 dataset by a large margin.

References

1. Anwer, R.M., Khan, F.S., van de Weijer, J., Laaksonen, J.: Combining holistic and part-based deep representations for computational painting categorization. In: Proceedings of the 2016 ACM on International Conference on Multimedia Retrieval, pp. 339–342. ACM (2016)
2. Bianco, S., Buzzelli, M., Mazzini, D., Schettini, R.: Deep learning for logo recognition. Neurocomputing (2017). http://dx.doi.org/10.1016/j.neucom.2017.03.051
3. Bianco, S., Mazzini, D., Pau, D., Schettini, R.: Local detectors and compact descriptors for visual search: a quantitative comparison. Digit. Sig. Proc. **44**, 1–13 (2015)
4. Carneiro, G., da Silva, N.P., Del Bue, A., Costeira, J.P.: Artistic image classification: an analysis on the PRINTART database. In: Fitzgibbon, A., Lazebnik, S., Perona, P., Sato, Y., Schmid, C. (eds.) ECCV 2012. LNCS, vol. 7575, pp. 143–157. Springer, Heidelberg (2012). doi:10.1007/978-3-642-33765-9_11
5. Cimpoi, M., Maji, S., Vedaldi, A.: Deep filter banks for texture recognition and segmentation. In: Proceedings of the IEEE Conference on Computer Vision and Pattern Recognition, pp. 3828–3836 (2015)
6. Gong, Y., Wang, L., Guo, R., Lazebnik, S.: Multi-scale orderless pooling of deep convolutional activation features. In: Fleet, D., Pajdla, T., Schiele, B., Tuytelaars, T. (eds.) ECCV 2014. LNCS, vol. 8695, pp. 392–407. Springer, Cham (2014). doi:10.1007/978-3-319-10584-0_26
7. He, K., Zhang, X., Ren, S., Sun, J.: Deep residual learning for image recognition. In: Proceedings of the IEEE Conference on Computer Vision and Pattern Recognition, pp. 770–778 (2016)
8. Hentschel, C., Wiradarma, T.P., Sack, H.: Fine tuning CNNS with scarce training data–adapting imagenet to art epoch classification. In: 2016 IEEE International Conference on Image Processing (ICIP), pp. 3693–3697. IEEE (2016)
9. Ioffe, S., Szegedy, C.: Batch normalization: accelerating deep network training by reducing internal covariate shift. In: Proceedings of the 32nd International Conference on Machine Learning, pp. 448–456 (2015)
10. Khan, F.S., Beigpour, S., Van de Weijer, J., Felsberg, M.: Painting-91: a large scale database for computational painting categorization. Mach. Vis. Appl. **25**(6), 1385–1397 (2014)
11. Krizhevsky, A., Sutskever, I., Hinton, G.E.: Imagenet classification with deep convolutional neural networks. In: Advances in Neural Information Processing Systems, pp. 1097–1105 (2012)

12. Mensink, T., Van Gemert, J.: The rijksmuseum challenge: museum-centered visual recognition. In: Proceedings of International Conference on Multimedia Retrieval, p. 451. ACM (2014)
13. Nair, V., Hinton, G.E.: Rectified linear units improve restricted boltzmann machines. In: Proceedings of the 27th International Conference on Machine Learning (ICML 2010), pp. 807–814 (2010)
14. Peng, K.C., Chen, T.: Cross-layer features in convolutional neural networks for generic classification tasks. In: 2015 IEEE International Conference on Image Processing (ICIP), pp. 3057–3061. IEEE (2015)
15. Peng, K.C., Chen, T.: A framework of extracting multi-scale features using multiple convolutional neural networks. In: 2015 IEEE International Conference on Multimedia and Expo (ICME), pp. 1–6. IEEE (2015)
16. Sharif Razavian, A., Azizpour, H., Sullivan, J., Carlsson, S.: CNN features off-the-shelf: an astounding baseline for recognition. In: Proceedings of the IEEE Conference on Computer Vision and Pattern Recognition Workshops, pp. 806–813 (2014)
17. Simonyan, K., Zisserman, A.: Very deep convolutional networks for large-scale image recognition. In: Proceedings of the International Conference on Learning Representations (ICLR) (2015)
18. Tan, W.R., Chan, C.S., Aguirre, H.E., Tanaka, K.: Ceci n'est pas une pipe: A deep convolutional network for fine-art paintings classification. In: 2016 IEEE International Conference on Image Processing (ICIP), pp. 3703–3707. IEEE (2016)
19. Westlake, N., Cai, H., Hall, P.: Detecting people in artwork with CNNs. In: Hua, G., Jégou, H. (eds.) ECCV 2016. LNCS, vol. 9913, pp. 825–841. Springer, Cham (2016). doi:10.1007/978-3-319-46604-0_57
20. Widjaja, I., Leow, W.K., Wu, F.C.: Identifying painters from color profiles of skin patches in painting images. In: Proceedings of 2003 International Conference on Image Processing, ICIP 2003, vol. 1, pp. I–845. IEEE (2003)
21. Zhao, L., Wang, K., Do, B.: Sherlocknet: exploring 400 years of western book illustrations with convolutional neural networks. Technical report, Stanford University (2016)

Weighty LBP: A New Selection Strategy of LBP Codes Depending on Their Information Content

Maria De Marsico[1] and Daniel Riccio[2]([✉])

[1] Sapienza University of Rome, Rome, Italy
demarsico@di.uniroma1.it
[2] University of Naples Federico II, Naples, Italy
daniel.riccio@unina.it

Abstract. This paper presents a novel variation of the use of LBP codes. Similarly to Uniform LBP and Local Salient Patterns (LSP), it aims at both obtaining an effective texture description, and decreasing the length of the feature vectors, i.e., of the chains of LBP histograms. Instead of considering uniform codes, we rather consider the codes providing the highest "representativeness" power with respect to texture features. We identify this subset of codes by a generalized notion of entropy. This allows determining the most informative items in an homogeneous set.

Keywords: Local binary patterns · Information content · Entropy

1 Introduction

Though quite simple and light to compute, Local Binary Patterns (LBP) represents a very efficient texture operator. The basic procedure uses the value of each image pixel in turn as a binarization threshold for the values in its neighborhood (originally a 3×3 window); afterward, the code assigned to each pixel is the binary number represented by the string of binary elements obtained in such neighborhood. While the gray-level value of a given pixel represents its spectral propriety, its LBP code represents the textural aspect of the given pixel. LBP has achieved a great and still increasing popularity since its introduction in [9], where it is presented as a simplification of texture units (TUs) [18] making up the texture spectrum of an image. Similarly to LBP, TUs are obtained from a neighborhood of 3×3 pixels, yet using three $(0, 1, 2)$ instead of two values, giving a much higher number of codes. The texture spectrum is defined as the histogram (frequency of occurrences) of texture units computed over a region. The work in [9] shows that LBP, when used together with a simple local contrast measure, achieves better performance in unsupervised texture segmentation than other methods for texture analysis quite popular at that time. Due to this descriptive power, since its introduction, LBP has been the object of extensive investigations and evaluations, as well as variations [10]. It has been applied to address many problems, in particular, in the field of biometrics. A few examples include face recognition [1], demographics classification [19], gender recognition [15,17],

© Springer International Publishing AG 2017
S. Battiato et al. (Eds.): ICIAP 2017, Part I, LNCS 10484, pp. 424–434, 2017.
https://doi.org/10.1007/978-3-319-68560-1_38

or even face expression recognition [14]. A comprehensive survey of the use of LBP in Computer Vision can be found in [11].

This work deals with a novel approach to reduce the size of the code set for LBP, along the line of Uniform LBP codes [16] and Local Salient Patterns (LSP) [2]. Therefore, these two are the reference techniques we will compare with. In other words, the contribution of this work is to propose a new code reduction technique, and to compare it with the previously proposed ones.

The new operator is denoted as Weighty LBP (W-LBP). The core idea is to choose the most relevant codes according to their informative content with respect to textural features. Though Uniform LBP codes can be perceptually appreciated as more relevant with respect to the remaining ones, there is no assessment of their individual "informative" power. The same holds for LSP. We propose to analyze the LBP codes by exploiting a generalized definition of entropy. This was introduced to identify relevant face images in a set, and used for image analysis [6], for template selection in video-surveillance tasks [4], for the construction of a difference space for face image classification [13], and also for clustering [8]. Then it was extended to analyze generic items in a set, e.g., to quantize colors for image segmentation [5]. We apply the underlying approach here to select the most "representative" LBP codes.

The application of the method based on set entropy requires to exploit a suited similarity measure able to capture the characteristic nature of the items at hand (see Sect. 3.1). In the case of LBP such items are binary strings. Despite many different similarity/distance measures have been proposed, each such measure captures different aspects. Therefore, in the following we will briefly present the ones that we chose to analyze in order to capture the possible "informative power" of LBP codes. Afterward, we will follow a double evaluation procedure. In the first place, for each similarity measure, the subset of the most representative codes will be identified, and then among the obtained subsets those achieving the best classification results will be selected.

2   Related Work

LBP can be used in two ways. It can be used to characterize an image by a feature vector built with the histogram of LBP codes from the image, or to produce a gray level *feature image* by substituting each pixel in the original image by its LBP code. Feature images then undergo further computer vision processing. In particular applications, e.g., face recognition, LBP robustness can be increased by processing the image divided in cells according to a grid, whose size depends on image resolution. In this case, LBP is applied separately to each grid cell, and, in the case of histograms, the final feature vector is obtained by chaining the single cell histograms. This causes possibly huge feature vectors. Methods requiring a training step may incur the curse of dimensionality problem. For this reason, an interesting research line investigates how to identify and use a reduced the number of LBP codes, achieving a possibly better texture characterization with shorter feature vectors. However, finding the optimal subset of patterns is a demanding combinatorial problem. The selection of a subset of NP or less patterns out of the total 256, requires to assess the performance of a number of possible solutions that even for moderate values of NP requires huge computing resources. Therefore a suboptimal yet satisfying solution is often searched for. The work in [16] compares two approaches to extract a relevant subset of LBP codes. The first one uses beam search and explores subsets of patterns minimizing the classification error. The method iteratively increases the size of the pattern subset up to dimension NP and updates a list of the best BS subsets identified. The classification at each iteration exploits a reduced LBP histogram that contains one bin for each pattern chosen so far. All the remaining patterns are collapsed into a single bin. After NP iterations, the procedure returns BS distinct pattern sets, from which the optimal patterns can be chosen.

The second approach proposes the nowadays popular Uniform LBP patterns. It first defines a measure of *nonuniformity* U(LBP), which corresponds to the number of transitions (from 0 to 1 or the inverse) in the circular bitwise representation of the code. The assumption is that the lower the number of transitions, the more robust the code to image distortions. Based on this, the authors propose using the nine uniform patterns and their circularly rotated versions (this allows some transformation invariance). In practice, this corresponds to use 58 out of the 256 original unrotated patterns . Even in this case, all the remaining patterns are compressed into a single bin, therefore obtaining a 59-bin histogram. The conclusions drawn in [16] underline that every application may have its optimal set of patterns, but uniform patterns appear to perform well in many situations.

An example of a different strategy to address the reduction of LBP codes is represented by Local Salient Patterns (LSP) [2]. This recent approach derived from the original formulation of LBP focuses on the location of the largest positive and negative differences within the pixel neighborhood. This is deemed to remove the noise influence. The coding takes into account the possible pairs of neighbor indexes $(p_{diffmax}, p_{diffmin})$ that provide the maximum and the minimum difference with the central value of the neighborhood (usually, a 3×3 window). Therefore there are 57 distinguished codes (the last one corresponds to equal differences for all neighbors). This descriptor has achieved good performances in different facial analysis tasks, and experiments reported in [2] show that in most cases, LSP can outperform Uniform LBP.

3   The Proposed LBP Reduction

3.1   Entropy to Select Representatives in a Set

In image analysis, entropy is usually exploited as a measure of randomness/homogeneity of image pixels. Each pixel x in image I is treated as a symbol in the alphabet emitted by a source S. As for gray scale images, the alphabet is the set of 8-bit integers in $[0, 255]$. After normalizing the image histogram values in the range $[0, 1]$, each bin represents the probability of occurrence of the corresponding symbol in I. Entropy H(I) is:

$$H(I) = -\sum_{k=0}^{255} p(k) log_2(k) \tag{1}$$

Equation 1 can be generalized to express the amount of homogeneity in a set of items of any kind, given a suited abstraction. We summarize here the basic notation. More details can be found in [6].

Given a set G of objects/elements/observations (items from here on), we first search for a suitable similarity measure s, which is used to associate a real scalar value to any pair of feature vectors (templates) used to characterize the items of interest according to a chosen set of characteristics. The choice of the similarity measure to exploit depends in the first place on the kind of items to compare, and on the extracted feature templates. Computational cost of measuring this similarity can provide a further criterion. Popular examples are Euclidean distance, if feature vectors are represented as points in a space, or Dynamic Time Warping (DTW) for time series. The noticeable property of the following definitions is that they hold whichever is the chosen similarity measure s. From here on, if not otherwise specified, the notation will identify templates with the items they describe. In a preliminary definition step, let us assume to compare the probe template v to classify with a set templates g_i (from now on denoted as *gallery*.) This produces a similarity measure $s(v, g_i)$, denoted as s_i. After score normalization, s_i is a real value in the interval $[0, 1]$. The score s_i can be interpreted as the probability that v conforms (adapts) to g_i, therefore obtaining a probability distribution over the set G, i.e., $s_{i,v} = p(v \approx g_i)$. In order to compute a total value for the entropy of the set G each of its elements is considered in turn as a probe v, to compute all-against-all similarities. Let's denote as Q the number of pairs $\langle q_i, q_j \rangle$ in G such that $s_{i,j} > 0$, used as a normalization factor; then entropy of G is denoted as $H(G)$ and computed as:

$$H(G) = -\frac{1}{log_2(|Q|)} \sum\nolimits_{q_{i,j} \in Q} s_{i,j} log_2(s_{i,j}) \qquad (2)$$

The value of $H(G)$ can be considered as a measure of heterogeneity for the items in G. It is possible to order all of them according to their *informative power* or *representativeness*, by computing their contribution to $H(G)$. Given G, the devised procedure computes the complete similarity matrix M and the value for $H(G)$. For each item $g_i \in G$, M is used to compute the value of $H(G \backslash g_i)$ obtained by ignoring g_i. The item g_i, achieving the minimum difference $f(G, g_i) = H(G) - H(G \backslash g_i)$, is selected; the matrix M is updated by deleting the $i - th$ row and column, and the process is repeated, until all elements of G have been selected. According to this procedure, we first select the most representative samples, i.e. those causing the lower entropy decrease. The algorithm progressively reduces the inhomogeneity of the set. We finally obtain an ordering of the elements as they are selected by the algorithm, with the corresponding value of $f(\cdot)$. The trend of the resulting curve presents local maxima and minima in a smooth saw tooth shape, that can be usually quite well approximated by a parabola (see [6]). The values obtained for $f(\cdot)$ can be used to cluster the set elements, in a way similar to one of the approaches in [13]. The first relative maximum becomes the representative element of the first cluster. The following elements along $f(\cdot)$ until the next relative maximum are included in this cluster. A new cluster is created when the next relative maximum is found, and cluster population is continued as before. This procedure is repeated till the end of $f(\cdot)$.

3.2 Binary Similarity and Distance Measures

In order to explore the information content of LBP codes with respect to different similarity measures, we refer to the survey presented in [3]. In that work, 76 binary similarity and distance measures are discussed that have been used over the last century, and their correlation is investigated through a hierarchical clustering technique. The interested reader can refer to that paper. For our purposes, we selected a subset of 65 out of the measures mentioned there, leaving out or merging duplicates. Similarly to [3], the definitions of measures are expressed by Operational Taxonomic Units (OTUs) [7]. Assume to have two binary vectors, i and j. Let n be their common dimension. The following notation is used:

- a = the number of vector entries where the values of i and j are both 1 (or *presence*, if the binary values are interpreted in this way), meaning positive matches: $a = i \bullet j$
- b = the number of entries where the value of i and j is $(0, 1)$ (or i *absence mismatch*): $b = \bar{i} \bullet j$
- c = the number of attributes where the value of i and j is $(1, 0)$ (or j *absence mismatch*): $c = i \bullet \bar{j}$
- d is the number of attributes where both i and j have 0 (or *absence*), meaning negative matches: $d = \bar{i} \bullet \bar{j}$.

The sum $a + d$ gives the total number of matches between i and j, while the sum $b + c$ gives the total number of mismatches between i and j. Measures defined as distances were transformed into similarities to obtain consistent measures.

4   Experimental Results

The experiments carried out for this work aimed at investigating a novel strategy to identify the most "informative" binary patterns produced by the LBP feature extractor, and how reducing the LBP code set to them can affect the performance of a classifier in terms of recognition accuracy. In the specific case, the experiments adopted a very simple classifier, namely Nearest Neighbor (NN) in order to avoid the dependence of the observed variations from factors not related to the aspect under study (the different sets of LBP codes). With the same idea in mind, the face database used as testbed is EGA [12]. This dataset is the result of the integration of subsets of a number of existing face datasets, that are quite different in nature in terms of ethnicity (E), gender (G), and age (A) of subjects. While EGA is expressly built to be quite balanced with respects to such demographic traits, it also offers a good variety in terms of image quality. It includes a total of 2345 images captured from 469 subjects, 5 images per subject. More details on source datasets and EGA organization can be found in

[12]. As for this work, it is important to underline that, since EGA collects face images extracted from datasets with different characteristics, both for the demographics of subjects, for capture setting and for capture devices, carrying out experiments on it is equivalent to carrying out experiments on the corresponding subsets of the source datasets. All experiments in this work considered all EGA subjects, with two out of the five images each: the first one in the dataset for the experiment gallery and the second one as the probe. Each image was pre-processed by Viola-Jones algorithm to detect the position of the face and of the center of the eyes. Faces were resized in order to have a constant inter-eye distance of 40 pixels, and cropped to 64 × 100 pixels. No pre-processing was performed regarding illumination, because LBP is in itself an operator quite robust to most illumination distortions. Firstly, for each similarity measure the subset of the most representative codes was identified, and then among the obtained subsets those achieving the best classification results were selected. This is quite different from beam search, that tries to add any missing code to a candidate subset, and from Uniform LBP, that selects LBP codes based on some code pattern (e.g., uniformity). We rather try to identify "weighty" LBP codes.

For sake of space, it is not possible to report the definition of all the 65 considered measures. Table 1 only reports those providing results worth mentioning, with the number indicating the ordering used here. Such numbering is maintained to preserve the relation with Figs. 2 and 3 below that report exper-

Table 1. The considered similarity (S) and distance (D) measures.

S	INTERSECTION	a	(10)
D	EUCLID	$\sqrt{b+c}$	(14)
D	HELLINGER or CHORD	$2 * \sqrt{1 - a/\sqrt{(a+b) * (a+c)}}$	(22)
S	MOUNTFORD	$a/(0.5 * (a*b + a*c) + b*c)$	(29)
S	JOHNSON	$a/(a+b) + a/(a+c)$	(34)
S	DENNIS	$(a*d - b*c)/\sqrt{n * (a+b) * (a+c)}$	(35)
S	SIMPSON	$a/min(a+b, a+c)$	(36)
S	FAGER & McGOWAN	$a/\sqrt{(a+b) * (a+c)} - max(a+b, a+c)/2$	(38)
S	BARONI-URBANI & BUSER-I	$(\sqrt{a*d} + a)/(\sqrt{a*d} + a + b + c)$	(60)

Fig. 1. LBP feature images produced with different strategies to reduce the set of codes: (a) All LBP (256 bins), (b) Uniform LBP (59 bins), (c) Entropy with MOUNTFORD measure (80 bins) and with (d) BARONI-URBANI & BUSER-I measure (41 bins).

imental results. The complete list can be found in [3] and a compacted one at the end of the paper (Table 3). Figure 1 shows some examples of LBP images produced for the same face image, but with sets of LBP codes obtained by computing representativeness according to two different measures, namely MOUNT-FORD measure (80 bins) and BARONI-URBANI & BUSER-I measure (41 bins), respectively entries indexed as (29) and (60) in Table 1. The discussion about experimental results will show that these two measures provide complementary yet orthogonal improvements. The first experiment aimed at verifying if and how a different identification of relevant LBP codes can affect the accuracy of a simple NN classifier. Classifier performances were measured in terms of Equal Error Rate (EER) in verification mode (1:1 matching with identity claim) and Recognition Rate (RR) in identification mode (1:N matching without identity claim). Chosen a similarity measure, the resulting $f(\cdot)$ function was computed and used for the clustering procedure as described in Sect. 3.1. When coding images, each code is substituted by the representative of the cluster to which it belongs. A further information provided by the used clustering algorithm is the number of returned clusters for the corresponding similarity measure. This helps evaluating also the efficiency of the produced coding (the lower the number of clusters, the lower the number of codes required), together with the obtained accuracy. Therefore, in the following figures, we have on the y axis three different items of information: the number of clusters produced, the EER and the RR value, for each of the 65 measures whose index is on x axis.

LBP feature vector is the chaining of histograms from a grid of image subregions, therefore a further element of interest is the size of such sub-regions (the smaller the size, the higher the number of histograms to chain, the higher the size of the final feature vector). Therefore the first experiment was repeated with four different sub-region dimensions: 16×16, 24×24, 32×32, and 36×36. Figures 2 and 3 show the plots obtained for the two extreme cases, where of course the plot of the number of clusters is always the same. The plots obtained by intermediate region sizes show a consistent trend.

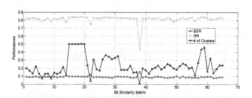

Fig. 2. Performance of NN with LBP over 16×16 sub-regions.

The results obtained in this experiment suggest that all similarity measures are affected in a generally similar way by sub-region sizes. It is possible to observe that with more information (smaller region size) both RR and EER are constantly better, and as the sub-region dimension increases, the accuracy decreases for almost all measures. RR is especially negatively affected by growing size, since

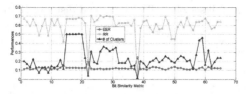

Fig. 3. Performance of NN with LBP over 36 × 36 sub-regions.

not only it decreases in general, but also becomes much more dependent from the exploited similarity measure. In this sense, some measures show a very different behavior from the others, either in positive or negative sense. Measures EUCLID, HELLINGER or CHORD, and FAGER & McGOWAN, respectively (14), (22) and (38) in Table 1, generate a number of bins that is too low, taking to an excessive performance decrease, which is further accentuated whit larger regions. On the contrary, BARONI-URBANI & BUSER-I, i.e., (60) in Table 1, though producing a very low number of bins, is able to provide an accuracy comparable with the others, and is also robust to region growing. A similar result holds, though with slightly lower performance, for INTERSECTION, JOHNSON, DENNIS and SIMPSON, respectively (10), (34), (35) and (36) in Table 1. Though dramatically decreasing the size of the feature vector, they maintain a sufficient discriminative power of extracted features.

The second experiment compared the proposed approach with LBP and with Uniform LBP. Table 2 shows the results, and reports the number of bins used by the different variations of LBP with the corresponding values of EER and RR,

Table 2. Performance of NN with different strategies to reduce the set of LBP codes.

16×16 pixels			
LBP code type	Histogram bins	EER	RR
Standard	256	0.089	0.840
Uniform	59	0.091	0.842
Entropy (29)	80	0.081	0.833
Entropy (60)	41	0.095	0.827
24×24 pixels			
LBP code type	Histogram bins	EER	RR
Standard	256	0.100	0.803
Uniform	59	0.097	0.805
Entropy (29)	80	0.092	0.798
Entropy (60)	41	0.094	0.792
32×32 pixels			
LBP code type	Histogram bins	EER	RR
Standard	256	0.093	0.842
Uniform	59	0.092	0.845
Entropy (29)	80	0.085	0.840
Entropy (60)	41	0.082	0.810
36×36 pixels			
LBP code type	Histogram bins	EER	RR
Standard	256	0.113	0.746
Uniform	59	0.116	0.739
Entropy (29)	80	0.114	0.696
Entropy (60)	41	0.104	0.671

Table 3. The full set of similarity (S) and distance (D) measures.

S\D	Name	OTU	
S	JACCARD or TANIMOTO	$a/(a+b+c)$	(1)
S	DICE or CZEKANOWSKI or NEI & LI	$2*a/(2*a+b+c)$	(2)
S	3W-JACCARD	$3*a/(3*a+b+c)$	(3)
S	SOKAL & SNEATH-I	$a/(a+2*b+2*c)$	(4)
S	SOKAL & MICHENER	$(a+d)/(a+b+c+d)$	(5)
S	SOKAL & SNEATH-II	$2*(a+d)/(2*a+b+c+2*d)$	(6)
S	ROGER & TANIMOTO	$(a+d)/(a+2*(b+c)+d)$	(7)
S	FAITH	$(a+0.5*d)/(a+b+c+d)$	(8)
S	GOWER & LEGENDRE	$(a+d)/(a+0.5*(b+c)+d)$	(9)
S	INTERSECTION	a	(10)
S	INNERPRODUCT	$a+d$	(11)
S	RUSSELL & RAO	$a/(a+b+c+d)$	(12)
D	HAMMING	$b+c$	(13)
D	EUCLID	$\sqrt{b+c}$	(14)
D	CANBERRA or MANHATTAN or CITYBLOCK or MINKOWSKI	$b+c$	(15)
D	MEAN MANHATTAN	$(b+c)/(a+b+c+d)$	(16)
D	VARI	$(b+c)/(4*(a+b+c+d))$	(17)
D	SIZEDIFFERENCE	$(b+c)^2/(a+b+c+d)^2)$	(18)
D	SHAPEDIFFERENCE	$(n*(b+c)-(b-c)^2)/(a+b+c+d)^2$	(19)
D	PATTERNDIFFERENCE	$4*b*c/(a+b+c+d)^2$	(20)
D	LANCE & WILLIAMS or BRAY & CURTIS	$(b+c)/(2*a+b+c)$	(21)
D	HELLINGER or CHORD	$2*\sqrt{1-a/\sqrt{(a+b)*(a+c)}}$	(22)
S	COSINE	$a/(a+b)*(a+c)$	(23)
S	GILBERT & WELLS	$log(a)-log(n)-log((a+b)/n)-log((a+c)/n)$	(24)
S	OCHIAI-I or OTSUKA	$a/\sqrt{(a+b)*(a+c)}$	(25)
S	FORBES-I	$n*a/(a+b)*(a+c)$	(26)
S	FOSSUM	$n*(a-0.5)^2/(a+b)*(a+c)$	(27)
S	SORGENFREI	$a^2/(a+b)*(a+c)$	(28)
S	MOUNTFORD	$a/(0.5*(a*b+a*c)+b*c)$	(29)
S	MCCONNAUGHEY	$(a^2-b*c)/(a+b)*(a+c)$	(30)
S	TARWID	$(n*a-(a+b)*(a+c))/(n*a+(a+b)*(a+c))$	(31)
S	KULCZYNSKI II	$(a/2*(2*a+b+c))/((a+b)*(a+c))$	(32)
S	DRIVER & KROEBER	$(a/2*(1/(a+b)+1/(a+c)))$	(33)
S	JOHNSON	$a/(a+b)+a/(a+c)$	(34)
S	DENNIS	$(a*d-b*c)/\sqrt{n*(a+b)*(a+c)}$	(35)
S	SIMPSON	$a/min(a+b,a+c)$	(36)
S	BRAUN & BANQUET	$a/max(a+b,a+c)$	(37)
S	FAGER & McGOWAN	$a/\sqrt{(a+b)*(a+c)}-max(a+b,a+c)/2$	(38)
S	FORBES-II	$(n*a-(a+b)*(a+c))/(n*min(a+b,a+c)-(a+b)*(a+c))$	(39)
S	SOKAL & SNEATH-IV	$1/4*(a/(a+b)+a/(a+c)+d/(b+d)+d/(c+d))$	(40)
S	GOWER	$(a+d)/\sqrt{(a+b)*(a+c)*(b+d)*(c+d)}$	(41)
S	PEARSON-I	χ^2 with $\chi^2=n*(ad-bc)^2/(a+b)(a+c)(c+d)(b+d)$	(42)
S	PEARSON-II	$\sqrt{\chi^2/(n+\chi^2)}$	(43)
S	PEARSON-III	$\sqrt{\rho/(n+\rho)}$ with $\rho=(ad-bc)/\sqrt{(a+b)(a+c)(c+d)(b+d)}$	(44)
S	PEARSON & HERON-I	ρ	(45)
S	PEARSON & HERON-II	$cos(\pi*\sqrt{b*c}/(\sqrt{a*d}+\sqrt{b*c}))$	(46)
S	SOKAL & SNEATH-III	$(a+d)/(b+c)$	(47)
S	SOKAL & SNEATH-V or OCHIAI-II	$a*d/\sqrt{(a+b)*(a+c)*(b+d)*(c+d)}$	(48)
S	COLE	$\sqrt{2}*(ad-bc)/\sqrt{a*d-b*c)^2-(a+b)*(a+c)*(b+d)*(c+d)}$	(49)
S	STILES	$log_{10}(n*(\|ad-bc\|-n/2)^2/((a+b)*(a+c)*(b+d)*(c+d)))$	(50)
S	YULE Q	$(a*d-b*c)/(a*d+b*c)$	(51)
D	YULE Q	$2bc/(a*d+b*c)$	(52)
S	YULE w	$(\sqrt{a*d}-\sqrt{b*c})/(\sqrt{a*d}+\sqrt{b*c})$	(53)
S	KULCZYNSKI-I	$a/(b+c)$	(54)
S	DISPERSON	$(a*d-b*c)/(a*d+b*c)^2$	(55)
S	HAMANN	$((a+d)-(b+c))/(a+b+c+d)$	(56)
S	MICHAEL	$4*(a*d-b*c)/((a+d)^2+(b+c)^2)$	(57)
S	GOODMAN & KRUSKAL	$(\sigma-\sigma')/(2*n-\sigma')$ with $\sigma=max(a,b)+max(c,d)+max(a,c)+max(b,d)$ and $\sigma'=max(a+c,b+d)+max(a+b,c+d)$	(58)
S	ANDERBERG	$(\sigma-\sigma')/(2*n)$	(59)
S	BARONI-URBANI & BUSER-I	$(\sqrt{a*d}+a)/(\sqrt{a*d}+a+b+c)$	(60)
S	BARONI-URBANI & BUSER-II	$(\sqrt{a*d}+a-(b+c))/(\sqrt{a*d}+a+b+c)$	(61)
S	PEIRCE	$(a*b+b*c)/(a*b+2*b*c+c*d)$	(62)
S	EYRAUD	$n^2*(n*a-(a+b)*(a+c))/((a+b)*(a+c)*(b+d)*(c+d))$	(63)
S	TARANTULA	$a*(c+d)/(c*(a+b))$	(64)
S	AMPLE	$\|(a*(c+d)/(c*(a+b)))\|$	(65)

having chosen the most representative measures according to the results of the first experiment. We can observe that using MOUNTFORD measure, i.e. (29) in Table 1, and carrying out our clustering/selection procedure, we obtain an LBP coding able to achieve better performance than Uniform LBP at the expense of a higher number of bins. On the contrary, BARONI-URBANI & BUSER-I is able to finally produce a lower number of bins, with a decrease of performance that might be negligible according to the accuracy requirements. In summary, it is possible to improve the performance over Uniform codes either in terms of feature vector length or in terms of EER.

5 Conclusions

This paper presented a novel approach to the selection of the most representative LBP codes in order to obtain smaller though sufficiently discriminative feature vectors. The proposed method neither performs an unaffordable exhaustive search nor relies on codes with special patterns. It rather exploits a clustering procedure based on a measure of representativeness of the different codes. Such measure is based in turn on a suitable similarity measure among binary codes. The obtained results show that it is possible to reduce the number of LBP codes to use in building feature vectors, without affecting the classification performance too much. The experiments aimed at analyzing 65 different similarity/distance measures. Though some common aspects of behavior were detected, some measures resulted better able to improve the selection of an appropriate subset of codes, by either reducing the size of the feature vectors without a dramatic decrease in performance, or obtaining a slightly better result than Uniform LBP at the expense of using some more codes. Our future work will be focused on testing the generality of these outcomes on different classes of images. In practice, LBP can be used in many applications based on texture analysis, and it will be interesting to evaluate our approach in a different context. In particular, it will be interesting to investigate if the same similarity measures produce equivalent results on different classes of images.

References

1. Ahonen, T., Hadid, A., Pietikainen, M.: Face description with local binary patterns: application to face recognition. IEEE Trans. Pattern Anal. Mach. Intell. **28**(12), 2037–2041 (2006)
2. Chai, Z., Sun, Z., Tan, T., Mendez-Vazquez, H.: Local salient patterns - a novel local descriptor for face recognition. In: International Conference on Biometrics (ICB) (2013)
3. Choi, S.S., Cha, S.H., Tappert, C.C.: A survey of binary similarity and distance measures. J. Syst. Cybern. Inf. **8**(1), 43–48 (2010)
4. De Marsico, M., Nappi, M., Riccio, D.: ES-RU: an entropy based rule to select representative templates in face surveillance. Multimedia Tools Appl. **73**(1), 109–128 (2014)

5. De Marsico, M., Nappi, M., Riccio, D.: Entropy-based automatic segmentation and extraction of tumors from brain MRI images. In: Azzopardi, G., Petkov, N. (eds.) CAIP 2015. LNCS, vol. 9257, pp. 195–206. Springer, Cham (2015). doi:10.1007/978-3-319-23117-4_17

6. De Marsico, M., Nappi, M., Riccio, D., Tortora, G.: Entropy-based template analysis in face biometric identification systems. SIViP 7(3), 493–505 (2013)

7. Dunn, G., Everitt, B.S.: An Introduction to Mathematical Taxonomy. Cambridge University Press, Cambridge (1982)

8. Nappi, M., Riccio, D., De Marsico, M.: Entropy based biometric template clustering. In: ICPRAM, pp. 560–563 (2013)

9. Ojala, T., Pietikäinen, M., Harwood, D.: A comparative study of texture measures with classification based on featured distributions. Pattern Recogn. 29(1), 51–59 (1996)

10. Ojala, T., Pietikäinen, M., Mäenpää, T.: Multiresolution gray-scale and rotation invariant texture classification with local binary patterns. IEEE Trans. Pattern Anal. Mach. Intell. 24(7), 971–987 (2002)

11. Pietikäinen, M., Hadid, A., Zhao, G., Ahonen, T.: Computer Vision Using Local Binary Patterns. Springer, London (2011). doi:10.1007/978-0-85729-748-8

12. Riccio, D., Tortora, G., De Marsico, M., Wechsler, H.: EGA - ethnicity, gender and age, a pre-annotated face database. In: 2012 IEEE Workshop on Biometric Measurements and Systems for Security and Medical Applications (BIOMS), pp. 1–8 (2012)

13. Riccio, D., De Marsico, M., Plasencia-Calaña, Y., Mendez-Vazquez, H.: GETSEL: gallery entropy for template selection on large datasets. In: International Joint Conference on Biometrics (IJCB 2014) (2014)

14. Shan, C., Gong, S., McOwan, P.W.: Facial expression recognition based on local binary patterns: a comprehensive study. Image Vis. Comput. 27(6), 803–816 (2009)

15. Sun, N., Zheng, W., Sun, C., Zou, C., Zhao, L.: Gender classification based on boosting local binary pattern. In: Wang, J., Yi, Z., Zurada, J.M., Lu, B.-L., Yin, H. (eds.) ISNN 2006. LNCS, vol. 3972, pp. 194–201. Springer, Heidelberg (2006). doi:10.1007/11760023_29

16. Topi, M., Timo, O., Matti, P., Maricor, S.: Robust texture classification by subsets of local binary patterns. In: Proceedings of 15th International Conference on Pattern Recognition, Barcelona, Spain, vol. 3, pp. 935–938 (2000)

17. Tapia, J.E., Perez, C.A., Bowyer, K.W.: Gender classification from iris images using fusion of uniform local binary patterns. In: Agapito, L., Bronstein, M.M., Rother, C. (eds.) ECCV 2014. LNCS, vol. 8926, pp. 751–763. Springer, Cham (2015). doi:10.1007/978-3-319-16181-5_57

18. Wang, L., He, D.C.: Texture classification using texture spectrum. Pattern Recogn. 23(8), 905–910 (1990)

19. Yang, Z., Ai, H.: Demographic classification with local binary patterns. In: Lee, S.-W., Li, S.Z. (eds.) ICB 2007. LNCS, vol. 4642, pp. 464–473. Springer, Heidelberg (2007). doi:10.1007/978-3-540-74549-5_49

Indoor Actions Classification Through Long Short Term Memory Neural Networks

Emanuele Cipolla$^{(\boxtimes)}$, Ignazio Infantino, Umberto Maniscalco, Giovanni Pilato, and Filippo Vella

ICAR, National Research Council of Italy, via Ugo la Malfa 153, Palermo, Italy
{emanuele.cipolla,ignazio.infantino,umberto.maniscalco,
giovanni.pilato,filippo.vella}@icar.cnr.it,
http://www.icar.cnr.it

Abstract. This work presents a system based on a recurrent deep neural network to classify actions performed in an indoor environment. RGBD and infrared sensors positioned in the rooms are used as data source. The smart environment the user lives in can be adapted to his/her needs.

Keywords: Deep learning · Human actions · LSTM · Indoor activities

1 Introduction

Ambient intelligence exploits smart sensors, pervasive computing and artificial intelligence techniques in order to make environments responsive, flexible and adaptive to the people living inside them to improve their daily life [5].

After the features of the users and their surroundings have been determined, a reasoning on perceived data takes place, followed by the selection of the more suitable actions aimed at assisting and improving the living conditions of users.

An effective Ambient Intelligence System can encompass hearing, vision, language, and knowledge. As a consequence, houses can nowadays be provided with sophisticated sensors networks like cameras, audio and pressure sensors, motion detectors as well as wearable technologies in order to realize an intelligent system which proactively perceives and analyzes the activities occurring in an apartment, setting up actions to provide help in the execution of tasks and optimizing the resources for the efficiency and the well being of the people living in it [17].

In this context a proper recognition of meaningful patterns in data is the crucial point towards the realization of such a system capable of detecting and recognizing user actions and activities on the environment [4,6,10,11].

An accurate classification of the user's actions allows the effective understanding of user's habits and preferences [1,20]. In a domestic environment the detection of the user activities makes it possible the optimization of home resources in compliance with the distribution of the activities [14]. It also can be exploited to assist home's residents in their daily activities, possibly monitoring also their health conditions [12,15].

© Springer International Publishing AG 2017
S. Battiato et al. (Eds.): ICIAP 2017, Part I, LNCS 10484, pp. 435–444, 2017.
https://doi.org/10.1007/978-3-319-68560-1_39

In this work we illustrate the evolution of an action detection module we have developed [7,19] that can be embedded into an ambient assisted living system to properly classify patterns of measures according to a specific set of actions.

The proposed system is based on a deep learning approach, and in particular relies on a deep recurrent neural networks [13]. As widely shown in different applicative contexts, a deep learning approach allows for more detailed feature representations compared to conventional neural networks.

Approaches of this kind have been widely exploited even in the field action classification and recognition: see, for example, [2].

Our has been tested using the dataset of the SPHERE (Sensor Platform for HEalthcare in Residential Environments) project [18]. The dataset consists in collection of measures from RGB-d cameras, worn accelerometers and passive environmental sensors, collected asking a set of trained people to perform a set of action in an indoor environment.

We have compared the approach with a previous methodology aimed at indoor action detection through probabilistic induction model [19].

The paper is organized as follows: Deep Learning Neural Networks are presented, with a focus on Long Short Term Memory Neural Networks in Sect. 2. The proposed classification approach and the pre-processing operations on the SPHERE dataset are shown in Sect. 3, while Sect. 3 discusses some experimental results, with a comparison with a baseline classifier employing a conventional Multi Layer Perceptron Network. Section 4 contains some conclusions and a discussion on future developments.

2   Neural Networks for Indoor Activity Classification

The new trend in machine learning aims at overcoming the limits of conventional techniques and approaches in the processing of data in raw form. They typically require a careful engineering and domain expertise to design the set of feature to transform the original data in suitable vectors.

Deep-learning methods usually employ a set of non-linear modules that automatically extract a set of features from the input and transfer them to the next module [13]. The weights of the layers of features are learned directly from data, allowing to discover intricate structures in high-dimensional data, regardless of their domain (science, business, etc.). With this mechanism very complex functions can be learned combining these modules: the resulting networks are often very sensitive to minute details and insensitive to large irrelevant variations.

2.1   MLP Multilayer Perceptron

A multilayer perceptron (MLP) is a feedforward network that maps sets of input data onto a set of desiderata outputs; it consists of at least three layers - an input layer, a hidden layer and an output layer - of fully connected nodes in a directed graph. Except for the input nodes, each node is a neuron (or processing element) with a nonlinear activation function - usually a sigmoid, or the hyperbolic tangent, chosen to model the bioelectrical behaviour of biological neurons in a natural brain. Learning occurs through backpropagation algorithm that modifies connections weights in order to minimize the difference between the actual network output and the expected result.

2.2 Deep Neural Networks

The new techniques for sub-symbolic representation has given a strong impulse for the machine learning algorithms.

Several advantages in using Recurrent Neural Networks for sequence labeling are listed in [8], the most important of which are the flexible use of context information to choose what information are to be stored, and the ability to give a reliable output even in the presence of sequential distorsions. In particular, Long-Short Term Memory units have been successfully employed for time series classification, as they are able to better retain the influence of past inputs decays quickly over time with respect to other recurrent networks, thus mitigating the so-called *vanishing error problem*. A subsequent section will describe the internal structure of a LSTM unit in detail.

Anyway, a single LSTM unit is unlikely to learn a satisfactorily meaningful, low-dimensional, and somewhat invariant feature space with anything but trivial datasets, because of the inherent complexity of many of them. More complex models with multiple layers may be used to represent multiple level of abstractions.

While a detailed theoretical explanation of the reasons behind the advantages of using more complex networks is offered in [16], an analogy that is suitable for the field of visual pattern recognition can be considered: if we had a network made of multiple layers, the neurons in the first layer might learn to recognize edges, while those in the second layer could learn to recognize more complex shapes built connecting some of these edges, such as triangles or rectangles.

In our scenario, as there is not a clear hierarchy of features, we chose to gradually stack LSTM layers and measure the trend of the F1-score to determine what the correct number of strata can be. Each LSTM layer is separated from the next one by a ReLU function. In addition, given a sequence length, we strived to determine how many neurons are needed for the representation to be of good quality.

After the LSTM layers have determined the boundaries of the representation space, several fully-connected layers are used to learn a function in that space. In these layers every input is connected to every output by a set of trained weights. Its output is fed to an activation function, which is usually a non-linear operation. We chose to use two deeply connected strata; while the first one having twice the number of classes we use is connected to a ReLU activation function, the last one feeds its output to a softmax stratum, thus generating a probability distribution over classes, the most probable of which is chosen as output [3].

2.3   LSTM

LSTMs have been designed by Hochreiter and Schmidhuber [9] to avoid the long-term dependency problem, at the price of a more complex cell structure.

The key feature of LSTMs is the "cell state" that is propagated from a cell to another. State modifications are regulated by three structures called gates, composed out of a sigmoid neural net layer and a pointwise multiplication operation.

The first gate, called "forget gate layer", considers both the input x_t and the output from the previous step h_{t-1}, and returns values between 0 and 1, describing how much of each component of the old cell state C_{t-1} should be left unaltered: if the output is 0, no modification is made; if the output is one, the component is completely replaced.

New information to be stored in the state is processed afterwards. The second sigmoid layer, called the input gate layer, decides which values will be updated. Next, a *tanh* layer creates a vector of new candidate values, $\tilde{C}_t$, that could be added to the state.

$$f_t = \sigma(W_f \cdot [h_{t-1}, x_t] + b_f)$$

$$i_t = \sigma(W_i \cdot [h_{t-1}, x_t] + b_i)$$

$$\tilde{C}_t = \tanh(W_C \cdot [h_{t-1}, x_t] + b_c)$$

$$C_t = f_t * C_{t-1} + i_t * \tilde{C}_t$$

To perform a state update, C_{t-1} is first multiplied by the output of the forget gate f_t, and the result is added to the pointwise multiplication of the input gate output i_t and $\tilde{C}_t$.

$$o_t = \sigma(W_o \cdot [h_{t-1}, x_t] + b_o)$$

$$h_t = o_t * \tanh(C_t)$$

Finally, the output h_t can be generated. First, a sigmoid is applied, taking into account both h_{t-1} and h_{t-1}; its output is then multiplied by a constrained version of C_t, so that we only output the parts we decided to.

3   Action Detection Through Classification

The action detection task exploits and regards a set of sampled human body joint configurations coming from different kinds of sensors. Each and every one of these samples will have a label describing the behaviour of the action the subject was carrying out.

Each sample has an inherent temporal relationship with its predecessors and successors: the complete dataset set of samples is thus a time series, and fixed-length sequences of joints can display interesting regularities.

Data has been pre-processed to make it homogeneous and let the computation proceed smoothly.

First of all, the input sensors have different sampling frequencies. Cameras acquire data with 30 fps, while accelerometers have a frequency of 20 Hz. We supposed that human actions are much slower, so we downsampled the data to 2 Hz considering the time frame detailed enough to capture human actions. We also considered a time window to capture information and associate a label to a set of movements. The size of the time window depends on the data and on the task, and has been determined after several batches of experiments.

Our experiments have been performed on the SPHERE dataset, available online[1] and a detailed outline of the dataset is available in [18]. The 2787 samples have been manually annotated with one of the given labels. The values sampled by accelerometer, RGB-D and environmental data for a vector of eighteen values. In particular the value are referred to

- x, y, z: acceleration along the x, y, z axes
- centre_2d_x, y: the coordinated of the center of the bounding box along x and y axes
- bb_2d_br_x, y: The x and y coordinates of the bottom right (br) corner of the 2D bounding box
- bb_2d_tl_x, y: The x and y coordinates of the top left (tl) corner of the 2D bounding box
- centre 3d x, y, z: the x, y and z coordinates for the centre of the 3D bounding box
- bb_3d_brb_x, y, z: the x, y, and z coordinates for the bottom right back (brb) corner of the 3D bounding box
- bb_3d_flt_x, y, z: the x, y, and z coordinates of the front left top (flt) corner of the 3D bounding box.

Twenty activity labels have been used to annotate the dataset, as follows. There are three main categories:

- action
- position
- transitions

All the activities requiring movements are called actions; they are *ascent stairs, descent stairs, jump, walk with load* and *walk*;

Position is referred to a still person; its labels are *bending, kneeling; jump, lying, sitting, squatting, standing.*

The transition are the intermediate steps between two positions and are: *stand-to-bend; kneel-to-stand; lie-to-sit; sit-to-lie; sit-to-stand; stand-to-kneel; stand-to-sit; bend-to-stand; turn.*

To increase training accuracy and have a richer set of examples to use, a set of more generic labels is compiled out of the original ones. All the transition labels have been clustered together in a simple label *transition*. The classes according the walking have been merged together in a single class. The final labels are:

---

[1] http://irc-sphere.ac.uk/sphere-challenge/home.

– bending
– standing
– lying
– sitting
– transition

The dataset has been divided in two parts. Four fifths have been used for the training set, while the remaining one fifth has been used as test set.

Every deep network has been trained using a batch size of 32, which has been found to give a good throughput while keeping variability low.

Classification performance is evaluated using standard functions from Information retrieval. The value of the True Positive (TP) counts the number of samples that have been correctly detected. False Positive (FP) is the number of times a wrong label has been assigned to a sample. False Negative (FN) is the number of samples that have not been correctly classified. The values of True Negative (TN) is referred to the wrong labels that have not been assigned to a sample. For these experiments it has always been set to zero. The accuracy of is defined as:

$$Acc = \frac{TP + TN}{TP + TN + FP + FN}$$

Precision and recall are instead defined as:

$$Prec = \frac{TP}{TP + FP} \qquad Rec = \frac{TP}{TP + FN}$$

The harmonic mean of precision and recall is called F_1-score:

$$F_1 = 2 \times \frac{Prec \times Rec}{Prec + Rec}$$

In our experiments timespans 2 to 20 samples long were used to ascertain the importance of using longer sequences to have better results. The F_1-score has been chosen as reference metric as it balances precision and recall.

First of all, we wondered if our more complex deep networks really fared better against simpler networks: in Fig. 1 we see that the latter generally has a worse performance even at its peak at timespan $= 19$.

Let us first consider networks having 2 stacked, hidden LSTM layers. While sharp changes in the F_1 score can be detected as the time span increases, an increasing trend may be recognized in the network having 128 neurons per LSTM layer as you use longer sequences, so we can deem it the best configuration in this subset (Fig. 2).

Using 4 hidden layers sharp changes in the F_1 score are still present, and the same tendency evaluation we applied before can suggest the use of 64 neurons, but the increase in F1 slows down as sequences longer than 15 frames are used (Fig. 3).

Besides, having enough hidden layers to employ, choosing the right number of neurons per layer is crucial to balance feature expressiveness and training time. Considering precision and recall curves might offer some advice in this discernment.

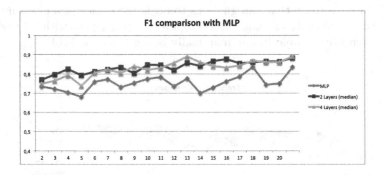

Fig. 1. F1 score: deep networks have better performance than MLP

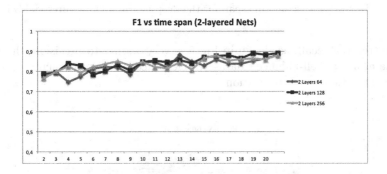

Fig. 2. F1 versus timespan for 2-layered LSTM

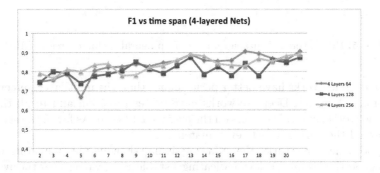

Fig. 3. F1 versus timespan for 4-layered LSTM

Considering precision alone as we have done in Fig. 4, we see that there is not a distinctive trend that can guide us towards the choice of a given number of frames; we can only exclude the configuration of 4 layers with 64 and 128 neurons each, that is able to fare marginally better than the MLP.

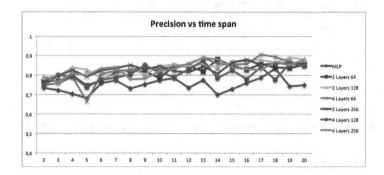

Fig. 4. Precision versus timespan for the deep neural architectures vs MLP

The trend of recall, as shown in Fig. 5 is very similar to precision: this may indicate either a well-balanced dataset, or the need of improvements in either feature extraction or regularization.

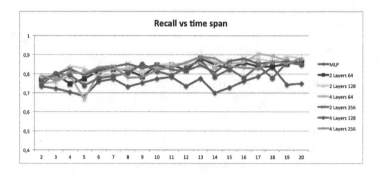

Fig. 5. Recall versus timespan for the deep neural architectures vs MLP

Both deep networks have better performance than the multilayer perceptron. On average, 2-layered Deep Networks have better precision and recall than 4-layered networks, and this reflects on the average F1 Score. As far as the accuracy is concerned they are a mostly even match.

For both deep networks the value of sigma, showing the spreading of the values across the average, is low, indicating a stable set of values near the average.

Anyway, if we take into account the very tiny differences between the shape of these curves, it seems that using 2 layers having 128 neurons each is the best course of action.

Table 1. Results for the test of labelling with the two deep neural networks and the multilayer perceptron

	MLP		2-layered DeepNet		4-layered DeepNet	
	μ	σ	μ	σ	μ	σ
Precision	0.75	0.001	0.831	1.2e−4	0.822	1.9e−4
Recall	0.74	0.001	0.833	7.46e−5	0.822	2.77e−2
Accuracy	0.75	0.001	0.832	7.46e−5	0.831	3.29e−6
F1 score	0.75	0.001	0.832	7.46e−5	0.822	1.9e−4

Taking into account both the metrics and the training time the 2-layered network must be preferred. The proposed system outperforms an analogous system based on a probabilistic model that is tuned with evaluation of the principal component analysis [19]. The accuracy is comparable but the F1 measure shows the neural system to be more robust (Table 1).

4 Conclusions

Within the framework of ambient intelligence, we have presented an approach to classify human indoor actions using deep neural network. We have shown that the use of networks having several stacked LSTM hidden layer have a good performance in classifying both short and long sequences of frames. To find an amenable number of hidden layers and neurons per layer a set of experiments have been carried on comparing different metrics varying the geometry the neural network.

Possible future works include:

- the use of different, richer datasets;
- the enrichment of the current dataset;
- a broader experimentation with different types of hidden layers and activation functions.

References

1. Augello, A., Ortolani, M., Re, G.L., Gaglio, S.: Sensor mining for user behavior profiling in intelligent environments. In: Pallotta, V., Soro, A., Vargiu, E. (eds.) Advances in Distributed Agent-Based Retrieval Tools, pp. 143–158. Springer, Heidelberg (2011). doi:10.1007/978-3-642-21384-7_10
2. Baccouche, M., Mamalet, F., Wolf, C., Garcia, C., Baskurt, A.: Sequential deep learning for human action recognition. In: Salah, A.A., Lepri, B. (eds.) HBU 2011. LNCS, vol. 7065, pp. 29–39. Springer, Heidelberg (2011). doi:10.1007/978-3-642-25446-8_4
3. Bengio, Y., Courville, A., Vincent, P.: Representation learning: a review and new perspectives. IEEE Trans. Pattern Anal. Mach. Intell. **35**(8), 1798–1828 (2013)

4. Castillo, J.C., Carneiro, D., Serrano-Cuerda, J., Novais, P., Fernández-Caballero, A., Neves, J.: A multi-modal approach for activity classification and fall detection. Int. J. Syst. Sci. **45**(4), 810–824 (2014)
5. Cook, D.J., Augusto, J.C., Jakkula, V.R.: Ambient intelligence: technologies, applications, and opportunities. Pervasive Mob. Comput. **5**(4), 277–298 (2009)
6. Donahue, J., Hendricks, L.A., Guadarrama, S., Rohrbach, M., Venugopalan, S., Saenko, K., Darrell, T.: Long-term recurrent convolutional networks for visual recognition and description. CoRR abs/1411.4389 (2014). http://arxiv.org/abs/1411.4389
7. Filippo, V., Agnese, A., Umberto, M., Vincenzo, B., Salvatore, G.: Classification of indoor actions through deep neural networks. In: 2016 International Conference on Signal-Image Technology & Internet-Based Systems (SITIS). IEEE (2016)
8. Graves, A.: Supervised Sequence Labelling with Recurrent Neural Networks. Studies in Computational Intelligence, vol. 385. Springer, Heidelberg (2012). doi:10.1007/978-3-642-24797-2
9. Hochreiter, S., Schmidhuber, J.: Long short-term memory. Neural Comput. **9**(8), 1735–1780 (1997)
10. Krishnan, K., Prabhu, N., Babu, R.V.: ARRNET: action recognition through recurrent neural networks. In: 2016 International Conference on Signal Processing and Communications (SPCOM), pp. 1–5, June 2016
11. Krishnan, N.C., Cook, D.J.: Activity recognition on streaming sensor data. Pervasive Mob. Comput. Part B **10**, 138–154 (2014)
12. Kyriazakos, S., Mihaylov, M., Anggorojati, B., Mihovska, A., Craciunescu, R., Fratu, O., Prasad, R.: eWALL: an intelligent caring home environment offering personalized context-aware applications based on advanced sensing. Wirel. Pers. Commun. **87**(3), 1093–1111 (2016)
13. LeCun, Y., Bengio, Y., Hinton, G.: Deep learning. Nature **521**(7553), 436–444 (2015)
14. Lima, W.S., Souto, E., Rocha, T., Pazzi, R.W., Pramudianto, F.: User activity recognition for energy saving in smart home environment. In: 2015 IEEE Symposium on Computers and Communication (ISCC), pp. 751–757. IEEE (2015)
15. Lowe, S.A., ÓLaighin, G.: Monitoring human health behaviour in one's living environment: a technological review. Med. Eng. Phys. **36**(2), 147–168 (2014)
16. Pascanu, R., Gülçehre, Ç., Cho, K., Bengio, Y.: How to construct deep recurrent neural networks. CoRR abs/1312.6026 (2013). http://arxiv.org/abs/1312.6026
17. Remagnino, P., Foresti, G.L.: Ambient intelligence: a new multidisciplinary paradigm. IEEE Trans. Syst. Man Cybern. Part A Syst. Hum. **35**(1), 1–6 (2005)
18. Twomey, N., Diethe, T., Kull, M., Song, H., Camplani, M., Hannuna, S., Fafoutis, X., Zhu, N., Woznowski, P., Flach, P., Craddock, I.: The SPHERE challenge: activity recognition with multimodal sensor data. arXiv preprint arXiv:1603.00797 (2016)
19. Maniscalco, U., Pilato, G., Vella, F.: Detection of indoor actions through probabilistic induction model. In: De Pietro, G., Gallo, L., Howlett, R.J., Jain, L.C. (eds.) KES-IIMSS 2017. SIST, vol. 76, pp. 129–138. Springer, Cham (2018). doi:10.1007/978-3-319-59480-4_14
20. Vella, F., Infantino, I., Scardino, G.: Person identification through entropy oriented mean shift clustering of human gaze patterns. Multimedia Tools Appl. **76**(2), 1–25 (2016)

Feature Clustering with Fading Affect Bias: Building Visual Vocabularies on the Fly

Ziyin Wang and Gavriil Tsechpenakis[(✉)]

Computer and Information Science Department,
Indiana University-Purdue University Indianapolis, Indianapolis, USA
gtsechpe@indiana.edu

Abstract. We present a fast and accurate center-based, single-pass approach for data clustering in a non-parametric fashion, with main focus on features from large image collections and streaming videos. We use a dictionary of clusters and a list ('short memory') of centers temporarily stored during parsing the data. The latter is used to determine emerging clusters, not previously observed, or outliers that are discarded. Our method assigns features to existing or newly created clusters with constant-time computations, and it can be used for more general static datasets or sequential (streaming) data. In our experiments, we make extensive comparisons with approaches commonly used in feature clustering, with respect to accuracy and efficiency.

Keywords: Online feature clustering · Visual vocabularies · Single-pass clustering

1 Introduction

A typical representative of building cluster center dictionaries is K-means [12] and its variants [2,15]. The basic K-means algorithm is known to strongly depend on initialization, and the results can vary arbitrarily [13]. Some more advanced versions, such as K-means++ [2] and Biset K-means [20], provide better initializations but are still sensitive to noise and usually do not yield the desired results for complex data. Other center-based algorithms, such as EM-clustering [17] and Fuzzy c-means [3], suffer from similar drawbacks. In terms of complexity, K-means is linear with respect to the data size, yet the algorithm iterates many times to converge, which requires multiple passes over the entire dataset. Also, like many center-based methods, it is an in-memory algorithm, which raises the issue of appropriate data handling/pre-processing when the dataset cannot fit into the main memory. Another issue of center-based greedily iterative algorithms is that they require the user to specify the number of centers. In most practical situations this corresponds to domain knowledge that may not be available. In the problem of finding feature similarities in large image collections, such information is *a priori* unknown. A rough approximation of the number of clusters can be obtained by first executing the algorithm many times for parameter

© Springer International Publishing AG 2017
S. Battiato et al. (Eds.): ICIAP 2017, Part I, LNCS 10484, pp. 445–456, 2017.
https://doi.org/10.1007/978-3-319-68560-1_40

tuning, which increases the overall running time significantly. Some algorithms, such as Affinity Propagation [10] and Density Peaks [24], are able to detect the number of clusters automatically, but usually either perform well only when the data are simple (with limited noise and clutter between clusters), or are inefficient for large datasets.

Improving Efficiency. Some hierarchy-based clustering algorithms, like BIRCH [29], are reported to provide increased efficiency. BIRCH builds a CF (clustering feature) tree and uses an agglomerative clustering algorithm to merge leaves towards a specific number of clusters. Since agglomerative clustering is very expensive with $O(N^2 lgN)$ time complexity, the efficiency is guaranteed only if the user chooses the correct parameters to generate a reasonable amount of leaves. Although it can be fast with careful parameter selection, it does not provide natural clusters and its performance is usually sensitive to the permutation of the data [29]. On the other hand, reducing the complexity of iterative methods has been a major motivation for developing single-pass algorithms, i.e., methods that parse the data once [25]. Despite its aforementioned limitations, BIRCH [29] is a popular single-pass framework. [9] describes a single-pass K-means that yields results similar to the iterative version. StreamSL [19] performs better than BIRCH, yet with higher complexity. StreamKM++ [1] approximates the performance of K-means++, is faster than StreamSL, yet is still slower than BIRCH. Overall, fast, stream clustering algorithms have reported accuracy lower than or equal to the accuracy of K-means++.

Visual Vocabularies. The greedily iterative paradigm is still very popular in Computer Vision, despite its drawbacks, due to its simplicity and its acceptable efficiency for some applications. Specifically, despite the advances of deep neural networks in image classification, building visual vocabularies [7,14,22,26] can still provide significant benefits for various tasks, including unsupervised object detection in image collections [5] or in streaming data where new categories may emerge. Coates *et al.* [6] use simple K-means clustering and a triangle metric to learn small image blocks, and use these learned features to encode an image. In [14,21,22], K-means and EM clustering are used to encode SIFT [16] features detected from an image, known as VLAD (Vector of Locally Aggregated Descriptors) and Fisher Vector encoding, respectively. For object retrieval, [23] uses randomized k-d forest when matching between centers and points to boost the speed of simple K-means, and reports better results than the vocabulary tree method in [18].

In this paper we present a center-based approach that improves the trade-off between accuracy and efficiency. Specifically, our method: (i) is able to detect accurately the number of natural clusters in a non-parametric fashion, (ii) requires only a single-pass through the dataset, while its efficiency can be further improved using hierarchy, and (iii) can be used for building visual vocabularies and/or object proposals from streaming images, where new clusters may emerge.

2   Method Overview

Consider a dataset $D = \{x_1; x_2; \ldots; x_N\}$, where x_i is $1 \times d$ feature vector. We build assignments to an *a priori* unknown number, K, of clusters $\{\pi_1; \pi_2; \ldots; \pi_K\}$, $\pi_k \bigcap \pi_l = \emptyset$, $\forall k \neq l$, such that $\bigcup_{k=1}^{K} \pi_k \subseteq \{x_1; x_2; \ldots; x_N\}$. Data not assigned to clusters are considered as outliers/noise. For x_i and x_j, $\forall i \neq j \leq N$, a threshold θ, and a similarity measure $s(x_i, x_j)$, x_j is matched to x_i if $s(x_i, x_j) > \theta$; here we consider s as the negative Euclidean distance. If x_i is a cluster center, then x_j is assigned to π_i.

In our method, illustrated in Fig. 1, we build a list of centers dynamically, what we call 'Dictionary', which is initially empty and then enriched by frequently matched features while parsing the given dataset. At any given instance of Dictionary, features near the cluster centers are in higher density, which means that these are the most representative samples of the formed patterns in the parsed subset of the data, though not representative of the entire dataset. Therefore, instead of parsing the data sequentially, we perform random sampling without replacement, which improves accuracy as we show below. A practical way to do so is shuffling: consecutive features in the shuffled order are actually random samples from the original dataset. We match each feature x_i in shuffled order with the closest existing center in the Dictionary. If the match succeeds with respect to a similarity threshold θ, then we assign the feature

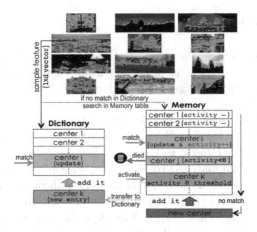

Fig. 1. Feature clustering with fading affect bias [28]: clustering while 'forgetting'. We cluster features from large image collections or streaming videos, with *a priori* unknown number of clusters. Dictionary is a dynamically populated list of formed cluster centers, while Memory is a temporary list of unmatched or rarely matched features. When a feature cannot be matched with any existing center in Dictionary, it moves to Memory; similar to that features also move to Memory, where they form a cluster, which is then transferred to Dictionary as a new cluster. The 'activity' counter is increased when a Memory entry is populated, and is reduced for every feature matched in Dictionary or moved to Memory but not assigned to the corresponding temporary center. This way, a center in Memory is 'activated' (transferred to Dictionary), or 'dies' (diminishes).

to the corresponding cluster π_k and update its center c_k in constant time as, $c_k^{(new)} = \frac{\|\pi_k\| c_k + x_i}{\|\pi_k\| + 1}$. If no matching center is found, we temporarily store the feature in what we call (short) memory with '*fading affect bias*' [28], or for simplicity 'Memory', which is a list of centers initially empty, then dynamically enlarged, and progressively diminished. Temporarily stored features are either 'forgotten' as outliers or move to Dictionary as members of a new cluster.

Memory uses a variable 'activity', indicating how frequently an entry (temporary center) is matched. The activity is set to an initial value, a_0, when a new entry is created and then varies during feature assignments. When a Memory entry is matched, its activity value is increased; when this value exceeds a threshold, ϕ, the corresponding entry is transferred to the Dictionary as a new cluster. The activity value of an entry is reduced when a feature is matched with either a center in the Dictionary or a different entry in Memory. When activity becomes negative, the corresponding entry 'dies', i.e., diminishes and is removed from Memory. This way we reject outliers, assuming that they are randomly assigned to Memory entries, or form new entries of small sizes that will diminish. This is why we consider this as a short memory: it keeps 'forgetting' data while receiving data, and the less persistent the assignments to a given entry (indicative to noise), the more likely it is for this entry to diminish. We borrow the term '*fading affect bias*' from cognitive psychology [28] to describe the fact that noise (negative memories) is discarded (fades) fast.

3   Parameter Estimation

In what is described above, the initial activity value a_0 and threshold ϕ essentially dictate how soon noise is discarded from Memory during parsing the dataset, while also determining whether informative features would be considered as noise.

In the shuffled list, each feature of the dataset has the same probability $1/N$ to appear in any cluster. Therefore the probability that a feature x_i in the shuffled list is from cluster π_k is $P(x_i \in \pi_k) = p_k = \frac{\|\pi_k\|}{N}$. If we assume all features from the same cluster can be matched to each other with respect to some similarity threshold θ, we can define noise as small clusters with population smaller than a certain number N_f, and then the probability of a feature appexaring in such a noise cluster is smaller than N_f/N. Therefore, in what follows, N_f/N and θ are related to each other: for smaller values of θ, higher values for N_f/N should be considered.

Consider a newly created entry in Memory with its activity value initialized to a_0. The activity value will decrease to 0 if no feature is matched with this entry during a_0 steps, i.e., during processing a_0 new features. These a_0 steps include sampling the dataset, matching and adding in Dictionary or Memory, and removing from Memory. Thus, ϕ indicates how frequently a Memory entry should be matched within a_0 steps to be considered 'informative' and not noise.

Our primary hypothesis $H_0 = $ '**informative**' for a feature x_i in the data is $\{H_0 : P(x_i \in \pi_k, \| \pi_k \| \geq N_f)\}$, and the alternative hypothesis, namely

$H_1 = $ 'noise', is $\{H_1 : P(\boldsymbol{x}_i \in \boldsymbol{\pi}_k, \| \boldsymbol{\pi}_k \| < N_f)\}$. This translates into calculating (a) a lower bound for a_0 that guarantees we have sufficient samples before rejecting a cluster, and (b) a lower bound for ϕ that indicates sufficient evidence that a temporary cluster is not noise. In other words, we must guarantee low probability of discarding real clusters as noise and low probability of permanently accepting noise, or respectively,

$$\begin{cases} P\{H_1 = 1 \mid H_0 = 1, H_1 = 0\} \le \alpha \ \textbf{Condition A} \\ P\{H_1 = 0 \mid H_0 = 0, H_1 = 1\} \le \beta \ \textbf{Condition B} \end{cases} \tag{1}$$

where α and β indicate probabilistic significance and are typically very small numbers (≤ 0.05). However, these two probabilities cannot be small simultaneously without increasing a_0, which is the sample size. A solution is to satisfy α first and then enlarge the sample size to satisfy β [11].

Assume an entry k in Memory has been matched X times during a_0 processing steps. Then, X follows binomial distribution $X \sim B(a_0, p_k)$. According to the Central Limit Theorem [11], when a_0 is large and $a_0 p_k \ge 5$, we can approximate binomial distribution $B(a_0, p_k)$ with normal distribution $N(a_0 p_k, \sqrt{a_0 p_k (1 - p_k)})$. Then, according to **Condition A** in Eq. (1), we expect that $P\{X \le X^*\} \le \alpha$, where X^* is the lowest bound of acceptance. If we consider the statistical variable, $Y = \frac{X - a_0 p_k}{\sqrt{a_0 p_k (1 - p_k)}} \sim N(0, 1)$,

$$P\{X \le X^*\} = P\left\{Y \le \frac{X^* - a_0 p_k}{\sqrt{a_0 p_k (1 - p_k)}}\right\} \le \alpha \tag{2}$$

If Φ is the cumulative distribution function of the standard normal distribution, i.e., $P\left\{Y \le \frac{X^* - a_0 p_k}{\sqrt{a_0 p_k (1 - p_k)}}\right\} = \Phi\left(\frac{X^* - a_0 p_k}{\sqrt{a_0 p_k (1 - p_k)}}\right)$, then the solution for X^* is,

$$X^* \ge a_0 p_k + \Phi^{-1}(\alpha) \sqrt{a_0 p_k (1 - p_k)} \ge X_f, \text{with} \tag{3}$$

$$X_f = a_0 \frac{N_f}{N} + \Phi^{-1}(\alpha) \sqrt{a_0 \frac{N_f}{N} \left(1 - \frac{N_f}{N}\right)}, \tag{4}$$

considering $p_k \ge \frac{N_f}{N}$ and $\frac{N_f}{N}, p_k \le 0.5$. Therefore, if an entry in Memory is matched more than X_f times within a_0 steps, we have confidence of $1 - \alpha$ that the features matched to this entry are from a real cluster and can be added to Dictionary, which satisfies **Condition A** in Eq. (1).

Once a Memory entry is matched with an input feature, its activity value, initialized with a_0, increases by a_0. If after m steps the entry is matched X times, $X \le m < a_0$, the activity value will be $a_0 + X a_0 - (m - X) = X(a_0 + 1) + a_0 - m$, and for $\{X = X_f, m = a_0\}$,

$$\phi = X_f(a_0 + 1), \tag{5}$$

which determines how long an entry is preserved in Memory, before being transferred to Dictionary as a permanent cluster.

Next, we examine how we can choose a_0 so that we do not discard real clusters from Memory, based on **Condition A** of Eq. (1). Let us assume an existing entry in Memory is from a real cluster, with population $N_\pi \geq N_f$. The probability of an informative feature, among a_0 samples, not being matched (e.g., if it appears only once) is $P = \left(\frac{N-N_\pi}{N}\right)^{a_0}$. If we need $1 - \alpha$ confidence that this will not happen, i.e.,

$$1 - \left(\frac{N - N_\pi}{N}\right)^{a_0} = 1 - P\{H_1 = 1 \mid H_0 = 1, H_1 = 0\} \geq 1 - \alpha, \text{then} \quad (6)$$

$$a_0 \geq \frac{\ln \alpha}{\ln(1 - \frac{N_\pi}{N})} \geq \frac{\ln \alpha}{\ln(1 - \frac{N_f}{N})} \quad (7)$$

Therefore, the probability of discarding a real cluster as noise is significantly low if we choose a_0 according to the condition above.

In **Condition B** of Eq. (1) we also require low probability of accepting noise as 'informative' features. Assume a noise cluster has been formed, π_z, where $\| \pi_z \| < N_f$, and it has been matched Z times during a_0 steps. Again, Z follows a binomial distribution $Z \sim B(a_0, p_z)$, with $p_z = \frac{\|\pi_z\|}{N}$, however, we cannot approximate it using a normal distribution since p_z is practically very small; instead, we use Poisson $P\{Z \geq X_f\} = \sum_{q=X^*}^{a_0} \frac{(a_0 p_z)^q}{q!} e^{-a_0 p_z}$. We cannot derive a closed form solution for this probability, however we can see with an example that it is insignificant: Recall $\frac{N_f}{N}$ is the minimum portion of the dataset that a real cluster can contain. For $\frac{N_f}{N} = 0.01$ and $\alpha = 0.01$, from Eq. (7) it is $a_0 \geq 458$. If we consider $a_0 = 458$ and $p_z = \frac{1}{2}\frac{N_f}{N}$, then $P\{Z \geq X_f\} = 1.3955 \cdot 10^{-4}$. In the worst-case where $p_z = \frac{N_f}{N}$, it is $P\{Z \geq X_f\} = 0.0191$, which determines the β-value in Eq. (1).

Sufficient Subset Size. Next, we show the portion of the dataset that needs to be processed for cluster centers to be calculated accurately. If we consider a cluster π_k and we sample N^* features from the dataset of size N, then the total number of features N_k expected to be from π_k follows the binomial distribution, $N_k \sim B(N^*, p_k)$. Let us consider the Chebyshev inequality,

$$P\{| N_k - E[N_k] | < \varepsilon\} \geq 1 - \frac{D[N_k]}{\varepsilon^2}, \quad (8)$$

with $E[N_k] = N^* p_k$ being the expectation and $D[N_k] = N^* p_k(1 - p_k)$ the variance of the random variable N_k, while ε is a positive constant. Then,

$$P\{| N_k - N^* p_k | < \varepsilon\} \geq 1 - \frac{N^* p_k(1 - p_k)}{\varepsilon^2} \quad (9)$$

If we consider that from the N^* samples it is expected that $N^*\frac{N_f}{N}$ outliers will emerge, we can assign $\varepsilon = \omega N^* \frac{N_f}{N}$, $0 < \omega < 1$. Then, Eq. (9) becomes

$$P\left\{| N_k - N^* p_k | < \omega N^* \frac{N_f}{N}\right\} \geq 1 - \frac{1 - \frac{N_f}{N}}{\omega^2 N^* \frac{N_f}{N}} \quad (10)$$

For a probability significance $1 - \gamma$, we expect,

$$P\left\{\mid N_k - N^* p_k \mid < \omega N^* \frac{N_f}{N}\right\} \geq 1 - \gamma \tag{11}$$

From Eqs. (10) and (11),

$$1 - \frac{1 - \frac{N_f}{N}}{\omega^2 N^* \frac{N_f}{N}} \geq 1 - \gamma \Rightarrow N^* \geq \frac{1 - \frac{N_f}{N}}{\omega^2 \gamma \frac{N_f}{N}}, \tag{12}$$

which is the condition for the size of the subset of the data that needs to be processed to generate accurate clusters.

Correctness. The analysis above involves a feature subset of size a_0. Since we have $T = \lfloor \frac{N}{a_0} \rfloor$ such subsets (sampling without replacement), the probability of failure to detect a cluster is (**Condition A** in Eq. (1)),

$$P_{fail} = (P\{H_1 = 1 \mid H_0 = 1, H_1 = 0\})^T = \alpha^T \tag{13}$$

and for all K clusters, the probability of success is

$$P_{success} = [1 - P_{fail}]^K = (1 - \alpha^T)^K \tag{14}$$

To illustrate the importance of this probability, let us consider a set of $N = 10^5$ features and $K = 100$ clusters (ground-truth). If we choose $\frac{N_f}{N} = 0.01$ $\alpha = 0.01$, then from Eq. (7) it is $a_0 \geq 458$. Considering the minimum number of features in each sample set, $a_0 = 458$, it is $P_{success} = \left(1 - 0.01^{\lfloor \frac{10^5}{458} \rfloor}\right)^{100} \approx 1$. Thus, under a distinct grouping pattern among the data, the algorithm is theoretically able to find a perfect clustering result. Note that $\frac{N_f}{N}$ determines the maximum portion of the data allowed to form a noise cluster, or equivalently, the minimum portion of the data that can be in a real cluster.

Clustering Streaming Features. Our method inherently carries the idea of sequential processing: it parses the dataset once, and each feature requires constant computation. However, we made the assumption that it shuffles the order at the beginning, in order to distribute noise evenly among the considered subsets of size a_0 and avoid noise accumulation. If we consider the problem of feature clustering in image collections, the average case is that noise is equally likely to appear in any image, and therefore shuffling the order of the features does not have significant effect. For the sequence paradigm, the worst case would be when successive images include more noise, or when noise is distributed spatially in an image in a non-uniform fashion. In such scenarios, if we do not shuffle the data, we can rely on the size of the formed clusters, and remove small ones transferred from Memory to Dictionary as statistically insignificant with respect to the content of the images.

Here we show that for a cluster π_k, $\| \pi_k \| \geq N_f$, no matter what the order of the input features is, there exists at least one consecutive subsequence of features $\mathbf{x}$, $\| \mathbf{x} \| = a_0$, such that,

$$\frac{\| \{\boldsymbol{x}_i \mid \boldsymbol{x}_i \in \pi_k, \boldsymbol{x}_i \in \mathbf{x}\} \|}{a_0} \geq \frac{N_f}{N} \tag{15}$$

In what follows, we consider binary variables ν to describe the feature membership to a specific cluster.

Lemma: For any permutation of a binary set $\{\nu_i \mid \nu_i = 0 \ or \ \nu_i = 1\}_N$ and any positive integer $a_0 \leq N$, there exists at least one consecutive subsequence ν of length a_0, such that,

$$\frac{\| \{\nu_i \mid \nu_i = 1, \nu_i \in \nu\} \|}{a_0} \geq \frac{\sum_{n=1}^{N} \nu_n}{N}, \tag{16}$$

where ν can include tail-head permutations, i.e., $[\nu_i, \nu_{i+1}, \ldots, \nu_N, \nu_1, \nu_2, \ldots]$.

Note: This condition means that there is at least one consecutive subsequence of length a_0 where the density of the cluster members is greater than or equal to the average density of the cluster members in the entire dataset.

Proof. We prove this lemma by contradiction. Assume for any a_0-length consecutive subsequence of an arbitrary permutation $[\nu_1, \ldots, \nu_N]$,

$$\sum_{j=i}^{i+a_0-1} \nu_{mod(j,N)+1} < a_0 \frac{\sum_{n=1}^{N} \nu_n}{N}, \quad \forall i = 0, 1, \ldots, N-1 \tag{17}$$

We consider the modulo index $mod(j, N) + 1$ to account for tail-head permutations (see above). In total, there are N distinct consecutive subsequences. Then,

$$\sum_{i=0}^{N-1} \left(\sum_{j=i}^{i+a_0-1} \nu_{mod(j,N)+1} \right) = a_0 \sum_{n=1}^{N} \nu_n, \tag{18}$$

i.e., each element in the dataset is added a_0 times (consider an a_0-length window 'sliding' N times along the dataset/sequence). However, according to the assumption in Eq. (17), we have,

$$\sum_{i=0}^{N-1} \left(\sum_{j=i}^{i+a_0-1} \nu_{mod(j,N)+1} \right) < \sum_{i=0}^{N-1} \left(a_0 \frac{\sum_{n=1}^{N} \nu_n}{N} \right) = a_0 \sum_{n=1}^{N} \nu_n, \tag{19}$$

which contradicts Eq. (18). □

4   Experimental Results

Image categorization involves, in general, three steps: (a) building visual vocabularies from image features, (b) image encoding using the vocabularies, and (c) classification. We used our method to cluster SIFT features [16] and create visual vocabularies. To show-case the benefit of using our method in such problems, namely improving the trade-off between accuracy and efficiency, we adopted three image classification methods: Vector of Locally Aggregated

Descriptors (**VLAD**) [14], Bag-of-Visual-Words (**BoVW**) [7], and Fisher Vector (**FV**) [22]. We make comparisons between our approach, ANN K-means [15], and Naive EM [17], when used in these three methods. Note that variations of K-means and Naive EM are among the most popular clustering approaches utilized in such Computer Vision tasks.

We used three publicly available image collections: (a) Object Discovery 100 (**Obj. Disc**); (b) Caltech 101 (**Caltech101**); and (c) PASCAL VOC 2007: we first used a subset of 6 randomly chosen categories (**PASC(6)**) and then the entire collection (**PASC(all)**).

In the experiments with K-means and EM, we follow the approach in [27] and randomly chose K·1000 features for VLAD and FV, and K·200 for BoVW. Since our algorithm is much more efficient than K-means and Naive EM, it allows us to mine a larger dataset and still be much faster than those two. For the experiments with our method, we used: for FV, 0.8 million features (N) from each dataset; for VLAD, 1 million features from Obj. Disc. dataset, 5 million from PASC(6), and 5 million from PASC(all); for BoVW, 2 million features from Obj. Disc. and 5 million from each of PASC(6) and PASC(all).

Table 1 summarizes the results of our method and the competition when used in each image classification method (VLAD, BoVW, FV) and for each dataset. Each row corresponds to a different vocabulary size K, as set by the competing

Table 1. Comparisons in building visual vocabularies for three popular image categorization methods. K = number of clusters; mAP = mean Average Precision [23]; DB = Davies-Bouldin index [8]; CH = Calinski-Harabasz index [4]; LL = Log-likelihood

VLAD

	Our method					ANN K-means				
	K	mAP	time	DB	CH	K	mAP	time	DB	CH
Obj.Disc.	131.8	94.60	6.21	4.17	1067.83	100	95.69	60.53	4.05	285.84
	265.7	95.03	7.11	3.99	639.23	200	95.55	149.43	3.82	350.48
	371.9	95.08	7.74	3.94	500.69	400	96.03	372.97	3.69	411.03
	587.9	94.51	9.08	3.88	356.59	600	95.65	579.98	3.63	445.56
Caltech101	252.6	79.18	56.23	3.90	1432.02	200	80.47	138.20	3.92	340.31
	420.0	80.85	90.48	3.79	972.06	400	80.57	331.60	3.72	411.67
	671.7	81.02	139.63	3.74	674.71	600	80.82	597.46	3.68	456.40
	839.6	**81.27**	**165.42**	3.71	559.55	800	**81.22**	847.33	3.66	482.09
PASC(6)	254.5	66.48	70.19	4.20	5995.89	200	66.06	168.91	4.09	325.35
	393.5	68.36	77.06	4.07	4336.50	400	69.32	387.05	3.85	389.80
	567.0	70.23	86.33	4.05	3229.48	600	70.28	705.26	3.78	420.65
	809.0	**71.55**	**104.60**	3.98	2439.39	800	**70.37**	953.47	3.77	443.26
PASC(all)	204.2	56.87	64.45	4.48	7351.11	200	53.25	147.50	4.08	324.90
	394.4	55.82	77.11	4.11	4262.41	400	57.43	365.70	3.87	389.66
	654.9	57.51	101.03	3.99	2883.65	600	57.18	587.63	3.76	420.72
	782.9	**58.77**	**101.44**	4.00	2499.83	800	**57.21**	883.62	3.76	443.91

FV

	Our method				Naive EM			
	K	mAP	time	LL(10^8)	K	mAP	time	LL(10^8)
Obj.Disc.	103.0	95.41	80.46	0.948	100	95.06	63.37	0.135
	206.5	96.50	151.74	0.957	200	95.82	334.31	0.214
	385.5	**97.03**	282.61	0.963	400	**96.05**	819.12	0.544
	629.0	5.46	527.34	0.954	600	95.80	2012.20	0.806
Caltech101	231.6	74.55	220.91	1.090	200	74.59	197.63	0.274
	414.3	76.26	423.07	1.100	400	75.96	879.18	0.551
	617.7	77.56	610.64	1.110	600	77.55	2146.14	8.350
	795.5	**77.62**	**677.01**	1.110	800	**77.08**	3654.57	1.120
PASC(6)	224.0	72.65	233.46	1.100	200	72.46	197.36	0.276
	390.0	74.49	410.76	1.111	400	74.17	806.97	0.557
	627.0	75.85	521.74	1.117	600	75.43	1718.42	0.840
	813.0	**76.34**	819.28	1.121	800	**76.30**	3636.91	1.125
PASC(all)	223.7	60.68	227.45	1.500	200	60.03	190.56	0.374
	419.4	63.08	512.09	1.500	400	62.26	782.23	7.560
	609.5	63.99	621.58	1.510	600	63.44	1551.14	1.140
	784.1	**64.68**	801.59	1.529	800	**63.82**	3523.66	1.520

BoVW

	Our method					ANN K-means				
	K	mAP	time	DB	CH	K	mAP	time	DB	CH
Obj.Disc.	2666.7	95.23	26.93	3.64	157.64	2000	95.25	729.31	3.41	118.73
	4227.8	95.21	38.46	3.51	110.36	4000	95.44	1772.24	3.25	124.34
	5977.2	**95.63**	**53.54**	3.39	85.03	6000	**95.30**	**2799.17**	3.10	93.09
Caltech101	4188.8	75.50	152.47	3.95	283.49	4000	75.55	1692.51	3.62	129.60
	6548.7	76.81	228.11	3.96	194.44	6000	76.65	2639.86	3.63	137.44
	8207.4	77.03	286.63	3.97	160.44	8000	77.36	4037.31	3.63	142.54
	9939.8	**77.66**	**342.77**	3.97	136.17	10000	**77.86**	**4945.88**	3.61	146.64
PASC(6)	4154.2	62.49	128.33	4.03	256.75	4000	60.90	1739.89	3.80	117.40
	6481.7	63.41	188.67	4.08	175.94	6000	62.75	2586.90	3.80	122.84
	8016.5	63.88	233.61	4.10	146.82	8000	63.07	4010.56	3.80	127.57
	9946.0	**64.95**	**290.25**	4.11	122.14	10000	**63.41**	**5034.19**	3.80	130.93
PASC(all)	4098.8	49.85	260.65	4.01	650.94	4000	48.06	1770.43	3.80	117.26
	7037.3	50.85	406.49	4.06	412.28	6000	49.10	2825.60	3.81	122.85
	8655.0	51.42	486.84	4.10	344.18	8000	49.91	3933.78	3.82	127.49
	10958.1	**51.67**	**594.20**	4.11	280.88	10000	**50.15**	**5028.55**	3.82	130.89

method. For each experiment (row), we ran our method and the competition 25 times, and we report average results. To evaluate clustering itself, we used the Davies-Bouldin (**DB**) index [8], Calinski-Harabasz (**CH**) index [4], and Log-likelihood (**LL**). To evaluate the overall accuracy of VLAD, BoVW, and FV, we used the mean Average Precision (**mAP**) [23]. The reported times are for clustering only. These results illustrate that our method and the competition yield, on average, comparable accuracy, with our method being significantly faster: in Table 1, the boldface numbers correspond to indicative comparison instances and examples where our method is 4 to 42 times faster than the competition. Note that SIFT features from natural images are usually very cluttered, therefore there are no natural groups in the clustered data. However, our method still generates competitive results while boosting the clustering efficiency.

To test the performance of our approach in sequential data (clustering on the fly), we used videos captured by an onboard camera of a quadrotor during flight. Figure 2 illustrates an indicative example of clustering detected SIFT features in eight non-consecutive 752×480 frames of a video (frame numbers are shown on top left of each image). The detected features are marked in different shapes and colors, indicating different cluster assignments. The magenta-yellow arrows in frames #250 and #375 show indicative examples of newly emerged clusters (orange and green square categories), while the long double arrow indicates correspondence between features (features in the same cluster) across frames. In this experiment we used $N_f/N = 0.005$ and $\theta = -275$.

Fig. 2. Building visual vocabularies on the fly. Colors and shapes indicate cluster assignments of the detected features, while the arrows in frames #250 and #375 indicate emerged clusters (orange/green squares). (Color figure online)

Finally, we also validated the efficiency of our method using synthetic datasets generated by different Mix Gaussian models with random means and covariances, large numbers of clusters, and uniformly distributed noise. Compared to K-means++ [2], BIRCH [29], and EM clustering [17], our approach was on average 3–7 times faster than the first two, while EM was the slowest among the competition. Note that BIRCH builds a CF-tree, where grouping the leaves into the desired number of clusters is computationally expensive with $O(N^2 log N)$.

5   Conclusions

We described a fast and accurate center-based clustering method suitable for large datasets with a high number of natural clusters. It produces clustering centers with a single pass through the data, by using a Dictionary and a (short) Memory list for building and enriching a global (sparse) histogram of the data: dense entries in Dictionary and Memory correspond to frequently matched features, thus indicating formed clusters. Input features that are not matched in Dictionary move to Memory, where either they are assigned to an existing entry, or create a new one. Memory entries that are not populated sufficiently are discarded as noise, while dense entries are moved to Dictionary permanently. In our results we showed that the trade-off between accuracy and efficiency is improved, compared clustering approaches commonly used in Computer Vision.

References

1. Ackermann, M.R., Lammersen, C., Martens, M., Raupach, C., Sohler, C., Swierkot, K.: StreamKM++: a clustering algorithm for data streams. J. Exp. Algorithmics **17**(article no. 2.4), 51–57 (2012)
2. Arthur, D., Vassilvitskii, S.: K-means++: the advantages of careful seeding. In: Proceedings of ACM-SIAM Symposium on Discrete Algorithms, pp. 1027–1035 (2007)
3. Bezdek, J.C., Ehrlich, R., Full, W.: FCM: the fuzzy c-means clustering algorithm. Comput. Geosci. **10**(2), 191–203 (1984)
4. Calinski, T., Harabasz, J.: A dendrite method for cluster analysis. Commun. Stat.-Theory Methods **3**(1), 1–27 (1974)
5. Cho, M., Kwak, S., Schmid, C., Ponce, J.: Unsupervised object discovery and localization in the wild: part-based matching with bottom-up region proposals. In: Proceedings of Computer Vision, Pattern Recognition, pp. 1201–1210 (2015)
6. Coates, A., Ng, A.Y., Lee, H.: An analysis of single-layer networks in unsupervised feature learning. In: Proceedings of International Conference on Artificial Intelligence and Statistics, pp. 215–223 (2011)
7. Csurka, G., Dance, C.R., Fan, L., Willamowski, J., Bray, C.: Visual categorization with bags of keypoints. In: Proceedings of Statistical Learning in Computer Vision, in European Conference on Computer Vision, pp. 1–22 (2004)
8. Davies, D.L., Bouldin, D.W.: A cluster separation measure. IEEE Trans. PAMI **1**(2), 224–227 (1979)
9. Farnstrom, F., Lewis, J., Elkan, C.: Scalability for clustering algorithms revisited. ACM SIGKDD Explor. Newsl. **2**(1), 51–57 (2000)
10. Frey, B.J., Dueck, D.: Clustering by passing messages between data points. Science **315**(5814), 972–976 (2007)
11. Ghahramani, S.: Fundamentals of Probability with Stochastic Processes, pp. 466–475. CRC Press, Boca Raton (2015)
12. Hartigan, J.A., Wong, M.A.: Algorithm AS 136: a k-means clustering algorithm. Appl. Stat. **28**(1), 100–108 (1979)
13. Jain, A.K.: Data clustering: 50 years beyond k-means. Pattern Recogn. Lett. **31**(8), 651–666 (2010)

14. Jégou, H., Douze, M., Schmid, C., Pérez, P.: Aggregating local descriptors into a compact image representation. In: Proceedings of Computer Vision, Pattern Recognition, pp. 3304–3311 (2010)
15. Kanungo, T., Mount, D.M., Netanyahu, N.S., Piatko, C.D., Silverman, R., Wu, A.Y.: An efficient k-means clustering algorithm: analysis and implementation. IEEE Trans. PAMI **24**(7), 881–892 (2002)
16. Lowe, D.G.: Distinctive image features from scale-invariant keypoints. Int. J. Comput. Vis. **60**(2), 91–110 (2004)
17. Moon, T.K.: The expectation-maximization algorithm. IEEE Sig. Proc. Mag. **13**(6), 47–60 (1996)
18. Nister, D., Stewenius, H.: Scalable recognition with a vocabulary tree. In: Proceedings of Computer Vision Pattern Recognition, vol. 2, pp. 2161–2168 (2006)
19. O'Callaghan, L., Mishra, N., Meyerson, A., Guha, S., Motwani, R.: Streaming-data algorithms for high-quality clustering. In: Proceedings of IEEE International Conference on Data, Engineering, pp. 685–694 (2002)
20. Pang-Ning, T., Steinbach, M., Kumar, V.: Introduction to Data Mining, pp. 508–509. Pearson, Boston (2006)
21. Perronnin, F., Dance, C.: Fisher kernels on visual vocabularies for image categorization. In: Proceedings of Computer Vision, Pattern Recognition, pp. 1–8 (2007)
22. Perronnin, F., Sánchez, J., Mensink, T.: Improving the fisher kernel for large-scale image classification. In: Proceedings European Conference on Computer Vision, pp. 143–156 (2010)
23. Philbin, J., Chum, O., Isard, M., Sivic, J., Zisserman, A.: Object retrieval with large vocabularies and fast spatial matching. In: Proceedings of Computer Vision, Pattern Recognition, pp. 1–8 (2007)
24. Rodriguez, A., Laio, A.: Clustering by fast search and find of density peaks. Science **344**(6191), 1492–1496 (2014)
25. Silva, J.A., Faria, E.R., Barros, R.C., Hruschka, E.R., De Carvalho, A.C.: Data stream clustering: a survey. ACM Comput. Surv. **46**(1), 13 (2013)
26. Sivic, J., Zisserman, A.: Video Google: a text retrieval approach to object matching in video. In: Proceedings of IEEE International Conference on Computer Vision, pp. 1470–1477 (2003)
27. Vedaldi, A., Fulkerson, B.: An open and portable library of computer vision algorithms. In: Proceedings of ACM International Conference on Multimedia, pp. 1469–1472 (2010)
28. Walker, W.R., Skowronski, J.: The fading affect bias: but what the hell is it for? Appl. Cognit. Psychol. **23**, 1122–1136 (2009)
29. Zhang, T., Ramakrishnan, R., Livny, M.: BIRCH: an efficient data clustering method for very large databases. Proc. ACM SIGMOD **25**(2), 103–114 (1996)

HoP: Histogram of Patterns for Human Action Representation

Vito Monteleone[(✉)], Liliana Lo Presti, and Marco La Cascia

Universita' degli Studi di Palermo, Palermo, Italy
vito.monteleone@unipa.it

Abstract. This paper presents a novel method for representing actions in terms of multinomial distributions of frequent sequential patterns of different length. Frequent sequential patterns are series of data descriptors that occur many times in the data. This paper proposes to learn a codebook of frequent sequential patterns by means of an apriori-like algorithm, and to represent an action with a *Bag-of-Frequent-Sequential-Patterns* approach. Preliminary experiments of the proposed method have been conducted for action classification on skeletal data. The method achieves state-of-the-art accuracy value in cross-subject validation.

Keywords: Action classification · Apriori algorithm · Frequent pattern

1 Introduction

In this work we propose to represent time series of descriptors by means of distributions of frequent sequential patterns of different length for action classification. We define a sequential pattern as a series of data descriptors indexed in time order, and a frequent pattern is one that occurs many times in the data [10].

A classical approach to represent actions is *Bag Of Visual Words (BoVW)* [5, 8,13,14,16]. In BoVW, an action is represented as a distribution of image/video patches (visual words). The codebook of visual words is generally computed by clustering algorithms, i.e. k-means [9,12,15,17]. To consider the dynamics of visual information in a time series within BoVW, spatio-temporal descriptors extracted from fixed-length cuboids [13,14,16] or multi-scale time windows [4] have been used. Visual feature dynamics are especially useful for discriminating actions that share similar body poses but show different temporal evolution; as an example, *sit down* and *get up* are actions sharing similar body poses, but these poses appear in different time order.

In contrast to the classical BoVW approach, we describe an action by means of frequent sequences of visual descriptors, thus focusing more on the body motion dynamics rather than actual body poses. Figure 1 gives an overview of the proposed *Bag-of-Frequent-Sequential-Patterns* approach. In our approach, the codebook of frequent sequential patterns is computed by means of a modified apriori algorithm [1,6]. Our implementation of the apriori algorithm allows

© Springer International Publishing AG 2017
S. Battiato et al. (Eds.): ICIAP 2017, Part I, LNCS 10484, pp. 457–468, 2017.
https://doi.org/10.1007/978-3-319-68560-1_41

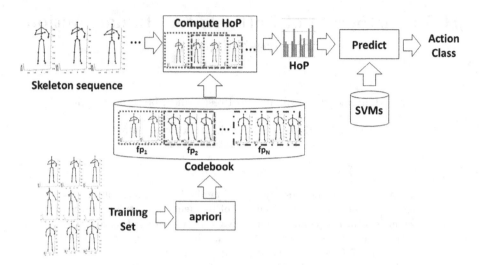

Fig. 1. *Bag-of-Frequent-Sequential-Patterns:* a test sequence is encoded in terms of frequent sequential patterns (fp₁, fp₂, ..., fp_N) by means of vector quantization; hence, a histogram of frequent sequential patterns is computed and used to predict the action class based on 1-vs-1 SVMs. In the proposed approach, the codebook is learned by a modified apriori algorithm on the training set.

us to calculate frequent patterns of different lengths, which represent different levels of body motion details. While in general clustering algorithms group elements based only on pairwise element similarities, our technique considers both similarity and frequency of the elements when learning a codebook of frequent sequential patterns. This allows us to ignore infrequent patterns that might be less informative or even confusing for classification purposes.

To summarize, our contribution in this paper is twofold:

1. we represent actions by multinomial distributions of frequent sequential patterns;
2. we propose an apriori algorithm-based learning approach for codebook of frequent sequential patterns.

We demonstrate our approach in the context of 3D skeleton-based action classification [11]. The proposed framework can be easily extended to other kinds of visual descriptors such as histograms of STIP features [16] or HOG [3]. We present preliminary experimental results on the Microsoft Research Cambridge-12 (MSRC-12) gesture dataset [18] in cross-subject validation. Our technique achieves state-of-the-art accuracy values.

The paper is organized as follows. Section 2 discusses related work; Sect. 3 explains both our modified apriori algorithm for learning a codebook of frequent sequential patterns, and how to represent an action in terms of histogram of frequent patterns (HoP); Sect. 4 presents experimental results and, finally, Sect. 5 discusses conclusions and future work.

2   Related Work

Two of the most successful approaches for representing visual content in images or videos, dictionary-based representations and Bag of Visual Words (BoVW), are based on dictionary/codebook learning. In dictionary-based representation approaches, the signal is represented as a linear combination of elements of a dictionary [23]. In Bag-of-Visual-Words (BoVW) approaches [14], introduced first for visual categorization in [5], visual content of images/videos is represented in terms of distributions of elements (codewords) in a codebook. Whilst [5] adopts a Bayesian hierarchical model to learn such kind of distributions, in practice the most commonly used pipeline requires the following steps [14]: local feature extraction, learning of a codebook by means of clustering techniques (e.g., k-means), vector quantization (for discretization of the analyzed signal in terms of codewords), codewords-based histogram computation. Such kind of paradigm has been adopted for action representation in several former works [4,8,9,12,13,15–17] In particular, in [4], sequences are represented as a distribution of local temporal texture descriptors estimated at different time scales. A codebook of multi-scale local representations is learned via k-means, and classification is performed via SVM. In [22], a codebook of temporal windows is learned via spectral clustering of data subsequences. Similarly to [4,22], we represent an action as a distribution of temporal windows of different lengths, but we adopt a data mining technique rather than a clustering technique to learn a codebook.

In the context of 3D Action Representation from skeletal data [11], the work in [20] represents actions in terms of co-occurring spatial and/or temporal configurations (poses) of specific body parts. A bag-of-words approach is adopted to represent an action where the codebook comprises co-occurring body-parts and is learned by contrast mining technique. In this sense, the codebook represents emerging patterns, that is patterns whose supports change significantly from one class to another. The work in [21] applies the apriori algorithm to find discriminative actionlet. An actionlet is defined as a subset of joints in the skeleton, and an action is represented by a linear combination of actionlets whose weights are learned via a multiple kernel learning approach. In contrast to this approach, our method aims at mining frequent sequential patterns and representing actions with a Bag-of-Frequent-Patterns approach. Our modified apriori algorithm is inspired by the work in [6]. The work in [6] focuses on detecting reduplications in a video of American Sign Language (ASL). The method detects frequent sequential patterns of increasing length by combining smaller frequent sequential patterns, and relies on approximate matching of the discovered sequential patterns with data. In counting frequencies of patterns, a waiting mechanism is used to account for poor matching arising in presence of small misalignments between patterns and data sequence. In this sense, [6] finds gapped sequential patterns. The focus of our paper is action classification; we use a method similar in spirit to [6] for mining sequential patterns to be added to our codebook. We apply our technique to a set of data streams rather than a single stream and look for non-gapped sequential patterns. During the pattern discovery process, all frequent patterns that do not contribute to the generation of longer patterns are added to our codebook. In contrast to [6], we learn frequent pattern models by averaging over matched data windows. In practice, this strategy proved to account for noise in data.

3   Representation by Histogram of Frequent Patterns

As shown in Fig. 1, a time series is represented as a histogram of frequent patterns by matching subsequences with the patterns stored in the codebook.

Frequent patterns may be found by data mining techniques, such as the apriori algorithm proposed for transactional databases [1]. In such kind of applications, a pattern $C^{(k)}$ is a set of k items from an alphabet $\mathscr{A}$, and the problem is that of finding the longest frequent patterns in the database.

Since in transactional databases there is no need of considering the order of the items within the patterns, the method is not appropriate for sequential data, such as time series, and requires some modifications in order to calculate frequent ordered item-sets. Modified apriori-algorithm for sequential data have been proposed in [2, 6, 10]. In particular, the method in [6] deals with the discovery of reduplication of ASL within a single data stream. As we will detail next, we borrow some of the ideas in [6] and adapt them to the learning (rather than discovering) of sequential patterns from a set of time series.

3.1   Codebook of Frequent Patterns

The main idea behind apriori-like algorithms is that a pattern $C^{(k)}$ is frequent if and only if each pattern $C^{(k-1)} \subset C^{(k)}$ is frequent as well. Therefore a frequent pattern $C^{(k)}$ may be generated iteratively by extending a pattern $C^{(k-1)}$ with an item $i \in \mathscr{A}$, and ensuring that the generated pattern is composed of only frequent sub-patterns.

At the k-th iteration, apriori-like algorithms consist mainly of three steps:

- Generation of candidates of length k by frequent patterns of length $k-1$;
- Counting of candidate frequencies;
- Removal of infrequent patterns.

Infrequent patterns have a frequency count lower than a predefined threshold ψ.

We modified these steps to adapt them to the processing of sequential data. Algorithm 1 shows the work-flow required to discover frequent patterns from training data $\mathscr{D}$. The algorithm generates frequent sequential patterns $\underline{C}^{(K_M)}$ of maximal length K_M. At the k-th iteration, $\underline{C}^{(k)}$ is a set of patterns $C_i^{(k)}$ with $i \in [1, \ldots, N_k]$, where N_k represents the number of frequent sequential patterns of length k that have been found in data $\mathscr{D}$. Each $C_i^{(k)}$ is an ordered sequence of feature descriptors $c_{i,j}$, i.e. $C_i^{(k)} = [c_{i,1}, c_{i,2}, \ldots, c_{i,k}]$. The set $codebook$ stores frequent sequential patterns of different-length. The set $\underline{fp}^{(k-1)}$ stores frequent sequential patterns of length $k-1$ that cannot be used to generate longer patterns.

Algorithm 1. Learning a codebook of frequent sequential patterns

1: **function** CODEBOOK = CODEBOOKLEARNING($\mathscr{D}, K_M$)
2:     $k \leftarrow \tau$
3:     codebook $\leftarrow \emptyset$
4:     $\underline{C}^{(k)} \leftarrow$ generateCandidatePatterns($\mathscr{D}$, k)
5:     **while** $k < K_M$ **do**
6:         $k \leftarrow k + 1$
7:         $[\underline{C}^{(k)}, \underline{fp}^{(k-1)}] \leftarrow$ newCandidatePatternGeneration($\underline{C}^{(k-1)}$)
8:         codebook $\leftarrow$ codebook $\cup \underline{fp}^{(k-1)}$
9:         $\underline{C}^{(k)} \leftarrow$ duplicatesRemoval($\underline{C}^{(k)}$)
10:        getFrequencies($\underline{C}^{(k)}, \mathscr{D}$)
11:        $\underline{C}^{(k)} \leftarrow$ infrequentPatternsRemoval($\underline{C}^{(k)}$)
12:    **end while**
13:    codebook $\leftarrow$ codebook $\cup \underline{C}^{(K_M)}$
14: **end function**

Candidate Pattern Generation: In the classical apriori algorithm [1], the initial set of items (alphabet $\mathscr{A}$) is known. In our application, this initial set is unknown and we start the algorithm with all possible windows of minimal length τ extracted from the data streams with a sliding window approach. We refine such initial set of candidate patterns $\underline{C}^{(\tau)}$ by pruning the duplicated and infrequent ones as detailed later.

Candidate Pattern Frequencies: Given a set of candidate patterns $\underline{C}^{(k)}$ and data $\mathscr{D}$, we need to count how many times each candidate pattern occurs in the data. In contrast to the classical apriori algorithm, our method entails the processing of non categorical data; therefore we need a strategy to establish approximate matches between candidate patterns and data. In particular, each candidate pattern $C_i^{(k)}$ has to be compared against temporal windows extracted from data and of the same length as the considered pattern. Let us assume for a moment that $\mathscr{D}$ contains only one sequence, i.e. $\mathscr{D} = [d_1, d_2, \ldots d_N]$, and consider a pattern $C_i^{(k)} = [c_{i,1}, c_{i,2}, \ldots, c_{i,k}]$. We consider a sliding window $W_t = [d_t, d_{t+1}, \ldots, d_{t+k-1}]$. The similarity between the candidate pattern and the temporal window W_t is measured by the following similarity score:

$$s(C_i^{(k)}, W_t) = \frac{1}{k} \cdot \sum_{j=1}^{k} e^{-\lambda \cdot ||c_{i,j} - d_{t+j-1}||_2} \tag{1}$$

where λ is a scaling parameter that multiplies the per-item squared Euclidean distance. When this score is greater than a threshold ϵ, it is possible to establish a match between the pattern and the window, and increment the candidate pattern frequency. For each pattern, we keep track of the matched temporal windows by considering the list $\underline{W}^{C_i} = \{W_j\}_{j \in J}$.

New Candidate Pattern Generation and Codebook Learning: Let us consider two frequent patterns $C_1^{(k-1)} = [c_{1,1}, c_{1,2}, \ldots, c_{1,k-1}]$ and $C_2^{(k-1)} = [c_{2,1}, c_{2,2}, \ldots, c_{2,k-1}]$ such that $c_{1,j} = c_{2,j-1}$ $\forall j \in [2, k-1]$. Following [6], a candidate frequent pattern of k items can be defined as $C^{(k)} = [C_1^{(k-1)}, c_{2,k-1}]$. Figure 2 sketches the new candidate pattern generation procedure.

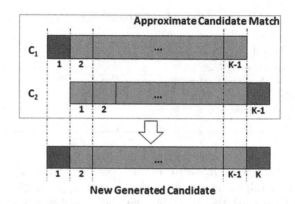

Fig. 2. The figure illustrates the idea behind the candidate pattern generation process. The new generated candidate is formed by concatenating the first item of C_1, the items shared by both C_1 and C_2, and the last item of C_2.

This candidate generation procedure would work in case of exact match of the items. In our implementation, we establish approximate matches between candidate patterns $C_1^{(k-1)}$ and $C_2^{(k-1)}$ when all corresponding items score a similarity greater than ϵ. By defining the following binary variable:

$$m(C_1^{(k-1)}, C_2^{(k-1)}) = \prod_{j=2}^{k-1} (e^{-\lambda \cdot ||c_{1,j} - c_{2,j-1}||} \geq \epsilon), \tag{2}$$

if $m(C_1^{(k-1)}, C_2^{(k-1)})$ is equal to 1 then an approximate match between the two candidate patterns can be established.

In contrast to [6], where the items of each frequent pattern comes from the data stream, we learn a pattern model by means of the lists of matched windows of the two candidate patterns, respectively $\underline{W}^{C_1}$ and $\underline{W}^{C_2}$. The new generated pattern will have the form $C^{(k)} = [\mu_1, \mu_{2:k-1}, \mu_k]$ where μ_1 is the expected value of the first item of $C^{(k)}$ and is computed by averaging the first elements of the windows in $\underline{W}^{C_1}$; $\mu_{2:k-1}$ are expected values of subsequent items in the pattern $C^{(k)}$ and are calculated by considering both the items of windows in $\underline{W}^{C_1}$ and windows in $\underline{W}^{C_2}$; finally, μ_k is the expected value of the last item in $C^{(k)}$ and is computed by averaging the last elements of the windows in $\underline{W}^{C_2}$.

Whenever a candidate pattern of length $k-1$ does not contribute to generate candidate patterns of length k, and its frequency is greater than a threshold ψ, then the pattern is stored into the codebook.

Removal of Duplicated and Infrequent Candidate Patterns: After the generation step, a pairwise comparison of candidate patterns is carried on. Each pair of candidates with a similarity score greater than ϵ is replaced by a new candidate generated averaging the lists of matched windows. Such kind of pruning is necessary to deal with approximate matches between data and patterns. To focus on frequent patterns, candidate patterns with a frequency count smaller than a threshold ψ are considered infrequent and, hence, pruned.

3.2 Histogram of Frequent Patterns

Provided with a codebook of N frequent sequential patterns $\{C_i\}_{i\in[1,N]}$ of different length, we aim at representing a time series $V = \{y_1, y_2, \ldots, y_v\}$ as a histogram of frequent patterns (HoP) by performing vector quantization (VQ) [14]. For each frame in V and for each pattern C_i in the codebook, we consider a subsequence of V that starts from the current frame, and of length equal to that of the considered pattern C_i. We compare each window to the patterns by the score in Eq. (1) and only increment the bin of the histogram that corresponds to the pattern achieving the highest similarity (i.e. we apply hard coding).

At the top of Fig. 3, a sample of the action class *Push-Right* is shown. The bar under the sequence indicates which patterns in the codebook have been detected in the sequence (each color corresponds to a different pattern); the patterns are represented under the bar while, at the bottom of the figure, the histogram of patterns is plotted.

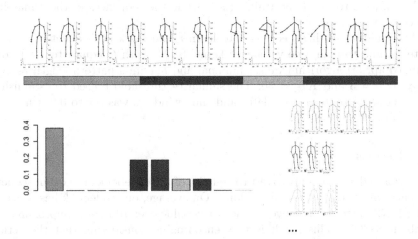

Fig. 3. The figure illustrates the HoP of a sample of the $Push - Right$ class in terms of frequent patterns learned by our apriori algorithm. In the figure, only few distinctive skeletons of the sequence and of the patterns are shown.

4   Experimental Results

We validated our technique on the Microsoft Research Cambridge-12 (MSRC-12) gesture dataset [18]. The dataset consists of sequences of skeletons described by means of the coordinates of 20 3D body joints. Skeletons were estimated by using the Kinect Pose Estimation pipeline [19]. The dataset includes 594 sequences representing the performances of 12 actions (*Start system* (SS), *Duck* (D), *Push Right* (PR), *Goggles* (G), *Wind it up* (W), *Shoot* (S), *Bow* (B), *Throw* (T), *Had enough* (H), *Change weapon* (C), *Beat both* (BB), *Kick* (K)) from 30 different subjects. Each sequence is a recording of one subject performing one gesture several times. Considering that the MSRC-12 dataset has been proposed for action detection, no temporal segmentation of the single performance is provided with the dataset but only the time when the action is considered recognizable. In order to test our method in action classification, we adopted the annotation made publicly available by [7]. Such annotation specifies the initial/final frame when each performance starts/ends. This annotation has produced 6243 different action sequences. In order to account for biometric differences, we preprocessed each action sequence by removing its average skeleton. In general, mining algorithms are used over a single sequence to discover repetitive patterns. In contrast, our algorithm learns frequent patterns over the entire training dataset, which includes segmented action sequences from different classes and performed by different subjects. Thus, our training approach allows us to learn more general frequent patterns. We repeated the experiment 10 times in cross-validation with a 50% subject split experimental protocol, that is we randomly select half of the subjects to build the training set, while the sequences of the remaining subjects are used for test.

The training set has been used to learn a codebook of frequent sequential patterns, and to train one vs one χ^2 kernel SVMs with C equals to 10. In our modified apriori algorithm we set minimal and maximal pattern length respectively to $\tau = 3$ and $K_M = 30$. The similarity threshold ϵ used to establish a match between pattern candidates and time windows was set to 0.9, while the threshold ψ was set to 75.

4.1   Results

We performed experiments to test the quality of the codebook of frequent sequential patterns generated via Algorithm 1. On average, our codebook has a size of 120 ± 14.72 patterns. The average accuracy value over 10 runs is approximately of about 88.32%. This result is very encouraging considering that the action representation is very compact.

As detailed in Sect. 3.1, the codebook stores all patterns with a frequency count greater than ψ that do not contribute to the generation of longer patterns. However, since we adopt an approximate matching strategy, the frequency count of the generated patterns is not a very reliable measure of the importance of the learned patterns. Hence, it is reasonable to wonder if patterns that are considered infrequent during the codebook learning procedure might actually improve

T vs P	SS	D	PR	G	W	S	B	T	H	C	BB	K
SS	**80.23**	0	0	0	0.43	4.18	0.04	0.90	v4.59	1.07	7.63	0.93
D	0	**99.96**	0	0	0	0	0.04	0	0	0	0	0
PR	0.04	0	**96.35**	0	0.73	1.42	0	0.24	0.12	1.09	0	0
G	0.12	0	0	**93.14**	1.00	1.66	0	0	3.12	0.48	0.48	0
W	0.42	0	1.24	0.09	**92.43**	1.18	0	0.10	0	2.00	2.54	0
S	0.59	0	0.07	0.11	0.30	**93.76**	0.04	0.04	0.12	2.28	2.67	0
B	0	4.38	0	0	0	0.20	**95.15**	0.04	0	0.03	0	0.19
T	0.04	0	0.08	0	0.04	0.81	0.44	**93.10**	0	1.42	0.04	4.03
H	2.74	0	0.04	5.35	0.12	1.28	0	0	**89.00**	0.04	1.42	0
C	0.08	0	0.04	0.08	0.28	3.31	0	0	0.04	**95.77**	0.40	0
BB	3.19	0.62	0.08	0.41	3.89	6.22	0	0.15	1.61	2.47	**81.33**	0.04
K	0.20	0.07	0	0	0.04	0.24	0.30	0.35	0	0.49	0	**98.30**

action classification. To validate our hypothesis, we also included in the codebook sequential patterns that are pruned in line 11 of Algorithm 1 and having a frequency count greater than a threshold ϕ. Then, we study how frequent a frequent pattern should be for being included in the codebook by studying how the average recall changes when varying ϕ in the range $[0, 100]$.

Figure 4(a) shows the trend of the average per-class recall over 10 runs when varying ϕ. Vertical bars represent standard deviations of recall values. Figure 4(b) shows the number of patterns in the codebook with a frequency greater than ϕ. As shown in the latter plot, the codebook size decreases exponentially; on average, the codebook size ranges between 50583 (when $\phi = 0$, i.e. all infrequent patterns are included in the codebook) and 44 (when $\phi = 100$).

On the other hand, as shown in Fig. 4(a), there is an increase of the recall values for growing values of ϕ. For value of ϕ in [20–70] there is a very limited variation of the average recall; what it really changes is the codebook size that affects the complexity of the vector quantization step. The best average per-class recall is obtained for $\phi = 40$ and is of about $92.38\% \pm 0.97$. The corresponding

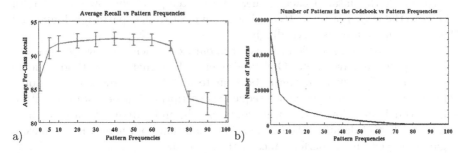

Fig. 4. Plots in (a) and (b) show how the average per-class recall and the number of patterns in the codebook, respectively, change by varying the minimal pattern frequency. Values are averaged over 10 runs, and vertical bars show standard deviations.

codebook size is of about 3086. For $\phi = 70$, the average recall is of about 91.31% and the codebook size is on average 400. For $\phi > 70$, the recall value decreases, however the information embedded in very frequent patterns is still very high considering that, with only 44 codewords (on average) with $\phi = 100$, the method achieves an average recall of about 82.26%.

Experimental results shows the confusion matrix obtained with our technique averaged over 10 runs when $\phi = 40$. Columns of the table represents predicted class labels while rows represent true class labels. As shown in the table, most of the confusion is between the action classes *Start System (SS)* and *Beat both (BB)*, *Had enough (H)* and *Goggles(G)*, *Beat both (BB)* and *Shoot (S)*. We stress here that our technique has been tested directly on the 3D joints coordinates and the only preprocessing of the sequences consists of making them zero mean. Since the method is very general, we believe that the use of more complex features extracted from skeletal data might result in higher value of the average recall.

We compare our method against the work in [7] on equal terms of experimental protocol. In [7], a pyramid of covariance matrices of 3D joints coordinates is used to represent a sequence of skeletons: the root node encode information about the entire sequence; at lower levels, sequences of covariance matrices calculated by a sliding window approach are considered. Action classification is performed by linear SVM. The work only reports the average correct classification rate or accuracy value averaged over 10 runs in different configuration, and achieves the best accuracy value of about 91.7%. Our accuracy value is of about 92.31% at $\phi = 40$, which is slightly superior to the one of [7].

5   Conclusions and Future Work

In this paper we demonstrate the idea of representing sequences of skeletons by means of distributions of frequent patterns. In our framework, frequent sequential patterns are computed by means of a modified version of the apriori algorithm. At each iteration, all frequent patterns that cannot be used for generating longer patterns are stored and used as codewords. This approach yields to a codebook of patterns of different length.

To encode the data, at each frame, we use a temporal window whose length adapts to the length of the pattern. Then, the most similar pattern is found and the histogram is updated accordingly.

One question we have tried to answer in our experiments is how frequent our frequent patterns have to be. Our experiments show that the method benefits from ignoring infrequent patterns both in terms of recall and computational complexity, since a more compact sequence description can be obtained with a smaller codebook. However, considering only the most frequent patterns may result in a lost of details of the action representation and, hence, might have a negative impact on the performance of the method.

We presented preliminary results by validating our method on skeletal data. On the MSRC-12 dataset our method achieves state-of-the-art accuracy values. In future work, we will extensively study the effect of varying some parameters, such as ϵ and ψ, on the performance of the method. The main limitation of our method is that it might not be able to cope with varying execution velocity of the action, which also depends on the subject. Therefore, we also plan to extend our formulation by accounting for the misalignments between patterns and matched temporal window in order to improve the learning of sequential patterns.

Acknowledgement. We are grateful to Mr. Giovanni Caruana for making available his implementation of the classic apriori algorithm, which he implemented in his Master thesis work at University of Palermo.

References

1. Agrawal, R., Imieliński, T., Swami, A.: Mining association rules between sets of items in large databases. In: ACM SIGMOD record, vol. 22. no. 2. ACM (1993)
2. Agrawal, R., Srikant, R.: Mining sequential patterns. In: Proceedings of the Eleventh International Conference on Data Engineering. IEEE (1995)
3. Dalal, N., Triggs, B.: Histograms of oriented gradients for human detection. In: IEEE Conference on Computer Vision and Pattern Recognition (CVPR), vol. 1. IEEE (2005)
4. Demirdjian, D., Wang, S.: Recognition of temporal events using multiscale bags of features. In: IEEE Workshop on Computational Intelligence for Visual Intelligence (CIVI). IEEE (2009)
5. Fei-Fei, L., Perona, P.: A bayesian hierarchical model for learning natural scene categories. In: IEEE Proceedings of Conference on Computer Vision and Pattern Recognition (CVPR), vol. 2. IEEE (2005)
6. Gavrilov, Z., Sclaroff, S., Neidle, C., Dickinson, S.: Detecting reduplication in videos of american sign language. In: Proceedings of Eighth International Conference on Language Resources and Evaluation (LREC), Instanbul, Turkey, May 2012
7. Hussein, M.E., et al.: Human action recognition using a temporal hierarchy of covariance descriptors on 3D joint locations. In: IJCAI, vol. 13 (2013)
8. Karaman, S., et al.: L1-regularized logistic regression stacking and transductive CRF smoothing for action recognition in video. In: ICCV workshop on action recognition with a large number of classes, vol. 13 (2013)
9. Laptev, I., et al.: Learning realistic human actions from movies. In: IEEE Conference on Computer Vision and Pattern Recognition (CVPR). IEEE (2008)
10. Laxman, S., Sastry, P.S.: A survey of temporal data mining. Sadhana **31**(2), 173–198 (2006)
11. Presti, L.L., La Cascia, M.: 3D skeleton-based human action classification: a survey. Pattern Recogn. **53**, 130–147 (2016)
12. Murthy, O.V., Goecke, R.: Ordered trajectories for large scale human action recognition. In: Proceedings of the IEEE International Conference on Computer Vision Workshops (2013)
13. Niebles, J.C., Wang, H., Fei-Fei, L.: Unsupervised learning of human action categories using spatial-temporal words. Int. J. Comput. Vis. **79**(3), 299–318 (2008)
14. Peng, X., et al.: Bag of visual words and fusion methods for action recognition: comprehensive study and good practice. Comput. Vis. Image Underst. **150**, 109–125 (2016)
15. Peng, X., et al.: Exploring motion boundary based sampling and spatial-temporal context descriptors for action recognition. In: British Machine Vision Conference (BMVC) (2013)

16. Schuldt, C., Laptev, I., Caputo, B.: Recognizing human actions: a local SVM approach. In: Proceedings of the 17th International Conference on Pattern Recognition (ICPR), vol. 3. IEEE (2004)
17. Wang, H., et al.: Dense trajectories and motion boundary descriptors for action recognition. Int. J. Comput. Vis. **103**(1), 60–79 (2013)
18. Fothergill, S., et al.: Instructing people for training gestural interactive systems. In: Proceedings of the SIGCHI Conference on Human Factors in Computing Systems. ACM (2012)
19. Shotton, J., et al.: Real-time human pose recognition in parts from single depth images. Commun. ACM **56**(1), 116–124 (2013)
20. Wang, C., Wang, Y., Yuille, A.L.: An approach to pose-based action recognition. In: Proceedings of the IEEE Conference on Computer Vision and Pattern Recognition (2013)
21. Wang, J., et al.: Mining actionlet ensemble for action recognition with depth cameras. In: IEEE Conference on Computer Vision and Pattern Recognition (CVPR). IEEE (2012)
22. Zhao, X., et al.: Online human gesture recognition from motion data streams. In: Proceedings of the 21st ACM international conference on Multimedia. ACM (2013)
23. Zhu, Y., Zhao, X., Fu, Y., Liu, Y.: Sparse coding on local spatial-temporal volumes for human action recognition. In: Kimmel, R., Klette, R., Sugimoto, A. (eds.) ACCV 2010. LNCS, vol. 6493, pp. 660–671. Springer, Heidelberg (2011). doi:10. 1007/978-3-642-19309-5_51

Revisiting Human Action Recognition: Personalization vs. Generalization

Andrea Zunino[1,2(✉)], Jacopo Cavazza[1,2], and Vittorio Murino[1,3]

[1] Pattern Analysis and Computer Vision (PAVIS),
Istituto Italiano di Tecnologia, 16163 Genova, Italy
{andrea.zunino,jacopo.cavazza,vittorio.murino}@iit.it
[2] Electrical, Electronics and Telecommunication Engineering and Naval Architecture
Department (DITEN), Università degli Studi di Genova, 16145 Genova, Italy
[3] Computer Science Department, Università di Verona, 37134 Verona, Italy

Abstract. By thoroughly revisiting the classic human action recognition paradigm, we analyzed different training/testing strategies, discovering that standard (cross-validating) testing strategies are not always the suitable validation procedures to assess an algorithm's performance. As a consequence, we design a novel action recognition architecture, applying a "personalized" strategy to learn how any subject performs any action. We discover that it is advantageous to customize (*i.e.*, personalize) the method to learn the actions carried out by each subject, rather than trying to generalize the actions executions across subjects. Leveraging on that, we propose an action recognition framework consisting of a two-stage classification approach where, given a test action, the subject is first identified before the actual recognition of the action takes place. Despite the basic, off-the-shelf descriptors and standard classifiers adopted, we score a favorable performance with respect to the state-of-the-art as to certify the soundness of our approach.

Keywords: Action recognition · Kinematic analysis · Generalization · Personalization strategy

1 Introduction

The video-based classification of human actions is a very complex task due to contextual clutter and noise, illumination variations, occlusions, and the implicit variability and complexity of actions. All these problems can be mitigated by the three-dimensional (3D) sensor technology, which allows to capture human motion at high spatial/temporal resolution (VICON), with good accuracy and low cost (Kinect). As a consequence, the development and improvement of computational approaches for 3D action recognition sharply rose in the recent year [12].

Within the context of 3D action recognition, this work undertakes a revisiting perspective, probing the principal evaluation strategies applied in the literature on the most common, publicly available, benchmark datasets. Thus, we aim at providing a deep understanding about the challenges that have to be faced when

© Springer International Publishing AG 2017
S. Battiato et al. (Eds.): ICIAP 2017, Part I, LNCS 10484, pp. 469–480, 2017.
https://doi.org/10.1007/978-3-319-68560-1_42

devising classification protocols: such awareness leads us to introduce a new effective, yet simple, approach for action recognition. The experimental testbed we have chosen consists of 3 public datasets, namely MSR-Action3D [11], MSRC-Kinect12 [6] and HDM-05 [13]. Each has own peculiar traits, *e.g.*, the amount and type of considered action classes or the number of skeletal joints. However, a common shared aspect is that a same action is performed by several subjects and a same subject actually performs each action more times. The variability of considered actions aim at reproducing real-world scenarios, while repeating actions and considering multiple actors allow to increase the learning methods in robustness and generalization, respectively. Usually, action recognition methods in the literature do not exploit the information associated to the subject identity, but they typically consider different splits of all action instances (*e.g.*, k-fold cross-validation) in the training/testing phases. Nevertheless, such information is quite relevant, indeed discriminant, for the actual recognition of the actions since *each* human being shows peculiar features which are reflected in the way an action is performed. The former aspects have been rarely investigated and seldom quantified by previous recognition system to date and, to this end, we focus on two main aspects:

- *Inter-subject variability*, which either refers to anthropometric differences of body parts or to incongruous personal styles in accomplishing the scheduled action. In practice, different subjects may perform the same (even very simple) action in different ways.
- *Intra-subject variability*, which represents the random nature of each single action class (*e.g.*, throwing a ball), which can also be dictated by pathological conditions or environmental factors. In other words, this reflects the fact that a subject never performs an action in the same exact way.

Both aspects lead to the fact that a same action could not be performed exactly equal to itself, either it is executed by the same or different human beings. In this line, the additional information of subject identity has empirically demonstrated to be effective in customizing the classification on a specific user for speech [15], handwriting [4], and gesture [10,19] recognition.

Among the few works which studied the variability within/across subjects, for instance, [1] did not register a strong impact of different subjects in daily activities classification, and [5] documented the stability of the performance on an *ad hoc* acquired dataset characterized by biometric homogeneity of the participants. Differently, in [16], the performance of checking the correct execution of gymnastics sharply falls when the subject under testing is excluded from the training phase. A similar trend was registered by [17,20] for computer assisted rehabilitation tasks, as well as by [2] which performed a theoretical dissertation about within-subject and across-subjects noise using wearable motion sensors. Globally, [1,2,5,16,17,20] did not mutually agree in their conclusions and, also, their investigation is actually limited by the use of private datasets explicitly designed for the considered application.

Despite some previous approaches grant in some way the importance of the knowledge of the human subject (especially for rehabilitation purposes, where the goal is directed to a specific subject), no study has been systematically reported to date on commonly used and publicly available datasets for general action/activity recognition. In other words, it is still an open problem to quantify how much those datasets are affected by *inter-* and *intra-subject variability*, and hence to figure out the impact of subjectiveness in action recognition to actually investigate the trade-off between personalization and generalization in the design of robots and automatic systems.

These arguments are investigated through the following main contributions.

(*i*) We analyze the role of the individual subject in human action recognition. By considering MSR-Action3D [11], MSRC-Kinect12 [6] and HDM-05 [13] benchmark datasets, we propose a novel testing strategy, called **Personalization**, where action classification is performed by considering the instances belonging to one specific subject at a time. We register a superior performance of **Personalization** while comparing it against **One-Subject-Out**, which left out the data of one subject as the test set, and **Cross-Validation**, where testing is performed on all subjects (which are also used for training).

(*ii*) In order to explain the latter performance and analyze the role of subjectiveness, we introduce a quantitative statistical analysis. This allows to evaluate the impact of retrieving in testing all the subjects used in the training phase, ultimately assessing the role played by either *inter-* or *intra-subject variability*.

(*iii*) Capitalizing on our improved understanding, we boost action recognition by learning the subject's identity. In particular, we propose a two-stage recognition pipeline (Fig. 1) where the preliminary estimation of the subject is followed by a subject-specific action classification. Overall, our new proposed pipeline shows a strong performance with respect to both *Cross-Validation* and *One-Subject-Out* strategies, also being superior to the state-of-the-art methods [18].

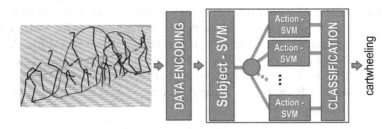

Fig. 1. As opposed to the generic recognition of an action performed by an unspecified human agent, we investigate a counterpart approach in which the action recognition accuracy is boosted by adopting a "personalization" 2-stage method, where the subject is first identified, followed by the actual classification of the action.

The rest of the paper is organized as follows. In Sect. 2, we present the considered datasets and the features adopted, and the evaluation strategies investigated are reported in Sect. 3. Section 4 presents and widely discusses the experimental results, and we illustrate the aforementioned two-stage classification pipeline in Sect. 5. Finally, Sect. 6 draws the conclusions of this study.

2   Datasets and Feature Encoding

Our investigation involves three publicly available MoCap datasets for activity recognition: MSR-Action3D, MSRC-Kinect12 and HDM-05. In all our experiments, we only used the 3D skeleton coordinates while the other data available (*e.g.*, depth maps or RGB videos) were not considered. For the sake of clarity, we briefly introduce each of them.

- **MSR-Action3D** [11] dataset has 20 action classes of mostly sport-related actions (*e.g.*, *jogging* or *tennis-serve*), performed by 10 subjects. $J = 20$ joints are extracted from the Kinect sensor data to model the human pose of the human agents. Each subject performs each action 2 or 3 times. In total, we used 544 sequences [8].
- **MSRC-Kinect12** [6] is a relatively large dataset of 3D skeleton data, recorded by means of a Kinect sensor. The dataset has 5881 sequences, containing 12 action classes performed by 30 different subjects. Each subject accomplishes each class of action 16 times, on average. The available motion files contain the trajectories estimated for $J = 20$ 3D skeleton joints.
- In **HDM-05** [13], the number of skeleton joints is $J = 31$, each action is repeated 5 times on average by each of the 5 subjects involved during the acquisition through a VICON system. We followed the 14-classes experimental protocol of [8,18].

For all the aforementioned datasets, each trial can be formalized as a collection $\mathbf{S}$ of τ different acquisitions $\mathbf{p}(1), \dots, \mathbf{p}(\tau)$. For any $t = 1, \dots, \tau$, we denote with $\mathbf{p}(t)$ the column vector which stacks $\mathbf{p}_1(t), \dots, \mathbf{p}_J(t) \in \mathbb{R}^3$, the three-dimensional x, y, z coordinates of the J skeletal joints. Using this notation, we now briefly introduce the two different representations for MoCap data.

First, we investigated the usage of dynamic time warping (DTW), a classical tool to quantify the similarity across two different time series by means of alignment [7,14]. In order to apply DTW, we evaluated the differences between any two joints collection $\mathbf{S} = [\mathbf{p}(1), \dots, \mathbf{p}(\tau)]$ and $\mathbf{S}' = [\mathbf{p}'(1), \dots, \mathbf{p}'(\tau')]$ through the following distance

$$d(\mathbf{p}(s), \mathbf{p}'(t)) = \frac{1}{J} \sum_{j=1}^{J} \|\mathbf{p}_j(s) - \mathbf{p}'_j(t)\|, \tag{1}$$

where $\| \cdot \|$ is the Euclidean norm, $s = 1, \dots, \tau$ and $t = 1, \dots, \tau'$. The final similarity measure, provided by DTW to compare $\mathbf{S}$ and $\mathbf{S}'$, is $\delta(\mathbf{S}, \mathbf{S}')$ which is the minimum value of (1) computed over all the sequences of timestamps which optimally align $\mathbf{S}$ with $\mathbf{S}'$ (see [14] for more details).

Second, we also estimated the $n \times n$ covariance matrix

$$\mathcal{C} = \frac{1}{\tau - 1} \sum_{t=1}^{\tau} (\mathbf{p}(t) - \overline{\mathbf{p}})(\mathbf{p}(t) - \overline{\mathbf{p}})^{\top}, \tag{2}$$

related to any trial $\mathbf{S}$, where $\overline{\mathbf{p}} = \frac{1}{\tau} \sum_{s=1}^{\tau} \mathbf{p}(s)$ averages all the τ coordinates and we denote $n = 3J$ for convenience. Since $\mathcal{C}$ is positive definite, we thus exploited the theory of the Riemannian manifold Sym_n^+ and projected (2) onto the tangent space to obtain $\widetilde{\mathcal{C}}$ [9]. Then, using the symmetry of $\widetilde{\mathcal{C}}$, we extracted its independent entries, yielding the following $n(n+1)/2$ vector

$$\text{COV} = [\widetilde{\mathcal{C}}_{11}, \ldots, \widetilde{\mathcal{C}}_{1n}, \widetilde{\mathcal{C}}_{21}, \ldots, \widetilde{\mathcal{C}}_{2n}, \ldots, \widetilde{\mathcal{C}}_{nn}]. \tag{3}$$

Note that the usage of covariance is inspired by [18], which set the new state-of-the-art performance for action recognition from MoCap data. Also, our approach is similar to the case $L = 1$ in [8], where a L-layered temporal hierarchy of covariance descriptors is proposed, but differently from us, the projection stage onto the tangent space is not considered.

For both representations, we used the support vector machine[1] (SVM) for classification: when fed with COV, we normalized the data imposing zero mean and unit variance and we then used a linear kernel. Instead, the negative dynamic time warping kernel function [7] produced the training and testing Gram matrices given in input to the SVM.

This will allow us to validate the testing strategies using the same basic classification approach with two different descriptors.

3   Evaluation Strategies

We compare the following three testing modalities.

For testing, *One-Subject-Out* considers any action instance belonging to one subject separately, the system being training on the remaining ones. The final classification results average all the subject-out intermediate scores. This is in line with the protocols of [3,11,18].

In the *Cross-Validation* strategy, we performed a subject-balanced shuffling of data. Precisely, for each subject $\frac{2}{3}$ of samples are used in training and the remaining $\frac{1}{3}$ in testing. To guarantee robustness, the final classification results are averaged over 20 random choices for the aforementioned partitions[2].

For the *Personalization* strategy, each model is trained on the action instances of a single subject at a time. To do this, we fix a subject and, for any action class, $\frac{2}{3}$ of samples are used in training, testing on the remaining $\frac{1}{3}$. Classification accuracies (in testing) are computed on each subject separately, finally fusing the single scores. As previously done, we average the classification results over 20 random splits of all the subject-specific instances.

---

[1] In all experiments, for the SVM cost parameter, we fixed $C = 10$.

[2] For the sake of clarity, please note that a test sample is never seen by the system in training.

4   Experimental Results and Discussion

In this Section, we compare *One-Subject-Out*, *Cross-Validation* and *Personalization*, using the descriptors of Sect. 2: the results related to DTW and COV are reported in Tables 1 and 2, respectively.

Table 1. DTW classification accuracies on the three MoCap datasets. Mean and standard deviation are reported in percentages for each testing strategies (best results are in bold).

Testing strategy	MSR-Action3D	MSRC-Kinect12	HDM-05
One-Subject-Out	28.42 ± 12.76	51.73 ± 17.58	92.39 ± 3.60
Cross-Validation	57.90 ± 3.07	66.93 ± 1.81	96.93 ± 1.72
Personalization	$\mathbf{81.75 \pm 2.71}$	$\mathbf{99.57 \pm 0.16}$	$\mathbf{97.59 \pm 0.85}$

Table 2. COV classification accuracies on the three MoCap datasets. Mean and standard deviation are reported in percentages for each testing strategies (best results are in bold).

Testing strategy	MSR-Action3D	MSRC-Kinect12	HDM-05
One-Subject-Out	70.49 ± 9.02	92.47 ± 6.01	87.78 ± 7.04
Cross-Validation	77.18 ± 3.59	98.57 ± 0.30	96.32 ± 1.97
Personalization	$\mathbf{92.46 \pm 1.09}$	$\mathbf{99.65 \pm 0.07}$	$\mathbf{99.02 \pm 0.98}$

In most case, the COV obtains higher performance with respect to DTW. We can observe a common trend: the action classification performance grows when switching from *One-Subject-Out* to *Cross-Validation*, reaching its peak with *Personalization*. Since common to both DTW and COV, such behavior is actually independent from the data representation.

It is worth noting that the *ranking in the accuracies obtained with the three different modalities is inversely depending on the number of the samples used in the training phase.*

Indeed, in both Tables 1 and 2, the lowest performance is always scored by *One-Subject-Out*, although such modality adopts the larger amount of training data if compared to either *Cross-Validation* or *Personalization*. The reason is that *One-Subject-Out* has to extrapolate more from the data, finding action-specific patterns which are also subject-invariant. Differently, the *Personalization* strategy is required to find action-specific patterns, totally neglecting intra-subject generalization. This helps explaining why *Personalization* obtains the best results for all datasets. Note that the latter fact occurs despite the *Personalization* strategy exploits the least number of samples within *One-Subject-Out* and *Cross-Validation*. In particular, by considering MSR-Action3D dataset (see Sect. 2), very few trials (and sometimes only one) are available per each action class and subject. In spite of that, *Personalization* scores 92.46% and 81.75% with COV and DTW respectively, and outperforms all the other two strategies. Indeed, MSRC-Kinect12 and HDM-05 are almost saturated by *Personalization*: *e.g.*, 99.57 ± 0.16 of DTW and 99.02 ± 0.98 of COV respectively.

Cross-Validation deserves an own discussion. Indeed, such strategy can be seen as a compromise between the two, since each subject is seen in both training and testing (as in *Personalization*) but is required to generalize across agents (as in *One-Subject-Out*). In terms of registered performance, *Cross-Validation* scores intermediately with respect to the other two strategies. Precisely, with respect to *One-Subject-Out*, *Cross-Validation* improves by margin: therefore, exploiting the same subject in both training and testing appears to be effective.

However, all *Cross-Validation* accuracies are always lower than the *Personalization* one, although the gap between them is sometimes very small (*e.g.*, *Cross-Validation* scores about 1% less with respect to *Personalization* on MSRC-Kinect12 dataset, see Table 2). Actually, this can be interpreted in the following manner: *adding many training samples belonging to different subjects does not always lead to an improvement, frequently confusing the (SVM) classifier.*

Evidently, the quality of the data is superior to quantity for the sake of performance. In the next Section, we will carry out a statistical analysis to characterize the concept of "quality" in terms of *inter-* and *intra-subject variability*.

4.1 Quantitative Statistical Analysis

Let us define the following statistics.

① $p_{subject}$ For all testing action instances $\bar{a}$, which are correctly classified in *Cross-Validation*, consider the training action instance $\bar{\bar{a}}$ which is closest to $\bar{a}$. We call $p_{subject}$ the (average) probability that both $\bar{a}$ and $\bar{\bar{a}}$ belongs to the same subject.

Clearly, $p_{subject}$ measures how often a good prediction is obtained by exploiting the information exactly coming from the same subject. Hence, high/low $p_{subject}$ values check if testing on the same subjects used for training gives a pros/cons for the classification, respectively.

② p_{inter} For each action class c, and for any instance a_c of that class, consider the instance a_c (still belonging to the same class) which is closest to a_c in the features space. While averaging on c, the frequency of that a_c and a_c belonging to the *same subject* is denoted by p_{inter}.
We can notice that $p_{inter} \approx 0$ when *inter-subject variability* is negligible.

③ p_{intra} For any subject s and for any instance a_s, consider a_s which is the closest to a_s within the ones in the dataset which belongs to the s-th subject. p_{intra} counts how frequently a_s and a_s belong to a *different action class*.
From the definition, if $p_{intra} = 0$, all the trials of a given action and a given subject are almost identical and *intra-subject variability* is totally absent.

④ Δ For each action class c, compute d_c as the maximal distance between two c-labelled elements in the dataset. Similarly, $d_{c,s}$ is the maximal distance of two c-labelled instances from the same subject s. Define $\Delta_{c,s} = \frac{|d_{c,s} - d_c|}{d_c}$.

We have $0 \leq \Delta_{c,s} \leq 1$, where the extremal case $\Delta_{c,s} = 0$ correspond to a null *inter-subject variability*: since $d_{c,s} = d_c$, within the trials of class c, subjects are maximally shuffled (Fig. 2, left). Also, $\Delta_{c,s} = 1$ implies $d_{c,s} = 0$ which minimizes the *intra-subject variability* since all instances of class c from subject s collapse to a point (Fig. 2, right). We define Δ as the average of all $\Delta_{c,s}$ over c and s. By construction, Δ quantifies the relative importance between *inter-* and *intra-subject variability*, being the latter or the former preponderant on the other in case of low or high Δ values, respectively.

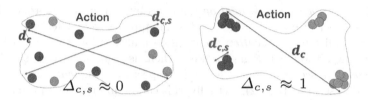

Fig. 2. In the feature space, we surround the region referring to a single action. Within, each point represents a trial and different colors relate to different subjects. Left: When $\Delta_{c,s} \approx 0$, *inter-subject variability* is minimized since, in general, trials from different subjects occupy nearby positions. Right: The case $\Delta_{c,s} \approx 1$ minimizes the *intra-subject variability* because all the instances of the same subject are compactly clustered.

In the definition of p_{subject}, p_{inter}, p_{intra} and Δ, a notion of "closeness" is involved. The latter depends on the exploited data representation. For COV, the distance is the Euclidean one, since induced by a linear kernel. Instead, for DTW, we use the dynamic time warping distance δ, as introduced in Sect. 2.

Discussion. Table 3 shows the values of our statistics in all the considered datasets. We only report the values related to COV since no remarkable differences are registered when moving to DTW[3].

Table 3. Quantitative evaluation of *inter* and *intra-subject variability*.

Dataset	p_{subject}	p_{inter}	p_{intra}	Δ
MSR-Action3D	0.78	0.86	0.19	0.71
MSRC-Kinect12	0.97	0.97	0.01	0.90
HDM-05	0.89	0.95	0.01	0.74

In all cases, p_{subject} is extremely high (*e.g.*, 0.89 for HDM-05). Therefore, in *Cross-Validation* testing strategy, the performance is actually boosted by leveraging on how each subject perform a given action. Therefore, the scored p_{subject} values attest that the role of the subject is crucial in 3D action recognition.

---

[3] For instance, the value of p_{subject} for MSR-Actio3D is 0.77, for MSRC-Kinect12 is 0.97 and for HDM-05 is 0.85.

Inter-subject variability is a problem ($p_{inter} > .85$). Thus, the same action is likely to be performed very differently by different subjects. This explains the difficulty of *One-Subject-Out* strategy.

On MSR-Action3D p_{intra} is low, being actually almost zero in the other cases. Especially in MSRC-Kinect12 and HDM-05, each subject identically repeats each action almost in the same way. As a consequence, *intra-subject variability* is not remarkably affecting the classification. Hence, even knowing one only action instance per subject can actually boost the recognition. This explains the favorable *Personalization* performance, despite the small data regime embraced.

Inter-subject variability is the actual burden to tackle, being totally overwhelming with respect to intra-subject one. The high values for Δ (*e.g.*, 0.9 for MSRC-Kinect12) certify that the gap to fill across subject is actually remarkable, where the challenges related benchmark datasets analyzed can be intuitively imagined as in Fig. 2, right.

Globally, if we can automatically recognize the subject's identity of a training/testing instance, we can cast action recognition as an easier subproblem: we do not have to fill huge inter-subjects gaps, but just learning how to discriminate different actions of the same subjects (which are likely to be more separable). As we will prove in the next Section, such divide et impera strategy is very effective.

5   Divide et Impera. Two-Stage Recognition Pipeline

In comparison to *Cross-Validation* and *One-Subject-Out*, the *Personalization* strategy always achieves the best scores (Tables 1 and 2). As explained, this happens because *inter-subject variability* is highly problematic, being *intra-subject variability* small as in MSR-Action3D and eventually absent in the other cases. However, *Personalization* leverage on the unfavorable assumption: it requires the subject's identity to be known in order to classify the action.

Actually, in this Section we tackle this issue, obtaining an equivalently effective action recognition system, which is now able to operate in real-world conditions. The key is *learning the subject's identity*.

Inspired by our findings (Sect. 4.1), we posit that we can proficiently apply features designed for action representation in order to recognize the subject's identity. This originate a divide et impera paradigm where, first the subject's identity is recognized and then action recognition is performed using a subject-specific classifier, trained on the instance of a single subject only. Despite the reduced amount of data, the task should be easier to train due to the better separability of action classes when the subject's identity is fixed. Precisely, we propose the following two-stage pipeline (Fig. 1).

Stage 1. A unique SVM model (*subject-SVM*) recognizes subject's identity.
Stage 2. Within many subject-specific action classifiers (called *action-SVMs*), the final action recognition step is performed by the one corresponding to the subject identified in **Stage 1**.

For training *subject-SVM* and *action-SVMs*, we performed a $\frac{2}{3}/\frac{1}{3}$ random splitting for training and testing data related to any subject and any action. Obviously, for each of the *action-SVMs*, we used only the training and testing examples belonging to one subject at a time. During testing, the *subject-SVM* scores is used to select one of the *action-SVMs* (actually the one corresponding to the recognized subject): this is the model exploited for action classification.

To validate our proposed pipeline, both *subject-SVM* and *action-SVMs* are fed with COV features, more powerful than DTW. The results in Tables 4, 5 provide the mean and standard deviation of the accuracies scored in the two steps separately, over 20 different random partitions of the data.

Discussion. Since COV is designed for action recognition, it is suboptimal for subjects' identification. In fact, despite the classification performance we registered is still reliable (Table 4), when a subject is misclassified, the action classifier corresponding to another subject is used and performance can deteriorate.

Nevertheless, we only registered a 2% the drop with respect to *Personalization* strategy, which can be considered as our two-stage pipeline with perfect subject recognition in the first stage. Such performance is remarkable since, after all, *Personalization* requires the subjects' identity to be known, whereas we are effectively able to automatically learn it[4].

Although a comparison of our simple approach with more sophisticated approaches [3,8,18] is challenging, we score a favorable performance with respect to the state-of-the-art. Despite the simplicity of our pipeline, we only pay 6% on MSR-Action3D (96.9%, [18]). This is coherent with the fact that *intra-subject variability* is not totally absent in such a case ($p_{intra} \approx 0.2$ in Table 3), therefore mining the underlying assumption of our approach. Differently, we are scoring

Table 4. Two-stage recognition pipeline - subject identification accuracies.

	MSR-Action3D	MSRC-Kinect12	HDM-05
subject-SVM	90.74 ± 2.41	85.18 ± 0.55	85.67 ± 3.18

Table 5. Two-stage recognition pipeline - action classification accuracies compared to SoA.

	MSR-Action3D	MSRC-Kinect12	HDM-05
action-SVMs	90.46 ± 1.17	**97.14 ± 0.39**	97.03 ± 1.36
SoA	**96.9** [18]	95.0 [3]	**98.1** [3]

---

[4] To have a better insight of the importance of the knowledge of the subject who is performing the action, we have conducted an experiment on MSRC-Kinect12 using COV features where we assume that the correct *action-SVM* is not available. Using the best *action-SVMs* belonging to all other subjects the performance drops from 97.14% to 80.68%.

almost on par with respect to [3] (98.1%) on HDM-05, also improving the state-of-the-art on MSRC-Kinect12 by about 2% (95.0%, [3]).

6 Conclusions

In this paper, we investigated the generalization capability of automatic activity recognition systems analyzing the proposed *Personalization* strategy in comparison with standard *Cross-Validation* and *One-Subject-Out* approaches. To this aim, we exploit classical representations (DTW and COV), with basic a classifier (linear SVM) on the MSR-Action3D, MSRC-Kinect12 and HDM-05 benchmark datasets.

From the experiments, *One-Subject-Out* resulted the more challenging strategy, although being able to ensure a better generalization. Differently, despite *Cross-Validation* was actually boosted from the usage of the same subject in both training and testing, the additional information relative to the other subjects could mislead. The *Personalization* strategy, gave the highest performance, despite the lowest number of instances used in training.

In addition, we also provided several quantitative statistics to measure *inter* and *intra-class variability* on the considered datasets: as a result, the latter is almost marginal, while the former is the actual burden that has to be tackled when devising new techniques.

Finally, we proposed a two-step classification pipeline by first identifying the subject and, second, by using subject-specific classifiers for action recognition. This paradigm can be applied to general surveillance tasks, by monitoring the activities of unknown subjects by means of the model corresponding to the most similar training subject. Additionally, this opens to the design of custom human-robotic systems and novel authentication procedures.

References

1. Bao, L., Intille, S.S.: Activity recognition from user-annotated acceleration data. In: Ferscha, A., Mattern, F. (eds.) Pervasive 2004. LNCS, vol. 3001, pp. 1–17. Springer, Heidelberg (2004). doi:10.1007/978-3-540-24646-6_1
2. Barshan, B., Yurtman, A.: Investigating inter-subject and inter-activity variations in activity recognition using wearable motion sensors. Comput. J. **59**, 1345–1362 (2015)
3. Cavazza, J., Zunino, A., San Biagio, M., Murino, V.: Kernelized covariance for action recognition. In: ICPR (2016)
4. Connell, S., Jain, A.: Writer adaptation for online handwriting recognition. PAMI **24**(3), 329–346 (2002)
5. Dalton, A., ÓLaighin, G.: Comparing supervised learning techniques on the task of physical activity recognition. IEEE J. Biomed. Health Inf. **17**(1), 46–52 (2013)
6. Fothergill, S., Mentis, H.M., Kohli, P., Nowozin, S.: Instructing people for training gestural interactive systems. In: ACM Conference on Computer-Human Interaction (2012)

7. Gudmundsson, S., Runarsson, T.P., Sigurdsson, S.: Support vector machines and dynamic time warping for time series. In: IJCNN (2008)
8. Hussein, M., Torki, M., Gowayyed, M., El-Saban, M.: Human action recognition using a temporal hierarchy of COV descriptors on 3d joint locations. In: IJCAI (2013)
9. Jayasumana, S., Hartley, R., Salzmann, M., Li, H., Harandi, M.: Kernel methods on the riemannian manifold of symmetric positive definite matrices. In: CVPR (2013)
10. Joshi, A., Ghosh, S., Betke, M., Pfister, H.: Hierarchical bayesian neural networks for personalized classification. In: NIPS workshop (2016)
11. Li, W., Zhang, Z., Liu, Z.: Action recognition based on a bag of 3d points. In: CVPR workshop (2010)
12. Lo Presti, L., La Cascia, M.: 3D skeleton-based human action classification: a survey. Pattern Recogn. 53, 130–147 (2016)
13. Müller, M., Röder, T., Clausen, M., Eberhardt, B., Krüger, B., Weber, A.: Documentation mocap database HDM-05. Technical report CG-07-2, June 2007
14. Albrecht, T.: Dynamic time warping. In: Information Retrieval for Music and Motion, pp. 69–84. Springer (2009)
15. Shinoda, K., Lee, C.H.: A structural bayes approach to speaker adaptation. IEEE Trans. Speech Audio Process. 9, 276–287 (2001)
16. Taylor, P.E., Almeida, G.J.M., Kanade, T., Hodgins, J.K.: Classifying human motion quality for knee osteoarthritis using accelerometers. In: EMBC (2010)
17. Tormene, P., Giorgino, T., Quaglini, S., Steanelli, M.: Matching incomplete time series with dynamic time warping: an algorithm and an application to post-stroke rehabilitation. Artif. Intell. Med. 45(1), 11–34 (2009)
18. Wang, L., Zhang, J., Zhou, L., Tang, C., Li, W.: Beyond covariance: feature representation with nonlinear kernel matrices. In: ICCV (2015)
19. Yao, A., Van Gool, L., Kohli, P.: Gesture recognition portfolios for personalization. In: CVPR (2014)
20. Yurtman, A., Barshan, B.: Automated evaluation of physical therapy exercises using multi-template dynamic time warping on wearable sensor signals. Comput. Methods Programs Biomed. 117(2), 189–207 (2014)

Multiview Geometry and 3D Computer Vision

Efficient Confidence Measures for Embedded Stereo

Matteo Poggi$^{(\boxtimes)}$, Fabio Tosi, and Stefano Mattoccia

Department of Computer Science and Engineering (DISI), University of Bologna,
Viale del Risorgimento 2, Bologna, Italy
{matteo.poggi8,fabio.tosi5,stefano.mattoccia}@unibo.it

Abstract. The advent of embedded stereo cameras based on low-power and compact devices such as FPGAs (Field Programmable Gate Arrays) has enabled to effectively address several computer vision problems. However, being the depth data generated by stereo algorithms affected by errors, reliable strategies to detect wrong disparity assignments by means of confidence measures are desirable. Recent works proved that confidence measures are also a powerful cue to improve the overall accuracy of stereo. Most approaches aimed at predicting match reliability rely on cost volume analysis, an information seldom available as output of most embedded depth sensors. Therefore, in this paper we analyze and evaluate strategies compatible with the constraints of embedded stereo cameras. In particular, we focus our attention on methods to infer match reliability inside depth sensors based on highly constrained computing architectures such as FPGAs. We quantitatively assess, on Middlebury 2014 and KITTI 2015 datasets, the impact of different design strategies for 16 confidence measures from the literature, suited for implementation on such embedded systems. Our evaluation shows that, compared to the confidence measures typically deployed in this context and based on storing intermediate results, other approaches yield much more accurate predictions with negligible computing requirements and memory footprint. This enables for their implementation even on highly constrained architectures.

1 Introduction

The recent availability of embedded depth sensors paved the way to a variety of computer vision applications for autonomous driving, robotics, 3D reconstruction and so on. In these application depth is crucial and several approaches have been proposed to tackle this problem following two main strategies. On one hand *Active* sensors infer depth by perturbing the sensed scene by means of structured light, laser projection and so on. On the other hand, *passive* depth sensors infer depth not altering at all the sensed environment. Although sensors based on active technologies are quite effective they have some limitations. In particular, some of them (e.g., Kinect) are not suited for outdoor environments during daytime while others (e.g., LIDAR) provide only sparse depth maps and are quite expensive, cumbersome and containing moving mechanical parts.

© Springer International Publishing AG 2017
S. Battiato et al. (Eds.): ICIAP 2017, Part I, LNCS 10484, pp. 483–494, 2017.
https://doi.org/10.1007/978-3-319-68560-1_43

Stereo vision is the most popular passive technique to infer dense depth data from two or more images. Many algorithms have been proposed to solve the stereo correspondence problem, some of them particularly suited for hardware implementation, thus enabling the design of compact, low-powered and real-time depth sensors [2,4,7,10,22,24,26]. Despite the vast literature in this field, challenging conditions found in most practical applications represent a major challenge for stereo algorithms. Popular benchmarks Middlebury 2014 [21] and KITTI 2015 [11] clearly highlighted this fact. Therefore, regardless of the stereo algorithm deployed, it is essential to detect its failures to filter-out wrong unreliable points that might lead to a wrong interpretation of the sensed scene. To this aim, confidence measures have become a popular topic on recent works concerning stereo. Some recent confidence measures combine multiple features within random forest frameworks to obtain more reliable confidence scores while an even more recent trend aims to infer confidence prediction leveraging on Convolutional Neural Networks (CNN) [19,23]. Despite their effectiveness, the latter strategies are often not compatible with the computing resources available inside the depth sensor, typically a low cost FPGA or a System-On-Chip (SoC) based on ARM CPU cores and an FPGA (e.g., Xilinx Zynq). Moreover, the features required by most of these machine-learning frameworks are not available as output of the embedded stereo cameras being in most cases computed from the cost volume (often referred to as disparity space image (DSI) [20]).

Therefore, in this paper we consider a subset of confidence measures compatible with embedded devices evaluating their effectiveness, on two popular challenging datasets and two algorithms typically deployed for real-time stereo for embedded systems, focusing our attention on issues related to their FPGA implementation. Our study highlights that some of the considered confidence measures, appropriately modified to fit with typical hardware constraints found in the target architectures, clearly outperform those currently deployed in most embedded stereo cameras.

2   Related Work

Stereo represents a popular and effective solution for depth estimation. It exploits epipolar geometry to find corresponding pixels on two or multiple synchronized frames, thus enabling to infer distance of the observed points by means of triangulation. According to the taxonomy by Scharstein and Szeliski [20], algorithms can be grouped into local and global methods. Algorithms belonging to the former group are usually very fast algorithms but typically less accurate than global ones. The Semi-Global Matching (SGM) [6] algorithm represents a very good trade-off between speed and accuracy and for this reason one of the most popular approach to infer depth even with embedded devices. The core of SGM algorithms consists of multiple and independent *scanline optimization* (SO) [20] along different directions. Each SO is fast, but affected by *streaking* artifacts near discontinuities. However, by combining multiple SOs as done by SGM significantly softens this issue. Moreover its computational structure allows for different optimization strategies and simplifications that enabled to implement it on almost any computing architecture (e.g., CPUs, GPUs, SoC, FPGAs). In particular, low power and massively parallel devices such as FPGAs represents a very good design choice for depth sensors with optimal performance/Watt. Examples of stereo pipeline based on SGM mapped on FPGAs are [2,4,7,10,22,24,26]. Some of them deploy hardware-friendly implementations, based on census transform [28] and 4 or 5 scanlines computed in a single image scan from top-left to bottom-right. On FPGAs a smart design is crucial in order to achieve accurate real-time results without violating the limited logic resource available.

Despite the good accuracy of SGM and state-of-the-art algorithms [29], stereo is still an open problem, as witnessed by recent, challenging datasets [11,21]. Thus, detecting failures of the stereo algorithm is a desirable property to achieve a more meaningful understanding of the sensed environments.

Several confidence measures have been proposed to tackle match reliability. In [8] the authors highlighted how different cues available inside the pipeline of general-purpose stereo algorithms implemented in software lead to different degrees of effectiveness on well-known ill-conditions of stereo such as occlusions, lack of texture and so on. Most recent proposals in this field proved that machine-learning can be effectively deployed to infer more accurate confidence measures, capable to better detect disparity errors. The very first work [5] trained a random forest classifier on multiple measures or features extracted from the DSI. More recent and effective proposals based on this strategy were proposed in [15,25], while in [18] was shown that a confidence measure could be effectively inferred by processing cues computed only from the disparity map. In [14] was proposed a data generation process based on multiple view points and contradictions, to select reliable labels to train confidence measures based on random forests. Latest works on confidence measures rely on deep-networks: [19,23] address confidence estimation by means of a CNN processing patches, respectively, from the left disparity map and from both left and right disparity maps.

Finally, we conclude this section observing that confidence measures have been deployed to detect occlusions [6,13] and sensors fusion [9,12]. Moreover, they were also plugged inside stereo pipeline to improve the overall accuracy by acting on the initial DSI [15,16,18,25].

3   Hardware Strategies for Confidence Implementation

When dealing with conventional CPU based systems confidence measures are generally implemented in C, C++ and to maintain the whole dynamic range single or double floating point data types are deployed. However, floating point arithmetic is sometimes not available in embedded CPU and generally unsuited to FPGAs. In particular, transcendental functions and divisions represent major issues when dealing with such devices. To overcome these limitations, fixed point arithmetic is usually deployed [1]. Fixed point represents an efficient and hardware-friendly way to express and manipulate fractional numbers with a fixed number of bits [1]. Indeed, fixed-point math can be represented with an integer number split into two distinct parts: the integer content (I), and the fractional content (F). Through the simple use of integer operations, the math can be efficiently performed with little loss of accuracy taking care to use a sufficient number of bits. The steps required to convert a floating point value to the corresponding fixed representation with F bits - the higher, the better in terms of accuracy - are the following:

1. Multiply the original value by 2^F
2. Round the result to the closest integer value
3. Assign this value into the fixed-point representation

Fixed point encoding greatly simplifies arithmetic operations with non-integer values, but integer divisions can be demanding - in particular on FPGAs - except when dealing with divisors which are powers of 2. In fact, in this case division requires almost negligibly hardware resources being carried out by means of a simple right shift. Thus, a simplified method to avoid integer divisions consists in rounding the dividing value to the closest power of 2, then shifting right according to its $\log_2$. This strategy will be referred to as *pow*.

Although fixed point increases the overall efficiency, some confidence measures rely on transcendental functions (in particular, exponentials and logarithms) which represent an a further major issue even when dealing with CPU based systems. An effective strategy to deal with such functions consists in deploying Look-Up Tables (LUTs) to store pre-computed results encoded with fixed point arithmetic. That is, given a function $\mathcal{F}(x)$, with x assuming n possible values, a LUT of size n can store all the possible outcome of such function. Of course, this approach is feasible only when the size of the LUT (proportional to n) is compatible with the memory available in the device.

4   Confidence Measures Suited for Hardware Implementation

In this section we describe the pool of confidence measures from the literature suited for implementation on target embedded devices. Figure 1 shows the matching cost curve for a pixel of the reference image. Given a pixel $\mathbf{p}(x, y)$, we will refer to its minimum cost as c_1, the second minimum as c_2 and the second local minimum as c_{2m}. The matching cost for any disparity hypothesis d will be referred to as c_d while the disparity corresponding to c_1 as d_1, the one corresponding to c_2 as d_2 and so on. If not specified otherwise, costs and disparities are referred to the reference left image (L) of the stereo pair. When dealing with right image (R), we introduce the ^{R} symbol on costs (e.g., c_1^R) and disparities. We denote as $\mathbf{p}'(x', y')$ the homologous point of $\mathbf{p}$ according to d_1 (i.e., $x' = x - d_1$, $y' = y$). It is worth to note that, assuming the right image as reference, the matching costs can be easily obtained by scanning in diagonal the cost volume computed with reference the left image without any further new computation. Nevertheless, adopting this strategy would require an additional buffering of $\frac{d_{max} \cdot (d_{max}+1)}{2}$ matching costs with d_{max} the disparity range deployed by the stereo algorithm.

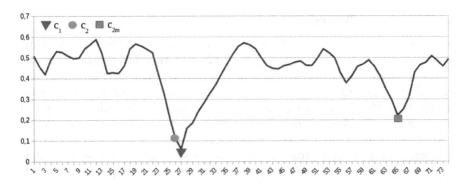

Fig. 1. Example of cost curve, showing the matching cost c_1, the second minimum c_2 and the second local minimum c_{2m}. On x axis the disparity range, on y magnitude of the costs.

We distinguish the considered pool of confidence measures in two, mutually exclusive, categories:

- *Hardware friendly*: confidence measures whose standard implementation is fully compliant with embedded systems.
- *Hardware challenging*: confidence measures involving transcendental functions and/or floating point divisions not well suited for embedded systems in their conventional formulation.

4.1   Hardware Friendly

This category groups confidence measures involving simple math operations that do not represent issues when dealing with implementation on embedded systems. The *matching score measure* (MSM) [8] negates the minimum cost c_1 assuming it related to the reliability of a disparity assignment. *Maximum margin* (MM) estimates match uncertainty by computing the difference between c_{2m} and c_1 while its variant *maximum margin naive* (MMN) [8] replaces c_{2m} with c_2. Given two disparity maps computed by a stereo algorithm assuming as reference L and R, the *left-right consistency* (LRC) [8] sets as confidence the negation of the absolute difference between the disparity of a point in L and its homologous point in R. This method represents one of the most widely adopted strategy by most algorithms even for those implemented on embedded devices. Another popular and more efficient strategy based on a single matching phase is the *uniqueness constraint* (UC) [3]: it assumes as poorly confident those pixels colliding on the same point of the target image (R) with the exception of the one having the lowest c_1. *Curvature* (CUR) [8] and *local curve* (LC) [27] analyze the behavior of the matching costs in proximity of the minimum c_1 and its two neighbors at $(d_1 - 1)$ and $(d_1 + 1)$ according to two similar strategies. Finally, *number of inflections* (NOI) [8] simply counts the number of local minima in the cost curve assuming that the lower, the more confident is the disparity assignment.

4.2   Hardware Challenging

Confidence measures belonging to this category can not be directly implemented in embedded systems following their original formulation. We consider *peak ratio* (PKR) [8] which computes the ratio between c_{2m} and c_1 and its variant *peak ratio naive* (PKRN) [8] which replaces c_{2m} with the second minimum c_2. According to the literature, these measures are quite effective but seldom deployed in embedded stereo cameras. Another popular measure is *winner margin measure* (WMN) [8] which normalizes the difference between c_{2m} and c_1 by the sum of all costs. Its variant *winner margin measure naive* (WMNN) [8] follows the same strategy replacing c_{2m} with c_2. The *left-right difference* measure (LRD) [8] computes the difference between c_2 and c_1 divided by the absolute difference between c_1 and the minimum cost of the homologous point in R (c_1^R). For these confidence measures the major implementation issue on embedded systems is represented by the division. For the remaining confidence measures the main problem is represented by transcendental functions: exponentials and logarithms. *Maximum likelihood measure* (MLM) [8] and *attainable maximum likelihood* (AML) [8] infer from the cost curve a *probability density function* (pdf) related to an ideal c_1, respectively, equal to zero for MLM and to c_1 for AML. A more recent and less computational demanding approach *perturbation* (PER) [5], encodes the deviation of the cost curve from a Gaussian function ant its implementation requires a division by a constant value suited for a LUT-based strategy. Finally, we also mention two very effective confidence measures based on distinctiveness, namely *distinctive similarity measure* (DSM) and *self-aware matching measure* (SAMM) and one *negative entropy measure* (NEM) [8] that infers the degree of uncertainty of each disparity assignment from the negative entropy of c_1. However, they require additional cues (e.g., self-matching costs on both reference and target images for SAMM) not well suited to embedded systems and thus not included in our evaluation.

5   Experimental Results

In this section we evaluate the 16 confidence measures previously reviewed and implemented following the design strategies outlined so far. We test their effectiveness with the output of two popular stereo algorithms well-suited for implementation on embedded systems:

- AD-CENSUS: aggregates matching costs according to the Hamming distance computed on 5×5 patches with census transform [28]. A further aggregation step is performed by a 5×5 box-filter. To reduce the amount of bits required by the single matching cost, we normalized aggregated costs by the dimension of the box-filter (to be more hardware-friendly, by 16), with negligible reduction of accuracy according to [17].
- SGM [6]: four scanline implementation using as data term the same AD-CENSUS aggregated costs and for parameters P1 and P2, respectively, 11 and 110. The four directions are those processed by scanning the image from top-left to bottom-right as suggested in [2,10,17].

We encode matching costs with, respectively, 6 and 8 bit integer values, being this amount enough to encode the entire ranges. Regarding parameters of the confidence measures: for LC, we set the normalization factor γ to 1 to avoid division, while for PER, MLM and AML we set s_{PER} to 1.2 and σ_{aml}, σ_{mlm} to 2 before initializing the LUTs. The other 12 confidence measures do not have parameters.

For CUR, LRC, LC, MM, MMN, MSM, NOI and UC we provide experimental results with the conventional implementation since their mapping on embedded devices is totally equivalent. Moreover, regarding PER, we do not report results concerned with division by the closest power of two being the divisor a constant value and thus such operation can be addressed with a LUT. Finally, it is worth observing that most embedded stereo vision systems rely on LRC [2,7] and UC [2,10] for confidence estimation.

In Sect. 5.1 we describe the evaluation protocol and in Sect. 5.2 we report experimental results on Middlebury 2014 (at quarter resolution) and KITTI 2015 datasets for AD-CENSUS and SGM algorithms.

5.1   Evaluation Protocol

The standard procedure to evaluate the effectiveness of a confidence measure is the ROC curve analysis, proposed by Hu and Mordohai [8] and adopted by all recent works [5,15,18,19,23,25] in this field. By extracting subsets of pixels from the disparity map, according to descending order of confidence, a ROC curve is depicted by computing the error rate, starting from a small subset of points (i.e., 5% most confident) and then increasing the pool of pixels iteratively, up to include all pixels. This leads to a non-monotonic ROC curve, whose area (AUC) is an indicator of the effectiveness of the confidence measure. Given a disparity map with $\varepsilon\%$ of pixels being erroneous, an optimal confidence measure should draw a curve which is zero until $\varepsilon\%$ pixels have been sampled. The area of this curve represents the optimal AUC achievable by a confidence measure and can be obtained, according to [8], as:

$$AUC_{opt} = \int_{1-\varepsilon}^{\varepsilon} \frac{p - (1 - \varepsilon)}{p} dp = \varepsilon + (1 - \varepsilon)\ln(1 - \varepsilon) \tag{1}$$

As reported on Middlebury 2014 and KITTI 2015 benchmarks, ε is obtained by fixing a threshold value on disparity error of, respectively 1 and 3 for the two datasets following the guidelines. Confidence measures achieving lower AUC values (closer to optimal) better identify wrong disparity assignments.

5.2   Experimental Evaluation on Middlebury 2014 and KITTI 2015

In this section we report results on Middlebury 2014 and KITTI 2015 datasets in terms of average AUC values achieved by confidence measures implemented in software. For hardware challenging measures of Sect. 4.2 we also report multiple AUC obtained with increasing number of bits dedicated to fixed point operations (i.e., from 6 to 16 for AD-CENSUS and from 8 to 16 for SGM, so as to handle the whole cost range). Moreover, for such measures, we also report the results obtained by rounding to the closest power of 2 and, then, shifting right (referred to as *pow* in the charts).

Table 1 shows for Middlebury 2014 that LRC and UC, confidence measures typically deployed in embedded stereo cameras, are less effective than MM, LRD, PKR, PKRN, WMN, WMNN with AD-CENSUS and MM, MSM, AML, MLM, PER, PKR, WMN, WMN with SGM. We can notice that LRC provides poor confidence estimation with SGM but achieves better results with AD-CENSUS while UC has average performance with both algorithms. Considering the more effective confidence measures in the table, we can notice that PKR and WMN, as well as their naive formulations, performs pretty well with both algorithms clearly providing much more accurate confidence estimation compared to LRC and UC. Moreover, we can notice that PER achieves the best performance with SGM but it does not perform as well with AD-CENSUS, yielding slightly better confidence predictions with respect to UC. Specularly, LRD provides very reliable predictions with AD-CENSUS but poor results with SGM. Finally, we point our that top-performing confidence measures always belong to the hardware challenging category.

Therefore, in Fig. 2 we report the performance of hardware challenging confidence measures, on Middlebury 2014 with AD-CENSUS and SGM, with multiple simplification settings. Observing the charts, PER is independent of the adopted strategy, being based on a LUT. Moreover, excluding PER, we can notice that the best performing ones (PKR, PKRN, WMN and WMNN at the right side of the figure) are those less affected by the number of bits deployed for fixed-point computations, thus resulting in reduced computational resources. In particular, we can observe that with only 8 bits, PKR and WMN achieve with both

Table 1. Experimental results, in terms of AUC, on Middlebury 2014 dataset with AD-CENSUS (a) and SGM (b) algorithms for the 16 confidence measures using a conventional software implementation. In red, top-performing measure. We also report the absolute ranking.

measure	standard	measure	standard	measure	standard	measure	standard
Opt.	0.08891	Opt.	0.08891	Opt.	0.04367	Opt.	0.04367
CUR	0.24377 (14)	AML	0.21173 (11)	CUR	0.11602 (11)	AML	0.08843 (3)
LRC	0.19933 (7)	LRD	0.17004 (3)	LRC	0.16853 (15)	LRD	0.11725 (13)
LC	0.24377 (15)	MLM	0.22413 (12)	LC	0.11602 (12)	MLM	0.09567 (6)
MM	0.17765 (6)	PER	0.20687 (9)	MM	0.09371 (5)	PER	0.08766 (1)
MMN	0.19933 (8)	PKR	0.16250 (1)	MMN	0.12920 (14)	PKR	0.08813 (2)
MSM	0.23182 (13)	PKRN	0.17185 (5)	MSM	0.10181 (7)	PKRN	0.10527 (10)
NOI	0.39053 (16)	WMN	0.16503 (2)	NOI	0.32028 (16)	WMN	0.08898 (4)
UC	0.20974 (10)	WMNN	0.17169 (4)	UC	0.10347 (9)	WMNN	0.10232 (8)

(a)   (b)

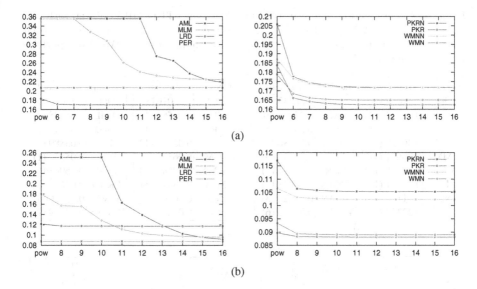

Fig. 2. Average AUC values on the Middlebury 2014 dataset for hardware challenging measures, varying the implementation settings (i.e., *pow* and number of bits of fixed-point arithmetic). (a) AD-CENSUS, (b) SGM algorithm.

algorithms results almost comparable to their conventional software implementation. A similar behavior can be observed, with slightly worse performance, for their naive formulation PKRN and WMNN and for LRD that, excluding PER, is the approach less dependent of the number of bits. On the other hand, AML e MLM with both algorithms are significantly affected by the number of bit deployed for their implementation achieving results comparable to their traditional software formulation, respectively, only with 13 and 16 bits. Finally, excluding PER, we can observe that dividing by a power of 2 always provides poor results with respect to other simplifications. However, we highlight that even with this very efficient implementation strategy, PKR, WMN outperform LRC and UC with both stereo algorithms. Thus, trading simplified computations with memory footprint leads to design better alternatives to standard confidence measures for embedded systems.

Table 2 reports the average AUCs for the two considered stereo algorithms on KITTI 2015 for software implementation of the 16 confidence measures. Compared to Table 1 we can notice a similar behavior with a notable difference. In fact, observing Table 2 we highlight that LRC achieves almost optimal results on AD-CENSUS but yields very poor performance with SGM. Looking at the behavior of the hardware challenging measures, reported in Fig. 3, we observe on KITTI 2015 a substantially similar behavior with respect to Fig. 2 concerned with Middlebury 2014.

Table 2. Experimental results, in terms of AUC, on KITTI 2015 dataset with AD-CENSUS (a) and SGM (b) algorithms for the 16 confidence measures using a conventional software implementation. In red, top-performing measure. We also report the absolute ranking.

measure	standard	measure	standard
Opt.	0.08055	Opt.	0.04367
CUR	0.30692 (14)	AML	0.23053 (10)
LRC	0.20018 (2)	LRD	0.20706 (5)
LC	0.30692 (15)	MLM	0.25180 (12)
MM	0.20601 (4)	PER	0.22575 (9)
MMN	0.24588 (11)	PKR	0.19821 (1)
MSM	0.25571 (13)	PKRN	0.20931 (7)
NOI	0.31160 (16)	WMN	0.20221 (3)
UC	0.22324 (8)	WMNN	0.20795 (6)

(a)

measure	standard	measure	standard
Opt.	0.01618	Opt.	0.01618
CUR	0.08585 (11)	AML	0.05738 (2)
LRC	0.10377 (15)	LRD	0.08744 (13)
LC	0.08585 (12)	MLM	0.05889 (3)
MM	0.06374 (8)	PER	0.05657 (1)
MMN	0.09549 (14)	PKR	0.06003 (6)
MSM	0.05999 (5)	PKRN	0.07611 (10)
NOI	0.16308 (16)	WMN	0.05970 (4)
UC	0.06310 (7)	WMNN	0.07149 (9)

(b)

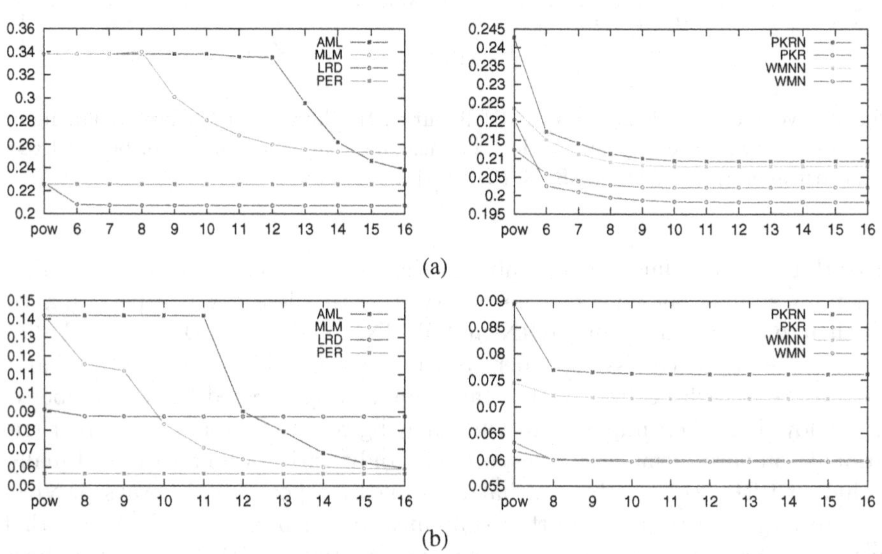

(a)

(b)

Fig. 3. Average AUC values on the KITTI 2015 dataset for hardware challenging measures, varying the implementation settings (i.e., *pow* and number of bits of fixed-point arithmetic). (a) AD-CENSUS, (b) SGM algorithm.

6   Conclusions

In this paper we have evaluated confidence measures suited for embedded stereo cameras. Our analysis shows that conventional approaches, LRC and UC, are outperformed by other considered solutions, whose implementation on embedded devices enables to achieve more accurate confidence predictions with a negligible amount of hardware resources and/or computations. In particular, according to our evaluation on Middlebury 2014 and KITTI 2015, PKR and WMN represent the overall best choice when dealing with two popular algorithms, AD-CENSUS and SGM, frequently deployed for embedded stereo systems.

References

1. Bailey, D.: Space efficient division on FPGAs. In: Electronics New Zealand Conference (ENZCon 2006), pp. 206–211 (2012)
2. Banz, C., Hesselbarth, S., Flatt, H., Blume, H., Pirsch, P.: Real-time stereo vision system using semi-global matching disparity estimation: architecture and FPGA-implementation. In: ICSAMOS, pp. 93–101 (2010)
3. Di Stefano, L., Marchionni, M., Mattoccia, S.: A fast area-based stereo matching algorithm. Image Vis. Comput. **22**(12), 983–1005 (2004)
4. Gehrig, S.K., Eberli, F., Meyer, T.: A real-time low-power stereo vision engine using semi-global matching. In: Fritz, M., Schiele, B., Piater, J.H. (eds.) ICVS 2009. LNCS, vol. 5815, pp. 134–143. Springer, Heidelberg (2009). doi:10.1007/978-3-642-04667-4_14
5. Haeusler, R., Nair, R., Kondermann, D.: Ensemble learning for confidence measures in stereo vision. In: CVPR (2013)
6. Hirschmuller, H.: Stereo processing by semiglobal matching and mutual information. PAMI **30**(2), 328–341 (2008)
7. Honegger, D., Oleynikova, H., Pollefeys, M.: Real-time and low latency embedded computer vision hardware based on a combination of FPGA and mobile CPU. In: IROS (2014)
8. Hu, X., Mordohai, P.: A quantitative evaluation of confidence measures for stereo vision. PAMI **34**, 2121–2133 (2012)
9. Marin, G., Zanuttigh, P., Mattoccia, S.: Reliable fusion of tof and stereo depth driven by confidence measures. In: Leibe, B., Matas, J., Sebe, N., Welling, M. (eds.) ECCV 2016. LNCS, vol. 9911, pp. 386–401. Springer, Cham (2016). doi:10.1007/978-3-319-46478-7_24
10. Mattoccia, S., Poggi, M.: A passive RGBD sensor for accurate and real-time depth sensing self-contained into an FPGA. In: 9th ICDSC (2015)
11. Menze, M., Geiger, A.: Object scene flow for autonomous vehicles. In: CVPR (2015)
12. Merrell, P., Akbarzadeh, A., Wang, L., Frahm, J., Nistér, R.Y.D.: Real-time visibility-based fusion of depth maps. In: CVPR (2007)
13. Min, D.B., Sohn, K.: An asymmetric post-processing for correspondence problem. Sig. Proc.: Image Commun. **25**(2), 130–142 (2010)
14. Mostegel, C., Rumpler, M., Fraundorfer, F., Bischof, H.: Using self-contradiction to learn confidence measures in stereo vision. In: CVPR (2016)
15. Park, M.G., Yoon, K.J.: Leveraging stereo matching with learning-based confidence measures. In: CVPR (2015)
16. Pfeiffer, D., Gehrig, S., Schneider, N.: Exploiting the power of stereo confidences. In: CVPR (2013)
17. Poggi, M., Mattoccia, S.: Evaluation of variants of the SGM algorithm aimed at implementation on embedded or reconfigurable devices. In: IC3D (2016)
18. Poggi, M., Mattoccia, S.: Learning a general-purpose confidence measure based on 0(1) features and a smarter aggregation strategy for semi global matching. In: 3DV (2016)
19. Poggi, M., Mattoccia, S.: Learning from scratch a confidence measure. In: BMVC (2016)

20. Scharstein, D., Szeliski, R.: A taxonomy and evaluation of dense two-frame stereo correspondence algorithms. Int. J. Comput. Vision **47**(1–3), 7–42 (2002)
21. Scharstein, D., Szeliski, R.: High-accuracy stereo depth maps using structured light. In: CVPR (2003)
22. Schmid, K., Hirschmuller, H.: Stereo vision and IMU based real-time ego-motion and depth image computation on a handheld device. In: ICRA (2013)
23. Seki, A., Pollefeys, M.: Patch based confidence prediction for dense disparity map. In: BMVC (2016)
24. Shan, Y., Hao, Y., Wang, W., Wang, Y., Chen, X., Yang, H., Luk, W.: Hardware acceleration for an accurate stereo vision system using mini-census adaptive support region. ACM Trans. Embed. Comput. Syst. **13**(4s), 132:1–132:24 (2014)
25. Spyropoulos, A., Komodakis, N., Mordohai, P.: Learning to detect ground control points for improving the accuracy of stereo matching. In: CVPR. IEEE (2014)
26. Ttofis, C., Theocharides, T.: Towards accurate hardware stereo correspondence: a real-time FPGA implementation of a segmentation-based adaptive support weight algorithm. In: DATE (2012)
27. Wedel, A., Meißner, A., Rabe, C., Franke, U., Cremers, D.: Detection and segmentation of independently moving objects from dense scene flow. In: Cremers, D., Boykov, Y., Blake, A., Schmidt, F.R. (eds.) EMMCVPR 2009. LNCS, vol. 5681, pp. 14–27. Springer, Heidelberg (2009). doi:10.1007/978-3-642-03641-5_2
28. Zabih, R., Woodfill, J.: Non-parametric local transforms for computing visual correspondence. In: Eklundh, J.-O. (ed.) ECCV 1994. LNCS, vol. 801, pp. 151–158. Springer, Heidelberg (1994). doi:10.1007/BFb0028345
29. Zbontar, J., LeCun, Y.: Stereo matching by training a convolutional neural network to compare image patches. J. Mach. Learn. Res. **17**, 1–32 (2016)

3D Reconstruction from Specialized Wide Field of View Camera System Using Unified Spherical Model

Ahmad Zawawi Jamaluddin$^{(\boxtimes)}$, Cansen Jiang, Olivier Morel, Ralph Seulin, and David Fofi

Université Bourgogne Franche-Comté, Laboratoire Le2i, UMR6306, FRE CNRS, 12 rue de la Fonderie, 71200 Le Creusot, France
{ahmad.jamaluddin,cansen.jiang,olivier.morel,ralph.seulin, david.fofi}@u-bourgogne.fr
http://le2i.cnrs.fr/

Abstract. This paper proposed a method of three dimensions (3D) reconstruction from a wide field of view(FoV) camera system. This camera system consists of two fisheye cameras each with 180° FoV. The fisheye cameras placed back to back to obtain a full 360° FoV. A stereo vision camera is placed to estimate the depth information of anterior view of the camera system. A novel calibration method using unified camera model representation has been proposed to calibrate the multiple camera systems. An effective fusion algorithm has been introduced to fuse multi-camera images by exploiting the overlapping area. Moreover, direct and fast 3D reconstruction of sparse feature matches based on the spherical representation are obtained using the proposed system.

Keywords: Fisheye camera calibration · Unified spherical model · 3D reconstruction · Interior Point Optimization algorithm

1 Introduction

Sensors that provide wider FoV are preferred in the robotic applications, as both navigation and localization can benefit from the wide FoV [1–3]. This paper presents a novel camera system which provides scenes of 360° FoV and the depth information concurrently. The system minimizes the use of equipment, image-data and it has the ability to acquire sufficient information of the scene. The robotic applications of wide FoV camera system are mainly mapping [4,5], object tracking [6], video surveillance [7–9], virtual reality [10] and structure from motion [11,12].

1.1 Proposed System

The proposed vision system consists of two fisheye cameras (omnidirectional camera) [13,14] with each has 185° FoV. The cameras are placed back to back

© Springer International Publishing AG 2017
S. Battiato et al. (Eds.): ICIAP 2017, Part I, LNCS 10484, pp. 495–506, 2017.
https://doi.org/10.1007/978-3-319-68560-1_44

so that they cover the whole 360° of the scene. A high-resolution stereo vision camera, named ZED [16] is placed in front of the rig so that its baseline is in parallel with the baseline of the fisheye cameras. Figure 1 shows the proposed system and the predicted FoV from camera rig.

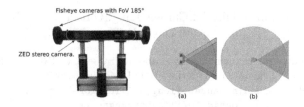

Fig. 1. The front view of the proposed vision system. The illustration (a) and (b) are the predicted FoV viewed from top and right side of the camera rig. The gray color are referred to the FoV of fisheye cameras. The red and green colour are referred to the FoV of stereo vision camera. (Color figure online)

The major contributions of this paper are:

1. We proposed an omni-vision system alongside with a stereo-camera, which offers immense information on 360° FoV of the environment as well as detailed depth information.
2. A new camera calibration method taking the advantages of Unified Camera Model representation has been proposed, which outperforms the state-of-the-art methods.
3. An Interior Point Optimization algorithm (IPO) based on pure rotation matrix estimation approaches has been proposed to fuse the two fisheye and ZED images, which offers seamless images stitching results.
4. A projective distortion has been proposed to be added to the ZED image before projecting onto the unit sphere, which results in enhancing the quality of the overlapping image.

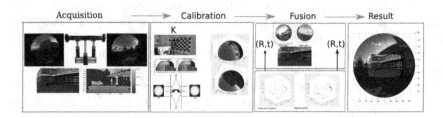

Fig. 2. Block acquisition contains the proposed camera system with the images of fisheye and ZED cameras. Block calibration consists of calibration method to estimate intrinsic and extrinsic parameters. Block fusion consists of the method to fuse all images onto unit sphere. Block result consist of final result with image from fisheye and ZED cameras fused together.

2   Methodology

2.1   Unified Spherical Camera Model

The unified spherical camera model has been proposed by Geyer [14] and Barreto [17]. The image formation of dioptric camera was effected by the radial distortion. Due to that, the point on the scene is nonlinear with the point in the dioptric image. The 3D points χ are projected to the image point $\mathbf{x}$ using the pin-hole camera model representation. It is also considered the representation of a linear and non linear transformation mapping function which is depending on the type of camera. The model was extended by Mei [18] and an omnidirectional camera calibration toolbox [23] has been developed. This model has been used as a reference to map the image on the unified spherical model. All points m are projected to the image plane using K, which is a generalized camera projection matrix. The value of f and η should be also generalized to the whole system (camera and lens).

$$p = Km = \begin{bmatrix} f_1\eta & f_1\eta\alpha & u_0 \\ 0 & f_2\eta & v_0 \\ 0 & 0 & 1 \end{bmatrix} m, \tag{1}$$

where the $[f_1, f_2]^T$ are the focal length, (u_0, v_0) are the principal point and α is the skew factor. By using the projection model, the point on the normalized camera plane can be lifted to the unit sphere by the following equation:

$$\hbar^{-1}(m) = \begin{bmatrix} \frac{\xi + \sqrt{1 + (1-\xi^2)(x^2+y^2)}}{x^2+y^2+1}x \\ \frac{\xi + \sqrt{1 + (1-\xi^2)(x^2+y^2)}}{x^2+y^2+1}y \\ \frac{\xi + \sqrt{1 + (1-\xi^2)(x^2+y^2)}}{x^2+y^2+1} - \xi \end{bmatrix}, \tag{2}$$

where the parameter ξ quantifies the amount of radial distortion of the dioptric camera.

2.2   Camera Calibration Using Zero-Degree Overlapping Constraint

A new multi-camera setup is proposed, where two 185° fisheye cameras are rigidly attached in opposite direction to each other. Since the fisheye camera has more than 180° of FoV, the proposed setup contributes an overlapping area along the periphery of the two fisheye cameras. Taking the advantage of overlapping FoV of the two fisheye cameras, we propose a new fisheye camera calibration using the constraint on overlapping zero-degree with Unified Camera Model under the following assumption:

- If ξ is estimated correctly, the 180° line of the fisheye camera should ideally lay on the zero degree plane of the unit sphere. See Figs. 3 and 4.
- A correct calibration (registration) of multi-fisheye camera setup contributes to a correct overlapping area.

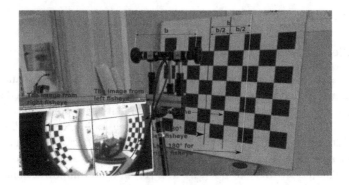

Fig. 3. Experimental setup to calibrate the value of ξ. The baseline of the camera rig, noted as b (from left fisheye lens to right fisheye lens) is measured and two parallel lines with the same distance to each other as well as a centre line is drawn on a pattern. The rig is faced and aligned in front of the checkerboard pattern such that the centre line touches the edges of both fisheye camera images.

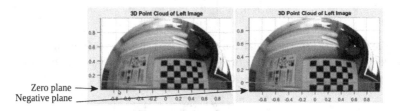

Fig. 4. The left image was projected with initial estimate of ξ. The 180° lines should ideally lay on the zero plane. After the iterative estimation of ξ, the 180° linear now lay on the zero plane.

Pure Rotation Registration. One of the major objective of our setup is to produce a high quality 360° FoV unit sphere which ables to handle the visualization. A common way to do this is to calibrate the camera setup such that the relative poses between the cameras are known. Let the features from the left and right fisheye cameras (projected onto the unit sphere) be denoted as $\mathbf{x}^{Lf}$ and $\mathbf{x}^{Rf}$, respectively. The transformation between the two fisheye cameras is noted as $T \in R^{4 \times 4}$, such that:

$$\mathbf{x}^{Lf} = R\mathbf{x}^{Rf}, \tag{3}$$

The estimation of the transformation matrix, T as discussed in [19,21,22] contains the rotation R and also the translation t.

In our method, we used a pure rotation matrix to solve this problem by enforcing the transformation matrix contains zero translation [20], which represented as:

$$\min_{R} \sum_{i=1}^{n} \Psi(\|\mathbf{x}^{Lf} - R\mathbf{x}^{Rf}\|), \quad \text{s.t.} \quad RR^{\mathsf{T}} = 1, \det(R) = 1, \tag{4}$$

where R is the desired pure rotation matrix, $\Psi(\cdot)$ is the Huber-Loss function for robust estimation. By solving the above equation, a pure rotation matrix that minimizes the registration errors between fusion of the two fisheye cameras. Here, we adopt the Interior Point Optimization algorithm (IPO) to solve the system.

Fusion of Perspective Camera onto Unit Sphere. A new method has been proposed to fuse perspective image onto the unit sphere. A major difference between perspective and spherical images is the existence of distortion which deforms the object on the scene. The direct matching point from the perspective is deficient to match the features on the unit sphere. This due to the characteristic of unit spheres which has several levels of distortion. The projective distortion parameters have been proposed to be added to the perspective image plane before projecting onto the unit sphere.

$$P = K^{-1} \cdot H \cdot I, \qquad \begin{bmatrix} x \\ y \\ 1 \end{bmatrix} = K^{-1} \cdot H \cdot I, \tag{5}$$

where, H is a projective distortion parameter. K is a camera matrix. I is an image frame.

$$\begin{bmatrix} x \\ y \\ 1 \end{bmatrix} = \begin{bmatrix} f_x & \delta & v_o \\ 0 & f_y & \nu_o \\ 0 & 0 & 1 \end{bmatrix} \cdot \begin{bmatrix} h_{11} & h_{12} & h_{13} \\ h_{21} & h_{22} & h_{23} \\ h_{31} & h_{32} & h_{33} \end{bmatrix} \cdot \begin{bmatrix} v \\ \nu \\ 1 \end{bmatrix}, \tag{6}$$

the value x, y and ξ ($\xi = 0$ for the perspective camera) are replaced into mapping function (in Eq. (2)) for projecting onto the unit sphere.

Fusion of Multi-camera Images. In our multi-camera setup, the fusion of fisheye cameras alongside with the ZED camera based on a unified model representation can be achieved in a similar manner.

Let $\mathbf{x}^z$ and $\mathbf{x}^{Lf}$ be the feature correspondences (mapped from χ^z and χ^{Lf}) on a Unified Sphere. The fusion of the ZED camera and the fisheye camera can be framed as a minimization problem of the feature correspondences on a unified sphere, which is defined as:

$$\underset{R}{\mathrm{argmin}} \sum_{i=1}^{n} \Psi \left(\left\| \mathbf{x}^{Lf} - \mathbf{x}^Z(R) \right\|_2 \right), \tag{7}$$

where $\Psi(\cdot)$ is the Loss function for the purpose of robust estimation, while

$$\chi(\theta_{x,y,z}) = R(\theta_{x,y,z}) \begin{bmatrix} x_s & \cdots & x_s^n \\ y_s & \cdots & y_s^n \\ z_s & \cdots & z_s^n \end{bmatrix}, \tag{8}$$

stands for the registration of ZED camera sphere points to the left fisheye camera (the reference), where $R(\theta_{x,y,z})$ is the desired pure rotation matrix with estimated rotation angles $\theta_{x,y,z}$. This can be solved on a similar manner solving Eq. (4), by applying an IPO algorithm.

2.3   Epipolar Geometry of Omnidirectional Camera

The epipolar geometry for an omnidirectional camera has been studied [17] and it was originally used as a model for catadioptric camera. The study was then extended to the dioptric or fisheye camera system. Figure 5 shows the epipolar geometry of fisheye camera. Lets consider the two positions of a fisheye camera observed from point P in the space. The points P_1 and P_2 are the projection point P onto the unit sphere image in two different fisheye's positions. The points P, P_1, P_2, O_1 and O_2 are coplanar, such that:

$$\overline{O_1 O_2} \times \overline{O_2 P_1} \cdot \overline{O_2 P_2} = 0, \qquad O_1^2 \times P_1^2 \cdot P_2 = 0, \tag{9}$$

where, O_1^2 and P_1^2 are the coordinates of O_1 and P_1 in coordinate system O_2. The transformation between system X_1, Y_1, Z_1 and X_2, Y_2, Z_2 can be described by rotation R and translation t. The transformation equations are:

$$O_1^2 = R \cdot O_1 + t = t, \qquad P_1^2 = R \cdot O_1 + t, \tag{10}$$

O_1^2 is the pure translation. By substituting (10) in (9) we get:

$$P_2^T E P_1 = 0, \tag{11}$$

where $E = [t]_\times R$ is the essential matrix which consists of rotation and translation. In order to estimate the essential matrix, the points correspondence pairs on the fisheye images are stacked into the linear system, thus the overall epipolar constraint becomes:

$$U f = 0, \; where \; U = [u_1, u_2, \ldots, u_n]^T, \tag{12}$$

and u_i and f are vectors constructed by stacking column of matrices P_i and E respectively.

$$P_i = P_i P_i'^T, \tag{13}$$

$$E = \begin{bmatrix} f_1 & f_4 & f_7 \\ f_2 & f_5 & f_8 \\ f_3 & f_6 & f_9 \end{bmatrix}. \tag{14}$$

The essential matrix can be estimated with linear least square by solving Eqs. (12) and (13), where P_i' is the projected point which corresponds to P_2 of the Fig. 5, U is $n \times 9$ matrix and f is 9×1 vector containing the 9 elements of E. The initial estimated essential matrix is then utilized for the robust estimation of essential matrix. An iterative reweighted least square method [15] is proposed to re-estimate the essential matrix of omnivision camera. This assigns minimal weight to the outliers and noisy correspondences. The weight assignment is performed by the residual r_i for each point.

$$r_i = f_1 x_i' x_i + f_4 x_i' y_i + f_7 x_i' z_i + f_2 x_i y_i' + f_5 y_i y_i' + f_8 y_i' z_i + f_3 x_i z_i' + f_6 y_i z_i' + f_9 z_i z_i', \tag{15}$$

$$err \rightarrow \min_f \sum_{i=1}^{n} \left(w_{Si} f^T u_i \right)^2, \tag{16}$$

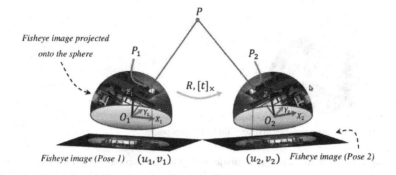

Fig. 5. The diagram of epipolar geometry of fisheye camera for 3D reconstruction.

$$w_{Si} = \frac{1}{\nabla r_i}, \; where \; r_i = (r_{xi}^2 + r_{yi}^2 + r_{zi}^2 + r_{xi'}^2 + r_{yi'}^2 + r_{zi'}^2)^{\frac{1}{2}}, \qquad (17)$$

where w_{Si} is the weight (known as Sampson's weighting) that will be assigned to each set of corresponding point and ∇r_i is the gradient; r_{xi} and so on are the partial derivatives found from Eq. (15), as $r_{xi} = f_1 x_i' + f_2 y_i' + f_3 y_i'$.

Once all the weights are computed, U matrix is updated as follow: $U = WU$.

where W is a diagonal matrix of the weights computed using Eq. (16). The essential matrix is estimated at each step and forced to be of rank 2 in each iteration. The procrustean approach is adopted here and singular value decomposition is used for this purpose.

3 Experimental Results

3.1 Estimation of Intrinsic Parameters

The unknown parameters f_1, f_2, u_0, v_0 and ξ of fisheye cameras are estimated using the camera calibration toolbox provided by Mei [23]. The fisheye images are projected onto the unit sphere using the Inverse Mapping Function defined in Mei's projection model. Figure 6(a) shows the two dimension(2D) images from a fisheye camera is projected onto the unit sphere.

3.2 Estimation of Extrinsic Parameters

Rigid Transformation Between Two Fisheyes Image. The overlapping features are taken along the periphery on the left and right fisheye images. The selected points are projected onto the unit sphere. The rigid 3D transformation matrix are estimated using the selected overlapping features. The IPO algorithm is used to estimate the rotation between the set of projected points. Figure 6(b) shows the set of projected points (green-left fisheye and red-right fisheye) aren't aligned. After using IPO algorithm, the selected points are aligned together as shown in Fig. 6(c).

The rotation matrix is parameterized in terms of Euler angles and cost function is developed that minimize the Euclidean distance between the reference (point projections of left camera image) and the three dimensional points from the right camera

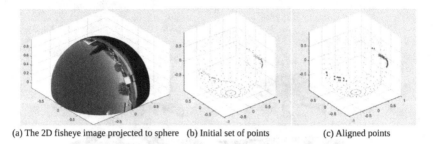

(a) The 2D fisheye image projected to sphere   (b) Initial set of points   (c) Aligned points

Fig. 6. (a) The 2D fisheye image is projected onto unit sphere. (b) The selected points (green and red point) aren't aligned together. (c) After use IPO algorithm, the green and red points are aligned with their respective point. (Color figure online)

image. The transformation results using Singular Value Decomposition (SVD) though are very close to pure rotation. It assumed that translation is also as a parameter to align the set of points. Figure 7 shows the fusion result. The points on the hemispheres that are beyond the zero plane are first eliminated. Then the transformation is applied on the hemisphere of the right fisheye camera and the point matrices are concatenated to get a full unit sphere.

(a)   (b)

Fig. 7. (a) The image from fisheye and ZED cameras are fused together. (b) The fusion result after applying the projective distortion on the ZED image. Focusing to the border between ZED and unit sphere, the ZED image is perfectly over-lapped on the unit sphere.

Rigid Transformation Between a Fisheyes and ZED Camera. The same procedures are used to estimate the transformation matrix between the image from ZED camera and two hemispheres.

As shown in Fig. 7(a), the RGB images from ZED camera are overlapped onto the unit sphere. It is also recovered the scale between the ZED and fisheye cameras.

The fusion has been enhanced by adding the projective distortion to the ZED image. Figure 7(b) shows that the result is much better after handling the distortion on the fisheye images.

3.3   Estimation the Three Dimensional Registration Error

The computation of registration error during mapping on the unit sphere is done to prove the registration method. The Root Means Square Error (RMSE) is used to

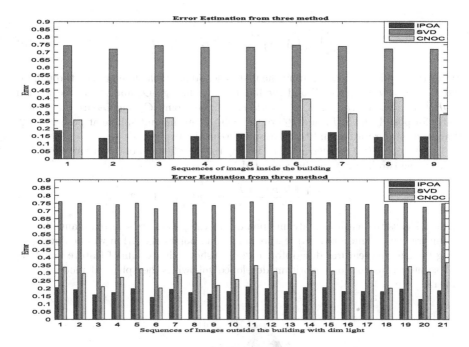

Fig. 8. The registration error estimation used three different methods. The images sequences have been taken inside and outside of the building. The average registration error using proposed method is 0.1612 (inside building) and 0.1812 (outside building).

calculate the error. The rigid 3D transformation matrix and parameter ξ which was obtained from calibration are used to determine the residual error of the point pairs registration on the unit sphere. Three methods have been compared:

1. IPO: Our method - The pure rotation estimated using feature matches and IPO algorithm.
2. SVD: The transformation matrix is estimated using features matches with SVD [21].
3. CNOC: Calibration Non Overlapping Cameras, Lébraly [19].

The image sequences were taken in several different environments. The feature points were selected on the overlapping area. The same data set are used in all three methods. Figure 8 shows that the proposed method has the lowest registration errors.

3.4   3D Reconstruction Using the Camera Rig

The goal of triangulation is to minimize the distance between the two lines toward point P in 3D the space. This problem can be expressed as a least square problem.

$$\min_{a,b} \|aP_1 - bRP_2 - t\|, \tag{18}$$

$$\begin{bmatrix} a^* \\ b_* \end{bmatrix} = \left(A^T A \right)^{-1} A^T t, A = [P_1 - RP_2], \tag{19}$$

By referring to Fig. 5, looking from the first pose, point P, the line passing through O_1 and P_1 can be written as aP and the line passing through O_2 and P_2 can be written as $bRP_2 + t$, where $a, b \in \mathbb{R}$, P is a world coordinate point. O_1 and O_2 are the camera center for pose 1 and 2. P_1 and P_2 are the point P on the unit sphere at pose 1 and 2. R and t are the rotation and translation between the two poses.

The 3D point P is reconstructed by finding the middle point of the minimal distance between the two lines. It can be computed by;

$$P_k = \frac{a^* P_1 + b^* R P_2 + t}{2} \ , \ where, \ k = 1, 2, 3, 4 \tag{20}$$

Figure 9 shows the features matching points. All the points are selected manually. For the future works, an automated features matching points algorithm will be developed using the existing features descriptor. Figure 10 shows the results of feature matching and scene reconstruction algorithm developed following the spherical model of the camera.

Fig. 9. The features matching points between two different poses of fisheye cameras

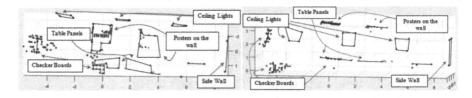

Fig. 10. The front view (left) and top view (right) of the three dimensional reconstruction scenes

4   Conclusions

This paper proposed a new camera system which has 360° FoV and detail depth information at anterior. The two fisheye cameras each 180° FoV are placed back to back to obtain 360° FoV. A stereo vison camera is placed perpendicular to obtain depth information at anterior. A novel camera calibration method taking advantages the Unified

Spherical Model has been introduced to calibrate multi camera system. A pure rotation matrix based-on IPO algorithm has been used to fuse images from multi camera setup by exploiting the overlapping area. The result are reduced the registration error and enhance the quality of image fusion. The 3D reconstruction based on the spherical representation has been estimated using the proposed system.

References

1. Li, S.: Full-view spherical image camera. In: ICPR 2006, pp. 386–390. IEEE (2006)
2. Zhang, C., Xu, J.: Development of an omni-directional 3D camera for robot navigation. In: AIM, pp. 262–267. IEEE (2012)
3. Kim, S.-H., Park, J.-H., Jung, I.-K.: Global localization of mobile robot using an omni-directional camera. In: IPCV, p. 1 (2014)
4. Liu, M., Siegwart, R.: Topological mapping and scene recognition with lightweight color descriptors for an omnidirectional camera. IEEE Trans. Robot. 30(2), 310–324 (2014)
5. Lukierski, R., Leutenegger, S., Davison, A.J.: Rapid free-space mapping from a single omnidirectional camera. In: ECMR, pp. 1–8. IEEE (2015)
6. Marković, I.: Moving object detection, tracking and following using an omnidirectional camera on a mobile robot. In: ICRA, pp. 5630–5635. IEEE (2014)
7. Cogal, O., Akin, A., Seyid, K., et al.: A new omni-directional multi-camera system for high resolution surveillance. In: SPIE, pp. 91200N–91200N-9 (2014)
8. Depraz, F., Popovic, V.: Real-time object detection and tracking in omnidirectional surveillance using GPU. In: SPIE, pp. 96530N–96530N-13 (2015)
9. Sablak, S.: Omni-directional intelligent autotour and situational aware dome surveillance camera system and method. U.S. Patent No 9,215,358, 15 December 2015
10. Li, D., et al.: Motion interactive system with omni-directional display. In: 2013 International Conference on Virtual Reality and Visualization (ICVRV). IEEE (2013)
11. Chang, P., Hebert, M.: Omni-directional structure from motion. In: Workshop on Omnidirectional Vision, pp. 127–133. IEEE (2000)
12. Micusik, B., Pajdla, T.: Structure from motion with wide circular field of view cameras. In: PAMI, vol. 28, no 7, pp. 1135–1149 (2006)
13. Nayar, S.K.: Catadioptric omnidirectional camera. In: CVPR, pp. 482–488. IEEE (1997)
14. Geyer, C., Daniilidis, K.: A unifying theory for central panoramic systems and practical implications. In: ECCV, pp. 445–461 (2000)
15. Hartley, R.I., Sturm, P.: Triangulation. Comput. Vis. Image Underst. 68(2), 146–157 (1997)
16. ZED Documentation. https://www.stereolabs.com/documentation/overview/getting-started/introduction.html. Accessed 15 June 2017
17. Barreto, J.P.: A unifying geometric representation for central projection systems. Comput. Vis. Image Underst. 103(3), 208–217 (2006)
18. Mei, C., Rives, P.: Single view point omnidirectional camera calibration from planar grids. In: ICRA, pp. 3945–3950. IEEE (2007)
19. Lébraly, P., Ait-Aider, O., Royer, E., et al.: Calibration of non-overlapping cameras-application to vision-based robotics. In: BMVC, pp. 10.1-10.12 (2010)
20. Othmani, A.A., Jiang, C.: A novel computer-aided tree species identification method based on burst wind segmentation of 3D bark textures. MVA 27(5), 751–766 (2016)

21. Llourakis, M.I.A., Deriche, R.: Camera self-calibration using the singular value decomposition of the fundamental matrix: from point correspondences to 3D measurements. Ph.D. thesis. INRIA (1999)
22. Jamaluddin, A.Z., Mazhar, O., Morel, O., et al.: Design and calibration of an omni-RGB + D camera. In: URAI, pp. 386–387. IEEE (2016)
23. Mei, C.: Active Vision Group. http://www.robots.ox.ac.uk/cmei/Toolbox.html. Accessed 15 June 2017

A Matrix Decomposition Perspective on Calibrated Photometric Stereo

Luca Magri[1]([✉]), Roberto Toldo[2], Umberto Castellani[1], and Andrea Fusiello[3]

[1] Dip. Informatica, Università di Verona, Strada Le Grazie 15, Verona, Italy
magri.luca.1@gmail.com
[2] 3Dflow SRL, c/o Computer Science Park, Strada Le Grazie 15, Verona, Italy
[3] DPIA, Università di Udine, Via Delle Scienze, 208, Udine, Italy

Abstract. Leveraging on recent advances in robust matrix decomposition, we revisit Lambertian photometric stereo as a robust low-rank matrix recovery problem with both missing and corrupted entries, tailoring Grasta and R-GoDec to normal surface estimation. A method to automatically detect shadows is proposed. The performance of different robust matrix completion techniques are analyzed on the challenging DiLiGenT datasets.

Keywords: Calibrated photometric stereo · Robust matrix factorization

1 Introduction

Robust matrix decomposition and completion has been an active research topic in recent years, and many methods exploiting low rank and sparsity constraints have sprouted out in several fields of applications, such as pattern recognition, machine learning, and signal processing just to name a few. In this work, we explore the performances of these techniques on calibrated Lambertian photometric stereo [9], *i.e.* the problem of estimating the surface normals of an object by observing several intensity images captured by a fixed camera under different known lighting conditions. In particular, we offer an overview on robust matrix decomposition methods tailored to photometric stereo – using for the fist time Grasta and R-GoDec for this scope – and a quantitative experimental evaluation on the recently proposed DiLiGenT dataset. A simple yet effective shadow detection method is also presented.

Notation: Matrix will be indicated in sans serif font $\mathsf{A} = [a^i_j]$, the i-th row of A is denoted by A^i, while the j-th column of A is indicated by A_j.

2 The Geometry of Single-Light Images

Let $I \in \mathbb{R}^p$ be an image composed by p pixels stacked by column. Following [3], under Lambertian assumption the proprieties of interest of an object Y can

© Springer International Publishing AG 2017
S. Battiato et al. (Eds.): ICIAP 2017, Part I, LNCS 10484, pp. 507–517, 2017.
https://doi.org/10.1007/978-3-319-68560-1_45

be encoded in matrix form as $\text{diag}(R)\mathsf{N}^\top \in \mathbb{R}^{p \times 3}$, where $R = (\rho_1, \ldots, \rho_p)^\top$ is the vector of pixels albedos, and $\mathsf{N} = [\mathsf{N}_1, \ldots, \mathsf{N}_p] \in \mathbb{R}^{3 \times p}$ collects the unitary normals of the object. Thus the image of Y illuminated by a distant point-light source $L \in \mathbb{R}^3$, is given by:

$$I = \max(\text{diag}(R)\mathsf{N}^\top L, 0). \tag{1}$$

Varying L, one obtains the so-called *illumination space* of Y defined as $\mathcal{L} = \{I : I = \text{diag}(R)\mathsf{N}L \colon L \in \mathbb{R}^3\} \subset \mathbb{R}^p$. Clearly $\dim(\mathcal{L}) = \text{rank}(\mathsf{N}^\top L) \leq 3$, therefore, if the normals span $\mathbb{R}^3$, the dimension of $\mathcal{L}$ is 3.

Belhumeur and Kriegman also observe that $\mathcal{L}$ intersect at most[1] $p(p-1)+2$ orthants of $\mathbb{R}^p$. Let $\mathcal{L}_0 = \mathcal{L} \cap \mathbb{R}^p_+$, where $\mathbb{R}^p_+ = \{x \in \mathbb{R}^p \colon x_i > 0 \,\forall i\}$, and $\mathcal{L}_i$ be the intersection of $\mathcal{L}$ with the other orthants. By construction, $\mathcal{L}_i$ are convex cones, and correspond to different shading configuration of pixels. As instance, $\mathcal{L}_0$ corresponds to images having all the pixels illuminated by a lighting source. The space of all possible images of Y is obtained by adding to $\mathcal{L}_0$ the images where not all the pixels are simultaneously illuminated, i.e. the projection of the cones $\mathcal{L}_i$, $i \neq 0$, on the boundary of $\mathbb{R}^p_+$ via the map $P \colon I \mapsto \max(I, 0)$. Therefore, the space of all the images of a convex Lambertian object, varying the direction of a single light source is given by the union of at most $p(p-1)+2$ convex cones.

$$\bigcup_{i=0}^{\nu(\nu-1)+2} P(\mathcal{L}_i). \tag{2}$$

Experimentally, it was demonstrated that this union of cones is "flat" and can be approximated by a linear subspace of dimension 3.

3   Robust Matrix Completion and Decomposition

The linear property of light superposition inspired the use of matrix completion and robust decomposition techniques to tackle the photometric stereo problem [10,11]. Given f images of the same object organized as a $p \times f$ matrix $\mathsf{X} = [I_1, \ldots, I_f]$, with images stacked as columns, the main intuition is to recover the illumination space as a low-rank matrix A that models the diffusive Lambertian observations, and to handle the non-Lambertian measurements as outliers. In particular, shadows, i.e. pixels outside $\mathcal{L}_0$, are treated as missing entries, whereas a *sparse* error matrix S accounts for the corruptions produced by strong specularities (highlights).

More formally, the image formation model can be rephrased as

$$\mathsf{X} = P_\Omega(\mathsf{A}) + \mathsf{S} \tag{3}$$

where $\mathsf{A} = \text{diag}(R)\mathsf{N}^\top L$ is low rank, $\mathsf{L} = [\mathsf{L}_1, \ldots, \mathsf{L}_f]$ collects the known light source vectors, $\Omega = \{(i,j) \colon \text{where } \mathsf{N}^\top L \text{ is nonzero}\}$ is the set of observed

---

[1] More precisely, $\mathcal{L}$ intersects $\nu(\nu-1)+2$ orthants, where ν is the number of distinct normal in B.

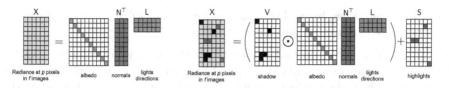

Fig. 1. The factorization of the intensities of an ideal Lambertian object (left), the same factorization in presence of shadow and highlights

entries, P_Ω indicates a linear projection of matrices defined component-wise as $[P_\Omega(\mathsf{A})]^i_j = \mathsf{A}^i_j$ if $(i,j) \in \Omega$ and 0 otherwise, and S is the matrix of sparse error. A graphical representation of this model is depicted in Fig. 1, where the operator P_Ω – which is the matrix counterpart of the projection P onto the positive orthant – is represented in an equivalent fashion as the Hadamard (element-wise) product between A and a matrix V defined component-wise as

$$\mathsf{V}^i_j = \begin{cases} 1 \text{ if } (i,j) \in \Omega, \\ 0 \text{ otherwise.} \end{cases} \tag{4}$$

In this way, photometric stereo becomes the problem of recovering a low-rank matrix with both missing entries – the shadows – and corrupted entries corresponding the unmodelled phenomena (e.g., non Lambertian). The rank of A may vary according to the image formation model adopted, and for Lambertian photometric stereo rank(A) = 3. Once the low rank matrix have been recovered, in the calibrated scenario, the normals can be easily estimated in closed form by normalizing the row of $\mathsf{L}^+\mathsf{A}$, where L^+ denotes the pseudo-inverse of L.

Decomposition into low-rank and sparse matrices has been developed in different formulation problems, hereinafter, we briefly review some of them that can be profitably adopted to tackle the Problem (3), namely: Robust Principal Component Analysis and L1-ALM (that have already been tailored to photometric stereo problem), together with Grasta and R-GoDec that we are going to apply to this scenario for the first time.

Robust Principal Component Analysis (RPCA) decomposes X into a low rank and sparse terms, without being given rank(A). The cost function is:

$$\arg\min_{\mathsf{A},\mathsf{S}} \text{rank}(\mathsf{A}) + \lambda\|\mathsf{S}\|_0 \text{ s.t. } X = P_\Omega(\mathsf{A}) + \mathsf{S}. \tag{5}$$

Unfortunately this problem turns to be intractable, therefore, instead of directly minimizing the discontinuous rank function and the ℓ_0 norm, the above objective function is relaxed to its convex surrogate; the rank of A is replaced with the nuclear norm $\|\mathsf{A}\|_*$ – i.e. the sum of the singular values of A – and the ℓ_0 norm is substituted for the ℓ_1 norm:

$$\arg\min_{\mathsf{A},\mathsf{S}} \|\mathsf{A}\|_* + \lambda\|\mathsf{S}\|_1 \text{ s.t. } X = P_\Omega(\mathsf{A}) + \mathsf{S}. \tag{6}$$

Several technique can be used to minimize Eq. (6); in [10], e.g., an adaptation of the augmented Lagrange multiplier method is used.

L1-ALM [11] proposes to find a solution to Problem (3) by enforcing exactly the low rank constraint $\text{rank}(A) = r$, and leverages on the factorization of the matrix $A = HK$ as the product of a $p \times r$ matrix H and a $r \times f$ matrix K. As the factorization is defined up to an invertible matrix, in order to shrink the solution space, the matrix H is enforced to be column-orthogonal, i.e. $H^\top H = I_r$, where I_r denotes the $r \times r$ identity matrix. The objective L1-ALM tries to minimize is

$$\arg\min_{H,K} \|P_\Omega(X - HK)\|_1 + \lambda\|K\|_* \text{ s.t. } H^\top H = I_r. \tag{7}$$

where $P_\Omega(X - HK)$ is the ℓ_1 norm of the sparse error matrix $S = X - P_\Omega(A)$ and $\|K\|_* = \|HK\|_* = \|A\|_*$ is a trace-norm regularization term, which, due to the orthogonality of H, is equivalent to the nuclear norm of A. This optimization problem is resolved via inexact augmented Lagrange multiplier and Gauss-Seidel iterations.

Grassmannian Robust Adaptive Subspace Tracking Algorithm (Grasta) [5] is an online robust subspace tracking algorithm, that works in the presence of corrupted and missing data. Given a sequence of incomplete vectors $\{v_1, \ldots, v_t\}$ that lie on a r-dimensional subspace, Grasta estimates this subspace, by minimizing the ℓ_1 error between the recovered subspace and the observed partial vector. This formulation can be casted to the problem of Eq. (3) as

$$\min_S \|S\|_1 \text{ s.t. } X = P_\Omega(HK) + S \tag{8}$$

where, similarly to Eq. (7), A is expressed as the product of two factors H, K, the first being an element of the Grassmanian Manifold $G(r, p)$. The problem is iteratively solved for H and K separately: fixed H, K is update via ADMM, whereas, when K is fixed, H is updated performing incremental gradient descent on the Grassmanian manifold. Even if the partial measurements of the matrix X are required to be exactly fixed, nevertheless, in practice, it was demonstrated that the algorithm is robust to small non sparse additive noise.

Robust Go Decomposition (R-GoDec) [2] proposes a robust approximate matrix completion and decomposition technique that improves GoDec [12]. An additional sparse term S' that has support on Ω^C – the complementary of Ω – is introduced to account for missing entries. In addition small sparse noise E is explicitly introduced in the decomposition:

$$X = A + S + S' + E. \tag{9}$$

The corresponding minimization problem is

$$\arg\min_{A,S} \|X - A - S - S'\|_F^2 \tag{10}$$

such that $\text{rank}(A) \le r$, S is sparse and S' has support in Ω^C. This problem is solved using a block-coordinate minimization scheme. At first, the rank-r

projection of the incomplete matrix given in input is computed through Bilateral Random Projection – faster than SVD – and assigned to A. Then, the two sparse terms S and S' are updated separately. The outlier term S is computed via soft-thresholding operator, and S' is updated as $-P_{\Omega^C}(A)$.

4   Detecting Shadows

This section is aimed at estimating the set Ω^C of shadowed pixels, in order to treat them as missing data and to reduce their influence in the low-rank matrix recovery. For this purpose, it becomes necessary to reason about the *visibility* of light source with respect to each image pixel in order to recognize which lights shine on which points and to discard the pixels in shadow.

To this end, a commonly employed solution is intensity-based thresholding: Pixels whose intensity lies below a certain threshold are considered in shadow. While this heuristic in some cases is enough to recover the light-visibility information, in general, the intensity of individual pixels depends on the variations of the *unknown* albedo of the object, thus, the brightness alone turns to be an unreliable cue.

In order to overcome this pitfall, other techniques have been proposed. For example, [4] adopts a graph cuts based method to estimate light visibility in a Markov Random Field formulation, where a per-pixels error, based on photometric stereo, is balanced by a smoothness constraints on shadows, aimed at promoting spatial coherence. Sunkavalli et al. in [8] avoid to enforce spatial coherence on shadows and present a method that works both in the calibrated and uncalibrated scenario leveraging on subspace clustering. Pixels sharing the same visibility configuration lie on linear subspaces, termed visibility subspaces, that are extracted using Sequential RANSAC. Once these subspaces are recovered and the object surface is segmented accordingly, the set of lights that shine on each region are identified analysing the magnitude of the subspace lighting obtained via SVD.

The visibility information can be encoded in the $n \times f$ *visibility matrix* V defined as in Eq. (4). Each row of V can be seen as the indicator function of the subset of lights visible by each pixel.

In our calibrated scenario, we want to recover V given the intensity matrix X and the lighting directions L. To this end, assuming that there are at leas $f \geq 4$ images, we propose a simple approach based on LMEDS [6].

The main idea is to approximate at first the space of the possible visibility configuration by randomly sampling triplets ω of lights. Fixed a pixel i, a tentative normal vector is estimated via least square regression for every lighting triplets. Hence, the normal $\widehat{N}_i$ which minimise the median of squared residuals is retained as a solution. By scrutinising the residual vector $I^i - \max(0, \widehat{N}_i^\top L)$, a binary weighting vector w_i is defined setting its j-th entry equals 1, if the j-th error is smaller of $2.5\hat{\sigma}$, and 0 otherwise, where $\hat{\sigma}$ is a robust estimate of the variance of the per pixels residuals defined by:

$$\hat{\sigma} = 1.4826(1 + 5/(f - 3))\sqrt{\operatorname{median} r_{\omega}^2} \qquad (11)$$

At the end, the normal estimate $\widehat{N}_i$ is refined using iteratively reweighted least squares (IRLS) on the set of lights $\{L_j : w_i^j = 1\}$.

The matrix $W = [w_1, \ldots, w_p]^\top$, composed by the weight-vectors arranged by row, could be used as a proxy for the visibility matrix, however here we take light directions into account, and we obtain a visibility matrix $\widehat{V}$ setting:

$$\widehat{V}_j^i = \begin{cases} 1 \text{ if } \widehat{N}_i^\top L_j > 0, \\ 0 \text{ otherwise.} \end{cases} \tag{12}$$

5   Evaluation the DiLiGenT Dataset

The methods presented in the previous section are here challenged on the *Di*rectional *Li*ghtings, objects of *Gen*eral reflectance, and ground *T*ruth shapes datasets (DiLiGenT) [7], a recently proposed benchmark of ten objects shown in Fig. 2.

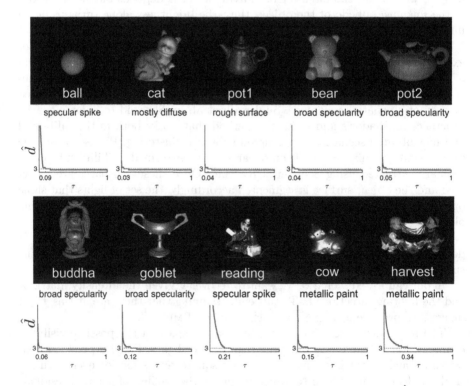

Fig. 2. The ten object of the DiLiGenT datasets with the respective $\hat{d}$ index.

This collection offers a great variety in terms of materials, appearances, geometries and type of deviations from the Lambertian model – from sparse specular spikes to broad specular lobes. This miscellany of non-diffusive phenomena can be captured analyzing the behavior of the index

Fig. 3. Example of visibility masks on the *ball* and on the *pot1* dataset. Visibility patterns are color coded: same colors correspond to the same shadowing configurations. (Color figure online)

$$\hat{d} = \min_{d} \left\{ \frac{\sum_{i=d+1}^{f} \sigma_i^2}{\sum_{i=1}^{f} \sigma_i^2} < \tau \right\} \tag{13}$$

with respect to τ which represents the smallest number such that the fraction of information discarded by the corresponding rank approximation is less than a threshold τ: the last three objects are the ones that deviates more from the rank-3 Lambertian approximation.

Visibility Mask. Sample results attained by this method on the DiLiGenT dataset are shown in Fig. 3, where it can be appreciated that $\hat{V}$ well approximates the ground truth visibility – computed as in Eq. 4 using the ground truth normals. As a reference, we also compare the LMEDS approach with the one based on visibility subspace [8]. Some differences can be pointed out. First, the extraction of visibility subspaces requires two parameters, namely the inlier threshold of RANSAC and a threshold on the magnitude of light. The inlier threshold is not always an educated guess, as the subspace estimation may be strained by the presence of highlighted pixels whose intensity profiles follow a different distribution with respect to shadowed points. LMEDS, on the contrary, is parameter-free and avoids this difficulties.

Second, LMEDS estimates the visibility configuration locally per pixels, visibility subspace, on the other hand, are estimated globally and pixels that lie in the intersections of multiple subspaces are not properly handled. Third, the random sampling performed to extract the visibility subspace acts on pixels, therefore the dimension of possible samples is $\binom{p}{3}$, which, as usually $p > f$ is higher than the upper bound on the number of samples of LMEDS $\binom{f}{3}$. Finally, LMEDS procedure can be parallelized in a straightforward way.

The visibility masks estimated by LMEDS are fed to the matrix completion algorithms to reduce the influence of shadowed pixels on the low rank estimation step.

Normal Estimation. We randomly chose 9 different lighting configurations for each dataset, and we compare the estimated normals with the ground truth ones, averaging the results on 10 trials per dataset. The rank was fixed to 3 for Grasta and R-GoDec, and the regualarization parameter to $\lambda = 1/\sqrt{p}$, whereas for L1-ALM we used $\lambda = 10^{-3}$ as suggested in the authors implementation [1]. The performances of the matrix factorization methods are recorded in Table 1 where the mean, the median and the standard deviation of angular errors were

Table 1. Angular error on the DiLiGenT dataset

	Least Square			Lmeds			RPCA			L1-ALM			Grasta			R-Godec		
	mean	med	std	mean	med	std	mean	med	std	mean	med	std	mean	med	std	mean	med	std
ball	4.38	2.53	0.74	3.59	2.40	0.86	5.23	3.77	0.53	5.20	2.53	0.87	2.70	2.28	0.92	2.86	2.35	0.80
cat	9.09	6.78	0.67	8.50	6.45	0.84	10.05	7.28	1.22	10.85	7.64	6.39	8.24	6.23	1.27	8.24	6.28	1.24
pot1	9.46	6.83	0.70	8.93	6.43	0.93	9.33	6.68	0.93	9.41	6.69	0.99	8.69	6.11	0.95	8.87	6.43	0.87
bear	10.49	8.16	1.15	9.90	7.77	1.21	10.13	8.28	0.99	10.32	8.08	1.31	9.39	7.50	1.50	9.36	7.55	1.48
pot2	15.89	12.02	1.12	15.46	11.35	1.13	12.37	10.21	0.88	15.88	11.77	1.03	15.38	10.88	1.05	15.57	11.41	1.10
buddha	15.45	10.25	0.87	14.34	9.46	1.00	15.37	10.96	1.30	14.28	9.33	1.35	14.37	9.20	1.39	14.19	9.24	1.28
goblet	19.43	15.72	0.71	18.65	14.76	0.75	17.26	13.96	0.91	20.82	16.21	5.51	18.43	14.03	0.81	18.39	14.05	0.83
reading	20.20	12.55	1.48	18.16	10.84	1.36	23.12	19.79	0.92	29.44	21.62	7.33	37.08	33.77	4.04	19.38	12.17	1.57
cow	26.48	26.87	0.75	25.63	25.75	0.84	15.00	13.95	1.37	33.02	28.16	8.78	31.83	31.15	2.43	26.66	26.58	0.61
harvest	31.19	25.59	0.61	30.34	24.16	0.59	27.74	22.37	1.61	35.75	29.57	3.18	33.88	27.59	2.55	32.31	24.73	0.96

reported for each method. As a reference we also detailed the errors attained by Least Square and LMEDS. When the accuracy of a method is worse than Least Square, we colored the corresponding cell with gray. Other colors are used to highlights the best results achieving the minimum error.

Grasta performed well on those datasets that manifest a clear diffusive component corrupted by local and sparse non-Lamberitan effects, whereas it worsened the results of Least Square estimation with respect to the last three sequences.

On the contrary RPCA achieved less accurate results on the first sequences and performed better on those challenging datasets characterized by board specularity and complex BRDF (pot2, goblet,cow and harvest have a metal appearance). R-GoDec behavior is similar to Grasta as can be sensed, looking at Fig. 4 – where the mean angular error is plotted for each sequence of the dataset.

Sample results of attained normals are shown in Fig. 5. One can also note that LMEDS always improved the performance of LS.

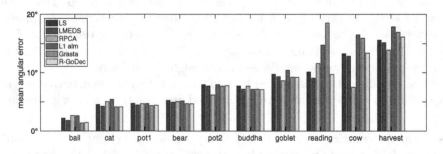

Fig. 4. Mean angular error on the diligent dataset (9 images, average on 10 trials)

Regularization Parameter. We recall that λ is a weighting parameter that is used by L1-ALM, R-GoDec and RPCA to balance between the low-rank and the

ground truth   RPCA   Grasta   R-GoDec

Fig. 5. Sample normal maps obtained on cat, buddha and harvest.

sparsity terms. In all the above experiments, this parameter was fixed. However, with better choice, it is possible to correct larger amount of outliers, enhancing the performance of the algorithms. Here we demonstrated the effect of λ on L1-ALM, R-GoDec and RPCA with respect to different number of input images we performed normal estimation on $4, 6, 12, 18, 24, 30, 36$ randomly drawn images using $C\frac{1}{\sqrt{p}}$ with $C \in \{0.05, 0.1, 0.2, 0.4, 0.8, 1, 1.2, 1.4, 1.6\}$. The corresponding mean angular errors are shown in Fig. 6, where it can be appreciated that L1-ALM and R-Godec benefit of the prior knowledge of rank being less sensitive to the number of images and the choice of λ. The minimum mean angular error per each datasets are reported in Table 2, where, for completeness, we also added the performance of Grasta varying only the number of images (λ is not required).

Table 2. Minimum mean error in degree varying λ and the number of images

	ball	cat	pot1	bear	pot2	buddha	goblet	reading	cow	harvest	mean	median
RPCA	2.68	7.41	7.42	6.38	10.40	11.99	14.56	15.04	10.80	25.58	11.23	10.60
L1-ALM	2.11	7.14	7.89	6.10	12.74	12.41	16.65	14.30	24.09	29.49	13.29	12.58
Grasta	2.11	7.13	7.95	6.11	12.78	12.45	16.64	20.81	25.03	29.60	14.06	12.61
R-godec	2.11	7.08	7.90	6.09	12.74	12.41	16.65	14.29	23.96	29.51	13.27	12.58

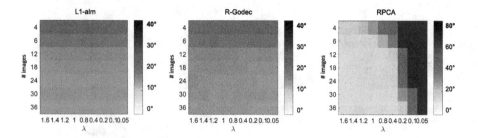

Fig. 6. Average angular error on the whole DiLiGenT dataset varying the number of images and the regularization parameter λ. (the scale of the colorbar is different for each methods) (Color figure online)

Distribution of Light Directions. In this experiment, we studied the effects of the distribution of light sources. We considered three different light configurations depicted in Fig. 7b: (A) 9 lights are randomly selected; (B) we choose a central light and the reaming 8 are those maximizing their distance from it; (C) we select 9 neighboring light sources. We run all the methods on the ball dataset, which is the only convex object and therefore results are less affected by the actual light orientations.

The summary of the experiment is that Grasta and R-GoDec preferred random and spread distribution, whereas RPCA and L1-ALM take advantage of the redundancy provided by the dense configuration.

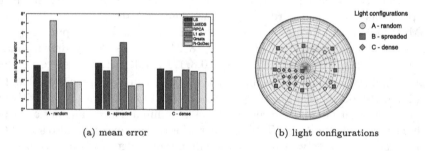

(a) mean error   (b) light configurations

Fig. 7. Varying the light configurations on the *ball* dataset

6   Conclusion

In this work, we tackle the problem of photometric stereo leveraging on robust matrix factorization techniques. We showed that the proposed shadow estimation based on LMEDS is able to produce accurate results, that, in turn, can be profitably fed to matrix completion algorithms. Experiments on a challenging datasets demonstrate that, if the object of interest is mostly Lambertian with strong and sparse non diffusive phenomena, it is advisable to adopt matrix approximation method with fixed rank. In this situation Grasta, followed by R-GoDec, performs better than L1-ALM. On the other side, if one is interested in recovering the normals of a surface that does not exhibit a strong diffusive behavior, all the methods suffer of low precision, but RPCA attains the more accurate results.

References

1. https://sites.google.com/site/yinqiangzheng/home/RegL1-ALM.zip?attredirects =0
2. Arrigoni, F., Magri, L., Rossi, B., Fragneto, P., Fusiello, A.: Robust absolute rotation estimation via low-rank and sparse matrix decomposition. In: 2014 2nd International Conference on 3D Vision (3DV), vol. 1, pp. 491–498. IEEE (2014)
3. Belhumeur, P.N., Kriegman, D.J.: What is the set of images of an object under all possible illumination conditions? Int. J. Comput. Vis. **28**(3), 245–260 (1998)
4. Chandraker, M., Agarwal, S., Kriegman, D.: ShadowCuts: photometric stereo with shadows. In: IEEE Conference on Computer Vision and Pattern Recognition CVPR 2007, pp. 1–8. IEEE (2007)
5. He, J., Balzano, L., Szlam, A.: Incremental gradient on the grassmannian for online foreground and background separation in subsampled video. In: 2012 IEEE Conference on Computer Vision and Pattern Recognition (CVPR), pp. 1568–1575. IEEE (2012)
6. Leroy, A.M., Rousseeuw, P.J.: Robust Regression and Outlier Detection. Wiley Series in Probability and Mathematical Statistics. Wiley, New York (1987)
7. Shi, B., Wu, Z., Mo, Z., Duan, D., Yeung, S.K., Tan, P.: A benchmark dataset and evaluation for non-Lambertian and uncalibrated photometric stereo. In: The IEEE Conference on Computer Vision and Pattern Recognition (CVPR), June 2016
8. Sunkavalli, K., Zickler, T., Pfister, H.: Visibility subspaces: uncalibrated photometric stereo with shadows. In: Daniilidis, K., Maragos, P., Paragios, N. (eds.) ECCV 2010 Part II. LNCS, vol. 6312, pp. 251–264. Springer, Heidelberg (2010). doi:10. 1007/978-3-642-15552-9_19
9. Woodham, R.J.: Photometric method for determining surface orientation from multiple images. Opt. Eng. **19**(1), 191139 (1980). http://dx.doi.org/10.1117/12. 7972479
10. Wu, L., Ganesh, A., Shi, B., Matsushita, Y., Wang, Y., Ma, Y.: Robust photometric stereo via low-rank matrix completion and recovery. In: Kimmel, R., Klette, R., Sugimoto, A. (eds.) ACCV 2010 Part III. LNCS, vol. 6494, pp. 703–717. Springer, Heidelberg (2011). doi:10.1007/978-3-642-19318-7_55
11. Zheng, Y., Liu, G., Sugimoto, S., Yan, S., Okutomi, M.: Practical low-rank matrix approximation under robust L1-norm. In: 2012 IEEE Conference on Computer Vision and Pattern Recognition (CVPR), pp. 1410–1417. IEEE (2012)
12. Zhou, T., Tao, D.: GoDec: randomized low-rank & sparse matrix decomposition in noisy case. In: Getoor, L., Scheffer, T. (eds.) Proceedings of the 28th International Conference on Machine Learning (ICML 2011), NY, USA, pp. 33–40. ACM, New York, June 2011

Dynamic 3D Scene Reconstruction and Enhancement

Cansen Jiang[(✉)], Yohan Fougerolle, David Fofi, and Cédric Demonceaux

Le2i, FRE CNRS 2005, Arts et Métiers,
University of Bourgogne Franche-Comté, Dijon, France
{cansen.jiang,yohan.fougerolle,david.fofi,
cedric.demonceaux}@u-bourgogne.fr

Abstract. In this paper, we present a 3D reconstruction and enhancement approach for high quality dynamic city scene reconstructions. We first detect and segment the moving objects using 3D Motion Segmentation approach by exploiting the feature trajectories' behaviours. Getting the segmentations of both the dynamic scene parts and the static scene parts, we propose an efficient point cloud registration approach which takes the advantages of 3-point RANSAC and Iterative Closest Points algorithms to produce precise point cloud alignment. Furthermore, we proposed a point cloud smoothing and texture mapping framework to enhance the results of reconstructions for both the static and the dynamic scene parts. The proposed algorithms are evaluated using the real-world challenging KITTI dataset with very satisfactory results.

Keywords: 3D reconstruction · 3D scene enhancement · Motion segmentation · Point cloud registration

1 Introduction

For the past decades, 3D scene reconstruction has been widely studied due to the need of many applications, such as city map modelling [1], robot navigation [2], autonomous driving [3], etc. Among numerous works in this context, the most representative approaches are: structure-from-motion of image sequence [4], RGB-D data fusion [5], and laser scans registration [6]. These approaches make use of the common assumptions that the environments are mostly static or contain very few moving objects. However, such assumptions do not hold for many practical scenarios, such as crowed campus and markets.

To address the problem of 3D reconstruction of dynamic environments, in our previous works [7,8], we proposed to detect and extract the moving objects prior to the scene reconstruction using a 2D-3D (RGB camera + 3D laser scanner) mobile camera system. Followed by, the static parts of the scene and the dynamic parts of the scene are independently reconstructed using a 3-point Random Sample Consensus (RANSAC) registration approach. Consequently, high quality static map and rigidly moving object reconstructions are achieved from

© Springer International Publishing AG 2017
S. Battiato et al. (Eds.): ICIAP 2017, Part I, LNCS 10484, pp. 518–529, 2017.
https://doi.org/10.1007/978-3-319-68560-1_46

highly dynamic environments. Since the 3-point RANSAC algorithm estimate the 3D-to-3D rigid transformation between two corresponding point sets, the accuracy of registration highly relies on the quality of corresponding sets. The 3D-to-3D feature correspondences are established by the tracking of their associated 2D (image) features which is sensitive to noise, as detailed in Sect. 4 [7]. Moreover, point cloud registration from long term observations inherently suffers from multi-layered problem due to the multiple scans of the same area. This problem can largely decrease the quality of the registration while increase the memory consumption. Building on top of [7,9], in this work, we propose a more robust and effective algorithm, call Dual-Weighted Iterative Closest Point (DW-ICP) algorithm, and a 3D reconstruction enhancement framework is presented to produce photographic quality results of real outdoor scenes.

Point Cloud Registration: Iterative Closest Point (ICP) is one of the most commonly used algorithm due to its simplicity and robustness. However, the convergence of ICP algorithm requires a good initialization and rich geometric structures of the point clouds. For instance, ICP registration of two planar objects can easily fall into a local minimum. To overcome these problems, we exploit that an initialization using 3-point RANSAC registration algorithm is very effective. Moreover, a DW-ICP algorithm is introduced to iteratively estimate the rigid transformation by assigning different weights to the RANSAC inlier point pairs and the ICP correspondences, as detailed in Sect. 4.

3D Reconstruction Enhancement: Due to the measurement noise of data, the 3D registration from multiple observations has multi-layered artefacts. To address this problem, we employ a 3D Thin Plane Spline algorithm which smooths the object surface to a single layer. Furthermore, a ball pivoting surface triangulation approach is applied to construct 3D meshes of the smoothed point clouds. Finally, the textures of the 3D meshes are mapped and refined using mutual information, as detailed in Sect. 5.

2   Related Work

State-of-the-art methods in 3D point cloud registration are categorized as: ICP-based point cloud alignment [10–12], RANSAC-based [1,7,13,14] point cloud registration, and volumetric representation-based point cloud fusion [5,15,16]. ICP-based methods are generally robust and accurate without prior knowledge of point-to-point correspondences. However, when the geometric structure of the point cloud is low, ICP registration yields to an ill-posed problem. RANSAC-based approaches are robust and efficient while they require sufficient number of precise 3D-to-3D matching pairs (at least 50% of them are inliers). Volumetric representation-based algorithms utilize the Signed Distance Function to describe the object surface using RGB-D camera. These methods work well for dense point cloud registration of large scene, but they suffer from over-smoothing problems.

3 Dynamic Scene 3D Reconstruction

In this section, we briefly revisit the principles of 3D reconstruction of dynamic scenes using 3D-based Sparse Subspace Clustering (3D-SSC) algorithm, see Fig. 1 Red Block. Given a mobile 2D-3D camera system, *i.e.* a car equipped with a 2D camera and a 3D laser scanner, our objective is to detect and extract the moving objects from a point cloud sequence, which yields to solve a Motion Segmentation (MS) problem. For this purpose, the 3D-SSC analyses the motion behaviours of the feature trajectories and segments them into independent motions. The principle of 3D-SSC is to construct an affinity matrix which encodes the similarity between the feature trajectories, followed by a spectral clustering algorithm to group the trajectories into their corresponding motion subspaces.

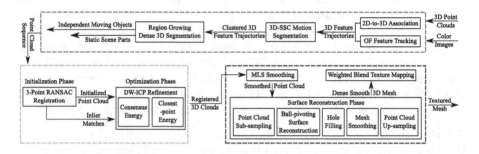

Fig. 1. Dynamic scene 3D reconstruction and enhancement framework: red block segments the point cloud into the dynamic and the static scene parts. Green block registers the point cloud sequence using our DW-ICP algorithm. Blue block refines the registered point cloud, followed by the texture mapping. (Color figure online)

Let $\mathbf{X} = [\mathbf{x}_1, \cdots, \mathbf{x}_F]^\mathsf{T}$ be a vectorized 3D feature trajectory of F frames, where $\mathbf{x}_i = [x, y, z] \in \mathbb{R}^3$ is a 3D feature point at frame i. Let $\mathbf{X} = [\mathsf{X}_1, \cdots, \mathsf{X}_P]$ be the assembly of P feature trajectories belonging to k different motions. Note that each independent motion determines a unique subspace. An element can be approximated by the linear combination of other elements from the same subspace, so called self-representation property. The self-representation model of MS problem is defined as a minimization problem:

$$\min \|\mathbf{C}\|_{1,1} \quad \text{s.t.} \quad \mathbf{X} = \mathbf{X}\mathbf{C}, \quad \text{diag}(\mathbf{C}) = 0, \tag{1}$$

where $\mathbf{C} = [\mathsf{C}_1, \cdots, \mathsf{C}_P]$ is a square-sized sparse permutation matrix, and operator $\|\cdot\|_{1,1}$ denotes the l_1-norm of each column of $\mathbf{C}$. The diagonal elements $\text{diag}(\mathbf{C})$ of $\mathbf{C}$ are constrained as zeros to avoid the trivial solution, so that X_i cannot be used to represent X_i itself. More specifically, the sparse vector $\mathsf{C}_i \in \mathbb{R}^P$ contains a few of non-zero elements such that $\mathsf{X}_i = \mathbf{X}\mathsf{C}_i$. The sparsity of C_i constrains that the least number of closest feature trajectories are selected, which contributes to its robustness to noise and outliers. By minimizing Eq. (1), the desired sparse permutation matrix $\mathbf{C}^*$ is obtained. Afterwards, a symmetric affinity matrix $\mathbf{A} = |\mathbf{C}^*| + |\mathbf{C}^*|^{\mathsf{T}}$ is constructed to perform K-means spectral clustering to separate the k independent motion subspaces. More details refer to [7].

4   Robust Point Cloud Registration

Point cloud registration from long term observations is a challenging problem. To tackle, we formulate an optimization problem that jointly minimizes both the feature matching energy and the nearest neighbour energy.

4.1   3-Point RANSAC Registration

Given a set of correspondences between two 3D point clouds, the rigid transformation parameters, *i.e.* R and t, can be estimated by solving a linear system. Let $\mathbf{x} = [x, y, z]^{\mathsf{T}}$ and $\mathbf{y} = [x', y', z']^{\mathsf{T}}$ be two corresponding points under rigid transformation, denoted as $\mathbf{x} = \mathsf{R}\mathbf{y} + \mathbf{t}$. In which, R is a 3×3 rotation matrix and t is a 3×1 translation vector. Let $\mathbf{g} = [g_x, g_y, g_z]^{\mathsf{T}}$ be the Gibbs representation [17] of the rotation matrix R, we have $\mathsf{R} = (\mathsf{I}_3 + \mathsf{G})^{-1}(\mathsf{I}_3 - \mathsf{G})$, where $\mathsf{G} = [\mathbf{g}]_\times$ is the skew-symmetric matrix form of $\mathbf{g}$ and I_3 is a 3×3 identity matrix.

By employing the Gibbs representation and the Cayley transform [18], the 3D registration problem is formulated as follow:

$$\left[-[\mathbf{x} + \mathbf{y}]_\times \ \ \mathsf{I}_3 \right] \begin{bmatrix} \mathbf{g} \\ \tilde{\mathbf{t}} \end{bmatrix} = \mathbf{x} - \mathbf{y}, \tag{2}$$

where $[\cdot]_\times$ denotes the skew-symmetric form of a vector and $\tilde{\mathbf{t}} = (\mathsf{I}_3 + \mathsf{G})\mathbf{t}$. Since each matching pair provides 2 independent equations, solving the 6 unknowns of Eq. (2) requires minimum 3 pairs of correspondences. For the sake of robustness to outliers, a RANSAC [19] framework is adopted, so called 3-Point RANSAC registration. The 3-point RANSAC point cloud registration algorithm is efficient and robust to outliers. However, in the presence of inaccurate correspondences, the quality of RANSAC registration is usually not very satisfactory. Therefore, we further propose to refine the registration by minimizing a dual-weighted closet-point energy taking into account both the RANSAC inlier matches as well as the full 3D point clouds.

4.2   Robust ICP Registration

When two overlapping point clouds of the same rigid object are given, the transformation between them is generally obtained by minimizing the energy derived from the closest-points distance. In most of the cases, this energy is minimized using an iterative method – also known as Iterative Closest Point (ICP) algorithm [20]. In each iteration, the ICP algorithm considers the closest points across two point clouds, say the reference and the model, as the corresponding ones. Let $\mathbf{X} = \{\mathbf{x}_1, \cdots, \mathbf{x}_n\}$ be the reference point cloud, and $\mathbf{Y} = \{\mathbf{y}_1, \cdots, \mathbf{y}_m\}$ be the new model, the robust method of ICP iteratively minimizes the following energy:

$$\mathcal{E}_I(\hat{\mathsf{T}}) = \min_{\mathsf{T}} \sum_{i=1}^{n} \rho(\min_{j \in \{1, \cdots, m\}} \|\mathbf{x}_i - \mathsf{T}\mathbf{y}_j\|_2), \tag{3}$$

where $\hat{\mathsf{T}}$ is the desired transformation matrix that relates the two point clouds. Note that the energy term $\mathcal{E}_I$ includes a robust cost function to handle noisy and partial data. Our choice of robust cost, say $\rho(x)$, is the Tukey's biweight function [21]:

$$\rho(x) = \begin{cases} (\tau^2/6)(1 - [1 - (x/\tau)^2]^3) & \text{if } |x| \leq \tau \\ (\tau^2/6) & \text{if } |x| > \tau \end{cases}, \tag{4}$$

and the weight of each corresponding pair is defined by:

$$w(x) = \frac{1}{x}\frac{\mathrm{d}\rho(x)}{\mathrm{d}x} = \begin{cases} [1 - (x/\tau)^2]^2 & \text{if } |x| \leq \tau \\ 0 & \text{if } |x| > \tau \end{cases}, \tag{5}$$

where τ is the inlier threshold, such that outliers $(|x| > \tau)$ are assigned with zero weights.

4.3   Dual-Weighted ICP Registration

While consensus-based registration method requires a subset of accurate correspondences, closest-point-based method requires rich structure of the point clouds. These prohibit us to make a choice of one method over another. Therefore, we propose to minimize a combined energy function – one from consensus, say $\mathcal{E}_R$, and the other from closest-point, say $\mathcal{E}_I$. We minimize the joint energy function in an iterative manner, named as dual-weighted ICP.

First, we define an energy function that measures the quality of the inlier set obtained from 3-point RANSAC. Note that due to the sparsity and noise, the inlier set obtained from RANSAC is not precise. Let $\{\mathbf{x}_i \leftrightarrow \mathbf{y}_i\}, i = 1, \ldots, k$ be the inlier correspondence set, the energy $\mathcal{E}_R$ for matching consensus is expressed as:

$$\mathcal{E}_R(\hat{\mathsf{T}}) = \min_{\mathsf{T}} \sum_{i=1}^{k} \tilde{\rho}(\|\mathbf{x}_i - \mathsf{T}\mathbf{y}_i\|_2), \tag{6}$$

where $k \leq m, n$, and $\tilde{\rho}(x)$ is the Huber's weight function denoted as:

$$\tilde{\rho}(x) = \begin{cases} (x^2/2) & \text{if } |x| \leq \tilde{\tau} \\ \tilde{\tau}[|x| - (\tilde{\tau}/2)] & \text{if } |x| > \tilde{\tau} \end{cases}, \tag{7}$$

$$\tilde{w}(x) = \frac{1}{x}\frac{\mathrm{d}\tilde{\rho}(x)}{\mathrm{d}x} = \begin{cases} 1 & \text{if } |x| \leq \tilde{\tau} \\ (\tilde{\tau}/|x|) & \text{if } |x| > \tilde{\tau} \end{cases}, \tag{8}$$

where $\tilde{\tau}$ is the threshold for inlier matches. The Huber loss function is selected under the assumption that the provided inlier set is noisy without severe outlier that needs to be completely discarded. In the spirit of Eqs. (3) and (6), we formulate our combined energy function as follows:

$$\mathcal{E}(\hat{\mathsf{T}}) = \min_{\hat{\mathsf{T}}} \left\{ \alpha \sqrt{\frac{1}{n} \sum_{i=1}^{n} \rho(\min_{j \in \{1, \cdots, m\}} \|\mathbf{x}_i - \mathsf{T}\mathbf{y}_j\|_2)} + \right.$$
$$\left. (1 - \alpha) \sqrt{\frac{1}{k} \sum_{i=1}^{k} \tilde{\rho}(\|\mathbf{y}_i - \mathsf{T}\mathbf{y}_i\|_2)} \right\}, \tag{9}$$

where α is the regularization term to control the influence of the $\mathcal{E}_I$ and $\mathcal{E}_R$ energy terms. Rather than optimizing the closest-point energy $\mathcal{E}_I$ or matching consensus energy $\mathcal{E}_R$ independently, the DW-ICP aims to iteratively and simultaneously optimize the joint energy $\mathcal{E}$ of Eq. (9).

4.4 Discussions

As summarized in Fig. 1 Green Block, our algorithm takes the 3-Point RANSAC registration as initialization. Afterwards, the DW-ICP is applied to refine the registration. Note that (also refer to Eq. (9)) the DW-ICP iteratively minimizes the combined energy of $\mathcal{E}_R$ and $\mathcal{E}_I$. On the one hand, $\mathcal{E}_I$ minimizes the overall registration error of the whole 3D point clouds. On the other hand, $\mathcal{E}_R$ minimizes the registration error of the inliers obtained form RANSAC. These two terms are usually complementary to each other, which is the key to the success of the proposed optimization framework. On top of the traditional ICP, there are two main advantages of our DW-ICP: (a) Feature matching constraint promises a proper registration regardless of the poor geometry structures of the point clouds. (b) Robust estimation framework is preserved such that the algorithm is generic and robust to outliers during a long term registration.

5   3D Reconstruction Enhancement

A complete pipeline for 3D reconstruction refinement is introduced to produce photo-realistic high quality 3D models, as shown in Fig. 1 Blue Block. There are three major steps involved, namely Moving Least Square (MLS) [22] point cloud smoothing, Surface Reconstruction [23], and Weighted Blend Texture Mapping [24]. Figure 2 depicts the evolutions of a car object from raw registered point cloud to high quality textured mesh.

Fig. 2. Illustration of 3D reconstruction enhancement: from left to right are raw registration, smoothed point cloud, surface reconstruction, textured mesh in side view, back view and top view, respectively.

Point Cloud Smoothing: The registered point cloud from long term observation suffers from outliers and multi-layered effects due to the measurement noise and imperfect registrations. Surface reconstruction using such point cloud suffers from many visual artefacts, such as spiky surfaces and holes. Therefore, a MLS algorithm, which smooths an unorganized point could using a polynomial fitting, is applied due to its simplicity and effectiveness.

Surface Reconstruction: To avoid the redundant (overlapped) points caused by multiple observations, a sub-sampling processing is performed based on the points' poisson-disk distribution [25]. Later on, a Ball Pivoting triangulation (or Poisson triangulation) algorithm is utilized to establish the neighbour-points relationships, followed by a dilation operation for hole closing. The Taubin Surface Smoothing [26] method is adopted to smooth the reconstructed surface while preserving the sharp edges. Finally, a Least Square Subdivision approach [27] is performed to refine and produce high quality meshes.

Texture Mapping: We make use of the 2D images for texture mapping. During this process, photographic alignment between the 3D mesh and the 2D images are required. Since the 2D-3D camera system is calibrated, and the motion of the camera is known, all the images are aligned with respect to the mesh reconstructed frame. The camera poses (between the cameras and the reconstructed mesh) are estimated by computing the inverse of the transformation matrices (obtained from registration) and using the camera calibration parameters. Furthermore, the blurring effect during the texture fusion from multiple images is reduced by using a Weighted Blending algorithm.

6 Experiments

We conducted experiments on both synthetic and real data (KITTI benchmark [28]). Since there is no ground truth data available for 3D reconstruction quantification, we generated three sets of synthetic data to quantify the robustness and accuracy of the proposed algorithms. Qualitative results of the proposed framework is presented using real data. All the experiments are conducted in a computer with Intel Quad Core i7-2640M, 2.80 GHz, 8 GB Memory. The algorithm parameters were set as: $\alpha = 0.8$, $\tau = 0.08$ m, $\tilde{\tau} = 0.03$ m, rotation tolerance $\epsilon_R = 10e-6$, translation tolerance $\epsilon_T = 10e-6$, and max iteration as 100.

Synthetic Datasets: The synthetic datasets were generated from three different objects, namely the Van, Red Car, and Cola Truck, see Fig. 4 for example. We simulate the motion behaviours the rigidly moving objects with smooth rotation and translation of 100 frames. Practical scenarios, such as partial overlaps, occlusions, and poor 3D geometric structures, are also taken into consideration. We applied 10 different levels of Gaussian noise, from 0.005 to 0.050 in meters. The maximum noise level is chosen as 2.5 times higher than the expected accuracy (0.02 m) of the Velodyne laser scanner. We compare the performances of the algorithms using the averaged absolute rotation and translation errors.

Fig. 3. High quality 3D reconstruction comparison: row 1 are selected images. Row 2 is the 3D reconstruction using [7]. Row 3 is the 3D reconstruction of the proposed method, which is more accurate than [7]. Last row is the textured reconstructed 3D mesh of static scene parts, where details of small objects are lost as shown in the zoom-in region.

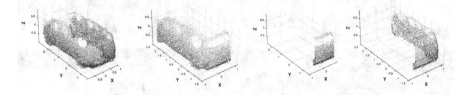

Fig. 4. Synthetic Van object with left, back and right side views.

Figure 5 shows the performances of 4 different algorithms, namely 3-Point RANSAC [7], RANSAC+ICP refinement [20], RANSAC+Robust-ICP [10] and RANSAC+DW-ICP. The overall performance of the algorithms are ranked (from top to down) as: DW-ICP, Robust-ICP, RANSAC+ICP and RANSAC. The Robust-ICP (using M-Estimator) has significantly better performance against that of traditional ICP. Most importantly, the proposed DW-ICP consistently outperforms the other approaches, regardless of rotation or translation.

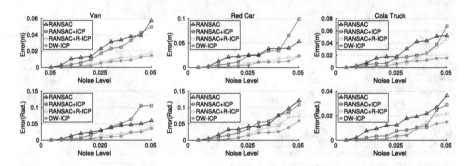

Fig. 5. Synthetic data quantification: top and bottom are averaged translation and rotation errors on Van, Red Car, and Cola Truck dataset, respectively. (Color figure online)

Real Datasets: Table 1 depicts the dataset information, where the 3D Error (averaged Leave-One-Out Error) metric was used to quantify the registration performance. The registration error of our method is consistently lower than [7], although we have slightly more computation time. Moreover, the high quality reconstructions of Figs. 2 and 6 were obtained using the proposed framework of Fig. 1. Note that the objects are reconstructed from long-term and faraway observations (see Table 1). The framework effectively overcomes the accumulation errors during the registration process and products very satisfactory results. Moreover, Figs. 3 and 6 demonstrates that significant better registration quality of our method is achieved compared to [7].

Table 1. Dataset Information: *Col. Sides* is number of object sides (left, right, back, and front) being captured. *Col. Dist.* is the averaged distance from the camera to the object. *Col. 3-Point RANSAC* [10] and *Col. Ours* show their respective averaged 3D error and computation time.

Object	# Frame	Sides	Dist. (m)	3-Point RANSAC [4]		Ours	
				Error (m)	Time (s)	Error (m)	Time (s)
Van	44	3	16.5	0.0150	3.1	**0.0131**	4.6
Red Car	60	3	10.8	0.0084	2.8	**0.0080**	4.3
Cola Truck	48	2	30.0	0.0234	3.7	**0.0229**	4.1

Fig. 6. Reconstructed Van and Cola truck: top are registered point clouds using [7]; bottom are our high quality meshes.

7   Conclusion and Future Work

We have proposed an effective high quality 3D reconstruction and enhancement framework which is evaluated using both synaesthetic and realistic outdoor dataset. The reconstructed 3D mesh of rigidly moving objects achieve photorealistic quality, while some small details of the large-scale 3D scene reconstruction are not well preserved. As future work, we expect to reconstruct the higher quality 3D mesh of the static scene parts.

References

1. Paudel, D.P., Demonceaux, C., Habed, A., Vasseur, P., Kweon, I.S.: 2D-3D camera fusion for visual odometry in outdoor environments. In: IROS (2014)
2. Castellanos, J.A., Tardos, J.D.: Mobile Robot Localization and Map Building: A Multisensor Fusion Approach. Springer Science and Business Media, Heidelberg (2012)
3. Berger, C., Rumpe, B.: Autonomous driving-5 years after the urban challenge: the anticipatory vehicle as a cyber-physical system. arXiv:1409.0413 (2014)
4. Pollefeys, M., Nistér, D., Frahm, J.M., Akbarzadeh, A., Mordohai, P., Clipp, B., Engels, C., Gallup, D., Kim, S.J., Merrell, P., Salmi, C.: Detailed real-time urban 3D reconstruction from video. IJCV **78**(2), 143–167 (2008)

5. Izadi, S., Kim, D., Hilliges, O., Molyneaux, D., Newcombe, R., Kohli, P., Shotton, J., Hodges, S., Freeman, D., Davison, A., Fitzgibbon, A.: KinectFusion: real-time 3D reconstruction and interaction using a moving depth camera. In: Proceedings of ACM Symposium on User Interface Software and Technology, pp. 559–568, October 2011

6. Zhang, J., Singh, S.: Low-drift and real-time lidar odometry and mapping. Auton. Robots **41**(2), 401–416 (2017). Springer

7. Jiang, C., Paudel, D.P., Fougerolle, Y., Fofi, D., Demonceaux, C.: Static-map and dynamic object reconstruction in outdoor scenes using 3-D motion segmentation. IEEE Robot. Autom. Lett. (RAL) **1**(1), 324–331 (2016)

8. Jiang, C., Paudel, D.P., Fougerolle, Y., Fofi, D., Demonceaux, C.: Reconstruction 3D de scènes dynamiques par segmentation au sens du mouvemen, In: Le 20èmé Congrès National sur la Reconnaissance des Formes et l'Intelligence Artificielle (RFIA), Clermont-Ferrand, France, June 2016

9. Jiang, C., Christie, D., Paudel, D.P., Demonceaux, C.: High quality recontruction of dynamic objects using 2D–3D camera fusion. In: Proceedings of International Conference on Image Processing (ICIP) (2017)

10. Fitzgibbon, A.W.: Robust registration of 2D and 3D point sets. Image Vis. Comput. **21**(13), 1145–1153 (2003)

11. Pomerleau, F., Colas, F., Siegwart, R., et al.: A review of point cloud registration algorithms for mobile robotics. Found. Trends® Robot. **4**(1), 1–104 (2015). Now Publishers, Inc

12. Attia, M., Slama, Y., Kamoun, M.A.: On performance evaluation of registration algorithms for 3D point clouds. In: Proceedings of Computer Graphics, Imaging and Visualization (CGiV) (2016)

13. Pankaj, D.S., Nidamanuri, R.R.: A robust estimation technique for 3D point cloud registration. Image Anal. Stereol. **35**(1), 15–28 (2016)

14. Christie, D., Jiang, C., Paudel, D.P., Demonceaux, C.: 3D reconstruction of dynamic vehicles using sparse 3D-laser-scanner and 2D image fusion. In: Proceedings of International Conference on Informatics and Computing (ICIC) (2016)

15. Yuheng Ren, C., Prisacariu, V., Murray, D., Reid, I.: STAR3D: simultaneous tracking and reconstruction of 3D objects using RGB-D data. In: ICCV (2013)

16. May, S., Koch, P., Koch, R., Merkl, C., Pfitzner, C., Nüchter, A.: A generalized 2D and 3D multi-sensor data integration approach based on signed distance functions for multi-modal robotic mapping. In: VMV, pp. 95–102 (2014)

17. Gibbs, J.W.: Elements of Vector Analysis: Arranged for the Use of Students in Physics. Tuttle Morehouse & Taylor, New Haven (1884)

18. Diele, F., Lopez, L., Peluso, R.: The Cayley transform in the numerical solution of unitary differential systems. Adv. Comput. Math. **8**(4), 317–334 (1998)

19. Fischler, M.A., Bolles, R.C.: Random sample consensus: a paradigm for model fitting with applications to image analysis and automated cartography. Commun. ACM **24**(6), 381–395 (1981)

20. Besl, P.J., McKay, N.D.: Method for registration of 3-D shapes. In: Robotics-DL Tentative, pp. 586–606 (1992)

21. Cressie, N., Hawkins, D.M.: Robust estimation of the variogram. J. Int. Assoc. Math. Geol. **12**(2), 115–125 (1980)

22. Lancaster, P., Salkauskas, K.: Surfaces generated by moving least squares methods. Math. Comput. **37**(155), 141–158 (1981)

23. Bernardini, F., Mittleman, J., Rushmeier, H., Silva, C., Taubin, G.: The ball-pivoting algorithm for surface reconstruction. IEEE Trans. Vis. Comput. Graph. **5**(4), 349–359 (1999)

24. Callieri, M., Cignoni, P., Corsini, M., Scopigno, R.: Masked photo blending: mapping dense photographic data set on high-resolution sampled 3D models. J. Comput. Graph. **32**(4), 464–473 (2008)
25. Corsini, M., Cignoni, P., Scopigno, R.: Efficient and flexible sampling with blue noise properties of triangular meshes. IEEE Trans. Vis. Comput. Graph. **18**(6), 914–924 (2012)
26. Taubin, G.: Curve and surface smoothing without shrinkage. In: ICCV (1995)
27. Boyé, S., Guennebaud, G., Schlick, C.: Least squares subdivision surfaces. Comput. Graph. Forum **29**(7), 2021–2028 (2010)
28. Geiger, A., Lenz, P., Stiller, C., Urtasun, R.: Vision meets robotics: the KITTI dataset. IJRR **32**(11), 1231–1237 (2013)

Feature Points Densification and Refinement

Andrey Bushnevskiy$^{(\boxtimes)}$, Lorenzo Sorgi, and Bodo Rosenhahn

Leibniz University of Hannover, Appelstraße 9a, 30167 Hannover, Germany
{andrey.bushnevskiy,bodo.rosenhahn}@tnt.uni-hannover.de

Abstract. A large part of computer vision algorithms and tools rely on feature points as an input data for the future computations. Given multiple views of the same scene, the features, extracted from each of the views can be matched, establishing correspondences between pairs of points and allowing their use in higher-level computer vision applications, such as 3D scene reconstruction, camera pose estimation and many others. Nevertheless, two matching features often do not represent the same physical 3D point in the scene, which may have a negative impact on the accuracy of all the further processing. In this work we suggest a feature refinement technique based on a Harris corner detector, which replaces a set of initially detected feature points with a more accurate and dense set of matching features.

Keywords: Feature points · Densification · Dense · Reconstruction · SIFT · SURF · GFTT · KLT · Harris corners · FREAK

1 Introduction

Feature points extraction is a powerful tool, which has found multiple applications in the field of computer vision. Features are descriptive points, which, being extracted from multiple views of the same scene, are to be matched and further applied in higher-level algorithms, i.e. 3D reconstruction, camera pose estimation, SfM and many others. The specific challenges while working with feature points are improving the performance of the extraction task, minimizing the number of incorrectly identified matches, ensuring localization accuracy of the points in detected matches with respect to the 3D points of the captured scene.

Among the most popular feature point detectors are Harris corner detector [1] and GFTT (Good Features to Track) [2], which, however, do not provide the scale and rotation invariance. Thus, often an additional data structure, called a descriptor, is used for feature points comparison and matching. One of the most well-known descriptor-based feature types is SIFT (Scale Invariant Feature Transform) [3], which is providing invariance to a uniform scale, rotation and partially to affine distortion. The SURF (Speeded up robust features) detector and descriptor based on a fast Hessian detector approximation and a gradient-based descriptor is presented in [4]. In the [5] FREAK (Fast Retina Keypoint) keypoint descriptor inspired by the human retina has been presented, which

© Springer International Publishing AG 2017
S. Battiato et al. (Eds.): ICIAP 2017, Part I, LNCS 10484, pp. 530–538, 2017.
https://doi.org/10.1007/978-3-319-68560-1_47

also provides rotation and scale invariance as well as an advantage in terms of performance. The performance of several types of feature point descriptors has been evaluated under different conditions in [6], confirming the advantages and robustness of SIFT descriptor. The performance of a number of feature detectors and descriptors has also been evaluated in [7] for the task of 3D object recognition. The results of the comparison suggest that SIFT and affine rectified [8] detectors the are the best choice for the task due to their robustness to change of viewpoint as well as changes in lighting and scale.

A new type of scale-invariant feature points is presented in [9]. There the Harris corner detector is combined with SIFT descriptor in order to obtain scale invariance and achieve real-time performance for the tasks of tracking and object recognition by skipping a time consuming scale space analysis. Recent works are applying a deep learning approach to the task of feature extraction. The LIFT (Learned Invariant Feature Transform) [10] presents a deep network architecture trained using a sparse multi-view 3D reconstruction of a scene, which implements three pipeline components, namely feature detection, feature orientation estimation and descriptor extraction.

In this paper we are presenting a novel approach for replacing an initial set of SIFT or other type of feature points with a new and more accurate set of Harris corner matches, extracted from the local neighbourhoods of the matching pairs of the initial set. We test the performance and demonstrate the efficiency of the proposed approach for the tasks of camera pose estimation and sparse point cloud reconstruction.

2   Feature Points Densification and Refinement

Typically, the task of scale-invariant feature extraction is performed on scaled-down versions of original images in order to improve the performance, ensure robustness of the algorithm and maximize the number of correctly detected feature matches [11,12]. Feature points in a correct match, however, often do not represent the same physical 3D point of the object. If the feature point in the first image is considered a reference, the matching feature in the second image may be displaced from a corresponding image point by a few pixels (Fig. 1), which affects the accuracy of the further processing. The number of extracted features may also be significantly reduced for the same reason. Moreover, for the tasks of 3D scene reconstruction and representation, the most descriptive and suitable points are corners, which may be naturally omitted by some of the feature detectors [13].

The approach presented in this paper is aimed at handling these factors by providing a new set of precise corner points, allowing for an accuracy improvement for all the further computer vision applications. The proposed feature densification pipeline is comprised of three steps, namely feature initialization, iterative feature patch warp and tracking of new refined feature points.

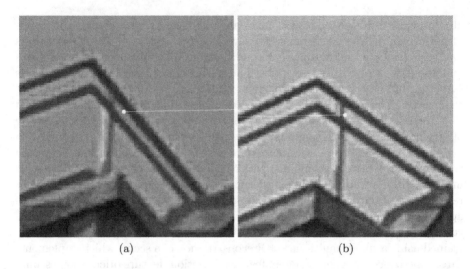

Fig. 1. SIFT features in the original images. Corresponding matching SIFT feature points in the left and right images are having a noticeable displacement.

2.1   Initialization

The proposed algorithm requires an initial set of conventional feature points and matches to be detected in the corresponding pairs of scaled-down images. In this paper we are considering SIFT feature points, however, the approach can also be adapted to the other types of features, such as SURF, FREAK or GFTT.

2.2   Feature Patches Warp

Each feature point depicts image content in its neighborhood, which can be described by an image patch with its center coinciding with the feature point location (Fig. 2(a)). Since two matching features represent the same 3D point of the

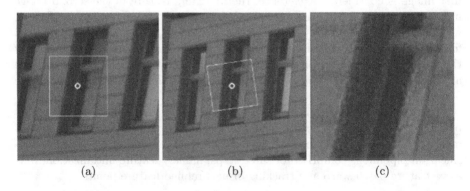

(a)   (b)   (c)

Fig. 2. A patch in the reference image (a) and the corresponding patch in the target image (b), warped using the estimated homography H (c).

captured scene, their corresponding image patches would represent the same area of the scene. Therefore, a new search for matching feature points can be performed locally in corresponding patches of each pair of initial matching features.

Nevertheless, in case of using scale and rotation invariant features (i.e. SIFT, SURF), two image patches have to be transformed before a local feature search can be performed in order to compensate the differences in the scale and orientation of their seed feature points. If one of the images is considered a reference and second a target, for each feature match, it is possible to define a homography, which is relating the reference image patch and the target image:

$$p_t = H \cdot p_{p_r}, \tag{1}$$

where $p_t = (x_t, y_t)$ and $p_{p_r} = (x_{p_r}, y_{p_r})$ are the points in the target image and the reference patch respectively. The size of the reference patch can be defined with respect to the scale of the seed feature point using a user-defined multiplication factor (1.3–2.7 in our experiments). The homography H can be approximated using the positions of two matching feature points together with their orientation, and scale parameters:

$$H = T_2 \cdot S_2 \cdot R_2 \cdot R_1^{-1} \cdot S_1^{-1} \cdot T_1^{-1}, \tag{2}$$

where

$$T_1^{-1} = \begin{bmatrix} 0 & 0 & -x_r + x_{p_{rc}} \\ 0 & 0 & -y_r + y_{p_{rc}} \\ 0 & 0 & 0 \end{bmatrix}, T_2 = \begin{bmatrix} 0 & 0 & -x_t \\ 0 & 0 & -y_t \\ 0 & 0 & 0 \end{bmatrix}, \tag{3}$$

$p_{p_{rc}} = (x_{p_{rc}}, y_{p_{rc}})$ is the top left corner point of the feature patch in the reference image, R_1 and R_2 are the rotation matrices built using orientation angles of the features, S_1 and S_2 are the corresponding feature scale matrices.

Once the homography H is known, the target image (Fig. 2(b)) can be warped and cropped to the target patch (Fig. 2(c)) representing the same part of the scene as the reference, allowing for extraction and tracking of a new feature set.

2.3 Feature Densification and Refinement

The new set of feature points is first extracted from the reference patch using the Harris corner detector [1]. The detected points are then tracked in the transformed target image patch using an iterative Lucas-Kanade tracker [14] (Fig. 3). It is important to mention, that the number of tracked point in the target patch depends on the patch content as well as the size of the reference patch and the quality of the initially detected feature match. Thus, one feature point in the initial set may produce multiple feature points within one patch in a refined set.

The newly extracted and tracked features are then brought back to the reference and target image domains using the homography H:

$$\begin{cases} x_r = x_{p_r} + x_{p_{rc}}, \\ y_r = y_{p_r} + y_{p_{rc}} \end{cases} \tag{4}$$

and

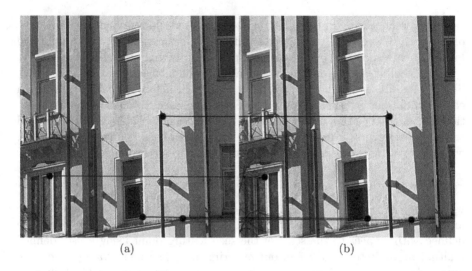

(a)   (b)

Fig. 3. A reference patch with extracted corners (a) and a target patch with the tracked points (b).

$$\begin{cases} x_t = \dfrac{h_{00} \cdot x_{p_t} + h_{01} \cdot y_{p_t} + h_{02}}{h_{20} \cdot x_{p_t} + h_{21} \cdot y_{p_t} + h_{22}}, \\ y_t = \dfrac{h_{10} \cdot x_{p_t} + h_{11} \cdot y_{p_t} + h_{12}}{h_{20} \cdot x_{p_t} + h_{21} \cdot y_{p_t} + h_{22}} \end{cases} \tag{5}$$

where (x_r, y_r) are the coordinates of the new feature, extracted from the reference patch, in the reference image and (x_t, y_t) are the coordinates of the matching feature in the target image.

The points extracted from the reference image and their matches tracked in the target image are then added to the new feature set and the next match from the initial set is processed (Fig. 4).

Fig. 4. A set of refined feature points matches.

3 Results

In order to provide a quantitative evaluation of the proposed approach, we have created a dataset comprised of 30 stereo image pairs taken at a 12 MP resolution, using a calibrated camera of a mobile device.

For each image pair, we have performed the tasks of SIFT and SURF features extraction and matching using the scaled-down versions of the original images with the maximum image width of 1024 px. The feature points of this initial set then have been refined using the proposed method (Fig. 4). Each of two feature sets has been used for estimation of the camera poses using the approach described in [15] and triangulation of a sparse point cloud. The set of 3D points has been reprojected back on the images using the corresponding camera poses and camera model parameters. The error then has been evaluated as a pairwise Euclidean distance in pixels between the originally detected feature points and the backprojected point cloud (Fig. 6).

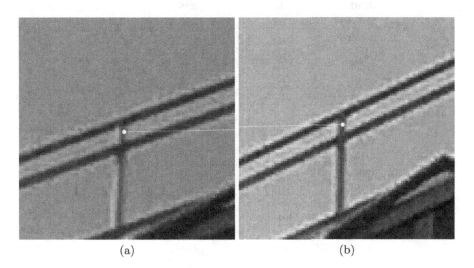

(a)   (b)

Fig. 5. A feature match from the refined set. Localization error is practically non-existent.

The results, presented in the Table 1, show that the proposed method allows for a significant increase in localization accuracy of the detected feature matches (Fig. 5). This accuracy improvement allows for a more precise estimation of the camera poses as well as a point cloud triangulation. A sample dataset image and the triangulated sparse point cloud, estimated using a refined feature set are shown in Fig. 7.

Table 1. Evaluation results. Number of detected feature matches and the 80th percentile of the backprojection error histogram for the initial and refined feature sets.

Type	Number of features	Backproj. error, [px]
Dataset 1		
SIFT	1214	0.60
Refined (SIFT)	1264	0.25
SURF	549	0.84
Refined (SURF)	1109	0.22
Dataset 2		
SIFT	702	1.84
Refined (SIFT)	675	1.06
SURF	215	3.57
Refined (SURF)	384	1.09
Dataset 3		
SIFT	290	0.84
Refined (SIFT)	256	0.67
SURF	65	2.36
Refined (SURF)	102	1.57
Dataset 4		
SIFT	319	2.23
Refined (SIFT)	305	1.02
SURF	103	4.59
Refined (SURF)	223	0.86
Dataset 5		
SIFT	841	0.94
Refined (SIFT)	891	0.54
SURF	178	2.20
Refined (SURF)	492	0.26
Dataset 6		
SIFT	1065	1.05
Refined (SIFT)	1051	0.92
SURF	212	2.42
Refined (SURF)	441	1.08

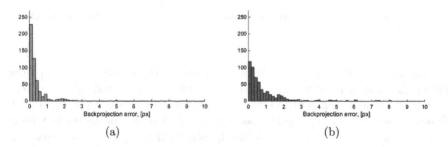

(a)   (b)

Fig. 6. Error histogram for the refined (a) and the initial SIFT features (b).

(a)   (b)

Fig. 7. A reference image and the corresponding sample sparse 3D reconstruction using the feature points from the refined set.

4   Conclusions

The paper presents a new approach for refinement of an initial set of SIFT, SURF or other types feature points. The initial set of matching features is replaced by a new set, obtained by performing a search for Harris corners in the corresponding patches, representing neighborhoods of the original feature points. In contrast to the original one, the new set features an improved localization accuracy as well as a smaller number of incorrectly identified matches. These two factors combined allow for a significant accuracy improvement for the computer vision applications, which are using feature points as an input.

The experimental results prove the efficiency of the proposed approach and demonstrate an accuracy improvement for the tasks of camera pose estimation and a 3D point cloud triangulation using a refined set of matching feature points.

References

1. Harris, C., Stephens, M.: A combined corner and edge detector. In: Proceedings of the 4th Alvey Vision Conference, pp. 147–151 (1988)
2. Shi, J., Tomasi, C.: Good features to track. Technical report, Ithaca, NY, USA (1993)
3. Lowe, D.G.: Distinctive image features from scale-invariant keypoints. Int. J. Comput. Vis. **60**(2), 91–110 (2004)
4. Bay, H., Ess, A., Tuytelaars, T., Van Gool, L.: Speeded-up robust features (SURF). Comput. Vis. Image Underst. **110**(3), 346–359 (2008)
5. Ortiz, R.: FREAK: fast retina keypoint. In: Proceedings of the 2012 IEEE Conference on Computer Vision and Pattern Recognition (CVPR), CVPR 2012, Washington, DC, pp. 510–517. IEEE Computer Society (2012)
6. Mikolajczyk, K., Schmid, C.: A performance evaluation of local descriptors. IEEE Trans. Pattern Anal. Mach. Intell. **27**(10), 1615–1630 (2005)
7. Moreels, P., Perona, P.: Evaluation of features detectors and descriptors based on 3D objects. Int. J. Comput. Vis. **73**(3), 263–284 (2007)

8. Mikolajczyk, K., Schmid, C.: An affine invariant interest point detector. In: Heyden, A., Sparr, G., Nielsen, M., Johansen, P. (eds.) ECCV 2002. LNCS, vol. 2350, pp. 128–142. Springer, Heidelberg (2002). doi:10.1007/3-540-47969-4_9

9. Azad, P., Asfour, T., Dillmann, R.: Combining Harris interest points and the SIFT descriptor for fast scale-invariant object recognition. In: 2009 IEEE/RSJ International Conference on Intelligent Robots and Systems, 11–15 October 2009, St. Louis, MO, USA, pp. 4275–4280 (2009)

10. Yi, K.M., Trulls, E., Lepetit, V., Fua, P.: LIFT: learned invariant feature transform. In: Leibe, B., Matas, J., Sebe, N., Welling, M. (eds.) ECCV 2016. LNCS, vol. 9910, pp. 467–483. Springer, Cham (2016). doi:10.1007/978-3-319-46466-4_28

11. Aly, M.: Face recognition using SIFT features (2006)

12. Yoshioka, M., Maeda, Y., Omatu, S.: Criterion for optimal image resolution using SIFT. Artif. Life Robot. 14(1), 24–28 (2009)

13. Peng, K., Chen, X., Zhou, D., Liu, Y.: 3D reconstruction based on SIFT and Harris feature points. In: IEEE International Conference on Robotics and Biomimetics, pp. 960–964 (2009)

14. Bouguet, J.Y.: Pyramidal implementation of the Lucas Kanade feature tracker description of the algorithm (2000)

15. Sorgi, L., Bushnevskiy, A.: Two view geometry estimation by determinant minimization. In: Magnenat-Thalmann, N., Richard, P., Linsen, L., Telea, A., Battiato, S., Imai, F.H., Braz, J. (eds.) VISIGRApp, vol. 3, pp. 592–596. SciTePress (2016)

Fast and Accurate Facial Landmark Localization in Depth Images for In-Car Applications

Elia Frigieri, Guido Borghi$^{(\boxtimes)}$, Roberto Vezzani, and Rita Cucchiara

University of Modena and Reggio Emilia, Via P. Vivarelli 10, 41125 Modena, Italy
{elia.frigieri,guido.borghi,roberto.vezzani,rita.cucchiara}@unimore.it

Abstract. A correct and reliable localization of facial landmark enables several applications in many fields, ranging from Human Computer Interaction to video surveillance. For instance, it can provide a valuable input to monitor the driver physical state and attention level in automotive context. In this paper, we tackle the problem of facial landmark localization through a deep approach. The developed system runs in real time and, in particular, is more reliable than state-of-the-art competitors specially in presence of light changes and poor illumination, thanks to the use of depth images as input. We also collected and shared a new realistic dataset inside a car, called *MotorMark*, to train and test the system. In addition, we exploited the public *Eurecom Kinect Face Dataset* for the evaluation phase, achieving promising results both in terms of accuracy and computational speed.

Keywords: Facial landmarks localization · Depth maps · Convolutional Neural Networks · Automotive

1 Introduction

The autonomous driving of on-road vehicles is one of the most challenging and actual problems for both research and industrial communities. In recent years, it is gathering the attention of numerous researchers from different disciplines, with a strong involvement of the ICT community. Among the others, Computer Vision is playing a leading role in two main aspects.

First, Computer Vision and Pattern Recognition disciplines are applied to assist or even replace traditional sensors in the perception of the surround context, *i.e.*, the outside world.

Second, the ability to monitor the behavior of passengers and drivers is fundamental, for example as a safety aid to enable full or semi-autonomous driving: the intervention of the driver or, at least, his/her attention can be requested by the automatic system in exceptional cases of need. In this case, vision-based systems must operate on images provided by internal cameras, installed and configured to monitor the passengers and the driver.

A reliable localization of facial landmarks – *i.e.*, the ability to infer the position of prominent face elements relative to the view of the acquisition

© Springer International Publishing AG 2017
S. Battiato et al. (Eds.): ICIAP 2017, Part I, LNCS 10484, pp. 539–549, 2017.
https://doi.org/10.1007/978-3-319-68560-1_48

device – is one of the basic component to conduct driver physical state investigation, through eyes or mouth direct monitoring [23], facial expressions recognition [19,27], head pose estimation [21], all fundamental elements also for driver attention analysis, as reported in literature [15].

Facial landmark localization is also an important task in Computer Vision, and a key element for many other fields, such as age estimation [14], sign language recognition [4] and various applications in biometrics [1].

Many solutions of facial landmark localization have been proposed in the last decades. However, the automotive context is characterized by specific issues such as strong occlusions, dramatic light changes, high head pose variability. Moreover, additional requirements like non-intrusivity of the acquisition device (no physiological signals, like EEG, ECG, EMG) and the avoidance of initialization or on-user training are preferable.

In this paper, we present a deep-based approach specifically designed for real time facial landmarks localization in the automotive context, through a regression manner approach. Proposed method is rely only on depth data to achieve a good reliability in presence of illumination changes. Moreover, a new challenging and deep-oriented dataset is collected to train and test the entire proposed system (Fig. 1).

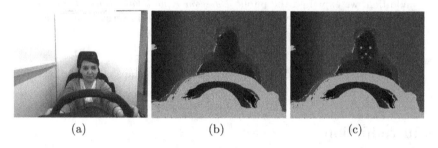

(a)   (b)   (c)

Fig. 1. Some visual samples of facial landmark localization on depth images. (a) input RGB frame, (b) input depth frame and (c) the depth frame with landmark annotation (green for eye pupils, blue for the nose tip and red for mouth corners). (Color figure online)

2   Related Work

Most of the systems for facial landmark localization in the literature are based on RGB images. However, these methods are prone to failure in case of illumination and pose changes. Only few works are based on depth images, even though the recent availability and diffusion of cheap and small sized depth acquisition devices.

Two main approaches are generally used in 3D landmark localization [10]: *heuristic* approaches and *statistical* methods.

Heuristic approaches rely on the properties of the face, like its symmetry or specific known shapes. In [1] the curvature information of the face is exploited to locate the nose tip and eye corners. In [35] facial features are exploited to build an automatic emotion annotation solution on depth maps and the solution consists of a facial landmarking method and a facial expression recognition technique. Also *Active Appearance Models* and their extensions have been proposed to tackle the problem of landmark localization task [3,29,33].

Statistical approaches exploit features directly extracted from the face and fed into a discriminative model. In [13] a *Linear Discriminant Analysis* and *AdaBoost* are exploited to perform landmarking on 3D facial models. In [25] landmark localization is performed in presence of occlusions and expression changes. A *Gappy Principal Component Analysis* is used in [2] to restore missing landmark coordinates. Fanelli *et al.* [16] used a *Random Forest* technique to place landmark on depth images. The *Supervised Descent Method* (SDM) is successfully exploited both with 2D and 3D data. For the case of 3D data, in [7,10] SDM achieves a good accuracy, relying on SIFT descriptor and HOG features. A global SDM is used in [32]. Besides, employing a cascade of regression functions good accuracy can be achieved [28,36].

Only recent works combine the use of *Convolutional Neural Networks* (CNN) and facial landmark localization task. In [22] a CNN is exploited for simultaneous face detection, landmark localization, pose estimation and gender recognition, starting from RGB images. In [34] landmark detection and localization is performed through a CNN in combination with logistic regressors. A cascade of CNN is used in [26] and requires a pre-partition of dataset faces in different parts. Each part is processed by a different network.

Very few works in the literature exploit on CNN and raw depth images as input. In the head pose estimation research field, there are some example [5, 6,30,31] in which normalized depth images acquired through *Microsoft Kinect* device are fed into a deep model that produces in output 3D head Euler angles.

The proposed method is one of the few works that combine depth images, also known as 2.5D images, and a deep architecture to localize facial landmark in a regression manner: this explains the lack of database and competitors for the presented work. This works aims to merge the CNN power in regression tasks, the use of depth maps and real time performance.

3   Proposed Method

The goal of the whole system is a reliable estimation of the facial landmark coordinates. Due to the limited spatial resolution of available depth images, we focus on a selection of five principal facial landmarks: eye pupils, mouth corners and the nose tip. Accordingly, the system outputs 10 coordinates, *i.e.*, the x and y values for each facial landmark.

The core of the method is a deep architecture that works in regression and receives a stream of depth images as input. The ground truth annotation of the landmark positions is required during the network training step and is used as a comparison during the test.

3.1   Image Pre-processing

Image pre-processing is an important step to reach high performance with deep approaches, as reported in [18]. All input images are equalized in an adaptive way, to enhance visual details: specifically, we have exploited the *Contrast Limited Adaptive Histogram Equalization* algorithm described in [37]. Besides, image values are scaled so that the mean and the variance of the values are 0 and 1, respectively. Head detection and localization are out of the scope of this paper; thus, we exploited the provided annotation during the experiments. However, the center of the driver head could be estimated with a face detector directly working on depth images [6,9,11] or locked to a predefined mean position. A fixed window containing the head as well as a small portion of background is cropped and, finally, all the cropped images are resized to 64 × 64 pixels. A visual example is provided in Fig. 2. Ground truth coordinates are normalized in the range $[-1, 1]$, accordingly to the specific activation function of the output network layer (Sect. 3.2).

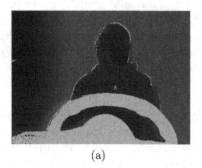 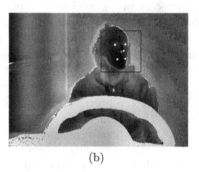

(a)   (b)

Fig. 2. Visual examples of the pre-processing step: (a) raw depth frame; (b) depth frame after the adaptive equalization. Facial landmarks are reported with green circles, the elaborated centroid in red and the square crop for face extraction is depicted with a red rectangle. (Color figure online)

3.2   Model Architecture

The model architecture is designed to deal with two main issues: low memory requirements and real time performance. The model is composed of 5 convolutional layers followed by three fully connected layers, composed by 120, 84 and 10 neurons, respectively. Between the fully connected layers are inserted 2 dropout layers (with $p = 0.5$). Due the limited input size, only the first three convolutional layers are followed by a max pooling layer of size (2×2). Hyperbolic tangent is exploited as activation function; as a consequence, the network outputs continuous values as required. We adopt *Adam* solver [17] with a initial learning rate set to 10^{-4}. Finally, a L_2 loss function is used:

$$L_2 = \sum_i^n ||y_i - f(x_i)||_2^2$$

where y_i is the $i - th$ coordinate from the ground-truth and $f(x_i)$ is the corresponding network output.

3.3 Training

The network has been trained with a decay value of 5^{-4} and a momentum value of 9^{-1}. The learning rate is automatically changed by the *Adam* solver.

A data augmentation procedure is implemented in order to increase the amount and the variety of input data. As outcome, the network is less prone to over fitting behaviors and its generalization capability are increased [18]. To this aim, random rotations, zoom-in and translations along x and y axis are applied to each input image. Moreover, a *Gaussian* noise is added (*jittering*).

4 Experimental Results

To train and test the proposed method and to collect experimental results, a new dataset has been defined, namely *MotorMark*, due to the lack of a public dataset containing high quality depth images, accurate landmark annotations and a sufficient amount of training data for a deep-based approach. We also exploit the publicly available *Eurecom Kinect Face Dataset* [20], that consists of multimodal facial images of 52 people (14 females, 38 males), obtained with the first version of the *Microsoft Kinect* device. Its limited size (around 700 frame in total) makes it not sufficient to train our deep architecture and thus we exploited it during the testing phase only.

Other existing datasets (*e.g.*, [8,12]) are not deep oriented or provide the annotation of facial landmarks only for the first frame or a subset of key frames.

4.1 MotorMark Dataset

We collected a dataset that includes RGB and the corresponding depth images, annotated with facial landmark coordinates on RGB and depth images. Frames are acquired through a *Microsoft Kinect One*. This *Time-of-Flight* depth device guarantees high quality images [24]. The dataset is publicly available[1]. The main features of *MotorMark* are:

- **Deep oriented**: is composed by more than 30k frames. A variety of subjects is guaranteed (35 subjects in total);
- **Automotive context**: we recreate an automotive context. The subject is standing in a real car dashboard (see Fig. 3) and performs real inside-car actions, like rotating steering wheel, shifting gears and so on;
- **Variety**: subjects are asked to follow a constrain path (4 led are placed in correspondence with the speedometer, the rev counter, infotainment system and the left wing mirror), to rotate their head in fixed position or to freely move their head. Besides, subjects can wear glasses, sun glasses and a scarf, to generate partial face and landmark occlusions;

---

[1] http://imagelab.ing.unimore.it/landmarkdepth.

- **Landmark annotations**: the annotation of 68 landmark positions on both RGB and depth frames is available, following the ISO MPEG-4 standard. The ground truth has been manually generated. The user was provided with an initial estimation done by means of the algorithm included in the *dLib* libraries[2], which gives landmark positions on RGB images. The projection of the landmark coordinates on the depth images is carried out exploiting the internal calibration tool of the *Microsoft Kinect SDK*.
- **High quality**: RGB and depth images are acquired with a spatial resolution of 1280 × 720 HD and 515 × 424, respectively;

Fig. 3. Sample frames from *MotorMark* dataset. Like in a real automotive context, subjects speak at the phone, drink, wear sunglasses or cap and perform different facial expressions.

4.2   Quantitative Evaluation

As previously described, input images are cropped around the head center. For our experiments, given the coordinates (x_i, y_i) of the $i - th$ facial landmark, we elaborate a face centroid of coordinates x_c, y_c computed as:

$$x_c = \frac{\sum_i^k x_i}{k}, \quad y_c = \frac{\sum_i^k y_i}{k} \tag{1}$$

where in our case $k = 5$. Based on x_c, y_c final square windows of 100×100 pixels are obtained. The dataset has been split in test (27 subjects) and train (all the remaining subjects) subsets.

Different tests, here referred as *baselines*, have been carried out and compared with the proposed pipeline:

1. A smaller window of 60×60 (instead of 100×100) is cropped in order to include less portions of background; all the extracted windows are then resized to 64×64 (baseline 1 in Fig. 4);

---

[2] www.dlib.net.

2. Final cropped input images are resized to 128×128, instead of 64×64, to enhance visual facial cues (baseline 2 in Fig. 4);

3. Background suppression is applied, through a threshold on depth values (baseline 3 in Fig. 4).

The results about previous three baselines and the presented method are reported in Fig. 4. In particular, we achieve a final average mean error of 0.97 pixel with a standard deviation of 0.84 pixel for all five facial landmarks on *MotorMark* dataset.

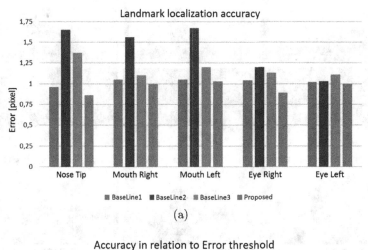

(a)

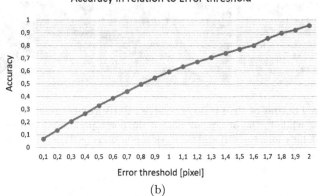

(b)

Fig. 4. (a) Mean and standard deviation of localization error on every facial landmark and for each test conducted; (b) landmark localization accuracy.

The experimental results show that the proposed system reaches good performance on *MotorMark* dataset. Moreover, we compare our system on *Eurecom Kinect Face Dataset* with the work of Zhao *et al.* [35]: Table 1 depicts the

Table 1. Results on *Eurecom Kinect Face* dataset, expressed as the mean error and the standard deviation in pixels w.r.t the ground truth, normalized by the interpupillary distance.

Method	Nose tip	Mouth right	Mouth left	Eye right	Eye left	Avg err
Zhao *et al.* [35]	4.4 ± 2.2	5.4 ± 3.2	5.4 ± 3.2	4.2 ± 2.1	4.2 ± 2.2	4.7 ± 2.6
Our	3.3 ± 4.5	3.5 ± 3.7	3.4 ± 3.9	3.5 ± 4.1	3.4 ± 4.0	3.4 ± 4.0

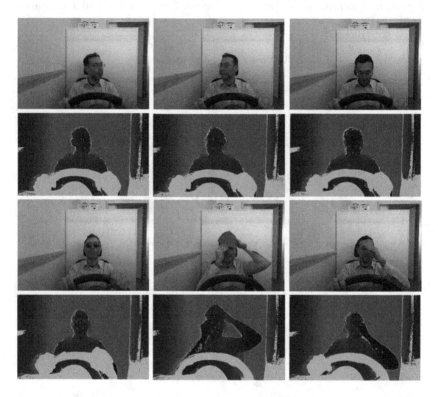

Fig. 5. The final output of the proposed system. RGB frames are reported in the first and third rows, while in the others are shown the corresponding depth maps. Ground truth landmark locations (green) and network predictions (red) are superimposed. [best on screen] (Color figure online)

mean and standard deviation errors, normalized by the interpupillary distance, in terms of the distance in pixels from the ground truth annotations. Results confirm a good accuracy and robustness also on expression variations contained in *Eurecom* dataset, sufficient for a real world monitoring system.

Real time performances are achieved with more than 30 fps in test phase. The system is tested on two platforms equipped with different GPUs, *Nvidia Quadro k2200* and *NVidia GTX 860m*, and requires less than 1 GB of video RAM.

All these elements allow to have an enabling technology for a real world monitoring system inside a car. Figure 5 shows an example of the working framework: the first and third rows contains the acquired RGB images, while in the second and fourth rows are reported the corresponding depth images with the landmark predictions from the described network.

5   Conclusions and Future Work

We have presented a CNN architecture to tackle the facial landmark localization task, with a good accuracy and real time performance. Due to the lack of deep oriented dataset for his dataset, a new and manually refined dataset is collected and publicly released. The proposed method deals with some requirements of the automotive context, like light changes in-variance, low computational load, no personal initialization and reliability to occlusions.

A variety of extensions and improvements are planned, to create a end-to-end pipeline for landmark localization, useful in an automotive context: for this goal, a embedded implementation of the deep model and a module for head detection are required. Finally, the output of the system can be exploited to elaborate indicators, like gaze direction, eyes tracking and so on.

Acknowledgements. This work has been carried out within the projects "Citta educante" (CTN01-00034-393801) of the National Technological Cluster on Smart Communities funded by MIUR and "FAR2015 - Monitoring the car drivers attention with multi-sensory systems, computer vision and machine learning" funded by the University of Modena and Reggio Emilia. We also acknowledge the CINECA award under the ISCRA initiative, for the availability of high performance computing resources and support.

References

1. Alyuz, N., Gokberk, B., Akarun, L.: Regional registration for expression resistant 3-D face recognition. IEEE Trans. Inf. Forensics Secur. 5(3), 425–440 (2010)
2. Alyüz, N., Gökberk, B., Spreeuwers, L., Veldhuis, R., Akarun, L.: Robust 3D face recognition in the presence of realistic occlusions. In: 2012 5th IAPR International Conference on Biometrics (ICB), pp. 111–118. IEEE (2012)
3. Antonakos, E., Alabort-i Medina, J., Tzimiropoulos, G., Zafeiriou, S.P.: Feature-based lucas-kanade and active appearance models. IEEE Trans. Image Process. 24(9), 2617–2632 (2015)
4. Ari, I., Uyar, A., Akarun, L.: Facial feature tracking and expression recognition for sign language. In: 2008 23rd International Symposium on Computer and Information Sciences, ISCIS 2008, pp. 1–6. IEEE (2008)
5. Borghi, G., Gasparini, R., Vezzani, R., Cucchiara, R.: Embedded recurrent network for head pose estimation in car. In: Proceedings of the 28th IEEE Intelligent Vehicles Symposium (2017)
6. Borghi, G., Venturelli, M., Vezzani, R., Cucchiara, R.: POSEidon: face-from-depth for driver pose estimation. In: IEEE Conference on Computer Vision and Pattern Recognition (CVPR) (2017)

7. Camgöz, N.C., Gökberk, B., Akarun, L.: Facial landmark localization in depth images using supervised descent method. In: 2015 23th Signal Processing and Communications Applications Conference (SIU), pp. 1997–2000. IEEE (2015)
8. Cao, C., Weng, Y., Zhou, S., Tong, Y., Zhou, K.: FaceWarehouse: a 3D facial expression database for visual computing. IEEE Trans. Vis. Comput. Graph. **20**(3), 413–425 (2014)
9. Chen, S., Bremond, F., Nguyen, H., Thomas, H.: Exploring depth information for head detection with depth images. In: 2016 13th IEEE International Conference on Advanced Video and Signal Based Surveillance (AVSS), pp. 228–234. IEEE (2016)
10. Cihan Camgoz, N., Struc, V., Gokberk, B., Akarun, L., Alp Kindiroglu, A.: Facial landmark localization in depth images using supervised ridge descent. In: Proceedings of the IEEE International Conference on Computer Vision Workshops, pp. 136–141 (2015)
11. Colombo, A., Cusano, C., Schettini, R.: 3D face detection using curvature analysis. Pattern Recogn. **39**(3), 444–455 (2006)
12. Colombo, A., Cusano, C., Schettini, R.: UMB-DB: a database of partially occluded 3D faces. In: 2011 IEEE International Conference on Computer Vision Workshops (ICCV Workshops), pp. 2113–2119. IEEE (2011)
13. Creusot, C., Pears, N., Austin, J.: A machine-learning approach to keypoint detection and landmarking on 3D meshes. Int. J. Comput. Vis. **102**(1–3), 146–179 (2013)
14. Dibeklioğlu, H., Alnajar, F., Salah, A.A., Gevers, T.: Combining facial dynamics with appearance for age estimation. IEEE Trans. Image Process. **24**(6), 1928–1943 (2015)
15. Dong, Y., Hu, Z., Uchimura, K., Murayama, N.: Driver inattention monitoring system for intelligent vehicles: a review. IEEE Trans. Intell. Transp. Syst. **12**(2), 596–614 (2011)
16. Fanelli, G., Dantone, M., Gall, J., Fossati, A., Van Gool, L.: Random forests for real time 3D face analysis. Int. J. Comput. Vis. **101**(3), 437–458 (2013)
17. Kingma, D.P., Ba, J.: Adam: a method for stochastic optimization. CoRR abs/1412.6980 (2014)
18. Krizhevsky, A., Sutskever, I., Hinton, G.E.: Imagenet classification with deep convolutional neural networks. In: Advances in neural information processing systems, pp. 1097–1105 (2012)
19. Medioni, G., Choi, J., Labeau, M., Leksut, J.T., Meng, L.: 3D facial landmark tracking and facial expression recognition. J. Inf. Commun. Converg. Eng. **11**(3), 207–215 (2013)
20. Min, R., Kose, N., Dugelay, J.L.: KinectFaceDB: a kinect database for face recognition. IEEE Trans. Syst. Man Cybern.: Syst. **44**(11), 1534–1548 (2014)
21. Murphy-Chutorian, E., Trivedi, M.M.: Head pose estimation in computer vision: a survey. IEEE Trans. Pattern Anal. Mach. Intell. **31**(4), 607–626 (2009)
22. Ranjan, R., Patel, V.M., Chellappa, R.: Hyperface: a deep multi-task learning framework for face detection, landmark localization, pose estimation, and gender recognition. arXiv preprint arXiv:1603.01249 (2016)
23. Reddy, K., Sikandar, A., Savant, P., Choudhary, A.: Driver drowsiness monitoring based on eye map and mouth contour. IJSTR **3**(5), 147–156 (2014)
24. Sarbolandi, H., Lefloch, D., Kolb, A.: Kinect range sensing: structured-light versus time-of-flight kinect. Comput. Vis. Image Underst. **139**, 1–20 (2015)
25. Sukno, F.M., Waddington, J.L., Whelan, P.F.: 3-D facial landmark localization with asymmetry patterns and shape regression from incomplete local features. IEEE Trans. Cybern. **45**(9), 1717–1730 (2015)

26. Sun, Y., Wang, X., Tang, X.: Deep convolutional network cascade for facial point detection. In: Proceedings of the IEEE Conference on Computer Vision and Pattern Recognition, pp. 3476–3483 (2013)
27. Tie, Y., Guan, L.: Automatic landmark point detection and tracking for human facial expressions. EURASIP J. Image Video Process. **2013**(1), 8 (2013)
28. Tzimiropoulos, G.: Project-out cascaded regression with an application to face alignment. In: Proceedings of the IEEE Conference on Computer Vision and Pattern Recognition, pp. 3659–3667 (2015)
29. Tzimiropoulos, G., Pantic, M.: Gauss-newton deformable part models for face alignment in-the-wild. In: Proceedings of the IEEE Conference on Computer Vision and Pattern Recognition, pp. 1851–1858 (2014)
30. Venturelli, M., Borghi, G., Vezzani, R., Cucchiara, R.: Deep head pose estimation from depth data for in-car automotive applications. In: Proceedings of the 2nd International Workshop on Understanding Human Activities through 3D Sensors (2016)
31. Venturelli, M., Borghi, G., Vezzani, R., Cucchiara, R.: From depth data to head pose estimation: a siamese approach. In: Proceedings of the 12th International Joint Conference on Computer Vision, Imaging and Computer Graphics Theory and Applications (2016)
32. Xiong, X., De la Torre, F.: Global supervised descent method. In: Proceedings of the IEEE Conference on Computer Vision and Pattern Recognition, pp. 2664–2673 (2015)
33. Zafeiriou, S., Zhang, C., Zhang, Z.: A survey on face detection in the wild. Comput. Vis. Image Underst. **138**(C), 1–24 (2015)
34. Zhang, Z., Luo, P., Loy, C.C., Tang, X.: Facial landmark detection by deep multi-task learning. In: Fleet, D., Pajdla, T., Schiele, B., Tuytelaars, T. (eds.) ECCV 2014. LNCS, vol. 8694, pp. 94–108. Springer, Cham (2014). doi:10.1007/978-3-319-10599-4_7
35. Zhao, X., Zou, J., Li, H., Dellandréa, E., Kakadiaris, I.A., Chen, L.: Automatic 2.5-D facial landmarking and emotion annotation for social interaction assistance. IEEE Trans. Cybern. **46**(9), 2042–2055 (2016)
36. Zhu, S., Li, C., Change Loy, C., Tang, X.: Face alignment by coarse-to-fine shape searching. In: Proceedings of the IEEE Conference on Computer Vision and Pattern Recognition, pp. 4998–5006 (2015)
37. Zuiderveld, K.: Contrast limited adaptive histogram equalization. In: Graphics Gems IV, pp. 474–485. Academic Press Professional, Inc. (1994)

Emotion Recognition by Body Movement Representation on the Manifold of Symmetric Positive Definite Matrices

Mohamed Daoudi[1], Stefano Berretti[2(✉)], Pietro Pala[2], Yvonne Delevoye[3], and Alberto Del Bimbo[2]

[1] IMT Lille Douai, University of Lille, CNRS, UMR 9189 – CRIStAL – Centre de Recherche en Informatique Signal et Automatique de Lille, 59000 Lille, France
[2] University of Florence, Florence, Italy
stefano.berretti@unifi.it
[3] University of Lille, CNRS, UMR 9193 SCALab, Lille, France

Abstract. Emotion recognition is attracting great interest for its potential application in a multitude of real-life situations. Much of the Computer Vision research in this field has focused on relating emotions to facial expressions, with investigations rarely including more than upper body. In this work, we propose a new scenario, for which emotional states are related to 3D dynamics of the whole body motion. To address the complexity of human body movement, we used covariance descriptors of the sequence of the 3D skeleton joints, and represented them in the non-linear Riemannian manifold of Symmetric Positive Definite matrices. In doing so, we exploited geodesic distances and geometric means on the manifold to perform emotion classification. Using sequences of spontaneous walking under the five primary emotional states, we report a method that succeeded in classifying the different emotions, with comparable performance to those observed in a human-based force-choice classification task.

Keywords: Emotion recognition · Symmetric Positive Definite matrices

1 Introduction

Automatic analysis of human motion has been an active research topic for several years, with outcomes that have been beneficial to a number of different applications, including security surveillance, health-care at home, athletes training and natural interfaces, to say a few. The variety in human body (size, height, corpulence), in the way different people perform an action, and even in the way a same person performs one action at different times, makes the task of human motion analysis very challenging. In the last decades, a consolidated line of research has analyzed the human motion from RGB and depth data enabling tasks such as action and gesture recognition [8]. However, body movements carry a multitude of information, also indicative of our intentions, inter-personal attitudes,

S. Battiato et al. (Eds.): ICIAP 2017, Part I, LNCS 10484, pp. 550–560, 2017.
https://doi.org/10.1007/978-3-319-68560-1_49

expectations and emotions. Of particular interest are basic emotions (i.e., *anger, disgust, fear, happiness, sadness,* and *surprise*) that are innate in all humans and are cross-culturally recognizable. These basic emotions can be further clustered in *active* (anger, happiness, surprise) and *passive* (fear, sadness, disgust). Recently, the study of computational models for human emotion recognition has gained increasing attention not only for commercial applications (to get feedback on the effect of advertising material), but also for gaming and monitoring of the emotional state of operators that act in risky contexts such as aviation. Most of these studies have focused on the analysis of facial expressions, but important clues can be derived by the analysis of the dynamics of body parts as well.

The first rigorous investigation on the expression of emotions through the body dates back to Darwin's seminal work on "The expression of the emotions in man and animals". Since then, research in the field of Emotional Body Language (EBL) has addressed this subject from both a bio-mechanical and a psychological perspective. The recognition of emotions from the analysis of body movements entails a higher level of complexity; indeed, since the body is primarily used to perform manipulative actions and enable motion, emotional clues can only be detected as secondary signatures on top of those ongoing actions. Hence, most EBL studies have addressed only the question of what aspects of 3D body kinematics are impacted by emotional states. Such studies have reported that body rhythmicity is slower for low energy emotions (*sadness* and *fright*) and faster for high-energy emotions (*anger*); these patterns have been confirmed across a variety of natural actions, e.g., door knocking, walking, and dancing. Nevertheless, such behavioral findings are not sufficient to tackle the difficult question of emotion classification through body motion observation.

Finding a compact and effective representation of body movement is a difficult task when considering the complexity of temporal dynamics. In addition, measuring the similarity between two temporal sequences for the purpose of classification is complicated in itself. In fact, the Euclidean distance is unsuitable for comparing temporal sequences, and Dynamic Time Warping is often used as an alternative [15]. To address these issues, there is a recent trend that investigates matrix based solutions. The idea of these methods is to embed the non-linearity of the sequence into a matrix representation, then exploit the geometric properties of the space (manifold) the matrices lay in to perform distance measurement and classification. Examples are the block Hankel matrix [3], and the Gram matrix [19]. Along this line of research, *covariance matrices* have found success in several computer vision applications, including activity recognition, visual surveillance and diffusion tensor imaging. Recently, several properties of the covariance matrices have been popularized by investigating the related Riemannian manifold of Symmetric Positive Definite matrices (SPD) [10].

Based on the above considerations, in this paper we propose a new solution to perform human emotion recognition from the analysis of the temporal dynamics of the joints of the body skeleton in the 3D space. Human motion is captured by the evolution across time of the 3D position of the joints in an appropriate reference system. Then, a covariance matrix descriptor is extracted from the features across the sequence frames. Exploiting the properties of the covariance matrix, this descriptor is mapped to the non-linear Riemannian manifold of SPD matrices. Finally, emotion classification is performed on the manifold by computing geodesic distances between test sequences and template emotions obtained as the average on the manifold of training examples. Experiments show the potential of the proposed solution, which obtains comparable results to those scored by human evaluators. In summary, the contributions of this work are: (i) Analysis of the dynamics of the full-body movement to understand human emotions over long sequences, while most of existing works use body-parts and short time; (ii) A representation of the body movement that uses the covariance descriptor to capture the dynamics of the skeleton joints, and analyzes these descriptors in the related Riemannian manifold of SPD matrices. This is obtained by the adoption of a suitable distance measure and mean computation to perform classification on the manifold.

The rest of the paper is organized as follows: Previous work related to the proposed method is summarized in Sect. 2; In Sect. 3, we present the mathematical background for the non-linear Riemannian manifold of SPD matrices; In Sect. 4, the adopted representation of the joints of the skeleton and its movement is presented; The classification approach on the manifold is discussed in Sect. 5; Results and a comparative evaluation are reported in Sect. 6; Finally, conclusions and future work directions are drawn in Sect. 7.

2   Related Work

The decreasing cost of whole-body sensing technology and its increasing reliability, make it possible to investigate the role played by body expressions as a powerful affective communication channel. Kapur et al. [11] were among the first to address these aspects in 3D. Using a Vicon Motion Capture system, they collected gestural sequence data depicting sadness, joy, anger, and fear emotions of five subjects. The 3D position of 14 markers, plus their velocity and acceleration were calculated, and the mean values of velocity and acceleration and the standard deviation values of position, velocity and acceleration across the sequence were considered as descriptors. Finally, classification was performed comparing five different classifiers. Gong et al. [6], addressed the problem of recognizing affect from non-stylized human body motion using 3D joints of the skeleton. Motion capture data were represented by a descriptor based on the shape of signal probability density function, and SVM were used for classification. Experiments were performed on a dataset of 30 individuals performing knocking, throwing, lifting and walking motions in four affective states (i.e., neutral, happy, angry and sad). Karg et al. [13] analyzed the human gait to reveal persons affective state, comparing inter-individual versus person dependent recognition. The dynamics of the body was captured by measuring features such as the stride length, cadence, velocity, minimum mean and maximum values of angles between body parts. Then, these features were reduced using PCA, kernel PCA, LDA and GDA techniques, while classification was performed with

NN, Naive-Bayes and SVM. Results showed that recognition is highly affected by individual walking styles and individual expressions of affect (accuracy of 69% and 95% were reported for the inter-individual and person dependent case, respectively, based on the observation of a single stride). They also observed that automatic recognition based on gait patterns tends to better recognize *active* than *passive* emotional states. For a comprehensive coverage of the topic, we refer to the survey by Kleinsmith and Bianchi-Berthouze [14] that reviewed the literature on affective body expression perception and recognition, and the survey by Karg et al. [12] that summarized methods to recognize affective expressions from body movements, and the converse problem of generating movements for virtual agents or robots, which convey affective expressions.

Several works used the special Riemannian manifold of SPD matrices. One typical case for which such matrices arise in practice is when covariance descriptors are used to model image sets or temporal frame sequences in videos. Covariance features were first introduced by Tuzel et al. [18] for texture matching and classification. Several studies have extended the use of covariance descriptors to the temporal dimension, with application to human action and gesture recognition. Sanin et al. [16], proposed an action and gesture recognition method from videos based on spatio-temporal covariance descriptors. Prior to classification, points on the manifold were mapped to an Euclidean space, through Riemannian Locality Preserving Projection [7]. Bhattacharya et al. [4] constructed covariance matrices, which capture joint statistics of both low-level motion and appearance features extracted from a video. To facilitate the classification task, matrices were mapped to an equivalent vector space obtained by the matrix logarithm operation, which approximates the tangent space of the original SPSD space of covariance matrices. Then, human action recognition was formulated as a sparse linear approximation problem, in which these mapped features are used to construct an overcomplete dictionary of the covariance based descriptors built from labeled training samples. In [5], Faraki et al. noted that when covariance descriptors are used to represent image sets, the result is often rank-deficient. Most of the existing methods solve this problem by accepting small perturbations to avoid null eigenvalues and thus, employ standard inference tools. What they proposed, instead, were novel similarity measures specifically designed for the particular case where symmetric matrices are not full-rank (i.e., Symmetric Positive Semi-Definite matrices, SPSD).

3   Manifold of Symmetric Positive Definite Matrices

Let f ($f \in \mathbb{R}^d$) be a d-dimensional feature vector of landmarks, and $D_{d \times n} = [f_1, \cdots, f_n]$ denote a set containing the d-dimensional feature descriptors of n images of an image set. The covariance matrix $\mathbf{C}$ of the set is defined by:

$$\mathbf{C} = \frac{1}{(n-1)} \sum_{i=1}^{n} (f_i - \mu)(f_i - \mu)^T, \tag{1}$$

where μ is the sample mean. A non-singular covariance matrix of size $d \times d$ belongs to the set of symmetric positive-definite (SPD) matrices. These do not form a vector space (the space is not closed under matrix subtraction), rather they form a connected Riemannian manifold Sym_d^+ [2]. As such, the distance between SPD matrices is not accurately captured by the Euclidean distance. Covariance matrix has recently received increasing attention in Computer Vision by leveraging Riemannian geometry of SPD matrices.

Indeed, several distance measures on Sym_d^+ have been proposed. The most widely used is the Log-Euclidean Riemannian Metric (LERM) [1]. Given two covariance matrices $\mathbf{C}_1$ and $\mathbf{C}_2$, their LERM is computed as:

$$d(\mathbf{C}_1, \mathbf{C}_2) = \| \log(\mathbf{C}_1) - \log(\mathbf{C}_2) \|_F, \tag{2}$$

where $\| \cdot \|_F$ is the Frobenius norm, and $\log(\mathbf{C})$ is the matrix logarithm of $\mathbf{C}$.

4   Representation of Body Movement

The dynamics of body movements is expressed by a sequence of observation vectors capturing the position of body joints across time. More specifically, the human body is approximated by a skeleton composed of N_J joints. Accordingly, the posture of the body at a generic observation time t is expressed by a vector $p \in \mathbb{R}^{3N_J}$ composed of the (X, Y, Z) coordinates of body joints at time t:

$$p(t) = [x_1, y_1, z_1, \ldots, x_{N_J}, y_{N_J}, z_{N_J}]. \tag{3}$$

In order to also keep track of the body dynamics at each observation time, the posture vector is augmented with the velocity vector that is composed of the (X, Y, Z) components of the velocity of body joints at time t:

$$v(t) = \left[v_{x_1}, v_{y_1}, v_{z_1}, \ldots, v_{x_{N_J}}, v_{y_{N_J}}, v_{z_{N_J}} \right]. \tag{4}$$

The velocity of a generic joint at time t is computed by finite difference of joint positions at time t and $t-1$, assuming zero velocity at $t=0$.

In order to make the position and velocity vectors invariant to the orientation of the body with respect to the camera, coordinate values (X, Y, Z) are normalized by expressing them in a skeleton centered coordinate system (X_S, Y_S, Z_S). This is computed as the orthonormal basis resulting from the PCA of the positions of the torso joints at $t = 0$. A compact yet representative description of the dynamics of body movements across a temporal observation window $[0, T]$ is extracted by computing the covariance matrix of the concatenated posture/velocity vectors. This results into a symmetric $6N_J \times 6N_J$ square matrix.

Figure 1 shows the idea of capturing the body movement in a sequence through a covariance matrix which, in turn, is a point on the SPD manifold.

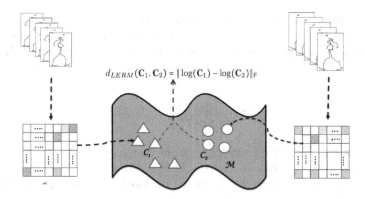

Fig. 1. Each body motion sequence is represented through a covariance matrix, which is a point on the Riemannian manifold of SPD matrices. Distance between sequences is then evaluated as the geodesic between points on the manifold

5 Emotion Classification

The covariance matrix computed from skeleton data observed across a temporal window $[0, T]$ retains a signature of the emotional state of the observed person. To perform emotion recognition, in the proposed approach, the covariance matrix computed from an unknown observation is compared with the Prototype Emotional Matrices (PEMs) representative of the target emotions (in the experiments reported in Sect. 6, five basic emotions are considered, namely, *anger, fear, joy, neutral, sadness*). The unknown sequence is classified according to the emotion associated with the closest PEM. Computation of PEMs relies on extraction of representative examples from training data. It should be noted that, according to what is described in Sect. 3, the computation of the distance to PEMs as well as the identification of PEMs from training data should both take into account the fact that covariance matrices lie on the Riemannian manifold of SPD matrices Sym_d^+. This prevents the use of common tools adopted in Euclidean spaces to compute distances between points and cluster them.

Let $\{\mathbf{C}_i, l_i\}_{i=1\ldots N}$ be a training set of labeled samples composed of covariance matrices $\mathbf{C}_i$ and corresponding emotion labels $l_i \in \{l_1, \ldots, l_E\}$. The emotion classification task acts like a function that associates with a generic element of Sym_d^+ its classification label $l \in \{l_1, \ldots, l_E\}$. A possible solution would be to adopt a nearest-neighbor (k-NN) approach by comparing the covariance matrix to be classified to all the labeled covariance matrices in the training, and assigning to it the same label of the closest matrix (for instance, the LERM distance in (2) can be used for the comparison). A better solution, both in terms of computation and of generalization of training examples is to extract some representative prototypes from the training examples. Then, it would be possible to compare the covariance matrix to classify to these prototypes, instead of using all training examples. Following this idea, we extract a PEM from each emotion class. This is achieved by computing, for each emotion class l_i the Riemannian

Center of Mass of all the training examples with label l_i. Given a set of covariance matrices $\{\mathbf{C}_i\}_{i=1...N}$ on the Riemannian manifold Sym_d^+, the Riemannian Center of Mass, also referred to as *Karcher mean* in the literature, is the point on Sym_d^+ that minimizes the sum of squared Riemannian distances:

$$\mu = \arg \min_{\mathbf{C} \in Sym_d^+} \sum_{i=1}^{N} d^2 (\mathbf{C}, \mathbf{C}_i), \tag{5}$$

being $d(\cdot)$ a suitable distance measure on the manifold.

It should be noted that, in case the LERM distance in (2) is used, the Riemannian Center of Mass can be computed in closed form through the following expression [19]:

$$\mu = \exp \left(\frac{1}{N} \sum_{i=1}^{N} \log (\mathbf{C}_i) \right), \tag{6}$$

being $\exp(\cdot)$ and $\log(\cdot)$ the matrix exponential and logarithm operators, respectively. In this way, for the emotion corresponding to label l_i, the Prototype Emotional Matrix $\mathbf{Pem}_{l_i}$ is computed as the Riemannian center of mass of all training samples $\{\mathbf{C}_k, l_k\}$, such that $l_k = l_i$. A generic covariance matrix to be classified is assigned the label corresponding to the closest $\mathbf{Pem}_{l_i}$. In doing so, the identification of Prototype Emotional Matrices as well as the classification of the emotion to be associated to a new covariance matrix rely on a measure of distance that preserves the inherent structure of the manifold.

6   Experiments

Experiments have been performed on the Body Motion-Emotion dataset (P-BME), that has been acquired at the Cognitive Neuroscience Laboratory (INSERM U960 - Ecole Normale Supérieure) in Paris [9]. It includes Motion Capture (MoCap) 3D data sequences recorded at a high frame rate (120 frames per second) by an Opto-electronic Vicon V8 MoCap system wired to 24 cameras. The body movement is captured by using 43 landmarks that are positioned at joints and other parts of the body as illustrated in Fig. 2. To create the dataset, 8 subjects (professional actors) were instructed to walk following a predefined "U" shaped path that includes forward-walking, turn, and coming back (Fig. 2). For each acquisition, actors move along the path performing one emotion out of a set of five different emotions, namely, *anger*, *fear*, *joy*, *neutral*, and *sadness*. So, each sequence is associated with one emotion label. In doing so, the emotional gait patterns show to be characterized by different walking velocity, wrist velocity and acceleration, body and head postures. Each actor performed at maximum five repetitions of a same emotional sequence for a total of 156 instances. Though there is some variation from subject to subject, the number of examples is well distributed across the different emotions: 29 *anger*, 31 *fear*, 33 *joy*, 28 *neutral*, 35 *sadness*.

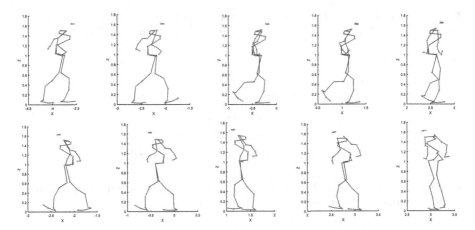

Fig. 2. Frames from a MoCap skeleton sequence of the P-BME dataset. In this example, an actor moves following a "U" shaped trajectory showing an *anger* emotion. In the top row, the subject advances towards the turning point (plots from left-to-right); in the bottom row, the subject moves away from the turning point (plots from right-to-left). The changes in the moving direction at the turning point can be observed in the rightmost frame of both top and bottom rows. In each frame, the skeleton is represented by 43 joints. Connections between joints are shown (except for the four joints of the head) to evidence the silhouette of the body and the limbs

6.1 Results and Comparative Evaluation

Experiments on the P-BME dataset were performed by using a *leave-one-subject-out* cross validation protocol. With this solution, iteratively, all the emotion sequences of a subject are used for test, while all the sequences of the remaining subjects are used for training. As discussed in Sect. 5, the training sequences are used to perform supervised clustering in the five emotional classes. This is obtained by first computing the Riemannian center of mass of each emotion class and retaining it as representative element of the class. Then, nearest-neighbor classification of the test sequence is performed by computing the LERM distance to these representative elements. A confusion matrix is thus computed for each fold. Averaging such matrices across the eight folds (also weighting each matrix according to the relative number of test examples, which is different from subject to subject) we obtain the overall results reported in Table 1. It can be observed the diagonal dominance of the matrix (average positive classification of about 71%), with the best results scored by *neutral* and *anger* (about 80%), followed by *sadness* and *fear* (about 68%), with the lowest accuracy for *joy* (about 58%).

We also performed experiments by using nearest-neighbor (NN) classification with respect to all the training sequences, without reducing them with any clustering operation. In addition to be much more computational demanding, this classification scores substantially lower results as reported in Table 2 (the average of the diagonal values decreases to about 51%). This confirms us the intuition that performing the Riemannian center of mass on the training sequences can

Table 1. P-BME dataset: Emotion recognition accuracy obtained using the Riemannian center of mass (results in percentage). Average accuracy is 71.12%

	Anger	Fear	Joy	Neutral	Sadness
Anger	**79.31**	3.45	13.79	0.00	3.45
Fear	3.57	**67.86**	10.71	0.00	17.86
Joy	3.23	6.45	**58.06**	9.68	22.58
Neutral	6.06	0.00	0.00	**81.82**	12.12
Sadness	2.86	20.00	2.86	5.71	**68.57**

Table 2. P-BME dataset: Emotion recognition accuracy obtained using a nearest-neighbor approach (results in percentage). Average accuracy is 50.74%

	Anger	Fear	Joy	Neutral	Sadness
Anger	**41.38**	0.00	3.45	31.03	24.14
Fear	0.00	**67.86**	7.14	3.57	21.43
Joy	0.00	3.23	**16.13**	32.26	48.39
Neutral	0.00	0.00	0.00	**45.45**	54.55
Sadness	2.86	11.43	0.00	2.86	**82.86**

reduce the effects induced by outliers included in the training examples that were provided for each emotion.

To also validate the importance of measuring distances between covariance matrices using geodesic distances on the manifold, compared to standard matrix norm computation, we performed NN-classification using the Frobenius norm of the difference between covariance matrices. This resulted in an average classification of 43.4% which is more than 7% less than the result obtained using LERM in Table 2.

Finally, we performed a user based test in order to evaluate the performance of the proposed classification method in comparison with a human-based judgment. In this test, thirty-two naive individuals (with heterogeneous age and no experience in human emotion classification) were asked to perform a force-choice task. Participants were seated in front of a computer screen, and videos were presented following a semi-randomized block design, with nature of emotion randomly presented for each actor. The order of the presentations of the video clips for each actor was also counter-balanced. Participants were required to categorize the observed motion sequences in one of the five emotional categories within 5*secs* after the end of the video presentation, using the Geneva Emotional Wheel (GEW) [17]. The task was a force choice situation in which the participants had to choose between one of five emotions: *anger, fear, joy, sadness* or *neutral*. Table 3 reports the scores obtained for emotion classification based on RGB videos by the human evaluators. The results reveal an average value of about 74%, which is just 3% over the average result found in Table 1.

It is relevant to note that the user based test being based on RGB videos provides to the users much more information for evaluation, including the actor's face. Notably, our method is capable to score comparable results based on the skeleton joints only.

Table 3. P-BME dataset: emotion recognition of body motion by human evaluator

Anger	Fear	Joy	Neutral	Sadness	Average
84.0	81.5	73.5	65.0	67.0	74.2

7   Conclusions

In this work, we focus on 3D dynamic sequences of the body skeleton and propose a new method to relate automatically human body movements to inner sensorial emotion. This is obtained by first representing the 3D evolution of the skeleton joints across time by using a covariance matrix. Then, we account for the fact that these matrices lay in the non-linear Riemannian manifold of SPD matrices. Exploiting geodesic distances and geometric average computation on the manifold, emotion classification is performed. Results obtained in the experiments show an average recognition of about 71% for the proposed method, which is comparable with the average score produced by human evaluation. Notably, our results have been obtained using only joints information, while humans evaluators exploited the richer RGB video channel. The covariance matrix captures the dependence of locations of different joints on one another during the performance of an human action. The covariance matrix does not capture the order of motion in time. Future work will address more advanced approaches for modeling the temporal evolution and machine learning and classification methods on a non-linear manifold. We will also investigate the generalization of the method by applying it to other types of voluntary motor actions besides walking (e.g., cycling, running, or cooking a meal).

Acknowledgements. This work has been partially supported by PIA, ANR (grant ANR-11-EQPX-0023), European Founds for the Regional Development (Grant FEDER-Presage 41779). We thank Julie Grèzes and Alain Berthoz (INSERM U960, France) and Halim Hicheur (University of Fribourg, Switzerland) for the possibility to use their data set. During the preparation of this paper (July 2016), M. Daoudi enjoyed excellent working conditions at the University of Florence, Italy.

References

1. Arsigny, V., Fillard, P., Pennec, X., Ayache, N.: Geometric means in a novel vector space structure on symmetric positive-definite matrices. SIAM J. Matrix Anal. Appl. **29**(1), 328–347 (2007)
2. Bhatia, R.: Positive Definite Matrices. Princeton University Press, Princeton (2007)

3. Bhattacharya, S., Kalayeh, M.M., Sukthankar, R., Shah, M.: Recognition of complex events: exploiting temporal dynamics between underlying concepts. In: IEEE Conference on Computer Vision and Pattern Recognition (CVPR), pp. 2243–2250 (2014)

4. Bhattacharya, S., Souly, N., Shah, M.: Covariance of motion and appearance features for spatio temporal recognition tasks. ArXiv e-prints, June 2016

5. Faraki, M., Harandi, M.T., Porikli, F.: Image set classification by symmetric positive semi-definite matrices. In: IEEE Winter Conference on Applications of Computer Vision (WACV), pp. 1–8 (2016)

6. Gong, L., Wang, T., Wang, C., Liu, F., Zhang, F., Yu, X.: Recognizing affect from non-stylized body motion using shape of Gaussian descriptors. In: ACM Symposium on Applied Computing (SAC), pp. 1203–1206 (2010)

7. Harandi, M.T., Sanderson, C., Wiliem, A., Lovell, B.C.: Kernel analysis over Riemannian manifolds for visual recognition of actions, pedestrians and textures. In: IEEE Workshop on Applications of Computer Vision (WACV), pp. 433–439 (2012)

8. Herath, S., Harandi, M., Porikli, F.: Going deeper into action recognition: a survey. Image Vis. Comput. **60**, 4–21 (2017)

9. Hicheur, H., Kadone, H., Grèzes, J., Berthoz, A.: The combined role of motion-related cues and upper body posture for the expression of emotions during human walking. In: Mombaur, K., Berns, K. (eds.) Modeling, Simulation and Optimization of Bipedal Walking. COSMOS, vol. 18, pp. 71–85. Springer, Heidelberg (2013). doi:10.1007/978-3-642-36368-9_6

10. Jayasumana, S., Hartley, R., Salzmann, M., Li, H., Harandi, M.: Kernel methods on Riemannian manifolds with Gaussian RBF kernels. IEEE Trans. Pattern Anal. Mach. Intell. **37**(12), 2464–2477 (2015)

11. Kapur, A., Kapur, A., Virji-Babul, N., Tzanetakis, G., Driessen, P.F.: Gesture-based affective computing on motion capture data. In: Tao, J., Tan, T., Picard, R.W. (eds.) ACII 2005. LNCS, vol. 3784, pp. 1–7. Springer, Heidelberg (2005). doi:10.1007/11573548_1

12. Karg, A., Samadani, A.A., Gorbet, R., Kühnlenz, K., Hoey, J., Kulić, D.: Body movements for affective expression: a survey of automatic recognition and generation. IEEE Trans. Affect. Comput. **4**(4), 341–359 (2013)

13. Karg, M., Kuhnlenz, K., Buss, M.: Recognition of affect based on gait patterns. IEEE Trans. Syst. Man Cybern. Part B **40**(4), 1050–1061 (2010)

14. Kleinsmith, A., Bianchi-Berthouze, N.: Affective body expression perception and recognition: a survey. IEEE Trans. Affect. Comput. **4**(1), 15–33 (2013)

15. Müller, M.: Information Retrieval for Music and Motion. Springer, Heidelberg (2007)

16. Sanin, A., Sanderson, C., Harandi, M.T., Lovell, B.C.: Spatio-temporal covariance descriptors for action and gesture recognition. In: IEEE Workshop on Applications of Computer Vision (WACV), pp. 103–110 (2013)

17. Scherer, K.R.: What are emotions? And how can they be measured? Soc. Sc. Inf. **44**(4), 693–727 (2005)

18. Tuzel, O., Porikli, F., Meer, P.: Region covariance: a fast descriptor for detection and classification. In: Leonardis, A., Bischof, H., Pinz, A. (eds.) ECCV 2006. LNCS, vol. 3952, pp. 589–600. Springer, Heidelberg (2006). doi:10.1007/11744047_45

19. Zhang, X., Wang, Y., Gou, M., Sznaier, M., Camps, O.: Efficient temporal sequence comparison and classification using Gram matrix embeddings on a Riemannian manifold. In: IEEE Conference on Computer Vision and Pattern Recognition (CVPR), pp. 4498–4507 (2016)

Lifting 2D Object Detections to 3D: A Geometric Approach in Multiple Views

Cosimo Rubino[1(✉)], Andrea Fusiello[2], and Alessio Del Bue[1]

[1] Visual Geometry and Modelling (VGM) Lab, Istituto Italiano di Tecnologia (IIT),
Via Morego 30, Genova, Italy
{cosimo.rubino,alessio.delbue}@iit.it
[2] DPIA, Università di Udine, Via Delle Scienze, 208, Udine, Italy

Abstract. We present two new methods based on Interval Analysis and Computational Geometry for estimating the 3D occupancy and position of objects from image sequences. Given a calibrated set of images, the proposed frameworks first detect objects using off-the-shelf object detectors and then match bounding boxes in multiple views. The 2D semantic information given by the bounding boxes are used to efficiently recover 3D object position and occupancy using solely geometrical constraints in multiple views. We also combine further constraints to obtain a solution even when few images are available. Experiments on three different realistic datasets show the applicability and the potentials of the approaches.

Keywords: Object localisation · Object detection · Interval Analysis

1 Introduction

Despite strong efforts in the Computer Vision community, object detection has been mostly restricted in 2D, even if multiple exposures of the same scene are present. In this paper we are trying to tackle instead this appealing question: "If multiple images of a rigid scene are available, is it possible to recover the 3D location and occupancy of the objects only having 2D bounding boxes returned by any object detectors?". This question is becoming impelling given the recent advancements in object detection, boosted by the advent of deep net architectures. Indeed, it is now possible to have accurate and repeatable 2D localisations of several object class instances in generic image scenes. An approach that would be able to leverage such 2D detections in 3D would make easier the geometrical interpretation of images, nowadays necessary for applications such as human robot interaction, visual question and answering, and navigation.

Object detections are, in general, represented as 2D bounding boxes containing the object image shape. This coarse representation was dictated by annotation easiness when tracing the box while labelling large datasets as in the Pascal VOC challenge [8]. Although some works leverage finer object shape annotations (e.g. [26]), only few methods can provide a detailed silhouette of the detected objects [14,27]. Object detections in 3D have been mainly tackled using RGBD

© Springer International Publishing AG 2017
S. Battiato et al. (Eds.): ICIAP 2017, Part I, LNCS 10484, pp. 561–572, 2017.
https://doi.org/10.1007/978-3-319-68560-1_50

data [16] in single images. This is a direct extension of the 2D case, where annotations are directly extracted from the depth data by using ellipsoids. Gupta et al. [13] use labelled 3D dataset as the NYUD2 dataset [20] to retrain a region-based convolutional network (R-CNN [24]), proposing candidate 2.5D regions.

Even if these works show that it is feasible to localise 3D objects just from a single RGBD image, there are less examples showing that object localisation is possible from just image sequences without any depth information. To this extent we present two new approaches that, using geometrical reasoning only, can extract the localisation of objects in a calibrated image sequence as a set of polyhedra in 3D. The first approach is inspired by [10] and it is based on a computational geometry (CG) method which has been applied to estimate, for each object, the polyhedra given by the intersections of all the pyramids, having the vertex on each camera centre and passing through the bounding boxes of the object detections. The second approach is based on Interval Analysis (IA), and following [9] solves a similar problem based on stereo triangulations. These two methods can be readily applied to any calibrated image sequence with matched bounding boxes detections. In particular, in Sect. 4 we show results on a subset of the ScanNet dataset [6] comprising more than 1250 image sequences in realistic indoor environments. To show further the flexibility of the proposed approach in different scenarios, we also show performance on two datasets (ACCV [15] and TUW [1]) with available ground truth.

2   Related Work

In this review we will restrict to single or multiple views methods for 3D object localisation, to which our approach is more closely related. As the most challenging scenario, strong efforts have been devoted to the study of single image pose estimation problems. This led to the necessity to learn image to object relations in order to generalise pose estimation in 3D to several classes of objects. In many cases a training phase is performed using images of a specific category of objects from different viewpoints. Many works have exploited 3D object models to get a 3D interpretation of the scene. Zia et al. [28] used the CAD models of cars to reconstruct the scene and the objects, including additional information about the ground plane. Pepik et al. [21] reformulated the model as a 3D deformable part model by learning the part appearances according to the CAD model. Recently, Mousavian et al. [19] used two networks to regress the orientation and the dimension of cars and bikes, then applied geometrical constraints to 2D detections to obtain the 3D bounding boxes.

When multiple images are available, recent works have tried to include geometrical reasoning to explicitly use constraints given by the multiple views. Bao et al. [3] tried to deduce both the viewpoint motion between multiple images and the pose of the objects using a part-based object detector. To reach the same goal a monocular SLAM approach was used by Dame et al. [7], combining it with shape priors-based 3D tracking and 3D reconstruction approaches, while Fidler et al. [11] reduced all the objects to 3D bounding boxes with each side being a planar approximation of the object.

Differently from these methods that use strong semantics and heuristics, our approaches are based exclusively on geometrical reasoning, using directly the 2D bounding boxes to define a polyhedra reconstruction problem, indicating where the objects are located in 3D. Unlike the Visual Hull of Laurentini [17] where the siluettes of the objects are used, we used bounding box detections as 2D input. An approach to infer the location of the objects was presented by Crocco et al. [5], estimating the occupancy of the objects through a quadric reconstruction problem. Differently to our work, they apply the simpler orthographic camera model. Furthermore, our approach is resilient if some of the detections are missing, since [5] solves the problem using the factorisation of a complete matrix containing the ellipses parametrisation for every object at every frame.

3   Lifting 2D Bounding Boxes to 3D

Our approach first extracts object detections from every frame of a generic image sequence. Given all the detections in each frame, we use a modified tracking-by-detection method [12] to associate the bounding boxes among different frames. This algorithm computes a distance matrix using patch appearance and associate detections using the Hungarian method for bipartite matching. We relaxed the part associated to the smoothness of the object trajectory because we might not have consistent camera motion among consecutive frames thus causing the corresponding consecutive bounding boxes to be far apart. Notice that, it is common that bounding boxes might not be precisely aligned with the true object centre and often they include a portion of background.

We then assume that the object is bounded by a rectangular region $\mathcal{B}_i$ in image i. In 3D space, each region $\mathcal{B}_i$ defines a semi-infinite pyramid $\mathcal{Q}_i$ with its apex in the camera center (see Fig. 1), which bounds the possible locus of the object. In the case of two views, assuming that the object's projections are bounded by rectangles $\mathcal{B}_1$ and $\mathcal{B}_2$ in the images respectively, the object in space must lie within a polyhedron $\mathcal{D}$ as in Fig. 1. Geometrically, $\mathcal{D}$ is obtained by intersecting the two semi-infinite pyramids defined by the two rectangles $\mathcal{B}_1$ and $\mathcal{B}_2$ and the respective centres of projection C_1 and C_2.

In the general case of n views, the object is localised inside the polyhedron formed by the intersection of the n semi-infinite pyramids generated by the rectangles $\mathcal{B}_1, \ldots, \mathcal{B}_n$:

$$\mathcal{D} = \mathcal{Q}_1 \cap \mathcal{Q}_2 \cdots \cap \mathcal{Q}_n. \tag{1}$$

Analytically, the polyhedron $\mathcal{D}$ is defined as the following set:

$$\mathcal{D} = \{X \in \mathbb{R}^3 : \exists x_i \in \mathcal{B}_i, i = 1 \ldots n \text{ s.t. } \forall i : x_i = \Pi_i(X)\} \tag{2}$$

where Π is the known perspective projection onto the i-th image.

3.1   Vertex Enumeration Solution

The semi-infinite pyramid $\mathcal{Q}_i$ can be written as the intersection of the four negative half-spaces $\mathcal{H}_1^i, \mathcal{H}_2^i, \mathcal{H}_3^i, \mathcal{H}_4^i$ defined by its supporting planes. Thus, the solution set D can be expressed as the intersection of $4n$ negative half-spaces:

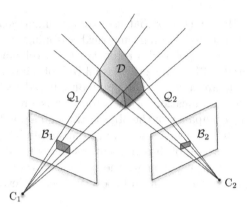

Fig. 1. Bounding the object in 3D from 2D detections. Here a graphical example with two images, where the semi-infinite pyramid is defined from the centre of projection and the bound $\mathcal{B}_i$.

$$\mathcal{D} = \bigcap_{\substack{i=1...n \\ \ell=1...4}} \mathcal{H}_\ell^i. \tag{3}$$

Implicitly these equations represent the polyhedron $\mathcal{D}$, and indeed this is also called the H-representation of $\mathcal{D}$. However, we aim at an explicit description of $\mathcal{D}$ in terms of vertices and edges, also called a V-representation. The problem of producing a V-representation from an H-representation is called the VertexEnumeration problem, in Computational Geometry. The vertices and the faces of $\mathcal{D}$ can be enumerated in $O(n \log n)$ time, being n the number of cameras [22]. In particular we used the implementation of the reverse search vertex enumeration algorithm described in [2] and available on the web[1].

In the following, this approach based on Computational Geometry (proposed in [10]) will be referred to as the "CG approach". In the next section, following [9], we shall describe how the solution set can be enclosed with an axis-aligned box using an approach based on *Interval Analysis*, henceforth dubbed "IA approach".

3.2   Bounded Computational Geometry Method

The polyhedron generated by the CG approach can approximate effectively the 3D volume occupied by a detected object if several images of the object with a large baseline between cameras are available. Otherwise, when there are few images with a narrow baseline between cameras, the computed polyhedron can easily overestimate the occupancy volume. To reduce this effect, we bounded the estimated volume by including a prior over its maximum elongation. This is done by first finding the centroid of the object using triangulation between the centres of the bounding boxes in different views [4]. Then, the final polyhedron

---

[1] http://cgm.cs.mcgill.ca/~avis/C/lrs.html.

is obtained by cutting the pyramid, generated by CG with two planes, with a distance before and after the object 3D centroid equal to half of the maximum size of the object[2], and with the normal aligned to the optical axis of the camera. We will henceforth refer to this variation as the CG_b method.

3.3 Interval Analysis

Interval Analysis [18] is an arithmetic defined on intervals, rather than on real numbers. It was firstly introduced for bounding the measurement errors of physical quantities for which no statistical distribution was known. In the sequel of this section we shall denote intervals with boldface. Underscores and overscores will represent respectively lower and upper bounds of intervals. $\mathbb{IR}$ stands for the set of real intervals. If $f(x)$ is a function defined over an interval $\boldsymbol{x}$ then range$(f, \boldsymbol{x})$ denotes the range of $f(x)$ over $\boldsymbol{x}$.

If $\boldsymbol{x} = [\underline{x}, \overline{x}]$ and $\boldsymbol{y} = [\underline{y}, \overline{y}]$, a binary operation between $\boldsymbol{x}$ and $\boldsymbol{y}$ is defined in interval arithmetic as:

$$\boldsymbol{x} \circ \boldsymbol{y} = \{x \circ y \mid x \in \boldsymbol{x} \wedge y \in \boldsymbol{y}\}, \forall \circ \in \{+, -, \times, \div\}.$$

Operationally, interval operations are defined by the min-max formula:

$$\boldsymbol{x} \circ \boldsymbol{y} = \left[\min\{\underline{x} \circ \underline{y}, \underline{x} \circ \overline{y}, \overline{x} \circ \underline{y}, \overline{x} \circ \overline{y}\}, \ \max\{\underline{x} \circ \underline{y}, \underline{x} \circ \overline{y}, \overline{x} \circ \underline{y}, \overline{x} \circ \overline{y}\}\right] \quad (4)$$

Interval division $\boldsymbol{x}/\boldsymbol{y}$ is undefined when $0 \in \boldsymbol{y}$.

In general, for arbitrary functions, interval computation cannot produce the exact range, but only approximate it.

Definition 1 (Interval extension [23]). *A function $\boldsymbol{f} : \mathbb{IR} \to \mathbb{IR}$ is said to be an interval extension of $f : \mathbb{R} \to \mathbb{R}$ provided that* range$(f, \boldsymbol{x}) \subseteq \boldsymbol{f}(\boldsymbol{x})$ *for all intervals $\boldsymbol{x} \subset \mathbb{IR}$ within the domain of $\boldsymbol{f}$.*

Such a function is also called an *inclusion function*. So, given a function f and a domain $\boldsymbol{x}$, the inclusion function yields a rigorous bound (or enclosure) on range$(f, \boldsymbol{x})$. This property is particularly suited for error propagation: If $\boldsymbol{x}$ bounds the input error on the variable x, $\boldsymbol{f}(\boldsymbol{x})$ bounds the output error. Therefore, if the exact value is contained in interval data, the exact value will be contained in the interval result.

Definition 2 (Natural interval extension [23]). *Let us consider a function f computable as an arithmetic expression* f, *composed of a finite sequence of operations applied to constants, argument variables or intermediate results. A natural interval extension of such a function, denoted by* f$(\boldsymbol{x})$, *is obtained by replacing variables with intervals and executing all arithmetic operations according to the rules above.*

---

[2] An upper bound for the size of several object classes has been extracted from the ShapeNet dataset: https://www.shapenet.org/.

Please note how different expressions for the same function yield different natural interval extensions. For instance, $f_1(x) = x^2 - x$, and $f_2(x) = x(x-1)$ are both natural interval extensions of the same function. For example, consider the expression $f(x) = x - x$ which is equivalent to 0. However evaluating the expression with the interval [1,2], gives $f([1,2]) = [1,2] - [1,2] = [-1,1]$, because the piece of information that the two intervals represent the same variable is lost. In general, although the ranges of interval arithmetic operations are exact, this is not so if operations are composed. For example, if $x = [0,1]$ we have $f_2(x) = [0,1]([0,1]-1) = [0,1][-1,0] = [-1,0]$, which strictly includes range($f, [0,1]$) = $[-1/4, 0]$.

It is well-known that Interval Analysis systematically overestimates the bound on the results of a computation: this is the price to pay for its simplicity.

3.4   Interval-Based Triangulation

Let us assume that we can write a closed form expression that relates the 3D point X to its projections $x_1 = \Pi_1(X)$ and $x_2 = \Pi_2(X)$ in two images (see [9]):

$$X = f(x_1, x_2) \tag{5}$$

If we let x_1 and x_2 in Eq. (5) vary in $\mathcal{B}_1$ and $\mathcal{B}_2$ respectively, then range($f, \mathcal{B}_1 \times \mathcal{B}_2$) describes the polyhedron $\mathcal{D}$ that contains the object. Interval Analysis gives us a way to compute an axis-aligned bounding box containing $\mathcal{D}$ by simply evaluating $f(x_1, x_2)$, the natural interval extension of f, with $\mathcal{B}_1 = x_1$ and $\mathcal{B}_2 = x_2$.

The 3D interval $f(x_1, x_2)$ encloses the polyhedron $\mathcal{D}$, and, in general, it is an overestimate. In fact, intervals can model only axis-aligned rectangular boxes; moreover, as seen in the examples, interval evaluation inevitably introduces overestimation.

The approach is easily extensible to the general n-views case. As defined in Sect. 3, the sought polyhedron $\mathcal{D}$ is formed by the intersection of the semi-infinite pyramids generated by back-projecting in space the sets $\mathcal{B}_1, \ldots, \mathcal{B}_n$. Thanks to the associativity of intersection, ($\mathcal{D}$) can be obtained by first intersecting pairs of such pyramids and then intersecting the results. Let $\mathcal{D}_{i,j}^2$ be the solution set of the triangulation between view i and view j. Then:

$$\mathcal{D} = \bigcap_{\substack{i=1,\ldots,n \\ j=i+1,\ldots,n}} \mathcal{D}_{i,j}^2. \tag{6}$$

An enclosure of the solution set $\mathcal{D}$ is obtained by intersecting the $n(n-1)/2$ enclosures of $\mathcal{D}_{i,j}^2$ computed with the IA method described above. Since each enclosure contains the respective solution set $\mathcal{D}_{i,j}^2$, their intersection contains $\mathcal{D}$. In summary, the IA approach yields a rectangular axis-aligned bounding box $f(x_1, x_2)$ that contains the polyhedron $\mathcal{D}$. This method is faster and easier to implement (basing on an interval arithmetic library, such as INTLAB [25]) than the CG one, but the enclosure is – in general – an overestimate.

4   Experiments

We tested our methods on three datasets: ACCV [15], TUW [1] and ScanNet [6]. These datasets present different imaging conditions related to camera motion, number of frames for each sequence, number of objects and their distance from the camera. In total we tested 1240 different image sequences with an overall number of 42,000 frames.

All the datasets provide the camera parameters and the annotated ground truth (GT) point clouds of the objects inside the scene. For each object, we evaluated the GT 3D bounding box by enclosing the given 3D point clouds. For each frame and each object we also generated a set of 2D bounding boxes to simulate the output of an object detector. This is done by fitting with a box the 2D reprojections of the labelled point clouds associated to each object. Additionally, we have also evaluated oriented bounding boxes, by aligning the box with respect to the orientation of the objects onto the 2D image frames. The alignment is performed by considering the orientation of an image mask associated to the reprojected points, returned by the function *regionprops* in MATLAB (Fig. 2).

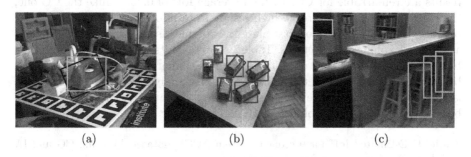

(a)   (b)   (c)

Fig. 2. A frame with oriented bounding boxes (red) and bounding boxes aligned to the axes (green) of Seq. "Iron" of the ACCV dataset, Seq. 7 of the TUW dataset and scene0000 of the ScanNet dataset. (Color figure online)

Results have been evaluated by computing the 3D Intersection over Union (IoU) between the bounding boxes associated to the GT and to the reconstruction. Our methods perform very well for the ACCV dataset since the sequences have a high number of images taken from a camera that performs a large rotation around the objects. Differently, the TUW and ScanNet datasets have a reduced number of frames and a limited motion of the camera, thus reducing drastically the performance of the proposed methods. The computational costs of both methods can be deduced by [9,10].

4.1   ACCV Dataset Evaluation

The ACCV dataset [15] contains 15 sequences, each of them depicting a single object laying on a table at different camera viewpoints (from 100 to 1000 per

Fig. 3. Results for 2 ACCV sequences. In the figures we show the GT point clouds of the objects and in green the estimated 3D bounding box. On the left are displayed the results by using the CG, where the red wire-frame represents the estimated polyhedron. On the right are shown the results of the IA method. (Color figure online)

sequence). We used 9 sequences for which the 3D point cloud of the object is provided, and limited the number of views to 100 for each sequence. The number of view and the motion affect positively the CG approach, as shown in the left image of Fig. 3: The larger the angle spanned by the viewpoints around the object, the better the performances of the method. As shown in Table 1, the results are remarkable for CG, with an average IoU of 0.85. Unlike the CG one, the IA approach does not reach high results in term of IoU (average IoU: 0.37) because of its tendency to overestimate the volume, as can be seen in Fig. 3 and as already explained in Sect. 3.3. Table 2 shows results using oriented bounding boxes with an average IoU of all the sequences similar to the average IoU given by bounding boxes aligned to the image axis. By analysing each sequence, there is a net improvement in the "Driller" and "Can" because the oriented bounding boxes can describe better objects with an anisotropic shape.

Table 1. Estimated IoU for 9 sequences from ACCV dataset, by using CG and IA with bounding boxes aligned to the image axis.

	Iron	Duck	Ape	Can	Driller	Vise	Glue	Cat	Lamp	Avg.
IA	0.34	0.14	0.27	0.39	0.33	0.63	0.50	0.18	0.53	**0.37**
CG	0.81	0.85	0.89	0.87	0.77	0.85	0.90	0.75	0.87	**0.84**

Table 2. Estimated IoU for 9 sequences from ACCV dataset, by using CG with bounding boxes aligned to the point cloud reprojections.

	Iron	Duck	Ape	Can	Driller	Vise	Glue	Cat	Lamp	Avg.
CG	0.80	0.84	0.73	0.92	0.90	0.82	0.93	0.70	0.87	**0.83**

4.2   TUW Dataset Evaluation

The TUW dataset [1] contains 15 annotated sequences showing a table with different sets of objects deployed on it. The number of frames per sequence ranges from 6 to 20, therefore fewer frames are available with respect to the ACCV dataset. Moreover, the objects in the images are not centred in the 3D scene as in the previous case.

We used both the CG and IA on these sequences, and the results are displayed in Table 4. In this case, it is clear a drop of performance for the CG approach, on average 0.27, while the IA approach fails to provide usable localisations by overestimating the volume when there are few frames available (Table 3).

We also performed an evaluation by considering the 2D bounding boxes aligned with the objects and we also evaluated the performance of the CG_b method. As expected, the results (reported in Table 4) outperform the original CG method in terms of IoU, reaching an average precision of 0.40. Indeed, if few frames are present, the constraint on the volume of the polyhedra is fundamental for not obtaining excessively overestimated volumes.

Table 3. Estimated IoU for 15 sequences from TUW dataset, by using CG and IA with bounding boxes aligned to the image axis.

	1	2	3	4	5	6	7	8	9	10	11	12	13	14	15	Avg.
IA	0.00	0.00	0.03	0.02	0.00	0.01	0.01	0.00	0.00	0.02	0.01	0.01	0.01	0.01	0.03	**0.01**
CG	0.17	0.05	0.53	0.25	0.32	0.29	0.23	0.25	0.12	0.41	0.24	0.23	0.23	0.38	0.33	**0.27**

Table 4. Estimated IoU for 15 sequences from TUW dataset, by using CG_b with bounding boxes oriented to the point cloud reprojections.

	1	2	3	4	5	6	7	8	9	10	11	12	13	14	15	Avg.
CG_b	0.45	0.09	0.34	0.37	0.39	0.33	0.42	0.57	0.28	0.57	0.43	0.45	0.39	0.33	0.52	**0.40**

4.3   ScanNet Dataset

ScanNet is a RGB-D dataset of real-world indoor environments proposed by [6] and it is the most challenging tested dataset. ScanNet main advantage is the high number of annotated sequences, 1513 in total. This dataset provides, for each sequence, all the camera parameters and a dense 3D reconstruction of the environment. Several objects and regions in the 3D point cloud are labelled, thereby providing ground truth for object localisation and occupancy estimation. We selected a subset of 1215 image sequences that have a minimum of 3 frames. We also did not consider all the sequences with a poor estimation of the motion of the camera, which can heavily affect objects localisation (Fig. 4).

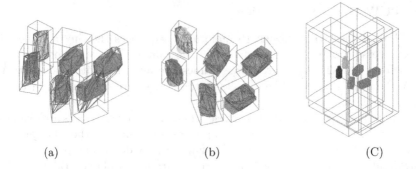

(a)   (b)   (C)

Fig. 4. Results for Seq. 7 of the TUW dataset. In the figures we show the GT point clouds of the objects and in green the estimated 3D bounding box. In (a) is displayed the result by using the CG, where the red wire-frame represents the estimated polyhedron. In (b) is shown the results of the CG_b with oriented bounding boxes, while in (c) the estimation performed by using the IA [9] method.

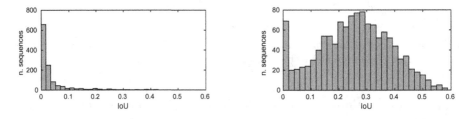

Fig. 5. Distribution of the IoU results for the 1215 selected sequences of the ScanNet dataset with the CG (left) and with the CG_b (right) approaches.

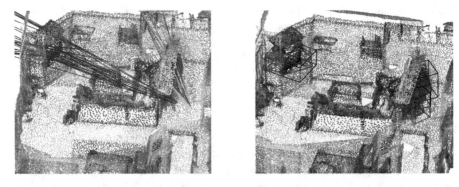

Fig. 6. Results for scene0000 of the ScanNet dataset. On the left is displayed the reconstruction by using only the CG approach, with the polyhedrons coloured differently to distinguish each reconstructed objects; on the right the estimation by using oriented bounding boxes and the CG_b approach, with the estimated polyhedrons in red and the associated bounding box in green. (Color figure online)

In this case the average results for the CG method is 0.04, while IA fails the reconstruction. The reason of this poor performance is mainly due the short baseline and small camera rotation. We also applied the CG_b to the ScanNet dataset by considering as input the oriented 2D bounding boxes. since the inclusion of two extra planes in the CG_b helps to limit the volume of the reconstruction as can be seen in Fig. 5(b), especially when motion of the camera is reduced and the polyhedron computed by the CG is unlimited, as in Fig. 5(a). In Fig. 5 we included some statistical information about the estimations, like the distribution among the sequences of the IoU results by using both the CG Fig. 6(a) and with the CG_b Fig. 6(b) approaches.

5 Conclusion

We have presented two approaches based on two already existing methods to perform the localisation (position and occupancy) of detected objects by using as input 2D bounding boxes associated to the objects and the camera parameters. Extensive experiments on real datasets confirm that the problem of estimating 3D localisation and occupancy from 2D bounding boxes is solvable. Between the two proposed approaches, IA tends to overestimate the enclosure with respect to CG. It is also clear that higher performance are obtained with higher number of frames and camera motion. Further improvements can still be obtained by including more data-driven priors about the surrounding environment and on the objects sizes and appearance. In particular, the ScanNet dataset performance can be further improved, representing a new challenge for the community.

References

1. Aldoma, A., Faulhammer, T., Vincze, M.: Automation of ground truth annotation for multi-view RGB-D object instance recognition datasets. In: IROS (2014)
2. Avis, D.: A revised implementation of the reverse search vertex enumeration algorithm. In: Kalai, G., Ziegler, G.M. (eds.) Polytopes Combinatorics and Computation. DMV Seminar, vol. 29, pp. 177–198. Birkhäuser, Basel (2000). doi:10.1007/978-3-0348-8438-9_9
3. Bao, S.Y., Xiang, Y., Savarese, S.: Object co-detection. In: Fitzgibbon, A., Lazebnik, S., Perona, P., Sato, Y., Schmid, C. (eds.) ECCV 2012. LNCS, vol. 7572, pp. 86–101. Springer, Heidelberg (2012). doi:10.1007/978-3-642-33718-5_7
4. Byröd, M., Josephson, K., Åström, K.: A Column-pivoting based strategy for monomial ordering in numerical gröbner basis calculations. In: Forsyth, D., Torr, P., Zisserman, A. (eds.) ECCV 2008. LNCS, vol. 5305, pp. 130–143. Springer, Heidelberg (2008). doi:10.1007/978-3-540-88693-8_10
5. Crocco, M., Rubino, C., Del Bue, A.: Structure from motion with objects. In: CVPR, pp. 4141–4149. IEEE (2016)
6. Dai, A., Chang, A.X., Savva, M., Halber, M., Funkhouser, T., Nießner, M.: ScanNet: richly-annotated 3D reconstructions of indoor scenes. arXiv (2017)
7. Dame, A., Prisacariu, V.A., Ren, C.Y., Reid, I.: Dense reconstruction using 3D object shape priors. In: CVPR, pp. 1288–1295. IEEE (2013)

8. Everingham, M., Van Gool, L., Williams, C.K.I., Winn, J., Zisserman, A.: The pascal visual object classes (VOC) challenge. IJCV **88**(2), 303–338 (2010)
9. Farenzena, M., Fusiello, A., Dovier, A.: Reconstruction with interval constraints propagation. In: CVPR, pp. 1185–1190 (2006)
10. Farenzena, M., Fusiello, A.: Stabilizing 3D modeling with geometric constraints propagation. Comput. Vis. Image Underst. **113**(11), 1147–1157 (2009)
11. Fidler, S., Dickinson, S., Urtasun, R.: 3D object detection and viewpoint estimation with a deformable 3D cuboid model. In: NIPS, pp. 611–619 (2012)
12. Geiger, A., Lauer, M., Wojek, C., Stiller, C., Urtasun, R.: 3D traffic scene understanding from movable platforms. PAMI **36**(5), 1012–1025 (2014)
13. Gupta, S., Girshick, R., Arbeláez, P., Malik, J.: Learning rich features from RGB-D images for object detection and segmentation. In: Fleet, D., Pajdla, T., Schiele, B., Tuytelaars, T. (eds.) ECCV 2014. LNCS, vol. 8695, pp. 345–360. Springer, Cham (2014). doi:10.1007/978-3-319-10584-0_23
14. He, K., Gkioxari, G., Dollár, P., Girshick, R.: Mask R-CNN. arXiv (2017)
15. Hinterstoisser, S., Lepetit, V., Ilic, S., Holzer, S., Bradski, G., Konolige, K., Navab, N.: Model based training, detection and pose estimation of texture-less 3D objects in heavily cluttered scenes. In: Lee, K.M., Matsushita, Y., Rehg, J.M., Hu, Z. (eds.) ACCV 2012. LNCS, vol. 7724, pp. 548–562. Springer, Heidelberg (2013). doi:10.1007/978-3-642-37331-2_42
16. Kim, B.S., Xu, S., Savarese, S.: Accurate localization of 3D objects from RGB-D data using segmentation hypotheses. In: CVPR, pp. 3182–3189 (2013)
17. Laurentini, A.: The visual hull concept for silhouette-based image understanding. TPAMI **16**(2), 150–162 (1994)
18. Moore, R.E.: Interval Analysis. Prentice-Hall, Upper Saddle River (1966)
19. Mousavian, A., Anguelov, D., Flynn, J., Kosecka, J.: 3D bounding box estimation using deep learning and geometry. arXiv (2016)
20. Silberman, N., Hoiem, D., Kohli, P., Fergus, R.: Indoor segmentation and support inference from RGBD images. In: Fitzgibbon, A., Lazebnik, S., Perona, P., Sato, Y., Schmid, C. (eds.) ECCV 2012. LNCS, vol. 7576, pp. 746–760. Springer, Heidelberg (2012). doi:10.1007/978-3-642-33715-4_54
21. Pepik, B., Gehler, P., Stark, M., Schiele, B.: 3D^{2}PM – 3D deformable part models. In: Fitzgibbon, A., Lazebnik, S., Perona, P., Sato, Y., Schmid, C. (eds.) ECCV 2012. LNCS, vol. 7577, pp. 356–370. Springer, Heidelberg (2012). doi:10.1007/978-3-642-33783-3_26
22. Preparata, F.P., Shamos, M.I.: Computational Geometry. An Introduction, Chap. 2. Springer, New York (1985). pp. 72–77
23. Kearfott, R.B.: Rigorous Global Search: Continuos Problems. Kluwer, Dordrecht (1996)
24. Ren, S., He, K., Girshick, R., Sun, J.: Faster R-CNN: towards real-time object detection with region proposal networks. In: NIPS, pp. 91–99 (2015)
25. Rump, S.: INTLAB - INTerval LABoratory. In: Developments in Reliable Computing, pp. 77–104. Kluwer Academic Publishers (1999)
26. Russell, B.C., Torralba, A., Murphy, K.P., Freeman, W.T.: LabelME: a database and web-based tool for image annotation. IJCV **77**(1), 157–173 (2008)
27. Zheng, S., Jayasumana, S., Romera-Paredes, B., Vineet, V., Su, Z., Du, D., Huang, C., Torr, P.: Conditional random fields as recurrent neural networks. In: ICCV (2015)
28. Zia, M.Z., Stark, M., Schindler, K.: Towards scene understanding with detailed 3D object representations. IJCV **112**(2), 188–203 (2015)

Image Analysis, Detection
and Recognition

A Computer Vision System for the Automatic Inventory of a Cooler

Marco Fiorucci[1]([✉]), Marco Fratton[2], Tinsae G. Dulecha[1,2], Marcello Pelillo[1,3],
Alberto Pravato[2], and Alessandro Roncato[1,2]

[1] DAIS, Ca' Foscari University, Via Torino 155, 30172 Venezia, Mestre, Italy
marco.fiorucci@unive.it
[2] PROSA S.r.l., via Dell'Elettricità 3/d, 30175 Venezia-Marghera, Italy
[3] ECLT, Ca' Foscari University, S. Marco 2940, 30124 Venezia, Italy

Abstract. In this paper we describe a system for beverage product recognition through the analysis of cooler shelf images. The extreme objects occlusion, the strong light influence and the poor quality of the images make this task a challenging one. To overcome these limitations, we rely on simple computer vision algorithms, like chamfer and color histogram matching and we introduce simple $3D$ modeling techniques. In our experiments, we demonstrate the effectiveness of our approach in terms of both detection accuracy and computational time.

Keywords: Cooler inventory · Image segmentation · Chamfer matching · Object recognition

1 Introduction

This paper describes a computer vision system for the automatic inventory of a commercial cooler. The goal is to count, for each brand, the number of beverage products (bottles and cans) contained in the cooler at any given moment in order to efficiently schedule a refill if necessary. This is done through the continuous analysis of the images of the cooler's shelves taken by (low-cost) wide-angle cameras.

Although at first glance the task looks trivial, as the objects to be recognized are clearly distinguishable, rigid and in a well-known static environment, it is in fact a challenging one due to a combination of several factors. In particular, a first difficulty arises from the severe occlusion conditions under which the system has to work. In fact, in a typical scenario involving densely packed shelves, visibility decreases row by row, the rear products being almost completely hidden from the front ones (see Fig. 1 for some typical examples). The items are also typically very close to each other and this makes segmentation and detection more difficult. Recognition is also complicated by the lighting conditions: light is not uniform in the images, not only due to the shadows generated by the shelves and by the products themselves, but also due to the influence of external light. As a result, our images have typically poorly defined edges and distorted

© Springer International Publishing AG 2017
S. Battiato et al. (Eds.): ICIAP 2017, Part I, LNCS 10484, pp. 575–585, 2017.
https://doi.org/10.1007/978-3-319-68560-1_51

color representation, thereby making segmentation and brand classification more difficult. Also, the system has to be flexible enough to recognize new products after software installation. These difficulties are exacerbated by the need to cut off production costs and by the consequent use of low-quality cameras and limited computational resources. Indeed the whole system has to run on an embedded low-performance computer and this poses serious limitations as to the kind of algorithms that can be used, as computationally intensive techniques are clearly not feasible.

Fig. 1. Typical images analyzed by our system.

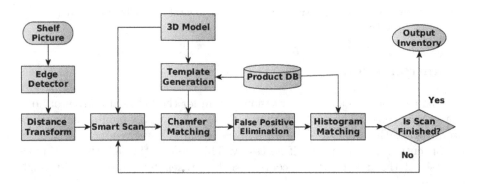

Fig. 2. Flow-chart of the proposed system.

The proposed system uses a combination of simple techniques to address these limitations. It is implemented into a pipeline of simple modules, as shown in Fig. 2. The pipeline begins with an edge detector which extracts the features that will be used by the distance transform module to construct a distance image. The next step in the pipeline is chamfer matching [1], which detects the shape of beverage products by shifting their templates at various locations of the distance image. A matching measure is used to detect a candidate beverage shape, which is then checked by a false positive elimination module. Finally, the brand of the beverage products is recognized using simple color histogram matching. The color histogram of the pixels which lie under a detected shape is compared with the color histograms build from the images of reference products. Despite the simplicity of the used techniques, preliminary results show the effectiveness of the proposed system in terms of both detection accuracy and computational time.

2   The Pipeline

The proposed pipeline is based on simple techniques applied in a cascaded way to enhance the recognition accuracy and to provide robustness. As previously mentioned, the pipeline begins with a learning-based edge detector [4] which extracts the most useful product edges that will be used to construct a distance image. This is used by the chamfer matching module [1] in order to detect candidate product shapes which will be checked by the false positive elimination module. The last module is the histogram matching that allows brand recognition. The algorithm is optimized by using 3D modeling techniques for template generation and by a space management system which allows faster image scan and avoids the need of a non-maximum suppression. Further accuracy is achieved by splitting a beverage into its main characterizing parts, processing them independently and considering the results as a whole. Occlusion is dealt by building an occlusion mask which keeps track of the image portions occupied by the detected beverages and masks the templates occluded parts. Figure 2 shows the flow chart of the proposed pipeline.

2.1   Edge Detection

Edge detection is the preprocessing stage of the pipeline. It relies on the OpenCV 3.2.0 [7] implementation of the fast edge detector proposed by Dollár and Zitnick [4], which is inspired by the work of Kontschieder et al. [8]. It exploits the high interdependence of the edges in a local image patch. In particular, edges exhibit well-known patterns that can be used to train a structured learning model. Dollár and Zitnick's algorithm segments an image into local patches used to train a structured random forest model. This model provides a local edge mask which is applied to extract edges in an accurate and efficient way. Figure 3 shows edge detection results obtained from Dollár and Zitnick's algorithm.

2.2   Shape Detection

Template matching is the first stage of the proposed system in which beverage candidates are evaluated and discarded if they do not satisfy the shape requirements. It relies on a chamfer template matching [6] for the shape detection, on a 3D modeling for the template generation, on a smart sliding window for the space management and on a simple yet essential mechanism for the occlusion management.

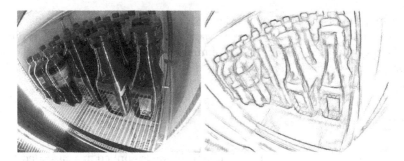

Fig. 3. Example of the edge detection results: the original shelf image on the left and the edge image on the right.

Chamfer matching is a simple template matching algorithm which offers high performance and a robust detection as it is very flexible and more tolerant to low quality edges than other algorithms of the same kind. First, a morphological transformation, known as distance transform [5], is applied to the previously extracted edges. The resulting picture will be a gray-scale image in which each pixel will have the value of the distance from that pixel to the nearest edge. Finally, a query template is slided onto the distance image. At each position, a matching measure is computed by summing the pixel values of the distance transform image which lie under the edge pixels of the template. If the computed matching measure lies below a certain threshold, the target beverage shape is considered detected. The template threshold should be chosen to achieve a desired trade-off between false positives and false negatives.

Chamfer matching is very inefficient as all beverage templates of varying shape and size have to be tested at each locations of the distance image. Thus, a $3D$ model of the shelf is introduced to speed-up the matching process. It allows to check only one template per product at each location of the distance image avoiding to check, for each products, a bunch of templates of varying shape and size. To achieve this aim, we exploit the available information related to the objects, the cooler and the camera in order to render the shelf and to build the template for the shape matching. In particular, each object is accurately measured as follows: first the bottom diameter is measured, then, going up, for each change in the shape the value of the height and the corresponding diameter are collected. In this way we sum up the product contour as a collection of diameter discontinuities and their relative heights. The beverage partition into contour and horizontal parts can reproduce well most of the bottles and cans, even those which are not circular based, with a little error. Furthermore, cameras intrinsic parameters are collected, while real position of the camera and rotation angles are measured. For this purpose we introduce an artificial reference points in the picture: a special sheet of paper with a printed grid is laid on the shelf, while the same grid is rendered in a $3D$ representation of that shelf, using the cooler information. At the beginning the virtual grid is in a random position but, using special buttons on the keyboard, a user is able to modify the camera position and the rotation angles in order to match as close as possible the virtual grid with the real grid. When the grids match, we obtain the camera position and the orientation with a good accuracy. This whole process should be done only once, when the camera is installed. Figure 4 shows the calibration process. Finally, after the calibration step, the template of each product is rendered at any desired point of the shelf (Fig. 5).

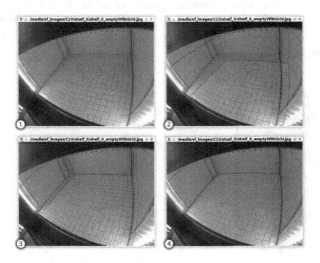

Fig. 4. Calibration procedure: the goal is to match the grid on the shelf. (1): Real grid in the shelf. (2): Starting virtual grid with predefined camera position and orientation. (3): Close match of the grids. (4): Good grid match; now the camera parameters are known.

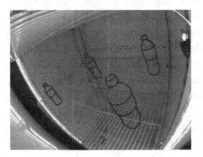

Fig. 5. Examples of templates generated by the 3D modeling.

To further speed-up the matching process, a smart sliding window for the space management, named smart scan module, is introduced. It relies on a $3D$ shelf model which allows to switch from virtual coordinates (pixels of the image) to physical ones (millimeters of the real shelf) (Fig. 6). The scan is then performed referring to the physical shelf position (x, z) so that the spatial information can be exploited to avoid points in which the template cannot fit due to the lack of space. In particular, the scan starts from the lowest right angle $(x = maxLength, z = 0)$ and goes up column-wise: at each detection step we keep the x fixed and we increase the z by $step_z$, until the innermost part is reached; then we reset z to 0, we shift left by $step_x$ $(x = x - step_x)$ and we start increasing the z again; this procedure goes over until the left highest corner is reached. Thus, the $3D$ model and the smart scan allow to check only one template per product at each permissible position (x, z) speeding up the template matching phase.

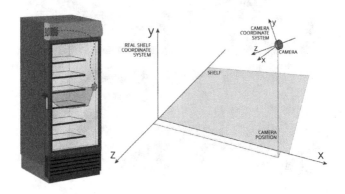

Fig. 6. Real shelf and camera coordinate systems.

To deal with the occlusion conditions, a binary image, called occlusion mask (see Fig. 7), keeps track of the detections found at every step. The occlusion mask has the same size of the shelf image, and it can be thought as a sort of shelf shadow doublet: each time a detection is confirmed in an image point, the occlusion mask is updated accordingly by setting to zero all the pixels belonging to the filled template shape at that same point. In this way the occlusion mask will be a binary image in which black pixels denote the scan image space occupied by the products found until that moment, while white pixels denote the free space left. We then update the query template by masking it with the occlusion mask, so that only the visible template portion is used in the subsequent matching. If the remaining template portion is under a certain threshold, it is discarded as not reliable enough. This solution offers good performance while keeping the problem at a very simple level, but it is not always accurate enough as it is based on a strong assumption which sometimes does not hold: products are considered to be picked in order from the visible ones to the most occluded.

Finally, to achieve better accuracy, a procedure known as false positive elimination is performed: each beverage part of a candidate detection is compared against the results achieved by the chamfer matching applied to a reference background image. If the results are too close to each other, the algorithm states that

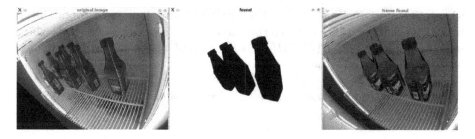

Fig. 7. Example of the mask image during an ongoing detection. The source image is on the left, the occlusion mask is in the middle and the objects found until that moment are on the right. (Color figure online)

the match is a fake one (the match is a part of the background which is wrongly detected as a real object).

2.3 Color Classification

Histogram matching is the second and last stage of the proposed pipeline in which the brand of a previously detected shape is recognized. In particular, the histogram matching module exploits all the elements defining a visual beverage, i.e. shape and color, to enhance the correctness of the shape detection and to recognize the brand of previously detected shapes.

This module relies on the same distinction between the product parts done in the template matching one: a product is split into its main components (cap, bottle liquid and logo for the bottles, the top part of the can and the can surface for the cans) so that it is possible to focus on simple algorithms while keeping the spatial color information (as an example, the cap should be blue while the liquid is green, and not the opposite). It is worth nothing that in the same product part the color is often uniform, so there is no need to split the objects further.

The color analysis is based on simple color histograms [2,3,9] guided by the $3D$ model: only the image portion under the filled template is used to build the histogram. The color space is divided into n sub-parts, called bins, covering specific color ranges. Three normalized color histograms, one for each channel, are then computed. Finally, the histogram of each product part is compared against histograms build from the products database in order to decide the fitness of the detection.

The product database contains reference photos of each product the algorithm should recognize. In particular, for each product, a series of photos are snapped in controlled conditions: the middle shelf of the reference fridge is divided into 9 zones and for each zone four pictures are snapped using 90° rotation.

Histogram comparison is based on the following measure:

$$d(H(I), H(I')) = d_{mode}(H(I), H(I'))(1 - H(I) \cap H(I')) \tag{1}$$

where $H(I)$ and $H(I')$ is a pair of normalized histograms, each containing n bins; d_{mode} is the distance between the bins of each histogram having the highest frequency indexes and $H(I) \cap H(I')$ is the sum of the smallest corresponding bins between two histograms, i.e. the histogram intersection.

The measure (1) is a weighted distance which is robust against color distortion because of the modes, while keeping a deeper histogram comparison because of the intersection.

3   Experimental Results

We have performed a series of experiments to verify the performances and the accuracy level that can be obtained by our system. All the module of the pipelines have been implemented using GNU C++ and have been run on dual core CPU with 1.6 GHz/core and 1 GB of RAM.

The results here presented are divided into two sections:

- the first section shows examples of products placed at random in the shelf;
- the second section shows examples of real cooler cases, where a shelf is filled by columns and each column will contain only bottles/cans of the same brand.

The experiments have been conducted in a 654×594 mm cooler shelf with 10 beverage brands. For each test it is shown: the original shelf image (on the left); the beverage edge image where detected caps are highlighted in red (in the middle) and, finally, the $3D$ rendering of the products detected by the pipeline (on the right).

3.1   Random Shelf Configurations

Figure 8 shows some examples of products randomly placed in the shelf and a few products placed at the rear. The recognition is high, even if some Lipton cans are seen as Kickstarter, since they are very similar; we can also note that the difference between the cans themselves is very little, as just a little part of the logo is different. It is worth to note that Gatorade are detected despite having a different shape from the one in our database: this highlights the algorithm is flexible enough to recognize even unknown products sharing similar properties to the known ones. As for the cans, the Lipton bottle brands (brown bottles) are so similar that it is almost impossible to distinguish between them. Finally, Pepsi and MtnDew (green bottles) have a distinctive color, hence we can achieve a good accuracy on them.

3.2   Ordered Shelf Configurations

Figure 9 shows some examples of real cooler cases, where a shelf is filled by columns and each column will contain only bottles/cans of the same brand. In the first row there are two tea bottles placed in the rear of an almost empty fridge which are correctly recognized, while the second row there are two Gatorade and

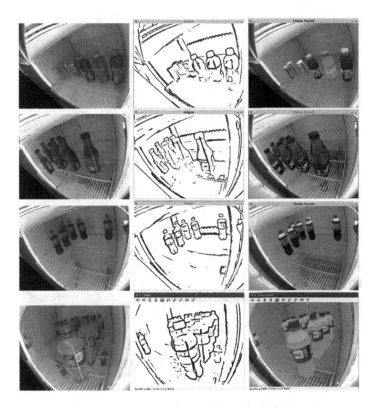

Fig. 8. Examples of products randomly placed in the shelf and a few products placed at the rear.

three Lipton cans which are correctly recognized too. The cooler is recognized to be almost empty in both cases. In the third row there are some missed Pepsi. This is due to weak edges which are not recognized by the template matching. In the last row, there is a shelf full of bottles and, in this case, some products are missed.

From the analysis of 100 experiments we can state that:

- the overall average accuracy level we have obtained is over 80%. In particular, an empty shelf can be identified with 100% precision, while the accuracy decreases to 70% if the shelf is almost full, because of the product occlusion that forces the algorithm to rely only on the top part of the product instead of considering it in its entirety.
- Since the system should send a cooler inventory every 10 min, the performances are quite satisfactory, as the whole scan of a 654 × 594 mm cooler shelf takes approximatively 100 s.
- Some products are more easily detectable than others since the colors of beverages like Pepsi, MtnDew, Gatorade create a well defined contrast with the background and are very different from the colors of other products. By

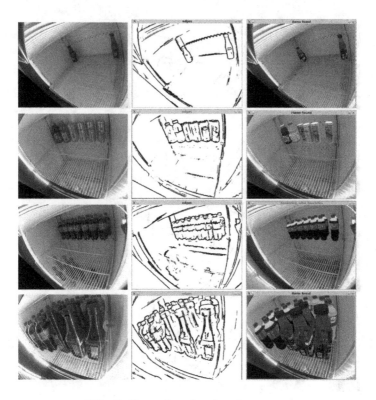

Fig. 9. Examples of real cooler cases.

contrast, Aquafina is very difficult to be identified because of its transparent bottle and its white cap which blends into the background.

4 Conclusions

We have described a simple yet effective system for monitoring the content of a commercial cooler through the visual analysis of the shelves' images taken with low-cost wide-angle cameras. The difficulty of this task lies mainly in the challenging set-up in which it has to be carried out, such as severe or almost complete occlusion, uneven lighting conditions, poor image quality, and low-cost hardware. The proposed solution combines simple techniques which effectively work under these challenging conditions.

Despite the simplicity of the used techniques, we achieved a satisfactory accuracy level, being able to detect from 70% to 95% of the whole shelf in 100 images. Since the system should send a cooler inventory every 10 min, the computational performances are acceptable as a full shelf scan takes approximately 100 s using limited computational resources. Finally, the system is very flexible, as it needs just a simple and quick learning phase to add new products.

In future, we are planning to better handle irregular light intensity and color distortion in order to improve the recognition accuracy.

References

1. Borgefors, G.: Hierarchical chamfer matching: a parametric edge matching algorithm. IEEE Trans. Pattern Anal. Mach. Intell. **10**(6), 849–865 (1988)
2. Chang, P., Krumm, J.: Object recognition with color cooccurrence histograms. In: IEEE Conference on Computer Vision and Pattern Recognition, pp. 2498–2504 (1999)
3. Dalal, N., Triggs, B.: Histograms of oriented gradients for human detection. In: IEEE Conference on Computer Vision and Pattern Recognition, pp. 886–893 (2005)
4. Dollr, P., Zitnick, C.L.: Fast edge detection using structured forests. IEEE Trans. Pattern Anal. Mach. Intell. **37**(8), 1558–1570 (2015)
5. Felzenszwalb, P.F., Huttenlocher, D.P.: Distance transforms of sampled functions. Theor. Comput. **8**(1), 415–428 (2012)
6. Barrow, H.G., Tenenbaum, J.M., Bolles, R.C., Wolf, H.C.: Parametric correspondence and chamfer matching: two new techniques for image matching. In: Proceedings of the 5th International Joint Conference on Artificial Intelligence, pp. 659–663 (1977)
7. Itseez: opencv_contrib (2016). https://github.com/opencv/opencv_contrib
8. Kontschieder, P., Bulò, S.R., Bischof, H., Pelillo, M.: Structured class-labels in random forests for semantic image labelling. In: 2011 International Conference on Computer Vision, pp. 2190–2197, November 2011
9. Ling, H., Okada, K.: Diffusion distance for histogram comparison. In: IEEE Conference on Computer Vision and Pattern Recognition, pp. 246–253 (2006)

A Convexity Measure for Gray-Scale Images Based on hv-Convexity

Péter Bodnár$^{(\boxtimes)}$, Péter Balázs, and László G. Nyúl

Department of Image Processing and Computer Graphics, University of Szeged,
Árpád tér 2., Szeged 6720, Hungary
{bodnaar,pbalazs,nyul}@inf.u-szeged.hu

Abstract. There are several measures for the convexity of digital images that extend the basic binary decision of the classic geometrical convexity. Some algorithms measure the convexity of a binary image using intensity profiles from horizontal and vertical directions. In this paper, we generalize the idea of binary, directional convexity and evaluate the proposed algorithm on gray-scale images. Furthermore, instead of a single convexity value, a vector can be formed using our approach, which provides a more prominent feature for various applications, such as computer vision, classification, retrieval, or medical image processing. The proposed feature can also be used locally on image parts, which makes that applicable as a shape descriptor.

Keywords: Convexity measure · Digital geometry · hv-convexity · Gray-scale convexity

1 Introduction

Convexity is a widely studied and applied shape descriptor in image analysis and classification. On digital shapes, there are various measures that approximate the continuous convexity, like area based [6,19,20] and boundary-based ones [22]. It shall be noted that many convexity measures produce continuous output [15,17,18], unlike the classic, geometrical approach, which gives a binary decision whether or not the observed shape is convex.

In case of digital images, directional convexity is a common alternative for the convexity for continuous shapes, due to the pixel-based representation of the image. Mostly horizontal and vertical convexity is used (shortly, hv-convexity), which means that the convexity measure is defined by the aggregation of the convexity degree along horizontal and vertical sweeping lines. The property of hv-convexity is deeply studied in Binary Tomography [14], where one problem in focus is to reconstruct binary images (matrices) from their row and column sums according to geometrical constraints. Several reconstruction methods utilize the preliminary information of hv-convexity about the binary image to be reconstructed [3,8,11]. Enforcing compactness of the image to reconstruct can also

---

The research was supported by the NKFIH OTKA [grant number K112998].

S. Battiato et al. (Eds.): ICIAP 2017, Part I, LNCS 10484, pp. 586–594, 2017.
https://doi.org/10.1007/978-3-319-68560-1_52

result in binary images which are (almost) hv-convex [12]. In [21] the authors introduced a measure of directional convexity and proved it to be useful in binary tomographic reconstruction. However, they also showed that a 2D extension of this measure is not straightforward [2]. Later, immediate 2D convexity measures were also proposed in [1,9], while in [5] an upgrade of the measure of [2] was published.

The aim of this paper is to generalize the directional convexity measure from binary to gray-scale images, that can be used with existing binary convexity measures [1,5,21]. The structure of the paper is the following. In Sect. 2 we describe the proposed gray-scale convexity measure. In Sect. 3 we present experimental results. Section 4 is for the conclusion.

2 The Proposed Gray-Scale Convexity Measure

2.1 Preliminaries

A digital image M is a matrix having m rows and n columns (where $m, n \in \mathbb{N}$). Numbering of rows and columns start with 1 from top to bottom and left to right, respectively. If M is a digital image then M^T is the image we get by interchanging the rows and columns of M. Let $I = \{i_0, \ldots, i_l\}$ be the set of possible intensity values of the image such that $i_k < i_{k+1}$ $(k = 0, \ldots, l-1)$ and $M(r, c) \in I$ denote the intensity value corresponding to the position (r, c). A typical choice is $I = \{0, \ldots, 255\}$ (8-bit images) or $I = \{0, \ldots, 65535\}$ (16-bit images).

For binary images $I = \{0, 1\}$. In this case, a *run* of object (background) points within a row or column is a sequence of consecutive pixels, all of them being object (resp. background) points, such that it cannot be expanded by further neighboring pixels of the same color. Obviously, each row and column of the image can be expressed by an alternating sequence of object and background runs. The length of an arbitrary run a will be denoted by $|a|$.

2.2 Measure of hv-Convexity for Binary Images

Originally, we follow the idea of hv-convexity measuring on binary images in [5] which is a modified version of [2]. According to that paper, first, the convexity defect $\varphi_h^{bin}(r)$ for each row $r = 1, \ldots, m$ is calculated in the following way (*bin* stands for "binary").

Let R be the pixel sequence of an arbitrary row. To compute the *non-convexity* of R, we split it into a list of object and background runs. If the first or last run is a background run then we omit them. Thus the rest of the row can be encoded as $R = b_1 w_1 b_2 w_2 \ldots w_{n-1} b_n$, where each b_i is an object run $(i = 1, \ldots, n)$ and each w_i $(i = 1, \ldots, n-1)$ is a background run.

Let O_R be the ordered set of object runs in row r, i.e., $O_R = \{b_1, b_2, \ldots, b_n\}$. The sum of object pixels in R is $N_R = |b_1| + |b_2| + \cdots + |b_n|$. Now, let $b_i, b_j \in O_R$ such that $i < j$. We select one random point from both, say, the k-th from

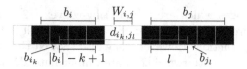

Fig. 1. Calculation of the non-convexity between two object points from different object runs, proposed in [5].

left in b_i denoted by b_{i_k} and the l-th from left in b_j denoted by b_{j_l}. The section connecting these two points is characterized by the non-convexity measure, which value depends on the number of background pixels between b_i and b_j. Let $W_{i,j} = \sum_{l=i}^{j-1} |w_l|$, $B_{i,j} = \sum_{l=i+1}^{j-1} |b_l|$ and d_{i_k,j_l} denote the distance of the two chosen points. This distance is partially made up of the points of b_i to the right of b_{i_k}, the points of b_j to the left of b_{j_l}. There are, $|b_i| - k + 1$ and l such points (including the chosen points, too), respectively. Additionally, the section contains the $W_{i,j}$ background points, and further object point runs $(B_{i,j})$, if $j > i + 1$. That is, $d_{i_k,j_l} = |b_i| - k + 1 + W_{i,j} + B_{i,j} + l$ (Fig. 1 illustrates the calculation). The normalized non-convexity measure for this section is

$$\frac{W_{i,j}}{d_{i_k,j_l}} , \qquad (1)$$

and the cumulated non-convexity of R is

$$\frac{\sum_{b_i,b_j \in O_R, i<j} \sum_{k=1}^{|b_i|} \sum_{l=1}^{|b_j|} \frac{W_{i,j}}{d_{i_k,j_l}}}{C_r} , \qquad (2)$$

where C_r is the number of combinations to select the two object points from different object point runs, computed as

$$C_r = \binom{N_R}{2} - \sum_{b \in O_R} \binom{|b|}{2} . \qquad (3)$$

The horizontal convexity of M is defined as

$$\Psi_h^{bin}(M) = 1 - \frac{\sum_{r=1}^{m} \varphi_h^{bin}(r)}{m} . \qquad (4)$$

The vertical convexity $\Psi_v^{bin}(M)$ can be calculated analogously by the observation that $\Psi_v^{bin}(M) = \Psi_h^{bin}(M^T)$. Finally, the hv-convexity is the algebraic mean of the horizontal and vertical convexity, i.e.,

$$\Psi_{hv}^{bin}(M) = \frac{\Psi_h^{bin}(M) + \Psi_h^{bin}(M^T)}{2} . \qquad (5)$$

2.3   Extension of the Convexity Measure to Gray-Scale Images

The aforementioned approach only measures convexity of binary images, since we need to define sequences of object and background pixels. In most cases, binarization is solved by thresholding (for example, with Otsu's method [16]), which leads to loss of information. To overcome this, we propose to aggregate the convexity using all possible thresholds. Let $T(M, t)$ denote the binary image we get by thresholding M at level t. For the continuous case, convexity is computed as

$$\Psi_{hv}(M) = \frac{1}{i_l - i_0} \int_{i_0}^{i_l} \Psi_{hv}^{bin}(T(M,t))dt \ . \tag{6}$$

This calculation theoretically takes infinite time, however, it collapses to a factor of $\mathcal{O}(|I|)$ when the input is quantized. Assuming that the image is in the positive intensity range, convexity is computed as

$$\Psi_{hv}(M) = \frac{1}{|I|} \sum_{t=i_0}^{i_l} \Psi_{hv}^{bin}(T(M,t)) \ . \tag{7}$$

The aforementioned approach calculates the convexity of the same binary image multiple times if the intensity value t does not occur within the original one. Exploiting this, the calculation of $T(M, t)$ is only necessary where $t \in J$ with $J = \{j_0, j_1, \ldots, j_{|J|-1}\} \subseteq I$ being the ordered set of distinct intensity values of I. For the sake of technical simplicity we assume that the maximal element i_l of I is always contained in J even if it is not present in the image. Each $\Psi_{hv}^{bin}(T(M,t))$ can be assigned a weight, reflecting how many times we could have calculated that. Let $W(t)$ be a weight corresponding to t. We perform thresholding at all intensity levels of J and aggregate the results of binary convexity measures (Algorithm 1) as

$$\Psi_{hv}(M) = \frac{1}{|I|} \sum_{t=0}^{|J|-1} \Psi_{hv}^{bin}(T(M,j_t))W(t) \tag{8}$$

with

$$W(t) = \begin{cases} j_0 + 1 & \text{if } t = 0 \\ j_t - j_{t-1} & \text{otherwise} \end{cases} \ . \tag{9}$$

The weight values $W(t)$ would be 1 for all t input, if all intensities occur in the image within the full intensity range. For an other example, if $I = \{0,1,2,3,4\}$ and the ordered set of intensities in the image is $J = \{0,1,4\}$, then the corresponding weights are $\{1,1,3\}$. It shall be noted that the sum of weights is always equal to the size of the intensity range in which the image is represented.

Algorithm 1. Convexity calculation for gray-scale images using binary hv-convexity measure $\Psi_{hv}^{bin}(M)$.

```
 1: function CONVEXITY(M)                        ▷ convexity for gray-scale image M
 2:     J ← i_l                                  ▷ Set of distinct intensity values
 3:     for i ← 1...m × j ← 1...n do
 4:         if M(i, j) ∉ J then J ← J ∪ M(i, j)
 5:     end for
 6:     W ← []                                   ▷ Array of weights
 7:     W[0] ← J[0]+1
 8:     for t ← 1...|J| − 1 do
 9:         W[t] ← J[t] − J[t − 1]
10:     end for
11:     c ← 0
12:     for t ← 0...|J| − 1 do
13:         M_t ← T(M, J[t])                     ▷ Thresholding at level t
14:         c ← c + Ψ_{hv}^{bin}(M_t) * W[t]     ▷ Convexity calculation with existing method
15:     end for
16:     c ← c / |I|                              ▷ Normalization
17:     return c
18: end function
```

3   Evaluation and Experiments

Our first experiment is about to show the basic difference between the original binary convexity [5] and the proposed one. In Fig. 2, binary thresholding leads to the same result for both squares. On the other hand, the proposed algorithm forms the weighted sum of thresholds on all occurring gray levels, and can differentiate between the two images. It gives a convexity value of 0.8940 for the gray-scale image and 0.5975 to its binarized version.

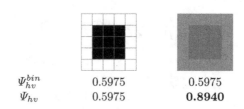

Ψ_{hv}^{bin}	0.5975	0.5975
Ψ_{hv}	0.5975	**0.8940**

Fig. 2. Images of two empty square objects and their corresponding convexity values. The gray square is intuitively more "full", which attribute is also supported by the proposed gray-scale convexity value.

We also examined the proposed algorithm on a real gray-scale image (Fig. 3). We thresholded the image at 50% of the intensity range, produced another image using 16 quantization levels, and finally, measured the hv-convexity of the original 8-bit image. According to this example, the quantization levels may be reduced for 8-bit images in order to achieve faster run-time of the algorithm, however, that only gives an approximation of the original convexity.

Levels	2	16	256
Ψ_{hv}^{bin}	0.6924	0.6924	0.6924
Ψ_{hv}	0.6924	0.7954	0.7864

Fig. 3. The proposed approach on a real 8-bit image.

It shall be noted that not only a scalar value can be derived from this approach. If desired, the convexity values can be used for each occurring intensity (Figs. 4 and 5). Thus, two vectors can be formed for each image, one containing the convexity values for each threshold level, and another with the corresponding weights. Both vectors have the same length (the number of distinct intensities of the source image). Those vectors can also be computed locally on image parts, which renders them applicable as a shape descriptor for computer vision, classification and object recognition.

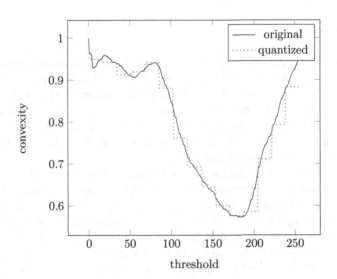

Fig. 4. Convexity values for the 8-bit real image (represented in Fig. 3) and its 16-level quantized version w.r.t. threshold level. The vector of individual convexities may give a more prominent feature for classification tasks than a single convexity value.

t	0	17	34	51	68	85	102	119
Ψ_h^{bin}	1.0000	0.9360	0.9363	0.9165	0.9219	0.9305	0.8872	0.8033
Ψ_v^{bin}	1.0000	0.9638	0.9500	0.9096	0.9178	0.9521	0.8741	0.7200
Ψ_{hv}^{bin}	1.0000	0.9499	0.9431	0.9130	0.9199	0.9413	0.8806	0.7616

t	136	153	170	187	204	221	238	255
Ψ_h^{bin}	0.7227	0.6640	0.5953	0.5822	0.5865	0.7172	0.7888	0.8600
Ψ_v^{bin}	0.6623	0.6250	0.6074	0.5712	0.5840	0.7053	0.7969	0.9063
Ψ_{hv}^{bin}	0.6925	0.6445	0.6014	0.5767	0.5853	0.7112	0.7928	0.8832

Fig. 5. The thresholded versions of the image represented on Fig. 3, quantized to 16 levels, and its corresponding values of h-, v-, and hv-convexity.

The proposed generalization of the binary convexity measure has the same behavior w.r.t. rotation and scale invariance than the original convexity measure we generalize. While this paper only evaluates the convexity measure of [5], the proposed idea can be used with other binary convexity measures as well [1].

4   Conclusion

In this paper, we presented a gray-scale generalization of an hv-convexity measure for binary images. Using this approach, the loss of information at the thresholding step is avoided, while all existing convexity measures that work on binary images [1,2,5] can be adapted to work on gray-scale images, too. Having only a few distinct intensity levels in an image, the calculation can be performed rapidly. If less precise calculation is acceptable and speed is more desired, intensity levels of the image may be further quantized.

The descriptor can be also computed locally to an image part, therefore it may also be used as an additional shape descriptor in applications, such as computer vision, classification, object recognition, image retrieval, or medical image processing. A further perspective is to use the single gray-scale convexity measure as prior information in multivalued discrete tomography. The reconstruction of multicolor images (i.e., containing at least 3 different gray intensity values) is in general an NP-hard problem, however, for certain image classes and/or with appropriate heuristics it can be effectively solved [4,7,10,13]. It needs a further investigation whether gray-level convexity measures can also facilitate such kind of reconstruction problems.

References

1. Balázs, P., Brunetti, S.: A measure of Q-convexity. In: Normand, N., Guédon, J., Autrusseau, F. (eds.) DGCI 2016. LNCS, vol. 9647, pp. 219–230. Springer, Cham (2016). doi:10.1007/978-3-319-32360-2_17
2. Balázs, P., Ozsvár, Z., Tasi, T.S., Nyúl, L.G.: A measure of directional convexity inspired by binary tomography. Fundam. Inform. 141(2–3), 151–167 (2015)
3. Barcucci, E., Del Lungo, A., Nivat, M., Pinzani, R.: Medians of polyominoes: a property for reconstruction. Int. J. Imaging Syst. Technol. 9(2–3), 69–77 (1998)
4. Barcucci, E., Brocchi, S., Frosini, A.: Solving the two color problem: an heuristic algorithm. In: Aggarwal, J.K., Barneva, R.P., Brimkov, V.E., Koroutchev, K.N., Korutcheva, E.R. (eds.) IWCIA 2011. LNCS, vol. 6636, pp. 298–310. Springer, Heidelberg (2011). doi:10.1007/978-3-642-21073-0_27
5. Bodnár, P., Balázs, P.: An improved directional convexity measure for binary images. In: Karray, F., Campilho, A., Cheriet, F. (eds.) ICIAR 2017. LNCS, vol. 10317, pp. 278–285. Springer, Cham (2017). doi:10.1007/978-3-319-59876-5_31
6. Boxer, L.: Computing deviations from convexity in polygons. Pattern Recogn. Lett. 14(3), 163–167 (1993)
7. Brocchi, S., Frosini, A., Rinaldi, S.: A reconstruction algorithm for a subclass of instances of the 2-color problem. Theor. Comput. Sci. 412(36), 4795–4804 (2011)
8. Brunetti, S., Lungo, A.D., Ristoro, F.D., Kuba, A., Nivat, M.: Reconstruction of 4- and 8-connected convex discrete sets from row and column projections. Linear Algebra Appl. 339(1), 37–57 (2001)
9. Brunetti, S., Balázs, P., Bodnár, P.: Extension of a one-dimensional convexity measure to two dimensions. In: Brimkov, V.E., Barneva, R.P. (eds.) IWCIA 2017. LNCS, vol. 10256, pp. 105–116. Springer, Cham (2017). doi:10.1007/978-3-319-59108-7_9
10. Chrobak, M., Dürr, C.: Reconstructing polyatomic structures from discrete X-rays: NP-completeness proof for three atoms. In: Brim, L., Gruska, J., Zlatuška, J. (eds.) MFCS 1998. LNCS, vol. 1450, pp. 185–193. Springer, Heidelberg (1998). doi:10.1007/BFb0055767
11. Chrobak, M., Durr, C.: Reconstructing hv-convex polyominoes from orthogonal projections. Inf. Proc. Lett. 69(6), 283–289 (1999)
12. Cipolla, M., Bosco, G.L., Millonzi, F., Valenti, C.: An island strategy for memetic discrete tomography reconstruction. Inf. Sci. 257, 357–368 (2014)
13. Dürr, C., Gunez, F., Matamala, M.: Reconstructing 3-colored grids from horizontal and vertical projections is NP-hard: a solution to the 2-atom problem in discrete tomography. SIAM J. Discret. Math. 26(1), 330–352 (2012)
14. Herman, G.T., Kuba, A.: Advances in Discrete Tomography and Its Applications. Applied and Numerical Harmonic Analysis. Birkhauser, Boston (2007)
15. Latecki, L.J., Lakamper, R.: Convexity rule for shape decomposition based on discrete contour evolution. Comput. Vis. Image Underst. 73(3), 441–454 (1999)
16. Otsu, N.: A threshold selection method from gray-level histograms. IEEE Trans. Syst. Man Cybern. 9(1), 62–66 (1979)
17. Rahtu, E., Salo, M., Heikkila, J.: A new convexity measure based on a probabilistic interpretation of images. IEEE Trans. Pattern Anal. Mach. Intell. 28(9), 1501–1512 (2006)
18. Rosin, P.L., Žunić, J.: Probabilistic convexity measure. IET Image Proc. 1(2), 182–188 (2007)

19. Sonka, M., Hlavac, V., Boyle, R.: Image Processing, Analysis, and Machine Vision. Cengage Learning, Boston (2014)
20. Stern, H.I.: Polygonal entropy: a convexity measure. Pattern Recogn. Lett. 10(4), 229–235 (1989)
21. Tasi, T.S., Nyúl, L.G., Balázs, P.: Directional convexity measure for binary tomography. In: Ruiz-Shulcloper, J., Sanniti di Baja, G. (eds.) CIARP 2013. LNCS, vol. 8259, pp. 9–16. Springer, Heidelberg (2013). doi:10.1007/978-3-642-41827-3_2
22. Zunic, J., Rosin, P.L.: A new convexity measure for polygons. IEEE Trans. Pattern Anal. Mach. Intell. 26(7), 923–934 (2004)

A Hough Voting Strategy for Registering Historical Aerial Images to Present-Day Satellite Imagery

Sebastian Zambanini$^{(\boxtimes)}$ and Robert Sablatnig

Computer Vision Lab, TU Wien, Vienna, Austria
{zamba,sab}@caa.tuwien.ac.at

Abstract. In this paper we present an approach for the georeferencing of historical World War II images by registering the images to present-day satellite imagery, with the aim of supporting the risk assessment of unexploded ordnances. We propose to exploit the local geometry of corresponding interest points in a Hough voting scheme to identify the most likely transformation parameters between the images. Our method combines the evidences from local as well as global correspondences and uses a spatial zoning rule to establish solutions with preferably uniformly distributed correspondences. An experimental evaluation is conducted on a set of 42 pairs of historical and present-day images and reveals the outstanding performance of our method compared to state-of-the-art image matching and registration algorithms, including commonly used hypothesize-and-verify and graph matching methods.

Keywords: Historical image registration · Georeferencing · Hough voting

1 Introduction

Assessing the risk of UneXploded Ordnances (UXOs) is an important concern for public safety [20]. Nowadays, a particular risk still comes from World War II bombing, as it is assumed that 10–30% of bombs remained unexploded [4]. UXO risk assessment involves the analysis of aerial photographs taken after bombing [17]. Indications of past bombardment such as craters can be used by analysts to derive such risk maps, but demands for a tedious prior georeferencing process by manually registering the images with modern satellite imagery.

In this paper we present an approach to register old World War II images with modern satellite images for automatic georeferencing. While being a typical image registration problem [25], this task poses specific challenges that demand for a well-adapted solution. First and foremost, the time spans of over 70 years lead to changes in image content of varying degrees, as shown in Fig. 1. Furthermore, the historical images are grayscale only and are affected by over- or underexposure, uneven illumination, low spatial resolution, blurring, sensor noise

© Springer International Publishing AG 2017
S. Battiato et al. (Eds.): ICIAP 2017, Part I, LNCS 10484, pp. 595–605, 2017.
https://doi.org/10.1007/978-3-319-68560-1_53

or cloud coverage. Due to this challenges, existing solutions for registering historical aerial photographs rely on manual interaction steps like line and point feature detection [18], prior rough alignment [7] or registration of a reference image [1,10].

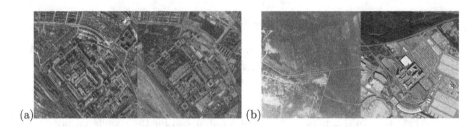

(a)   (b)

Fig. 1. Examples of image changes between World War II aerial images and present-day satellite imagery; (a) Vienna's 3rd district, (b) Vienna airport in Schwechat, Austria.

In image registration, commonly feature-based registration techniques are exploited due to their ability to ground the registration only on a few salient image parts, which makes them robust against occlusions and other image deterioration effects and allows to handle complex spatial transformation between the images [25]. These techniques consist of a feature matching step followed by a spatial verification step, where outlier correspondences are filtered out and the transformation with highest support from the set of putative matches is identified. A dominant example for spatial verification are hypothesize-and-verify methods such as RANSAC [8] and similar techniques [21], where transformation model hypotheses are created from randomly chosen correspondence samples and evaluated by means of congruence with the remaining correspondences. However, these estimators are only effective if the extracted local features are discriminative enough to deliver a certain inlier ratio in the set of putative matches. For weaker local features the feature matching step can be coupled with the spatial verification step to increase the inlier ratio, as done by graph matching methods [6,22]. Here feature matching is cast as an optimization problem involving both the similarities of the local feature vectors as well as pairwise or higher-order geometric consistencies of matches. However, graph matching is only computationally tractable for extracted feature sets in the order of 10^1–10^3 [6], thus allowing only a small set of initial candidate matches resulting in low recall.

The method presented in this paper builds upon the idea of using Hough voting in the transformation space, as used for object recognition [14], image retrieval [11] or image co-segmentation [5]. Due to the available local geometry of interest points, each single correspondence is able to cast a weighted vote based on feature similarity. Although a single correspondence gives only a weak evidence about the image transformation, stronger evidences are produced the higher the number of correspondences, and false votes can be effectively ruled out. As adding new correspondences has only linear costs, we can rely the transformation estimation on much more correspondences (in our case in the order of

10^5). This is especially effective for high-resolution remote sensing image data, as all small local structures in the comparatively large earth surface areas covered are possibly useful for image registration. Hence, we transfer the idea of Hough voting from single correspondences to our problem of old-to-new image aerial registration and propose two reasonable extensions to improve the performance: the combination of local and global image similarities and the use of a correspondence zoning scheme to favor solutions with spatially evenly distributed correspondences.

The remainder of this paper is organized as follows. Section 2 reviews related work in the registration of remote sensing images in general and aerial WWII images in particular. Our methodology is described in Sect. 3. In Sect. 4, quantitative results are reported and discussed. Concluding remarks are given in Sect. 5.

2 Related Work

Registration of Remote Sensing Images: Image registration plays a major role in various remote sensing applications, such as image fusion, change detection and georeferencing [12]. Depending on the specific scenario, the effectiveness of an algorithm is mainly determined by weather it uses the *global* image information or rather focuses on *local* parts of the image [24]. Global techniques aim at optimizing the transformation parameters based on a global similarity metric of pixel intensities, e.g. mutual information [13]. They are usually favored when the detection of salient structures in the image is not possible and accurate sub-pixel registration is privileged, but suffer from high computational load and local minima trapping and are thus limited to registration problems with a bounded search space of transformation parameters, e.g. the fine registration of roughly aligned image pairs [13]. Therefore, local techniques are more prominently used in general scenarios [15,16], as here the registration is based on a few salient features, which makes them also more robust against dissimilar, non-matchable image parts and other types of appearance changes. Typical appearance changes that are considered between remote sensing images are different modalities [16], illumination effects or disaster damages [3]. Automatic multitemporal image registration is also followed [9,19,23], but commonly not for such long time-spans as in our case of historical-to-modern image registration.

Registration of WWII Aerial Images to Present-Day Satellite Imagery: only a few works consider the problem addressed in our paper, with Murino et al. [18] being the first to provide a semi-automatic solution. In their approach, all possible matches between interactively selected line and point features are included for a RANSAC homography estimation. Automatic alignment of aerial images from WWII was also addressed in the GeoMemories project [1], but only between the historical images, whereas the actual georeferencing is achieved by selecting a reference historical image with known coordinates. The same principle is followed by Jao et al. [10], as historical-to-historical image registration has to deal with much less image variations and thus spatially verified SIFT feature matching is a suitable choice. Another solution to simplify the problem is presented by Cléry et al. [7]. Here a coarse initial registration is assumed, which can then be automatically refined by matching line features to a topographic map of the area. In contrast, our method does not make such assumption and represents, to the best of our knowledge, the first fully automatic approach to register aerial WWII images to present-day satellite imagery with a-priori unknown orientation and translation relation.

3   Methodology

In our application scenario, historical aerial images cover an area of 2.5–$16\,\text{km}^2$ and are registered to regions-of-interest in modern satellite imagery with a size of 1.5–10 times the size of the aerial image. We exploit the already known approximate image scale of the historical images derived from the recorded aircraft altitude and camera focal length to scale-normalize both images and limit the Hough parameter space to translation and rotation only. A preliminary inspection of our test data revealed an error of the estimated image scales of only 4.7%, with the maximum being at 30%. Hence, neglecting scale differences in local feature extraction and the Hough transformation space is a justified choice, whereas the small scale differences are respected in the final estimation of the image transformation from the correspondences responsible for the global peak in Hough space.

3.1   Hough Voting from Corresponding Local Interest Point Geometry

Extraction of local image features, e.g. SIFT features [14], from an image delivers descriptor vectors $\mathbf{d}_i$ as well as local feature frames $\mathbf{f}_i = (x_i, y_i, \sigma_i, \theta_i)$. Here, (x_i, y_i) is the feature location relative to the image center, σ_i is its scale and θ_i is the orientation. When registering image I' to image I'', we first compute similarities between all features $\mathbf{d}'_i$ and $\mathbf{d}''_j$ as $s_{i,j} = (\|\mathbf{d}'_i - \mathbf{d}''_j\|_2)^{-1}$. For computation of the Hough space, we take the subset $\mathcal{M}$ containing the N matches with highest similarity, $\mathbf{m}_{i,j} = (\mathbf{f}'_i, \mathbf{f}''_j) \in \mathcal{M}$. Each $\mathbf{m}_{i,j} \in \mathcal{M}$ votes for a rigid transformation in the 3D Hough Space $H(x_{\mathbf{m}_{i,j}}, y_{\mathbf{m}_{i,j}}, \theta_{\mathbf{m}_{i,j}})$, with

$$\theta_{\mathbf{m}_{i,j}} = \theta'_i - \theta''_j, \tag{1}$$

$$\begin{pmatrix} x_{\mathbf{m}_{i,j}} \\ y_{\mathbf{m}_{i,j}} \end{pmatrix} = \begin{pmatrix} x''_j \\ y''_j \end{pmatrix} - \begin{pmatrix} \cos\theta_{\mathbf{m}_{i,j}} & -\sin\theta_{\mathbf{m}_{i,j}} \\ \sin\theta_{\mathbf{m}_{i,j}} & \cos\theta_{\mathbf{m}_{i,j}} \end{pmatrix} \cdot \begin{pmatrix} x'_i \\ y'_i \end{pmatrix} \tag{2}$$

H is initialized with zeros and updated as

$$H(x_{\mathbf{m}_{i,j}}, y_{\mathbf{m}_{i,j}}, \theta_{\mathbf{m}_{i,j}}) = H(x_{\mathbf{m}_{i,j}}, y_{\mathbf{m}_{i,j}}, \theta_{\mathbf{m}_{i,j}}) + s_{i,j} \tag{3}$$

for each $\mathbf{m}_{i,j} = (\mathbf{f}'_i, \mathbf{f}''_j) \in \mathcal{M}$.

After the N votes are cast, the transformation parameters x^*, y^* and θ^* are identified at the maximum value in H. We then select the matches $\mathcal{M}^*$ that voted for a similar transformation by thresholding the translation and rotation differences, i.e.

$$\mathcal{M}^* = \{\mathbf{m} \mid \left\| \begin{pmatrix} x_{\mathbf{m}} \\ y_{\mathbf{m}} \end{pmatrix} - \begin{pmatrix} x^* \\ y^* \end{pmatrix} \right\|_2 < T_t \ \wedge \ \pi - ||\theta_{\mathbf{m}} - \theta^*| - \pi| < T_\theta \} \tag{4}$$

with T_t and T_θ being the thresholding values for translation and orientation, respectively. From all correspondences in $\mathcal{M}^*$, we estimate the final similarity transformation to account also for small scale differences between the images.

3.2 Combination of Votes from Local and Global Features

The strong appearance changes between old and new images not only decreases the discriminative power of the extracted features, but also affects the repeatability of interest point detection. Therefore, we rather apply a dense feature extraction scheme where local features are sampled on a regular grid with a fixed feature scale.

In this case, the selection of feature scale is influential to the performance of feature matching as it decides the level of image details to be compared. For multitemporal remote sensing data, both smaller and larger structures are potentially helpful: while smaller scales capture finer details like buildings, larger scales capture rougher structures like the courses of rivers and streets, which are likely less affected by changes over time. Therefore, we combine evidences from correspondences of both local and global image features by constructing two Hough spaces H_l and H_g and join them to the combined Hough space H_c as

$$H_c = \lambda \cdot H_l + (1 - \lambda) \cdot H_g, \tag{5}$$

where λ serves as weighting parameter.

3.3 Correspondence Zoning

Dense feature extraction ensures that features are uniformly distributed in both images, but the feature points belonging to correspondences responsible for a certain peak in Hough space might be concentrated on very few image parts with high appearance similarity. However, spatially uniform distributed correspondences should be preferred over clustered ones, since the estimation of the global image transformation is more robust and the final result is more trustable as it is grounded on evidences from different image parts.

Therefore, we utilize a zoning procedure that ensures that only one vote is allowed between two image regions, and consequently gives preference to solutions with spatially uniformly distributed correspondences. For this purpose, we set up an indicator function $Z(x_i', y_i', x_j'', y_j'') \in \{0, 1\}$ that defines if a correspondence between the points (x_i', y_i') and (x_j'', y_j'') has already been used. The Hough space update of Eq. 3 for H_l is applied only if $Z(x_i', y_i', x_j'', y_j'') = 0$ and after an update with the match $\mathbf{m}_{i,j}$, $Z(x', y', x'', y'')$ is set to 1 for all $(x', y') \in \mathcal{N}_{x_i', y_i'}$ and all $(x'', y'') \in \mathcal{N}_{x_j'', y_j''}$, with $\mathcal{N}_{x,y}$ specifying a neighborhood system of the point (x, y).

The effect of correspondence zoning is exemplarily shown in Fig. 2. When no correspondence zoning is applied, nearby erroneous matches produce a global peak in Hough space, resulting in a wrong registration result (Fig. 2a). However, with correspondence zoning the influence of nearby matches is reduced and the final matches of the global Hough peak are correct and more uniformly distributed over the image (Fig. 2b).

(a)   (b)

Fig. 2. Image matches chosen based on global Hough peak; (a) without correspondence zoning; (b) with correspondence zoning.

3.4 Implementation Details

In our implementation, we use SIFT features [14] to describe both the local and global image patches. For the local Hough space H_l, features are extracted on a regular grid with an interval of 40 m and a patch size of 120 m, based on empirical tests. For each feature the orientation θ_i is determined from the dominant gradient direction within the patch [14]. For the global Hough space H_g, we extract a SIFT feature over the whole historical image area for 18 regularly spaced orientations. Features of the same size are extracted from the present-day satellite image with a fixed orientation and an interval of 100 m.

For the discretization of the Hough space, we use a step size of one pixel for the translation parameters and $\frac{2\pi}{18}$ for the orientation. Due to the rougher discretization of the orientation parameter, we use bilinear interpolation to distribute the value of θ_i to adjacent bins. For the correspondence zoning, a circular neighborhood with a radius of 80 m is used. Other parameters of our method are empirically set as follows: we equally weight the contribution of the local and global Hough space ($\lambda = 0.5$ in Eq. 5), take the $N = 10^5$ best correspondences to fill the Hough space, and set the thresholding values for the final inlier correspondences in Eq. 4 to $T_t = 100$ m and $T_\theta = \frac{2\pi}{36}$.

4 Experiments

In this section we report quantitative results of our method on annotated test data and compare it to other image matching and registration algorithms proposed in literature.

Dataset and Evaluation Protocol: Our dataset consists of 8 reference satellite images from urban and non-urban areas in Austria. For each reference image, 3–11 historical aerial images are available, leading to a total of 42 image pairs. All images have been scale-normalized to a spatial resolution of 1 m prior to processing. Manually selected ground truth correspondences are used to measure the root mean squared error (RMSE) [2] of the image transformations determined by the different evaluated algorithms.

Algorithms: We compare our method to the following algorithms:

SIFT+RANSAC: standard SIFT feature matching [14] with RANSAC [8] for spatial verification serves as baseline performance for the comparison. For a fair comparison, we again use a dense feature extraction with fixed scale, set to 360 m in this case for best performance. Putative matches are achieved by SIFT matching with an inlier ratio threshold of 1.3 [14]. Additionally, RANSAC solutions of the similarity transform are validated only if the scale difference Δs is within the bounds $(1/T_{\Delta s}) \leq \Delta s \leq T_{\Delta s}$, with $T_{\Delta s}$ set to 1.4.

Locally Linear Transforming (LLT) [15]: similar to RANSAC, LLT is a method for the simultaneous transformation estimation and outlier removal from a set of putative matches, but embedded in a maximum-likelihood framework with a locally linear constraint. The method is included in the evaluation as it showed an outstanding performance compared to other robust estimators on remote sensing data [15]. In our evaluation, we applied the rigid transformation LLT version to the same set of putative matches as for SIFT+RANSAC. We used the parameter settings reported in [15], but changed the uniform distribution parameter from 10 to 5 due to a better performance (see [15] for details).

Position-Scale-Orientation-SIFT (PSO-SIFT) [16]: PSO-SIFT is another recently proposed method for remote sensing image registration. Like our method, it is also based on statistical evidences from the local geometry of correspondences. However, transformation parameters are treated individually and their modes are only used for an enhanced distance metric of local descriptors.

Progressive Graph Matching (PGM) [6]: PGM is a general image matching algo-rithm based on graph matching. Starting with a set of initial matches, graph matching results are iteratively enriched with correspondences and re-matched. We use the SIFT matching results from *SIFT+RANSAC* to initialize PGM and apply RANSAC to its final matching results for transformation estimation.

Additionally, we report results on various versions of our Hough voting (HV) method in order to demonstrate the effects of the proposed improvements of local-global feature combination and correspondence zoning: a version with local Hough voting only (HV_{local}), global Hough voting only (HV_{global}), combined Hough voting according to Eq. 5 ($HV_{local+global}$), and the full method with cor-respondence zoning ($HV_{local+global,CZ}$).

Results and Discussion: In Fig. 3 the number of correct registration results achieved with the various algorithms are plotted. Correctness of a result is deter-mined by comparing its RMSE with the threshold on the logarithmically scaled x-axis of the plot. It can be seen that the competing methods perform poorly on the test data set compared to the proposed methods. SIFT+RANSAC and LLT have a similar low correct registration rate which demonstrates the limits of outlier removal methods when weak feature descriptions produce too low inlier ratios in the set of putative matches. PGM has a slightly higher correct regis-tration rate for higher acceptable errors due to its correspondence enrichment, but still does not reach the performance of our method.

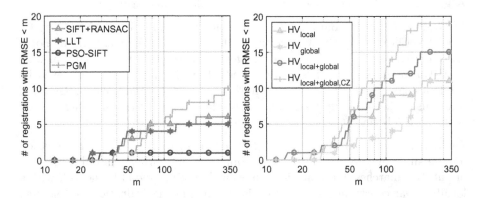

Fig. 3. Number of correct registration results as a function of max. RMSE in meters for the competing methods (left) and different versions of our method (right).

The results shown in Fig. 3 also verify the effectiveness of combining local and global correspondences as well as zoning the local correspondences. $HV_{local+global}$ shows a better performance than HV_{local} and HV_{global}, and the full method $HV_{local+global,CZ}$ with correspondence zoning gives another significant perfor-mance boost. Examples of correct registration results are shown in Fig. 4. For these examples, $HV_{local+global,CZ}$ is the only method able to achieve a correct

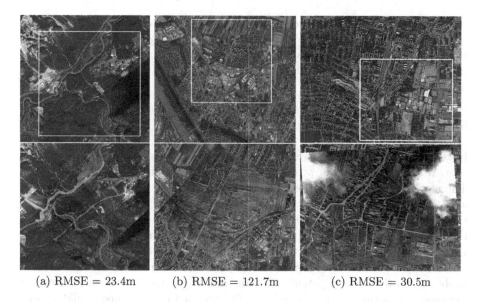

(a) RMSE = 23.4m   (b) RMSE = 121.7m   (c) RMSE = 30.5m

Fig. 4. Example results of image registration with $HV_{local+global,CZ}$; top: reference satellite image; bottom: registration result (zoom of red area shown in reference image). (Color figure online)

registration. Nevertheless, it is evident from the results that correctly registering images with such a high time distance is an enormously challenging problem, as even the best performing method $HV_{local+global,CZ}$ is only able to register around 45 % of the cases with an error of less than 350 m.

The generally low precision of our Hough voting methods is primarily a result of the regular feature sampling with a step size of 40 m. This is a necessary compromise as feature detection has shown to have very unreliable repeatability, but on the other hand regular sampling prevents a precise localization of corresponding features. As part of future research, we plan to investigate fine registration as postprocessing step with a flexible transformation model, also to account for the actually non-linear spatial transformation between the images.

Runtime Analysis: In Table 1 we compare the average runtimes of our MAT-LAB implementation of Hough voting to the runtimes of the competing methods on the same machine. The tests for the competing methods have been performed with the original MATLAB implementations provided by the authors. Due to the high number of matches to be processed ($N = 10^5$), the better registration performance of our method comes at the price of an considerably longer

Table 1. Comparison of average registration runtimes per image pair, including feature extraction.

SIFT+RANSAC	LLT	PSO-SIFT	PGM	$HV_{local+global}$	$HV_{local+global,CZ}$
12.7 s	11.0 s	55.7 s	156.7 s	32.8 s	30.2 s

runtime than the outlier removal methods SIFT+RANSAC and LLT. It can also be seen that correspondence zoning does not only give a boost in registration performance, but also saves around 8% of computation time as not all of the N matches have to be evaluated for their local transformation and included in the Hough space.

5 Conclusions

Registration of aerial images from the times of WWII to modern satellite imagery proves to be a challenging problem due to the severe changes between images. Therefore, previously published approaches rely on strong initial assumptions about the geometric relation of images or have to make use of a manual step in the processing pipeline. In this paper, we introduced a Hough voting strategy that allows for the fully automatic historical-to-modern aerial image registration with a-priori unknown translation and orientation differences.

Although our method outperforms state-of-the-art methods for this kind of problem, it offers much potential for further improving the performance. For instance, our voting strategy can be easily extended to combined Hough spaces leveraging multiple descriptors for matching. The encoded evidences about relative geometric relations between images can also be integrated to reason about the overall geometric relations of images in a groupwise registration scenario. Additionally, a final fine registration step can be used to obtain a more precise solution. These issues, besides adapting the methodology to other domains, will be investigated for future research.

Acknowledgements. This work is supported by the Austrian Research Promotion Agency (FFG) under project grant 850695. The authors wish to thank *Luftbilddatenbank Dr. Carls GmbH*.

Acquisition of historical aerial imagery: *Luftbilddatenbank Dr. Carls GmbH*; Sources of historical aerial imagery: *National Archives and Records Administration* (Washington, D.C.) and *Historic Environment Scotland* (Edinburgh).

References

1. Abrate, M., Bacciu, C., Hast, A., Marchetti, A., Minutoli, S., Tesconi, M.: Geomemories a platform for visualizing historical, environmental and geospatial changes in the italian landscape. ISPRS Int. J. Geo-Inf. **2**(2), 432–455 (2013)
2. Bouchiha, R., Besbes, K.: Comparison of local descriptors for automatic remote sensing image registration. Sig. Image Video Process. **9**(2), 463–469 (2015)
3. Bowen, F., Hu, J., Du, E.Y.: A multistage approach for image registration. IEEE Trans. Cybern. **46**(9), 2119–2131 (2016)
4. Busé, M.S.: WWII ordnance still haunts europe and the asia-pacific rim. J. Conv. Weapons Destr. **4**(2), 83–87 (2016)
5. Chen, H.Y., Lin, Y.Y., Chen, B.Y.: Co-segmentation guided hough transform for robust feature matching. IEEE PAMI **37**(12), 2388–2401 (2015)
6. Cho, M., Lee, K.M.: Progressive graph matching: making a move of graphs via probabilistic voting. In: CVPR, pp. 398–405 (2012)

7. Cléry, I., Pierrot-Deseilligny, M., Vallet, B.: Automatic georeferencing of a heritage of old analog aerial photographs. ISPRS Ann. 2(3), 33–40 (2014)

8. Fischler, M.A., Bolles, R.C.: Random sample consensus: a paradigm for model fitting with applications to image analysis and automated cartography. Commun. ACM 24(6), 381–395 (1981)

9. Gong, M., Zhao, S., Jiao, L., Tian, D., Wang, S.: A novel coarse-to-fine scheme for automatic image registration based on sift and mutual information. IEEE GRS 52(7), 4328–4338 (2014)

10. Jao, F.J., Chu, H.J., Tseng, Y.H.: Historical image registration and land-use land-cover change analysis. Environments 1(2), 181–189 (2014)

11. Jégou, H., Douze, M., Schmid, C.: Improving bag-of-features for large scale image search. IJCV 87(3), 316–336 (2010)

12. Le Moigne, J., Netanyahu, N.S., Eastman, R.D.: Image Registration for Remote Sensing. Cambridge University Press, Cambridge (2011)

13. Liang, J., Liu, X., Huang, K., Li, X., Wang, D., Wang, X.: Automatic registration of multisensor images using an integrated spatial and mutual information (SMI) metric. IEEE GRS 52(1), 603–615 (2014)

14. Lowe, D.: Distinctive image features from scale-invariant keypoints. IJCV 60(2), 91–110 (2004)

15. Ma, J., Zhou, H., Zhao, J., Gao, Y., Jiang, J., Tian, J.: Robust feature matching for remote sensing image registration via locally linear transforming. IEEE GRS 53(12), 6469–6481 (2015)

16. Ma, W., Wen, Z., Wu, Y., Jiao, L., Gong, M., Zheng, Y., Liu, L.: Remote sensing image registration with modified sift and enhanced feature matching. IEEE GRSL 14(1), 3–7 (2017)

17. Merler, S., Furlanello, C., Jurman, G.: Machine learning on historic air photographs for mapping risk of unexploded bombs. In: Roli, F., Vitulano, S. (eds.) ICIAP 2005. LNCS, vol. 3617, pp. 735–742. Springer, Heidelberg (2005). doi:10.1007/11553595_90

18. Murino, V., Castellani, U., Etrari, A., Fusiello, A.: Registration of very time-distant aerial images. In: ICIP, vol. 3, pp. 989–992. IEEE (2002)

19. Nagarajan, S., Schenk, T.: Feature-based registration of historical aerial images by area minimization. ISPRS J. Photogramm. Remote Sens. 116, 15–23 (2016)

20. Paunila, S.: Managing residual clearance: learning from Europe's past. J. Conv. Weapons Destr. 18(1), 22–25 (2015)

21. Raguram, R., Frahm, J.M., Pollefeys, M.: A comparative analysis of RANSAC techniques leading to adaptive real-time random sample consensus. In: Forsyth, D., Torr, P., Zisserman, A. (eds.) ECCV 2008. LNCS, vol. 5303, pp. 500–513. Springer, Heidelberg (2008). doi:10.1007/978-3-540-88688-4_37

22. Sanromà, G., Alquézar, R., Serratosa, F.: A new graph matching method for point-set correspondence using the EM algorithm and softassign. CVIU 116(2), 292–304 (2012)

23. Vakalopoulou, M., Karantzalos, K., Komodakis, N., Paragios, N.: Simultaneous registration and change detection in multitemporal, very high resolution remote sensing data. In: CVPR Workshops, pp. 61–69 (2015)

24. Wen, G.J., Lv, J.J., Yu, W.X.: A high-performance feature-matching method for image registration by combining spatial and similarity information. IEEE GRS 46(4), 1266–1277 (2008)

25. Zitova, B., Flusser, J.: Image registration methods: a survey. IVC 21(11), 977–1000 (2003)

Automatic Detection of Subretinal Fluid and Cyst in Retinal Images

Melinda Katona[1], Attila Kovács[2], Rózsa Dégi[2], and László G. Nyúl[1(✉)]

[1] Department of Image Processing and Computer Graphics, University of Szeged,
Árpád tér 2, Szeged 6720, Hungary
{mkatona,nyul}@inf.u-szeged.hu
[2] Department of Ophthalmology, University of Szeged, Korányi fasor 10-11,
Szeged 6720, Hungary
{kovacs.attila,degi.rozsa}@med.u-szeged.hu

Abstract. A modern tool for age-related macular degeneration (AMD) investigation is Optical Coherence Tomography (OCT) that can produce high resolution cross-sectional images about retinal layers. AMD is one of the most frequent reasons of blindness in economically advanced countries. AMD means degeneration of the macula which is responsible for central vision. Since AMD affects only this specific part of the retina, unattended patients lose their fine shape- and face recognition, reading ability, and central vision. We present a novel algorithm to localize subretinal fluid and cyst segments and extract quantitative measures thereof. Since, these algorithms are fully automated, the doctor does not need to perform extremely time-consuming manual contouring and human inaccuracies can be also eliminated.

Keywords: Optical Coherence Tomography · SD-OCT · Age-related macular degeneration · AMD · Subretinal fluid · Cyst

1 Introduction

Age-related macular degeneration is one of the most frequent reasons of acquired blindness in economically advanced countries. The constant growing of AMD patient population is more and more challenging. AMD means degeneration of the macula which is the region of the retina responsible for central vision. Since AMD affects only this specific part of the retina, unattended patients lose their fine shape- and face recognition, reading ability, and central vision [9].

Basically, AMD has two forms: dry and wet form, and the latter causes rapid and serious visual impairment in 10% of the cases [13]. In this type of the disease, abnormal angiogenesis starts from the choroid under the macula. Fluid and blood leak out of the neovascularized membrane into retina layers that ruins the photoreceptors.

Experiments have demonstrated that the vascular endothelial growth factor (VEGF) plays a vital role in the formation of choroidal neovascularization [4].

© Springer International Publishing AG 2017
S. Battiato et al. (Eds.): ICIAP 2017, Part I, LNCS 10484, pp. 606–616, 2017.
https://doi.org/10.1007/978-3-319-68560-1_54

Currently, the most common and effective clinical treatment for wet AMD is anti-VEGF therapy, which is a periodic intravitreal (into the eye) injection [11].

In the last decade, Optical Coherence Tomography (OCT) has been widely used in the diagnosis of AMD and follow-up therapy. Spectral domain OCT (SD-OCT) produces 3D volumes of data, which have been useful in clinical practice. Existing OCT systems are partially suited to monitoring the progress of the disease, but OCT shows many features about AMD such as hyper-reflective dots (HRD), subretinal fluid and cysts. Figure 1 illustrates an SD-OCT B-scan with biomarkers of AMD.

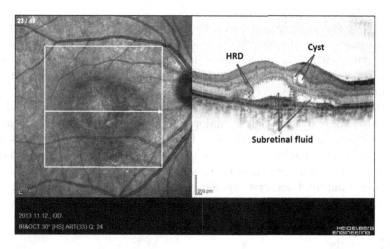

Fig. 1. Original Optical Coherence Tomography (SD-OCT) image with biomarkers of AMD in inverted display (as our medical experts use it in daily routine).

A large number of publications in the scientific literature deal with the problem of detecting retinal layers based on various techniques. One approach is the automatic segmentation procedure using graph theory [1, 2, 6]. In this approach, the graph nodes usually relate to image pixels, the graph edges are assigned to pairs of pixels, the edge weights depend on the intensity differences between the node pixels, and also may depend on the spatial distance between the pixels. Image segmentation then becomes a graph cutting problem, which can be solved via dynamic programming. These approaches are less tolerant to noise, that is a disadvantage, because real images are often very noisy. Another idea relies on the well-known energy-minimizing active contour method which, unfortunately, also has problems in handling low contrast and noise. Yazdanpanah et al. [18] suggested a multi-phase framework with a circular shape prior in order to model the boundaries of retinal layers and estimate the shape parameters. They used a contextual scheme to balance the weight of different terms in the energy functional. Machine learning is widely used in recent years, also for retinal image analysis. Lang et al. [12] used random forest classifier to segment retinal layers.

The random forest classifier learns the boundary pixels between layers and produces an accurate probability map for each boundary, which is further processed to finalize boundaries. Procedures based on active contour or machine learning provide effective solution, but these methods are too time-consuming. Hassan et al. [8] used a structure tensor approach combined with a nonlinear diffusion process for layer detection. A structure tensor is a second-moment matrix that shows similarities and prominent orientations of the image gradient. Some other approaches use optimized boundary tracking [5] or polynomial smoothing [14]. These algorithms are rather complex. We have developed an algorithm which uses simple operations to localize subretinal areas. It is based on vertical profile analysis [10].

Relatively few publications deal with the problem of automatically detecting cysts. Gonzalez et al. [7] described a method based on watershed segmentation and different machine learning classifiers. They focused on feature extraction which can help to eliminate false regions. Other approaches can also be found in the literature, two of which are discussed in Sect. 2.1.

In this study, we deal with the automatic localization of subretinal fluid areas and cysts and also analyze major retinal layers, since layer information can help localizing and distinguishing fluid and cyst regions. We present an algorithm that automatically delineates the ILM (inner-limiting membrane) and RPE (retinal pigment epithelium) retinal layers. We also describe a method to detect subretinal fluid and cyst segments and distinguish them from each other. We compare our results with some other algorithms from literature.

2   Methods

In this section, we present several algorithms that use different approaches to determine cyst and subretinal fluid. First, we briefly describe two existing approaches that we re-implemented according to the original papers for comparison. Then, we describe our novel approach in more detail. The procedures first delineate the inner and outer boundary layers (ILM and RPE, resp.) for easier determination of the important areas.

2.1   Literature Procedures

Firstly, we describe Wieclawek's [16] algorithm. First, the input image is normalized to the [0 1] interval, because images can be made with different settings, so that their intensity range may vary. The OCT images are affected by distortions like noise, so the authors used a non-linear filtering to reduce this effect. Next, they applied a spatial averaging filtering technique which is based on the real product of complex diffusion. The tools of mathematical morphology was used to delineate specific cystic areas, based on the observation that cysts appear as darker segments in the images. Among other operations, they used H-minima transform to highlight important regions. The single control parameter is a threshold value. This value has been fixed experimentally to 30% of the maximum brightness in the image. The next step was the binarization of the obtained image with a given threshold value. Since the result may still contain false regions, they filtered all objects that are above ILM and below RPE layers.

Wilkins et al. [17] investigated the problem from another point of view. They discarded color information in the first step of the algorithm and determined the major layer boundaries. To improve image quality and filter noise, they used the combination of a median filter and a bilateral filter during preprocessing. After binarization, the method determined the boundaries of the remaining possible cyst segments and they defined three conditions based on empirical studies. They investigated the extent of the objects, the degree of scattering between the intensity of the pixels in the segment, and whether the object is located between the ILM and RPE layers.

2.2 Proposed Method

OCT images are affected by distortions like "shadowing" by blood vessels, that may yield to false detections. In the first step of our proposed algorithm, we improve the image quality by noise filtering and contrast enhancement using a fuzzy operator [3]. This operation can highlight major retinal layers. We analyze vertical profiles of the filtered image and large intensity steps in pixel density are assumed to correspond to the change of tissue. The used fuzzy function is defined as

$$\kappa_\nu^* = \frac{1}{1 + \frac{1-\nu}{\nu}\left(\frac{1-x}{x}\frac{\nu}{1-\nu}\right)^\lambda} , \tag{1}$$

where ν is a threshold, x is pixel intensity and λ denotes the sharpness of the filtering. The function κ_ν^* can highlight boundary layers while suppressing noise. We determine dynamically the input parameter ν in a simple way. We sample from the top range of the image and calculated average intensity for this ROI. The parameter λ was set to 3 empirically. Figure 2 shows an example of applying the function κ_ν^*.

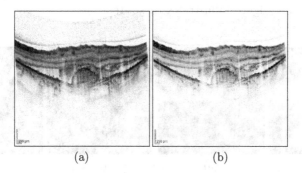

(a)   (b)

Fig. 2. Sample OCT image before (a) and after (b) applying the function κ_ν^*. It filters out the noise outside the retina and highlights boundary layers, so it is easier to delineate them.

After filtering, we divide the image into bars with fixed width. A bar consists of 10 contiguous pixel columns and we calculate horizontal projections of each bar to determine boundaries. One of the main steps of our proposed method is to analyze the vertical profiles. This signal is usually noisy, so there is a need for filtering the data. We use the Savitzky-Golay filter [15] which is a smoothing digital filter. This filter is effective in preserving the relevant high frequency components of the signal, which is an important aspect for our detection method.

Determining the outer layer boundary is harder than that of the inner boundary, because Choriocapillaris and Chorodoidal vessels are located under the RPE layer. The intensity of these regions vary, so several peaks appear in the projections. Fortunately, in most cases, these minimum points are not prominent, and do not cause problem in choosing the right locations. The algorithm chooses the most important local minimum from the projected data to identify the possible inner and outer layer. In the next step, we filter out the outliers and we fit a curve to the remaining points.

After we determined the boundary layers, the next step is the segmentation of fluid and cyst areas. It can be observed that these regions appear as spots with brighter intensity in the image. For processing, we use the inverse of the signals, because our medical colleagues used the inverted presentation of images for visual assessment and also exported the image data for us in this format. The zones of the disease and the intensity of the vitreous body of the eye are almost within the same range (if distortions are not considered). Anisotropic diffusion is used to eliminate various errors from imaging or blood shadows. Using the filter, it is more apparent that some parts of the retina are within a given intensity range, so we quantize the grayscale image into five intensity levels. Our observations showed that the layers of the retina are only in some intensity ranges, so this operation facilitates the separation of the 8–10 main retinal layers. During the binarization, we keep the brightest points, because we know that the reflectivity of the cyst similar to that of the vitreous body. By this step, we create a mask for the active contour process. To achieve the appropriate segmentation result, the input parameters of the model were given based on empirical studies. After that, there may be holes in some objects, so we use hole filling. Figure 3 illustrates these steps.

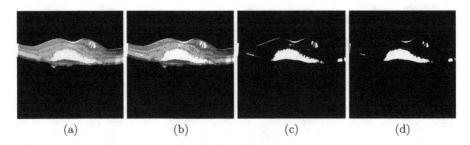

(a)   (b)   (c)   (d)

Fig. 3. Intermediate stages of the proposed algorithm. (a) Anisotropic diffusion, (b) Quantization, (c) Binarization, (d) Result of active contour.

So far we have identified possible important segments. Next, we need to separate cyst, fluid and false segments from each other. This step may be omitted in some cases, when there is no object in the picture that would be detected as a pathological mutation. We developed a condition set for filtering cysts and fluid regions. We test the fulfillment of the four criteria for classification at the object level.

The greater distortion of layers was observed in those parts where these symptoms appear, so firstly, we examine whether the actual layer is distorted or not. This plays an important role in distinguishing between cysts and fluid areas. To determine whether the layers are creased or not, we use the top layer boundary points. We calculate the minimal y coordinate of the top points. For this, the method does not take into account the left and right 25% of the image. The sides of the image may not contain information, because of the image registration, and large distortions also may appear in these parts of the images. In Fig. 4 we illustrate these mentioned effects. Next, we search the minimal y coordinate of the top points, divide this image into two parts along the established peak, and we investigate the maximal y point on both sides. We estimate the degree of creasing. Sometimes there is no change in the middle of the image, so we determined a threshold to decide if there is any crease in the slice or not. The threshold for the minimum y point and the given maximum y point was defined in 5 pixels experimentally.

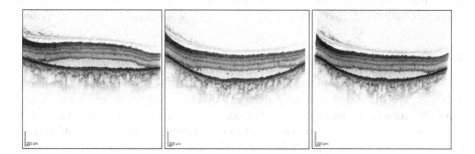

Fig. 4. Example of distortion in consecutive slices.

Various a priori information can be used to distinguish between cysts and fluid areas, and to filter out the false segments. We investigated a condition system and we considered where the object is located within the retina, what is the extent and the shape of the object, and whether the layer is distorted or not. Fluid areas have larger extent and they are located close to the bottom layer boundary or may appear in the distorted area. In the case of higher distortion of the retina, they may also appear on the left or right side. When determining fluid regions, it is also important to examine the cases where there is no creasing in the layer. In the case of cysts, we need to find objects with oval shapes and the observations show that these segments are in the increased zone. Contrary to fluid regions, cysts are found in higher layers. We distinguish the symptoms based on these characteristics. False segments may also appear in the image, but these are small objects, so we can remove them easily with an area-based filtering. The key stages of the procedure are summarized in Fig. 5.

Fig. 5. Flowchart of the proposed algorithm.

3 Evaluation

3.1 Image Data

Our evaluation dataset contained 11 Heidelberg Spectralis OCT scans of wet age related macular degeneration patients treated with anti-VEGF intravitreal injections. The scanning parameters are: 49 scan pattern, pattern size: 5.8×5.8 mm, distance between B-scans: $121\,\mu$m, size X: 512 pixel, size Z: 496 pixel, the pixel size is $11.44\,\mu$m and $3.87\,\mu$mm in X and Z direction, respectively.

Manual ILM and RPE layer segmentation was performed by ophthalmologists for 7 image sequences. This was considered as a ground truth for evaluating the boundary layer detection method on these 7 image volumes.

3.2 Results and Discussion

We implemented our proposed method in MATLAB, using the Image Processing Toolbox. We evaluated our retinal layer detection algorithm in two different ways. We compared the results of our algorithm against the manual delineations and we also compared the proposed method for automatic detection of subretinal fluid and cyst with some published methods from the literature.

Firstly, we consider the result of localization of major layers. We calculated the mean, maximum and standard deviation of boundary errors for every surface. The 7 curves shown in Fig. 6 depict the error histogram for those OCT volumes where manual annotation was available. Each curve aggregates the boundary errors in the 49 scans (slices) of a study. It can be observed that the largest error is between 1 and 4 pixels in most cases and Table 1 asserts to this statement. As presented in Table 1, the maximal distance between manually segmented and automatically detected layer boundary is 19 pixels (ca. $73.5\,\mu$m). This deflection comes from two sources, namely, the substantial jumping between B-scans and layer distortions due to the disease. Unfortunately, we could not exploit 3D information directly to segment the retina layers, because there are some anomalies among slices of the OCT volume, due to the image acquisition and registration process (within the device's software).

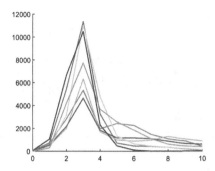

Fig. 6. Error histogram of 7 image sequences.

Table 1. Summary of mean, standard deviation and maximum error (in pixels) between manually segmented and automatically detected layers in 7 annotated OCT image sequence.

		Seq_02	Seq_03	Seq_04	Seq_05	Seq_06	Seq_07	Seq_08	All
Mean	Wilkins	1.90	1.86	2.90	1.88	9.00	1.22	10.91	4.24
	Wieclawek	6.86	10.34	11.20	17.82	6.88	5.42	7.11	9.38
	Proposed	**2.01**	**2.10**	**1.44**	**1.96**	**2.39**	**2.17**	**1.83**	**1.98**
Std. dev.	Wilkins	3.34	2.05	2.67	1.62	2.73	0.18	2.93	2.28
	Wieclawek	7.92	5.59	5.92	6.31	5.97	4.78	5.39	5.98
	Proposed	**1.56**	**0.69**	**0.65**	**0.80**	**1.63**	**0.64**	**0.65**	**0.94**
Maximum	Wilkins	26	19	17	17	25	11	25	20
	Wieclawek	35	31	24	22	21	20	19	24.57
	Proposed	**17**	**15**	**15**	**19**	**18**	**15**	**17**	**16.57**

In Sect. 2, we have presented two methods from the literature for the segmentation of cysts, as well as our proposed method, which is also suitable for delineating fluid areas. Unfortunately, expert annotation was not yet available to evaluate segmentation results, so we compare visually the outputs of the algorithms.

Figure 7 illustrates segmentation results by the algorithms in some slices. The method developed by Wieclawek detected fewer possible cyst regions, which may be due to the fact that the given threshold only keeps the actual light points. The disadvantage of this is, that important areas may be lost during processing. The other method from literature by Wilkins yields almost the same segmentation results, but in many cases it holds false objects, because the thresholds are not dynamically defined. In contrary, our method uses dynamic requirements based on a priori information.

We tested the re-implemented earlier published methods on images in which cysts and fluids may also appear. They can also detect these regions because these segments are also lighter object in the image. Our algorithm, however, as it can be seen in Fig. 7(c), can also distinguish these two types of structures from each other.

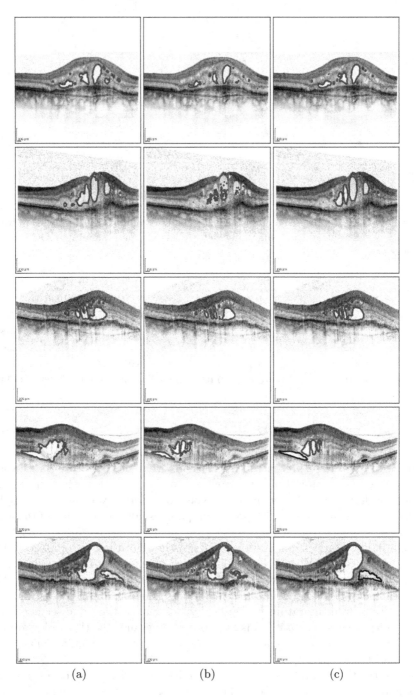

(a)   (b)   (c)

Fig. 7. Illustration of detected cyst (red curve) and subretinal fluid (blue curve) regions by the described algorithms. Columns: (a) Wilkins et al., (b) Wieclawek, (c) Proposed method. Top three rows contain only cysts, while the bottom two scans also have fluid areas. (Color figure online)

4 Conclusion

We presented a novel algorithm for the detection of subretinal fluid areas and cysts and we compared it with two methods from the literature for cyst localization. After having seen the results, the medical colleagues believe that digital image processing can help the quantitative assessment of the OCT features of AMD by providing automatic tools to detect abnormalities and to describe by objective metrics the current state and longitudinal changes during disease evolution and treatment. Using SD-OCT to follow up changes of subretinal fluid and cysts volume will become a useful tool in detecting subtle changes during the treatment process. Further studies are planned to evaluate these new tools in a cohort of AMD patients.

Acknowledgments. This work was supported by the NKFIH OTKA [grant number K112998]. The authors thank Dr. József Dombi for suggesting the use of fuzzy operators for the image preprocessing step.

References

1. Abhishek, A.M., Berendschot, T., Rao, S.V., Dabir, S.: Segmentation and analysis of retinal layers (ILM & RPE) in Optical Coherence Tomography images with Edema. In: 2014 IEEE Conference on Biomedical Engineering and Sciences (IECBES), pp. 204–209 (2014)
2. Chiu, S.J., Li, X.T., Nicholas, P., Toth, C.A., Izatt, J.A., Farsiu, S.: Automatic segmentation of seven retinal layers in SDOCT images congruent with expert manual segmentation. Opt. Express **18**(18), 19413–19428 (2010)
3. Dombi, J.: Modalities. In: Melo-Pinto, P., Couto, P., Serôdio, C., Fodor, J., De Baets, B. (eds.) Eurofuse 2011. Advances in Intelligent and Soft Computing, vol. 107, pp. 53–65. Springer, Heidelberg (2011). doi:10.1007/978-3-642-24001-0_7
4. Ferrara, N.: Vascular endothelial growth factor: basic science and clinical progress. Endocr. Rev. **25**(4), 581–611 (2004)
5. Fu, D., Tong, H., Luo, L., Gao, F.: Retinal automatic segmentation method based on prior information and optimized boundary tracking algorithm. In: Proceedings of SPIE, vol. 10033, p. 100331C: 1–100331C: 6 (2016)
6. Garvin, M.K., Abramoff, M.D., Wu, X., Russell, S.R., Burns, T.L., Sonka, M.: Automated 3-D intraretinal layer segmentation of macular spectral-domain optical coherence tomography images. IEEE Trans. Med. Imaging **28**(9), 1436–1447 (2009)
7. González, A., Remeseiro, B., Ortega, M., Penedo, M.G., Charlón, P.: Automatic cyst detection in oct retinal images combining region flooding and texture analysis. In: Proceedings of the 26th IEEE International Symposium on Computer-Based Medical Systems, pp. 397–400 (2013)
8. Hassan, B., Raja, G., Hassan, T., Akram, M.U.: Structure tensor based automated detection of macular edema and central serous retinopathy using optical coherence tomography images. J. Opt. Soc. Am. **33**(4), 455–463 (2016)
9. Hee, M.R., Baumal, C.R., Puliafito, C.A., Duker, J.S., Reichel, E., Wilkins, J.R., Coker, J.G., Schuman, J.S., Swanson, E.A., Fujimoto, J.G.: Optical coherence tomography of age-related macular degeneration and choroidal neovascularization. Ophthalmology **103**(8), 1260–1270 (1996)

10. Katona, M., Nyúl, L.G.: An approach to the quantitative assessment of retinal layer distortions and subretinal fluid in SD-OCT images. Acta Cybern. Accepted for publication (2017)
11. Kovach, J.L., Schwartz, S.G., Flynn Jr., H.W., Scott, I.U.: Anti-VEGF treatment strategies for wet AMD. J. Ophthalmol. **22**, 786870:1–786870:7 (2012)
12. Lang, A., Carass, A., Hauser, M., Sotirchos, E.S., Calabresi, P.A., Ying, H.S., Prince, J.L.: Retinal layer segmentation of macular OCT images using boundary classification. Biomed. Opt. Express **4**(7), 1133–1152 (2016)
13. Lim, J.: Age-Related Macular Degeneration. CRC Press, Boca Raton (2012)
14. Lu, S., Cheung, C.Y.L., Liu, J., Lim, J.H., Leung, C.K.S., Wong, T.Y.: Automated layer segmentation of optical coherence tomography images. IEEE Trans. Biomed. Eng. **57**(10), 2605–2608 (2010)
15. Schafer, R.W.: What is a Savitzky-Golay filter? IEEE Sig. Process. Mag. **28**(4), 111–117 (2011)
16. Wieclawek, W.: 'Automatic cysts detection in optical coherence tomography images. In: 2015 22nd International Conference on Mixed Design of Integrated Circuits Systems (MIXDES), pp. 79–82 (2015)
17. Wilkins, G.R., Houghton, O.M., Oldenburg, A.L.: Automated segmentation of intraretinal cystoid fluid in optical coherence tomography. IEEE Trans. Biomed. Eng. **59**(4), 1109–1114 (2012)
18. Yazdanpanah, A., Hamarneh, G., Smith, B., Sarunic, M.: Intra-retinal layer segmentation in optical coherence tomography using an active contour approach. In: Yang, G.Z., Hawkes, D., Rueckert, D., Noble, A., Taylor, C. (eds.) MICCAI 2009. LNCS, vol. 5762, pp. 649–656. Springer, Heidelberg (2009). doi:10.1007/978-3-642-04271-3_79

Computer Aided Diagnosis of Pleural Effusion in Tuberculosis Chest Radiographs

Utkarsh Sharma$^{(\boxtimes)}$ and Brejesh Lall

Indian Institute of Technology Delhi, New Delhi, India
{ee5120562,brejesh}@ee.iitd.ac.in

Abstract. Tuberculosis (TB) is one the leading killers in the world, and its early detection at scale is a challenge that remains. Computer Aided Detection of Tuberculosis is an important possibility for the world due to the mismatch in the incidences of this disease with the number of trained human readers for its identification. In this paper, we propose novel features for the detection of one of the symptoms observed in cases of TB, Pleural Effusion (PE). We begin by segmenting the lung regions, followed by creation of a novel feature set. We achieve an ROC of 0.961 on discriminating PE against Chest X-Rays (CXRs) without incidences of TB. To validate that our system discriminates against PE, we achieve an ROC of 0.864 against CXRs showing incidences of TB but a lack of PE. These features are then tested on two publicly available datasets (One collected from the United States, and the other from China). Due to the lack of other work for detection of PE on these datasets, a direct comparison is unfortunately not possible. However, the results obtained surpass those of work on PE detection on other private datasets.

Keywords: Tuberculosis · Pleural Effusion · Computer Aided Diagnosis

1 Introduction

Tuberculosis (TB) is one of the world's leading killers, with a mortality rate of 1.2 million people in 2010 [1]. While being curable, identifying it in its early stages on scale is a challenge that remains. While the gold-standard tests for TB are slow, Chest X-Rays (CXRs) form the first phase of test for TB, and work as an inexpensive preliminary screening method [2]. However, the interpretation of the CXR depends on the skill of the reader, and is subject to human error and biases. As a result, computer aided detection of tuberculosis through analysis of CXRs seems like a promising idea. This will result in a more easily accessible preliminary screening, and reduce the possibility of errors.

Over the years, much work has been done towards detection of abnormalities in CXRs, but only a few have been geared towards specifically detecting TB. Kuo et al. [3] propose a method for detection of abnormalities (including TB) in CXRs, through the use of novel roughness and symmetry measures for parts of the lung. For specifically discriminating against TB, Jaeger et al. [4] identify TB

© Springer International Publishing AG 2017
S. Battiato et al. (Eds.): ICIAP 2017, Part I, LNCS 10484, pp. 617–625, 2017.
https://doi.org/10.1007/978-3-319-68560-1_55

through a combination of shape and texture descriptors along with the features based on the Lucene Image Retrieval Library. Ginneken et al. [5] begin off by subdividing the lung into smaller regions followed by texture analysis. More information on TB screening systems can be found in the survey by Jaeger et al. [6].

In this paper, we use a different approach to work towards identifying TB in CXRs. We plan to look at the symptoms radiologists tend to look at while reading CXRs, and build classifiers for them individually. Here, we look at the challenge of identifying pleural effusion (PE). While PE by no means is a phenomenon exclusive to TB, it sits in line with our larger aim of building a system to automatically detect TB through analysis of CXRs. This approach would allow a more thorough diagnosis where not only will the overall system assign a score of TB, it will also diagnose the CXR with the correct symptoms.

PE is characterized by the buildup of pleural fluid in the lung regions. This fluid tends to accumulate near the bottom of the lungs and the chest cavity [7]. Refer to Fig. 1 to view the manifestations of PE at different severity levels. Limited research has been done on automated detection of PE. Avni et al. [8] proposed a bag of visual words approach to discriminate between various pathologies in CXRs, including a very small dataset of PE. Maduskar et al. [9] work towards identifying the costophrenic point with great accuracy followed by building features around that point. In this paper, we capture the signature of PE using novel features we define after consulting with radiologists on their methods of reading CXRs.

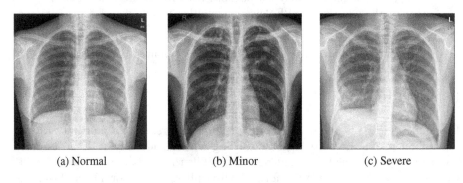

(a) Normal   (b) Minor   (c) Severe

Fig. 1. Chest Radiographs as labeled by a radiologist (a) normal CXR (b) minor PE in the left lung (c) severe PE in the right lung

The rest of the paper is structured as follows. In Sect. 2 we explain the process of segmentation of lungs, creation of the features, and details of the classifier. In Sect. 3, the details of the dataset being used are mentioned. Section 4 talks about the results of the experiments. Section 5 deals with the conclusion of the paper.

2   Method

This section talks about our implemented methods for lung segmentation, feature computation, and classification. The system uses a graph cut method to segment the lung, computes the novel features proposed in this work, and then feeds them to a random forest classifier which identifies thresholds around which to classify CXRs.

2.1   Lung Segmentation

The first task of PE detection method is to segment the region of lungs from the CXRs. Since the manifestations of PE occur near the lower boundaries of the lung, it places high requirements on the quality of the segmentation. We initially tried methods based on multilevel thresholding and region growing. Post this, we used a graph cut based segmentation method mentioned in [10]. The method consists of three main steps. It begins with content based image retrieval using a training set along with its defined masks. Post this, the initial patient specific anatomical model is created using SIFT-flow for deformable registration of training masks for the patient CXR. In the final step, a graph cuts optimization procedure with a custom energy function is used. The code made available by Candemir et al. [10] was used for the segmentation process, and hence further details may be inferred from their work. The results of the lung segmentation were good enough to allow us to proceed to the feature designing stage. Figure 2 shows an example of a CXR before and after segmentation.

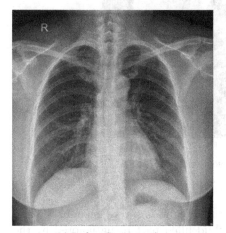
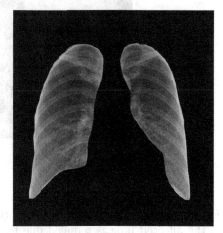

(a) Before Segmentation   (b) After Segmentation

Fig. 2. Chest Radiographs before and after segmentation

2.2   Features

The second stage involved crafting features which can discriminate PE. Through consultation with various radiologists, we arrived at a set of visual features they would look for and worked on translating them into their mathematical counterparts. The one valuable insight we gained was that radiologists tend to compare the left and right lung to identify anomalies. The effect of this insight influenced our design of the features.

Another important point to note is that PE is the accumulation of pleural fluid in the lungs, which accumulates in the bottom portion of the lung. This manifests itself as a white region in the CXR, which if large enough, is ignored by the segmentation process as not a part of the lung. We exploit this 'shortcoming' to design some of our features.

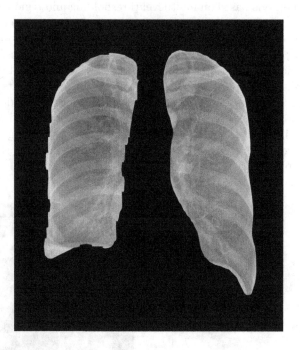

Fig. 3. Height difference post segmentation due to PE

Height Difference. If the accumulation of the fluid is severe enough, the size of the affected lung is reduced due to its fluid filled part being ignored by the segmentation process. This can be observed in Fig. 3. Hence we take the difference in the heights of the left and right lungs as one of the features. The height of a lung is defined as the distance between the top and bottom pixel of the lung after segmentation. Let H_l and H_r be defined as the heights of the left and right lung respectively. Then the feature H_{diff} is defined as:

$$H_{diff} = \frac{|H_l - H_r|}{max(H_l, H_r)} \tag{1}$$

We normalise by the maximum of the two heights to account for the fact that people have different sized lungs, and we don't want the difference to be magnified or reduced solely on the basis of lung size (or even the size of the CXR).

A possible issue which might arise is that the patient might have manifestations of PE in both of his lungs, which results in no height difference. However, such a case is extremely unlikely as a height reduction is observed in severe cases of PE, and such a manifestation in both lungs would be rare. Even if it did occur, such an occurrence of PE would also show up in the other features defined below.

Lung Bottom Curvature. The fluid accumulated at the bottom of the lung settles due to gravity with a flat horizontal surface. This manifests itself as a nearly horizontal white line in the CXR. This affects the segmentation of the lungs, and results in a less curved cut at the bottom after segmentation than normal.

Let B_r denote the row indexes of the bottom most pixels of columns containing the right lung. It should be noticed that while most of the bottommost pixels of each column will actually be the bottom of the lung, some of them might be the sides of the lung owing to the curvature of the lung. As a result, we disregard 10% of the values on each end of the B_r array. This figure of 10% was experimentally determined and it may need tuning for other datasets.

Let B_l denote the row indexes of the bottom most pixels of half the columns on the left side (i.e. only use the left half of the left lung to find row indexes) which contain the left lung. The reason we disregard the right half of the left lung is that a great variation is observed in the row indexes on this end during segmentation due to the presence of the heart, which results in a wide variance of the row indexes which may possibly hide the lack of variance which might be there due to pleural effusion. We also disregard 10% of the values on the other end for reasons similar to case of the right lung.

The feature V_r is defined as:

$$V_r = \frac{var(B_r)}{len(B_r)} \tag{2}$$

where the numerator is the variance of the elements in B_r and the denominator is the number of elements in B_r. The normalisation by the length of B_r prevents the variance from being increased or decreased due to individual variation in lung sizes. V_l is similarly defined for the left lung. These two features are fed to the classifier.

Lower Lung Intensity Variation. In non-severe cases of pleural effusion, a marginal deposition of the fluid in the bottom corners of the lungs is observed, which manifests itself into whiteness in the CXRs. A horizontal scan of the pixel values for each lung starting from the bottom is done, and the mean intensity of the first 7% pixels is calculated. The figure of 7% was arrived at by the desire to capture a large portion of the bottom part of lungs, but at the same time not include too much of the higher portions. The exact number was experimentally determined, and might needed to be tuned for other datasets. The mean intensity of the rest of the pixels is also calculated, and the ratio of the former with respect to the latter is taken. Let I_l and I_r denote these values for the right and left lung respectively.

I_m is one of the features fed to the classifier and is defined as:

$$I_m = max(I_l, I_r) \qquad (3)$$

I_d is the other feature used and is defined as:

$$I_d = |I_l - I_r| \qquad (4)$$

These two features don't need to be normalised unlike other features because the normalisation is inherent in the calculation of the I_l and I_r itself.

2.3   Classification

After constructing the features, we ended up with a 5 dimensional feature vector for each CXR. These feature vectors were then fed into a random forest classifier [11]. The standard MATLAB implementation of random forest was used for the purpose of the experiments [12].

A random forest classifier is an ensemble learning method which works by constructing multiple decision trees. Each of the decision trees is trained on a different subset on the whole training data, a method known as bootstrap aggregating. The output of the classifier system is taken as the mode of the output of the individual trees which constitute the random forest.

There are a few advantages of random forests which led to this being our classifier of choice. It can learn complex relationships between variables and requires minimal tuning, as opposed to some other classifiers such as SVM. It doesn't require too much data for learning, unlike neural networks. Additionally, the ensemble of trees help avoid the issue of over-fitting to the data.

While we currently look to classify CXRs on whether they have PE or not, it is also possible in future work to recognise the area where PE has occurred. Since PE mainly occurs near the bottom of the lungs, we need to identify which lungs are affected by PE. This can be done by designing similar features to those above but which aren't agnostic to the right and left lung. However, this would require a greater amount of data for learning due to increased dimensionality of the features, and hence not explored here.

3   Data

The datasets used in this study were two publicly available CXR dataset provided by the US National Library of Medicine [13]. The first dataset (Montgomery Dataset) contained 138 CXRs with 58 of them showing instances of TB.

Each CXR had a dimension of *4020* × *4892* pixels. The second dataset (China Dataset) contained 662 CXRs with 336 of them showing instances of TB. The CXRs in this set had varying dimensions, but were roughly in the range of *3000* × *3000* pixels.

While the sets came with information regarding the presence of TB or not in the CXRs, they did not specifically talk about PE. For the ground truth regarding PE, the dataset was curated by an eminent radiologist. The data was partitioned into three sets. The first set consisted of CXRs which showed instances of TB along with manifestations of PE. The second set consisted of normal CXRs which showed no instances of TB. The third set consisted of CXRs which showed instances of TB but not any manifestations of PE. The number of CXRs showing instances of PE were limited to 63 CXRs, which were a mix from the China and Montgomery dataset. The other two sets were also kept the same size to this to prevent unbalanced classes, which could affect the training of the classifier.

Owing to the less amount of data available, it would not be a good idea to partition the entire data into separate testing and training sets. However, at the same time, it is not feasible to have any common data in the testing and training sets. So, leave-one-out-cross-validation was the preferred method of choice to avoid this issue during classification. It has the advantage of effectively increasing the amount of data for testing purposes, preventing overfitting, and avoiding the excessive computation issue faced in leave-out-p-cross-validation.

4   Results

The performance of the proposed PE detection system was analysed in terms of area (AUC) under the receiver operating characteristics (ROC) curve. As mentioned before, random forests were the classifiers used in the evaluation of performance. The classification accuracy on the PE set is measured against two sets, normal CXRs and those CXRs with manifestations of TB but not PE.

The results of classifying normal CXRs v/s those with PE can be seen in Fig. 4a. We report and AUC of 0.961. Since it is important to not miss TB, we would err on the side of caution and aim for higher sensitivity. The optimal operating point is shown in the ROC curve with a red circle, suggesting we operate at 100% sensitivity and 80.95% specificity. Sensitivity, also known as recall, is the true positive rate of the classifier. Specificity is the true negative rate. This corresponds to a precision (positive predictive value) of 84%.

To ensure that the classifier was learning pleural effusion specific features and not just tuberculosis, we also tested the classifier on the dataset which contained CXRs with manifestations of TB without instances of pleural effusion and the dataset which contained CXRs with manifestations of TB resulting in symptoms of pleural effusion. Even in this case, the classifier achieved an AUC of 0.864 and the ROC characteristics can be seen in Fig. 4b. The optimum point for operation here is at a sensitivity (recall) of 80.95% with a specificity of 77.8%. This corresponds to a precision of 78.4%. If we were to look for 100% sensitivity so as not to miss any manifestation of PE, our precision would fall to around 65%.

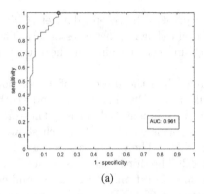 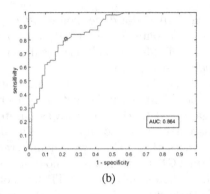

(a)   (b)

Fig. 4. ROC curves for Pleural Effusion detection. (a) For the dataset containing CXRs with no instance of TB v/s those showing PE. (b) For the dataset containing CXRs with TB but no instance of PE v/s those showing PE.

The latest work against which we evaluate our results is of Maduskar et al. [9]. They built a system to identify PE by creating features around the costrophrenic point. They evaluate the performance of the system independently for the left and right lungs. However, due to the dataset constraints on our side, this is not a feasible option for us. Their system returns an AUC of 0.84 and 0.90 for PE detection in the left and right lung respectively. This is lower than our AUC reported above. They also evaluate the performance of their system against cases of PE they classify as severe, a level above obvious. They report AUC of 0.88 and 0.94 for the left and right lungs respectively. This is again lower than the AUC of our system.

5   Conclusion

In this study, we have presented a system to detect PE in CXRs. We began by discussing with radiologists their methodology of reading CXRs, and looked to transform this into mathematical formulations. Before extracting features based on these formulations, the left and right lung were separated from the CXR based on a three step segmentation method. Post extracting the features, they were fed into a random forest classifier to arrive on decision boundaries to decide on whether a particular CXR shows instances of PE.

The AUCs of 0.961 and 0.864 are quite encouraging. They are better than the results we have seen in other papers. However, a direct comparison is not possible due to them being based on different datasets, and results being dependent on the datasets. In further work, one can look to build systems to identify other symptoms which manifest in the presence of TB, and then finally look to combine them into one main system.

Acknowledgement. The authors would like to express their thanks to Dr. Anurag Agrawal, CSIR Institute of Genomics and Integrative Biology, and Dr. Anjali Agrawal,

Teleradiology Solutions, for their help in describing the methods they use while reading CXRs and their help in annotating the data set with instances of PE.

References

1. World Health Organization: Global tuberculosis report (2012)
2. Toman, K., Frieden, T., World Health Organization: Toman's Tuberculosis: Case Detection, Treatment, and Monitoring: Questions and Answers, 2nd edn. World Health Organization, Geneva (2004)
3. Kao, E.-F., Kuo, Y.-T., Hsu, J.-S., Chou, M.-C., Liu, G.-C.: Zone-based analysis for automated detection of abnormalities in chest radiographs. Med. Phys. 8(7), 4241–4250 (2011)
4. Jaeger, S., Karargyris, A., Candemir, S., Folio, L., Siegelman, J., Callaghan, F., Xue, Z., Palaniappan, K., Singh, R.K., Antani, S., Thoma, G., Wang, Y.-X., Pu-Xuan, L., McDonald, C.J.: Automatic tuberculosis screening using chest radiographs. IEEE Trans. Med. Imaging 33(2), 233–245 (2014)
5. van Ginneken, B., Katsuragawa, S., ter Haar Romeny, B.M., Doi, K., Viergever, M.: Automatic detection of abnormalities in chest radiographs using local texture analysis. IEEE Trans. Med. Imag. 21(2), 139–149 (2002)
6. Jaeger, S., Karargyris, A., Candemir, S., Siegelman, J., Folio, L., Antani, S., Thoma, G., McDonald, C.J.: Automatic screening for tuberculosis in chest radiographs: a survey. Quant. Imaging Med. Surg. 3(2), 89 (2013)
7. Woodring, J.: Recognition of pleural effusion in supine radiographs: how much fluid is required? Am. Roentgen Ray Soc. (AJR) 142, 59–64 (1984)
8. Avni, U., Greenspan, H., Konen, E., Sharon, M., Goldberger, J.: X-ray categorization and retrieval on the organ and pathology level, using patch-based visual words. IEEE Trans. Med. Imaging 30(3), 733–746 (2011)
9. Maduskar, P., Philipsen, R.H.M.M., Melendez, J., Scholten, E., Chanda, D., Ayles, H., Snchez, C.I., van Ginneken, B.: Automatic detection of pleural effusion in chest radiographs. Med. Image Anal. 28, 22–32 (2016)
10. Candemir, S., Jaeger, S., Palaniappan, K., Musco, J.P., Singh, R.K., Xue, Z., Karargyris, A., Antani, S., Thoma, G., McDonald, C.J.: Lung segmentation in chest radiographs using anatomical atlases with nonrigid registration. IEEE Trans. Med. Imaging 33(2), 577–590 (2014)
11. Breiman, L.: Random forests. Mach. Learn. 45(1), 5–32 (2001)
12. MATLAB (R2016a), MathWorks, Natick, MA (2016)
13. Jaeger, S., Candemir, S., Antani, S., Wang, Y.-X.J., Pu-Xuan, L., Thoma, G.: Two public chest X-ray datasets for computer-aided screening of pulmonary diseases. Quant. Imaging Med. Surg. 4(6), 475–477 (2014)

Design of a Classification Strategy for Light Microscopy Images of the Human Liver

Luigi Cinque[1], Alberto De Santis[2], Paolo Di Giamberardino[2],
Daniela Iacoviello[2], Giuseppe Placidi[3($\boxtimes$)], Simona Pompili[4],
Roberta Sferra[4], Matteo Spezialetti[3], and Antonella Vetuschi[4]

[1] Department of Computer Science, Sapienza University of Rome,
Via Salaria 113, 00198 Rome, Italy
cinque@di.uniroma1.it

[2] Department of Computer, Control and Management Engineering Antonio
Ruberti, Sapienza University of Rome, Via Ariosto 25, 00185 Rome, Italy
{desantis,digiamberardino,iacoviello}@dis.uniroma1.it

[3] A2VI-Lab, Department of Life, Health and Environmental Sciences,
University of L'Aquila, Coppito 2, 67100 L'Aquila, Italy
giuseppe.placidi@univaq.it,
matteo.spezialetti@graduate.univaq.it

[4] Department of Biotechnological and Applied Clinical Sciences,
University of L'Aquila, L'Aquila, Italy
antonella.vetuschi@univaq.it

Abstract. Light Microscopy (LM) represents the method by which pathologists study histological sections; the observations by LM can be considered the gold standard for making diagnosis and for its diagnostic accuracy. The classes that can be defined through the observation of LM images of the liver are: normal, steatosis, fibrosis, cirrhosis and hepatocarcinoma (HCC). Normally, a pathologist has to examine by LM many histological sections to perform a complete and accurate diagnosis. For this reason, an automatic system for the analysis of LM images of the liver would be particularly useful. Goal of this paper is to propose an automatic multi-stage procedure to classify the normal tissue, and the pathologic ones from human liver microphotographs. Due to the articulated nature of the examined images, the analysis will first assess if steatosis is present, by using objects analysis, and then determine whether the image belongs to a normal tissue or to one of the other pathologic ones, by using a machine learning based technique. To this aim some texture features are calculated, and the Principal Component Analysis is applied to derive the best representation of the data. Four binary Support Vector Machines classifiers are trained, one for each kind the four classes of liver conditions to be identified. Experimental results show the classification capability of the proposed system, with promising theoretical and experimental basis for developing a fully automatic decision support system.

Keywords: Medical image processing · Liver tissues classification · Principal component analysis · Support vector machine

© Springer International Publishing AG 2017
S. Battiato et al. (Eds.): ICIAP 2017, Part I, LNCS 10484, pp. 626–636, 2017.
https://doi.org/10.1007/978-3-319-68560-1_56

1 Introduction

Liver diseases may be roughly divided into two categories, focal diseases, where the abnormality is concentrated in small area, and diffused diseases, where the abnormality is distributed all over the whole liver volume [1]. Different, noninvasive (in the sense that they do not require surgery for the patient), diagnostic imaging techniques, such as Magnetic Resonance Imaging (MRI), Computer Tomography (CT) or Ultrasound Tomography (UT), can be effectively used for preliminary diagnosis and for planning surgery interventions or pharmacological treatments. However, Light Microscopy (LM) represents the method by which pathologists study and review histological sections and the observations by LM can be considered the gold standard for making diagnosis and for its diagnostic accuracy, in particular regarding the possibility of defining the heaviness of a given pathology at a very high resolution. The classes that can be defined through the observation of LM images of the liver are: normal, steatosis, fibrosis, cirrhosis and hepatocarcinoma. Normally, a pathologist has to examine by LM many histological sections to perform a complete and accurate diagnosis. For this reason, an automatic system for the analysis of LM images of the liver would be particularly useful. Aim of this paper is to define a complete procedure for automatic classification of LM images presenting different pathologies affecting liver parenchyma. The problem considered in this paper has been addressed by many researchers [2]. Combination of methods from traditional image analysis and sophisticated machine learning and pattern recognition techniques has yielded interesting texture based information and effective quantitative characterization for a number of applications of practical interest, including medical image analysis [3–5]. Since the possible textures of interest may be very different, several methods can suit for different kind of medical images. Basically we may distinguish between statistical, spectral and structural analysis of textures; in particular, in texture analysis, one of the most difficult aspects is to define a set of features that adequately describe the characteristics of a texture [6].

Wavelet transform and Fisher Linear Discrimination Analysis are efficiently used in [7] in color medical images for liver fibrosis identification. A wavelet multi resolution analysis on the three color image components is applied to reduce the liver slice background noise, thus increasing the discrimination power of the Fisher algorithm in segmenting the liver fibrotic tissue from the other tissues on pathological section images. In [8] focal lesions in ultrasound images of the liver are automatically assigned to four classes (normal, cyst, benign and malignant masses). The texture features are extracted by four procedures (grey level co-occurrence and run length matrices for the statistical properties, Gabor wavelets and 2D Laws for the local spectral content). The two sets of textures features are reduced by either a manual or a PCA based selection. The former reduced set is classified by neural networks and the latter by k-means. The neural network achieves a higher correct classification rate than the k-means in this experiment. In [9] fractal dimension and M-band wavelet transform are used for composing the feature vector in the classification of ultrasonic liver images. Three conditions of normal, cirrhosis, and hepatoma are recognized with a high classification rate.

In [10] statistical methods of texture analysis are applied on microscopic liver images, in particular of liver fibrosis; the sensitivity of texture analysis is tested when fibrotic and normal tissues are stained with different fibrosis biomarkers. The texture analysis is performed by using the co-occurrence matrix and the run-length matrix; a classification using agglomerative hierarchical clustering and linear discriminant analysis with cross validation is applied on different biomarkers that in some cases influenced the results. In [11] an interesting review on machine learning techniques combined with image processing methods for automatic segmentation of liver CT and MRI images is presented; a particular attention is devoted to SVM based techniques [12, 13] that assumed as input texture descriptors. In [14] region-based shape descriptors, gray level and co-occurrence matrix (GLCM) features are adopted for automatic CT image classification, by SVM, of specific liver diseases like cysts, hepatoma, and cavernous hemangioma.

In this paper, LM liver images are analyzed to distinguish different tissue types: normal, steatosis, fibrosis, cirrhosis and hepatocarcinoma, [15–18]. A classification method is designed to assign a sample image to one of the five classes. The color images are first reduced to grey level scale and then a first level classification is accomplished to identify the steatosis liver tissue; to this aim a suitable segmentation algorithm is applied along with an object analysis to detect the roundish, smooth edge, fat droplets. The ratio of fat droplets area over the total image area determines a quite robust indicator to reliably distinguish the steatosis class from the others. Images of the remaining four classes are characterized by texture analysis by considering two groups of features: statistical properties of the grey level value (contrast, uniformity, entropy), and statistical features of the grey level spatial distribution as obtained by the co-occurrence matrix (contrast, correlation, homogeneity, energy). Any sample image of the given class is partitioned into tiles of suitable size, and the average and standard deviation of the texture descriptors are computed over the set of tiles of each image of the training set. Fourteen textures features are obtained with a good separation between classes (strong correlation within classes, and weak correlation between classes). It is worth noting that the tiling procedure strengthens the local character of the texture parameters in order to better capture the local parenchyma structure in the different tissues (that is sometimes very subtle as between fibrosis and cirrhosis tissue). The set of features is processed by PCA to obtain a more efficient representation of the information content used to train four SVMs binary classifiers. High correct classification rates are obtained for each class, and the ROC curves denote a quite satisfactory behavior of the classifiers over a repeated random selection of the training set. The result is a very flexible and general purpose approach for the classification of the LM images of the human liver. In a future work, by using a richer image dataset, the proposed approach will also be applied to images where multiple kinds of tissue are present.

The paper is organized as follows. In Sect. 2, the LM structure of the liver parenchyma, for the considered five classes, is described and the features extraction and classification procedure is proposed. In Sect. 3 the numerical results are presented and discussed. Conclusions and future developments are outlined in Sect. 4.

2 Materials and Methods

In this paper microscopic images of different kind of liver tissue are observed under light microscope. Two independent pathologists examined various histological sections of the hepatic parenchyma and, on the basis on specific structures, they placed samples in different groups and described the relevant and specific shapes they considered to define the allowance to different groups. The considered images may be grouped into five classes: normal (N), steatosis (S), fibrosis (F), cirrhosis (C), HCC (H) see Fig. 1, though the transition from one class to the other is often gradual and different states could be contemporary present (steatosis aspects are interleaved with normal tissue; fibrotic structures can be also present in an early cirrhosis; focal steatosis sis present in alcoholic cirrhosis; etc.). The automatic analysis of this kind of samples may present a number of technical issues due to the contemporary presence of different states. In fact, irregular regions that can be easily detected by LM may represent pathologies completely different and not so well distinguishable (for example, fibrotic tissue can be easily present in mainly cirrhotic images).

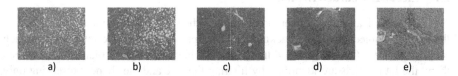

a)   b)   c)   d)   e)

Fig. 1. The microphotographs show different classes of liver parenchyma (Haematoxylin & Eosin, original magnification 4X). (a) normal liver; (b) steatosis; (c) fibrosis; (d) cirrhosis; (e) HCC.

For each class, a binary classifier simply labels the test image as belonging or not to a given class. Then, the overall classification process consists in the application of five binary classifiers, according to the block diagram of Fig. 2, one for each class (N, S, F, C and H), due to the fact that images belonging to different classes can be very different, even if, in some cases, different pathologic states could be contemporary present. The result is a binary string containing 1 where the answer for a specific class is positive and 0 in the case of negative answer. Each of the five binary classifiers is structured for the specificity of the tissue to be recognized.

As a matter of fact, a fatty liver tissue is mainly characterized by the presence of roundish fat droplets spread over the liver surface, therefore a steatosis can be easily classified by object segmentation and evaluating shape and size of bright items over the background; a fat presence indicator can be defined and a suitable threshold value determined to discriminate easily between a steatosis/non steatosis condition. The other kinds of tissue present diffused abnormalities that can be described by texture analysis: the grey level texture features and the grey level spatial distribution texture features computed by the co-occurrence matrix.

Therefore the first result that must be assessed is whether a tissue is a steatotic one or not. If the answer is negative a set of features, adequately transformed by the PCA, are used to train SVMs classifiers, as will be described in the following.

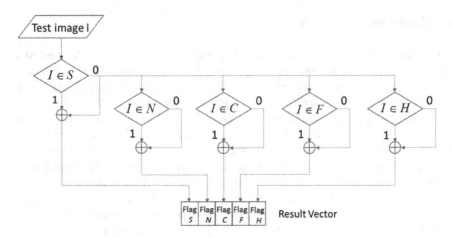

Fig. 2. Block diagram of the classification process.

2.1   Steatosis Characterization

An image binarization is sufficient to distinguish the fat droplets (typical of steatosis) as the brightest objects over the background. We applied the discrete level set approach proposed in [19]. On the binarized image, the brighter objects are isolated and, between them, the fat droplets are identified by filtering the size and the shape, preserving only the non-eccentric non-ragged objects with a significant area. On the selected set of items, the *Percentage of Fatness* (PoF) is computed as the total area of fat droplets (number of white pixels) over the image size.

An image is classified as a steatosis one if the PoF is above a chosen threshold; by analysing the data of all the classes it was noted that generally there is a difference of an order of magnitude between the PoF of an S image and the others.

2.2   Feature Extraction for Non-steatosis Aspect

The liver images not belonging to class S denote an appearance that is difficult to characterize as objects over a background. Even though some structures are detectable (as described above), the very difference between the classes N, C, F and H is mainly due to the texture structure. The considered texture features belong to two groups: the first group is related to the grey level, the second to the grey level spatial distribution as characterized by the co-occurrence matrix. The features are computed on a training set of N_{tr} images by the Matlab Image Processing Toolbox standard functions; each class contributes with the same number of sample images.

For the grey level texture features, each image of size $m \times n$ is partitioned assuming tiles T of size $\ell \times \ell$ (a part of each rectangular image is left out). The grey level texture features considered are the Contrast, the Uniformity and the Entropy. The contrast C_T is a measure of the variability of the grey level within the tile, the higher the contrast the better the details are identified over the background. The uniformity U_T and the entropy E_T describe the degree of regularity of the grey level values in a tile: if all the pixels have the same grey level it would be $U_T = 1$ and $E_T = 0$, meaning that the tile is maximally uniform (it has constant grey value indeed) and maximally ordered (all the pixels are equal). On the contrary an unstructured noise, would have the grey values all different and maximally disordered ($U_T = 1/L$ and $E_T = \log_2 L$). The image features are then computed as average and standard deviation over the set of tiles, obtaining six features.

The grey level *spatial distribution* texture features may be characterized by defining some relations between the grey level values of neighbouring pixels, and by computing the frequency of the occurrence of any such a relation in the whole image domain. These frequencies for all the pairs of grey level values $\{g_i, g_j\}$ define the co-occurrence matrix GCO. Quantities like Contrast, Correlation, Homogeneity and Energy are computed as averages over the whole image weighted with the entries of GCO, so that the local spatial distribution constraint at different scales $s = 1, 2, \ldots, \bar{s}$, and directions $d = 1, 2, \ldots, \bar{d}$ is taken into account. These quantities have similar meaning of the ones previously defined but their values depend of the given scale and direction. For each image of the training set, eight features are obtained by the average and standard deviation of the contrast, the correlation, the homogeneity and the energy for each pair of scale-direction values.

2.3   Feature Analysis and Principal Component Analysis

Fourteen features are computed for the N_{tr} images of the training set and are collected in the matrix Φ of dimension $N_{tr} \times 14$. A correlation analysis of these features among the N_{tr} images showed that the features are highly correlated within each class but substantially uncorrelated between classes. This in turn suggests that the selected set of features is suitable for classifying liver images belonging to the chosen four classes. Nevertheless, numerical experiments showed that in the space of these features the images are not linearly separable, therefore the use of the correlation analysis would deliver a classification system with poor performances. Therefore a more efficient representation of the data is advisable; this could obtained by principal components analysis, choosing, for any class training set of images, p^* principal components maintaining the $P\%$ of the information content.

2.4   Binary Classifier Training

So far, for any of the four classes N, F, C, and H, a set of p^* principal components is selected. For any class, a binary classifier is trained to recognize a test image either belonging to the class or not; the procedure is briefly outlined for a single class, being the same for all classes. Consider the class N and let $\tilde{E}_N$ be the selected p^* principal components. Compute now the coordinates $\tilde{L}_{N,N}$ of the features vectors M_{FN} of the $N_{tr,N}$ images of the training set of class N: for any row vector of $\tilde{L}_{N,N}$ the response variable of the class N classifier is set to 1. Now repeat the process for the training set images of the remaining classes determining $\tilde{L}_{N,C}$, $\tilde{L}_{N,F}$, $\tilde{L}_{N,H}$ and set the classifier response variable to 0 for these set of coordinates. The *perfect classifier* would separate the points with response equal to 1 from the points with response equal to 0. Such a task can be accomplished by training a SVM, [12]: it is a well-established method aiming at the determination of the best hyperplane (in general a manifold) able to separate a set of response points into two classes. The parameters of the SVM are determined by using the ten-fold cross validation [20] and the classification is performed by LIBSVM 3.18 [21].

Four classifiers C_i i = 1, 2, 3, 4 are trained to determine if the image belongs to the class N, F, C or H respectively; it means that, for example, the classifier C_1 is trained to identify the normal images, i.e. it is able to distinguish normal tissues versus F, C or H ones, assigning label 1 if the image is classified as belonging to the N class or label 0 if not. When a fibrotic image X is tested with the classifier C_1 trained to identify the normal images N versus all the other, the classifier should identify the X image as "not-normal" and assign label "0", whereas when one uses the "right classifier" C_2 (trained to identify the type of images like the X data), the classifier should assign label "1".

3 Numerical Results and Discussion

The classification procedure proposed in this paper considered a set of 120 images of size $m \times n$, pixels, $m = 543$, $n = 780$, 24 for each of the five classes N, S, F, C, H.

The set of 24 steatosis images is used to tune and validate the steatosis classifier. The remaining group of 96 images is divided into two sets, with images for each class in the proportion of 60% and 40%: the first set, N_{tr}, is used to train the classifiers whereas the latter, N_{test}, is used for testing the classifiers over data not used for the training.

The described classification procedure starts with the decision whether the image could be in the S class or not. The steatotic images are characterized by the presence of circular white elements; as already said, they can be easily determined by a binarization procedure that allows the recognition of white objects. To identify only the fat droplets, a morphological filtering is performed, preserving only objects with area larger than 6 pixels, with eccentricity lower than 0.6. Moreover, to avoid too ragged objects, only white objects whose ratio between their area and the area of the ellipse circumscribing the objects is greater than 0.5 are considered.

The analysis of the percentage of fat droplets in all the images yields an evident difference, see Table 1:

Table 1. Mean values and standard deviations of the percentage of fat in liver tissues images

Fat percentage	N	F	C	H	S
Mean value	0.27	0.29	0.05	0.32	4.85
Standard deviation	0.08	0.32	0.05	0.81	2.69

If a sample image has a percentage of fat less than 1% it can be assumed that the tissue is not in the S class. The fat droplets identification method has provided convincing results approved by pathologists that evaluated by themselves the fat percentage and compared their results with the ones obtained by applying the described automatic method.

To determine the grey level texture features each image is partitioned considering tiles of size $l = n/10$. This tiling resolution yields good results to the subsequent classifiers training phase. To determine the Uniformity and the Energy also the number of bins must be fixed and $L = 4$ appeared a good choice, allowing a simplification of the data and the preservation of interesting structures. Therefore, from this analysis six features are computed as average and standard deviation over the set of tiles.

As far as the grey level spatial distribution texture features is concerned, four scales $s = 1, 2, 3, 4$ and four directions $d = 0, 45°, 90°, 135°$ for $L = 4$ grey level values are considered, thus obtaining a co-occurrence matrix GCO of size $4 \times 4 \times 16$. From the co-occurrence matrix GCO the Contrast, Correlation, Homogeneity and Energy are evaluated, yielding the eight features obtained as their average and standard deviation.

To classify if an image belongs to one of the N, F, C or H class the analysis based on feature classification is performed and four different classifier C_i, i = 1, 2, 3, 4 are designed.

Once the fourteen features are calculated for all the images of the N_{tr} set, the principal component analysis is applied; to preserve the percentage $P = 99\%$ of information, after the evaluation of the eigenvalues of the covariance matrix, the first $p = 8$ principal components are retained. For the classification, the chosen kernel function is the radial basis function. Each classifier C_i is trained to assign the label "1" to the i-th class and the label "0" to all the others; more precisely, C_1 assigns 1 to the class of normal tissues and 0 to all the others, C_2 assigns 1 to the class of fibrotic tissues and 0 to all the others, C_3 assigns 1 to the class of cirrhotic tissues and 0 to all the others and C_4 assigns 1 to the class of HCC tissues and 0 to all the others. The accuracy in the training phase relies in the percentage of success in assigning the labels "1" and "0".

The obtained classifiers are tested on the test set of $N_{test} = 10$ images of each class N, F, C, H; in this case we assumed to ignore the nature of the data X to be classified, and by default we initially assign it the label "1". Therefore the generic classifier should confirm label "1" if it is the classifier trained to identify the class of the specific unknown image X, otherwise the classifier should assign label "0" (meaning that the image belongs to one of the other classes). The mean value of the results over the 10 test images of each class are reported in Table 2. The results reported are obtained as mean values of the results of each classifiers after randomly choosing the training set and the test set, in order to avoid lucky choices of the test images.

Table 2. Results of the test: mean value of the percentage of success of the classifiers.

	C1	C2	C3	C4
Test images of class N	95%	6.25%	16.25%	1.25%
Test images of class F	0%	96.67%	6.67%	10%
Test images of class C	7.5%	1.25%	95%	7.5%
Test images of class H	0%	13.75%	25%	86.25%

From Table 2 it could be noted that the percentages of identification are high on the diagonal of the table (default label "1" identified correctly as label "1"), whereas if the test image is tested with a classifier trained to identify with label "1" a different kind of image the percentage of success must be low (default label "1" identified correctly as label "0").

The results are encouraging; it could be noted in fact that only the 6.25% of test images of normal parenchyma could be confused with a fibrotic one, 16.25% could be confused with cirrhotic aspect and only 1.25% could be wrongly classified as HCC.

As far as test images of class F, when tested versus all the classifiers, they appear clearly identifiable; they are not confused with normal tissues, a percentage of 10% could be wrongly classified as HCC and only 6.67% could be confused with cirrhotic parenchyma. A cirrhotic test image could be confused (percentage of 7.5%) with a normal tissue or with an HCC one and only a 1.25% could be wrongly classified as fibrotic tissue.

The less robust results appear to be the ones connected with the HCC images; for example, a percentage of 25% could be identified as image of the cirrhotic class C. Trying to analyse the motivations of this result, one has to take into account that cirrhosis can be lead to HCC and the latter can appear as a multiple nodules that resemble cirrhotic nodules and for this reason the two histological aspects could be confused by the automatic system.

The classifiers have been further tested by choosing randomly the training and the test set, thus obtaining 10 trials for each classifier. In each trial the *true positive rate* (TRP, the rate of the images correctly classified), and the false positive rate (FPR, the rate of the images misclassified) were computed. All the trials show a score above the intercept (random classifier) and the most part of them have score between 0.8 and 1 (and therefore confirming the average score reported in Table 2), thus denoting good performances.

It is worth noting that the proposed algorithm is tuned to classify the unknown images into each of the 5 classes by considering the possibility of the contemporary presence of more than one aspects at once. This comes from the overall adopted procedure. However, in the images used therein each tissue can be assigned to a single specific class: in this way it has been possible to tune unambiguously the parameters of the different classifiers. Nevertheless it may happen that on a given sample image different classifiers yield a positive identification, i.e. the method classifies the image as belonging to different classes. These preliminary data of our investigation would only suggest the presence in the liver parenchyma of different pathologies. In this case, the image should be classified as referred to the heaviest pathology between those recognized and the other observed features should additional give information about the possible contemporary presence of different hepatopaties in the same sample.

4   Conclusions and Future Developments

In this paper the classification of different hepatopathies is addressed by proposing an automatic multi-stage procedure. We combine a textural based segmentation method with a support vector machine supervised pattern recognition procedure for automatic classification of microscopic images of liver in order to detect the presence of abnormal regions of a given family of pathologies, thus supporting medical diagnosis. The liver specimen is classified into one of the five following classes: normal, steatosis, fibrosis, cirrhosis and HCC, by considering both object analysis and a machine learning approach. More precisely, the former is used to determine first if the tissue is a steatosis one, by using the presence of the fat bright circular structures as a useful indicator; the machine learning approach is applied to determine if the tissue belongs to one of the other four classes. Suitable features are evaluated considering texture properties of the images and a principal component analysis is applied to derive the best representation of the data to be submitted to the support vector machine. Four distinct binary classifiers are trained providing promising results with good capability in separating the considered data. In this early investigation the selected texture features allowed the training of binary classifiers with encouraging performances that could be further improved by a better description of the spatial distribution of the grey level in the LM liver images; to this aim a richer set of scales and directions values to compute the co-occurrence matrix could be considered, along with some differential characteristics of the image signal. Moreover the overall classifying process will be applied on images containing more than a single pathology in order to establish the nature of the pathology and/or its heaviness. An effort will be done in order to indicate also the percentage of image occupied by different classes. This generalization will be investigated by enriching the data set with LM images containing also mixtures of the discussed five aspects of the human liver parenchyma.

References

1. Horng, M.H.: An ultrasonic image evaluation system for assessing the severity of chronic liver disease. Comput. Med. Imaging Graph. **31**, 85–491 (2007)
2. Oliveira, F.P.M., Tavares, J.M.R.S.: Medical image registration: a review. Comput. Methods Biomech. Biomed. Eng. **17**(2), 73–93 (2014). doi:10.1080/10255842.2012.670855
3. Placidi, G., Sotgiu, A.: A novel restoration algorithm for reduction of undersampling artifacts from magnetic resonance images. Magn. Reson. Imaging **22**, 1279–1287 (2004)
4. Franchi, D., Sotgiu, A., Placidi, G.: A novel acquisition–reconstruction algorithm for surface magnetic resonance imaging. Magn. Reson. Imaging **26**, 1303–1309 (2008)
5. Tsipouras, M.G., Giannakeas, N., Tzallas, T., Tsianou, Z.E., Manousou, P., Hall, A., Tsoulos, I., Tsianos, E.: A methodology for automated CPA extraction using liver biopsy image analysis and machine learning techniques. Comput. Methods Progr. Biomed. **140**, 61–68 (2017)
6. Ahmadian, A., Mostafa, A., Abolhassani, M., Salimpour,Y.: A texture classification method for diffused liver diseases using Gabor wavelets. In: Proceedings of IEEE Engineering in Medicine and Biology 27th Annual Conference, Shanghai, China. vol. 70, pp. 1567–1570 (2005)
7. Lu, Z., Song, E., Wang, Q., Wang, X.: The liver fibrosis identification based on color 2D wavelet transform for the medical image. In: 2008 IEEE International Conference on Wavelet Analysis and Pattern Recognition, pp. 205–208 (2008)
8. Balasubramanian, D., Srinivasan, P., Gurupatham, R.: Automatic classification of focal lesions in ultrasound liver images using principal component analysis and neural networks. In: 29th Annual International Conference of the IEEE Engineering in Medicine and Biology Society, pp. 2134–2137. IEEE (2007)

9. Lee, W.L., Hsieh, K.S.: A robust algorithm for the fractal dimension of images and its applications to the classification of natural images and ultrasonic liver images. Sig. Process. **90**, 1894–1904 (2010)

10. Amin, A., Mahmoud-Ghoneim, D.: Texture analysis of liver fibrosis microscopic images: a study on the effect of biomarkers. Acta Biochim. Biophys. Sin. **43**(3), 193–203 (2011)

11. Punia, R., Singh, S.: Review on machine learning techniques for automatic segmentation of liver images. Int. J. Adv. Res. Comput. Sci. Softw. Eng. **3**, 666–670 (2013)

12. Cristianini, N., Shawe-Taylor, J.: An Introduction to Support Vector Machines and Other Kernel-based Learning Methods. Cambridge University Press, New York (2000)

13. Schölkopf, B., Smola, A.J.: Learning with Kernels. MIT Press, Cambridge, (2002)

14. Lee, C.C., Chen, S.H., Chiang, Y.C.: Classification of liver disease from CT images using a support vector machine. J. Adv. Comput. Intell. Intell. Inform. **11**, 396–402 (2007)

15. Burkitt, G.H., Stevens, A., Lowe, J.S., Young, B.: Wheather's Basic Histopathology: A Colour Atlas and Text. Pearson Professional Limited, London (2013)

16. Friedman, V.: Mechanisms of disease: mechanisms of hepatic fibrosis and therapeutic implications. Nat. Clin. Pract. Gastroenterol. Hepatol. **1**, 98–105 (2004)

17. Pinzani, M., Rombouts, K., Colagrande, S.: Fibrosis in chronic liver disease: diagnosis and management. J. Hepatol. **42**, S22–S36 (2005)

18. Latella, G., Vetuschi, A., Sferra, R., Catitti, V., D'Angelo, A., Zanninelli, G., Flanders, K.C., Gaudio, E.: Target disruption of Smad3 confers resistance to the development of dimethylnitrosamine-induced hepatic fibrosis in mice. Liver Int. 997–1009 (2009). doi:10. 1111/j.1478-3231.2009.02011.x

19. De Santis, A., Iacoviello, D.: A discrete level set approach for image segmentation. Sig. Image Video Process. **1**, 303–320 (2007)

20. Efron, B.: Estimating the error rate of a prediction rule: improvement on cross-validation. J. Am. Stat. Assoc. **78**(382), 316–331 (1983)

21. Chang, C.C., Lin, C.J.: LIBSVM: A library for support vector machines. ACM Trans. Intell. Syst. Technol. **2**(3), 1–27 (2011). Article no. 27

Improving Face Recognition in Low Quality Video Sequences: Single Frame vs Multi-frame Super-Resolution

Andrea Apicella, Francesco Isgrò, and Daniel Riccio$^{(\boxtimes)}$

Università degli Studi di Napoli Federico II, Naples, Italy
and.api87@gmail.com, {francesco.isgro,daniel.riccio}@unina.it

Abstract. Re-Identification aims to detect the presence of a subject spotted in one video in other videos. Traditional methods use information extracted from single frames like color, clothes, etc. A sequence in time domain of consecutive subject images could contain a greater amount of information compared with a single image of the same subject. Typically, these sequences are taken from surveillance cameras at very poor resolution. Even with modern cameras the resolution can be a problem when dealing with a subject who is far from the camera. A possible way of handling low resolution images is by using a multi-frame super-resolution algorithm. Multi-frame super-resolution image reconstruction aims at obtaining a high-resolution image by fusing a set of low-resolution images. Low-resolution images are usually subject to some degradation which causes substantial information loss. Therefore, contiguous images in a sequence could be viewed as a degraded version (SR image) of an image at higher resolution (HR image). Using a multi-frame SR algorithm could achieve a restoration of the HR image. This work aims to investigate the possibility of using a multi-frame super-resolution algorithm to enhance the performance of a classic re-identification system by exploiting information provided by video sequences made available by a video surveillance system. In the case that the SR technique employed results in an effective performance enhancement, we intend to show empirically how many match frames are required to have an effective improvement.

1 Introduction

The creating, broadcasting and archiving of information in a video format is a growing phenomenon that is a direct consequence of the reduction in the cost of technology and the increase of the available network bandwidth. The availability of adsl/vdsl home connections with a large bandwidth has opened up a new class of services, such as IPTV, with the consequence that a large number of data streams need to be managed and organised. This is particularly true for video-surveillance systems, where a large amount of video data needs to be analysed, a requirement which has in the last few years stimulated research in the area of video analytics. In the field of video-surveillance a very important topic is

© Springer International Publishing AG 2017
S. Battiato et al. (Eds.): ICIAP 2017, Part I, LNCS 10484, pp. 637–647, 2017.
https://doi.org/10.1007/978-3-319-68560-1_57

face recognition, both for the identification of the person in the scene, and for the re-identification of subjects from among different video footage captured at different times and/or places. The two problems have very different objectives. The former aims at assigning an identity to a subject detected in the scene. The target of the latter is to decide whether it is the same subject appearing in different videos, without considering the subject's identity. Despite the increase in camera resolution and video quality which has occurred over the last few years, the problem of image resolution in the context of face recognition is still an open topic of research. Many video surveillance systems do not use state-of-the-art technology and even the most modern systems are unable to recognize a face when the distance of the subject from the camera is great, with respect to the camera resolution. In this case, the face can occupy just a very small portion of the acquired image, with an effective resolution that is not sufficient for the recognition task.

Super-resolution methods can give a valid support to face-recognition systems that use low resolution video equipment, and can solve this problem fully, or partially. In the literature many different methods have been proposed [5,8,17], covering a wide range of activities. We can divide the different algorithms into two main macro-categories: single image methods, and multi-frame methods. The techniques falling into the first group try to increase the image resolution by using structures contained in the image itself; conversely, the techniques in the second group aim at obtaining an image with greater informative content by combining many observations of the same scene taken at different moments. A natural benchmark for all these algorithms is the classic technique of zooming (e.g., linear interpolation), which does not need any assumption on the image content. It is worth mentioning that a zoomed image obtained from an interpolation method has merely a larger number of pixels than the original, but this does not correspond necessarily to a greater informative content. However, it is the obtaining of better informative content that can be a key element for the improvement of the performance of a pattern recognition task, such as, in this case, face recognition. This paper introduces a super-resolution method in the pipeline of a biometric face-recognition task. In particular, it provides various innovative contributions compared to similar proposals already present in literature. First and foremost, two different super-resolution methods are compared: the first operating on a single image, the second based on the combination of consecutive frames. The second contribution is the analysis of the performance of the two super-resolution methods on two different face-recognition frameworks: the first is based on local feature extraction computed at pixel level, while the second works on patches of greater size with respect to a 3×3 mask.

The goal is to show how super-resolution techniques can have a better performance when using global recognition methods. Finally, the super-resolution method [14] adopted in this work does not require any face registration, which is a typical limitation of the majority of the techniques proposed in the literature.

The paper is structured as follows. Related works are briefly reviwed in the next section, Sect. 2. The architecture proposed is described in Sect. 3, and the experimental results are reported in Sect. 4. Section 5 concludes the paper with the final remarks.

2   Related Work

Over the last few years Person Re-identification has posed a significant challenge. One of the main difficulties is the low resolution of old cameras that can make every traditional technique to improve image quality unworkable. Person re-identification methods can be divided into two main groups:

- *single frame methods*, that aim to extract information about a person by analysing a single image;
- *multi-frame methods*, that use multiple images of the same person (usually obtained from one or more sequences) to build his/her signature.

For the first class, color and histogram-like methods have proved to be well suited for the retrieval of images with similar content, as in [6, 8, 11]; the main drawback of histogram based methods is the lack of any geometric or spatial information. In [5] the silhouettes of people are segmented into multiple horizontal stripes, and then color features are computed to characterize each segment. In [16] color features, together with a set of SURF points of interest, are extracted from the images and used to build a *person's descriptor*. Other techniques exploit the availability of other sources of information, such as the color of the clothes the subject is wearing, biometrics or collateral features, such as gait [15]. Such methods suffer from several drawbacks, like enlightenment sensitivity or pose changes, in addition to possible occlusions in the field of view.

On the other hand, multi-frame methods, like [10], collect several views from different cameras and build feature based on a variant of the SURF points. Authors in [1] adopt a cascade of grids of common region descriptors (e.g., SURF, SIFT). In [4] the use of a sequence of frames from a video instead of single still frames provides a significant increase in the performance.

In recent years the progress in camera technology, that can now record videos at a high resolution, has opened up the possibility of exploring unexploited paths in this research field, like the adoption of super-resolution algorithms in the face recognition [3, 9, 18] and re-identification pipeline. For the latter, previous works that try to take advantage of the SR algorithm include [2], that proposes a procedure for the recognition of low-resolution faces by using the features extracted from a high-resolution training employed as prior information in a super-resolution algorithm, and [13] that learns a pair of HR and LR dictionaries to generate a mapping function from the features of HR and LR training images. With the learned dictionary pair and mapping function, the features of LR images can be converted into discriminating HR features.

3   System Architecture

In a traditional video-surveillance system, different sequences taken from cameras are used to verify which identities declared for a set of people are true. Usually, the system is based on the individual biometric keys used to identify a single person; the key signatures are not invariant to sequence conditions and the image quality can be a factor that can affect the correctness of the identification. We aim to show how the introduction of a super-resolution algorithm in a classic face-based recognition framework can improve the performance. For this purpose, we add a super-resolution algorithm to a classic recognition pipeline (see Fig. 1), obtaining the following configuration:

1. two (or more) low-resolution cameras, each observing different not overlapping areas;
2. a super-resolution algorithm;
3. a bio-metric key producer (in our case a feature extractor);
4. a score function $S(k_1, k_2) \rightarrow \mathbb{R}$ used to indicate if the identification proposed is accepted or rejected by the system.

So, instead of using raw images taken from image sequences produced by cameras, we compute a higher quality image constructed by sequence fragments. More formally, given a gallery set G taken from image sequences of a set of people at a given resolution r_G, and a probe set P taken from image sequences at a given resolution r_P, our approach is based on extrapolating n contiguous frames from every sequence in G and generating a higher resolution image for each of them by using a given multi-frame super resolution algorithm. We indicate as G_{SR}^n the image set obtained. Subsequently, the same process is applied to P taking m contiguous frames from each sequence, and we indicate the resulting set as P_{SR}^m.

This process can be repeated varying m and n in order to compare performances. The resulting G_{SR} and P_{SR} sets are then used in a re-identification task with the two different matching schemes described in Sect. 3.2. The objective of this work is to show not only how SR can improve the performances of

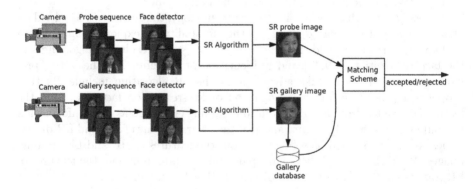

Fig. 1. Architecture of the system proposed.

a generic re-identification system, but also how the temporal information taken from different frames of the same sequence can affect the results.

3.1 Super-Resolution Algorithm

Our framework requires a multi-frame SR algorithm to synthesize the information provided by multiple frames of the same sequence in a single shot at a higher resolution. Super-resolution addresses the problem of reconstructing high-resolution data from a single or multiple low resolution observations. The key idea is based on the observation that every low-resolution image can contain different information on the same subject and that the fusion of these images can make it possible to extract subpixel information from the low-resolution image. Although the main focus of super-resolution methods is to obtain higher resolution images from low-resolution sequences, techniques of image restoration and image enhancement are also under consideration. SR techniques can be classified by using two parameters:

- methods that work in different domains (spatial/frequency)
- methods that work on the number of frames used in the restoration process (single/multiple)

Single frame methods can be considered equivalent to image interpolation, where the information in the image can be taken from the image itself. In our opinion, the state-of-the-art has been achieved by [12]; briefly, this study aims to search for similar patches in one image and computes the best homography which can generate the same image at a higher resolution. The following sequence of steps can clarify how it works:

Given an image I_O, to obtain an image I_H of a higher resolution of a k factor;

1. downsample the image of a k factor obtaining I_L image at a lower resolution
2. split images I_O and I_L in patches
3. for every patch P_O of I_O, compute a homography T s.t. $Q_L = T(P_O)$ where Q_L is the best patch matching I_L using a distance measure.
4. extract from I_O the Q_O patch that matches to Q_L in I_L
5. compute the inverse transformation T^{-1}
6. use T^{-1} to obtain P_H from Q_O, which is the patch in I_H that matches the position of P_O.

This first method is used to obtain SR single frame images.

Instead, multi-frame resolution enhancement is used when the information is taken from multiple frames. We can suppose that every image of the same subject owns inside it a certain information load; by contrast, multi.frame algorithms require a more complex pipeline to obtain significant results consisting of crucial intermediate steps like the registration between images; multi-frame methods often involve auxiliary algorithms taken from object detection or image registration. The authors in [14] propose a Bayesian method that seems to obtain good results, for which reason we have used it in our experiments. It aims to estimate the best high resolution image that can generate a set of low-resolution images of the same scene using a probabilistic approach. In our experiments, we have compared the performances using a single-frame SR algorithm (i.e. [12]) and a multi-frame SR algorithm (i.e. [14]).

In Fig. 2 we show the output from both algorithms relative to a frame of a video sequence.

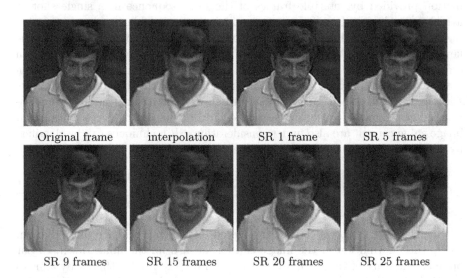

Original frame   interpolation   SR 1 frame   SR 5 frames

SR 9 frames   SR 15 frames   SR 20 frames   SR 25 frames

Fig. 2. Results of the SR algorithms considered. Top left to bottom right: orginal full resolution frame, result obtained using bi-linear interpolation, single frame SR, and multi-frame results using from 5 to 25 frames.

3.2 Feature Extraction and Matching Scheme

Once we have the SR images, we need a biometric key in order to discriminate in a unique manner a single face; our experiments have been performed with two different set-ups: the former using as the picture key-signature *LBP features* [2], a well-known type of global descriptor used in computer vision and face recognition, and the latter using the *spatial correlation index* [7].

- **LBP**: given an image I, the LBP operator assigns a label to every pixel by thresholding the 3×3-neighbourhood of each pixel with the central pixel value and considering the result as a binary number. Then, the labels histogram can be used as a descriptor. The facial image is divided into local regions and each descriptor is extracted from each region independently. All the descriptors are then concatenated to form a unique global face description $LBP(I)$.

    Once we have the keys for every SR_P and SR_G set, we need a matching scheme to compare the extracted bio-metric key; The similarity between two images I_1 and I_2 can be computed as the cosine similarity between $LBP(I_1)$ and $LBP(I_2)$.

- **Spatial Correlation**: the matching is performed by a localized version of the Spatial Correlation index. Given two images I_1, I_2 and the respective means $\mathbb{E}(I_1), \mathbb{E}(I_2)$, defining S as

$$s(I_1, I_2) = \frac{\sum_i \sum_j (I_1(i,j) - \mathbb{E}(I_1))(I_2(i,j) - \mathbb{E}(I_2))}{\sqrt{\sum_i \sum_j (I_1(i,j) - \mathbb{E}(I_1))^2 (I_2(i,j) - \mathbb{E}(I_2))^2}}$$

In our case, we divide all the images into subregions; for each subregion $r_1 \in I_1$ we search, in a subwindow around the same position in I_2, the region $r_2 \in I_2$ that maximizes $S(r_1, r_2)$. The global correlation is then obtained as

$$S_g = \sum_{r_1} s(r_1, (\arg\max_{r_2} s(r_1, r_2))).$$

The similarity between the two images I_1 and I_2 can be computed as $S_g(I_1, I_2)$.

4 Experimental Assessment

We decided to conduct an experimental evaluation of our proposed method on the publicly available ChokePoint video dataset[1]: this consists of videos of 29 subjects (23 male and 6 female) taken from different cameras. The videos in the dataset have a frame rate of 30 fps, and the frame resolution is 800×600 pixels. In total, the dataset consists of 48 video sequences, and $64,204$ face images. In all the sequences, only one subject is present in every image at any one time.

4.1 Description of Experiments

The ChokePoint dataset gives us a set of contiguous frame sequences taken from a camera at a fixed resolution of 800×600; every sequence contains images of a moving person the face of whom can be enclosed in a bounding box of about 80×80; this resolution is too high for our purposes, so we resize every face to 40×40, obtaining what we consider our face-gold or Original Set $S_{original}$. Next, we randomly select a subset of 25 subjects and, for each of them, we take two distinct sequences, forming respectively the Gallery set $G_{original}$ and the Probe set $P_{original}$. We perform two sets of experiments, the former using the SR single-frame approach, the latter using the SR multi-frame approach. The performances are evaluated with the Genuine Acceptance Rate vs False Acceptance Rate (GAR-FAR) curves, and Cumulative Matching Curves (CMC). We subsample $G_{original}$ of a given factor d obtaining a set of low resolution sequences $G_{LR}^{\downarrow d}$. Next, we aim to restore the original resolution by applying an up-sample of the same factor using a standard interpolation algorithm obtaining the $G_{HR}^{\uparrow d}$ set. So, we can now use the SR algorithms to obtain our test-cases.

---

[1] Available from http://arma.sourceforge.net/chokepoint/.

The SR Single-Frame Approach: For every sequence in $G_{LR}^{\downarrow d}$, we select a frame where the face appears close to the camera and in a frontal pose, and choose this as a single frame. We then use the SR algorithm proposed by [12] to obtain a first set of super-resolved images G_{SR}^1. Analogously, we build P_{SR}^1.

The SR Multi-frame Approach: For every sequence in $G_{LR}^{\downarrow d}$, we select a sub-sequence (close to the conditions used for selecting the frame for the single-frame case) of the n_G contiguous frame that we want to combine together using the SR algorithm proposed by [14]; n_G is taken alternatively with values of $5, 9, 15$ and 20, obtaining $G_{SR}^5, G_{SR}^9, G_{SR}^{15}$ and G_{SR}^{20} sets. Analogously, we build $P_{SR}^5, P_{SR}^9, P_{SR}^{15}$ and P_{SR}^{20} sets.

Our objective is to evaluate how the identification performance changes when using

- SR images (P_{SR}, G_{SR});
- images at low resolution (P_{LR}, G_{LR});
- direct high-resolution images ($P_{original}$, $G_{original}$).

We use the distances discussed in Sect. 3.2, and we compare the G_{SR} sets versus the P_{SR} sets.

4.2 Results

As a first experiment we compared the performance when using the multi-frame SR with a different number of frames. Figures 3 and 4 show the results for the Spatial Correlation and LBP, respectively. The graphs show that using 5 frames gives, in general, the best performance. This is reasonable if we consider that the more frames we use, the more the pose of the subject can change, making the information *fusion* between the frames less coherent.

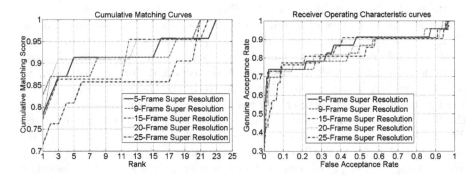

Fig. 3. Performance when using spatial correlation and the multi-frame SR algorithm with a different number of frames. Left: CMC curves. Right: GAR-FAR curves.

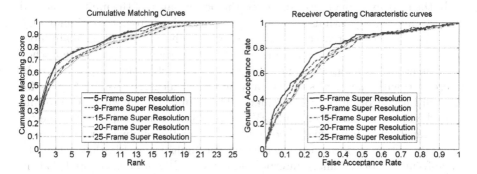

Fig. 4. Performance when using LBP and the multi-frame SR algorithm with a different number of frames. Left: CMC curves. Right: GAR-FAR curves.

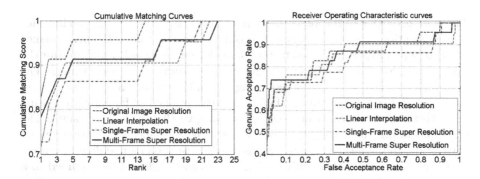

Fig. 5. Performance of the two SR algorithms against original HR images and interpolated images. Left: CMC curves. Right: GAR-FAR curves. Spatial correlation was used for this experiment.

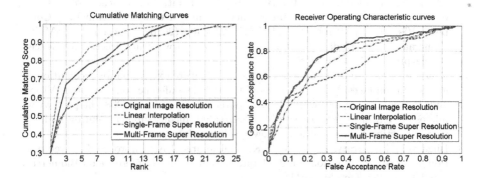

Fig. 6. Performance of the two SR algorithms against original HR images and interpolated images. Left: CMC curves. Right: GAR-FAR curves. LBP was used for this experiment.

In the next experiment we compared the performance of the multi-frame and single frame SR algorithms against the ones obtained from the original high resolution images. Moreover, in order to verify if the SR algorithm really does produce a positive result, we also measured the performance of the system when using images with a super-resolution with a simple interpolation procedure. In accordance with the results of the previous experiment, we used 5 frames for the multi-frame SR in this experiment. The results are shown in Figs. 5 and 6 for the Spatial Correlation and LBP, respectively. The results show a better performance when using spatial correlation. As for the resolution we can notice a slightly better performance when using the multi-frame SR than when the single-frame SR or the image interpolation is used. It can be also noted that, in general, the single-frame SR gives worse results than the image interpolation.

5   Conclusions

The work presented in this paper investigates the use of super-resolution in a face re-identification system. To achieve this objective we compared the performance of two different super-resolution methods, the former using only one frame for the super-resolution, the latter fusing the information from a sequence of frames. The performances have been measured using two different face-recognition frameworks: spatial correlation and LBP. The results obtained so far show that using a multi-frame super-resolution algorithm can give slightly better results. The experiments presented here also show a better performance when using spatial correlation. In the future our system needs to be tested on more difficult datasets than the one used for this study, and the experiments should be extended to a wider range of features.

References

1. Alahi, A., Vandergheynst, P., Bierlaire, M., Kunt, M.: Cascade of descriptors to detect and track objects across any network of cameras. Comput. Vis. Image Underst. **114**(6), 624–640 (2010). SpecialIssueonMulti-CameraandMulti-ModalSensorFusion. http://www.sciencedirect.com/science/article/pii/S1077314210000275
2. Baker, S.: Simultaneous super-resolution and feature extraction for recognition of low resolution faces. In: Proceedings of the IEEE Conference on Computer Vision and Pattern Recognition. IEEE Computer Society, June 2008. https://www.microsoft.com/en-us/research/publication/simultaneous-super-resolution-and-feature-extraction-for-recognition-of-low-resolution-faces/
3. Bilgazyev, E., Efraty, B., Shah, S.K., Kakadiaris, I.A.: Improved face recognition using super-resolution. In: 2011 International Joint Conference on Biometrics (IJCB), pp. 1–7. October 2011
4. Buml, M., Stiefelhagen, R.: Evaluation of local features for person re-identification in image sequences. In: 2011 8th IEEE International Conference on Advanced Video and Signal Based Surveillance (AVSS), pp. 291–296. August 2011

5. Cong, D.N.T., Achard, C., Khoudour, L.: People re-identification by classification of silhouettes based on sparse representation. In: 2010 2nd International Conference on Image Processing Theory, Tools and Applications, pp. 60–65. July 2010
6. D'angelo, A., Dugelay, J.L.: People re-identification in camera networks based on probabilistic color histograms. In: 3DIP 2011 Electronic Imaging Conference on 3D Image Processing and Applications, January 2011, San Francisco, CA, USA, vol. 7882, pp. 23–27 (2011). http://www.eurecom.fr/publication/3274
7. De Marsico, M., Nappi, M., Riccio, D.: Face: face analysis for commercial entities. In: 2010 IEEE International Conference on Image Processing, pp. 1597–1600. September 2010
8. Forssen, P.E.: Maximally stable colour regions for recognition and matching. In: 2007 IEEE Conference on Computer Vision and Pattern Recognition, pp. 1–8. June 2007
9. Gunturk, B.K., Batur, A.U., Altunbasak, Y., Hayes, M.H., Mersereau, R.M.: Eigenface-domain super-resolution for face recognition. IEEE Trans. Image Process. **12**(5), 597–606 (2003)
10. Hamdoun, O., Moutarde, F., Stanciulescu, B., Steux, B.: Person re-identification in multi-camera system by signature based on interest point descriptors collected on short video sequences. In: 2008 Second ACM/IEEE International Conference on Distributed Smart Cameras. pp. 1–6. September 2008
11. Hatakeyama, Y., Makino, M., Mitsuta, A., Hirota, K.: Detection algorithm for color image by multiple surveillance camera under low illumination based-on fuzzy corresponding map. In: 2007 IEEE International Fuzzy Systems Conference, pp. 1–6. July 2007
12. Huang, J.B., Singh, A., Ahuja, N.: Single image super-resolution from transformed self-exemplars. In: CVPR (2015)
13. Jing, X.Y., Zhu, X., Wu, F., You, X., Liu, Q., Yue, D., Hu, R., Xu, B.: Super-resolution person re-identification with semi-coupled low-rank discriminant dictionary learning. In: The IEEE Conference on Computer Vision and Pattern Recognition (CVPR). June 2015
14. Liu, C., Sun, D.: On Bayesian adaptive video super resolution. IEEE Trans. Pattern Anal. Mach. Intell. **36**(2), 346–360 (2014)
15. Man, J., Bhanu, B.: Individual recognition using gait energy image. IEEE Trans. Pattern Anal. Mach. Intell. **28**(2), 316–322 (2006)
16. de Oliveira, I.O., de Souza Pio, J.L.: People reidentification in a camera network. In: 2009 Eighth IEEE International Conference on Dependable, Autonomic and Secure Computing, pp. 461–466. December 2009
17. Park, S.C., Park, M.K., Kang, M.G.: Super-resolution image reconstruction: a technical overview. IEEE Signal Process. Mag. **20**(3), 21–36 (2003)
18. Wheeler, F.W., Liu, X., Tu, P.H.: Multi-frame super-resolution for face recognition. In: 2007 First IEEE International Conference on Biometrics: Theory, Applications, and Systems, pp. 1–6. September 2007

Learning to Weight Color and Depth for RGB-D Visual Search

Alioscia Petrelli[(⊠)] and Luigi Di Stefano

University of Bologna, Bologna, Italy
{alioscia.petrelli,luigi.distefano}@unibo.it
http://vision.deis.unibo.it

Abstract. Both color and depth information may be deployed to seek by content through RGB-D imagery. Previous works dealing with global descriptors for RGB-D images advocate a decision level fusion whereby independently computed color and depth representations are juxtaposed to pursue similarity search. Differently, in this paper we propose a *learning-to-rank* paradigm aimed at weighting the two information channels according to the specific traits of the task and data at hand, thereby effortlessly addressing the potential diversity across applications. In particular, we propose a novel method, referred to as *kNN-rank*, which can learn the regularities among the outputs yielded by similarity-based queries. A further novel contribution of this paper concerns the *Hyper-RGBD* framework, a set of tools conceived to enable seamless aggregation of existing RGB-D datasets in order to obtain new data featuring desired peculiarities and cardinality.

Keywords: RGB-D image search · Compact descriptors · Learning-to-rank

1 Introduction

Encoding image content into compact though distinctive representations is key to retrieval performance in large-scale visual search. To pursue visual search one would typically match the query image against those stored in a database by comparing global image representations, so as to receive the digital content linked to the most similar one. In this realm, numerous works, such as [4,14,18], address how to represent images by short binary codes conducive to efficient matching and storage when dealing with large-size databases.

Reliance on compact binary representations is an essential trait in mobile visual search alike. Here, the image acquired by a mobile device's camera is encoded and transmitted via a wireless network to a remote server undertaking database search. Therefore, bandwidth constraints mandate the images sent to the server to be represented as compactly as possible. How to design an effective mobile visual search architecture leveraging on compact image representations has been addressed in several research papers [4,7,9] as well as in the recently defined *Compact descriptors for visual search* (CDVS) standard by the MPEG group.

© Springer International Publishing AG 2017
S. Battiato et al. (Eds.): ICIAP 2017, Part I, LNCS 10484, pp. 648–659, 2017.
https://doi.org/10.1007/978-3-319-68560-1_58

Similar technology trends and research challenges are likely to become increasingly relevant in the field of RGB-D imagery. Indeed, broad diffusion of consumer depth cameras has enabled the creation of a few relatively large-size RGB-D datasets comprising thousands or tens of thousands images. Moreover, mobile devices start being endowed with the ability to sense depths, either by mountable cameras, like *Structure* by *Occipital*, or fully integrated sensors, e.g. as provided by Google's *Project Tango* technology which, in particular, is on the verge of deployment in off-the-shelf smartphones. Hence, one might be lead to foresee more and more large RGD-D datasets to become available as well as the emergence of applications performing Visual Search via RGB-D images taken by mobile devices. The above trends, thus, are likely to foster considerable research efforts towards the novel topic of compact binary representations for RGB-D visual search.

The work described in [15,16] proposes the first investigation on how to globally represent RGB-D images by compact binary codes. The experimental analyses reveal that encoding of depths is key to recognize object categories, whereas object instances are mainly told apart based on RGB information. More interestingly, though, the authors highlight how different tasks and datasets exhibit different peculiarities, so that, in general, naively chaining together the binary codes associated with color and depth yields sub-optimal performance. Rather, an effective approach to RGB-D visual search should pursue automatic learning of the relative prominence of color and depth in the addressed scenario.

In information retrieval, the *learning-to-rank* paradigm provides a sound framework to combine different strategies by learning a model that fuses into a joint ranking the individual rankings yielded independently by the different strategies. Learning-to-rank approaches perform a supervised learning aimed at discovering which strategies produce better rankings in the addressed scenario and, accordingly, learn how to weight properly the individual rankings into the final one. Such paradigm has been deployed in Content-Based Image Retrieval[1] to weight the contributions of different feature kinds extracted from RGB images.

In this paper, we propose the first investigation dealing with application of the learning-to-rank paradigm to RGB-D visual search by binary codes. In particular, we propose and apply to the architecture described in [16] a novel learning-to-rank approach, dubbed *kNN-rank*. This approach tries to obtain a joint ranking for the given query by learning the regularities within the k-NNs retrieved by matching color and depth codes, such regularities concerning both the types of object found as neighbors as well as the associated distances. Intuitively, if we query by a yellow cup we might retrieve cups based on depth and bananas based on color, so that we would wish to learn to ignore the color channel when aiming at category recognition while positively weighting it when willing to recognize that specific cup.

Although a few relatively large RGB-D datasets are available nowadays, their size is far smaller than that of state-of-the-art RGB datasets. To facilitate

---

[1] Here, unlike visual search, the task is to provide the user several images similar to the query.

experimentation with larger and diverse datasets, a second novel contribution of this paper concerns a software framework, referred to as *HyperRGBD*, that allows researchers to create straightforwardly new data with desired traits and peculiarities by mixing arbitrarily and seamlessly images drawn from different RGB-D datasets.

2 Previous Work

In the last few years, many papers have addressed the task of object recognition from RGB-D images. Most of them [2,8,17] fuse color and depth data at feature level through either hand-crafted descriptors or deep learning approaches. Such rich representations are then fed to a classifier (e.g. a SVM) trained to recognize the content of the query image. Differently, in [3] depth and color information are fused at decision level. Indeed, both color and depth are represented by eight different descriptors and a specific SVM is trained for each feature type and object category. The final decision is taken by a neural network presented with the output of all these SVMs.

Other works are focused on how to weight the contribution of color and depth as well as of diverse shape cues. [11] adopts an AdaBoost learning procedure to weight color and depth for the task of face recognition, whereas [13] analyzes different strategies for weighting five different 3D descriptors on the *Princeton Shape Benchmark*. In [1], Bar-Hillel et al. propose the O^2NBNN framework that describes images through multiple channels encoding intensity, depth information or a feature level fusion of the two contributions. At training time, an optimization allows for learning the proper weights for each class and channel that are, then, used to predict the object class from the query image. However, all the above mentioned methods rely on rich, high-dimensional descriptors and leverage on classifiers, while in the realm of visual search one would typically rely on compact representations and perform a similarity search across the database.

Learning-to-rank has been effectively applied in RGB-based image retrieval. [12] quantitatively compares three different approaches (pointwise, pairwise and listwise) on four datasets. The work in [6] applies and compare *Ranking SVM*, Genetic algorithms and Association Rules for ranking eighteen types of descriptors (color, texture and shape based) on two RGB datasets. To the best of our knowledge, the only work that exploits a learning-to-rank paradigm to fuse color and depth data has been recently described in [5]. The method measures the similarity between a query and a reference image by means of an ensemble of dense matchings that weight differently the features extracted from color and depth data. Then, the scores obtained by dense matchings are ranked through *Ranking SVM* [10]. However, this approach is not conceived for large-scale visual search but to re-rank a set of candidates priorly identified by a classifier, such as the algorithm proposed in [17]. Moreover, it would not be applicable to mobile scenarios due to the requirement of sending to the remote server the full RGB-D image rather than just a compact binary code.

3   Visual Search Architecture

In this section we outline the visual search architecture proposed in [16] and deployed in this paper to apply learning-to-rank methods for RGB-D image search. First, a set of patches are extracted densely from the query RGB-D image and described through Kernel Descriptors. In particular, the appearance information associated with each patch is represented by kernels dealing with intensity gradients (KD_I) and color (KD_C), while 3D shape information is captured by kernels encoding depth gradients (KD_D) and *Spin Images* descriptors (KD_S). Then, these local features are aggregated into a global image description by *Fisher Kernel*. Finally, the *Spherical hashing* algorithm provides the compact binary code used to carry out similarity search within the image database. The experimental analysis in [16] highlights that the information extracted from the depth and color images should better be aggregated at decision rather than feature level. Accordingly, the four Kernel Descriptors (KD_I, KD_C, KD_D, KD_S) are computed, aggregated and hashed separately, so as to end up with four binary codes, referred to as B_I, B_C, B_D and B_S, that are simply juxtaposed to create the final tag, B, deployed to seek for the most similar image within the database. Then, an object instance (category) gets recognized correctly if the most similar database image retrieved by matching the binary tag comes from the same instance (category) as the query image. Comparison between binary tags is achieved by the fast Hamming distance and the search performed efficiently by indexing the database through the *multi-probe LSH* scheme. Finally, the matching process is robustified by the weighted k-NN classifier ($k = 9$).

However, as highlighted in [15,16], simple juxtaposition of the binary codes hardly succeeds in capturing the diverse distinctiveness that the deployed feature channels may convey in different tasks and datasets. Accordingly, the next section describes an approach aimed at learning to weight the relative contributions of the individual binary codes in order to seamlessly adapt the pipeline to the peculiarities of the addressed scenario.

4   The kNN-rank Approach

Figure 1 allows for visualizing the results of a query carried out on the *Washington* dataset by the visual search architecture proposed in [16]. It can be observed that the binary codes dealing with depth information (B_D, B_S) succeed in identifying the correct category, whilst this is not the case of those extracted from the RGB image (B_I, B_C). In particular, matching based on color (B_C) mistakes the *bowl* for a *cup* due to the very similar texture patterns. This, in turn, hinders the final matching based on the juxtaposed codes, B, which returns a wrong category (i.e. *cup*).

However, had we be presented with these results and been told to trust depth much more than color, we would have been able to pick the correct category. Similar observations drawn from analyzing the results of several queries lead us to the intuition that the information conveyed by retrieved images contains

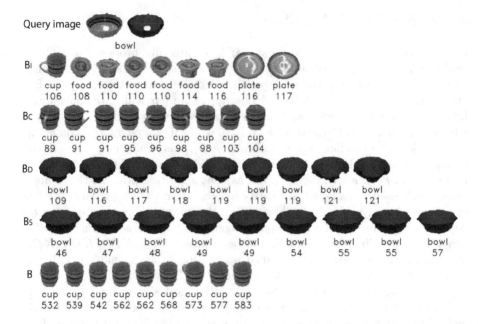

Fig. 1. Result of a query on the *Washington* dataset. The first row reports the query RGB-D image together with its associated category ("bowl"). The next four rows show the $k = 9$ most similar images according to the four binary codes (B_I, B_C, B_D and B_S). The last row shows the $k = 9$ images retrieved by the binary tag, B. The category and Hamming distance from the query image are shown below each retrieved image.

regularities that may be exploited in order to learn how to make decisions aware of the specific scenario and data.

Accordingly, this section describes a novel learning-to-rank method, dubbed *kNN-rank*, that, given the results of a query, defines a set of feature vectors based on both the labels (either instance or category labels, depending on the recognition task) and distances of retrieved images. In particular, a feature vector is created for each different retrieved label, such feature vectors used at training time to learn a ranking function while at test time to rank the label with respect to the query.

More in detail, the labels relevant to a query are those retrieved either by each of the individual binary codes or by juxtaposing them. For example, for the query illustrated in Fig. 1, the relevant labels are "cup", "food", "plate" and "bowl". Then, given a relevant label, l_i, an associated feature vector, x_i, is assembled by computing a pair of features for each retrieved image. The first feature in the pair encodes whether the corresponding image is labeled as l_i or not: in the former case, it is "fired" and equal to the measured Hamming distance, in the latter it is set to zero. Considering the exemplar query of Fig. 1 and label "cup", the first feature of the pair for each retrieved image is shown in blue on the left side of Fig. 2. Conversely, the second feature is fired, i.e. equal

106	0	0	0	0	0	0	0	0	0	108	110	110	110	114	116	116	117
89	91	91	95	96	98	98	103	104	0	0	0	0	0	0	0	0	0
0	0	0	0	0	0	0	0	0	109	116	117	118	119	119	119	121	121
0	0	0	0	0	0	0	0	0	46	47	48	49	49	54	55	55	57
532	539	542	562	562	568	573	577	583	0	0	0	0	0	0	0	0	0

Fig. 2. Feature vector produced by the *kNN-rank* method for label "cup" according to the query results depicted in Fig. 1. Each row deals with the images retrieved based on a different code (i.e. B_I, B_C, B_D, B_S, B) and consists of $2 \times k$ elements. "Blue" features encode the Hamming distance for "cup" images whereas "green" features the Hamming distance for "non-cup" images. (Color figure online)

to the measured Hamming distance, for all the retrieved images showing labels other than l_i. Considering again the query of Fig. 1 and label "cup", the second feature of the pair for each retrieved image is shown in green on the right side of Fig. 2.

Similarly to the Ranking SVM approach [10], we solve a binary classification problem. More precisely, at training time we randomly select N images from the database to be treated as queries. For each query, we apply a k-nn search in the database based on B_I, B_C, B_D, B_S and B. As described, we create a feature vector for each relevant label, l_i, and then assign either $+1$ or -1 to each feature vector based on whether l_i is correct or wrong for the query. These samples are normalized to similarity scores in the interval $[0,1]$ and used to train a linear SVM. In particular, denoted as $x_i(d)$, $d \in \{I, C, D, S\}$, the Hamming distances associated with the four binary codes, the corresponding normalized features are given by

$$\tilde{x}_i(d) = \frac{\max_i x_i(d) - x_i(d)}{\max_i x_i(d)} \qquad (1)$$

At test time, given a query, each relevant label l_i is ranked with respect to the query according to the score computed by the trained SVM:

$$f(l_i) = <w, x_i> \qquad (2)$$

5   The HyperRGBD Framework

This section outlines a C++ software framework, referred to as *HyperRGBD*, devised to enable researchers and practitioners to build effortlessly new datasets by aggregating images from different existing RGB-D datasets. For example, one might wish to experiment with datasets larger than existing ones, which would seamlessly be attainable by deploying *HyperRGBD* to aggregate the images belonging to existing datasets into a larger data corpus. Furthermore, should a dataset be biased towards certain abundant categories with others featuring a few samples only, it would be just as seamless to build a more balanced dataset by using *HyperRGBD* to draw samples for the rare categories from other datasets. Another example may deal with changing the granularity of categories, e.g. aggregating "chair", "table" and "couch" into a broader "furniture" category or splitting "fruit" into more specific categories like "apple", "orange" and "banana". At present, we have integrated in the framework the main existing RGB-D datasets for object recognition, i.e. *Washington*, *CIN 2D+3D*, *BigBIRD* and *MV-RED*, that are briefly described in a project page[2] we make available, along with the source code of the framework, so to foster research activity on perception from RGB-D imagery and enable researchers to integrate their data.

We exploited *HyperRGBD* to obtain two new RGB-D datasets used in our experiments besides the main existing ones. We aggregated the above four datasets to create two new datasets and tested both in instance as well as category recognition scenarios. The former, *HyperRGBD*, merges all the available images. The latter, *HyperRGBD - Balanced*, addresses the wide differences in size between existing datasets by balancing them upon aggregation. More precisely, for instance recognition scenario, we identify the dataset with the fewest instances (*BigBIRD* comprising 114 instances) and level down the others by randomly selecting 114 instances per dataset. In the case of category recognition, instead, for each of the categories of the aggregated dataset, we search for the dataset providing the smallest amount of instances and, accordingly, populate the category by randomly selecting that amount of instances from each dataset. Once the datasets are gathered, both for category and instance recognition, a tenth of the dataset is used as test set and the remaining to perform the training. The procedure is repeated 10 times on different randomly generated test sets so to obtain 10 different trials.

6   Experimental Evaluation

To assess the ability of the novel *kNN-rank* method to properly weight color and depth channels across different tasks and data, our experimental evaluation compares it against the *SVMrank* approach proposed in [10], a *Ranking SVM* formulation that has proved to be effective in a variety of real settings. As a baseline, we also include in the evaluation the matching of juxtaposed binary codes encoding depth and color information, as delineated in previous work dealing with RGB-D visual search [15,16].

In the experiments reported in this section, queries and database images are encoded by allocating 512 bits to each of the four binary codes (B_I, B_C, B_D, B_S), so that the final tag, B, gets as large as 2048 bits. Indeed, extensive experimental investigation showed that longer descriptions would not provide significant improvement in the recognition capability of the architecture. Furthermore, even though recognition rates decrease as the description length decreases, the ranking between the approaches considered in this section remains identical. We also report the recognition performance achieved by individually matching binary

---

[2] http://www.vision.disi.unibo.it/research/78-cvlab/107-hyperrgbd.

codes B_I, B_C, B_D, B_S, which in these kinds of experiment are given the same length (2048 bits) as B.

In the *SVMrank* approach settings, the feature vectors x_i are four-dimensional and consist of the four Hamming distances between the binary codes B_I, B_C, B_D, B_S computed from query and database images as described in Sect. 3. To perform the training we randomly select N images from the database to be treated as queries; then, for each query, we randomly pick 500 relevant images (i.e. for which the category/instance is the same of the query image) and equally many irrelevant ones, so as to create pairs of feature vectors x_i, x_j dealing with the same query in which one is associated to a relevant image and the other to an irrelevant one. Thereby, the binary classifier can be provided with training samples according to the standard formulation of the *Ranking SVM* approach. At query time we avoid the computation of the ranking score for all database images and instead rank only a subset of candidates. Purposely, we individually match the four binary codes B_I, B_C, B_D, B_S to identify, for each, the k most similar images. Moreover, we match the tag given by juxtaposing the four binary codes, B, to retrieve equally many images. The final set of candidates is the union of these retrieved images (i.e., at most $k \times 5$ images).

Both for *SVMrank* and *kNN-rank*, to perform training, we extract $N = 2000$ images treated as queries and, as suggested in [16], similarity searches have been performed by setting $k = 9$ for all the methods.

6.1 Results

Table 1 summarizes all the results obtained by our quantitative evaluation on all the available datasets in both category and instance recognition tasks. We evaluate performance based on the recognition rate, i.e. top-1 accuracy, as this is the standard metric concerning visual search scenarios, where one would wish

Table 1. Recognition rates obtained on the considered datasets by matching the binary codes B_I, B_C, B_D, B_S, B, by a learning-to-rank approach (*SVMrank*) and by our novel proposal (*kNN-rank*).

	Dataset	B_D	B_S	B_I	B_C	B	SVMrank	kNN-rank
Category recognition	Washington	0.568	0.550	0.531	0.450	0.791	0.778	0.801
	CIN 2D+3D	0.666	0.597	0.616	0.562	0.768	0.764	0.771
	MV-RED	0.494	0.586	0.550	0.578	0.760	0.743	0.769
	HyperRGBD - Balanced	0.518	0.521	0.500	0.462	0.732	0.719	0.748
	HyperRGBD	0.539	0.557	0.525	0.505	0.773	0.751	0.785
Instance recognition	Washington	0.342	0.393	0.489	0.845	0.843	0.882	0.879
	CIN 2D+3D	0.606	0.503	0.723	0.795	0.834	0.840	0.841
	MV-RED	0.434	0.597	0.686	0.957	0.929	0.960	0.960
	BigBIRD	0.281	0.361	0.434	0.822	0.727	0.814	0.820
	HyperRGBD - Balanced	0.434	0.493	0.619	0.896	0.831	0.884	0.894
	HyperRGBD	0.413	0.482	0.612	0.916	0.863	0.908	0.912
	Average:	0.481	0.513	0.571	0.708	0.805	0.822	0.834

to receive information linked to image content[3]. This is the metric adopted in [15,16] as well as in most previous work related to instance/category recognition from RGB-D imagery [2,5,8,17]. Each row reports the recognition rates obtained by the considered approaches on a different dataset and type of experiment (i.e. either category or instance recognition). The adopted color code allows for perceiving clearly the differences in performance as higher recognition rates are denoted by darker background colors within cells.

The comparison between the results obtained by separate deployment of the different cues (B_I, B_C, B_D, B_S) and concatenation of descriptors, B, confirms the findings already discussed in Sect. 3. As a matter of fact, fusing descriptions is clearly beneficial for category recognition, whereas, in general, much less effective to tell apart specific object instances. In the latter task, indeed, performance depends quite significantly on the specific type of data, with juxtaposition providing higher recognition rate in the *CIN 2D+3D* dataset and turning out useless with the type of objects included in the *Washington* dataset, where description based on color (B_C) suffices in delivering the highest performance. On the remaining datasets, juxtaposing representations (B) is even detrimental with respect to allocating all the available bits to color (B_C). Thus, although the simple recognition strategy based on matching juxtaposed descriptors delineated in [15,16] is overall effective, as vouched by the average figures across the first five columns (B_I, B_C, B_D, B_S, B) reported in the last row of Table 1, it turns out clearly sub-optimal in many relevant settings.

The *SVMrank* approach partly addresses such issue by providing, generally, higher recognition rates, as reported by the average recognition rate in the last row. Nonetheless, even if the method properly deals with instance recognition tasks by providing top recognition rates on all the datasets, a comparison limited to the category recognition task between *SVMrank* and B shows slightly better results in favor of the latter. Such behaviour could be ascribed to the large intra-class variability of the objects belonging to a category which renders the task more challenging than telling apart a specific object from others. *SVMrank* may not be powerful enough to learn the regularities that tie the objects of a same category.

That is not the case of the novel *kNN-rank* method introduced in this paper, that, as vouched by the last column of Table 1, can yield recognition rates higher than B also in category recognition experiments, behaves effectively on both the tasks and correctly adapts to all the datasets. The background color code permits to catch at a glance that our proposal provides the highest recognition rates on most of the datasets and ties on the others. Again, the average figures on the last row show the overall superiority of *kNN-rank*. Hence, we can conclude that learning the regularities underlying retrieved images is an effective strategy for obtaining correct rankings.

Table 1 reports also the results on the two new datasets created through the *HyperRGBD* framework. The results are coherent with those obtained on the

---

[3] Differently, in image retrieval, one is interested in receiving several images and therefore top-n accuracy is adopted.

individual datasets and highlight the good scalability of learning-to-rank methods to larger datasets. It is worth showing that *HyperRGBD* and *HyperRGBD - Balanced* are genuinely new datasets and not plain aggregations of the constituent datasets. As evidence of that, Fig. 3 reports three examples of queries performed on the *HyperRGBD - Balanced* dataset by matching the binary tag, B. The retrieved images belong to different datasets. Furthermore, in the first two examples, even though the query images belong to the *Washington* dataset, two images from the *CIN 2D+3D* are returned as top-1 result. These examples show that the *HyperRGBD* framework mixes datasets effectively and prove that the recognition rates reported in Table 1 for the *HyperRGBD* and *HyperRGBD - Balanced* are not the mere averages of the results already obtained on the other datasets.

Fig. 3. Result of three queries on the *HyperRGBD - Balanced* dataset. On the left we show the query images, whereas on the right the $k = 9$ images retrieved by matching the binary tag, B. Each image is labeled with the dataset it comes from.

7   Final Remarks

This paper shows that applying the learning-to-rank paradigm for weighting color and depth cues in RGB-D visual search does improve performance significantly and, in particular, allows for handling seamlessly diverse datasets and tasks. This is achieved by applying the novel *kNN-rank* method, that analyses the regularities in the retrieved images so as to learn the contribution conveyed by the different cues. The approach provides top performance on all the experiments we performed, both on the main existing RGB-D datasets as well as on two new datasets we created by means of the proposed *HyperRGBD* framework.

Although the *kNN-rank* method has been applied to Hamming distances of binary codes encoding color and depth cues, nothing indicates that the approach could not be successfully deployed in other contexts. So far, learning to rank methods have been applied in Content-Based Image Retrieval wherein large-scale RGB databases are encoded by numerous color, shape and texture features. Thus, we plan to test and evaluate our proposal in these settings so as to assess the ability of *kNN-rank* to scale to databases comprising million of images and to properly weight a larger number of cues.

References

1. Bar-Hillel, A., Hanukaev, D., Levi, D.: Fusing visual and range imaging for object class recognition. In: International Conference on Computer Vision, pp. 65–72 (2011)
2. Blum, M., Wulfing, J., Riedmiller, M.: A learned feature descriptor for object recognition in RGB-D data. In: International Conference on Robotics and Automation, pp. 1298–1303 (2012)
3. Browatzki, B., Fischer, J.: Going into depth: evaluating 2D and 3D cues for object classification on a new, large-scale object dataset. In: International Conference on Computer Vision Workshops (2011)
4. Chandrasekhar, V., Lin, J., Morere, O., Veillard, A., Goh, H.: Compact global descriptors for visual search. In: Data Compression Conference, pp. 333–342 (2015)
5. Cheng, Y., Cai, R., Zhang, C., Li, Z., Zhao, X., Huang, K., Rui, Y.: Query adaptive similarity measure for RGB-D object recognition. In: International Conference on Computer Vision, pp. 145–153 (2015)
6. Faria, F.F., Veloso, A., Almeida, H.M., Valle, E., Torres, R.D.S., Gonçalves, M.A., Meira, W.: Learning to rank for content-based image retrieval. In: International Conference on Multimedia Information Retrieval (2010)
7. Guan, T.A.O., Wang, Y., Duan, L., Ji, R.: On-device mobile landmark recognition using binarized descriptor with multifeature fusion. Trans. Intell. Syst. Technol. **7**(1), 12–29 (2015)
8. Gupta, S., Girshick, R., Arbeláez, P., Malik, J.: Learning rich features from RGB-D images for object detection and segmentation. In: Fleet, D., Pajdla, T., Schiele, B., Tuytelaars, T. (eds.) ECCV 2014. LNCS, vol. 8695, pp. 345–360. Springer, Cham (2014). doi:10.1007/978-3-319-10584-0_23
9. He, J., Feng, J., Liu, X., Cheng, T., Lin, T.H., Chung, H., Chang, S.F.: Mobile product search with bag of hash bits and boundary reranking. In: Conference on Computer Vision and Pattern Recognition, pp. 3005–3012 (2012)
10. Joachims, T.: Training linear SVMs in linear time. In: International Conference on Knowledge Discovery and Data Mining (2006)
11. Li, S.Z., Zhao, C.S., Ao, M., Lei, Z.: Learning to fuse 3D+2D based face recognition at both feature and decision levels. In: Zhao, W., Gong, S., Tang, X. (eds.) AMFG 2005. LNCS, vol. 3723, pp. 44–54. Springer, Heidelberg (2005). doi:10.1007/11564386_5
12. Li, Y., Zhou, C., Geng, B., Xu, C., Liu, H.: A comprehensive study on learning to rank for content-based image retrieval. Sig. Process. **93**(6), 1426–1434 (2013)
13. Lv, T., Liu, G., Huang, S.B., Wang, Z.X.: Selective feature combination and automatic shape categorization of 3D models. In: International Conference on Fuzzy Systems and Knowledge Discovery, pp. 447–451 (2009)
14. Perronnin, F., Liu, Y., Jorge, S.: Large-scale image retrieval with compressed fisher vectors. In: Conference on Computer Vision and Pattern Recognition, pp. 3384–3391 (2010)
15. Petrelli, A., Pau, D., Stefano, L.: Analysis of compact features for RGB-D visual search. In: Murino, V., Puppo, E. (eds.) ICIAP 2015. LNCS, vol. 9280, pp. 14–24. Springer, Cham (2015). doi:10.1007/978-3-319-23234-8_2

16. Petrelli, A., Pau, D., Plebani, E., Di Stefano, L.: RGB-D visual search with compact binary codes. In: International Conference on 3D Vision, pp. 82–90 (2015)
17. Socher, R., Huval, B., Bhat, B., Manning, C.D., Ng, A.Y.: Convolutional-recursive deep learning for 3D object classification. In: Advances in Neural Information Processing Systems, pp. 1–9 (2012)
18. Song, D., Liu, W., Ji, R., Meyer, D.A., Smith, J.R.: Top rank supervised binary coding for visual search. In: International Conference on Computer Vision, pp. 1922–1930 (2015)

Mine Detection Based on Adaboost and Polynomial Image Decomposition

Redouane El Moubtahij$^{(\boxtimes)}$, Djamal Merad, Jean-Luc Damoisaux, and Pierre Drap

LSIS-UMR CNRS 7296, Aix-Marseille University,
163 Avenue of Luminy, Cedex 9, 13288 Marseille, France
redouane.el-moubtahij@univ-amu.fr

Abstract. In this paper, we propose a new method for underwater mine detection. This detection strategy is based on the use of the Adaboost algorithm with a Polynomial Image Decomposition (PID). PID splits a given image into two components the geometrical component (cartoon) and the textural one (small scale). This decomposition is based on the use of a polynomial transform. The use of PID reduces the noise and turbidity of underwater images, which results a consequent improvements on the visibility of underwater objects. As a result, our detector achieves a high detection rate and good efficiency. It also shows better performance against the use of a simple adaboost algorithm for underwater mine detection.

Keywords: Underwater mine · Object recognition · Polynomial transform · Anisotropic diffusion

1 Introduction

The high resolution photogrammetric surveys in underwater environment has taken an operational turn with the automation of processing and their integration in a tele-operated machines. This new type of data opens numerous perspectives for operational exploitation in various sectors of activities, from ecology to civil engineering, to oil-related applications and, of course, to the specific needs of mine warfare. For the mine detection application, the underwater images sequence that is captured with remotely operated vehicles could have some specificities. It could have more or less accentuated, such as light absorption and diffusion which leads to noisy images, less contrast or unusable color information. For these reasons, and because detection is sensitive to these problems, the images need pre-processing before any recognition of underwater mine. After this step, the object recognition is intended to extract automatically and efficiently interesting content. Our research focuses on a system that recognizes and locates mines.

The rest of this paper is organized as follows: Sect. 2 gives a summary of related works. Section 3 gives brief introduction to image decomposition with a polynomial transform, and describes our proposed adaboost techniques. Section 4

© Springer International Publishing AG 2017
S. Battiato et al. (Eds.): ICIAP 2017, Part I, LNCS 10484, pp. 660–670, 2017.
https://doi.org/10.1007/978-3-319-68560-1_59

presents the detection results that obtained from the samples of underwater images and shows the superior performance of our underwater mine detection method.

2   Related Work

Object detection is a central problem in computer vision domain. Most object detection systems are based on the use one of the two main approaches, either the use of global or local image features.

For the local features approaches, we can cite LBP [1], Haar [2,3] and HOG [4]. For underwater object detection, the most commonly used methods are LBP and HOG due to the satisfactory detection results [1,5,6].

The LBP was introduced in 1996 by Ojala et al. to characterize the neighborhood of a point within a given image. This is done by calculating the gray level difference between a given pixel and its neighbors. This feature is widely used for recognition purposes such as by [2], and in object recognition by [1]. The disadvantage of this descriptor is the number of parameters to be fixed according to the fixed value N corresponding to the number of neighboring pixels and the chosen radius R. The other commonly used feature is the HOG which is used to calculate the occurrences of the orientations of gradients orientations in a localized portion of the image. They were introduced by Dalal and Triggs [4] to recognize pedestrians in a given image. This descriptor is also used for underwater object detection such as underwater fish [5] or underwater plan images [6].

For global features, the image is represented by one feature vector which describe the information in the whole image. In other words, global feature representation produces a single multidimensional vector with values that measure various aspects of the image, such as texture or shape. These feature vectors may be used with different classification strategies for object detection. Our objective is the detection of underwater mines, where images resulting from the acquisition step are very noisy, with non uniform lighting, muted colors, and low contrast. For this case, we found that image decomposition is effective to separate the geometrical component of the image. The use of geometric component enable the reduction of noise to obtain a usable image. In our context, we used the image decomposition approach as pre-processing for mine detection.

The literature on image decomposition is very rich. Many methods were proposed among which two categories are closely related to this work: those based on the study of image by variational methods and partial differential equations (PDE) and those based on signal processing that perform frequency analysis of images, such as morphological component analysis (MCA), and polynomial transforms. For the first category, we can mention [7–9].

In Buades et al. [7] work, they derived a non-linear filter pair to decompose image into cartoon and texture parts based on the theory given by Meyer [8]. Each image can be decomposed into geometrical component (u) and texture component (v). For each pixel, a decision is made whether it belongs to the geometric part or to the texture part. This decision is made by computing a local total variation of the image around the point and comparing it to the local total variation after the application of a low-pass filter. Zhang et al. [9] proposed a method that is easy to implement and had low computational complexity. This method is based on Rudin-Osher-Fatemi (ROF) model by using split Bregman algorithm [10] as a first step. Then, the salient features are selected from the cartoon and texture components respectively to form a composite feature space. Finally, the salient information presented by the local features of the source images are integrated to construct the fused image.

Another common approach consists of decomposing image by using the PDE. This method demonstrated powerful tools to decompose a given image into its structure, texture and noise components [11,12]. In this context, Hiremath et al. [11] proposed a novel method that uses PDE and local directional binary patterns for texture analysis. Based on their approach [13], they introduced a method for texture analysis using wavelet transform. This method is based on four basic steps. The first step consists in calculating the information about direction from the image by using a Haar wavelet transform. Then, in the second step, the texture is approximated by applying an anisotropic diffusion on horizontal, vertical and diagonal components resulting from the first step. The third step consists on the extraction of statistical features from approximation texture image and the optimization of feature sets by using a linear discriminant analysis. The fourth and last step is the classification of feature sets from textural images using KNN classifier [11].

For the second category, many approaches are dealing with the problem of image decomposition by using MCA. This method was very successful in separating various components in many practical applications [13,14]. The cartoon and texture components are represented by well-chosen dictionaries, such as dictionaries corresponding to represent Discrete Cosine Transform (DCT) or the Discrete Sine Transform (DST) for texture component representation, and dictionaries corresponding to wavelet, curvelet or shearlet for geometric component representation. However, one of the limitations of this approach is that some textures found in many practical applications cannot be modeled by DCT or DST dictionaries which tend to produce a poor decomposition.

For the same purpose, some new methods were proposed to extract and describe the texture by using a Polynomials transform. Among these methods, we may mention Bordei et al. [15] who proposed a method based on the projections of images on a complete polynomial basis. The proposed method consists of replacing the texture representation model by polynomial projections on a complete orthonormal basis. In the same context, El Moubtahij et al. [16,17] proposed a method that describe the texture, and decompose an image into geometric and texture component. Their method is based on the projection of the image on a complete polynomial basis, while considering anisotropic diffusion in their decomposition equation. In this paper, we use this method of image decomposition to reduce the noise in order to obtain a usable image. As a result, we will have a cartoon image (using the equation proposed in [16,17]), where only the image contrasted shapes appear (without texture).

3   Adaboost Based Polynomial Image Decomposition

3.1   Complete Basis

In our previous work [16,17], we proposed to decompose image as a linear combination of polynomials from orthogonal basis, and we make anisotropic diffusion to find texture approximation from directional information.

Let a *Bivariate Polynomial* (BP) of degree d is a function of $x = (x_1, x_2) \in \mathbb{R}^2$ defined as:

$$P(x) = \sum_{\substack{(i,j)\in[0;d]^2 \\ i+j\leq d}} a_{i,j}\, x_1^i\, x_2^j \tag{1}$$

where $i,j \in \mathbb{R}^+$ are the maximum degrees of variables x_1, x_2 and $\{a_{i,j}\} \in \mathbb{R}$ are the coefficients of the polynomial. The overall degree of the polynomials is $d = i + j$.

Considering a finite set of pairs $D = \{(i,j)\} \subset \mathbb{N}^2$, we represent by $\mathbb{E}_D$ the space of all BP such as $a_{i,j} \equiv 0$ if $(i,j) \notin D$ and by $\mathcal{K}_D$ the subset of monomials

$$\mathcal{K}_D = \left\{ K_{d_1,d_2}(x) = x_1^i\, x_2^j \right\}_{(i,j)\in D} \tag{2}$$

Obviously $\mathcal{K}_D$ satisfies the linear independence and spanning conditions and so, $\mathcal{K}_D$ is a basis of $\mathbb{E}_D$, the canonical basis. In our context of color image decomposition, we look for bases with more suitable properties such as orthogonality or normality. So, to construct a discrete orthonormal BP finite basis we first have to consider the underlying discrete domain

$$\Omega = \left\{ x_{(i,j)} = \left(x_{1,(i,j)}, x_{2,(i,j)} \right) \right\}_{(i,j)\in D} \tag{3}$$

where D will now represent the set of pairs associated to Ω.

Starting from $\mathcal{K}_D$ we intend to construct a new orthonormal basis applying the Gram-Schmidt process. This implies that we need some product and norm for functions defined on Ω. Given two bivariate functions, F and G, their *discrete extended scalar product* is defined by

$$\langle F|G \rangle = \sum_{\omega} F(x)\, G(x)\, \omega(x) \tag{4}$$

with ω a real positive function over Ω (Legendre, Chebichev, Hermite, ...). Then, the actual construction process of an orthonormal basis

$$\mathcal{B}_{D,\omega} = \{B_{i,j}\}_{(d_1,d_2)\in D} \tag{5}$$

is a recurrence upon (i,j)

$$T_{i,j}(x) = K_{i,j}(x) - \sum_{(l_1,l_2)\prec_2(i,j)} \langle K_{i,j} \,|\, B_{l_1,l_2} \rangle_\omega B_{l_1,l_2}(x) \tag{6}$$

$$B_{i,j}(x) = \frac{T_{i,j}(x)}{|T_{i,j}|_\omega} \tag{7}$$

where $\prec_2$ is the lexicographical order and $|\;|_\omega$ the norm induced by $\langle\;|\;\rangle_\omega$. The resulting set of B polynomials verifies

$$\langle B_{i,j}|B_{l_1,l_2}\rangle_\omega = \begin{cases} 0 & \text{if } (i,j) \neq (l_1,l_2) \\ 1 & \text{if } (i,j) = (l_1,l_2) \end{cases} \tag{8}$$

and so $\mathcal{B}_{D,\omega}$ is effectively an orthonormal basis with respect to a weighting function ω. A complete basis is the orthonormal basis whose domain Ω and related to the discrete extended inner product (4) is defined by the family:

$$\{\mathcal{B}_{i,j}(x)\}_{i=0\cdots n_1\,j=0\cdots n_2} \tag{9}$$

with the number of polynomials in the complete polynomial basis is given by the size $(n_1 + 1) \times (n_2 + 1)$.

3.2   Polynomial Image Decomposition (PID)

Given an image I, the geometrical component is given by a partial reconstruction $\tilde{I}$ of I in an overlapped polynomial transform context (El Moubtahij et al. [16, 17]). The procedure of image decomposition follows this scheme:

1. Construction of a complete polynomial basis of degree d and sub-domain $M = S_1 \times S_2$.
2. Performing a polynomial approximation of image I.
3. Performing a partial reconstruction with brutal restriction or restriction based on energies. For example, by using the normality of the basis to assimilate the absolute value of its coefficients to a part of the energy of a subdomain, then sort the coefficients, and finally retain a fixed number of these coefficients or those satisfying a certain condition (c.f., principal component analysis).
4. Applying the redundancy equation:

$$\tilde{I}(x) = \frac{1}{c(x)} \sum_{\{\Omega_M \ni x\}} \left(\Psi(\Omega_M)\, \omega(x_M) \sum_{(i,j)\in\mathcal{P}_M} b_{i,j}(I_M)\, B_{i,j}(x_M) \right) \tag{10}$$

with:

- $\tilde{I}$: reconstructed image, in our case, it is the geometric image.
- X: a point referring to Ω, X_M same point referring to sub-domain Ω_M.
- I_M: restriction of I to sub-domain Ω_M.
- P_M: selected polynomials for I_M approximation.
- $m(x)$: sum of contributions from point X.
- $\Psi(\Omega_M)$: degree of anisotropy assign to sub-domain Ω_M.

The texture component I^T is simply deduced from the partial reconstruction of the redundancy equation as: $I^T = I - \tilde{I}$.

The function $\Psi(\Omega_M)$ plays a very important role in this equation, thanks to it, one can control the amount of contour extraction in the image. It is evaluated as follow:

$$\Psi(\Omega_M) = \frac{1}{1 + \lambda^r} \tag{11}$$

with λ is the largest eigenvalue of a tensor structure composed with the approximations of partial derivatives (El Moubtahij et al. [16,17]). The balance between isotropic and anisotropic diffusion is adjusted by the parameter r that controls the degree of anisotropy. A small value of r induces a uniform blur effect on the image, this is known as the isotropic diffusion effect. However, if r increases, rather than blur, the contours will become very contrasted. The complexity of this approach in terms of time/memory is cited in [17].

We present in Fig. 1 the underwater mine image decomposition result from this process with a varied value of r. From this figure, image representing the geometric component is very smooth, while all that is texture, contour and noise are in the texture component.

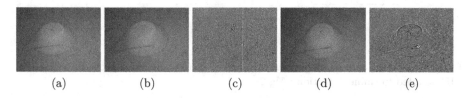

(a)   (b)   (c)   (d)   (e)

Fig. 1. Mine image decomposition by polynomial approach: (a) original image, (b) and (c) are respectively geometrical component and texture component with $r = 0.75$. (d) and (e) are respectively geometrical component and texture component with $r = 4$.

3.3 Adaboost Based PID

In this paper, a novel classification scheme based on the use of adaboost with the geometrical component of images is proposed. This classification scheme is detailed in Fig. 2. The traditional adaboost algorithm introduced by Viola and Jones [2] is composed of two stages that are the training and the detection process. The basic idea is to integrate the geometrical component extracted by polynomial transform as input for adaboost algorithm in a combination with LBP and HOG to train underwater mine detector. The noise and texture are treated as high frequency components. However, there is a probability of having noise contents in the geometrical image. To adaptively attenuate noise, the generated geometrical components are extracted based on the isotropic and anisotropic diffusions.

From Fig. 2(b), it can be seen that the algorithm starts by decomposing original images into geometrical positive and negative samples, extracting the LBP and HOG features from these samples, and then, training the adaboost in order to generate our mine classifier. Once the classifier is trained, the detection process will be launched. For this purpose, the LBP and HOG features are calculated for all geometrical test images and then, the adaboost detection process is applied on these images to detect the mines, the adaboost algorithm can distinguish objects that correspond to mines by separating the high frequencies (noise and texture) from underwater images.

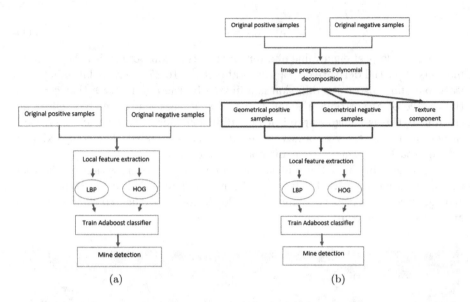

Fig. 2. (a) Traditional adaboost algorithm (Alg.1), (b) AdaBoost procedure based on PID adapted for mine detection (Alg.2)

4   Experimental Result and Analysis

4.1   Parameters Setting

In order to evaluate our mine detection, we used different values of r ranging from isotropic to anisotropic diffusions. However, to achieve a stable and reliable detection, the parameters are set as: Scale factor $= 1.1$, Mine size $= 30 \times 30$, $r = 0.75$, 3 and 4, where scale factor is the ratio of the sliding window of adaboost algorithm in the two successive scans, mine size is the minimum size region containing an object of adaboost algorithm, and r is the parameter that adjust isotropic and anisotropic diffusion.

4.2   Mine Detection

After decomposing images with our polynomial method, we used the adaboost algorithm to train a binary classifier with both LBP and HOG features that have been extracted from positive and negative samples. These samples are obtained using Remotely Operating Underwater Vehicle (ROUV). They were taken on various maritime sites close to Marseille, France. These samples were taken with different view angles, illuminations and noise densities. Many unnecessary images are provided by ROUV. The number of images that contain mines is minimal. These samples (a few examples are shown in Fig. 3) constitute the images data set that is used for mine detection. The images data set contains 184 positive samples and 398 negative samples. The positive samples are mines in different angles and the negative samples are the non-mines areas selected from background. On these high-resolution images 1936 × 1456, we selected the regions of interest to build a well-varied set of positive and negative samples. To perform training and evaluate the detection performance, two sub-sets of positive and negative samples were selected. For this, we took 80% of images as a training set and the remaining 20% as a test set for both positive and negative samples. After decomposing images and using the geometrical components for the positive and negative samples, we use the LBP and HOG features with a cascade classifier. The used classifier is a 20 stages strong classifiers, where each stage is an set of weak classifiers. For detection, the trained classifier is used with a sliding window on test images to localize windows containing the object of interest. The size of the window varies to detect objects at different scales, but its aspect ratio remains fixed.

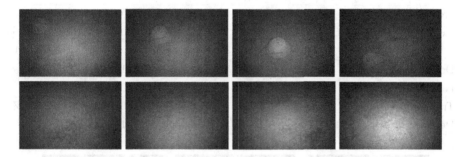

Fig. 3. The positive and negative samples: the first row show the positive samples, and the second row show the negative samples.

4.3 Results

To evaluate our algorithm, we used three criteria which are the detection rate, the false detection rate, and the true negative rate. The detection rate is the ratio between the number of true detection and the total number of positive and negative samples, and the false detection rate is the ratio between the number of negative samples wrongly classified as positive (false positives) and the total number of actual negative samples. Whereas the true negative rate is the proportion of negatives that are correctly identified. Table 1 shows the results of mine detection using $Alg.1$, and Table 2 shows the same results but this time using $Alg.2$. $Alg.1$ uses original images without any pre-processing whereas $Alg.2$ uses pre-processed images. From Table 2, we can see an improvement in performance with our proposed algorithm, the detection rate has achieved up to 96% with LBP feature and $r = 0.75$ much better than $Alg.1$ where the detection rate has achieved 51% with LBP features.

Example of mine detection results are shown in Fig. 4.

From the above results, we can see the geometrical information based polynomial image decomposition helps our descriptor to correctly detect the mine. The application of LBP features on the geometrical images are more effective than

Table 1. The mine detection results with *Alg*.1

Method	Alg.1	
Feature	LBP	HOG
Number of tested mines/non-mines	40/78	40/78
Detected mines	32	30
False detection rate	21.6%	8%
True negative rate	78%	87%
Detected rate	51%	87%

Table 2. The mine detection results with *Alg*.2

Method	Alg.2					
Feature	LBP			HOF		
Number of tested mines/non-mines	40/78			40/78		
	$r = 0.75$	$r = 3$	$r = 4$	$r = 0.75$	$r = 3$	$r = 4$
Detected mines	38	34	37	21	15	16
False detection rate	2.5%	8%	4%	19%	24%	23%
True negative rate	97.4%	91.6%	95.1%	80.4%	75.7%	76.4%
Detected rate	95.7%	84.7%	81%	81%	78.8%	79%

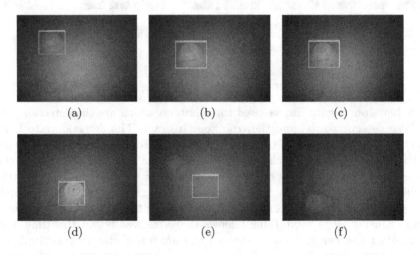

(a)   (b)   (c)

(d)   (e)   (f)

Fig. 4. Detection results with our proposed adaboost algorithm: by comparing the results shown on (a–d) and those shown on (e–f), we can see that most are detected except those positioned in dark regions. There are also some other mines are not detected due to hard noise on images

HOG features. It implies that characterizing textures of mines with LBP features on geometric images gives better detection compared to the use of traditional features on original images.

5   Conclusion

In this paper, a novel method for mine detection in underwater images has been proposed. This method is based on the use of Viola and Jones cascade classifier that is widely used in computer vision for object detection. In addition to the local features such as LBP and HOG, a geometrical component extracted by polynomial transform is proposed and used as input for adaboost algorithm to train underwater mine detector. Although the input underwater images are noisy; with our approach, the noise is reduced from the geometrical component, and our classifier can properly detect mines by using LBP and HOG features. The proposed approach has been tested on a set of underwater real images. Experimental results show that the proposed method has a high accuracy in detection and identification of underwater mine compared with Viola and Jones classifier cascade, which is indicate the effectiveness of our method.

Acknowledgments. This work has been supported by French Government Defense procurement and technology agency (DGA: Direction générale de l'armement) within the RAPID program and in the framework LORI project (Localisation and Recognition of Submerged Objects).

References

1. Satpathy, A., Jiang, X., Eng, H.-L.: Human detection by quadratic classification on subspace of extended histogram of gradients. IEEE Trans. Image Process. **23**(1), 287–297 (2014)
2. Viola, P., Jones, M.: Rapid object detection using a boosted cascade of simple features. In: CVPR (2001)
3. Yan, J., Zhang, X., Lei, Z., Liao, S., Li, S.Z.: Robust multiresolution pedestrian detection in traffic scenes. In: CVPR (2013)
4. Dalal, N., Triggs, B.: Histograms of oriented gradients for human detection. In: IEEE Conference on Computer Vision and Pattern Recognition, San Diego, USA, pp. 886–893, 20 June 2005
5. Chuang, M.-C., Hwang, J.-N., Ye, J.-H., Huang, S.-C., Williams, K.: Underwater fish tracking for moving cameras based on deformable multiple kernels. IEEE Trans. Syst. Man Cybern.: Syst. (2016)
6. Rajasekar, M., Celine Kavida, A., Anto Bennet, M.: Underwater k-means clustering segmentation using SVM classification. Middle-East J. Sci. Res. **23**(9), 2166–2172 (2015)
7. Buades, A., Le, T.M., Morel, J.M., Vese, L.A.: Fast cartoon + texture image filters. IEEE Trans. Image Process. **19**(8), 1978–1986 (2010)
8. Meyer, Y.: Oscillating Patterns in Image Processing and Nonlinear Evolution Equations. American Mathematical Society, Boston (2001)
9. Zhang, Y., Li, H., Zhao, Z.: Multi-focus image fusion with cartoon-texture image decomposition. Int. J. Sig. Process. Image Process. Pattern Recogn. **8**(1), 213–224 (2015). http://dx.doi.org/10.14257/ijsip.2015.8.1.19
10. Wang, Y., Yang, J., Yin, W., Zhang, Y.: A new alternating minimization algorithm for total variation image reconstruction. SIAM J. Imaging Sci. **1**(3), 248–272 (2008)

11. Hiremath, P.S., Bhusnurmath, R.A.: PDE based features for texture analysis using wavelet transform. Int. J. Cybern. Inform. (IJCI) **5**(1), 143–155 (2016)
12. Hiremath, P.S., Bhusnurmath, R.A.: Diffusion approach for texture analysis based on LDBP. Int. J. Comput. Eng. Appl. Part I **9**(7), 108–121 (2015)
13. Vishal, M.P., Glenn, R.E., Rama, C., Nasser, M.N.: Separated component-based restoration of speckled SAR images. IEEE Trans. Geosci. Remote Sens. **52**(2), 1019–1029 (2014)
14. Taheri, S., Patel, V.M., Chellappa, R.: Component-based recognition of faces and facial expressions. IEEE Trans. Affect. Comput. **4**(4), 360–371 (2013)
15. Bordel, B., Pascal, B., Augereau, B., Carre, P.: Polynomial based texture representation for facial expression recognition. In: IEEE International Conference on Acoustic, Speech and Signal Processing ICASSP (2014)
16. El Moubtahij, R., Augereau, B., Tairi, H., Fernandez-Maloigne, C.: Spatial image polynomial decomposition with application to video classification. J. Electron. Imaging **24**(6), 061114 (2015)
17. El Moubtahij, R., Augereau, B., Tairi, H., Fernandez-Maloigne, C.: A polynomial texture extraction with application in dynamic texture classification. In: QCAV, Le Creusot, France (2015)

Perceptual-Based Color Quantization

Vittoria Bruni[1,2], Giuliana Ramella[2(✉)], and Domenico Vitulano[2]

[1] Department of SBAI, University of Rome La Sapienza, Rome, Italy
vittoria.bruni@sbai.uniroma1.it
[2] Institute for the Applications of Calculus, CNR, Rome, Italy
{giuliana.ramella,d.vitulano}@iac.cnr.it

Abstract. The paper presents a method for color quantization (CQ) which uses visual contrast for determining an image-dependent color palette. The proposed method selects image regions in a hierarchical way, according to the visual importance of their colors with respect to the whole image. The method is automatic, image dependent and requires a moderate computational effort. Preliminary results show that the quality of quantized images, measured in terms of Mean Square Error, Color Loss and SSIM, is competitive with some existing CQ approaches.

Keywords: Human Visual System · Visual contrast · Color quantization · RGB color space

1 Introduction

Although the Human Visual System (HVS) is able to distinguish a large numbers of colors, it behaves as an imperfect sensor. It tends to group colors with similar tonality since few colors are generally enough for image representation and understanding. A color quantization (CQ) method attempts to emulate this perceptual behavior by selecting a suitable reduced number of representative colors and by producing a quantized image which still is visually similar to the original one with minimum distortion.

A number of CQ methods are available in the literature [1,3,13]. The standard approach is based on the interpretation of CQ as a clustering problem in the 3D color space. Colors are grouped into clusters, by using any clustering technique, and the representative color for each cluster is generally obtained as the average of the colors in the cluster. Most CQ methods belong to the category of image dependent clustering methods. Usually, they can be categorized into two families: preclustering methods [3,8,9,18] and postclustering methods [6]. Methods in the former class are based on a hierarchical structure and recursively find nested cluster either in a top-down or bottom-up manner; on the contrary, methods in the second class find all clusters simultaneously as a partition of the data.

Visual perception is mediated by a collection of individual mechanisms in the visual cortex due to the neuron response to stimuli above a certain contrast. Hence, to integrate the properties of the HVS in the quantization step,

© Springer International Publishing AG 2017
S. Battiato et al. (Eds.): ICIAP 2017, Part I, LNCS 10484, pp. 671–681, 2017.
https://doi.org/10.1007/978-3-319-68560-1_60

a perceptual-based method should exploit the spatio-temporal masking properties and establish thresholds based on psychophysical contrast phenomena. This contrast sensitivity varies with spatial frequency, temporal frequency and orientation and can be used to indicate the threshold at which a spatial frequency just becomes visible under certain viewing conditions. Some perceptual-based methods based on contrast sensitivity have been proposed in the literature [5,12,15], especially for image compression purposes. However, a contrast-based analysis, which allows an integration of the perceptual mechanisms of the HVS in the quantization step to achieve the best possible visual quality, has still not received the adequate attention.

In this paper we propose a CQ method which selects quantization bins according to measures related to contrast sensitivity in order to reach a good visual quality. The proposed model, named perception-based color quantization (PCQ), aims at applying some basic rules which guide human perception in the selection of the most K representative colors in an image, when K is given. It mainly consists of a 3D extension of the model proposed by the same authors in [2] for dermoscopic images processing. Specifically, the quantities used for measuring contrast variations have been generalized to the color space. They allow an automatic selection of the threshold to use for selecting those image pixels which contribute to the definition of representative image colors. PCQ is automatic since perceptive thresholds are automatically tuned according to the analyzed image. It can be framed in the preclustering method category and can be considered as context adaptable, since the resulting CQ is image-dependent.

PCQ has been compared with some representative methods belonging to the same class in terms of some well known objective measures. Experimental results show that the simple use of basic quantities related to human vision allows us to reach results that are comparable to some reference methods in the literature, with a very good subjective visual quality.

The outline of the paper is the following. Section 2 gives a detailed description of the general perceptual model extended to the three color channels. As well as a through description of the main steps of the whole quantization procedure. Experimental results, discussions and concluding remarks are in Sect. 3.

2   The Proposed CQ Model

Color contrast is one of the main property of vision and plays a key role in object detection and discrimination. It has a direct connection with two of the main rules of primary vision, like chromatic adaptation and color constancy. Chromatic adaptation is the ability of the HVS to discount the color of a light source and to approximately preserve the appearance of an object. Color constancy is the property by which objects tend to appear with the same color under changes in illumination. The strength of chromatic contrast is influenced by several factors including relative illumination, spatial scale, spatial configuration and context as well as object dimension and background variability. More precisely, the perception of an object with a given color (foreground) depends on the color of its background as well as on the chromatic variability of the same background. Based on this consideration, we are interested in quantifying: *(i)* how the visual contrast of the foreground changes if its background is gradually modified and *(ii)* how the perception of the same object changes if its color is modified while its background is leaved unchanged. The combination of these two quantities provides a sort of visual distortion curve where the optimal quantization bin can be determined.

More precisely, by denoting with $I_{ij}(k)$ the image I at point with coordinates $(i, j) \in \Omega$, where Ω is the image domain, and color channel k (for example, in the RGB color space, $k = 1, 2, 3$ respectively denote red, green and blue components), with R the reference color (R is a vector having three components) and with B the color which represents the background, it is possible to define the following quantity

$$D_1(i, j) = \left| \frac{\|I_{ij} - B\|_2^2}{\|B\|_2^2} - \frac{\|R - B\|_2^2}{\|B\|_2^2} \right|, \quad \forall\, (i, j) \in \Omega \tag{1}$$

where

$$C_{ij} = \frac{\|I_{ij} - B\|_2^2}{\|B\|_2^2} \tag{2}$$

is the square of the contrast of the object having color I_{ij} with respect to a background whose color is $B - \frac{\|R-B\|_2^2}{\|B\|_2^2}$ has a similar meaning; $\| * \|_2$ denotes the euclidean distance. D_1 quantifies the variation of the contrast of an object with respect to a fixed background having average color B if the object changes its color (from $I_{i,j}$ to R). It is worth noticing that Eq. (2) is a generalization of the classical Weber's contrast for monochromatic images [17].

Similarly, it is possible to define a quantity which works in the opposite way: the color of the object is fixed (I_{ij}), while its background changes (from B to B_R). It is defined as follows

$$D_2(i, j) = \left| \frac{\|I_{ij} - B\|_2^2}{\|B\|_2^2} - \frac{\|I_{ij} - B_R\|_2^2}{\|B_R\|_2^2} \right|. \tag{3}$$

D_1 and D_2 can be then combined to define a pointwise distortion as follows

$$D(i, j) = \sqrt{D_1(i, j) D_2(i, j)}, \tag{4}$$

which accounts for the two competing phenomena. In order to use D for determining the optimal detection threshold, it is necessary to define the spatial domain where those measures have to be defined. The latter depends on the rule used for the estimation of R and B_R in Eqs. (1) and (3). This rule can depend of the specific kind of application and purposes and it will be presented in the following section.

2.1   Representative Color Selection

The aim of this section is to separate image foreground and background in an iterative manner. At each iteration, the foreground represents the object of interest, while the background consists of the remaining part of the image. The object of interest is a region of the image whose color is perceived as homogeneous. Since we are interested in finding perceptual representative colors in the image, in this paper the foreground is determined starting from the color which occurs more in the image and enlarging the color region by including tones having increasing distance from it. The chromatic region growing process stops when the variation of contrast becomes clearly visible to a human observer—this contrast threshold determines the amplitude of the bin which gives a color in the final palette as well as the region of interest to which assign this color. This process is then iterated on the remaining part of the image. The number of iterations is the number of colors K to be used for image quantization, which is an input value.

More precisely, if $\mathbf{c} = [c_1, c_2, c_3]$ is the color having more occurrences in the image I, we define the domain

$$\Omega_m = \{(i,j) \in \Omega : |I_{ij}(k) - c_k| \leq m\delta_k, \quad k = 1, 2, 3\}, \quad m \geq 1, \ m \in \mathbf{N}, \quad (5)$$

where δ_k is the minimum allowed bin for the k-th color channel and it is estimated separately from each color component, as explained in the next subsection. Ω_m contains pixels having colors close to $\mathbf{c}$. R is then defined as the average color of I in the region Ω_m, B_R as the average color of I in $\Omega - \Omega_m$ while B as the average color of I in Ω.

The extension of Eq. (4) to the domain Ω_m is then

$$D(\Omega_m) = \frac{1}{|\Omega_m|} \sum_{(i,j) \in \Omega_m} D(i,j), \quad (6)$$

where $|\Omega_m|$ is the cardinality of Ω_m.

Regions of interest in I are selected using a threshold value that has to correspond to the point of maximum visibility of the foreground with respect to its background, which represents an optimal point of $D(\Omega_m)$ as a function of $|\Omega_m|$—see Fig. 1. More precisely, the region of interest is selected as the one which realizes the maximum curvature of D. This point can be approximated as follows

$$\bar{m} : \frac{\delta^2 D}{\delta |\Omega_m|^2} \Big|_{m=\bar{m}} = 0 \quad (7)$$

with $\frac{\delta^3 D}{\delta |\Omega_m|^3} \Big|_{m=\bar{m}} < 0$. This optimal point represents the frontier between image foreground and background, i.e. from that point on pixels of the background would be confused with foreground.

Finally, the mean value of the colors (in the RGB color space) of points belonging to $\Omega_{\bar{m}}$ is considered as the dominant color of the region and represents the first value c_1 of the color palette to be used in the quantization step.

The procedure is iterated by considering only the remaining image domain, i.e. $\Omega - \Omega_{\bar{m}}$, till the number of desired colors is reached.

2.2   Estimation of the Least Allowed Bin Size

In the preattentive phase, human eye acts as a low pass filter [17] since it is not interested in the detection of image details in this phase. As a result non-homogeneous colored image regions are usually perceived at the first glance as

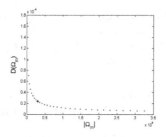

Fig. 1. Original Parrots image (*left*); distortion curve $D(\Omega_m)$ versus the size of Ω_m, as in Eq. (6) (*middle*) (the optimal point is marked); selected region in the original image which is estimated from the optimal point of the distortion curve (*right*).

uniform areas. This visual resolution also gives the minimum allowed bin width (i.e. the one to which human eye is almost insensitive at first glance). This visual resolution, namely δ, corresponds to a precise scale level of a pyramid decomposition of the image. For example, in the dyadic case, $\delta = 2^J$, where J is a fixed positive integer number. It means that a variation h in color components, reduces to $\frac{h}{2^{J-1}}$ at level J of the pyramid. In particular, if $h = 2^{J-1}$, it vanishes (less than 1) at level J—in other words, differences in amplitude greater than 2^{J-1} are hard to be perceived and then a bin size greater than 2^{J-1} can be considered. For the estimation of the "visual resolution", the method in [2] has been adopted. It computes the contrast between two successive low-pass filtered versions of the analysed image (where filters have increasing support) and selects δ as the resolution which gives the minimum perceivable contrast. This procedure is independently applied to the three color channels in this paper.

2.3   PCQ Algorithm

1. Compute the 3D histogram $\mathcal{H}(r, g, b)$ of the RGB image I.
2. For each color channel $I(k)$ ($k = 1, 2, 3$), estimate the minimum bandwidth (respectively $\delta_1, \delta_2, \delta_3$[1]) as in Sect. 2.2 as well as the mean value $M_{av}(k)$ and the mode $Mo(k)$. Let $\mathbf{Mav}, \mathbf{Mo}$ and δ be the corresponding 3D vectors.
3. Compute the correction parameter $\sigma = \frac{1}{3K} \| \frac{|\mathbf{Mav} - \mathbf{Mo}|}{\mathbf{Mav}} \|_1$ and the parameter $\Delta = \frac{128}{\|\delta\|_\infty}$.
4. Repeat the following steps K times (for $l = 1, 2, \ldots, K$)
   - Compute the average color B of I in the domain Ω and correct it using the following rule: $B = B(1 - l\sigma)$.
   - Set $\mathbf{c} = argmax_{r,g,b}\mathcal{H}(r, g, b)$ and $m = 1$.
   - For each integer $m \in [1, \Delta]$:
     • Find Ω_m using in Eq. (5).
     • Compute the average color R in Ω_m and the average color B_R in $\Omega - \Omega_m$.
     • Evaluate $D(\Omega_m)$ using in Eq. (6).

---

[1] They are given as power of 2.

- Extract the optimal $\bar{m}$ as in Eq. (7) and the corresponding region $\Omega_{\bar{m}}$.
- Compute the average color $\mathbf{c}_l$ of I in $\Omega_{\bar{m}}$ and put it in the palette and set $\mathcal{H}(I_{ij}(1), I_{ij}(2), I_{ij}(3)) = -1, \quad \forall\,(i,j) \in \Omega_{\bar{m}}$.
- Set $\Omega = \Omega - \Omega_{\bar{m}}$ and $I = I(\Omega)$ (the latter denotes I restricted to the domain Ω).
5. Assign to each pixel in the original image the closest color in the selected color palette $\{\mathbf{c}_1, \mathbf{c}_2, \ldots, \mathbf{c}_K,\}$ and let I_Q the quantized image.

The correction parameter σ is used for adapting the algorithm to the number of desired colors. In fact, the detection algorithm can be less sensitive to some details as K decreases; while it is the opposite as K increases. That is why, the value B, which represents the image background, is defined as a correction of the actual average value of the image to be analysed. It is also worth observing that for $K \le 16$ the algorithm is applied to the low pass filtered version of I at resolution $log_2(min\{\delta_r, \delta_g, \delta_b\})$.

3 Experimental Results and Concluding Remarks

PCQ has been tested on several color images having different features. In order to perform a comparative study, in this section results achieved on 21 images taken from some public available databases (such as [19–23]) and the 8 images used in [3] will be shown and discussed. The first dataset has been used for a direct comparison with some standard CQ methods. Specifically, the following methods have been considered: *(i)* the Median-cut (MC) [9], which recursively split boxes obtained using a uniformly quantized image along the longest axis at the median point. At each step, the split is applied to the box that contains the greatest number of colors; *(ii)* the Octree (OCT) [8], which merges colors represented in a tree data structure by pruning the tree until the desired number of colors is obtained; *(iii)* the greedy orthogonal bipartitioning (WU) [18], which uses the minimum SSE (sum of squared error) for boxes splitting; *(iv)* self-organizing map (SOM) [6], which uses a one-dimensional self-organizing map with K neurons and the weights of the final neurons define the color palette.

Table 1 contains the results achieved using no more than 16 colors ($K = 16$). They have been measured in terms of Mean Square Error (MSE) and Color Loss (CL), since commonly used measures for the evaluation of color image quality. We have also evaluated the structural similarity index (SSIM) as a measure which is more consistent with visual perception, even though it has not been specifically defined for color images. For two images v and w of dimension $H \times K$,

- MSE [14] is computed as: $MSE(v,w) = \frac{1}{HK} \sum_{i=1}^{H} \sum_{j=1}^{K} \sum_{k=1}^{3} (v_{ij}(k) - w_{ij}(k))^2$, where i, j denote pixel location while k is the color channel;
- CL [4,10,11] is the average color loss between v and w, i.e. $CL(v,w) = \frac{1}{HK} \sum_{i=1}^{H} \sum_{k=1}^{K} \sqrt{\sum_{k=1}^{3} (v_{ij}(k) - w_{ij}(k))^2}$;
- SSIM [7,16], for two gray-level images v and w is defined as: $SSIM(v,w) = \frac{(2\mu_v \mu_w + c_1)(2\sigma_{vw} + c_2)}{(\mu_v^2 + \mu_w^2 + c_1)(\sigma_v^2 + \sigma_w^2 + c_2)}$, where μ_* is the average of $*$; σ_*^2 is the variance of $*$;

Table 1. SSIM, MSE and CL results achieved on the images in Fig. 2 by Median Cut (MC) [9], the Octree (OCT) [8], Greedy orthogonal bipartitioning (WU) [18], Self-organizing map (SOM) [6] and the proposed perception-based color quantization method (PCQ) using 16 colors. For each metric, the average values (**Avg**) computed on the whole dataset are also given. Finally, for each method, the number of used colors K is provided.

img		MC	OCT	Wu	SOM	PCQ	img	MC	OCT	Wu	SOM	PCQ
1	K	16	16	16	16	16	12	16	16	16	13	16
	SSIM	0.84	0.83	0.90	**0.92**	0.87		0.90	0.90	**0.95**	0.93	0.85
	MSE	282.4	544.7	**179.5**	203.4	224.9		69.2	91.2	**43.9**	84.3	120.5
	CL	13.34	19.88	10.25	**9.77**	12.22		6.98	7.07	**5.42**	6.16	9.00
2	K	16	16	16	16	16	13	16	16	16	14	16
	SSIM	0.76	0.74	0.76	**0.79**	0.78		0.92	0.92	**0.93**	0.92	**0.93**
	MSE	905.2	1154.2	791.1	**681.7**	709.7		217.5	288.8	**162.1**	287.1	222.0
	CL	27.39	30.52	25.41	**23.28**	23.78		11.93	12.28	**9.40**	10.57	11.83
3	K	16	15	16	16	16	14	16	16	16	15	16
	SSIM	0.88	0.86	0.89	**0.90**	0.87		0.85	0.87	0.88	**0.89**	**0.89**
	MSE	153.88	198.71	125.87	127.13	286.03		178.6	952.7	130.6	137.3	**121.4**
	CL	10.73	11.65	9.58	**9.09**	14.24		12.14	24.84	9.72	9.52	**9.16**
4	K	16	15	16	16	16	15	16	16	16	16	16
	SSIM	0.71	**0.68**	0.76	0.79	0.78		0.76	0.75	0.81	**0.83**	0.78
	MSE	828.1	522.7	**363.3**	391.5	395.1		530.7	670.8	**366.2**	431.3	390.6
	CL	25.64	19.57	15.48	14.79	14.70		20.25	22.37	**15.47**	15.73	17.72
5	K	16	16	16	15	16	16	16	16	16	16	16
	SSIM	0.95	0.96	**0.97**	**0.97**	0.96		0.83	0.68	**0.85**	**0.85**	0.83
	MSE	92.6	68.2	**48.0**	52.6	88.2		283.1	958.9	257.7	**224.9**	268.4
	CL	8.40	6.57	5.48	**4.80**	8.07		14.96	25.07	14.30	**13.50**	14.70
6	K	16	16	16	16	16	17	16	16	16	16	16
	SSIM	0.80	0.74	**0.83**	**0.83**	0.82		0.81	0.80	0.82	**0.86**	0.81
	MSE	435.2	1048.7	354.2	345.4	**335.4**		270.4	353.8	227.8	**216.4**	243.5
	CL	17.84	25.71	15.66	**15.11**	15.50		14.51	16.48	12.83	**12.16**	14.28
7	K	16	16	16	13	16	18	16	16	16	16	16
	SSIM	0.96	0.95	0.96	**0.97**	0.96		0.84	0.85	0.86	**0.87**	0.85
	MSE	47.3	60.7	**46.3**	48.6	65.2		475.8	444.8	**356.5**	383.6	390.4
	CL	6.06	5.76	5.62	**4.86**	7.12		18.61	18.26	16.16	**15.91**	17.58
8	K	16	16	16	16	16	19	16	16	16	16	16
	SSIM	0.77	0.81	**0.84**	**0.84**	**0.84**		0.76	0.73	0.74	**0.77**	**0.77**
	MSE	412.8	408.3	206.8	204.4	**187.2**		776.8	964.8	621.2	566.3	**522.7**
	CL	18.08	16.70	12.40	12.10	**12.01**		25.22	25.99	21.55	**19.75**	19.85
9	K	16	16	16	15	16	20	16	12	16	16	16
	SSIM	**0.91**	0.89	**0.91**	0.90	0.89		0.72	0.74	0.77	0.76	**0.78**
	MSE	186.0	346.1	**169.0**	228.6	215.8		773.6	747.4	**461.5**	482.6	472.3
	CL	11.82	14.08	**10.86**	11.14	12.26		3.70	24.57	19.03	18.81	**18.69**
10	K	16	16	16	15	16	21	16	16	16	16	16
	SSIM	0.94	0.93	0.94	**0.95**	0.93		0.84	0.82	0.85	**0.87**	0.84
	MSE	129.1	281.2	**109.0**	138.2	215.7		451.2	1399.9	**313.9**	386.2	334.8
	CL	8.96	13.40	8.28	**7.90**	11.82		18.29	32.45	**14.14**	14.41	15.36
11	K	16	16	16	16	16	Avg					
	SSIM	0.85	**0.74**	0.86	0.87	0.86		0.84	0.82	0.86	**0.87**	0.85
	MSE	294.7	995.8	229.2	230.7	**216.1**		375.6	597.9	**269.2**	282.5	290.1
	CL	15.26	25.26	12.62	**12.00**	12.59		15.84	18.93	12.97	**12.58**	14.01

Fig. 2. Images used for the comparative studies in Table 1.

Table 2. MSE and MAE results achieved by PCQ and the two methods proposed in [3] (VC and VCL). The set of images is the same used in [3].

Image	Method ($K = 32$)	MSE	MAE	Image	Method ($K = 32$)	MSE	MAE
Fish	PCQ	179.8	17.6	Goldhill	PCQ	199.9	19.7
	VC	168.1	17.2		VC	174.8	17.8
	VCL	169.9	17.1		VCL	169.3	17.3
Motocross	PCQ	283.4	22.8	Lena	PCQ	156.4	16.9
	VC	253.2	20.5		VC	145.6	16.5
	VCL	240.6	19.4		VCL	146.3	16.5
Parrots	PCQ	287.4	22.0	Peppers	PCQ	292.0	22.9
	VC	290.6	22.4		VC	294.8	22.9
	VCL	263.7	21.6		VCL	261.1	22.9
Baboon	PCQ	452.3	29.1	Pills	PCQ	251.7	21.3
	VC	450.6	29.4		VC	234.4	20.9
	VCL	425.6	28.5		VCL	229.8	20.5

σ_{vw} is the covariance of v and w; c_1 and c_2 are two stabilizing constants. For RGB images, SSIM is computed for the three color channels independently and the quality value is obtained by averaging the three indexes.

As it can be observed in Table 1, PCQ provides, on average, results close to Wu and SOM, while it outperforms MC and OCT. It is worth observing that PCQ does not start from a rigid and prefixed uniform quantization of image colors. It adaptively quantizes the image according to the estimated resolution of each color channel; in addition, each bin is determined by evaluating the visibility of image regions having the assigned representative color with respect to a changing image background and fixes the size of the bins as the ones which provides a not negligible contrast. However, the computation of the optimal point of the distortion curve, as defined in Eq. (7), suffers from some numerical instability

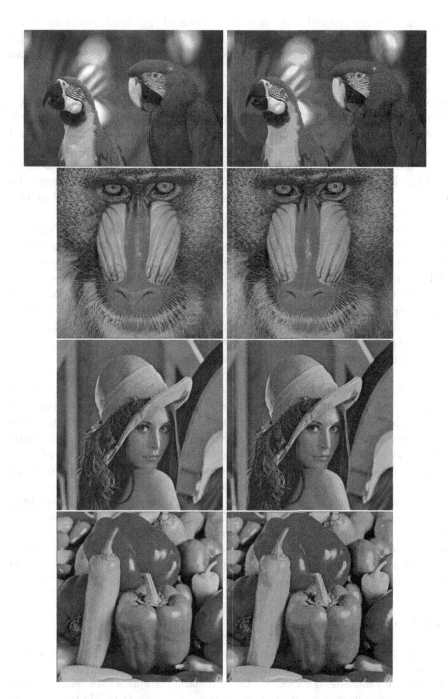

Fig. 3. *(Left)* Original image; *(Right)* quantized image by the proposed PCQ method with $K = 32$. (Color figure online)

that can influence the right selection of the optimal threshold, especially if K is low. Even though the correction of the numerical instability is able to further improve results in Table 1, this refinement has not been considered here, since out of the scope of the paper.

The second dataset, has been used for a direct comparison with the variance cut (VC) method in [3] and its refined version VCL. VC is a divisive CQ method which employs a binary splitting strategy. It starts from a $32 \times 32 \times 32$ color histogram obtained from a 5 bits/channel uniform quantization. At each iteration, the method splits the partition with the greatest SSE along the coordinate axis with the greatest variance at the mean point. The centroids of the resulting K sub-partitions define the color palette. VCL uses a few Lloyd-Max iterations for a local optimization of the two sub-partitions obtained at each step. In the same paper the authors compare their method with the ones considered in the first dataset and then they will be not reported in Table 2. MSE and MAE (Mean Absolute Error) are the two metrics used for comparing quantized image quality, as in [3]. The Mean Absolute Error MAE [14] is computed as: $MAE(v, w) = \frac{1}{HK} \sum_{i=1}^{H} \sum_{j=1}^{K} \sum_{k=1}^{3} |v_{i,j}(k) - w_{ij}(k)|$. As it can be observed in Table 2, PCQ, in its present and not optimized version, approaches and sometimes outperforms VC method. In addition the quality of quantized image is good, as it is shown in Fig. 3. Textured regions are well recovered, as for example the plumage of the Parrots, or in Lena hat or in Baboon. In addition, there is a good match between image region and assigned representative color.

These results show that PCQ is promising and the use of simple rules of human vision allows us to reach the results of some optimized methods which are based on statistical image features. In addition, using this kind of approach, some of the adopted measures and criteria could be embedded and interpreted in this new way of facing the problem. For example, the SSE is strictly related to the variability of the background which is used in the computation of image contrast. In addition, the definition of contrast measures allows us to simply embed some locality and spatial constraints which definitely would contribute to improve CQ, enabling the method to be more image content and perception dependent. Finally, the computational effort of PCQ is moderate since few simple operations are required. In fact, the most expensive step of the method is the iterative construction of the distortion curve. Future research will be devoted to define a more robust numerical scheme able to detect the optimal threshold, without constructing the whole curve.

References

1. Brun, L., Tremeau, A.: Chapter 9: Color quantization. In: Digital Color Imaging Handbook. Electrical and Applied Signal Processing. CRC Press (2002)
2. Bruni, V., Ramella, G., Vitulano, D.: Automatic perceptual color quantization of dermoscopic images. In: Braz, J., et al. (eds.) Proceedings of VISAPP 2015, pp. 323–330. SciTePress Science and Technology Publications, Berlin (2015)

3. Celebi, M.E., Wen, Q., Hwang, S.: An effective real-time color quantization method based on divisive hierarchical clustering. J. Real-Time Image Process. 10(2), 329–344 (2015)
4. Chan, H.C.: Perceived image similarity and quantization resolution. Displays 29, 451–457 (2008)
5. Chandler, D.M., Hemami, S.S.: Dynamic contrast-based quantization for lossy wavelet image compression. IEEE Trans. Image Process. 14(4), 397–410 (2005)
6. Dekker, A.: Kohonen neural networks for optimal colour quantization. Netw.: Comput. Neural Syst. 5(3), 351–367 (1994)
7. De Simone, F., Ticca, D., Dufaux, F., Ansorge, M., Ebrahimi, T.: A comparative study of color image compression standards using perceptually driven quality metrics. In: Proceedings of SPIE Conference on Optics and Photonics, Applications of Digital Image Processing XXXI (2008)
8. Gervautz, M., Purgathofer, W.: Simple method for color quantization: octree quantization. New Trends Comput. Graph. 219–231 (1988). Springer, Berlin
9. Heckbert, P.: Color image quantization for frame buffer display. ACM SIGGRAPH Comput. Graph. 16(3), 297–307 (1982)
10. Hsieh, I.S., Fan, K.C.: An adaptive clustering algorithm for color quantization. Pattern Recogn. Lett. 21, 337–346 (2000)
11. Kim, N., Kehtarnavaz, N.: DWT-based scene-adaptive color quantization. Real-Time Imaging 11, 443–453 (2005)
12. Nadenau, M.J., Reichel, J., Kunt, M.: Wavelet-based color image compression: exploiting the contrast sensitivity function. IEEE Trans. Image Process. 12(1), 58–70 (2003)
13. Ozturk, C., Hancer, C.E., Karaboga, D.: Color image quantization: a short review and an application with artificial bee colony algorithm. Informatica 25(3), 485–503 (2014)
14. Salomon, D.: Data Compression: The Complete Reference. Springer, London (2007)
15. Yao, J., Liu, G.: A novel color image compression algorithm using the human visual contrast sensitivity characteristics. Photonic Sens. 7(1), 72–81 (2017)
16. Wang, Z., Lu, L., Bovik, A.C.: Video quality assessment based on structural distortion measurement. Sig. Process. Image Commun. 19(2), 121–132 (2004)
17. Winkler, S.: Digital Video Quality. Vision Models and Metrics. Wiley, Hoboken (2005)
18. Wu, X.: Efficient statistical computations for optimal color quantization. In: Arvo, J. (ed.) Graphics Gems, vol. II, pp. 126–133. Academic Press, London (1991)
19. http://www.hlevkin.com/TestImages/
20. http://sipi.usc.edu/database/
21. http://r0k.us/graphics/kodak/
22. http://www.eecs.berkeley.edu/Research/Projects/CS/vision/bsds/
23. http://decsai.ugr.es/cvg/dbimagenes/

Product Recognition in Store Shelves as a Sub-Graph Isomorphism Problem

Alessio Tonioni$^{(\boxtimes)}$ and Luigi Di Stefano

University of Bologna, Bologna, Italy
{alessio.tonioni,luigi.stefano}@unibo.it
http://vision.disi.unibo.it

Abstract. The arrangement of products in store shelves is carefully planned to maximize sales and keep customers happy. Verifying compliance of real shelves to the ideal layout, however, is a costly task currently routinely performed by the store personnel. In this paper, we propose a computer vision pipeline to recognize products on shelves and verify compliance to the planned layout. We deploy local invariant features together with a novel formulation of the product recognition problem as a sub-graph isomorphism between the items appearing in the given image and the ideal layout. This allows for auto-localizing the given image within aisles of the store and improves recognition dramatically.

1 Introduction

Management of a grocery store or supermarket is a challenging task entailing personnel busy in supervising shelves and the whole sale point. In order to coordinate human resources more effectively, technology advances may be deployed to provide more reliable information in real time to the store manager. Such novelties, however, should turn out viable from a cost perspective, modify current practices moderately and not affect customer experience adversely. Computer vision techniques may fulfill the above requirements due to potential reliance on cheap cameras either mounted non-invasively in the store or embedded within the hand-held computers routinely used by sales clerks.

The problem addressed in this paper is **visual shelf monitoring** through computer vision techniques. The arrangement of products on supermarket shelves is planned very carefully in order to maximize sales and keep customers happy. The planned layout of products within shelves is called *planogram*: it specifies where each product should be placed within shelves and how many *facings* it should cover, that is how many packages of the same product should be visible in the front row of the shelf. A synthetic visual representation of a planogram can be observed in the middle column of Fig. 1. Thus far, planogram compliance is pursued by having sales clerks visually inspecting aisles during the quieter hours of the day, we propose, instead, to solve the task using computer vision techniques as planogram compliance can be seen as a very challenging object recognition task. As vouched by recently published patents [8,13], journal articles [11] and emerging companies (such as *Planorama*, *Vispera*, *Simble*

© Springer International Publishing AG 2017
S. Battiato et al. (Eds.): ICIAP 2017, Part I, LNCS 10484, pp. 682–693, 2017.
https://doi.org/10.1007/978-3-319-68560-1_61

Robotics)[1], major corporations are investigating as well on how to solve this task.

In this paper, we propose a computer vision pipeline that, given the planogram, an image of the observed shelve and one reference image for each item, can correctly localize each product, check whether the real arrangement is compliant to the planned one and detect missing or misplaced items. Key to our approach is a novel formulation of the problem as a sub-graph isomorphism between the product detected in the given image and those that should ideally be found therein given the planogram. Accordingly, our pipeline relies on a standard feature-based object recognition step, followed by our novel graph-based consistency check and a final localized image search to improve the overall product recognition rate.

2 Related Work

As pointed out by Merler et al. [12], dealing with grocery products on shelves exhibits peculiarities that render the task particularly challenging. Firstly to model each product one can rely only on a single or a few ideal views (renderings or taken in ideal studio-like condition) making it awkward to deploy directly object recognition methods that demand a large corpus of labelled training examples, such as deep convolutional neural networks. Moreover, as noticeable in Fig. 1, verifying planogram compliance calls for detecting and localizing each individual product instance within a shelves image crowded with lots of remarkably similar objects, distracting elements and nuisance. Merler et al. [12] in their seminal work, propose a public dataset and develop an assistive tool for visually impaired customers, pursuing products recognition without additional information concerning products layout. However, the performance of the proposed systems turned out quite unsatisfactory in terms of both precision and efficiency.

Further research has then been undertaken to ameliorate the performance of automatic visual recognition of grocery products [4,17,19]. In particular, Cotter et al. [4] report significant performance improvements deploying a machine learning based detector for each product. Such solution, however, requires many training images for each item and makes the system slow both at test and training time. The approach proposed in [4] was then extended in [1] through a contextual correlation graph between products that can be queried at test time to predict the products more likely to be seen given the last k detections, thereby reducing the number of detectors computed at test time and speeding up the whole system. Similarly in [3] the authors exploit machine learning classifier and statistically computed context information with good results in some product categories. In contrast with those works our solution deploys context constraints as well but does not require a huge corpus of annotated data as training.

---

[1] http://www.planorama.com/, http://vispera.co/, http://www.simberobotics.com/.

Another relevant work is due to George and Floerkemeier [7]. They carry out an initial classification to infer the categories of observed items, then, following detection, they run an optimization step based on a genetic algorithm to detect the most likely products from a series of proposals obtaining precision below 30%. The paper proposes also a publicly available dataset, referred to as *Grocery Products*, comprising 8350 product images classified into 80 hierarchical categories together with 680 high resolution images of shelves. We have used part of this public dataset as the main test bench for our method.

Marder et al. in [11] addressed the planogram compliance problem through detecting and matching SURF features [2] followed by visual and logical disambiguation between similar products. The paper reports a good 87.4% product recognition rate on a publicly unavailable dataset of cereal boxes and hair care products, though precision figures are not highlighted. To improve product recognition the authors deploy information dealing with the known product arrangement through specific hand-crafted rules, such as e.g. 'conditioners are placed on the right of shampoos'. Differently, we propose to deploy automatically these kinds of constraints by modelling the problem as a sub-graph isomorphism between the items detected in the given image and the planogram.

Systems to tackle the planogram compliance problem are described also in [6] and [10]. These papers delineate solutions relying either on large sensor/camera networks or mobile robots monitoring shelves while patrolling aisles. In contrast, our proposal would require just an off-the-shelf device, such as a smartphone, tablet or hand-held computer.

3   Proposed Pipeline

As depicted in Fig. 1, we propose to accomplish the planogram compliance check by a visual analysis pipeline consisting of three steps. We provide here an overview of the functions performed by the three steps.

The first step operates only on model images and the given shelves image without any additional information (e.g. location of acquisition or portion of the aisle pictured). Accordingly, the first step cannot deploy any constraint dealing with the expected product disposition, and is thus referred to as **Unconstrained Product Recognition**. As most product packages consist of richly textured piecewise planar surfaces, we obtained promising result through a standard object recognition pipeline based on local invariant features (as described, e.g., in [9]). Yet, the previously highlighted nuisances cause both missing product items as well as false detections due to similar products. In the following, using the set of detections (see Fig. 1) produced as output from the first step and despite the mistakes, our second step will identify the observed portion of the aisle to deploy constraint based on the expected product layout.

From the second step, dubbed **Graph-based Consistency Check**, we start leveraging on the information about products and their relative disposition contained in planograms. We choose to represent a planogram as a *grid-like fully connected graph* where each node corresponds to a product facing and is linked

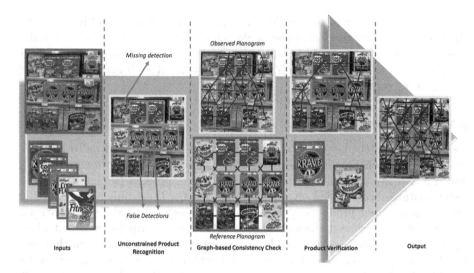

Fig. 1. Overview of our pipeline. For each step we highlight the inputs and outputs through red and yellow boxes. Product detections throughout stages are highlighted by green boxes, while blue lines show the edges between nodes in both the *Reference* and *Observed* planograms. (Color figure online)

to at most 8 neighbours at 1-edge distance, i.e. the closest facings along the cardinal directions. We rely on a graph instead of a rigid grid to allow for a more flexible representation; an edge between two nodes does not represent a perfect alignment between them but just proximity along that direction. This abstract representation, referred to as *Reference Planogram*, encodes information about the relative positions of products on shelves and can be visualized in Fig. 1. The detections provided by the first step are used in the second to build automatically another *grid-like graph* having the same structure as the *Reference Planogram* and referred to as *Observed Planogram*. Then, we find the *sub-graph isomorphism* between the *Observed* and *Reference* planograms, so as to identify sets of products placed in the same relative position in both graphs. As a result, the second step ablates away inconsistent nodes from the *Observed Planogram*, i.e. false detections yielded by the first step, and localizes the observed scene within the aisles by matching the *Observed Planogram* that concerns the shelves seen in the current image whit the *Reference Planogram* that models the whole aisle.[2]

After the second step the *Observed Planogram* should contain true detections only. Hence, those nodes that are missing compared to the *Reference Planogram* highlight items that appear to be missing wrt the planned product layout. The task of the third step, referred to as **Product Verification**, is to verify whether these items are really missing in the scene or not. Should the verification process

---

[2] More generally, matching the *Observed* to a set of *Reference* planograms does localize seamlessly the scene within a set of aisles or, even, the whole store.

highlight the presence of the product, a corresponding node would be added to the *Observed Planogram*; otherwise, a planogram compliance issue related to the checked node is reported (i.e. missing/misplaced item). The process is iterated till all the facings in observed shelves are either associated with detected instances or flagged as compliance issues.

3.1   Unconstrained Product Recognition

As already mentioned, we rely on the classical multi-object and multi-instance object recognition pipeline based on local invariant features presented in [9], which is effective with planar textured surfaces (e.g. product facings) and scales well to database comprising several hundreds or a few thousands models, i.e. the number of different products typically sold in grocery stores and supermarkets. Accordingly, we proceed through feature detection, description and matching, then cast votes into a pose space by a Generalized Hough Transform that can handle multiple peaks associated with different instances of the same model in order to cluster correspondences and filter out outliers. In our settings, it turns out reasonable to assume the input image to represent an approximately frontal view of shelves, so that both in-plane and out-of-plane image rotations are small. Therefore, we estimate a 3 DOF pose (image translation and scale change). Being the object recognition pipeline able to be deployed seamlessly with different local feature it turns out just as straightforward to rely on multiple types of features jointly to pursue higher sensitivity. Purposely, our implementation of the standard object recognition pipeline can run in parallel several detection/description/matching processes based on different features and have them eventually cast vote altogether within the same pose space. As reported in Sect. 4, we have carried out an extensive experimental investigation to establish which features would yield the best performance on the supermarket scenario.

3.2   Graph-Based Consistency Check

To build the *Observed Planogram* we instantiate a node for each item detected in the previous step and perform a search along 8 cardinal directions (N, S, E, W, NW, NE, SW, SE); if another bounding box is found at a distance less than a dynamically determined threshold, an edge is created between the two nodes. In the given node the edge is labelled according to the search direction (e.g. N), oppositely in the found neighbour node (i.e. S). The graph is kept self-coherent, e.g. if node B is the *North* node of A, then A must be the *South* node of B. In case of ambiguity, e.g. both A and C found to be the *South* node of B, we retain the edge between the two closest bounding boxes only.

Once built, we formulate our problem as follows: given I (*Reference Planogram*) and O (*Observed Planogram*), find an isomorphism between a subset of nodes in I and a subset of nodes in O such that the former subset has the maximum feasible cardinality given product placements constraints. Each node in I can be associated with a node in O only if they both refer to the same product and exhibit coherent neighbours. In theoretical computer science this problem is referred to as *subgraph isomorphism* and known to be NP-complete [18]. A general formulation may read as follows: given two graphs G an H, determine whether G contains a sub-graph for which does exist a bijection between the vertex sets of G and H. We choose not to rely on one of the many general algorithms, like e.g. [16], and, devised an ad-hoc heuristic algorithm that produces in a single run the isomorphism between observed and reference planograms while discarding wrongly detected products. Conversely, a general algorithm like [16] would yield either the isomorphism or a failure in case of wrong detection (i.e. no isomorphism) requiring an extensive search eventually speedable with some ad-hoc heuristic.

Algorithm 1. Find *sub-graph isomorphism* between I and O

$C_{max} \leftarrow 0$
$S_{best} \leftarrow \emptyset$
$\mathcal{H} \leftarrow CreateHypotheses(I, O)$
while $\mathcal{H} \neq \emptyset$ **do**
    $C, S, h_0 \leftarrow FindSolution(\mathcal{H}, C_{max}, \tau)$
    if $C > C_{max}$ **then**
        $S_{best}, C_{max} \leftarrow S, C$
    end if
    $\mathcal{H} \leftarrow \mathcal{H} - h_0$
end while
return S_{best}, C_{max}

Algorithm 1 outline our method. We starts with procedure *CreateHypotheses*, which establishes the initial set of hypotheses, $\mathcal{H} = \{\ldots h_i \ldots\}$, $h_i = \{n_I, n_O, c(n_I, n_O)\}$, with n_I and n_O denoting, respectively, a node in I and O related to the same product and $c(n_I, n_O) = \frac{nn_c}{nn_t}$ with nn_c number of coherent neighbours (e.g. referring to the same product both in O and I) and nn_t number of neighbours for that node in I. *CreateHypotheses* iterates over all $n_I \in I$ so to instantiate all possible hypotheses. An example of the hypotheses set is shown in the first row of Fig. 2. Then, procedure *FindSolution* finds a solution, S, by iteratively picking the hypothesis featuring the highest score (e.g. Fig. 2a). Successively, $\mathcal{H}$ is updated by removing the hypotheses containing either of the two nodes in the best hypothesis and increasing the scores of hypotheses associated with coherent neighbours (Fig. 2b). Procedure *FindSolution* returns also a confidence score for the current solution, C, which takes into account the cardinality of S, together with a factor which penalizes the presence in O of disconnected sub-graphs that exhibit relative distances different than those expected given the structure of I^3 which instead is always fully connected. *FindSolution* takes as input the score of the current best solution, C_{max}, and relies on a branch-and-bound scheme to accelerate the computation. In particular, as illustrated in Fig. 2c), after updating $\mathcal{H}$ (Fig. 2b), *FindSolution* calculates an upper-bound

---

[3] In the toy example in Fig. 2, O does not contain disconnected sub-graphs.

for the score, B_C, by adding to the cardinality of S the number of hypotheses in $\mathcal{H}$ that are not mutually exclusive, so as to early terminate the computation when the current solution can not improve C_{max}. The iterative process continues with picking the new best hypothesis until $\mathcal{H}$ is found empty or containing only hypotheses with confidence lower then a certain threshold τ (Fig. 2d). In the last step (Fig. 2e), the procedure computes C and returns also the first hypothesis, h_0, that was added into S. Upon returning from *FindSolution*, the algorithm checks whether or not the new solution S improves the best one found so far and removes h_0 from $\mathcal{H}$ (see Algorithm 1) to allow evaluation of another solution based on a set of different initial hypotheses.

Ideal Planogram (I)   Observed Planogram (O)

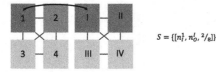

$H = \{ [n_I^1, n_O^I, {}^2/_8], [n_I^3, n_O^{III}, {}^2/_8], [n_I^4, n_O^{IV}, {}^2/_8], [n_I^1, n_O^{II}, {}^1/_8], [n_I^3, n_O^{IV}, {}^1/_8], [n_I^4, n_O^{III}, {}^0/_8] \}$

---

a) Pick the best hypothesis and add it to the solution. In case more hypotheses have equal score randomly pick one.

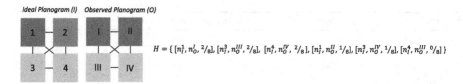

$S = \{ [n_I^1, n_O^I, {}^2/_8] \}$

---

b) Remove hypotheses and increase scores.
$H = \{ [n_I^1, n_O^I, {}^2/_8], [n_I^3, n_O^{III}, 1 + {}^2/_8], [n_I^4, n_O^{IV}, 1 + {}^2/_8], [n_I^1, n_O^{II}, {}^1/_8], [n_I^3, n_O^{IV}, {}^1/_8], [n_I^4, n_O^{III}, {}^0/_8] \}$

c) Compute B_c. If $B_c < C_{max}$ return $C = B_c$.
$H = \{ [n_I^3, n_O^{III}, {}^{10}/_8], [n_I^4, n_O^{IV}, {}^{10}/_8], [n_I^3, n_O^{IV}, {}^1/_8], [n_I^4, n_O^{III}, {}^0/_8] \}$
$S = \{ [n_I^1, n_O^I, {}^2/_8] \}$
$B_c = B(S, H) = 3$

d) Restart from step (a) until $H = \emptyset$ or $c(n_I, n_O) < \tau$ for all remaining hypotheses.

e) Compute C and return.

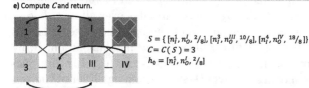

$S = \{ [n_I^1, n_O^I, {}^2/_8], [n_I^3, n_O^{III}, {}^{10}/_8], [n_I^4, n_O^{IV}, {}^{18}/_8] \}$
$C = C(S) = 3$
$h_0 = [n_I^1, n_O^I, {}^2/_8]$

Fig. 2. Toy example concerning two small graphs with 4 nodes used to describe procedure *FindSolution*. The colour of each node denotes the product the numbers within squares identify the different nodes in the text. (Color figure online)

Algorithm 1 finds self-consistent nodes in O given I, thereby removing inconsistent (i.e. likely false) detections and localizing the observed image wrt to the planogram. Accordingly, the output of the second steps contains information about which items appear to be missing given the planned product layout and where they ought to be located within the image.

3.3   Product Verification

We use an iterative procedure whereby each iteration tries to fill the observed planogram with one seemingly missing object through three stages. We start with the missing element featuring the highest number of already detected neighbours. The position and size of each neighbour, together with the average edge length in the *Observed Planogram*, provide an estimation of both the center and dimension[4] of the missing element: averaging estimations across the neighbours yields a good approximation. Thus, we define a coarse image ROI, allowing for some margin on account of possible localization inaccuracies, to find and localize the specific missing product therein. To accomplish this task we have experimented with template matching techniques as well as with a similar pipeline based on local features as deployed in Subsect. 3.1. Both approaches would provide a series of Detection Proposals filtered in the last stage of an iteration by first discarding those featuring bounding boxes that overlap with already detected items and then scoring the remaining ones according to the coherence of the position within the *Observed Planogram* and the detection confidence. Based on such a score, we pick the best proposal and add it to the *Observed Planogram*, so as to enforce new constraints that may be deployed throughout successive iterations. If either all detection proposals are discarded due to the overlap check or the best one exhibits too low a score, our pipeline reports a planogram compliance issue related to the currently analysed missing product.

4   Experimental Results

To assess the performance of our pipeline we rely on a subset of shelve pictures from the *Grocery Products* dataset [7] that we enrich with item-specific bounding boxes, the originals concerning only product categories (see Fig. 3(a) and (b)). For each image we also create an ideal planograms that encode in our graph like representation the perfect disposition of items on shelve (e.g. if the actual image contains voids or misplaced products they will not be encoded on the

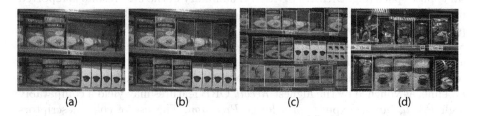

(a) (b) (c) (d)

Fig. 3. Qualitative samples: (a) ground-truth annotations provided with the *Grocery Products* dataset, (b) our instance-specific annotations, (c) (d) qualitative results obtained by our methods.

---

[4] Store databases contain product sizes: the image size of a missing product can be estimated from those of the detected neighbors and the known metric sizes.

ideal planogram that instead will model only the correct disposition). Our chosen subset consists of 70 images featuring box-like packages and dealing with 181 different products belonging to different categories such as rice, coffee, cereals, tea...; each image depicts ≈12 visible products. We judge a detection as correct if the intersection over union between the detected and ground-truth bounding boxes is >0.5. For each image we compute *Precision* (number of correct detections over total detections), *Recall* (number of correctly detected products over products visible in the image) and *F-Measure* (harmonic mean of *Precision* and *Recall*); in the following we report average figures across the dataset.

As regards comparative evaluation with respect to previous work, it is worth highlighting that the only work addressing exactly the same task as ours is [11], but neither their dataset nor their implementation are publicly available. Indeed, their system is quite complex and tailored for their specific use case so we judge that it would have been unfair to reproduce their results on our dataset by our own implementation. We have instead investigated on the use of region proposals, such as [15], followed by classification (e.g. by a CNN). Unfortunately, we found that this approach does not suit to the addressed task because in such a highly textured environment proposals tend to isolate logos and very colourful details from the underlying boxes while joining similarly colored regions belonging to different nearby products. We think that the most reasonable baseline to compare with is given by the first step of our pipeline, i.e. the standard object instance recognition approach based on local invariant features that has been proven to work effectively in a variety of diverse premises.

We will follow the processing flow along our pipeline starting with evaluating the **Unconstrained Product Recognition** step. To find the best suitable local features to be used in this scenario we have tested all the detectors and descriptors available in OpenCV (SIFT, SURF, ORB, BRISK, KAZE, AKAZE, STAR, MSD, FREAK, DAISY, LATCH, Opponent Color Space Descriptors), as well as the line segments features BOLD [14] (original code distributed by the authors for research purposes). We have considered features providing both the detector and descriptor (e.g. SIFT) as well as many different detector/descriptor pairs (e.g. MSD/FREAK) and multiple feature processes voting altogether in the same pose space (e.g. BRISK+SURF). A summary of the best results is reported in Fig. 4. As it can be observed, binary descriptors, such as BRISK and FREAK perform fairly well yielding the highest *Precision* and best *F-Measure* scores. SURF features provide good results alike, in particular as concerns *Recall*. It is also worth noticing how the use of multiple features, such as BRISK+SURF, to capture different image structures may help increasing the sensitivity of the pipeline, as vouched by the highest *Recall*. ORB features may yield a comparably high *Recall*, but at expense of a lower *Precision*. The use of color descriptors (Opponent SURF), instead, does not seem to provide significant benefits. As the second step is meant to prune out the false detections provided by the first, one would be lead to prefer those features yielding higher *Recall*. Yet, it may turn out hard for the second step to solve the *sub-graph isomorphism* problem in presence of too many false positives. A good balance between the two types of detection errors turns out preferable, therefore we will consider both BRISK and BRISK+SURF features within the first step in order to further evaluate the results provided by our pipeline after the **Graph-based Consistency Check**.

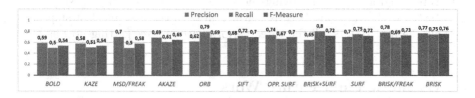

Fig. 4. Evaluation of different features for **Unconstrained Product Recognition**. Results ordered from left to right with increasing *F-Measure* scores.

For the second step we fixed $\tau = 0.25$ and deployed the algorithm proposed in Subsect. 3.2, the results are displayed in the left part of Fig. 5. First, the boost in *Precision* attained with both types of features compared to the output provided by the first step (Fig. 4) proves the effectiveness of the *sub-graph isomorphism* to remove false detections arising in unconstrained settings. In particular, when using BRISK features, *Precision* raises from ≈78 to ≈98% and with BRISK+SURF from ≈66 to ≈97%. Alongside, though, we observe a decrease in *Recall*, such as from ≈75 to ≈74% with BRISK and from ≈81 to ≈74% with BRISK+SURF. This is mostly due to items that, although detected correctly in the first step, cannot rely on enough self-coherent neighbours to be validated (i.e. $c(n_I, n_O) < \tau$). Overall, the **Graph-based Consistency Check** does improves performance significantly, as vouched by the increase of *F-Measure*.

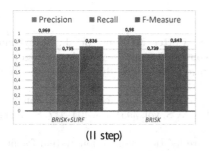

(II step)

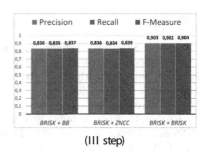

(III step)

Fig. 5. Left: results after **Graph-based Consistency Check** using either BRISK or BRISK+SURF in the first step. Right: different choices for the final **Product Verification** step, using BRISK features in the first step.

Given that BRISK slightly outperforms BRISK+SURF according to all the performance indexes and requires less computation, we pick the former features for the fist step and evaluate different design choices as regards the final **Product**

Verification. (i.e. different template matching and feature-based approaches). The best results, summarized in the right part of Fig. 5, concern template matching by the ZNCC (Zero-mean Normalized Cross Correlation) in the HSV color space, the *Best-buddies Similarity* method (BB) [5] in the RGB color space and a feature-based approach deploying the same features as in the first step, that is BRISK. As shown, using BRISK features in both the first and last step does provide the best results, all the three performance indexes getting now as high as $\approx 90\%$.

5   Conclusion and Future Work

We have shown that using our pipeline with BRISK features both for step one and three and deploying product arrangement constraints formulating the product recognition problem as a sub-graph isomorphism deals satisfactory performance on our test set. Accordingly, our proposed pipeline can work effectively in realistic scenarios in which just one model image per product and the planogram are available and the given image is not a priory localized with respect to the aisle. Our pipeline works quite well when applied to textured piece-wise planar products, however, grocery stores and supermarkets usually sells many different categories of products (e.g. bottles, deformable bags...) where local invariant features are likely to fail. To address this more challenging scenario, we plan to devise a preliminary product categorization step based on machine (deep) learning to segment the image into regions corresponding to different categories (e.g. piece-wise planar packages, bottles, jars, cans kitchenware...). Then, each detected segment may be handled by a specific approach to establish upon planogram compliance, the method described in this paper being applicable within segments labeled as piece-wise planar products.

References

1. Advani, S., Smith, B., Tanabe, Y., Irick, K., Cotter, M., Sampson, J., Narayanan, V.: Visual co-occurrence network: using context for large-scale object recognition in retail. In: 2015 13th IEEE Symposium on Embedded Systems for Real-time Multimedia (ESTIMedia), pp. 1–10. IEEE (2015)
2. Bay, H., Tuytelaars, T., Gool, L.: SURF: speeded up robust features. In: Leonardis, A., Bischof, H., Pinz, A. (eds.) ECCV 2006. LNCS, vol. 3951, pp. 404–417. Springer, Heidelberg (2006). doi:10.1007/11744023_32
3. Baz, I., Yoruk, E., Cetin, M.: Context-aware hybrid classification system for fine-grained retail product recognition. In: 2016 IEEE 12th Image, Video, and Multi-dimensional Signal Processing Workshop (IVMSP), pp. 1–5. IEEE (2016)
4. Cotter, M., Advani, S., Sampson, J., Irick, K., Narayanan, V.: A hardware accelerated multilevel visual classifier for embedded visual-assist systems. In: Proceedings of 2014 IEEE/ACM International Conference on Computer-Aided Design, pp. 96–100. IEEE Press (2014)
5. Dekel, T., Oron, S., Rubinstein, M., Avidan, S., Freeman, W.T.: Best-buddies similarity for robust template matching. In: 2015 IEEE Conference on Computer Vision and Pattern Recognition (CVPR), pp. 2021–2029. IEEE (2015)

6. Frontoni, E., Mancini, A., Zingaretti, P.: Embedded vision sensor network for planogram maintenance in retail environments. Sensors **15**(9), 21114–21133 (2015)
7. George, M., Floerkemeier, C.: Recognizing products: a per-exemplar multi-label image classification approach. In: Fleet, D., Pajdla, T., Schiele, B., Tuytelaars, T. (eds.) ECCV 2014. LNCS, vol. 8690, pp. 440–455. Springer, Cham (2014). doi:10. 1007/978-3-319-10605-2_29
8. Larsen, B.J.: Automated generation of a three-dimensional space representation and planogram verification. US Patent Ap. 14/073,231, 6 November 2013
9. Lowe, D.G.: Distinctive image features from scale-invariant keypoints. Int. J. Comput. Vis. **60**(2), 91–110 (2004)
10. Mankodiya, K., Gandhi, R., Narasimhan, P.: Challenges and opportunities for embedded computing in retail environments. In: Martins, F., Lopes, L., Paulino, H. (eds.) S-CUBE 2012. LNICSSITE, vol. 102, pp. 121–136. Springer, Heidelberg (2012). doi:10.1007/978-3-642-32778-0_10
11. Marder, M., Harary, S., Ribak, A., Tzur, Y., Alpert, S., Tzadok, A.: Using image analytics to monitor retail store shelves. IBM J. Res. Dev. **59**(2/3), 3:1–3:11 (2015)
12. Merler, M., Galleguillos, C., Belongie, S.: Recognizing groceries in situ using in vitro training data. In: . IEEE Conference on Computer Vision and Pattern Recognition, CVPR 2007, pp. 1–8. IEEE (2007)
13. Opalach, A., Fano, A., Linaker, F., Groenevelt, R.B.R.: Planogram extraction based on image processing. US Patent 8,189,855, 29 May 2012
14. Tombari, F., Franchi, A., Stefano, L.: Bold features to detect texture-less objects. In: Proceedings of IEEE International Conference on Computer Vision, pp. 1265–1272 (2013)
15. Uijlings, J.R., van de Sande, K.E., Gevers, T., Smeulders, A.W.: Selective search for object recognition. Int. J. Comput. Vis. **104**(2), 154–171 (2013)
16. Ullmann, J.R.: Bit-vector algorithms for binary constraint satisfaction and subgraph isomorphism. J. Exp. Algorithmics (JEA) **15**, 1–6 (2010)
17. Varol, G., Kuzu, R.S.: Toward retail product recognition on grocery shelves. In: Sixth International Conference on Graphic and Image Processing (ICGIP 2014), p. 944309. International Society for Optics and Photonics (2015)
18. Wegener, I.: Complexity Theory: Exploring the Limits of Efficient Algorithms. Springer Science & Business Media, Heidelberg (2005)
19. Winlock, T., Christiansen, E., Belongie, S.: Toward real-time grocery detection for the visually impaired. In: 2010 IEEE Computer Society Conference on Computer Vision and Pattern Recognition Workshops (CVPRW), pp. 49–56. IEEE (2010)

Real-Time Incremental and Geo-Referenced Mosaicking by Small-Scale UAVs

Danilo Avola[1], Gian Luca Foresti[1(✉)], Niki Martinel[1], Christian Micheloni[1],
Daniele Pannone[2], and Claudio Piciarelli[1]

[1] Department of Mathematics, Computer Science, and Physics, University of Udine,
Via delle Scienze 206, 33100 Udine, Italy
{danilo.avola,ginaluca.foresti,niki.martinel,christian.micheloni,
claudio.piciarelli}@uniud.it
[2] Department of Computer Science, Sapienza University,
Via Salaria 113, 00198 Rome, Italy
pannone@di.uniroma1.it

Abstract. In the last decade, the use of small-scale Unmanned Aerial Vehicles (UAVs) is increased considerably to support a wide range of tasks, such as vehicle tracking, object recognition, and land monitoring. A prerequisite of many of these systems is the construction of a comprehensive view of an area of interest. This paper proposes a small-scale UAV based system for real-time creation of incremental and geo-referenced mosaics of video streams acquired at low-altitude. The system presents several innovative contributions, including the use of A-KAZE feature extractor in aerial images, a Region Of Interest (ROI) to speed-up the stitching stage, as well as the use of the rigid transformation to build a mosaic at low-altitude mitigating in part the artifacts due to the parallax error. To prove the correctness of the proposed system at low-altitude, the public UMCD dataset and a simple metric based on the difference between image regions are presented. Instead, to show the overall effectiveness of the system, the public NPU Drone-Map dataset and a correlation measure are used. The latter metric evaluates the similarity between mosaics generated by the proposed method and those provided by a reference work of the current literature. Finally, the performance of the system compared with that of different modern solutions is also discussed.

Keywords: UAVs · Incremental mosaicking · Real-time mosaicking · ROI · UMCD dataset · NPU dataset · Rigid transformation · A-KAZE

1 Introduction

In recent years, the use of small-scale UAV based systems to support a wide range of application domains has increased considerably. In particular, these systems are especially useful in all those tasks in which a frequent, or even continuous, monitoring of an area of interest is required [16]. In military field, for

© Springer International Publishing AG 2017
S. Battiato et al. (Eds.): ICIAP 2017, Part I, LNCS 10484, pp. 694–705, 2017.
https://doi.org/10.1007/978-3-319-68560-1_62

example, a typical use of these systems regards the land monitoring for security purposes [14]. In fact, in many operative contexts, a large number of interesting areas has to be continually checked to detect the unexpected presence of people and vehicles, which can represent a possible danger [19,22]. Another typical example regards the frequent monitoring of strategic areas near military bases, refugee camps, and connecting roads, to detect the presence of objects (e.g., Improvised Explosive Devices, IEDs) that can threaten the crossing of humanitarian and military convoys. In civilian field, these systems can be suitably used for monitoring restricted areas after catastrophic events, such as earthquakes, tsunamis, damage to nuclear power plants, and so on. A prerequisite of many of the above introduced systems is the construction of a comprehensive view of an area of interest. The video sequence that represents the area is often acquired at low-altitude for several reasons, including the need to have a high spatial resolution for classifying objects [8,15], to safeguard the UAV, to hide the UAV, and many others.

This paper presents a small-scale UAV based system for real-time creation of incremental and geo-referenced mosaics of areas of interest acquired at low-altitude. The only input required by the system is a set of GPS coordinates that specifies one or more areas that have to be mosaicked. The proposed mosaicking algorithm presents several innovative contributions compared to the current state-of-the-art. First, to speed-up the feature extraction and matching processes, it adopts the A-KAZE extractor [1]. The recent literature [1–3] has shown that A-KAZE features are faster to compute than SIFT [13] and SURF [4], moreover they exhibit much better performance in detection and description than ORB [24]. Second, the mosaicking algorithm implements an automatic method to optimize the acquisition rate of the RGB camera based on the telemetry (i.e., speed and height). Third, to speed-up all steps involved in the stitching process, the mosaicking algorithm implements a ROI through which the computation required for the stitching of each new frame on the mosaic is reduced. Fourth, unlike the majority of the mosaicking algorithms known in literature that use RANSAC [7] to perform the geometric transformation stage, the proposed algorithm adopts the rigid transformation [18] that allows the building of mosaics at low-altitude mitigating in part the artifacts due to the parallax error [9]. Currently, public datasets for testing mosaicking algorithms contain video sequences acquired at high-altitude, for this reason we have implemented and made available the UAV Mosaicking and Change Detection (UMCD) dataset[1]. Instead, to test the algorithm at high-altitudes we have used the NPU Drone-Map dataset[2].

The rest of the paper is structured as follows. Section 2 presents some selected works near to that proposed. Section 3 introduces the architecture of the proposed mosaicking algorithm and discusses the different algorithmic choices, including the feature extraction by the A-KAZE extractor, the stitching process by the rigid transformation, and the implementation of the ROI. Section 4 reports

---

[1] http://www.umcd-dataset.net/.

[2] http://zhaoyong.adv-ci.com/downloads/npu-dronemap-dataset/.

the experimental results obtained by using both UMCD and NPU Drone-Map datasets. Finally, Sect. 5 concludes the paper.

2 Related Work

Regardless the specific size of the UAVs (e.g., small, medium, large), in the last years a wide range of tasks has been supported by their use, such as urban monitoring [23], vegetation analysis [17], surveillance [10], and others. Anyway, the pipeline of these systems is similar and includes specific main stages: extraction of salient points from frames (i.e., feature extraction), find image transformation values (i.e., rotation, scale, and translation), and merge frames together.

Several works in the literature produce a mosaic in off-line mode, i.e., when all the frames are available for the processing. Two examples are reported in [20] and [11], respectively, where the authors present a robust system that uses SIFT extractor and homography transformation based on RANSAC. Similar steps are used in [27], where the authors, first, utilize a transformation between frames based on an iterative threshold to find the edges and, subsequently, apply a correlation phase to merge them. From a performance point of view, the mosaicking of a high number of frames is a time-consuming duty that requires a wide availability of resources. A possible solution to this issue is reported in [21], where a fast algorithm using little amount of resources is presented. In particular, the proposed algorithm works by doing pairwise image registration, then it projects the resulting points to the ground and produces a new set of control points by moving these points closer to each other. Then, it fits image parameters to these new control points and repeats the process to convergence. Regarding the real-time processing, in [12] the authors use ORB as feature extractor and provide a spatial and temporal filter for removing the majority of the outlier points. In [28], the authors use SIFT as feature extractor and Euclidean distance with a threshold for matching the frames. Finally, in [26], the authors adopt an incremental technique and more UAVs to cover an area of interest and to build a qualitative mosaic. Inspired by several of these works, but unlike them, the proposed mosaicking algorithm uses A-KAZE and ROI to speed-up the stitching process. Moreover, the use of the rigid transformation allows to obtain mosaics whose video sequences are acquired at low-altitude.

3 The Mosaicking Algorithm

The logical architecture of the small-scale UAV based system and the pipeline of the proposed mosaicking algorithm are shown in Fig. 1. The algorithm consists of four main stages each of which is discussed below. The system is designed to work with standalone and client-server architectures. However, the latter is used to explain properly how the system works.

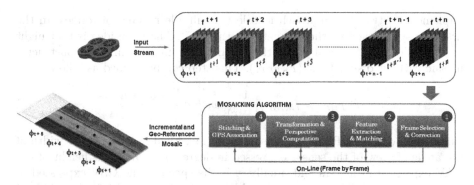

Fig. 1. The proposed mosaicking algorithm. f^t and ϕ_t are the frames and the linked GPS coordinates provided to the algorithm at each second t, respectively.

3.1 Background

In the following, let:

$$UAV_{path} = \{\phi_{t+i}(x_{t+i}, y_{t+i}) \mid t \in \mathbb{N} \wedge i \in [1, \ldots, n] \subset \mathbb{N}\} \tag{1}$$

be the set of GPS coordinates that defines the area of interest that needs to be mosaicked, where, t is the amount of seconds required by the UAV to reach the area, and n is the seconds of flight duration within the area. Besides, for each $i \in [1, \ldots, n]$, $\phi_{t+i}(x_{t+i}, y_{t+i})$ is the $t + i^{th}$ coordinate and (x_{t+i}, y_{t+i}) is the pair $(latitude, longitude)$. Without loss of generality, we can define ϕ_{start} and ϕ_{end} when $i = t + 1$ and $i = t + n$, respectively. In addition, let:

$$F_{TRS} = \{f^{t+i} \mid t \in \mathbb{N} \wedge i \in [1, \ldots, n] \subset \mathbb{N}\} \tag{2}$$

be the set of frames transmitted from the UAV to the processing unit (local or remote) within the UAV_{path}, where t and n are defined as above. For each $i \in [1, \ldots, n]$, $f^{t+i} = \{f_1^{t+i}, f_2^{t+i}, \ldots, f_{FPS}^{t+i}\}$ is the set of frames transmitted by the UAV at the second i. The set depends on frame per second (FPS) of the RGB camera. The UAV starts the transmission to the processing unit from the take-off up to the landing. In general, each second $k \in \mathbb{N}$ of transmission is composed of a GPS coordinate, $\phi_k(x_k, y_k)$, and a set of frames, $f^k = \{f_1^k, f_2^k, \ldots, f_{FPS}^k\}$.

3.2 Frame Selection and Correction

Since the aim of the algorithm is to build the mosaic of the area of interest defined by the UAV_{path}, all the frames transmitted outside of this path (i.e., $f^k \notin F_{TRS}$ for each $k \in \mathbb{N}$) are discarded by the processing unit. The rest of the frames transmitted by the UAV (i.e., $f^k \in F_{TRS}$ for each $k \in [t+1, \ldots, t+n] \subset \mathbb{N}$) are used in part to create the mosaic, while the remaining are discarded again. This is due to the fact that at each second the UAV tends to transmit more frames than ones necessary to create a proper mosaic. The proposed algorithm implements a two-step approach to select a suitable number of frames. In the first, the system adopts the telemetry of the UAV. The main idea is that hight and speed of the UAV can derive the number of frames required to construct a mosaic without disjunctions (F_{STEP_1}). This step can be defined as follows:

$$F_{STEP_1} = f_{MAX} \frac{1}{(1+c)} \; with \; c = \frac{FoV * h}{v} \tag{3}$$

where, f_{MAX} is the FPS of the sensor, FoV (i.e., Field of View) is the width of the angle of view of the sensor expressed in degree, h is the flight height of the UAV expressed in meters and, finally, v is the speed of the UAV expressed in meters per second. The frames selected by the first step can be further thinned out by means of a user parameter (F_{STEP_2}) that defines the amount of overlap between a current frame and the linked part of the mosaic. Although it is not a focus of the paper, it should be observed that the system implements also a layer in which the calibration parameters of the camera can be stored to correct the possible lens-distortion introduced by the RGB camera.

3.3   Feature Extraction and Matching

Let M_j be the mosaic built up to the second j and let $\hat{f}_s^{j+1}$, with $j + 1 \in [t + 1, \ldots, t + n] \subset \mathbb{N}$ and $s \in [1, \ldots, FPS] \subset \mathbb{N}$, the current selected frame, at the second $j + 1$, to be added to the mosaic. The main steps to built the new mosaic, $M_j \cup \hat{f}_s^{j+1}$, are the feature extraction and matching processes. In general, the features extracted from each current frame should be compared with those extracted from the whole mosaic to establish where the current frame has to be placed. Since the size of the mosaic grows over time, the comparison stage tends to become unmanageable after a certain period of time. With the aim to avoid such a issue, the proposed system uses a ROI to extract the features from the mosaic. The ROI tracks the last frame added to the mosaic and delimits, to a region surrounding it, the feature extraction process. A ROI centred on the last frame and sized three times than the size of a frame is sufficient to ensure the proper execution of the mosaicking algorithm. By the ROI the adding of a new frame takes a constant-time, no more dependent on the increasing size of the mosaic. Notice that the ROI concept is not new, but it is worth describing it due to the its effectiveness in increasing the system performance. The proposed algorithm uses A-KAZE, instead of the most popular extractors, such as SIFT, SURF or ORB. This is due to the fact that A-KAZE adopts both the Fast Explicit Diffusion (FED) embedded in a pyramidal framework and the Modified-Local Difference Binary (M-LDB) descriptor in order to speed-up feature detection in non-linear scale space and to exploit gradient information from the non-linear scale space, respectively. These aspects make A-KAZE an optimal compromise between speed and performance with respect to the current literature [1].

The keypoints extracted from M_j and $\hat{f}_s^{j+1}$ are used to detect the overlapping region between them. Let $\mathcal{X}_{M_j} = \{\alpha_1, \ldots, \alpha_h\}$ and $\mathcal{X}_{\hat{f}_s^{j+1}} = \{\beta_1, \ldots, \beta_t\}$ be the set of keypoints extracted by A-KAZE from M_j and $\hat{f}_s^{j+1}$, respectively. With the aim of finding the correspondence between the keypoints in $\mathcal{X}_{M_j}$ with those in $\mathcal{X}_{\hat{f}_s^{j+1}}$ a simple Brute Force Matcher (BFM) algorithm is used [25]. This algorithm performs an exhaustive search between the two sets of keypoints and matches only those keypoints that have an identical pattern (i.e., local structure of the pixels). Formally, at the end of the process, the algorithm generates two sub-sets $\hat{\mathcal{X}}_{M_j} = \{\alpha_{h_1}, \ldots, \alpha_{h_m}\} \subseteq \mathcal{X}_{M_j}$ and $\hat{\mathcal{X}}_{\hat{f}_s^{j+1}} = \{\beta_{t_1}, \ldots, \beta_{t_m}\} \subseteq \mathcal{X}_{\hat{f}_s^{j+1}}$ where for each $k \in \{h_1, \ldots, h_m\}$ exists a single $j \in \{t_1, \ldots, t_m\}$ such that $\alpha_k \equiv \beta_j$. As well-known, the two sub-sets have the same cardinality.

3.4 Transformation and Perspective Computation

Once obtained the corresponding keypoints (i.e., $\hat{\mathcal{X}}_{M_j}$ and $\hat{\mathcal{X}}_{\hat{f}_s^{j+1}}$) between the two frames, the system must compute the geometrical transformation by which the keypoints of the current frame, $\hat{f}_s^{j+1}$, are collimated with ones of the mosaic, $\mathcal{X}_{M_j}$, within the reference system of the latter. This transformation is subsequently used on each pixel of the frame to stitch it over the mosaic. In literature, the RANSAC algorithm to calculate the homography transformation is considered the reference approach. It consists in using the corresponding keypoints to iteratively estimate the parameters of a mathematical model by which to perform the geometric projection of each pixel between the two images. Despite this, as shown in Fig. 2a, the homography transformation can produce a high level of distortions especially when it is applied on images acquired a low-altitude. In particular, the mosaic can present an unreal curvature. This is due to the fact that the homography transformation matrix has 8 degrees of freedom, hence at least 4 corrected correspondences are required to build a proper mosaic. In the proposed mosaicking algorithm, the acquired images can be considered as a linear scanning of the ground surface, therefore a transformation with less degrees of freedom can be adopted. For this reason, the rigid transformation matrix that has only 4 degrees of freedom is implemented [18]. The reference example reported in Fig. 2b shows the goodness of the obtained results. The majority of the UAV based systems treat video sequences acquired at high altitude, or propose a orthorectification pre-processing step at the expense of the real-time

(a)   (b)

Fig. 2. Geometric transformation: (a) homography transformation by RANSAC algorithm, (b) rigid transformation.

processing [29] thus avoiding this type of issue. The last step of the module is to merge the pixels of the mosaic, $\mathcal{X}_{M_j}$, with the transformed pixels of the frame, $\Gamma(\hat{f}_1^{j+s})$, to obtain a new pixel matrix, $\mathcal{X}_{M_j} \cup \Gamma(\hat{f}_s^{j+1})$.

3.5   Stitching and GPS Association

The acquisition of the GPS coordinates is performed following the NMEA[3] format, one of the most widespread standards for the transmission of position data. Current commercial GPS transmitters provide one or more position data per second, however in the latter case a good practice is to derive a single information per second to reduce the intrinsic error due to the acquisition process. Since the construction of a mosaic can require more frames per second, this means that only the first of the n frames for second acquired by the RGB camera is associated to a GPS coordinate, the rest of the $n - 1$ frames, if added to the mosaic, has to be associated to coordinates inferred by ones previously acquired. Actually, once obtained two coordinates of the first frame of two consecutive seconds, then the coordinates of the remaining frames of the first second can be derived by adopting a simple linear interpolation. Let $\phi_j(x_j, y_j)$ and $\phi_{j+1}(x_{j+1}, y_{j+1})$ be the GPS coordinates acquired and associated with the frames $\hat{f}_1^j$ and $\hat{f}_1^{j+1}$, respectively ($s = 1$ in both cases since they are the first frames of each second). In addition, considering $\hat{f}_1^j$ belonging to the mosaic M_j and $\hat{f}_1^{j+1}$ the current frame. Then, the coordinate of any frame added to the mosaic between them can be derived as follows:

$$x_k = x_j + \frac{k}{FPS}(x_{j+1} - x_j), \ y_k = y_j + \frac{k}{FPS}(y_{j+1} - y_j) \qquad (4)$$

where, x_k and y_k are the interpolated *latitude* and *longitude*, respectively, of the new GPS coordinate $\phi_k(x_k, y_k)$ associated to the frame $\hat{f}_k^j$. Moreover, k specifies the coordinate of which frame needs to be computed, finally, FPS is the frames per second of the sensor. The current version of the system performs the mosaicking algorithm in on-line mode. This means that when the system acquires a new GPS coordinate, it also considers the previous acquired one, computes the interpolation process and associates the interpolated coordinates to the linked frames within the mosaic. Each GPS coordinate (acquired or interpolated) is anchored to the barycentre of the linked frame. This last is a main aspect to enable the system with a wide range of tasks. Once that the GPS coordinate has been linked to the new frame, the gain compensation between this latter and the mosaic is performed by using the multi-band blending [5]. This assures that there will be no seams when the new frame is added to the current mosaic.

4   Experimental Results and Discussion

For testing the mosaicking algorithm, two recent public datasets were used. The first is the UMCD dataset, that contains a collection of aerial video sequences

---

[3] http://www.nmea.org/.

acquired at low-altitudes. The second is the NPU Drone-Map dataset, that contains a collection of aerial video sequences acquired at high-altitude. In both cases, the sequences are acquired by small-scale UAVs. Regarding the first dataset, we tested the algorithm on 40 challenging video sequences and measured the quality of the obtained mosaics by a simple metric based on the difference between image regions. Regarding the second dataset, we compared the proposed mosaicking algorithm with that presented in [6]. The latter is one of the few works in the literature that makes available source code, video sequences (i.e., the NPU Drone-Map dataset), and obtained mosaics to support a concrete comparison with other approaches. In particular, 4 challenging video sequences were selected from the second dataset and a correlation measure was adopted to quantify the similarity between mosaics pairs.

4.1 Low-Altitude and High-Altitude Mosaicking

In this sub-section, key considerations about the quality of the obtained mosaics are reported and discussed. Regarding the low-altitude, the adopted 40 video sequences had an average acquisition height of about 15 meters. In Fig. 3a an example is shown. In order to measure the quality of the mosaics derived by these video sequences the image difference process presented in [3] is adopted. The main idea is that each part of the mosaic must have the same spatial and colour resolution with respect to the original frames that have generated it. For

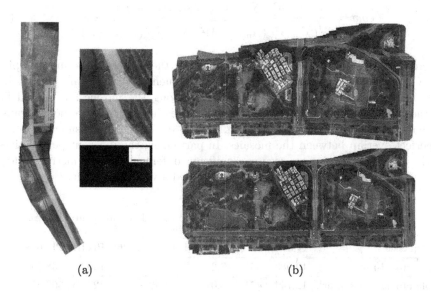

(a)   (b)

Fig. 3. Experimental results: (a) example of mosaic at low-altitude. The three miniatures are the frame extracted from the mosaic (up), one of the original frames used to build the mosaic (middle), the difference between the overlapped regions (bottom), (b) examples of mosaics at high-altitude by the proposed method (up), the method proposed in [6] (bottom).

this reason, the difference between each portion of the mosaic and the linked original frames is computed. Subsequently, a simple histogram is calculated on each image difference to evaluate the degree of deviation. Anyway, this simple but effective process has shown that each part of the mosaic generated by the proposed method is quite similar to the linked frames. On average, the difference images shown a deviation of about 15%. This can be considered a real good result taking into account all the geometrical distortion and error propagations that occur during the complex mosaicking process. Moreover, it should be considered that the incremental real-time mosaicking process at low-altitude is a topic that needs to be further investigated. By the implemented UMCD dataset and the provided results, the aim is to provide a concrete first contribute for the comparison of these algorithms. In Fig. 3b, examples of high-altitude mosaics are shown. In particular, the mosaic on the top of the Fig. 3b is generated with the proposed approach, while the mosaic on the bottom is generated with the method proposed in [6]. Both mosaics were created by using the same video sequence contained in the NPU Drone-Map dataset (named: phantom3-centralPark). How it is possible to observe, some visual differences are present. This is due to the fact that the proposed method applies only basic transformations, such as translation, rotation, and scale change, while the method with which we compare performs the orthorectification of the frames. Despite this, the degree of correlation between the two types of mosaic is impressive. To verify the similarity between them the following metric was adopted:

$$corr = \frac{\sum_m \sum_n (A_{mn} - \bar{A})(B_{mn} - \bar{B})}{\sqrt{(\sum_m \sum_n (A_{mn} - \bar{A})^2)(\sum_m \sum_n (B_{mn} - \bar{B})^2)}} \tag{5}$$

where A, B are the two mosaics, and $\bar{A}$, $\bar{B}$ are the means of the mosaics pixels. On average, considering all the 4 video sequences reported in Table 1, we obtained a correlation value of about 80% among the mosaics. It should be considered that due to the different image processing, such as geometric transformation, orthorectification, stitching, and so on, it is not possible to obtain a perfect overlap between the mosaics. In particular, the different perspectives of the obtained mosaics are seen as significant differences by the metric. In any case, the degree of correlation can be considered a very high value.

Table 1. Time needed for generating the mosaics. The unit is in minutes.

Sequence	Frames	KFs	Proposed	Bu et al. [6]	Pix4D	Photoscan
phantom3-npu	19,983	457	7.2	9.32	140.08	538.38
phantom3-centralPark	12,744	471	6.01	8.49	127.73	563.57
phantom3-village	16,969	406	10.4	11.31	132.07	360.70
phantom3-huangqi	14,776	393	8	10.36	102.83	462.32

4.2   Mosaicking Performance

In this sub-section, the performance of the proposed method is presented. All the experiments were performed on a laptop equipped with an Intel i7 6700HQ CPU, 16 GB DDR3 RAM and a nVidia GTX960 GPU. In Table 1, the time needed for generating the mosaics is reported. More specifically, we compared the proposed method with the algorithm reported in [6] and with two commercial software, Pix4D[4] and Photoscan[5], also reported in the same work. The proposed method stitches 1 frame per second, while the method proposed in [6] requires the stitching of 10 frames per second. Both Pix4D and Photoscan, instead, use only the keyframes to produce the final mosaic (i.e., similar to the proposed algorithm). Since all methods, with the exception of that proposed, use the GPU, a resize to the half of HD resolution (i.e., the original size of the frames) to be stitched is performed. In Table 1, the comparison is shown. As it is possible to observe, both the proposed and [6] algorithms take much less time than the commercial software. The proposed method show low processing times even with respect to the work proposed [6] and the generated mosaics by the two approaches result quite similar. Anyway, we are currently developing an approach to perform the orthorectification frame by frame.

5   Conclusions

This paper propose a small-scale UAV based system for the real-time creation of incremental and geo-referenced mosaics of video streams acquired at low-altitude. The system presents several innovative contributions, including the use of A-KAZE feature extractor in aerial images, a ROI to speed-up the stitching stage, as well as the use of the rigid transformation to build a mosaic at low-altitude mitigating in part the artifacts due to the parallax error. We implemented the UMCD dataset and used the NPU Drone-Map dataset to test the algorithm at low-altitude and high-altitude, respectively. The adopted metrics have shown remarkable results, in time and quality, compared with selected solutions of the current state-of-the-art.

Acknowledgments. This work was partially supported by both the "Proactive Vision for advanced UAV systems for the protection of mobile units, control of territory and environmental prevention (SUPReME)" FVG L.R. 20/2015 project and the "Augmented Reality for Mobile Applications: advanced visualization of points of interest in touristic areas and intelligent recognition of people and vehicles in complex areas (RA2M)" project.

References

1. Alcantarilla, P., Nuevo, J., Bartoli, A.: Fast explicit diffusion for accelerated features in nonlinear scale spaces. In: BMVC, pp. 1–11 (2013)

---

[4] https://pix4d.com/.
[5] http://www.agisoft.com/.

2. Avola, D., Foresti, G.L., Cinque, L., Massaroni, C., Vitale, G., Lombardi, L.: A multipurpose autonomous robot for target recognition in unknown environments. In: INDIN, pp. 766–771 (2016)
3. Avola, D., Cinque, L., Foresti, G.L., Massaroni, C., Pannone, D.: A keypoint-based method for background modeling and foreground detection using a PTZ camera. In: PRL (2016)
4. Bay, H., Ess, A., Tuytelaars, T., Gool, L.V.: Speeded-up robust features (SURF). CVIU 110(3), 346–359 (2008)
5. Brown, M., Lowe, D.G.: Automatic panoramic image stitching using invariant features. IJCV 74(1), 59–73 (2007)
6. Bu, S., Zhao, Y., Wan, G., Liu, Z.: Map2dfusion: real-time incremental UAV image mosaicing based on monocular slam. In: IROS, pp. 4564–4571 (2016)
7. Fischler, M.A., Bolles, R.C.: Random sample consensus: a paradigm for model fitting with applications to image analysis and automated cartography. Commun. ACM 24(6), 381–395 (1981)
8. García, J., Martinel, N., Gardel, A., Bravo, I., Foresti, G.L., Micheloni, C.: Modeling feature distances by orientation driven classifiers for person re-identification. JVCIR 38, 115–129 (2016)
9. He, B., Yu, S.: Parallax-robust surveillance video stitching. Sensors 16(1), 1–12 (2015)
10. Heikkilä, M., Pietikäinen, M.: An image mosaicing module for wide-area surveillance. In: VSSN, pp. 11–18 (2005)
11. Javadi, M.S., Kadim, Z., Woon, H.H.: Design and implementation of automatic aerial mapping system using unmanned aerial vehicle imagery. In: CGMIP, pp. 91–100 (2014)
12. Li, J., Yang, T., Yu, J., Lu, Z., Lu, P., Jia, X., Chen, W.: Fast aerial video stitching. IJARS 11, 1–11 (2014)
13. Lowe, D.G.: Distinctive image features from scale-invariant keypoints. IJCV 60(2), 91–110 (2004)
14. Martinel, N., Avola, D., Piciarelli, C., Micheloni, C., Vernier, M., Cinque, L., Foresti, G.L.: Selection of temporal features for event detection in smart security. In: Murino, V., Puppo, E. (eds.) ICIAP 2015. LNCS, vol. 9280, pp. 609–619. Springer, Cham (2015). doi:10.1007/978-3-319-23234-8_56
15. Martinel, N., Micheloni, C., Foresti, G.L.: The evolution of neural learning systems: a novel architecture combining the strengths of NTs, CNNs, and ELMs. IEEE Syst. Man Cybern. Mag. 1(3), 17–26 (2015)
16. Martinel, N., Micheloni, C., Piciarelli, C.: Pre-emptive camera activation for video surveillance HCI. In: ICIAP, pp. 189–198 (2011)
17. Michener, W.K., Houhoulis, P.F.: Detection of vegetation changes associated with extensive flooding in a forested ecosystem. Photogram. Eng. Remote Sens. 63(12), 1363–1374 (1997)
18. Ngo, P., Passat, N., Kenmochi, Y., Talbot, H.: Topology-preserving rigid transformation of 2D digital images. IEEE TIP 23(2), 885–897 (2014)
19. Piciarelli, C., Micheloni, C., Foresti, G.L.: Occlusion-aware multiple camera reconfiguration. In: ICDSC, pp. 88–94 (2010)
20. Piciarelli, C., Micheloni, C., Martinel, N., Vernier, M., Foresti, G.L.: Outdoor environment monitoring with unmanned aerial vehicles. In: Petrosino, A. (ed.) ICIAP 2013. LNCS, vol. 8157, pp. 279–287. Springer, Heidelberg (2013). doi:10.1007/978-3-642-41184-7_29
21. Pritt, M.: Fast orthorectified mosaics of thousands of aerial photographs from small UAVs. In: AIPR, pp. 1–8 (2014)

22. Remagnino, P., Velastin, S.A., Foresti, G.L., Trivedi, M.: Novel concepts and challenges for the next generation of video surveillance systems. MVA **18**(3), 135–137 (2007)

23. Ridd, M.K., Liu, J.: A comparison of four algorithms for change detection in an urban environment. Remote Sens. Environ. **63**(2), 95–100 (1998)

24. Rublee, E., Rabaud, V., Konolige, K., Bradski, G.: ORB: an efficient alternative to SIFT or SURF. In: ICCV, pp. 2564–2571 (2011)

25. Song, B.C., Kim, M.J., Ra, J.B.: A fast multiresolution feature matching algorithm for exhaustive search in large image databases. IEEE TCSVT **11**(5), 673–678 (2001)

26. Wischounig-Strucl, D., Rinner, B.: Resource aware and incremental mosaics of wide areas from small-scale uavs. MVA **26**(7), 885–904 (2015)

27. Yang, X.H., Xin, S.X., Jie, Z.Q., Jing, X., Dan, Z.D.: The UAV image mosaic method based on phase correlation. In: ICCA, pp. 1387–1392 (2014)

28. Yang, Y., Sun, G., Zhao, D., Peng, B.: A real time mosaic method for remote sensing video images from UAV. J. Sig. Inf. Process. **4**(3B), 168–172 (2013)

29. Zhou, G.: Near real-time orthorectification and mosaic of small UAV video flow for time-critical event response. IEEE TGRS **47**(3), 739–747 (2009)

Automatic Multi-seed Detection for MR Breast Image Segmentation

Albert Comelli[1], Alessandro Bruno[1(✉)], Maria Laura Di Vittorio[2],
Federica Ienzi[2], Roberto Lagalla[2], Salvatore Vitabile[2], and Edoardo Ardizzone[1]

[1] Dipartimento dell'Innovazione Industriale e Digitale (DIID),
Università di Palermo, Palermo, PA, Italy
alessandro.bruno15@unipa.it
[2] Dipartimento di Biopatologia e Biotecnologie Mediche,
Università di Palermo, Palermo, PA, Italy

Abstract. In this paper an automatic multi-seed detection method for magnetic resonance (MR) breast image segmentation is presented. The proposed method consists of three steps: (1) pre-processing step to locate three regions of interest (axillary and sternal regions); (2) processing step to detect maximum concavity points for each region of interest; (3) breast image segmentation step. Traditional manual segmentation methods require radiological expertise and they usually are very tiring and time-consuming. The approach is fast because the multi-seed detection is based on geometric properties of the ROI. When the maximum concavity points of the breast regions have been detected, region growing and morphological transforms complete the segmentation of breast MR image. In order to create a Gold Standard for method effectiveness and comparison, a dataset composed of 18 patients is selected, accordingly to three expert radiologists of University of Palermo Policlinico Hospital (UPPH). Each patient has been manually segmented. The proposed method shows very encouraging results in terms of statistical metrics (Sensitivity: 95.22%; Specificity: 80.36%; Precision: 98.05%; Accuracy: 97.76%; Overlap: 77.01%) and execution time (4.23 s for each slice).

Keywords: Automatic segmentation · Breast MR · Maximum concavity points · Seed detection

1 Introduction

Nowadays, medical research focuses on the optimization of the workflow from the acquisition to final report of detected lesions [1] and for volumetric measurements from preoperative staging and evaluation after neo-adjuvant chemotherapy [2]. In last decades the scientific community has shown a growing interest towards the analysis of breast images. As a matter of fact, nowadays the CAD (computer aided diagnostic) systems are representing a second reader to help radiologist in interpretation task. Breast imaging is an effective tool for detection of suspicious regions in breast tissue, contributing to a noticeable decrease of mortality

© Springer International Publishing AG 2017
S. Battiato et al. (Eds.): ICIAP 2017, Part I, LNCS 10484, pp. 706–717, 2017.
https://doi.org/10.1007/978-3-319-68560-1_63

associated with breast cancer [3,4]. Breast MRI is a medical imaging technique used for the analysis of breast tissue. MRI acquires a set of volumetric data with a high contrast between fatty tissue and fibroglandular tissue [5]. This imaging modality is excellent for measurement of volumetric breast density. In last decade several CAD (computer aided diagnostic) methods have been developed to allow the radiologists and physicians in advanced analysis and inspection of breast tissue properties such as breast density [6]. Furthermore CAD methods allow to detect breast lesions useful for prevent breast cancer. Breast MRI is increasing its popularity as a screening modality for high-risk patients or patients with dense breasts. A fundamental role in Computer Aided Diagnostic methods is played by segmentation of the breast. A good breast segmentation allows to avoid processing irrelevant features, such as background and the tissue not belonging to the breast regions. A good breast segmentation on MRI confines the CAD systems to focus on the breast tissue, improving the specificity by eliminating false positives outside the ROI (region of interest) i.e. the breast tissue. Generally, the first step in CAD systems is the segmentation of the breast area on MR images, then, the second step is the detection and the exclusion of the chest wall muscle. A key challenge for medical imaging is to measure the volume of fibroglandular tissue and the density in MRI by normalizing to the breast volume. It is not easy to design approaches for automatic breast segmentation because of the large variety of breast in shape and pattern. In more detail, the major issue is to delineate the lateral posterior and the chest wall muscle boundaries. The radiologists observed that in MRI of dense breasts, the visual properties of fibroglandular tissue can be quite similar to the chest wall muscle [7]. Thus, as consequence of the aforementioned remarks, it increases the technical difficulty to delineate and exclude the muscle while preserving the dense breast tissue.

In this paper an automatic multi-seed detection method for magnetic resonance (MR) breast image segmentation is proposed. The method shows very encouraging results compared to the gold standard in terms of statistical metrics (Sensitivity: 95.22%; Specificity: 80.36%; Precision: 98.05%; Accuracy: 97.76%; Overlap: 77.01%) and execution time (4.23 s for each slice). The implementation of the proposed method has been running on a general purpose PC with a 2,3 GHz Intel Core i5 processor, 8 GB 1333 Mhz DDR3 memory, and Mac OS x 10.8.5 version. The article is organized as follows: In Sect. 2 the meaningful works on images segmentation are described; In Sect. 3 the breast MRI dataset used for development, test, and evaluation of the proposed system is described; In Sect. 4, the proposed system is discussed; In Sect. 5 depicts results and discussion; The final considerations are treated in Sect. 6.

2   Related Works

In breast MRI, several elements are required to perform automatic analysis. Many examples of medical imaging require an initial segmentation phase: multimodal breast image registration, computer aided analysis of DCE (dynamic contrast enhanced) MRI [8], and breast density assessment [5,6]. In [8] the authors detected the left side and the right side of the breast and the center of mass in each side is used as the seed points for region growing. The region detection and extraction from the anatomical regions are very difficult tasks. Complicating factors are the large shape variations of pectoral muscles across different patients, the similarity between intensity distributions and texture descriptors of the breast MR in muscle and fibroglandular tissues.

In last few years, many researches appears in medical imaging and precise segmentations of relevant anatomical structures such as breast region and fibroglandular tissue are required. Most of state of the art methods for breast segmentation on MRI are semi or fully automated, furthermore they can be grouped in contour-based, region-based and atlas-based approaches [9]. Generally, on Breast MRI, the following operations precede the breast segmentation task: Pectoralis muscle boundary segmentation, breast-air boundary segmentation, In [10] the authors proposed a method based on the observation that the pectoralis muscle and breast-air boundaries exhibit smooth sheetlike surfaces in 3D. This surfaces which can be simultaneously enhanced by a Hessian-based sheetness filter. The authors in [11] proposed a method for breast segmentation, but it needs manual intervention. In [12] breast segmentation was based on a semiautomated model that accounting for partial volume effects.

In [13] the authors proposed an automatic segmentation method based on the second derivative information represented by the Hessian matrix. Koenig et al. [14] performed a method to detect the most important strutctural elements of the breast by using BI-RADS criteria. Nie et al. [15] proposed a method for the analysis of breast density based on three-dimensional breast MRI: they first performed breast segmentation including an initial segmentation based on body landmarks of each individual woman, then they used fuzzy C-mean classification to exclude air and lung tissue, last they performed B-spline curve fitting to exclude chest wall muscle. Xiaoua et al. [16] proposed a method within a Bayesian framework, based on a maximum a posteriori estimation method. In [17] Guberna-Mérida et al. performed breast segmentation by using a framework based on Atlas (a technique for automatic delineation of anatomical structures in different 3D image modalities). In [18] the authors extended the method [18] by adding a combination of image processing techniques such as signal intensity inhomogeneities correction and probabilistic analysis as Expectation Maximization. In [19] Gallego-Ortiz et al. performed breast segmentation on MRI by combining a 3-D edge detection method with a probabilistic atlas of the breast. In [20] the authors focused the attention on fibroglandular tissue segmentation on Breast MRI: a fully automated segmentation algorithm, to estimate the volumetric amount of fibroglandular tissue in breast MRI. To optimize the computational cost of image segmentations, a lot of approaches are applied. PCA [21], unsupervised classification [22], and fuzzy c-means [23] reduces the dimensionality of the data therefore reduces the computational cost of analyzing new data [21].

3   Materials

The dataset consists of 18 patients from UPPH. The patient were divided in two groups according to their age: group 1 (25/35 years old, glandular/fibroglandular tissue) and group 2 (45/55 years old, fibrofatty/fatty tissue). A GE signa excite1.5 T HD 23 scanner was used to acquire T1 FSE axial sequences with the following technical parameters: 4 channels coil; $TR/TE = 525$; echo train $= 2$; image slices $= 40$; slice thickness $= 5$ mm; slice gap $= 0$; $FOV = 160 \times 320$; bandwidth $= 41.67$ Hz; imaging matrix $= 512 \times 256$.

3.1   Gold Standard

Three medical doctors, one resident and two radiologists, with progressively increasing knowledge level of breast imaging, performed the manual segmentation by using DICOM viewer Osirix [24] and following these criteria: breast parenchyma and cutaneous surface were isolated from external air basing on its lower intensity; lower boundary of breast region was delimited by using pectoral muscle as landmark; lateral bounds were represented by axillary cavities. The radiologists usually do not agree with each other, then the results from several observers are used to define a consolidated reference to compare the inter-observer variance, as in [25].

4   Methods

The proposed method consists of three steps: (1) pre-processing step to locate three regions of interest (axillary and sternal regions); (2) processing step to detect maximum concavity points for each region of interest; (3) breast image segmentation step. Eighteen patients have been manually segmented accordingly to three expert Radiologists to generate Gold Standard ground-truth used to evaluate the effectiveness of the proposed method. The acquisition parameters and characteristics are depicted in the next section. The algorithms used in the proposed segmentation method are briefly described in the next sections.

The proposed system consists of three main steps, as depicted in Fig. 1:

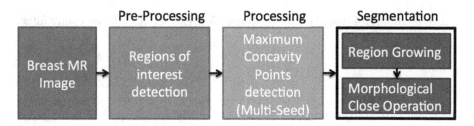

Fig. 1. Block scheme of the proposed method: pre-processing, processing, and segmentation.

- The pre-processing output as depicted in Fig. 2;
- The processing output as depicted in Fig. 3;
- The segmentation output as depicted in Fig. 4.

4.1 Pre-processing

The pre-processing step is as follows:

1. A breast MRI study is loaded. It contains, more or less 30 slices with thickness 5.00 mm (a single slice as shown in Fig. 2a);
2. A first binarization step is required to extract the boundary of the breast as shown in Fig. 2b, an adaptive thresholding is applied to the image by analyzing the trimodal distribution of intensity histogram;
3. An image crop containing the breast regions is considered (approximatively 2/3 of the whole image) as shown in Fig. 2c;
4. The holes are filled in the image, as shown in Fig. 2d;
5. The Largest Connected Component is found in the MRI and all other components are removed, as shown in Fig. 2e;
6. The objective is to find the coordinates of three pairs of points (A, B, C, D, E, F as shown in Fig. 2f), each pair of points includes a concave region of the breast boundary. The first and the third pair of points correspond to the vertices of axillary regions, the second pair of points correspond to the sternal region. The task is find a n-by-2 matrix that specifies the convex hull including the Breast Region and each row includes the coordinates of the convex hull corners. We notice that the larger side of the convex hull correspond to our regions of interest i.e. the axillary and stern regions (as suggested by the radiologists). In few words we find the three pairs of points by detecting and sorting the larger sides of the aforementioned convex hull. The first three sides of the convex polygon respectively correspond to the axillary (A, B, E, F see in Fig. 2f) and the sternal regions (C, D see in Fig. 2f). We sort the vector including the distances between the consecutive vertices of the convex polygon, in descend order, then we select the first three pairs of coordinates as the vertices of our regions of interest (axillary and sternal regions), as shown in Fig. 2f;
7. The convex hull of the binary regions found in step 5 is computed, as shown in Fig. 2g;
8. The boundary of image obtained in item 5 is extracted with the canny filter, as shown in Fig. 2h;

4.2 Processing

The coordinates of the points A–F (see in Fig. 2f) are grouped in three pairs: A and B belong to the first side of the convex hull (axillary region), C and D belong to the second side of the convex hull (stern region), E and F belong to the third side of the convex hull (axillary region). To detect the maximum concavity

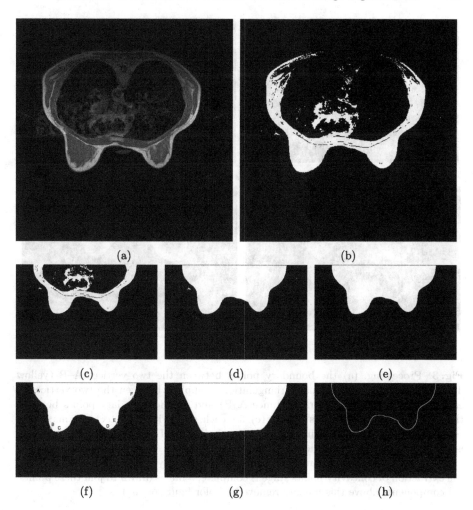

Fig. 2. Pre-processing: (a) a breast MRI is loaded; (b) the adaptive thresholding is applied in the MRI to emphasize the contours of the breast; (c) image crop containing the breast regions is considered; (d) the holes are filled in the image obtained in (c); (e) the Largest Connected Component is found in the MRI and all other components are removed; (f) the coordinates of three pairs of green points (A, B, C, D, E, F) are found; (g) the convex hull is computed in the image obtained in Figure e and it is returned a binary convex hull image; (h) the boundary of image obtained in Figure e is extracted with the canny filter. (Color figure online)

points (in breast image, see green points G, H, I in Fig. 3c) we process three regions of interest i.e. the axillary regions and the stern region. We highlight that the maximum concavity points correspond to the landmarks identified by the radiologists. To accomplish the detection of the maximum concavity points, each region of interest is processed as it follows:

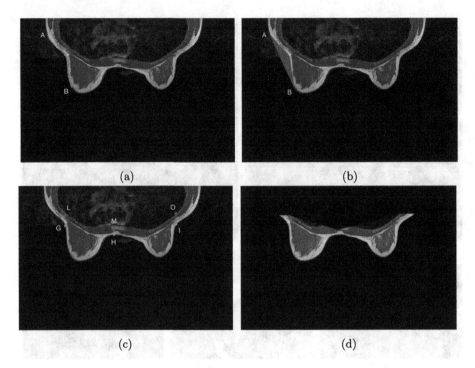

Fig. 3. Processing: (a) the boundary points between the two vertices A–B (yellow dots) are plotted; (b) Delaunay Triangulation is applied between the two vertices of each region of interest (in first instance A–B) and all the boundary points between the two vertices (yellow dots in (a); (c) for each concavity to detect the maximum concavity points (in breast image, see green points G, H, I); an additional margin is computed by measuring the vertical distance between G and the inner boundary extracted by filtering with canny algorithm, as depicted by red points in (c); (d) when the extraction of concavity points stage is complete, a line is drawn to join these points. All components above this line are removed. (Color figure online)

1. To detect the maximum concavity, we first apply Delaunay Triangulation between the two vertices of each region of interest (in first instance A–B in Fig. 2f) and all the boundary points between the two vertices (yellow dots in Fig. 3a);
2. The area of each triangles is computed;
3. The triangles are sort in descend order with respect to area value;
4. The first triangle is selected, it includes the larger area in the concave region of interest;
5. The maximum concavity point of the region is the third vertex of the triangle selected in the previous step (see green point G in Fig. 3c);
6. To avoid the exclusion of some region of interest including important features such as Skin, subcutaneous fat pad, and chest fat pad we add and additional margin to the G coordinates;

7. The additional margin is computed by measuring the vertical distance between G and the inner boundary extracted by filtering with canny algorithm. The same technique is applied to the other maximum concavity points, as depicted in Fig. 3c by red points;

8. When the maximum concavity points are detected, a line is drawn to join these points. All the elements located above the line are deleted, as shown in Fig. 3d.

4.3 Segmentation

The segmentation phase consists of three steps:

1. First, a region growing algorithm [26] with standard parameters (threshold) is applied to the image processed as described in the previous section. Region Growing algorithm needs a seed point to be executed. The maximum concavity points are then used as seed points for region growing. The result is shown in Fig. 4a;

2. In second step, morphological close operation has been used to fill the holes emerged from region growing. The structuring element of morphological operations is a disk with radius of 20 pixels so that the largest hole gets filled. The disk structuring element is used to preserve the circular nature of the object. The result is shown in Fig. 4b;

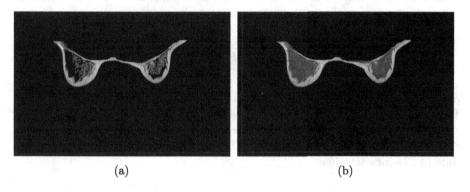

(a)   (b)

Fig. 4. Segmentation: (a) the region growing is applied to the image obtained in the end of processing step Fig. 3d; (b) the holes emerged form region growing are filled by applying the morphological close operations.

5 Results and Discussion

The proposed method has been tested on the dataset described in the materials section. The results showed that the proposed method achieves excellent results, as depicted in Table 1. Performance measures are then calculated regarding correct/incorrect segmentation.

The following measures are computed: Sensitivity, Specificity, Negative Predictive Value, Precision, Accuracy, and Error scores:

- Sensitivity: It is defined as the percentage of effective positives that are correctly identified as such:

$$Sensitivity : \frac{TP}{TP + FN} \tag{1}$$

- Specificity: It is defined as the percentage of effective negatives that are not classified as such:

$$Specificity : \frac{TN}{TN + FP} \tag{2}$$

- Precision: It is defined as, related to reproducibility and repeatability, the degree to which repeated segmentations under unchanged conditions show the same results:

$$Precision : \frac{TP}{TP + FP} \tag{3}$$

- Accuracy: It is defined as the degree of closeness of unsupervised segmentations of a breast to that manual segmentation:

$$Accurancy : \frac{TP + TN}{(TP + TN + FP + FN)} \tag{4}$$

- Overlap: It is defined as the index to quantify agreement between the unsupervised segmentation and manual segmentation:

$$Overlap : \frac{TP}{(TP + FP + FN)} \tag{5}$$

Furthermore, our method has been compared with a state of the art approach [8] to evaluate the effectiveness and the accuracy in breast segmentation. The results are depicted in Table 1. The proposed method shows very encouraging results in terms of statistical metrics (Sensitivity: 95.22%; Specificity: 80.36%; Precision: 98.05%; Accuracy: 97.76%; Overlap: 77.01%) and execution time (4.23 s for each slice).

Table 1. Experimental results

Segmentation	Methods	
	Proposed method	Template-based method [6]
Sensitivity Eq. (1)	95,22%	88,47%
Specificity Eq. (2)	80,36%	78,12%
Precision Eq. (3)	98,05%	92,99%
Accuracy Eq. (4)	97,76%	92,54%
Overlap Eq. (5)	77,01%	73,84%

6   Conclusions

In this paper an automatic multi-seed detection method for magnetic resonance (MR) breast image segmentation is presented. The multi-seed detection has been focused because of its importance in regional segmentation technique as the region growing. The maximum concavity points have been proposed as the seed points for the breast MR image segmentation. The detection of this points is based on the identification of three ROI of the breast MR image: the axillary regions and the sternal region. The Gold Standard, described in materials section, is used to compute the effectiveness and the performance of the proposed method. The preliminary results are very encouraging in terms of statistical metrics and execution time. In future works we are interested to extend the number of cases study and to develop a CAD (computer aided diagnostic) to detect suspicious regions on breast MRI: the first step is to detect the region of interest by using our proposed method (segmentation phase), than a further analysis and investigation should be conducted to detect suspicious regions by analyzing several features such as texture descriptors, statistical descriptors, histogram of gradients and others state of the art techniques.

References

1. Giannini, V., Vignati, A., Morra, L., Persano, D., Brizzi, D., Carbonaro, L., Bert, A., Sardanelli, F., Regge, D.: A fully automatic algorithm for segmentation of the breasts in DCE-MR images. In: 2010 Annual International Conference of the IEEE Engineering in Medicine and Biology Society (EMBC), pp. 3146–3149. IEEE (2010)
2. Mann, R.M., Balleyguier, C., Baltzer, P.A., Bick, U., Colin, C., Cornford, E., Evans, A., Fallenberg, E., Forrai, G., Fuchsjäger, M.H., et al.: Breast MRI: EUSOBI recommendations for women's information. Eur. Radiol. 25(12), 3669–3678 (2015)
3. Giger, M.L., Karssemeijer, N., Schnabel, J.A.: Breast image analysis for risk assessment, detection, diagnosis, and treatment of cancer. Annu. Rev. Biomed. Eng. 15, 327–357 (2013)
4. Roxanis, I., Colling, R.: Use of digital image analysis for outcome prediction in breast cancer. Diagn. Pathol. 1(8) (2016)
5. Brooksby, B., Pogue, B.W., Jiang, S., Dehghani, H., Srinivasan, S., Kogel, C., Tosteson, T.D., Weaver, J., Poplack, S.P., Paulsen, K.D.: Imaging breast adipose and fibroglandular tissue molecular signatures by using hybrid MRI-guided near-infrared spectral tomography. Proc. Natl. Acad. Sci. 103(23), 8828–8833 (2006)
6. Patel, B.C., Sinha, G.: An adaptive k-means clustering algorithm for breast image segmentation. Int. J. Comput. Appl. 10(4), 35–38 (2010)
7. Wu, S., Weinstein, S.P., Conant, E.F., Schnall, M.D., Kontos, D.: Automated chest wall line detection for whole-breast segmentation in sagittal breast MR images. Med. Phys. 40(4), 042301 (2013)
8. Lin, M., Chen, J.H., Wang, X., Chan, S., Chen, S., Su, M.Y.: Template-based automatic breast segmentation on MRI by excluding the chest region. Med. Phys. 40(12), 122301 (2013)

9. Wang, L., Chitiboi, T., Meine, H., Günther, M., Hahn, H.K.: Principles and methods for automatic and semi-automatic tissue segmentation in MRI data. Magn. Reson. Mater. Phys. Biol. Med. **29**(2), 95–110 (2016)

10. Wang, L., Platel, B., Ivanovskaya, T., Harz, M., Hahn, H.K.: Fully automatic breast segmentation in 3D breast MRI. In: 2012 9th IEEE International Symposium on Biomedical Imaging (ISBI), pp. 1024–1027. IEEE (2012)

11. Khazen, M., Warren, R.M., Boggis, C.R., Bryant, E.C., Reed, S., Warsi, I., Pointon, L.J., Kwan-Lim, G.E., Thompson, D., Eeles, R., et al.: A pilot study of compositional analysis of the breast and estimation of breast mammographic density using three-dimensional T1-weighted magnetic resonance imaging. Cancer Epidemiol. Prev. Biomark. **17**(9), 2268–2274 (2008)

12. Lee, N.A., Rusinek, H., Weinreb, J., Chandra, R., Toth, H., Singer, C., Newstead, G.: Fatty and fibroglandular tissue volumes in the breasts of women 20–83 years old: comparison of X-ray mammography and computer-assisted MR imaging. AJR. Am. J. Roentgenol. **168**(2), 501–506 (1997)

13. Wang, L., Filippatos, K., Friman, O., Hahn, H.K.: Fully automated segmentation of the pectoralis muscle boundary in breast MR images. In: SPIE Medical Imaging, International Society for Optics and Photonics, p. 796309 (2011)

14. Koenig, M., Laue, H., Boehler, T., Peitgen, H.O.: Automatic segmentation of relevant structures in DCE MR mammograms. In: Medical Imaging, International Society for Optics and Photonics, p. 65141S (2007)

15. Nie, K., Chen, J.H., Chan, S., Chau, M.K.I., Yu, H.J., Bahri, S., Tseng, T., Nalcioglu, O., Su, M.Y.: Development of a quantitative method for analysis of breast density based on three-dimensional breast MRI. Med. Phys. **35**(12), 5253–5262 (2008)

16. Xiaohua, C., Brady, M., Lo, J.L.-C., Moore, N.: Simultaneous segmentation and registration of contrast-enhanced breast MRI. In: Christensen, G.E., Sonka, M. (eds.) IPMI 2005. LNCS, vol. 3565, pp. 126–137. Springer, Heidelberg (2005). doi:10.1007/11505730_11

17. Gubern-Mérida, A., Kallenberg, M., Martí, R., Karssemeijer, N.: Segmentation of the pectoral muscle in breast MRI using atlas-based approaches. In: Ayache, N., Delingette, H., Golland, P., Mori, K. (eds.) MICCAI 2012. LNCS, vol. 7511, pp. 371–378. Springer, Heidelberg (2012). doi:10.1007/978-3-642-33418-4_46

18. Gubern-Mérida, A., Kallenberg, M., Mann, R.M., Marti, R., Karssemeijer, N.: Breast segmentation and density estimation in breast MRI: a fully automatic framework. IEEE J. Biomed. Health Inform. **19**(1), 349–357 (2015)

19. Ortiz, C.G., Martel, A.: Automatic atlas-based segmentation of the breast in MRI for 3D breast volume computation. Med. Phys. **39**(10), 5835–5848 (2012)

20. Wu, S., Weinstein, S.P., Conant, E.F., Kontos, D.: Automated fibroglandular tissue segmentation and volumetric density estimation in breast MRI using an atlas-aided fuzzy C-means method. Medical physics **40**(12), 122302 (2013)

21. Agnello, L., Comelli, A., Vitabile, S.: Feature dimensionality reduction for mammographic report classification. In: Pop, F., Kołodziej, J., Di Martino, B. (eds.) Resource Management for Big Data Platforms. CCN, pp. 311–337. Springer, Cham (2016). doi:10.1007/978-3-319-44881-7_15

22. Comelli, A., Agnello, L., Vitabile, S.: An ontology-based retrieval system for mammographic reports. In: 2015 IEEE Symposium on Computers and Communication (ISCC), pp. 1001–1006. IEEE (2015)

23. Agnello, L., Comelli, A., Ardizzone, E., Vitabile, S.: Unsupervised tissue classification of brain MR images for voxel-based morphometry analysis. Int. J. Imaging Syst. Technol. **26**(2), 136–150 (2016)

24. Rosset, A., Spadola, L., Ratib, O.: Osirix: an open-source software for navigating in multidimensional DICOM images. J. digit. Imaging **17**(3), 205–216 (2004)
25. Warfield, S.K., Zou, K.H., Wells, W.M.: Simultaneous truth and performance level estimation (STAPLE): an algorithm for the validation of image segmentation. IEEE Trans. Med. Imaging **23**(7), 903–921 (2004)
26. Adams, R., Bischof, L.: Seeded region growing. IEEE Trans. Pattern Anal. Mach. Intell. **16**(6), 641–647 (1994)

Efficient Image Segmentation in Graphs with Localized Curvilinear Features

Hans H.C. Bejar[1], Fábio A.M. Cappabianco[2], and Paulo A.V. Miranda[1(✉)]

[1] Institute of Mathematics and Statistics, University of São Paulo,
São Paulo, SP 05508-090, Brazil
{hans,pmiranda}@ime.usp.br
[2] Instituto de Ciência e Tecnologia, Universidade Federal de São Paulo,
São José dos Campos, SP, Brazil
cappabianco@unifesp.br

Abstract. In graph-based image segmentation, the arc weights are given by a local edge indicator function based on image attributes and prior object information. In boundary tracking methods, an edge integration process combines local edges into meaningful long edge curves, interconnecting a set of anchor points, such that a closed contour is computed for segmentation. In this work, we show that multiple short-range edge integrations can extract curvilinear features all over the image to improve seeded region-based segmentation. We demonstrate these results using edge integration by Live Wire (LW), combined with Oriented Image Foresting Transform (OIFT), due to their complementary strengths. As result, we have a globally optimal segmentation, that can be tailored to a given target object, according to its localized curvilinear features.

Keywords: Live wire · Image foresting transform · Boundary tracking

1 Introduction

Image segmentation can be interpreted as a graph partition problem subject to hard constraints, given by seed pixels selected in the image domain [3,9,12]. The *min-cut/max-flow* algorithm, also known simply as *Graph Cut* (GC) [3], and some methods that can be described according to *Image Foresting Transform* (IFT) [13], such as *Watersheds* [12] and *Fuzzy Connectedness* [9], correspond to the ε_1- and ε_∞-minimization problems, respectively, within a common framework, sometimes referred to as Generalized Graph Cut (GGC) [4,8].

Each class of methods has its own drawbacks. While methods from the ε_∞ family have problems related to irregular boundaries and "leaking" through

---

P.A.V. Miranda—Thanks to CNPq (308985/2015-0, 486083/2013-6, 486988/2013-9, FINEP 1266/13), FAPESP (2011/50761-2, 2014/12236-1, 2016/21591-5), CAPES, and NAP eScience - PRP - USP for funding, and Dr. J.K. Udupa (MIPG-UPENN) for the images.

© Springer International Publishing AG 2017
S. Battiato et al. (Eds.): ICIAP 2017, Part I, LNCS 10484, pp. 718–728, 2017.
https://doi.org/10.1007/978-3-319-68560-1_64

badly defined borders, the ε_1 methods suffer from metrication error ("blockiness"), shrinking bias, and higher computational time[1] [8].

Methods based on multiple energy minimizations, using iterations of different energy classes to address distinct image parts [2,10] or iterated merging the results of a single energy class over recomputed subgraphs [28], tend to alleviate the above mentioned problems, but they lose global optimality, since they do not perform a single energy minimization over all image elements. As consequence, it is harder to incorporate high-level priors for object segmentation in these approaches. On the other hand, methods like *Oriented Image Foresting Transform* (OIFT) [20,26] are very versatile, supporting several high-level priors, including global properties such as connectedness of the segmented object [19,22], shape constraints [21,27] and boundary polarity [20,26], which allow the customization of the segmentation to a given target object [18]. However, as an extension of the ε_∞-minimization problem to directed graphs, OIFT inherits its drawbacks.

In graph-based image segmentation, including the ε_1- and ε_∞-minimization problems, arc weights are given by a local edge indicator function from image attributes and prior object information [5,6,25]. In boundary tracking, an edge integration process, usually based on a path-cost function, combines local edges (arc weights) into meaningful long edge curves, interconnecting a set of anchor points, such that a closed contour is computed for segmentation [14,17].

Seeded region-based segmentation and boundary tracking methods are usually presented as different competing approaches, with the former being easier to extend to multidimensional images, while the latter is more sensitive to seed positioning errors [3]. In this work, we show that multiple short-range edge integrations can be used to extract curvilinear features all over the image in order to improve seeded region-based segmentation.

Our proposed method differs from [33]'s work and from hybrid approaches, such as the *Live Markers* paradigm [31], because our method does not require the specification of boundary constraints (anchor points or other point-based soft constraints [16]), while Live Markers requires the selection of anchor points over the object boundary to compute optimum boundary segments, which are turned into internal and external markers for region-based delineation. In our approach, the boundary tracking method is applied locally over the image inside circular regions to extract curvilinear features without the need of any user intervention. It improves the arc-weight assignment from a local edge indicator function to a more general short-range edge integration function. As advantages we have:

- Segments by local boundary tracking with inconsistent boundary polarity can be penalized.
- We can favor the segmentation of objects with more regular and smoothed forms by penalizing arcs in segments with accentuated curvature.

---

[1] The ε_∞-minimization methods have complexity $O(N \cdot logN)$ with respect to the image size N (linear-time implementations $O(N)$ can be achieved for some instances, depending on the data structure of the priority queue [7]), while the run time for the ε_1-minimization problem is $O(N^{2.5})$ for sparse graphs [3].

– We can fill boundary gaps at weak edge points to avoid leaking problems in the segmentation.
– We keep the user interface simple to use without the need for multiple types of user input.

The short-range edge integration helps to circumvent the main problems of the ε_∞ family (irregular boundaries and "leaking"). We demonstrate these results using edge integration by *Live Wire* (LW) [14], combined with OIFT [20,26], due to their complementary strengths, since live wire can be seen as a boundary-based version of the ε_1-minimization problem on a dual graph [3,24]. As result, we have a globally optimal segmentation by OIFT, that can be tailored to a given target object, according to its localized curvilinear features, and other high-level priors already supported by OIFT [19,21,22,27].

For the sake of completeness in presentation, Sect. 2 includes an overview of concepts on image graph and a revision of OIFT and live wire. Section 3 shows the proposed algorithm to compute the localized curvilinear features. In Sect. 4, we evaluate the proposed method, named *OIFT with Localized Curvilinear Features* (OIFT-LCF) and our conclusions are stated in Sect. 5.

2   Background

An image can be interpreted as a weighted digraph $G = \langle \mathcal{V}, \mathcal{A}, \omega \rangle$, whose nodes $\mathcal{V}$ are the image pixels in its image domain $\mathcal{V} \subset \mathbb{Z}^n$, and whose arcs are the ordered pixel pairs $\langle s, t \rangle \in \mathcal{A}$. For example, one can take $\mathcal{A}$ to consist of all pairs of ordered pixels $\langle s, t \rangle$ in the Cartesian product $\mathcal{V} \times \mathcal{V}$ such that $\|s - t\| \le \rho$ and $s \ne t$, where ρ is a specified constant (e.g., 4-neighborhood, when $\rho = 1$, and 8-neighborhood, when $\rho = \sqrt{2}$, in case of 2D images). Each arc $\langle s, t \rangle \in \mathcal{A}$ has a weight $\omega(s, t) \ge 0$. The digraph G is symmetric if for any of its arcs $\langle s, t \rangle \in \mathcal{A}$, the pair $\langle t, s \rangle$ is also an arc of G.

For a given image graph $G = \langle \mathcal{V}, \mathcal{A}, \omega \rangle$, a path $\pi = \langle t_1, t_2, \ldots, t_n \rangle$ is a sequence of adjacent pixels (i.e., $\langle t_i, t_{i+1} \rangle \in \mathcal{A}$, $i = 1, 2, \ldots, n - 1$) with no repeated vertices ($t_i \ne t_j$ for $i \ne j$). A path $\pi_t = \langle t_1, t_2, \ldots, t_n = t \rangle$ is a path with terminus at a pixel t. When we want to explicitly indicate the origin of a path, the notation $\pi_{s \rightsquigarrow t} = \langle t_1 = s, t_2, \ldots, t_n = t \rangle$ may also be used, where s stands for the origin and t for the destination node. A path is *trivial* when $\pi_t = \langle t \rangle$. A path $\pi_t = \pi_s \cdot \langle s, t \rangle$ indicates the extension of a path π_s by an arc $\langle s, t \rangle$, and $\pi_{r \rightsquigarrow t} = \pi_{r \rightsquigarrow s} \cdot \pi_{s \rightsquigarrow t}$ indicates the concatenation of two paths.

2.1   Live Wire (LW)

A *connectivity function* computes a value $f(\pi_t)$ for any path π_t, usually based on arc weights. A path π_t is *optimum* if $f(\pi_t) \le f(\tau_t)$ for any other path τ_t in G. By selecting to each pixel $t \in \mathcal{V}$ one optimum path with terminus at t, we obtain the optimum-path value $V_{opt}(t)$, which is uniquely defined by $V_{opt}(t) = \min_{\forall \pi_t \text{ in } G} \{f(\pi_t)\}$.

The live-wire function is given by:

$$f_{LW}(\langle t \rangle) = \begin{cases} 0 & \text{if } t \in \mathcal{S} \\ +\infty & \text{otherwise} \end{cases}$$

$$f_{LW}(\pi_s \cdot \langle s, t \rangle) = f_{LW}(\pi_s) + \omega(s, t) \tag{1}$$

where $\mathcal{S}$ is a seeds set, usually composed by a single starting anchor point.

The optimum path interconnecting two consecutive anchor points can be computed by a generalization of Dijkstra's algorithm to more general path-cost functions, known as *Image Foresting Transform* (IFT) [13]. In the IFT framework, the paths are stored in backward order in a *predecessor map* $P : \mathcal{V} \to \mathcal{V} \cup \{nil\}$, such that for any pixel $t \in \mathcal{V}$, a path π_t^P is recursively defined as $\langle t \rangle$ if $P(t) = nil$, and $\pi_s^P \cdot \langle s, t \rangle$ if $P(t) = s \neq nil$, according to the following algorithm:

Algorithm 1 – IFT ALGORITHM

INPUT: Image graph $G = \langle \mathcal{V}, \mathcal{A}, \omega \rangle$, and function f.
OUTPUT: Predecessor map P and the path-cost map V, which may converge to
             V_{opt} depending on f.
AUXILIARY: Priority queue $\mathcal{Q}$, variable tmp, and set $\mathcal{F}$.

1. **For each** $t \in \mathcal{V}$, **do**
2. │   *Set $P(t) \leftarrow nil$, $V(t) \leftarrow f(\langle t \rangle)$ and $\mathcal{F} \leftarrow \varnothing$.*
3. └   **If** $V(t) \neq +\infty$, **then** *insert t in $\mathcal{Q}$.*
4. **While** $\mathcal{Q} \neq \varnothing$, **do**
5. │   *Remove s from $\mathcal{Q}$ such that $V(s)$ is minimum.*
6. │   *Add s to $\mathcal{F}$.*
7. │   **For each** *pixel t such that $\langle s, t \rangle \in \mathcal{A}$ and $t \notin \mathcal{F}$*, **do**
8. │   │   *Compute $tmp \leftarrow f(\pi_s^P \cdot \langle s, t \rangle)$.*
9. │   │   **If** $tmp < V(t)$, **then**
10. │   │   │   **If** $V(t) \neq +\infty$, **then** *remove t from $\mathcal{Q}$.*
11. │   │   │   *Set $P(t) \leftarrow s$, $V(t) \leftarrow tmp$.*
12. └   └   └   *Insert t in $\mathcal{Q}$.*

In user-steered image segmentation [14,17,24], the computed path from the previous anchor point to the current mouse position is shown to the user, as he moves the cursor, so that the user can interactively select the desired path, that best matches the object boundary, and start a new path search from that point. All previous selected paths are kept unchanged (frozen) during the algorithm, so that their nodes cannot be revisited.

2.2   Oriented Image Foresting Transform (OIFT)

Let $\mathcal{X} = \{\mathcal{O} : \mathcal{O} \subseteq \mathcal{V}\}$ be the space of all possible binary segmented objects $\mathcal{O}$. A seed-based segmentation uses *seeds* $\mathcal{S} = \mathcal{S}_o \cup \mathcal{S}_b \subseteq \mathcal{V}$, where $\mathcal{S}_o$ and $\mathcal{S}_b$ are *object* ($\mathcal{S}_o \subseteq \mathcal{O}$) and *background* ($\mathcal{S}_b \subseteq \mathcal{V} \setminus \mathcal{O}$) seed sets, respectively. They restrict $\mathcal{X}$ to $\mathcal{X}(\mathcal{S}_o, \mathcal{S}_b) = \{\mathcal{O} \in \mathcal{X} : \mathcal{S}_o \subseteq \mathcal{O} \subseteq \mathcal{V} \setminus \mathcal{S}_b\}$. A *cut* is defined as $\mathcal{C}(\mathcal{O}) = \{\langle s, t \rangle \in \mathcal{A} : s \in \mathcal{O} \text{ and } t \notin \mathcal{O}\}$. We can associate an energy value $\varepsilon(\mathcal{O})$ to an object (and its cut), and restrict the set of solutions to those which minimizes it. Let energy $\varepsilon_q(\mathcal{O}) = (\sum_{\langle s,t \rangle \in \mathcal{C}(\mathcal{O})} \omega(s,t)^q)^{\frac{1}{q}}$. The original Graph Cut algorithm minimizes $\varepsilon_1(\mathcal{O})$, while OIFT minimizes $\varepsilon_\infty(\mathcal{O}) = \max_{\langle s,t \rangle \in \mathcal{C}(\mathcal{O})} \omega(s,t)$ [8,20].

In this work, we will present OIFT in its equivalent dual form as a maximization problem of the energy $\bar{\varepsilon}_\infty(\mathcal{O}) = \min_{\langle s,t \rangle \in \mathcal{C}(\mathcal{O})} \omega'(s,t)$, in a strongly connected and symmetric digraph $G = \langle \mathcal{V}, \mathcal{A}, \omega' \rangle$, where the weights $\omega'(s,t)$ are a combination of an undirected dissimilarity measure $\delta(s,t)$ between neighboring pixels s and t, multiplied by an orientation factor, as follows:

$$\omega'(s,t) = \begin{cases} \delta(s,t) \times (1 + \alpha) & \text{if } I(s) > I(t) \\ \delta(s,t) \times (1 - \alpha) & \text{if } I(s) < I(t) \\ \delta(s,t) & \text{otherwise} \end{cases} \tag{2}$$

where $\alpha \in [-1, 1]$ and $I(t)$ is the image intensity at pixel t. In this work, we consider $\alpha = 50\%$ in order to get a more balanced solution. Different procedures can be adopted for $\delta(s,t)$, as discussed in [5,25], such as the mean gradient magnitude (i.e., $\delta(s,t) = \frac{\|\nabla I(s)\| + \|\nabla I(t)\|}{2}$). Note that we usually have $\omega'(s,t) \neq \omega'(t,s)$ when $\alpha \neq 0$. For colored images, a reference map should be considered for $I(t)$ in Eq. 2, or α must be set to zero [20]. OIFT is build upon the IFT framework by considering the following path function [20]:

$$f^{\vec{\sigma}}(\langle t \rangle) = \begin{cases} -1 & \text{if } t \in \mathcal{S}_o \cup \mathcal{S}_b \\ +\infty & \text{otherwise} \end{cases}$$

$$f^{\vec{\sigma}}(\pi_{r \rightsquigarrow s} \cdot \langle s,t \rangle) = \begin{cases} \omega'(s,t) & \text{if } r \in \mathcal{S}_o \\ \omega'(t,s) & \text{otherwise} \end{cases} \tag{3}$$

The segmented object $\mathcal{O}$ by OIFT is defined from the forest P computed by Algorithm 1, with $f^{\vec{\sigma}}$, by taking as object pixels the set of pixels that were conquered by paths rooted in $\mathcal{S}_o$. For $\alpha > 0$, the segmentation by OIFT favors transitions from bright to dark pixels, and $\alpha < 0$ favors the opposite orientation [20,26].

3   OIFT with Localized Curvilinear Features (OIFT-LCF)

In *OIFT with Localized Curvilinear Features*, the live-wire method for boundary tracking is applied locally over the image inside circular regions to extract curvilinear features without the need of any user intervention. For each pixel c of the image, we consider a circular disc $\mathcal{D}(c) = \{t \in \mathbb{Z}^n \mid \|t - c\| \leq R\}$ of radius R centered at c. The optimum path $\pi_{a \rightsquigarrow c}$ from a pixel a in the disc boundary $\mathcal{B}(c) = \{s \in \mathcal{D}(c) \mid \exists t \notin \mathcal{D}(c) \text{ such that } \|s - t\| \leq 1\}$ to the central pixel c is computed by Algorithm 1 with the live-wire function f_{LW}, using $\mathcal{S} = \mathcal{B}(c)$ and the local edge indicator function $\omega(s,t)$ defined by:

$$\omega(s,t) = \begin{cases} (\frac{\bar{G}(s) + \bar{G}(t)}{2} \times (1 + \gamma))^\beta + \|s - t\| & \text{if } I(r) > I(l) \\ (\frac{\bar{G}(s) + \bar{G}(t)}{2} \times (1 - \gamma))^\beta + \|s - t\| & \text{if } I(r) < I(l) \\ (\frac{\bar{G}(s) + \bar{G}(t)}{2})^\beta + \|s - t\| & \text{otherwise} \end{cases} \tag{4}$$

where $\bar{G}(t)$ is the complement of the magnitude of some gradient like image $G(t)$ at pixel t, l and r are the neighboring left and right pixels of the arc $\langle s, t \rangle$ (Fig. 1) and we usually have $\gamma = 50\%$. The parameter γ is used to penalize segments with inconsistent boundary polarity, by favoring a particular boundary orientation. For example, $\gamma = 50\%$ improves arcs with the right pixel being darker than its left pixel (Fig. 2).

We then compute the optimum path from c to the disc boundary $\mathcal{B}(c)$, such that we end with a composite path $\pi_{a \leadsto b} = \pi_{a \leadsto c} \cdot \pi_{c \leadsto b} = \langle p_1 = a, p_2, \ldots, p_k = c, \ldots, p_m = b \rangle$, connecting two boundary points of the disc and passing through its center. In this short-range edge integration, the live-wire segments attach to the objects' boundaries, so that we can extract important contour information from the underlying objects all over the image (Fig. 3a).

For each pixel c, we can extract curvilinear features from the composite path $\pi_{a \leadsto c} \cdot \pi_{c \leadsto b}$, such as the following mean curvature measure:

$$\text{Curv}(\langle p_1, .., p_m \rangle) = \frac{1}{(L/2)} \cdot \sum_{i=L/2}^{L-1} \frac{\|p_{k-i} - 2 \cdot c + p_{k+i}\|}{\|p_{k-i} - c\| + \|c - p_{k+i}\|} \quad (5)$$

where $L = \min\{\text{Length}(\pi_{a \leadsto c}), \text{Length}(\pi_{c \leadsto b})\} + 1 = \min\{k - 1, m - k\} + 1$.

We can then improve the arc-weight assignment for the OIFT method from a local edge indicator function $\delta(s, t) = \frac{G(s) + G(t)}{2}$ to a more general short-range edge integration function $\delta(s, t) = \frac{G_{LCF}(s) + G_{LCF}(t)}{2}$, where $\bar{G}_{LCF}(t) = \bar{G}(t) \times (1 + \text{Curv}^2(\langle p_1, \ldots, p_m \rangle))$. This OIFT with localized curvilinear features (OIFT-LCF) helps to circumvent the irregular boundaries of the original OIFT (Figs. 3 and 4). Curvature regularity for region-based image segmentation usually results in an NP-hard problem, and linear programming relaxation with thresholding is used to obtain an approximate solution [30]. In our method, the curvilinear features can be fast computed using localized live-wire executions, which can be calculated in parallel for different regions of the image. Since both live wire and OIFT take linearithmic time in worst-case scenarios and live wire is computed only in circular discs of fixed size, the complexity of OIT-LCF is linearithmic.

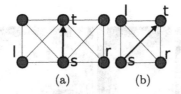

(a)   (b)

Fig. 1. The neighboring left and right pixels for each arc.

Another possibility is to consider in OIFT-LCF:

$$\bar{G}_{LCF}(t) = \left(\frac{f_{LW}(\pi_{a \leadsto c}) + f_{LW}(\pi_{c \leadsto b})}{\text{Length}(\pi_{a \leadsto c} \cdot \pi_{c \leadsto b})} \right)^{1/\beta} \times (1 + \text{Curv}^2) \quad (6)$$

which helps to circumvent "leaking" problems of the original OIFT (Fig. 5).

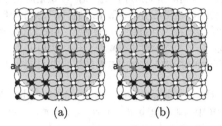

Fig. 2. Boundary tracking with boundary polarity. The arcs with the desired orientation (right pixels darker than their left pixels) are shown with thicker lines, while arcs with incorrect orientation are shown with dashed lines. (a) Segments by live wire with the correct boundary orientation. (b) Segments with inconsistent polarity.

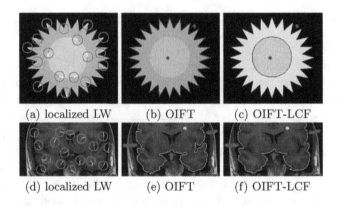

(a) localized LW   (b) OIFT   (c) OIFT-LCF

(d) localized LW   (e) OIFT   (f) OIFT-LCF

Fig. 3. (a) Curvilinear features can be extract by computing localized live wire segments. (b–c) We can favor the segmentation of objects with more regular and smoothed forms by penalizing arcs in segments with accentuated curvature. (d–f) An example using MR images of the brain.

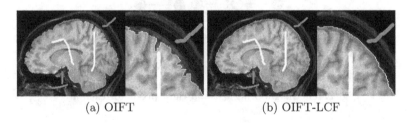

(a) OIFT   (b) OIFT-LCF

Fig. 4. The segmentation of the brain external surface in a MR-T1 image. OIFT-LCF gives the most regular contour, following more closely the dura mater, due to its curvature analysis.

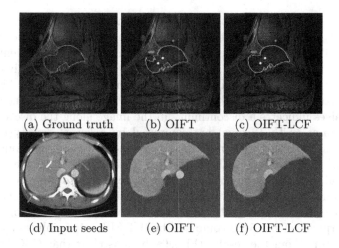

(a) Ground truth   (b) OIFT   (c) OIFT-LCF

(d) Input seeds   (e) OIFT   (f) OIFT-LCF

Fig. 5. The segmentation of: (a–c) A bone of the human foot and (d–f) the liver.

4   Experimental Results

We conducted quantitative experiments, using a total of 40 image slices of 10 thoracic CT studies to segment the liver, and 40 slice images from real MR images of the foot to segment the calcaneus bone. Several different gradients with and without the usage of localized curvilinear features, were assessed for accuracy employing the mean performance curve (Dice coefficient) and ground truth data obtained from an expert of the radiology department at the University of Pennsylvania for different seed sets.

In the first experiment, we used the second version of OIFT-LCF to avoid "leaking" problems, with $R = 3$ and $\beta = 2$, and considered different seed sets automatically obtained by eroding and dilating the ground truth at different radius values. By varying the radius value, we can compute the segmentation for different seed sets and trace accuracy curves using the Dice coefficient of similarity. However, in order to generate a more challenging situation, we considered a larger radius of dilation for the external seeds (twice the value of the inner radius), resulting in an asymmetrical arrangement of seeds. In order to demonstrate the versatility of OIFT-LCF and OIFT, which support several high-level priors, we considered these methods with shape constraints by *Geodesic Star Convexity* (GSC) [21], leading to the GSC-OIFT-LCF and GSC-OIFT methods.

Figures 6a–b show the mean accuracy curves for all the images of the first experiment, using different input gradients indicated by a superscript index, for each method. For the sake of simplicity, we only considered gradient like images from local image attributes without resorting to more sophisticated techniques by supervised learning [32]. The superscript index 1 is used to indicate $G(t)$ as the Sobel gradient magnitude. The superscript index 2 denotes the image-based weight by Miranda et al. [25], which is based on image smoothing at four different scales by a sequences of opening by reconstruction and closing by reconstruction.

The superscript index 3 corresponds to the morphological gradient with a radius of 1.5 pixels, that is, the difference between the dilation and the erosion of the image. The superscript index 4 describes the brightness gradient features from [23], where for each pixel, a circle of radius r is drawn and divided along the diameter at orientation θ. The half-disc regions are described by histograms, which are compared by the chi-squared distance. A large difference between the disc halves indicates a discontinuity in the image along the disc's diameter. The histograms (with 12 bins) are computed by a kernel density estimation using a Gaussian kernel with $\sigma = 10.0$. We considered $r = 4$ and four different orientations. The gradient with superscript 5 was inspired by the work from Rauber et al. [29], where superpixel graphs were shown to improve interactive segmentation, by exploring the mean color/intensity inside superpixels. We tried to reproduce similar results at the pixel level, by using the average of brightness inside superpixels of size 5×5 computed by IFT-SLIC [1] as input image for the Sobel operator. From the results (Figs. 6a–b) it is clear that GSC-OIFT-LCF outperformed GSC-OIFT for all corresponding indexes (i.e., GSC-OIFT-LCF^i better than GSC-OIFT^i, for $i = 1, \ldots, 5$).

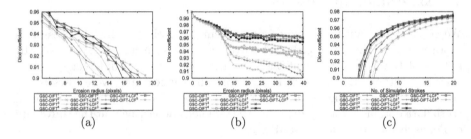

(a)   (b)   (c)

Fig. 6. The mean accuracy curves (Dice) using different gradients with and without the usage of localized curvilinear features. (a–b) First experiment for the segmentation of (a) calcaneus bone and (b) liver. (c) Calcaneus mean accuracy curve by a robot user.

In the second experiment, we used the first version of OIFT-LCF to circumvent irregular boundaries, with $R = 4$ and $\beta = 5$. Figure 6c shows the experimental curve using a robot user [15], which confirms similar results.

5   Conclusions

In conclusion, we developed extensions to the OIFT algorithm [26], by incorporating localized curvilinear constraints in its formulation, helping to circumvent irregular boundaries of OIFT. As future work, we intend to extend OIFT-LCF to 3D and test it with other boundary-based methods [11].

References

1. Alexandre, E.B., Chowdhury, A.S., Falcão, A.X., Miranda, P.A.V.: IFT-SLIC: a general framework for superpixel generation based on simple linear iterative clustering and image foresting transform. In: 2015 28th SIBGRAPI Conference on Graphics, Patterns and Images, pp. 337–344, August 2015
2. Bejar, H.H., Miranda, P.A.: Oriented relative fuzzy connectedness: theory, algorithms, and its applications in hybrid image segmentation methods. EURASIP J. Image Video Process. **2015**(1), 21 (2015)
3. Boykov, Y., Funka-Lea, G.: Graph cuts and efficient N-D image segmentation. Int. J. Comput. Vis. **70**(2), 109–131 (2006)
4. Couprie, C., Grady, L., Najman, L., Talbot, H.: Power watersheds: a unifying graph-based optimization framework. IEEE Trans. Pattern Anal. Mach. Intell. **99**(7), 1384–1399 (2010)
5. Ciesielski, K., Udupa, J.: Affinity functions in fuzzy connectedness based image segmentation I: equivalence of affinities. Comput. Vis. Image Underst. **114**(1), 146–154 (2010)
6. Ciesielski, K., Udupa, J.: Affinity functions in fuzzy connectedness based image segmentation II: defining and recognizing truly novel affinities. Comput. Vis. Image Underst. **114**(1), 155–166 (2010)
7. Ciesielski, K., Udupa, J., Falcão, A., Miranda, P.: Fuzzy connectedness image segmentation in graph cut formulation: a linear-time algorithm and a comparative analysis. J. Math. Imaging Vis. **44**(3), 375–398 (2012)
8. Ciesielski, K., Udupa, J., Falcão, A., Miranda, P.: A unifying graph-cut image segmentation framework: algorithms it encompasses and equivalences among them. In: Proceedings of SPIE on Medical Imaging: Image Processing, vol. 8314 (2012)
9. Ciesielski, K., Udupa, J., Saha, P., Zhuge, Y.: Iterative relative fuzzy connectedness for multiple objects with multiple seeds. Comput. Vis. Image Underst. **107**(3), 160–182 (2007)
10. Ciesielski, K.C., Miranda, P., Falcão, A., Udupa, J.K.: Joint graph cut and relative fuzzy connectedness image segmentation algorithm. Med. Image Anal. (MEDIA) **17**(8), 1046–1057 (2013)
11. Condori, M.A.T., Mansilla, L.A.C., Miranda, P.A.V.: Bandeirantes: a graph-based approach for curve tracing and boundary tracking. In: Angulo, J., Velasco-Forero, S., Meyer, F. (eds.) ISMM 2017. LNCS, vol. 10225, pp. 95–106. Springer, Cham (2017). doi:10.1007/978-3-319-57240-6_8
12. Cousty, J., Bertrand, G., Najman, L., Couprie, M.: Watershed cuts: thinnings, shortest path forests, and topological watersheds. Trans. Pattern Anal. Mach. Intell. **32**, 925–939 (2010)
13. Falcão, A., Stolfi, J., Lotufo, R.: The image foresting transform: theory, algorithms, and applications. IEEE Trans. Pattern Anal. Mach. Intell. **26**(1), 19–29 (2004)
14. Falcão, A., Udupa, J., Miyazawa, F.: An ultra-fast user-steered image segmentation paradigm: live-wire-on-the-fly. Trans. Med. Imaging **19**(1), 55–62 (2000)
15. Gulshan, V., Rother, C., Criminisi, A., Blake, A., Zisserman, A.: Geodesic star convexity for interactive image segmentation. In: Proceedings of Computer Vision and Pattern Recognition, pp. 3129–3136 (2010)
16. Jones, J.L., Xie, X., Essa, E.: Biomedical Image Segmentation: Advances and Trends. Taylor & Francis Group, CRC Press, Boca Raton (2017)
17. Kang, H.: G-wire: a livewire segmentation algorithm based on a generalized graph formulation. Pattern Recogn. Lett. **26**(13), 2042–2051 (2005)

18. Lézoray, O., Grady, L.: Image Processing and Analysis with Graphs: Theory and Practice. CRC Press, California (2012)
19. Mansilla, L.A.C., Miranda, P.A.V., Cappabianco, F.A.M.: Oriented image foresting transform segmentation with connectivity constraints. In: 2016 IEEE International Conference on Image Processing (ICIP), pp. 2554–2558, September 2016
20. Mansilla, L., Miranda, P.: Image segmentation by oriented image foresting transform: handling ties and colored images. In: 18th International Conference on Digital Signal Processing, Greece, pp. 1–6, July 2013
21. Mansilla, L.A.C., Miranda, P.A.V.: Image segmentation by oriented image foresting transform with geodesic star convexity. In: Wilson, R., Hancock, E., Bors, A., Smith, W. (eds.) CAIP 2013. LNCS, vol. 8047, pp. 572–579. Springer, Heidelberg (2013). doi:10.1007/978-3-642-40261-6_69
22. Mansilla, L.A., Miranda, P.A.: Oriented image foresting transform segmentation: connectivity constraints with adjustable width. In: XXIX Conference on Graphics, Patterns and Images (SIBGRAPI), So José Dos Campos, SP, Brazil, October 2016
23. Martin, D.R., Fowlkes, C.C., Malik, J.: Learning to detect natural image boundaries using local brightness, color, and texture cues. IEEE Trans. Pattern Anal. Mach. Intell. 26(5), 530–549 (2004)
24. Miranda, P., Falcao, A., Spina, T.: Riverbed: a novel user-steered image segmentation method based on optimum boundary tracking. IEEE Trans. Image Process. 21(6), 3042–3052 (2012)
25. Miranda, P., Falcão, A., Udupa, J.: Synergistic arc-weight estimation for interactive image segmentation using graphs. Comput. Vis. Image Underst. 114(1), 85–99 (2010)
26. Miranda, P., Mansilla, L.: Oriented image foresting transform segmentation by seed competition. IEEE Trans. Image Process. 23(1), 389–398 (2014)
27. de Moraes Braz, C., Miranda, P.: Image segmentation by image foresting transform with geodesic band constraints. In: 2014 IEEE International Conference on Image Processing (ICIP), pp. 4333–4337, October 2014
28. Peng, B., Zhang, L., Zhang, D., Yang, J.: Image segmentation by iterated region merging with localized graph cuts. Pattern Recogn. 44(1011), 2527–2538 (2011). semi-Supervised Learning for Visual Content Analysis and Understanding
29. Rauber, P.E., Falcão, A.X., Spina, T.V., de Rezende, P.J.: Interactive segmentation by image foresting transform on superpixel graphs. In: 2013 XXVI Conference on Graphics, Patterns and Images, pp. 131–138, August 2013
30. Schoenemann, T., Kahl, F., Cremers, D.: Curvature regularity for region-based image segmentation and inpainting: A linear programming relaxation. In: 2009 IEEE 12th International Conference on Computer Vision, pp. 17–23, September 2009
31. Spina, T.V., de Miranda, P.A.V., Falcão, A.X.: Hybrid approaches for interactive image segmentation using the live markers paradigm. IEEE Trans. Image Process. 23(12), 5756–5769 (2014)
32. Wolf, S., Schott, L., Köthe, U., Hamprecht, F.: Learned watershed: end-to-end learning of seeded segmentation. Technical report, Cornell University Library, April 2017. arXiv:1704.02249
33. Yang, W., Cai, J., Zheng, J., Luo, J.: User-friendly interactive image segmentation through unified combinatorial user inputs. IEEE Trans. Image Process. 19(9), 2470–2479 (2010)

Historical Handwritten Text Images Word Spotting Through Sliding Window HOG Features

Federico Bolelli[(✉)], Guido Borghi, and Costantino Grana

Dipartimento di Ingegneria "Enzo Ferrari",
Università degli Studi di Modena e Reggio Emilia,
Via Vivarelli 10, 41125 Modena, MO, Italy
{federico.bolelli,guido.borghi,costantino.grana}@unimore.it

Abstract. In this paper we present an innovative technique to semi-automatically index handwritten word images. The proposed method is based on HOG descriptors and exploits *Dynamic Time Warping* technique to compare feature vectors elaborated from single handwritten words. Our strategy is applied to a new challenging dataset extracted from Italian civil registries of the XIX century. Experimental results, compared with some previously developed word spotting strategies, confirmed that our method outperforms competitors.

Keywords: Word spotting · Handwriting recognition · Indexing

1 Introduction

The transition from handwritten to digitalized historical documents establishes a great challenge, due to the huge amount of documents, the peculiarity of this kind of data, and the noise on manuscripts: generally, automatic handwriting recognizers, also called *Optical Character Recognizers* (OCRs), or standard text analyzers fail.

In this context, we develop a new word spotting technique, or rather the ability to create word collections grouped into clusters containing all instances of the same word. The creation of these clusters is based on image matching results [10]. In this way, it is possible to semi-automatically index the content of handwritten historical documents. Manual transcription and index generation is extremely expensive and time-consuming in these cases and thus not always feasible for voluminous manuscripts.

In this paper we propose a method to extract features from historical document words and to match them exploiting the *Dynamic Time Warping* (DTW) technique, which compares and aligns feature vectors elaborated from single handwritten words. We collect a new dataset that is publicly available and it is acquired through a previously developed system of image dewarping [1]. This system starts from a curled page, usually taken by a digital scanner or digital

© Springer International Publishing AG 2017
S. Battiato et al. (Eds.): ICIAP 2017, Part I, LNCS 10484, pp. 729–738, 2017.
https://doi.org/10.1007/978-3-319-68560-1_65

camera, and outputs an image constituted only of horizontal straight text lines, without any distortion due to perspective, lenses and page warping. In particular, through this approach, a great amount of documents from Italian civil registries of the XIX century are available for our scope.

The paper is organized as follows. Section 2 presents an overall description of related literature works. In Sect. 3, the proposed method is detailed. The proposed dataset is described in Sect. 4. Section 5 reports experimental results. Finally, in Sect. 6 conclusions are drawn.

2   Related Work

The original idea of word spotting for handwritten manuscripts was initially presented in [8,9]. In these works, matching techniques and pruning methods are described: given a word's bounding box, unlikely matches are quickly discarded and similar words are clustered.

Generally, word spotting methods can be divided in two main classes: *line-segmentation* and *word-segmentation* based approaches.

Line-segmentation based methods rely on the hypothesis that each line in the document is separated and word segmentation techniques are not strictly required. Terasawa *et al.* [13,14] presented a word spotting method based on line segmentation, sliding window, continuous dynamic programming and a gradient-distribution-based feature with overlapping normalization and redundant expressions, also known as "slit style HOG features". In [6] a line-oriented process is applied to avoid the problem of segmenting cursive script into individual words. This approach exploits pattern matching techniques and dynamic programming algorithms. The presented system is tested on old Spanish manuscripts, showing a high recognition rate. Even the adoption of a number of heuristic to limit the search along document lines, this approach is expensive since words have to be searched for every possible position. Besides, DTW is separately applied on each feature vector and results are heuristically merged, producing different alignment for the same word-line pair.

On the other hand, *word-segmentation* approaches are based on the hypothesis that each word in the document images is separately clipped. A word-by-word mapping between a scanned document and a manual transcript is proposed in [15]: in this way, it is possible to exactly locate words in document pages. This method relies on a OCR used as a recognizer for multiple word segmentation hypothesis generated for each line of the document. Results shown that OCR is not a useful and feasible solution for historical manuscript recognition. In [11] a local descriptor, inspired by the SIFT [7] key-point descriptor, is proposed. Significant improvements are achieved exploiting two different word spotting systems, based on the well-known *Hidden Markov Models* and DTW.

In [10] a range of features suitable for DTW has been analyzed: this work is described in detail because we use it as a touchstone. Speed and precision have achieved as result of combining different text features which are extracted from pre-processed rectangular word images and that do not contain ascenders from other words. Moreover, inter-word variations such as *skew* and *slant* angles are detected and normalized. Investigated features include projection profile, partial projection profile, upper and lower word profile, background to ink transitions, gray scale variance, and feature sets containing horizontal and vertical partial derivatives applied through a Gaussian kernel. Best performance in terms of average precision are achieved by the combination of projection profile, upper and lower profile and ink transitions. In order to compare this strategy with our proposal, we produce an implementation of this algorithm, maintaining all details described in the corresponding paper.

3 Proposed Word Spotting Method

The method proposed in this paper is *word-oriented*, thus we describe it starting from single word image as the one reported in Fig. 1a (see Sect. 4 for extraction details). Before proceeding with feature extraction all word images are pre-processed as described in the following section.

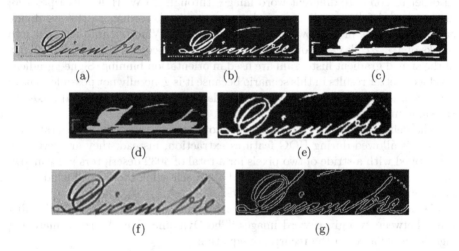

Fig. 1. Example of pre-processing steps applied to an handwritten word image. (a) is the raw input grayscale word image, (b) the result after binarization process, (c) the graphical output of the RLSA algorithm. Connected components are then labeled (d) and the bounding box of the biggest component is extracted (e) and (f). In (g) is reported the output of *Canny* algorithm applied on (e).

3.1 Word Image Pre-processing

All input images are binarized through an adaptive threshold [12] which deal with the light changes that occur in the original manuscripts (Fig. 1b). Then, we exploit the horizontal *Run Length Smoothing Algorithm* (RLSA) [16] to ensure that all pixels belonging to the word contained in the binary image are connected (Fig. 1c). The threshold used for RLSA is equal to text height that is calculated as described in [4]. Thanks to the *Connected Components Labeling* (CCL) [5] we are able to extract the word using the bounding box of the biggest component (see Fig. 1d, e and f for instance). We aim, through the combination of RLSA and CCL algorithms, to filter out all the graphical contents that do not belong to the handwritten word which are represented by remainders of other words in the original document. After these steps, it is possible to remove background from the image.

Moreover, images are vertically and horizontally resized to a fixed window of 352 × 90 pixels. This operation could be viewed as a normalization of handwritten word's width and height, and it is also a fundamental step for next elaborations.

Finally, *Canny* [2] algorithm is applied on the binarized and resized image in order to make our algorithm invariant to ink thickness.

Graphical result is reported in Fig. 1g.

3.2   Feature Extraction and Word Matching

In order to compare different word images through DTW, HOG descriptors [3] are computed follow a sliding windows approach. According to [14] we divide each input image in windows of fixed size 16 × 90. Windows are then split into blocks of 4 × 2 cells each composed by 4 × 4 pixels (see Fig. 2b). Finally, 12 bins of the signed gradient histogram are used in orientation binning. Signed gradient produces better results in this scenario because it is generally not possible to have some characters brighter and some other darker than the background mixed in the same manuscript.

The defined block has the same width of the window so no horizontal overlapping is allowed during HOG features extraction, instead, they are vertically overlapped with a stride of two pixels for a total of 4032 descriptors per window. In our experiments we test the proposed approach using overlapped windows in horizontal direction with different strides.

As said before, DTW is exploited to compute and align the similarity distance between two given word images. The Dynamic Time Warping matching algorithm is based on the recurrence equation

$$DTW(i,j) = min \left\{ \begin{array}{c} DTW(i-1,j) \\ DTW(i,j) \\ DTW(i,j-1) \end{array} \right\} + d(i,j) \qquad (1)$$

where $d(i,j)$ is the distance between the $i-th$ and $j-th$ feature vectors (respectively called x and y and both of length N) of the two images to match:

$$d(i,j) = \sum_{k=1}^{N} |x_{ik} - y_{jk}| \qquad (2)$$

In Fig. 2c is reported an example of DTW distance matrix calculated with Formula 1. In this example, HOG feature vectors are obtained with window stride

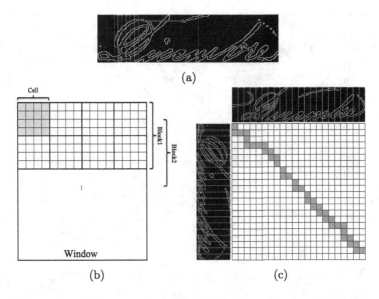

(a)

(b)   (c)

Fig. 2. (a) example of sliding window on a XDOCS's word image for HOG descriptors calculation; (b) details of the adopted windows; (c) example of DTW matrix obtained for two XDOCS's word images.

equal to window width (*i.e.* non overlapped windows). The green path, usually called *warping path*, represents the optimal match between the two words. This approach let us to determine a measure of word similarity independently from certain non-linear variations in the time dimension.

4 Dataset

As mentioned above, we collect a new challenging dataset. The dataset consists of a collection of handwritten month names extracted from Italian civil registries of the XIX century. We extract word images employing a template approach: given a rectified image of a whole page of the historical document, we directly extract month names placed in fixed position of the page. In this way, we can automatically collect a number of high quality and easy to annotate samples.

Specifically, the obtained dataset consists of around 1200 words and all 12 months are available. Moreover, the variety of handwritten words is guaranteed by three different official state writers. The dataset is publicly available[1].

The dataset creation approach relies on the assumption that rectified pages are available. These are obtained by the use of the dewarping technique described in [1]. The entire pipeline could be summarized in these steps:

- *Image pre-processing*: in this step document and page noise is filtered out; this is mainly due to the digitization process and to the intrinsic nature of the original images;

---

[1] http://imagelab.ing.unimore.it/XDOCS.

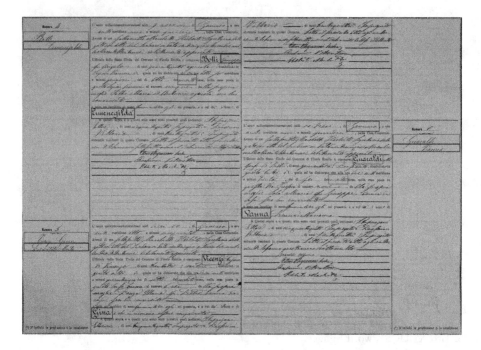

Fig. 3. Example of dewarped document and word extraction leaded by template. A template approach for word extraction is possible due to the prearranged structure in these particular historical registries.

- *Projection extraction*: this step aims to find the curved surface projection represented by two almost vertical straight lines and by two third degree polynomial curves surrounding the document page. This is required by the implemented dewarping method.
- *Dewarping phase*: this step is the core of the image rectification and dataset creation phases. During this phase, the projection of the curved surface is mapped into a rectangular normalized area.

At the end of this process, input images are correctly rectified and they do not suffer of any distortion effects. The result is depicted in Fig. 3 where colored bounding boxes show word of interest automatically extracted by image coordinates. In the following we will refer to the described dataset as XDOCS dataset.

5   Experimental Results

We test our system on the XDOCS dataset divided into three different group, one for each handwriting style. We refer to each group with the name of the municipality from which original documents belong, *i.e. Vignola, Carpi*, and *Formigine*. This approach let us to perform both *intra* and *inter* dataset evaluation.

Following a common practice for word spotting task, we exploit the *Mean Averages Precision* (MAP) with *cut-off* at $C = \{5, 10, 15\}$ (*i.e.* MAP@5, MAP@10, and MAP@15) to evaluate and compare different algorithms. Given a couple of datasets the first (with a word elements) is used to create queries that are then performed on the second one (with b word elements). For each query the average precision is calculated using the following formula:

$$ap@n = \frac{\sum_{k=1}^{n} P(k)}{\min(m, n)} \tag{3}$$

In Eq. 3, $P(k)$ is the precision at cut-off k in the item list, *i.e.*, the ratio of the number of correct word matches, up to position k, over the number k; n is the cut-off chosen from C and m is the total number of word images that match with the query in the second dataset.

Table 1. Results (on the XDOCS dataset) of the proposed method with different window strides (a), (b), and (c) and of the algorithm described by Rath *et al.* (d).

		Vignola	Carpi	Formig.		Vignola	Carpi	Formig.
Vignola	MAP@05	0.528	0.100	0.181	MAP@05	0.634	0.108	0.206
	MAP@10	0.380	0.086	0.144	MAP@10	0.465	0.098	0.168
	MAP@15	0.306	0.093	0.132	MAP@15	0.376	0.101	0.156
	CMF	75.25%	17.82%	26.73%	CMF	83.17%	18.32%	25.25%
Carpi	MAP@05	0.135	0.466	0.095	MAP@05	0.145	0.534	0.112
	MAP@10	0.101	0.434	0.078	MAP@10	0.110	0.489	0.093
	MAP@15	0.079	0.414	0.072	MAP@15	0.086	0.485	0.086
	CMF	14.53%	63.25%	15.38%	CMF	17.10%	66.67%	17.95%
Formig.	MAP@05	0.192	0.127	0.644	MAP@05	0.268	0.150	0.775
	MAP@10	0.156	0.114	0.541	MAP@10	0.209	0.133	0.662
	MAP@15	0.135	0.121	0.476	MAP@15	0.173	0.138	0.582
	CMF	24.69%	19.25%	77.82%	CMF	36.82%	21.34%	89.96%

(a) Our - 16 pixels stride.   (b) Our - 8 pixels stride.

		Vignola	Carpi	Formig.		Vignola	Carpi	Formig.
Vignola	MAP@05	0.665	0.102	0.222	MAP@05	0.468	0.042	0.077
	MAP@10	0.493	0.093	0.189	MAP@10	0.347	0.034	0.057
	MAP@15	0.400	0.098	0.170	MAP@15	0.276	0.028	0.050
	CMF	87.13%	14.85%	27.22%	CMF	68.32%	9.90%	13.37%
Carpi	MAP@05	0.159	0.578	0.125	MAP@05	0.086	0.445	0.087
	MAP@10	0.117	0.536	0.101	MAP@10	0.060	0.411	0.067
	MAP@15	0.091	0.527	0.096	MAP@15	0.050	0.382	0.058
	CMF	19.66%	73.50%	17.95%	CMF	13.78%	51.70%	15.34%
Formig.	MAP@05	0.309	0.177	0.823	MAP@05	0.097	0.053	0.557
	MAP@10	0.235	0.152	0.708	MAP@10	0.071	0.045	0.413
	MAP@15	0.194	0.153	0.621	MAP@15	0.060	0.042	0.342
	CMF	40.59%	26.77%	94.14%	CMF	19.25%	9.21%	80.33%

(c) Our - 2 pixels stride.   (d) Rath *et al.* [10].

Once the $ap@n$ is calculated for every query the $MAP@n$ is given by Eq. 4.

$$MAP@n = \frac{\sum_{i=1}^{Q} ap@n_i}{N} \qquad (4)$$

where $Q = a$ is the number of queries and $ap@n_i$ is the average precision for the $i\text{-}th$ query.

Moreover, an additional metric is included in our evaluation: for every test among two different datasets we store the number of queries that return a correct match as first. This value is reported in tables with the acronyms CMF (Correct Match First) and in percentage with respect off the total number of queries.

We compare our method with the one described by Rath and Manmatha [10] that is one of the state-of-the-art algorithms for word spotting task. This method is chosen because it uses a comparable approach based on DTW which exploits different features to perform matching between different words. As we said in Sect. 2, we implemented Rath's algorithm because, from our knowledge, a public implementation is not available. According to [10], we included in our code only the combination of features which achieve better performance in the original paper.

In Table 1a, b, and c the performance of the proposed method, using a stride ranging from 2 to 16 for HOG windows, is presented. The MAP increase when the stride size decrease and as consequence the best results are obtained with stride 2. Unfortunately, the computational cost of the process increase with lower strides. According to the application in which word spotting has to be applied, a compromise between accuracy and computation time can be selected. Even thought HOG descriptors can be calculated off-line during words extraction process, the DTW suffers with very long feature vectors such as the one obtained by HOG.

On the other hand, in Table 1d results achieved by our designed competitors on the same datasets are reported. As depicted, our method achieves better

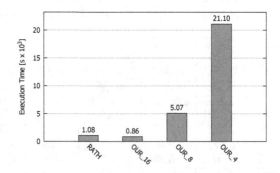

Fig. 4. Average query execution time for the algorithm by Rath and Manmatha [10] and our proposal with different window strides (*i.e.* 16, 8, 4). The time required by windows stride 2 (139.2×10^3 s) is not reported to facilitate the readability of the chart. Query search space is composed by almost 1200 samples (*i.e.* the entire XDOCS dataset).

accuracy in all tests also with 8 pixels stride. Moreover, CMF reveals that our proposal can be used as enabling technology for real word applications.

In Fig. 4 average query execution time required by the experimented word spotting algorithms is reported. All average times are computed considering the entire XDOCS dataset, thus the search space counts almost 1200 word images. No code optimizations are involved neither in the word spotting algorithm nor in the DTW implementation, so Fig. 4 serve only as comparison between time performance of the described algorithms.

6 Conclusions

In this paper two main contributions are described. Firstly, a new and challenging dataset of handwritten historical documents from Italian civil registries is publicly released. Secondly, a novel method to tackle the problem of word spotting task is presented: it is based on HOG descriptors and exploits the Dynamic Time Warping technique to compare feature vectors elaborated from single handwritten words. The system is able to achieve a good accuracy in terms of MAP and overcomes a literature competitor.

References

1. Bolelli, F.: Indexing of historical document images: ad hoc dewarping technique for handwritten text. In: 13th Italian Research Conference on Digital Libraries, IRCDL, February 2017
2. Canny, J.: A computational approach to edge detection. IEEE Trans. Pattern Anal. Mach. Intell. **6**, 679–698 (1986)
3. Dalal, N., Triggs, B.: Histograms of oriented gradients for human detection. In: IEEE Computer Society Conference on Computer Vision and Pattern Recognition, CVPR 2005, vol. 1, pp. 886–893. IEEE (2005)
4. Gatos, B., Pratikakis, I., Ntirogiannis, K.: Segmentation based recovery of arbitrarily warped document images. In: Ninth International Conference on Document Analysis and Recognition (ICDAR 2007), vol. 2, pp. 989–993. IEEE (2007)
5. Grana, C., Baraldi, L., Bolelli, F.: Optimized connected components labeling with pixel prediction. In: Blanc-Talon, J., Distante, C., Philips, W., Popescu, D., Scheunders, P. (eds.) ACIVS 2016. LNCS, vol. 10016, pp. 431–440. Springer, Cham (2016). doi:10.1007/978-3-319-48680-2_38
6. Kolcz, A., Alspector, J., Augusteijn, M., Carlson, R., Popescu, G.V.: A line-oriented approach to word spotting in handwritten documents. Pattern Anal. Appl. **3**(2), 153–168 (2000)
7. Lowe, D.G.: Distinctive image features from scale-invariant keypoints. Int. J. Comput. Vis. **60**(2), 91–110 (2004)
8. Manmatha, R., Croft, W.: Word spotting: Indexing handwritten archives. In: Intelligent Multimedia Information Retrieval Collection, pp. 43–64 (1997)
9. Manmatha, R., Han, C., Riseman, E.M., Croft, W.B.: Indexing handwriting using word matching. In: Proceedings of the first ACM International Conference on Digital Libraries, pp. 151–159. ACM (1996)

10. Rath, T.M., Manmatha, R.: Features for word spotting in historical manuscripts. In: Proceedings of Seventh International Conference on Document Analysis and Recognition, pp. 218–222. IEEE (2003)
11. Rodriguez, J.A., Perronnin, F.: Local gradient histogram features for word spotting in unconstrained handwritten documents. In: Proceedings of 1st ICFHR, pp. 7–12 (2008)
12. Sauvola, J., Pietikäinen, M.: Adaptive document image binarization. Pattern Recogn. 33(2), 225–236 (2000)
13. Terasawa, K., Nagasaki, T., Kawashima, T.: Eigenspace method for text retrieval in historical document images. In: Proceedings of Eighth International Conference on Document Analysis and Recognition, pp. 437–441. IEEE (2005)
14. Terasawa, K., Tanaka, Y.: Slit style hog feature for document image word spotting. In: 10th International Conference on Document Analysis and Recognition, ICDAR 2009, pp. 116–120. IEEE (2009)
15. Tomai, C.I., Zhang, B., Govindaraju, V.: Transcript mapping for historic handwritten document images. In: Proceedings of Eighth International Workshop on Frontiers in Handwriting Recognition, pp. 413–418. IEEE (2002)
16. Wahl, F.M., Wong, K.Y., Casey, R.G.: Block segmentation and text extraction in mixed text/image documents. Comput. Graph. Image Process. 20(4), 375–390 (1982)

Incremental Support Vector Machine for Self-updating Fingerprint Presentation Attack Detection Systems

Pierluigi Tuveri[(✉)], Mikel Zurutuza, and Gian Luca Marcialis

Department of Electrical and Electronic Engineering,
University of Cagliari, Cagliari, Italy
{pierluigi.tuveri,marcialis}@diee.unica.it, mikelzuru@gmail.com

Abstract. In this years Fingerprint Presentation Attack Detection (FPAD) had an increasing interest and the performances became acceptable, especially thanks to the LivDet protocols into the International Fingerprint Liveness Detection competition. A security issue arose from LivDet2015: the FPAD systems are not invariant towards the materials for fabricating spoofs. In other words, some previous works pointed out the vulnerability of these systems when an attackers uses unexpected materials. In this paper, we proposed a solution that exploit the self-update abilities of the classifier to adapt itself to never-seen-before attacks over the time. Experimental results on four LivDet data sets showed that the proposed method allowed to manage this vulnerability.

1 Introduction

In real world scenarios, one can attempt to circumvent a biometric sensor by using a copy of a certain required biometry. The artifact that is used as a counterfeit biometric is called *spoof*. Presentation attack detection (PAD) is the method which distinguishes genuine living biometric traits from spoof ones [4].

Despite the fact that currently several mature methods exist to distinguish impostors from genuine in fingerprint verification, the fingerprint PAD (FPAD) remains unsolved owing to the continuous evolution in the technology, materials for artificial fingerprint fabrication and cleverness of the attacker, as a sort of cops and robbers game. The use of novel (never-seen-before) materials, unknown when the PAD system was designed, is a constant challenge. A survey on FPAD is given in [6,7]. Attempting to propose a taxonomy for the present field, Coli *et al.* [8] distinguished two main categories for detection methods: hardware and software-based. Hardware-based detection of presentation attacks in the fingerprint can be made by using temperature, blood pressure, pulse or other methods. Conversely, PAD capabilities can also be added to the system by using software-based approaches. The procedure consists on extracting features from the fingerprint images acquired by the given sensors [13]. These measurements embed the information to determine the liveness degree of the targeted fingerprint. For example, a method for quantifying the perspiration phenomenon in a

© Springer International Publishing AG 2017
S. Battiato et al. (Eds.): ICIAP 2017, Part I, LNCS 10484, pp. 739–749, 2017.
https://doi.org/10.1007/978-3-319-68560-1_66

single image was developed by Tan e Schuckers in [14] through wavelet transform. Additionally, a curvelet transform approach for FPAD was proposed in [9,13], which allows representing singularities along curves in a more efficient way.

Besides the PAD method itself, it is also important to know the *modus operandi* for generating a spoof, and usually classified as cooperative and non-cooperative approaches. Within the cooperative method, the finger of the target individual must be placed into certain ductile material. Instead, the non-cooperative approach is the process to be performed when the subject left a latent fingerprint on a surface and it needs to be enhanced.

Given the even finer fabrication techniques and the discovering of high quality materials, a presentation attack attempted by using never-seen-before spoof materials becomes an endless challenge. Based on the LivDet2015 competition results, in [11], Ghiani *et al.* remark that it is still difficult being able to generalize against unexpected spoof attacks. On the basis of what they reported, the need arises for a PAD system able to *self-update* its recognition capacity.

In this sense, Rattani *et al.* highlighted in [3] the problem of encountering spoofs that were not considered or unobserved within the training stage. Based on the open set theory, they present a W-SVM-based novel material detector in order to cope with this scenario.

In accordance with the same working line, we present an alternative, incremental SVM system which adapts to unknown materials using the Stochastic Gradient Descent (Sect. 2). We compared the performance of our system against the W-SVM one, used as a sort of ground truth. Very competitive results for most of the scanner/feature method/spoof material combinations were attained. The experimental protocol is discussed in Sect. 3 as well as obtained results in Sect. 3.3. Section 4 draws conclusions and suggests future research works.

2   The Proposed Method

One of the key researches related to facing presentation attacks with novel materials was accomplished by Tan *et al.* [1] where the spoof material dependency was investigated. This pointed out that when the classifier is trained with specific spoof materials, it can misclassify spoofs made with materials that are not present in the training set. Marasco and Sansone proposed a comparison among FPAD state-of-the-art algorithms in order to find the most robust algorithm towards the introduction of new materials in the test set [2].

Since very different algorithms were used in the works above, we can claim that the problem seems to be independent of the algorithm type; on the other hand, all these algorithms cannot generalize the never-seen-before materials problem. The International Fingerprint Liveness (LivDet) competition, in the 2015 edition [11], took into consideration the problem of bias related to the material. As a matter of fact, the conclusion reported in [11] was that the performance decays when attacking the system with a material that is not present in the train set (never-seen-before); in a few of cases, this can be argued from the visual inspection of obtained images. This issue was partially addressed in the previously cited and other papers, and could be an interesting matter of research for the future.

Therefore, updating the PAD system is a solution which appears as necessary to avoid the performance decay. Updating could be done manually and offline, but it is a not trivial problem to detect a good spoof by visual inspection. An alternative is to add samples derived from novel materials when available, but this did not assure that a better performance will be obtained over never-seen-before materials.

In this paper, we focused on the semi-supervised update concept. It is common to refer to them as "self update" systems [12]. These use the data collected while they are in operation in order to re-train the classification parameters. Usually, these systems tend to become more robust over time as they adapt in the new working conditions (this have been observed in other applications [12]). In this case, the goal is to re-train the system especially if a novel material is used to attack it.

Therefore, we propose a statistical classifier that updates itself in order to adapt the estimated conditional distributions when there are novel materials. In particular, we explore the so-called incremental Support Vector Machine (SVM) [12]. This choice is motivated by the fact that the majority of PAD systems achieved the highest performance, given the feature set, when the classifier is a SVM, as reported in [12].

2.1   Incremental SVM for Fingerprint Presentation Attacks Detection

A linear SVM learns the function $f(x) = \omega^T x + b$ using the train set where the samples are a tuple in the form (x_i, l_i). $x_i \in \Re^n$ and $l_i = \pm 1$. In the test phase, if $f(x) > 0$ then the label value is $+1$ otherwise the label value is -1.

The parameter ω is the slope of the hyperplane and b is the intercept. In this way, two subspaces, one for the patterns with positive labels, and one for negative labels, are computed.

In order to find the best function $f(x)$, that is, the parameters ω e b, we use the following error function:

$$E(\omega, b) = \frac{1}{n} \sum_{i=1}^{n} L(y_i, f(x_i)) + \alpha \frac{1}{2} \sum_{i=1}^{n} \omega_i^2 \tag{1}$$

Where L is the hinge loss function, and $\alpha > 0$ is the hyperparameter.

In order to minimize the error function, we use a Stochastic Gradient Descent [15][1].

Let us introduce the sets $B_{t_0}, ..., B_{t_m}$, each B_{t_i} is composed by a collection of *Live* and *Fake* fingerprint representations. Each component (x_i, l_i) of B_{t_i} is made up by two parts, one is x_i, that is the feature vector, and other one l_i is the label of the sample.

---

[1] We use the python library http://scikit-learn.org/stable/modules/sgd.html.

For updating, a novel set of data B_{t_i} is collected, the gradient continues seeking the minimum from the previous point.In our approach, updating is performed off-line, that is, when the system is not operating. Basically, we train the SVM with initial data B_{t_0}, and we update the classifier with new data B_{t_i} in order to adapt the system's parameters.

The updating algorithm is given in Algorithm 1. The prediction function $f(x) \propto P(c \mid X)$ outputs an estimation of $P(c = Live \mid X)$, the so-called "liveness score" S. It represents the liveness degree of the pattern. We can estimate the probability density $p(S \mid c = Live)$, that is, the distribution of $Live$ samples based on the liveness score. We introduce $FRR(s) = \int_0^s p(S \mid Live)dS$ as the percentage of the misclassified $Live$ samples at liveness score value s. In order to reduce the probability of using misclassified patterns, we select $FRR(s) = 0.1$, in fact we want to insert in the update stage only 10% of misclassified patterns. The value s ranges in $[0,1]$ and defines threshold such that $FRR(s) = thrFRR_{t_i}$ related to B_{t_i}. When we update the SVM with new data, if a certain sample follows the $thrFRR_{t_i}$ condition, it is used to update the classifier.

Algorithm 1. Incremental SVM

- Let us indicate our classifier with C.
- Let B_{t_0}, B_{t_1}, ..., B_{t_m} be $m + 1$ sets of samples collected over time, and called batches.
- $B_{t_0} = \{(x_1, l_1), ..., (x_{N_0}, l_{N_0})\}$, where x_i is a feature vector and l_i is the related class $(Live/Fake)$. The first is labeled, that means, the class are known for each sample.
- For $t \in [t_1, ..., t_m]$, $B_t = \{x_1, ..., x_{N_t}\}$, where x_i is a feature vector.
- Let s be the estimated value of $P(c = Live|X = x)$, as output of C, being x a feature vector. s is also called "liveness score".
- Let $p(S|c = Live)$ be the so-called live fingerprints distribution of liveness scores.
$B \leftarrow B_{t_0}$
$C \leftarrow TrainSVM(B)$, that is, train the SVM classifier by estimating the gradients on the data set B.
Estimate $P(c = Live|X = x)$, $p(S|c = Fake)$ over B.
for $t \in [t_1, ..., t_m]$ **do**
  $thrFRR_t \leftarrow argmax_s(P(s \mid c = Live) \leq 0.1)$.
  $H \leftarrow \emptyset$
  for $x \in B_t$ **do**
    if $s \leq thrFRR_t$ **then**
      $l \leftarrow Fake$
      $H \leftarrow H \cup \{(x, l)\}$
    end if
  end for
  $C \leftarrow UpdateSVM(C, H)$
  $B \leftarrow B \cup H$
  Estimate $P(c = Live|X)$, $p(S|c = Live)$ over B.
end for

2.2 Prior Work

Rattani et $al.$ [3] proposed a self update system for facing with the same problem. They used the open set theory in order to re-train the system. Let $M = \{m_0, m_1, ..., m_k\}$ where m_0 is the class of $Live$ samples and m_i $i \neq 0$ is a generic spoof material. In order to find the open set, they used 1-Class RBF SVM, thus instantiating a 1-Class classifier f_i^0 for each material $m_i \in M$. A 1-Class RBF SVM constructs a decision boundary that encloses all the positive elements. It is needed a calibration stage in order to calculate the $P_0(l \mid f^0(x))$, that is the inclusion probability of a sample to that class (the updating stage). If the $P_0(l \mid f^0(x)) \leq \delta_\tau{}^2$ for a pattern, it is classified as a never-seen-before material. Basically a 1-Class classifier might suffer from the overfitting problem because it was only trained with positive samples. Therefore, a binary SVM in order to generalize the classification was used. In this case, the binary RBF classifier was trained using the positive and negative patterns. There was not a division per material, but there is only the class $Fake$. Let $s_i = f(x_i)$ be the output (score) of the binary RBF SVM where x_i is a sample of the train set. $Fake$ and $Live$ samples were fitted by the Weibull distribution. In this way, the value $P_\eta(l \mid f(x))$ is obtained which describes the probability that the score belongs to the $Live$ class. Equally, $P_\psi(l \mid f(x))$ describes the distribution of $Fake$ scores. In order to detect the never-seen-before material, they used:

$$ly^* = argmax_{l \in Y} \ P_{\eta,l} \times P_{\psi,l} \times t_l$$
$$subject \ to \ P_{\eta,l*} \times P_{\psi,l*} \geq \delta_R \tag{2}$$
$$t_l = 1 \ if \ P_0(l \mid x) \geq \delta_\tau \ otherwise \ t_l = 0$$

In order to detect spoof materials already known:

$$l = +1 \ \Leftrightarrow \ P_{\eta,+1} \times P_{\psi,+1} \times t_{+1} \geq \delta_R$$
$$t_{+1} = 1 \ if \ P_0(y \mid x) \geq \delta_\tau \ otherwise \ t_{+1} = 0 \tag{3}$$

The δ_R parameter is the threshold used to find the Equal Error Rate (EER) point, where FAR=FRR. δ_R is not a constant but varied. The details of Rattani's algorithm is in Algorithm 2.

We compared our approach with Ref. [3] because it is the only one, to the best of our knowledge, that proposed a self updating system to deal with never-seen-before materials, thus treating the FPAD problem as a cops and robbers game, where a "patch" is constantly applied to the existing system in order to adapt itself over time.

The method in [3] relied on the number of materials (one SVM per material), and every time all of them need a complete re-training because they are considered as a unique system (therefore it is not possible to use incremental SVM for each of them). Our proposed method relies on the linear SVM that self-updates by the stochastic gradient descent theory. The system is more scalable due to the fact that its size in terms of classifiers is independent of the number of materials involved. Only one linear SVM is necessary, without the need of both old and new data.

---

^{2} δ_τ is a threshold and in all experiments is set to 0.001.

Algorithm 2. Rattani (W-SVM)

- Let us indicate our classifiers with $C1$, $C2$.
- Let $B_{t_0}, B_{t_1}, ..., B_{t_m}$ be $m + 1$ sets of samples collected over time.
- Let $M = \{m_0, m_1, ..., m_k\}$ where m_0 is the class of *Live* samples and m_i $i \neq 0$ is a generic spoof material that are present in B_{t_0}
- $B_{t_0} = \{(x_1, l_1), ..., (x_{N_0}, l_{N_0})\}$, where x_i is a feature vector and l_i is the related class (*Live/Fake*). The first is labeled, that means, the class are known for each sample.
- Let $P_0(l \mid f^0(x))$, that is the inclusion probability of the class.
- Let $P_\eta(l \mid f(x))$, that is the Live distribution
- Let $P_\psi(l \mid f(x))$, that is the Fake distribution
- For $t \in [t_1, ..., t_m]$, $B_t = \{x_1, ..., x_{N_t}\}$, where x_i is a feature vector.

$B \leftarrow B_{t_0}$
$C1 \leftarrow 1 - Class\ RBF\ SVM(B)$, is an ensemble a SVM for each m_i where is
$i = 1, .., k$
$C2 \leftarrow RBF\ SVM(B)$
Estimate $P_0(l \mid f^0(x))$, $P_\eta(l \mid f(x))$, $P_\psi(l \mid f(x))$ over B.
for $t \in [t_1, ..., t_m]$ **do**
   $H = B$
   for $x \in B_t$ **do**
      Estimate the novel material using Eq. 2 as Known Negative $KnownN$.
      Estimate the spoof material using Eq. 3 as Known Negative $KnownN$.
      $H = H \cup KnownN$
   end for
   $C2 \leftarrow RBF\ SVM(C2, H)$
   Estimate $P_\eta(l \mid f(x))$, $P_\psi(l \mid f(x))$ over B.
end for

---

3 Experimental Results

3.1 Dataset

We tested our approach on the datasets publicly available and related to the second Liveness Detection Competition 2011 (LivDet 2011) [11]. They were collected in order to ascertain the current state of the art in FPAD systems. These datasets are based on images captured from four different sensors: Biometrika, Digital Persona, Italdata and Sagem. For each of these scanners, 4000 images were acquired in total, being 2000 *Live* and 2000 *Fake* images. Each dataset has a set M of n materials $M = \{m_1, ..., m_n\}$; without loss of generality we indicate m_0 the *Live* patterns. Remaining artificial materials are distributed as follows: for Digital Persona and Sagem spoofs materials are Latex, PlayDoh, Gelatine, Silicone and Wood Glue. Instead, for Biometrika and ItalData datasets there are image samples of Latex, Ecoflex (platinum-catalysed silicone), Gelatine, Silgum and Wood Glue.

3.2 Experimental Protocol

In the LivDet competition, each data set is splitted into two parts, training set for computing the system's parameter, and test set for evaluating its performance. In order to verify the performance under never-seen-before materials, we followed the experimental protocol described by Rattani *et al.* [3]. It basically consists of three steps: training, adaptation and testing.

1. Training: Each train set is splitted in 1,000 *Live* images and 2 sets of 200 *Fake* images each one, that is, B_{t_0}. These sets are grouped into the following $M_{t_0} = \{m_0, m_l, m_k\}$ combinations, where $l, k = \{1, .., n\}$; $l \neq k$ are considered as known spoof materials. We must split the train set with all combinations of $m_1, .. m_n$ materials according to their corresponding sensor described above.
2. Adaptation: Each test set is divided into two non-overlapping partitions called T_1 and T_2. Both partitions require 500 *Live* and 500 *Fake* samples. T_i $i = \{1, 2\}$ is composed by 200 *Fake* samples with the same materials of M_{t_0}, and 300 samples of 3 novel materials that are not present in M_{t_0}, called unknown materials. Those partitions are separately used to either adapt the system with novel spoofs as well as for testing purposes. When using for updating purposes, we call them B_{t_i} in order to incrementally adapt the system as previously described in Sect. 2.
3. Testing: T_1 and T_2 have also the role of testing the performance of the system. In particular, we update the system with T_i, $i \in \{1, 2\}$ and test with T_j, $j \in \{1, 2\}$, $j \neq i$. In order to see the benefits of the self-update system, we also compare its performance without the adaptation stage.

3.3 Results

The goal of the experiment is to compare our algorithm with results reported in Rattani *et al.* Thus we used the same textural features, namely, LBP, LPQ and BSIF [10,16,17].

Reported results are averaged over the cross-test using T_1 and T_2 data sets.

Figures 1(a), (b), (c) and (d) showed the performance of the system in terms of Equal Error Rate (EER), defined as the error rate where the false positive and false negative rates are equal. The x-axis represents the materials present in the training set. Thus, the remaining artificial materials according to each sensor are considered as never-seen-before spoofs. The y-axis is divided in two parts, where the upper part represents the results of Rattani *et al.*'s system and the one in the bottom represent results of the proposed system. For each system the contour plot represents the performance for LPQ, LBP and BSIF feature extraction methods. As indicated above, for each method, averaged EER values for non adapted and adapted protocols are shown. The not adapted system is trained using the same algorithms (W-SVM and incremental SVM), and there is not the adaptation phase, but only the testing step. We compare the performance in order to evaluate the benefits of the adapting strategy.

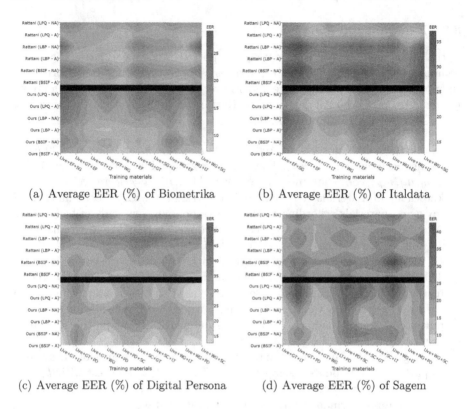

(a) Average EER (%) of Biometrika   (b) Average EER (%) of Italdata

(c) Average EER (%) of Digital Persona   (d) Average EER (%) of Sagem

Fig. 1. EER plots. Each chart is divided being the approach of Rattani et al. in the upper part and our approach below. x-axis represents materials used in training stage. Their acronyms are EF (Ecoflex), SG (Silgum), GT (Gelatine), LT (Latex), WG (WoodGlue), SC (Silicone), PD (Playdoh), that indicate (Color figure online).

We can observe the influence of each artificial material depending on whether they are known or unknown materials, that is, depending on whether the displayed material was present during the initial training stage or not. In these plots, the averaged EER value scale appears as a legend on the right column of the image, going from red (high) to green (low) values of EER. In order to keep certain coherence among the 4 plots, we maintain the same red-green scale: colors represent equal EER values within different plots. The highest existing EER value (52.05% when using LPQ with DigitalPersona samples and being Gelatine and Playdoh known materials, achieved by Rattani *et al.*) among all scanners and material combinations is represented as absolute red color whereas the lowest value (8.45% when using BSIF with Biometrika samples and being Latex and EcoFlex known materials, achieved by our method) represents the absolute green color.

Overall, the obtained results support the idea that through adapting the system with novel materials we may achieve a better performance against never-seen-before scenarios, for both W-SVM and incremental SVM. By following the above described convention, Fig. 1(a) clearly shows a poorer performance of Biometrika samples when avoiding adaptation through LPQ features for almost every material combinations. On the other hand, avoiding adaptation with LBP features may lead to differences in performance depending on the material combinations. This means that the role of the features set adopted may help in detecting never-seven-before materialis, due to "generalization" abilities that should be better investigated in future works. The W-SVM system performs worse when EcoFlex+Silgum, Silgum+Gelatine and WoodGlue+Silgum material combinations appear within the training set. Conversely, the incremental SVM seems to behave better with previous materials but performs worse when Gelatine+Latex, Latex+EcoFlex or Silgum+Latex pairs are known spoofs. Furthermore, BSIF features are generally less effective for the non adapted version of Rattani et al. than for our approach. Only WoodGlue+EcoFlex materials characterize a worse performance for our non adapted procedure. However, we also observe that once having adapted our classifier, it has a better performance than adapted BSIF from Rattani et al.

Table 1. EER values averaged from T1 and T2 (both Rattani and our approach)

Average EER	LBP			
	Rattani		Ours	
	(not adapted)	(adapted)	(not adapted)	(adapted)
Biometrika	16.45	10.69	15.97	12.78
Italdata	29.23	24.53	24.81	22.56
Digital persona	37.49	26.08	26.62	25.05
Sagem	21.79	17.63	17.22	15.46
Average EER	LPQ			
	Rattani		Ours	
	(not adapted)	(adapted)	(not adapted)	(adapted)
Biometrika	16.41	12.32	20.32	17.25
Italdata	21.81	17.43	19.50	17.22
Digital Persona	38.86	15.22	21.75	19.74
Sagem	26.315	17.03	26.34	23.06
Average EER	BISF			
	Rattani		Ours	
	(not adapted)	(adapted)	(not adapted)	(adapted)
Biometrika	17.27	11.82	13.9	11.34
Italdata	29.82	27.34	19.73	18.87
Digital persona	25.43	21.68	19.94	16.65
Sagem	25.43	18.46	20	15.17

Summarizing, our approach gives similar or even better results compared to Rattani's work. In fact, all the results achieved with BSIF features are better for our method, as well as almost every scanner with LBP. By focusing on the lowest performance obtained by incremental SVM, the deficiency generally lies on the LPQ features for Biometrika, Sagem and Digital Persona sensors and also on LBP features for Biometrika sensor. Similarly, special attention must be paid to which materials are present within the training set or novel spoofs set. Overall Biometrika and Italdata systems perform worse when Latex and WoodGlue are considered as never-seen-before materials. Instead, for DigitalPersona and Sagem scanners WoodGlue and Playdoh seem to cause the poorest classification rate.

In order to complement the plots with numerical results, in Table 1 we can observe that after adaptation, our system overcomes the W-SVM-based novel material detector in 8 out of 12 different cases (4 sensors times 3 feature extraction methods). Specifically, the samples obtained through Italdata, Digital Persona and Sagem scanners are better classified with our approach by using LBP method (22.56%, 25.05% and 15.46% respectively). Same thing occurs for average EER values from Italdata samples by using LPQ (17.22%) and also with every sensor when using BSIF features (11.34%, 18.87%, 16.65% and 15.17% respectively), thus suggesting the important role of the adopted feature set to distinguish among materials characteristics.

4   Conclusions

The obtained results confirm that a self-adaptation algorithms may greatly help in dealing with the never-seen-before materials problem in FPAD. Both W-SVM and our proposed system appear to be effective from the point of view of classification improvement through adaptive methods. It is remarkable that our system achieves a very competitive performance through a less complex and easier-to-implement approach. Nevertheless, generally speaking, it obtains weaker results with LPQ features in comparison to the W-SVM method. This fact suggests a different behaviour of LPQ with respect to LBP and BSIF. Regarding to the difference in performance when exchanging spoof materials between train and test datasets, obtained results support the existence of a general trend: facing against Latex or Gelatine as novel materials results in poorer performance of the system before having adapted it. Again, this can be a limitation of the feature sets adopted. Consequently, it means that facing high quality artificial materials leads to a big challenge. In spite of this, our method has demonstrated to be flexible through a simpler and scalable algorithm. In the future, we will continue investigating the differences related to feature extraction methods and algorithms. Additionally, we will extend these experiments to already available LivDet 2013 and LivDet 2015 datasets in order to deepen into the image quality (scanners) and spoof quality (materials) tradeoff, which was explicitly dealt with in those editions of LivDet.

References

1. Tan, B., Lewicke, A., Yambay D., Schuckers, S.: The effect of environmental conditions and novel spoofing methods on fingerprint anti-spoofing algorithms. In: IEEE WIFS, pp. 1–6 (2010)
2. Marasco, E., Sansone, C.: On the robustness of fingerprint liveness detection algorithms against new materials used for spoofing. In: BIOSTEC 2011, vol. 1, pp. 553–558 (2011)
3. Rattani, A., Scheirer, W.J., Ross, A.: Open set fingerprint spoof detection across novel fabrication materials. IEEE TIFS **10**(11), 2447–2460 (2015)
4. Jain, A.K., Flynn, P., Ross, A.: Handbook of Biometrics. Springer, USA (2007)
5. Schuckers, S.A.: Spoofing and anti-spoofing measures. Inf. Secur. Tech. Rep. **7**(4), 56–62 (2002)
6. Marasco, E., Ross, A.: A survey on antispoofing schemes for fingerprint recognition systems. ACM CSUR **47**(2), 28 (2015)
7. Sousedik, C., Busch, C.: Presentation attack detection methods for fingerprint recognition systems: a survey. IET Biom. **3**(4), 219–233 (2014)
8. Coli, P., Marcialis, G.L., Roli, F.: Vitality detection from fingerprint images: a critical survey. In: Lee, S.-W., Li, S.Z. (eds.) ICB 2007. LNCS, vol. 4642, pp. 722–731. Springer, Heidelberg (2007). doi:10.1007/978-3-540-74549-5_76
9. Nikam, S., Agarwal, S.: Curvelet-based fingerprint anti-spoofing. SIVP **4**(1), 75–87 (2010)
10. Nikam, S.B., Agarwal, S.: Texture and wavelet-based spoof fingerprint detection for fingerprint biometric systems. In: ICETET 2008, pp. 675–680. IEEE (2008)
11. Ghiani, L., Yambay, D.A., Mura, V., Marcialis, G.L., Roli, F., Schuckers, S.A.: Review of the Fingerprint Liveness Detection (LivDet) competition series: 2009 to 2015. IVC J. **58**, 110–128 (2017)
12. Rattani, A., Roli, F., Granger, E. (eds.): Adaptive Biometric Systems: Recent Advances and Challenges. Advances in Computer Vision and Pattern Recognition, 1st edn. Springer, Cham (2015)
13. Nikam, S.B., Agarwal, S.: Fingerprint liveness detection using curvelet energy and co-occurrence signatures. In: CGIV 2008, pp. 217–222. IEEE (2008)
14. Tan, B., Schuckers, S.: Liveness detection for fingerprint scanners based on the statistics of wavelet signal processing. In: 2006 Computer Vision and Pattern Recognition Workshop, CVPRW 2006, p. 26. IEEE, June 2006
15. Bottou, L.: Large-scale machine learning with stochastic gradient descent. In: Lechevallier, Y., Saporta, G. (eds.) COMPSTAT 2010, pp. 177–187. Springer, Heidelberg (2010)
16. Ghiani, L., Marcialis, G.L., Roli, F.: Fingerprint liveness detection by local phase quantization. In: ICPR 2012, pp. 537–540. IEEE (2012)
17. Ghiani, L., Hadid, A., Marcialis, G.L., Roli, F.: Fingerprint liveness detection using binarized statistical image features. In: BTAS 2013, pp. 1–6. IEEE (2013)

Tampering Detection and Localization in Images from Social Networks: A CBIR Approach

Cedric Maigrot[1]([✉]), Ewa Kijak[1], Ronan Sicre[2], and Vincent Claveau[2]

[1] Univ. Rennes I, UMR 6074 IRISA, Rennes, France
[2] CNRS, UMR 6074 IRISA, Rennes, France
{cedric.maigrot,ewa.kijak,ronan.sicre,vincent.claveau}@irisa.fr

Abstract. Verifying the authenticity of an image on social networks is crucial to limit the dissemination of false information. In this paper, we propose a system that provides information about tampering localization on such images, in order to help either the user or automatic methods to discriminate truth from falsehood. These images may be subjected to a large number of possible forgeries, which calls for the use of generic methods. Image forensics methods based on local features proved to be effective for the specific case of copy-move forgery. By taking advantage of the number of images available on the internet, we propose a generic system based on image retrieval, followed by image comparison based on local features to localize any kind of tampering in images from social networks. We also propose a large and challenging adapted database of real case images for evaluation.

Keywords: Tampering detection and localization · Tweet image analysis · Image forgery · Copy-move and splicing detection · Matching

1 Introduction

Massive amounts of information are spread over social networks, and among them a large quantity of fake information is conveyed. Messages are often composed of images or videos associated with text. Cases of misinformation take many forms: images can be modified for malicious purpose, or original images can be reused in a wrong context. Detecting such manipulations is now a key issue, and such process usually requires to examine the several modalities to get some contextual information about the transmission channel as well as information from the web. In this work, we focus on the visual aspect of this problem, and we are interested in automatically providing clues about images exchanged on the social networks.

Images may have undergone different types of modifications: some of them are malicious, like duplication of some parts of the image (known as copy-move attack), inserting a region from another image (copy-paste or splicing attack), or deleting some regions (thanks to techniques as in painting or seam carving);

© Springer International Publishing AG 2017
S. Battiato et al. (Eds.): ICIAP 2017, Part I, LNCS 10484, pp. 750–761, 2017.
https://doi.org/10.1007/978-3-319-68560-1_67

Fig. 1. Examples of images in social networks

but images posted on social networks can also typically be submitted to editing process, such as combination of several images into one, adding of text or shapes (arrows, circles, etc.), aesthetic filters, or simply cropped or re-compressed, see Fig. 1. Rather than only classifying an image as modified or pristine, we are interested in detecting and localizing any type of modifications.

Many studies in the image forensics field tackle the problem of assessing the authenticity of digital images. In the traditional forensics paradigm, no external information but the image is available. This is a difficult task, and forensics methods can usually only cope with copy-move attacks, and are evaluated on clean dedicated databases. We adopt a different paradigm as we rely on the access to external information such as image databases, or Web reverse image search. Indeed, one of the first step in manual checking of image integrity is to search it (or modified versions) on the Web[1], and there's no reason to refuse this information, in particular in the context of social network use. The problem is thus assimilated to a comparison task between pairs of images, which can handle various tampering operations, at a lower cost and faster than tampering detection methods based on a single image. These previous methods can be seen as an alternative approach, when no similar images are retrieved.

Difficulties lie in the wide variety of possible modifications. In this work, we propose a unified framework to detect and localize a large variety of forgeries in an image, by detecting inconsistencies between two images. The image to analyze is compared to the most similar images retrieved by a Content-Based Image Retrieval (CBIR) system. Such a system could be a reverse image search tool, but in our work we query our own database. Thus, we can evaluate the performance of our CBIR system when dealing with the particular class of images considered here, where strong editing process may trouble the recognition. Once similar images are retrieved, a local descriptor based approach is used to identify and localize differences. We also build two datasets containing various types of forgeries to evaluate our system.

In the next section, we discuss related studies on image forensics, image retrieval and social networks analysis. Our approach is described in Sect. 3, while datasets for evaluation and results are detailed in Sect. 4. Concluding remarks are presented in Sect. 5.

---

[1] http://www.stopfake.org/en/13-online-tools-that-help-to-verify-the-authenticity-of-a-photo/.

2   Related Work

Image forensics. The identification of tampered images has been largely studied in the field of image forensics. Various forms of image manipulation exist such as objects deletion, retouching objects, copy-moving parts of an image, or inserting elements taken from a different source, *i.e.* splicing or copy-paste. Such diverse scenarios require specific approaches and techniques. Traditionally in image forensics scenarios, the decision (tampered or not) must be made solely on the basis of the image to be analyzed, without using any external information. Most passive forgery detection techniques aim at revealing alteration of the underlying statistics of the forged image. However, almost all existing forensics methods detect only one type of image processing operations or are based on some assumptions regarding the image format or the camera used. Among these techniques, pixel-based approaches are the most related to our context. Indeed, for images transmitted on social networks, we have neither information about camera (as EXIF informations are erased), nor prior about format.

Pixel-based methods widely address the problem of copy-move forgery detection (CMFD) [24]. These methods, also called Local Descriptor-based forgery detection techniques, are typically based on feature matching. Block-based approaches split the image into overlapping blocks and extract features, such as DCT, DWT, histogram of co-occurrences on the image residual [11], Zernike moments, or Local Binary Pattern (LBP) [9]. Keypoints-based approaches compute features, usually SIFT or SURF [1,10], on local regions characterized by a high entropy. Features are then matched to detect similar regions, as a cue for copy-move forgery. Generally, it is shown that techniques based on dense fields provide a higher accuracy [7]. Also, some methods propose not only the detection but also the localization of the modified regions. We note that deep Convolutional Neural Networks (CNN) have been recently introduced in image forensics [5,17]. The general idea is to restrict the first convolutional layer to a set of high-pass filters in order to suppress image content. However, the CNNs are used either only for image binary classification (authentic/forged), without localization [17], or to identify some manipulations such as median filtering or Gaussian blurring, excluding copy-move or splicing attacks [5].

Content-Based Image Retrieval (CBIR). For several years state-of-the-art methods in image retrieval consisted in aggregating local descriptors, such as SIFT, into a global representation. These last years, the use of pre-trained CNN [13] became the new reference for global descriptors. [4] first showed that using fully connected layers of a pre-trained deep network as global descriptors can outperform descriptors based on SIFT features, even without fine-tuning. Similar conclusions were shared by [18,23] with region-based descriptors. Also, [3] proposed to aggregate deep local features, while [20] proposed new fusing schemes for compact descriptions.

Social networks information analysis. Analysis of information on social networks raises a growing interest, in particular detecting false information. This is illustrated by an increasing number of projects on this topic[2], and the emergence of a task dedicated to tweet classification on true or false at the *Mediaeval* benchmark, named *Verifying Multimedia Use*[3]. Usually the methods are interested in the multimodal nature of the messages to make a decision (text, social networks, image). It was also shown that the use of external knowledge is of great importance in the success of the proposed methods [15,16].

3 Proposed Method

We propose a unified approach to detected a large variety of forgeries, which is composed of two main steps. First, the image to analyze is used to query a database. The system searches for the most similar image. If an image is retrieved, it is then compared to the query image to detect and localize the forged areas; Otherwise, the process ends.

3.1 Content-Based Image Retrieval System

Initial Ranking. A CBIR system is used to retrieve candidate images, sufficiently similar to a query (the image to be analyzed), even if the images are different one from another due to tampering operations.

First, images are described using CNN-based representations. Following the recent works of [21,23], we choose to build descriptors using the seventh fully connected layer $fc7$ of the VGG vd19 CNN [22] trained on ImageNet. Images are first scaled to the standard 224×224 input size. Then, ℓ_2-normalization is performed and we obtain a 4096-dimensional vector.

Once all images descriptors are obtained, cosine similarity is computed between the query and images from the database. The nearest neighbors are retrieved using a KD-Tree to accelerate the search. Only images whose similarity exceeds a given threshold T, which is further evaluated in the experimental section, are considered as relevant. Otherwise no image is considered similar.

Filtering. A geometric verification step, *i.e.* filtering, is then employed to filter the false positives from the short list of top ranked images returned by the CBIR. Filtering is based on the number of inlier matches after estimating the spatial transformation between the query and each candidate images. Finally, only the image with the highest similarity is considered for further processing.

The proposed approach is based on SURF features matching, similarly to several reranking process used in CBIR systems. Specifically, dense SURF features are first extracted in both images and matched [19]. RANSAC algorithm is then applied to estimate the affine transformation H between the two images. To further decrease the number of false matches, only a subset S of points in

---

[2] See for example Reveal project (https://revealproject.eu/), InVID project (http://www.invid-project.eu/), or Pheme project (https://www.pheme.eu/).
[3] http://www.multimediaeval.org/mediaeval2016/verifyingmultimediause.

the query are kept as candidate matches for the RANSAC algorithm. These are points that match another point with a distance $d \leq 2 \times d_{min}$, where d_{min} is the minimum distance found between 2 descriptors of the pair of images.

After RANSAC estimation, we further apply H to each point of $\mathcal{S}$ and classify them as inlier if the distance d^* between the projected position and its match is lower than $0.15 \times diag$, where $diag$ is the length of the image diagonal in terms of pixels. Images with a majority of outliers in the set $\mathcal{S}$ are discarded as false positives. Among the remaining images, the one with the highest ratio of inliers over outliers of the set $\mathcal{S}$ is selected and given to the following localization part.

3.2   Tampering Localization

Once a pair of images is given by the CBIR system, the tampering localization step consists in identifying potential inconsistencies between them. The process should be robust to various transformations, such as rotation, illumination changes, crop, or translation, and is then based on local descriptors. In our case, we are interested in detecting outlier matches spatially close to one another, as a cue of tampering.

Having the homography H computed previously, we apply H to all keypoints of the query to identify inliers and outliers, as detailed in the previous section. Note that the matching criteria considered at this step (1-nn) is weaker than the one used to estimate the homography, in order to enforce a one-to-one matching of keypoints. Since this process is not symmetric, both images are used in turn as query. The image containing the most outliers is selected for the localization step, see Fig. 2(d).

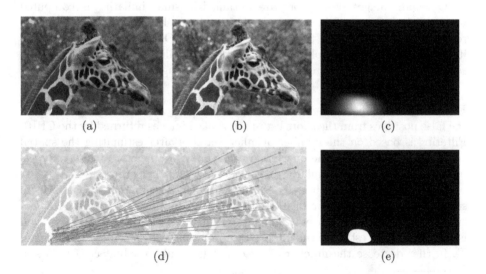

Fig. 2. (a) Query image; (b) candidate image returned by the CBIR (d) outliers computed from query to candidate image; (c) density map; (e) binary mask.

Finally, we identify the areas with high density of outliers and remove the isolated points. These two operations are carried out by a Kernel Density Estimation (KDE) technique. We compute a density map $\mathcal{D}$ on the set of outliers by applying a Gaussian kernel with bandwidth selected by Scott's Rule of thumb, see Fig. 2(c). This density map is then thresholded to obtain a binary mask $\mathcal{B}$ of the suspicious regions. Only points $\mathbf{p}$ of the density map verifying $\mathcal{D}(\mathbf{p}) \geq 1/2 \, max_{\mathbf{p} \in \mathcal{D}}(\mathcal{D}(\mathbf{p}))$ are retained in the final segmentation, see Fig. 2(e).

4 Experiments

We evaluate our approach on challenging datasets exhibiting a large variety of modifications. We first give an overview of the datasets involved and describe the different characteristics of the data. The CBIR is further evaluated using all these datasets and the tampering localization is finally tested.

4.1 Datasets

Many datasets of various size and difficulty have been proposed in image forensics to evaluate forgery detection methods. They differ by the realism of their construction (from simple artificial insertion to realistic complex objects with post-processing), by the types of attacks they address, and by the presence of the modification masks allowing the evaluation of the tampering localization.

Most existing datasets focus on copy-move attacks, thus we build two new datasets. *Reddit* is built from real data with every type of forgery especially copy-paste, which are almost not occurring in the other datasets. Similarly, *Synthetic* is artificially built with various and precise forgeries to better understand how our system copes with each type of attack. Also, we are interested in datasets allowing tampering localization and for which the original images are available.

MICC-F600 [1] is a dataset from image forensics. It contains 600 images: 440 original images from the 1,300 images of the MICC-F2000 dataset [2], and 160 forged images from the SATS-130 dataset [6]. Forged images contain realistic and challenging multiple copy-move attacks.

MediaEval (ME) is composed of 316 images associated to the tweets used in the *Verifying Multimedia Use* task of *Mediaeval 2016*. We use 40 images as queries: 17 fake images particularly challenging, which have their original image in the database, and 23 images with typical collage, cropping, or insertion of text and geometrical shapes (see Fig. 3). These last modifications are generally not achieved in a malicious purpose, but are challenging for the CBIR system. The groundtruth maps were manually constructed for these queries.

Reddit is a collection of 129 original images and their photoshopped versions from the Photoshop challenge on the *Reddit* website[4], totalling 383 images. 106 images are used as query and were manually annotated by up to three annotators,

---

[4] http://www.reddit.com.

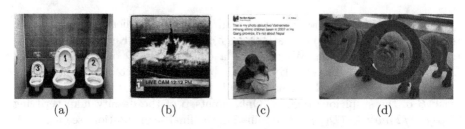

(a)   (b)   (c)   (d)

Fig. 3. Some examples of challenging images from the *ME* dataset.

(a)   (b)   (c)

Fig. 4. Examples of different kinds of attacks in *Reddit*: (a) copy-paste; (b) text insertion; (c) copy-move. Blue: unmodified regions; Red: tampered regions. (Color figure online)

with an inter-annotator agreement of 75.12% in terms of Jaccard's score. The tampering operations are mainly splicing of various size, which is not addressed by *MICC-F600*. Some examples are given in Fig. 4.

Synthetic is an artificially generated dataset of 3,500 forged images, including both copy-move and copy-paste attacks and different processing of the alien. For each 7 original images, we generate 500 forged versions. Each forged image is created by combining a random selection of different parameters among the number of modifications (between 0 and 3), the size of the alien (10, 20, 30, 40, or 50% of the host image), the rotation applied (0, 45, 90, 135, 180, 225, 270, or 315 degrees), a blurring or not of the alien, and the type of attack, *i.e.* copy-move or copy-paste. Note that we can find both copy-move and copy-paste attacks in a forged image, and that a blur attack can be applied on the whole host image (even without any attack). This dataset is not evaluated with the CBIR.

Distractors. Additionally, we collect distractors when evaluating the CBIR system. We use 8,035 images collected from 5 websites dedicated to hoax detection[5]. We further add 82,543 unique images from Twitter, corresponding to the top tweets during January and February 2017, for a total of 170 different topics.

4.2 CBIR System

Most CBIR systems are evaluated on benchmark databases composed of several views of a same object. However, we want to test whether our system is capable

---

[5] hoaxbuster.com, hoax-busters.org, urbanlegends.about.com, snopes.com, and hoax-slayer.com.

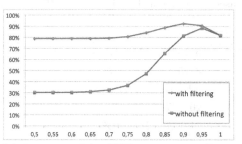
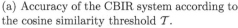

(a) Accuracy of the CBIR system according to the cosine similarity threshold T.

(b) Example of true positive

(c) Example of false positive

Fig. 5. CBIR evaluation results.

of returning a quasi-copy of a query at first rank and none if no copy exists. We further evaluate the behaviour of our system with tampered and noisy images.

The query set is composed of diverse tampered and pristine images and the database contains original images as well as distractors. Specifically, the database to query is composed of 93,121 images: 82,543 images from Twitter, 8,035 images from hoax websites, 316 images from *ME*, 129 original images from *Reddit*, 98 images from the *SATS-130* dataset and 2,000 images from the *MICC-F2000* dataset which contain the original images of *MICC-F600*.

Then, we use a set of 2,151 queries, both positives and negatives (meaning having or not a correspondence in the database): 600 images from *MICC-F600*, 106 photoshopped images from *Reddit*, and 40 tampered images from *ME* are positive examples. Amongst them, 440 images from *MICC-F600* are not tampered. 1,405 images from *Holidays* dataset [12] are used as negative queries.

Results. Unlike most CBIR measuring ranking performance in terms of precision (P@k, mAP, etc.), we evaluate our system in terms of mean accuracy, computed over all the queries. Indeed, we wish our CBIR-based system to output either the most similar image or no image, if no quasi-copy is found in the database.

Figure 5(a) shows the accuracy of the CBIR system for various threshold values T. We observe that the best threshold is $T = 0.9$ with an accuracy of 91.91% with filtering and 81.08% without filtering. The value $T = 0.9$ is kept for the tampering localization step.

Table 1 shows the performance with respect to each set of queries for given thresholds T. We observe a gain in accuracy for lower T on *Reddit*, *MICC-F600*, and *ME* (positive queries). However, *Holidays* performs best for a high T, as it only contains negative queries. Indeed, a low threshold allows to list all relevant images, while generating a lot of false positives.

As an insight, we observe that the CBIR mainly fails when the forged area is very large with respect to the image. This is particularly illustrated by poor

Table 1. CBIR accuracy per datasets for different threshold values T

T	Reddit	MICC-F600	ME	Holidays
0.75	**73.62%**	**99.83%**	**32.50%**	74.68%
0.80	73.23%	**99.83%**	**32.50%**	80.58%
0.85	71.65%	99.50%	**32.50%**	88.41%
0.90	64.57%	98.50%	20.00%	96.09%
0.95	37.80%	94.00%	15.00%	**100.00%**

Table 2. Tampering localization results per datasets.

Dataset	Synthetic	Synthetic unblurred	MICC-F600	Reddit	ME
F_P	12.41%	0.93%	10.82%	37.11%	24.37%
F_N	15.05%	9.90%	20.93%	24.82%	29.82%
TPR	100.00%	100.00%	95.61%	100.00%	100.00%
FPR	49.64%	0.00%	9.10%	0.00%	0.00%

performances on ME. This small set of queries was specially chosen to challenge the CBIR system, which is disturbed by overly large insertions (more than 50% of the image size), or border/banners insertions. Figure 3 shows such queries whose original image has not been retrieved. Examples of successful match despite a quite large forgery and false positive are given in Fig. 5(b) and 5(c).

4.3   Tampering Localization

We evaluate the tampering localization on $Synthetic$, and on the pairs of real forged images returned by the CBIR, from MICC-F600 (copy-move attacks), $Reddit$ (various attacks, mainly copy-paste), and ME (various modifications). For the $Synthetic$ dataset, image pairs are directly given.

The performance on patch localization is computed at the pixel level as the percentage of erroneously detected pixels F_P (i.e. false positives) and erroneously missed pixels F_N (i.e. false negatives). To compare with other methods, we also measure the detection performance at the image level in terms of True Positive Rate (TPR) and False Positive Rate (FPR), where TPR is the fraction of tampered images correctly identified, while FPR is the fraction of original images that are not correctly identified.

Results. Table 2 shows the localization results per datasets. We observe on $Synthetic$ that the localization method is robust to the size, rotation or number of inserted aliens, but unsurprisingly sensitive to blurring of the whole image. The high FPR corresponds to blurred original images classified as forged. Discarding the blurred images ($Synthetic\ Unblurred$), attacks are precisely detected.

Table 3. Results on MICC-F600 (best settings for each method, in %)

Method	Fan2016 [10]	Cozzolino2015 [8]	Li2016 [14]	Ours
TPR	88.13%	96.25%	96.25%	95.00%
FPR	6.82%	5.91%	4.77%	9.10%

Generally, the pixel-level localization is altered by two factors: (i) our predicted area is often smaller than the alien, which increases F_N. However, we do not focus on having the most accurate localization at the pixel level but rather precisely detecting whether a tampering is detected or not; (ii) when the image is wrongly matched by the CBIR with a false positive, the tampering localization failed, resulting in an increase of F_P. This doesn't concern *MICC-F600*, which offer cleaner and smaller attacks, and for which the accuracy of the CBIR is the highest, with no false positives.

At the image level, the detection of tampering in *Reddit*, and *ME* offers perfect results. The null FPR is due to the fact all queries are forged for these datasets. When not all queries are forged, as in *MICC-F600*, performance remains very high. In fact, we compute FPR and TPR for the sake of comparison with the state of the art on *MICC-F600*, as most of methods (except [8]) only deal with detection. Comparison with the state of the art is given in Table 3. We note that the CBIR is not applied there (whole images of *MICC-F600* are processed) to allow the comparison. Our system performs on par with recent state of the art methods, with a higher FPR.

Regarding the entire process and all the datasets (including Holidays as negative examples), we measure a TPR of 81.37% and a FPR of 5.14%. Errors are mainly due to the CBIR performance, as false positives at the retrieval step generate false positives for the tampering detection, while false negatives result in missed tampering detections.

5   Conclusion

In this paper, we address the problem of verifying the authenticity of images from social networks. Moreover, we built two complete dataset for the evaluation. We propose a system that detect and localize tampering on such images, based on image retrieval, followed by image comparison based on local features. Unlike methods from the literature, our system is generic and can handle a large variety of modifications. We evaluated our system on diverse datasets, and shown that the proposed method performs on par with the state of the art for copy-move. We also observed that images from social networks are challenging for state of the art CBIR, and there is room for improvement to deal with this particular type of images. Future work will be directed in this direction.

References

1. Amerini, I., Ballan, L., Caldelli, R., Bimbo, A.D., Tongo, L.D., Serra, G.: Copy-move forgery detection and localization by means of robust clustering with J-linkage. Signal Process.: Image Commun. (SPIC) **28**, 659–669 (2013)
2. Amerini, I., Ballan, L., Caldelli, R., Del Bimbo, A., Serra, G.: A SIFT-based forensic method for copy-move attack detection and transformation recovery. TIFS **6**, 1099–1110 (2011)
3. Babenko, A., Lempitsky, V.S.: Aggregating local deep features for image retrieval. In: International Conference on Computer Vision (ICCV) (2015)
4. Babenko, A., Slesarev, A., Chigorin, A., Lempitsky, V.: Neural codes for image retrieval. In: Fleet, D., Pajdla, T., Schiele, B., Tuytelaars, T. (eds.) ECCV 2014. LNCS, vol. 8689, pp. 584–599. Springer, Cham (2014). doi:10.1007/978-3-319-10590-1_38
5. Bayar, B., Stamm, M.C.: A deep learning approach to universal image manipulation detection using a new convolutional layer. In: Workshop on IHMS (2016)
6. Christlein, V., Riess, C., Angelopoulou, E.: On rotation invariance in copy-move forgery detection. In: Workshop on Information Forensics and Security (2010)
7. Christlein, V., Riess, C., Jordan, J., Riess, C., Angelopoulou, E.: An evaluation of popular copy-move forgery detection approaches. Trans. Inf. Forensics Secur. (TIFS) **7**, 1841–1854 (2012)
8. Cozzolino, D., Poggi, G., Verdoliva, L.: Efficient dense-field copy–move forgery detection. Trans. Info. Forensics Secur. (TIFS) **10**, 2284–2297 (2015)
9. Dixit, A., Dixit, R., Gupta, R.K.: Detection of copy-move forgery exploiting LBP features with discrete wavelet transform. Int. J. Comput. Appl. (3), 1–10 (2016)
10. Fan, Y., Zhu, Y.S., Liu, Z.: An improved sift-based copy-move forgery detection method using T-linkage and multi-scale analysis. IHMSP **7**, 399–408 (2016)
11. Fridrich, J., Kodovsky, J.: Rich models for steganalysis of digital images. Trans. Inf. Forensics Secur. (TIFS) **7**, 868–882 (2012)
12. Jegou, H., Douze, M., Schmid, C.: Hamming embedding and weak geometric consistency for large scale image search. In: Forsyth, D., Torr, P., Zisserman, A. (eds.) ECCV 2008. LNCS, vol. 5302, pp. 304–317. Springer, Heidelberg (2008). doi:10.1007/978-3-540-88682-2_24
13. Krizhevsky, A., Sutskever, I., Hinton, G.E.: Imagenet classification with deep convolutional neural networks. In: Neural Information Processing Systems 25 (2012)
14. Li, Y., Zhou, J.: Image copy-move forgery detection using hierarchical feature point matching. In: APSIPA Transactions on Signal and Information Processing (2016)
15. Maigrot, C., Claveau, V., Kijak, E., Sicre, R.: MediaEval 2016: a multimodal system for the verifying multimedia use task. In: MediaEval Workshop (2016)
16. Phan, Q.T., Budroni, A., Pasquini, C., De Natale, F.G.: A hybrid approach for multimedia use verification. In: MediaEval Workshop (2016)
17. Rao, Y., Ni, J.: A deep learning approach to detection of splicing and copy-move forgeries in images. In: Workshop on Information Forensics and Security (2016)
18. Razavian, A.S., Azizpour, H., Sullivan, J., Carlsson, S.: CNN features off-the-shelf: an astounding baseline for recognition. In: CVPR Workshops (2014)
19. Sicre, R., Gevers, T.: Dense sampling of features for image retrieval. In: ICIP, pp. 3057–3061. IEEE (2014)
20. Sicre, R., Jégou, H.: Memory vectors for particular object retrieval with multiple queries. In: ICMR. ACM (2015)

21. Sicre, R., Jurie, F.: Discriminative part model for visual recognition. CVIU **141**, 28–37 (2015)
22. Simonyan, K., Zisserman, A.: Very deep convolutional networks for large-scale image recognition. In: ICLR (2014)
23. Tolias, G., Sicre, R., Jégou, H.: Particular object retrieval with integral max-pooling of CNN activations. In: ICLR (2016)
24. Warbhe, A.D., Dharaskar, R., Thakare, V.: A survey on keypoint based copy-paste forgery detection techniques. Proced. Comput. Sci. **78**, 61–67 (2016)

Author Index

Printed in the United States
By Bookmasters